Progress in Drug Research
Fortschritte der Arzneimittelforschung
Progrès des recherches pharmaceutiques
Vol. 35

Progress in Drug Research
Fortschritte der Arzneimittelforschung
Progrès des recherches pharmaceutiques
Vol. 35

Edited by / Herausgegeben von / Rédigé par
Ernst Jucker, Basel

Authors / Autoren / Auteurs
C. R. Pfaltz · Robert B. McCall · Ray W. Fuller · Richard M. Schultz ·
David A. Solomon, Edison Miyawaki and Carl Salzman ·
Philip J. Podrid, Lisa Mendes, Scott L. Beau and John S. Wilson ·
Donald T. Witiak and Youg Wei · Satyavan Sharma ·
Alberto Mantovani and Annunciata Vecchi · Pushkar N. Kaul

1990

Birkhäuser Verlag
Basel · Boston · Berlin

ISBN 3-7643-2499-6
ISBN 0-8176-2499-6

Contents · Inhalt · Sommaire

The impact of state and society on medical research 9
By C. R. Pfaltz

Role of neurotransmitters in the central regulation of the cardiovascular system 25
By Robert McCall

Drugs affecting serotonin neurons 85
By Ray W. Fuller

The role of cytokines in macrophage activation 109
By Richard M. Schultz

Benzodiazepine augmentation of the treatment of disruptive psychotic behavior 139
By David A. Solomon, Edison Miyawaki and Carl Salzman

The oral antiarrhythmic drugs 151
By Philip J. Podrid, Lisa Mendes, Scott L. Beau and John S. Wilson

Dioxopiperazines: Chemistry and biology 249
By Donald T. Witiak

Vector-borne diseases 365
By Satyavan Sharma

Interaction of cancer chemotherapy agents with the mononuclear phagocyte system 487
By Alberto Mantovani and Annunciata Vecchi

Drug molecules of marine origin 521
By Pushkar N. Kaul

Index · Sachverzeichnis · Table des matières, Vol. 35 559

Index of titles · Verzeichnis der Titel · Index des titres · Vol. 1–35 567

Author and paper index · Autoren und Artikelindex · Index des auteurs et des articles, Vol. 1–35 577

Foreword

Volume 35 of "Progress in Drug Research" contains 10 articles and the various indexes which facilitate the use of this monograph and establish the connection with the previous volumes. While most of the articles deal with some of the topical aspects of drug research, the first review is an effort to analyze and to characterize the ongoing controversy between the mass media on the one hand and research organizations on the other. Professor C. R. Pfaltz, the Chancelor of the University of Basel, outlines in this article some of the misunderstandings and criticisms, particularly those related to gene technology and current practices in pharmacology and toxicology in use in medical research today, as expressed by state and society. The remaining 9 reviews provide an overview of the work involved in the search for new and better medicines, with a focus on chemical, pharmacological, toxicological, biological and biochemical studies. In the 30 years this series of monographs has existed, the Editor has enjoyed the help and advice of many colleagues: Readers, the authors of the individual articles, and, last but not least, the reviewers have all contributed greatly to the success of PDR. Although many comments received have been favorable, it is nevertheless necessary to analyze and reconsider the current position and the direction of such a series. So far, it has been the Editor's aim to help spread information on the vast domain of drug research, and to provide the reader with a tool helping him or her to keep abreast of the latest developments and trends. The reviews in PDR are useful to the non-specialists who can obtain an overview of a particular research field in a relatively short time. The specialist readers of PDR will appreciate the reviews' comprehensive bibliographies. Moreover they might even get fresh impulses for their own studies. Finally, all scientists interested in drug research can use the 35 volumes of PDR as an encyclopedic source of information.
It gives me great pleasure to present this new volume to our readers. At the same time I would like to express my gratitude to Birkhäuser Verlag and, in particular, to Mrs. L. Koechlin and Mssrs. H.-P. Thür and A. Gomm. Without their personal commitment and assistance, editing PDR would be a nearly impossible task.

Basel, November 1990 Dr. E. JUCKER

Vorwort

Der 35. Band der «Fortschritte der Arzneimittelforschung» enthält zehn Artikel sowie die verschiedenen Register, welche das Arbeiten mit diesen Monographien erleichtern. Mit einer Ausnahme behandeln die Referate aktuelle Gebiete der pharmazeutischen Forschung; die Ausnahme betrifft den Beitrag von Professor C. R. Pfaltz, Rektor der Universität Basel, und ist den zunehmenden Problemen und Spannungen gewidmet, die sich in beängstigender Weise zwischen Staat und Gesellschaft auf der einen Seite und der medizinischen Forschung auf der anderen manifestieren. Die übrigen neun Artikel behandeln in gewohnter Weise aktuelle Aspekte der Chemie, Pharmakologie, Biologie, Toxikologie, Biochemie und Medizin im Rahmen der heutigen Arzneimittelforschung.

Seit der Gründung der Reihe sind 30 Jahre vergangen, und der Herausgeber konnte stets auf den Rat seiner Fachkollegen, der Leser und der Autoren der Reihe zählen; ihnen allen möchte ich auch hier meinen Dank abstatten. In diesen Dank seien auch die Rezensenten eingeschlossen; sie haben mit ihrer Kritik und mit ihren Empfehlungen wesentlich zum guten Gedeihen der Reihe beigetragen. Viele Kommentare waren positiv und lobend. Trotzdem ist es aber angebracht, die Frage nach dem Sinn und Zweck der «Fortschritte» zu stellen und zu überprüfen.

Nach wie vor ist es unser Ziel, neueste Forschungsergebnisse in Form von Übersichten darzustellen und dem Leser auf diese Weise zu ermöglichen, sich rasch und mühelos über bestimmte Gebiete und Richtungen der pharmazeutischen Forschung zu informieren. Es wird ihm somit die Möglichkeit gegeben, sich in diesem komplexen Forschungsgebiet auf dem laufenden zu halten und den Kontakt zum Fortschritt in diversen Teilbereichen nicht zu verlieren. Die Übersichtsarbeiten bieten einerseits dem aktiven Forscher, der ja meistens in einem relativ engen Gebiet tätig ist, nützliche Vergleichsmöglichkeiten, andererseits stellen sie für den Spezialisten eine wertvolle Quelle der Originalliteratur dar. So kann dieser rasch feststellen, ob er in seiner Literaturrecherche alle wichtigen Publikationen erfaßt hat.

Zum Gelingen dieses Werkes haben nicht zuletzt auch die Mitarbeiter des Birkhäuser Verlages, vor allem Frau L. Koechlin und die Herren H.-P. Thür und A. Gomm, wesentlich beigetragen; auch ihnen möchte ich an dieser Stelle meinen Dank abstatten.

Basel, November 1990 Dr. E. JUCKER

The impact of state and society on medical research

By C. R. Pfaltz
Chancellor of the University of Basel, Switzerland

1 Prerequisites and goals of clinical research 11
2 Should clinical research be controlled or limited by state regulations? 14
3 The public "image" of clinical research 17
4 Priorities in medical research 17
5 Conclusions 21
References 23

Among sciences, medicine, in particular medical research, takes on an exceptional position because the objective of its scientific activity is man. Biology may be regarded as "real science" because its progress is based on the application of ways of thinking consistent with the knowledge or understanding of research in natural science. Medical research, however, is restricted by a borderline which cannot be surmounted by physical, chemical or biological methods of research (C. F. von Weizsäcker). The latter may provide the scientist with exact quantitative data, which can be used for a scientific analysis resulting in a few indications concerning the function of some human organs. But despite the enormous progress in scientific methodology, these data do not suffice for the explanation of the complex feedback control of the human organism. Its underlying mechanisms are based on regulation processes which in majority are no longer quantitatively measurable but are predominantly controlled by central-nervous and psychological factors. The theory of "Cybernetics" or feedback control is a scientific-causal way of thinking. If it were integrated in a more human, understanding and intuitive frame of mind, another dilemma would arise: Are we justified in setting limits to attempts of interpretation which are entirely based on scientific cognition, or do we really do justice to man if we reduce his path through life to a sequence of causally explicable processes? (C. F. von Weizsäcker).

Max Planck has pointed out that the gap between the metaphysical world and the phenomenologic world cannot be bridged by the exact sciences. We have to accept a border which cannot be crossed by scientific methods because we have reached a terminal point where science and ethics can no longer be separated from each other. W. Heitler, from the point of view of the nuclear physicist, has come to the same conclusion: If man sees himself only as a sophisticated chemico-physical laboratory he becomes the mere objective of chemico-physical experiments and therefore he loses the right to lay claim to an ethical behavior on behalf of his fellow creatures.

Medical research today finds itself in a situation in which ethical and legal conflicts are pre-programmed, because in current medical practice most diagnostic, therapeutic or prophylactic procedures involve hazards. This applies especially to biomedical research. Thus, the estimate of its scientific quality is considerably reduced through the eyes of state and society, and this judgement may even call in question the freedom of research and the scientific mind. In the field of medical research a fundamental distinction must be recognized between medical research, in which the aim is essentially diagnostic or therapeutic for the patient, and medical research, the essential object of which is purely scientific and does not imply direct diagnostic or therapeutic value to the person subjected to the research project. In medicine both basic and applied research are linked together very closely, but there is a gap between basic science and its application to human disease which has to be bridged and this task is only feasible by promotion and further development of *clinical research.*

1 Prerequisites and goals of clinical research

The point of departure and the goal of any clinical research is the human being, the patient.

In the office, at the hospital ward and in the operating theater the scientifically minded physician is continuously confronted with problems exceeding the limits of a merely practically oriented medicine, both diagnostically and therapeutically. These problems concern the cause, i. e. etiology and pathogenesis of a disease, the efficacy of a medical or surgical treatment, the causal factors of success or failure in therapy. Clinical research is a continuing process integrated in the professional, i. e. clinical activity of a physician or surgeon.

Prerequisites are an analytic mind, a scientific approach to methodology and, above all, a very high ethical standard as well as a strong sense of responsibility. We have to bear in mind that the scientist who is involved in clinical research may be confronted daily with the decision to transfer the knowledge gained experimentally in the laboratory to the human species. This transfer is part of an applied research methodology which is not without risk for the patient. The logical conclusion, that a series of medical problems may be solved in the best possible way by studies on human subjects, is rejected by our society because it is considered an unethical and illegal act to make a human subject the objective of a medical research project and to encroach upon his human rights by abusing him physically and mentally in a clinical study which he cannot or does not fully understand with respect to its purpose and consequences.

This complex problem demonstrates the malaise existing in our society as far as the necessity and the risks of clinical research are concerned.

Another important factor influencing the future of clinical research is the *scientific standard of the physician* involved in this particular field of medical research.

Increasingly, society demands that longstanding disease-related problems be solved. But who will solve these problems? The opportunities are greater than in the past because of the eminent conceptional and technical advances in basic biomedical science. There is every indication that such advances will continue to develop at an increasing rate. However, as science becomes more complicated, clinical scientists have greater difficulty in applying these advances to the treatment of disease and basic scientists are needed. The complexity of disease-related problems forms a spectrum. Some of the problems exceed the ability of the traditional clinical scientist to deal with them; others are less complex but necessitate collaboration between clinical and basic scientists. Unfortunately, most basic scientists have little knowledge of patho-biology or clinical medicine. Because most clinical scientists can no longer be completely competent as basic scientists, they need their assistance. A close and efficient cooperation between clinical and basic scientists necessitates mutual understanding of the problems and also the capability to speak the same scientific language. Hence basic scientists must learn enough pathobiology to attack any disease-related problems that are waiting to be solved together with the clinicians and the latter should have some fundamental knowledge of basic biological sciences.

Concern about the supply of research physicians is reaching crisis proportions. The factors responsible for the steady decline in the number of physician-scientists include the financial indebtedness of medical graduates and increasing financial pressures to earn money in practice, lengthened periods of post-graduate training for specialty certification, decreased emulation of the physician-scientist professor as a role model, uncertain long-term institutional support and increasing competition for research funding.

One reason clinical investigators find it difficult to remain scientifically competitive is the inadequacy of their post-graduate training in biomedical research. Many programs have been created to bridge the widening gap between advances in basic biology and their application in the treatment of human disease. Most notable are combined degree (MD – PhD) programs, the NIH program for physician scientists and foundation programs that are directed at students at every academic level (I. M. Arias, 1989).

The impact of clinical activities on medical research is great and clinical services are a major impediment to successful research programs. The development of networks for scientific communication between clinicians and basic scientists are essential to promote research. Particular attention must be given to the recruitment of the very best clinicians in academic positions who are scientifically active and who must be committed to a prospective and ongoing planning process. Research planning cannot be static if it is to be effective. It is equally important that the rising generation of clinicians interested in medical research be encouraged and promoted by:

- a goal-directed post-graduate training program providing the future clinical scientist with a fundamental knowledge of basic biological science
- the establishment of a hospital infrastructure allowing physicians to become independent clinical scientists (i. e. research facilities should be provided as well as partial exemption from routine work guaranteed).

These conditions can only be satisfied if the state, i. e. the ministries responsible for public health, education and science, pay enough attention to the needs and requirements of the subordinated universities and their medical faculties, particularly with respect to the promotion of research and the support of the future generation of clinical scientists. The interrelations between clinical research and medical care as well as those between research and medical technology are well known. Both medical care and medical technology are linked to rather high financial expenditures.

Politicians are primarily interested in keeping the continuously rising costs of medical care as low as possible. Restrictions of medical services are in general rejected by society, and also the demands made on a highly developed medical technology cannot be denied easily without a certain loss of political prestige. For those reasons the obvious solution is the application of economic measures on research. It would be unwise to cut the budget of medical research in order to lower the expenses of medical care and technology, because the quality of medical service depends entirely on the quality of medical research and teaching. As soon as the quality of research is lowered, the quality of medical services declines. The same holds true for medical teaching because the clinician holding the position of an academic teacher is not only bound to keep his professional knowledge updated but has to revise it continuously by analyzing critically the valid views and theories in medicine. Teaching and research should not be separated from professional practice but regarded as a unit in the sense of a mandate given by the university. This is not a problem of our time; already 65 years ago, A. Flexner, the reformer of medical education in the USA, emphasized the view that university hospitals should be equally concerned with teaching and research.

This infers that university hospitals must be centers of excellence in clinical research. This goal can only be achieved by providing the official departments with adequate personnel and a technical infrastructure which allows clinical research at a high scientific level. The other condition which should be met is the training of physicians in basic biomedical science which will enable them to communicate and cooperate with basic scientists at the same level, in order to solve disease-related problems.

2 Should clinical research be controlled or limited by state regulations?

One of the most important problems which turns up repeatedly in general discussions between the representatives of science and society is the question, . . ."what can we do to avoid the potential harmful effects of research on mankind? There are those who speak of the possible necessity of setting limits to research. Should research be restricted in areas in which incalculable potential dangers to the human race could arise, as for example in the manipulation of the human genetic code or in the breeding of certain viruses? Can science tolerate restrictions in the dissemination of research findings or inhibitions on the publication of new discoveries? Do men of such stature exist that they could accept the responsibility of sitting as judges, evaluating the potentials for good or bad of a particular discovery? Can such prohibitions be effective in a species whose outstanding members have been characterised by a probing intelligence and a never-ending search for knowledge? . . ." (A. Pletscher, 1972)

We have to bear in mind that it is practically impossible to prevent a scientist from ultimately carrying out a project he is capable of conceiving. However, there are limits based on moral principles which have to be respected. There are rules of ethical conduct which have to be followed, otherwise we run the risk of severe conflicts.

Ethical problems arise in the biomedical field as in all other areas of medicine, when decisions are influenced by conflicting sets of values. Such conflicts are often unrecognized or at least not explicitly formulated, especially when one set of values is so prevalent in society that it is taken as an absolute, while conflicting values are considered as "deviant". This is often observed in socio-political affairs: "For example some of the values of liberal democracy such as free speech and due process, proclaimed originally by a revolutionary bourgeoisie and widely assumed since then to be intrinsically good, are some times experienced as artificial restraints by radical groups bent on changing again the structure of society and – more frequently – are violated by the instruments of a now dominant class ... Another example, this one closer to medicine, is the conflict between the demand for optimal care for individual patients, including costly major surgery and the demand for adequate care on a broad social front – a conflict that may persist as long as medical services continue to be available in scarcity rather than in plenty. Both examples illustrate the simple truth that, in political as well as in medical affairs, values change as the situation changes. Conflicts arise, at first implicit and unrecognized, when seen as opposition between "valid" and "deviant" values, and finally as the expression of new human needs or aspirations or opportunities. Nowhere is this more evident than in the influence of modern technology on ethical values. Can this view of ethics, that is the awareness of the existence of conflicting sets of values, provide guidance for the practical resolution of specific conflicts, especially those that may face the biologist in his experimentation, or the pharmaceutical chemist in drug testing, or the physician in his daily practice? Are there any grounds for choice between guidelines, except one's personal convictions or prejudices or interests?" (S. E. Luria, 1972).

If we examine all the codes, regulations and the discussions of the ethical principles when human subjects serve as objectives in clinical research, the two key issues are *informed voluntary consent* and *favorable risk-benefit ratios*.

From a legal point of view, i. e. in the opinion of the former judge of the royal high court of justice, Sir Roger Ormrod, the phrase "informed consent" is wholly redundant and unfortunate. If the word *informed* is introduced, there seems to be an implication that there is something like an "un-informed consent". Its existence cannot be recognized from the point of view of the law. R. Ormrod regards the question of informed consent as a very strange situation because in the second half of the 20th century the medical profession is poised on this dilemma of informed consent. The jurists of the world have been debating this question of when consent is *"consent"* and when not for decennia. They have yet to come up with an answer. Over the years, lawyers have evolved a pragmatic solution to this question. In one sentence one could say: "The degree of information required to make a consent valid varies with the gravity of the matter in issue . . ." (R. Ormrod, 1972).

The second key issue of moral principles in clinical research is the favorable *risk-benefit ratio*. The guidelines of the Swiss Academy of Medical Sciences, which are based on the "declaration of Helsinki", issued in 1983 by the World Medical Association, emphasize: "Biomedical research on human subjects may only be carried out if the risks of a biomedical study in comparison with its clinical relevance and importance can be justified from the physician's point of view". We are, therefore, committed to inform the patient but the information has to be put into the patient's language. This can be done although it may be difficult. What the patient needs to know is whether there is any risk to him and if so, what kind of risk. It is very important to outline risk and possible benefit of a clinical study in the patient's language so that he can have a final say on whether or not he wants to participate.

If we try to answer the initial question "Should clinical research be controlled by state regulations?", we should keep in mind that limitations set to research by state regulations are only significant if the political power is strong enough to carry through these legal proceedings under strict control. According to H. Ruh (1985), ethical principles, particularly if they are very differentiated, cannot easily be transformed into legal paragraphs. Regulation should be restricted to limitations which are clearly aimed at the prevention of malpractice and its negative consequences, i. e. regulations which correspond with fundamental moral principles.

Freedom of research is closely related to *freedom of communication*. Any research program would become a sterile venture if it were not possible to inform society of the results. If the state tries to suppress certain scientific queries or problems, or prohibits the publication of research results, it interferes with our freedom of information and communication. This kind of state interference would correspond to a political censorship which is not compatible with the civil rights set up by a pluralistic democracy. (Ch. Brückner, 1989).

3 The public "image" of clinical research

Why does much of society today show so little confidence in science and scientists, and what is the reason for a great deal of antagonism to science and to scientists, particularly in the field of medical research?

The argument that science is an extremely complicated and highly developed subject, which cannot be properly explained in a language understood by the layman, can no longer be accepted. It is the duty of the scientist to provide society with clear and fundamental information about the results of his research. In the field of clinical research there is a great risk that unsatisfactory information by the scientist will lead to disinformation in the mass media. Very often sensational reports are published about medical research projects which are completely misunderstood and misinterpreted because of a lack of pertinent information on the part of the scientist. A network of communication between the layman and the scientist should be set up which guarantees objective, unbiased and clear information with respect to the risk-benefit and the cost-benefit ratio of the research project involved.

4 Priorities in medical research

The problem of setting priorities in medical research, particularly in clinical research, is a very complex one. It does not only touch upon economic questions but also and to a higher extent on ethical principles and finally the crucial problem of freedom of research.

At the beginning of this century, priorities in medical research were set by society demanding a lowering of the high mortality due to infectious diseases. The rate of infant mortality amounted to 147:1000 live births and the mean life expectancy for both sexes was 49,3 years. Nowadays, infant mortality is less than 1 % and the mean life expectancy has reached 79,1 years for the female sex, 72,4 for the male sex. In 1910, mortality due to tuberculosis came to 22 cases per 100 000 subjects and in 1986 there were less than 1,4 cases per 100 000 subjects.

This therapeutic breakthrough in the management of infectious diseases is entirely due to the enormous progress in basic and clinical research, particularly drug research. However, this evolution compels medical research to change and reset once again its order of priorities. It is society which takes again a leading role in selecting and imposing the priorities in biomedical research in general and in clinical research in particular. The spectacular rise in mean life expectancy results in a rapidly progressing change of the age structure of our population. This process in return leads to a fundamental change of the distribution of diseases within the various age categories. Between 1940–1986 the mortality rate due to cardiac and vascular diseases rose from 30 % to 40 %; within the same period the mortality rate due to malignancies increased from 13 % to 25 %. But also morbidity is changing because diseases linked to the process of aging are gradually increasing. These ailments are due to degenerative lesions of the central nervous system, of the sensory organs and of the musculo-skeletal system. Treatment of these geriatric diseases necessitates not only drug therapy and surgery, but also extensive rehabilitation measures.

Whereas nearly 100 years ago *longevity* was the top priority among the goals medical research was aiming at, we are now confronted with the top priority problem of *improvement of the quality of life.*

According to the statement of the Swiss Academy of Medical Sciences, the definition of the *"quality of life"* reads as follows:

The liberty to satisfy the following desires or needs:

- Physiological functions (nutrition, locomotion, sleep, reproduction)
- Exchange of information
- Mental activity
- Economic security
- Social rank
- Self-esteem and self-realization.

Medicine is concerned with prevention, diagnosis and treatment of physiological dysfunction. One of its major duties is the surveillance of the quality of life. On the one hand quality control means regular medical check-ups, aiming for an early diagnosis of severe ailments. On the other hand, quality control entails the confrontation of the individual with the ideal version of himself which he has formed unconsciously under the permanent influence of society and his environment, particularly through advertisement and the media. This confrontation leads to a continuous subconscious comparison of physical reality and idealistic concepts resulting in a progressive awareness of one's anatomical and physiological shortcomings. These are overestimated and perceived as a loss of quality of life. The individual develops a defensive attitude towards his physical insufficiencies in general and the aging process in particular.
The claim to improve the quality of life with respect to the satisfaction of its physiological needs and functions seems to increase with growing prosperity. This social demand has now become a highly important factor influencing the setting of priorities in research.
The Board of the Swiss Council of Science has recently entrusted the Swiss Academy of Medical Sciences with the launching of a research project on the problem of *life quality control* in medicine.

4 subjects have been selected which serve as representative models for the investigation of this complex *clinical research* problem:
1. Senile dementia
2. Non-inflammatory rheumatic diseases
3. Breast cancer
4. Prenatal screening.

The first two subjects – senile dementia and rheumatism – concern a steadily growing part of our overage society. Within a short time they will rank as top priority research projects, due to the increasing pressure of society on biomedical research.
The third research project concerns an ailment which to date takes the role of a *sacred disease:* Cancer. The importance of the psychological factors involved in the diagnosis of a malignant growth is clearly manifested in the case of breast cancer, the most frequent carcinoma of the female sex. Decisions which are made with respect to the adequate (and radical) treatment of the disease do not only depend on the findings of a regular gynecological check-up or the tho-

rough investigation of a suspicious palpable little lump, but also on the mental process which is necessary to overcome the initial psychological shock and the subsequent change in the patient's attitude towards the disease. The acceptance of this particular malignant disease is highly influenced by the fear of physical disfigurement leading to a loss of self-esteem, and by the fear of losing feminine desirability.

The management of these serious psychosomatic problems will be the determining factor with regard to the future quality of life of a woman with breast cancer. A great number of questions are still open and waiting for scientific investigation, above all psycho-social problems; but at the same time clinical research must still make great efforts to improve surgical and radio-oncologic treatment of the malignant disease. This means an interdisciplinary approach to the problem by a team of basic research scientists and physicians with a special training and fundamental knowledge in biomedical sciences.

The fourth model of the complex clinical research project on *"life quality control" concerns prenatal diagnosis* or *screening*. No other project shows the manifold ethical and legal aspects and the ambiguity of a recently opened field of clinical research.

On the one hand biomedical research enters here a domain in which – on behalf of society – ways and means should be found to control and improve the quality of life. On the other hand medical research will encounter a growing resistance from some social groups which are sensitized and reject research and practice of any prenatal screening for reasons of moral scruples or religious belief. Contrary to *postnatal diagnostic procedures prenatal screening* does not open the way to alternative therapeutic measures. The result of prenatal genetic diagnosis comprises only one decision: abortion or carrying the child to full term.

In this situation we are facing the dilemma of *knowing* on the one hand and of *not being able to help* on the other. From an ethical point of view we may be justified in questioning the moral value of prenatal diagnostic investigation. We may even go so far as to emphasize that ignorance is better than knowledge combined with inability to help. F. Boeckle (1988) refutes this notion from a philisophic-ethical standpoint. In his opinion it is inadmissible to judge genetic diagnostic methods without arranging them adequately in the context of consultation in human genetics and prenatal care.

Prenatal screening can eliminate the suspicion of a fetal malformation in more than 97 % of high-risk pregnancies; this diagnostic procedure is a great psychological relief for the majority of parents, who are afraid of a genetic defect in their progeny. However, F. Boeckle also points out that prenatal diagnostic procedures revealing the presence of a malformation can bring on the risk of artificial abortion because adequate therapy is not yet available, and the only measure to save both fetus and parents from a life of suffering is the termination of the pregnancy. In this situation an ethical conflict arises and we have to approve of the Boeckle's conclusion: From an ethical standpoint the diagnosis of a fetal malformation should and can *never* be accepted as an argument indicating the necessity of artificial abortion.

On the other hand the reasoning of the individual to accept or reject prenatal genetic diagnosis on ethical grounds should never lead to a situation, in which conflicting moral principles stir up emotions in society demanding the intervention of the state to prohibit entirely any research in human genetics.

Conclusions

Medical research is exposed to increasing pressure caused by the rising demands of our society with respect to the strict observation of human and civil rights on the one side and by the political targets and concepts of the state on the other. Both pressure-groups exert a continuous, albeit inconspicuous influence on medical research. On the other hand a certain feedback effect of biomedical research on those two groups cannot be denied. We may speak of a constant interaction influencing the structure of society and state as well as the direction and the points of main effort of biomedical research. This is demonstrated by the fact that medical research is no longer aiming merely at prolonging life but also at improving its quality.

Society, however, shows an ambiguous attitude to biomedical research in general and clinical research in particular. The same society which demands an increase in the quality of life by progress in medicine, condemns certain research projects for ethical reasons and postulates legal control of research. The main arguments concern the violation of human and civil rights. This situation is understandable because the layman cannot judge the risk-benefit-ratio of a research

project properly. He feels himself being manipulated and reacts emotionally. This condition should be improved by more transparency on behalf of the scientist, i. e. better information and communication between the scientist and the public is an absolute necessity. Only adequate information will help society to understand and approve the purpose of medical research and to motivate the individual to lose his mystic fear of manipulation by the scientist.
In the field of biomedical research the *state* makes more and more attempts to legally enforce the human and civil rights of its subjects. At the same time the state tries to get medical research gradually under control by financial measures. The funding of projects by the state is also influenced by political motifs related to public health problems, political economy and the need for new technologies developed in the interest of the state. Intervention by state regulations may become dangerous as soon as legal measures impose restrictions of freedom upon scientific research in medicine.
We find ourselves entangled in some sort of a dilemma. On the one hand we claim the liberty of research and on the other we have to accept the fact that neither the state nor private financial sources are able to subsidize research through unlimited funds. For those reasons *research planning* necessitates *priority setting*. Political and social demands must be considered but priorities in research should be set only by competent scientists who bear the responsibility for the scientific projects. This does not exclude the fact that biomedical and clinical research in general should be planned with the consent of state and society. Scientists are responsible for their research projects and should also consider what will be made of the results of their scientific activities; but they cannot be made responsible for a later misuse of their experimental results because in a distant future the final outcome of their discoveries is unpredictable.
In the field of clinical research the physician himself has to take full responsibility for the planning and the success of his scientific project. This responsibility is absolutely personal and cannot be taken over or shared by a collective panel such as a *committee on ethics*.
In conclusion, medical progress is based on research which ultimately must rely in part on experimentation involving human subjects. Biomedical research involving human subjects must lead to the improvement of diagnostic, therapeutic or prophylactic procedures and the understanding of the etiology and pathogenesis of a disease.

According to the "recommendations guiding physicians in biomedical research involving human subjects", adopted by the 18th World Medical Assembly in Helsinki 1964, the physician ... "can combine medical research with professional care, the objective being the acquisition of new medical knowledge, only to the extent that medical research is justified by its potential diagnostic or therapeutic value for the patient". In other words, the physician engaged in clinical research should always remember the maxim of François Rabelais (1532) – a physician himself – that: "Science without conscience is the ruin of mind".

References

1 I. M. Arias: Training basic scientists to bridge the gap between basic science and its application to human disease. The New Engl. J. Med. *321,* 972–973 (1989).

2 F. Böckle: Biotechnik und Menschenwürde. Schweiz. Ärztezeitung *69,* 2107–2110 (1988).

3 C. Brückner: Fortpflanzungsmedizin und Humangenetik auf dem Prüfstand des Gesetzgebers. Schweiz. Ärztezeitung *70,* 93–99 (1989).

4 A. Flexner: Medical Education 1925, p. 6–18. The MacMillan Co., New York.

5 W. Heitler: Naturphilosophische Streifzüge, p. 10. Vieweg, Braunschweig 1970.

6 S. E. Luria: Ethical aspects of the new perspectives in biomedical research, in: The Challenge of Life, p. 224–30. Birkhäuser, Basel und Stuttgart 1972.

7 Sir Roger Ormrod: Biomedical progress and human values, in: The Challenge of Life, p. 143. Birkhäuser, Basel und Stuttgart 1972.

8 M. Planck: Sinn und Grenzen der exakten Wissenschaft, p. 23. Kindler, München 1971.

9 A. Pletscher: Roche and the human problems of biomedical progress, in: The Challenge of Life, p. 23–28. Birkhäuser, Basel und Stuttgart 1972.

10 F. Rabelais: Pantagruel, chap. 8, 1532.

11 Freiheit und Begrenzung für den Zugriff zum Leben. Wissenschaftspolitik Beiheft 33, 1986.

12 C. F. von Weizsäcker: Die Rolle der Wissenschaft, in: Die Einheit der Natur, p. 26., München 1971.

Role of neurotransmitters in the central regulation of the cardiovascular system

By Robert B. McCall
Cardiovascular Diseases Research, The Upjohn Company, Kalamazoo, Michigan 49001, USA

1	Introduction	26
2	General organization of central sympathetic pathways	27
3	Rostral ventrolateral medulla	29
3.1	Function in cardiovascular regulation	29
3.2	Neurotransmitters contained within the descending pathway from the rostral ventrolateral medulla to sympathetic preganglionic neurons	30
3.2.1	Epinephrine	30
3.2.2	Substance P	33
3.2.3	Serotonin	35
3.2.4	Glutamate	35
4	Caudal ventrolateral medulla	36
4.1	Function in cardiovascular regulation	36
4.2	Regulation of rostral ventrolateral medulla	37
4.3	Regulation of vasopressin release	38
5	A5 Noradrenergic cell group	40
5.1	Norepinephrine regulation of SPN firing	41
6	Medullary raphe nuclei	43
6.1	GABA mediation of midline depressor response	44
6.2	Serotonin mediation of midline pressor response	45
7	Dorsal raphe nucleus	49
8	Paraventricular nucleus	50
9	Nucleus of the solitary tract	51
9.1	Putative neurotransmitters in baroreceptor afferent nerves	52
9.2	Putative neurotransmitters in NTS which modulate baroflexes	54
9.3	Neurotransmitters mediating baroreceptor inhibition of sympathetic neurons	57
10	Neurohumoral control	60
11	Centrally-acting antihypertensive agents	62
11.1	α-Receptor agonists and antagonists	62
11.2	Serotonin$_{1A}$ agonists	64
12	Summary and conclusions	66
	References	68

1 Introduction

Tremendous progress has been made over the last decade in identifying mechanisms which regulate sympathetic outflow from the central nervous system. This review describes recent developments which have had a major influence on our understanding of the central neurotransmitters involved in the regulation of sympathetic nerve discharge. A critical factor responsible for advances in this field is the development of neuroanatomical techniques which have allowed for the identification of interconnections between areas important in the central regulation of blood pressure. These techniques include the use of fluorescent dyes which exhibit retrograde transport properties and other markers such as the conjugated derivatives of horseradish peroxidase that allow tracing of pathways by both anterograde and retrograde transport. Immunochemical techniques have provided biochemical information indicating a wide variety of putative neurotransmitters are contained within these central pathways. Classical physiologic techniques have been employed to elucidate the role of these pathways in central autonomic regulation and to determine the functional significance of putative neurotransmitters contained within these pathways. Correlative techniques (e.g. single unit spike triggered average of sympathetic nerve activity [see ref. 1 for review]) have been developed to identify neurons contained within central sympathetic pathways. The techniques of microiontophoresis and microinjection of putative transmitters or their antagonists are widely used and have provided valuable information regarding the function of putative neurotransmitters. Similar information has been obtained in studies in which the cardiovascular effects of chemical or electrical micro-stimulation of central sympathetic pathways is altered by pharmacological manipulations. Thus, the complexity of the central autonomic nervous system has begun to yield to the combined use of the techniques described above. This multidisciplinary approach has provided the most complete understanding of the central sympathetic nervous system summarized in this review.

Portions of this review are taken with permission from a review written by the author and appearing in Cardiovascular Pharmacology, ed. MJ Antonaccio, Raven Press, New York.

2 General organization of central pathways

Arterial blood pressure is maintained by the tonic vasomotor activity of sympathetic preganglionic neurons (SPN) which are located in the intermediolateral cell column (IML) of the thoracic and lumbar spinal cord. Supraspinal inputs are critical in maintaining the activity of SPN. The medulla and pons have been recognized as essential for maintaining tonic sympathetic activity since the late 1800's when experiments showed that transection of the brain stem caudal to the inferior colliculus failed to lower blood pressure. Transections more caudally produced increasingly greater falls in blood pressure. Reductions in blood pressure reached a maximum with transections at the level of the obex [see ref. 2 for historical review]. A major goal of central autonomic research has been to identify the descending neuronal pathways which project to SPN and to determine the neurotransmitters contained within these pathways [see refs. 3–5 for review]. Supraspinal structures which project to the IML of the spinal cord include the rostral ventrolateral medulla, medullary and pontine raphe nuclei (obscurus, pallidus and magnus), the nucleus interfascicularis hypoglossi, the A5 noradrenergic cell group, the Kolliker-Fuse nucleus located in the lateral portion of the parabrachial nucleus complex, the paraventricular hypothalamic nucleus and the lateral hypothalamic area. Indeed, many of these brain stem structures are interconnected and appear to play a critical role in the elaboration of central sympathetic outflow (Figure 1). Medullary and pontine nuclei involved in autonomic control project to hypothalamic structures which are involved in the regulation of fluid balance (e.g. paraventricular hypothalamic nucleus and the organ of the lamina terminalis). These hypothalamic nuclei send descending projects to autonomic areas of the lower brain stem and spinal cord [6–10]. Thus reciprocal innervations exist between the ventrolateral medulla and 1) the parabrachial nucleus, 2) the paraventricular nucleus and 3) raphe nuclei. The parabrachial nucleus is also reciprocally innervated by the paraventricular nucleus. The A5 catecholamine cell group projects to all these areas. The baroreceptor reflex plays a critical role in the regulation of arterial blood pressure. Baroreceptors arising in the aortic arch and the carotid sinuses project centrally within the aortic depressor and carotid sinus branches of the IXth (glossopharyngeal) and Xth (vagus) cranial nerves. The nucleus trac-

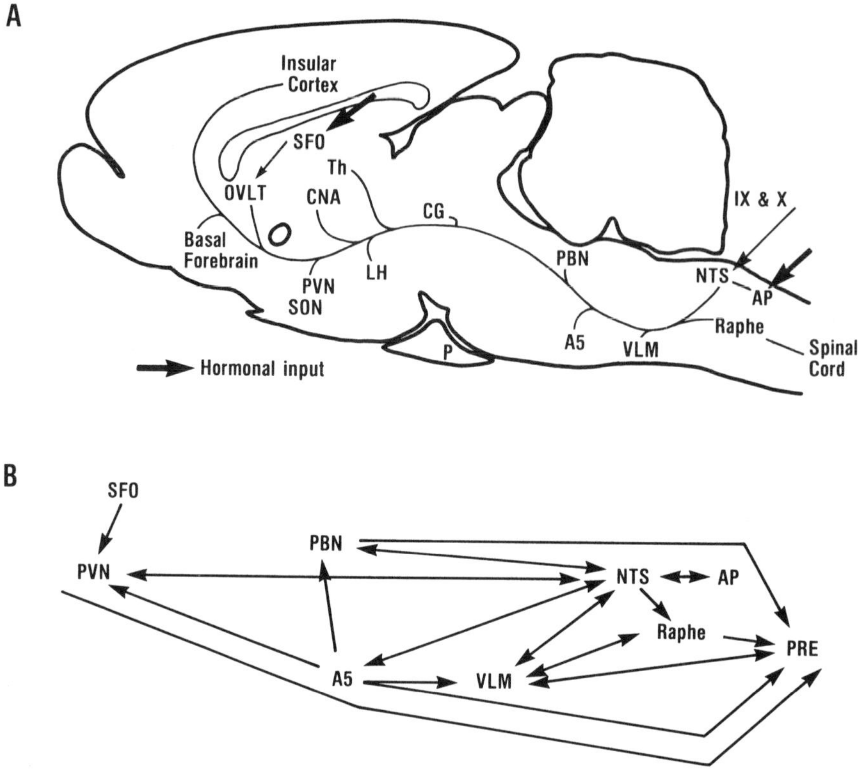

Figure 1
Brain areas involved in the control of cardiovascular function. Panel *A* illustrates locations of important nuclei involved in cardiovascular regulation. Panel *B* is a wiring diagram depicting connections between these nuclei. Abbreviations are: CG, central grey area; CNA, central nucleus of the amygdala; LH, lateral hypothalamus; NTS, nucleus tractus solitarius; OVLT, organum vasculosum of the lamina terminalis; P, pituitary; PBN, parabrachial nucleus; PVN, paraventricular nucleus; SFO, subfornical organ; SON, supraoptic nucleus; Th, midline thalamic nuclei, VLM, ventrolateral medulla. Reprinted from *Cardiovascular Pharmacology,* ed. M. J. Antonaccio, Raven Press, New York.

tus solitarius (NTS) and in particular its dorsomedial subdivision is the location of the first synapse in the baroreceptor afferent pathway [see ref. 10 for review]. Thus, efferent projections from this area define probable nuclei involved in the central regulation of the cardiovascular system. Afferents into this region may function to modulate baroreceptor reflexes. The NTS has been shown to be reciprocally innervated by the ventrolateral medulla, the A5 catecholamine cell group, the parabrachial nucleus and the paraventricular nucleus.

These pathways are summarized in figure 1 and undoubtedly play an important role in the regulation of the cardiovascular system. In addition, the NTS projects to cardiac vagal motoneurons located in the dorsal motor nucleus of the vagus and in the nucleus ambiguus complex [10]. This pathway represents the neuronal basis for the vagal-mediated baroreceptor control of heart rate.

Finally, it is now appreciated that humoral substances in the blood can directly influence the central integration of sympathetic nerve activity. Sites of potential interaction between circulating substances and central structures include the circumventricular organs (i.e. subfornical organ [SFO], organum vasculosum of the lamina terminalis [OVLT], median eminence [ME] and the area postrema [AP]). Circumventricular organs lack an effective blood brain barrier and thus provide sites through which humorally derived substances can penetrate into the brain and activate neural and humoral mechanisms. Circumventricular organs have been shown to project to central nuclei which are involved in cardiovascular regulation. For example, the SFO projects to the paraventricular nucleus and the median preoptic area. SFO neurons which project to paraventricular neurons receive inputs from the NTS [10–12]. The AP shares reciprocal projections with the NTS [10, 11]. These pathways represent the basis by which blood-borne substances can regulate central sympathetic nerve activity.

The above discussion is a brief overview of the areas of the central nervous system that are critical in the regulation of the cardiovascular system. Subsequent sections of this review will discuss these areas in greater detail. The nature and function of neurotransmitters which help to regulate activity within and between central autonomic nuclei will be emphasized.

3 Rostral ventrolateral medulla

3.1 Function in cardiovascular regulation

Transection at the pontomedullary border has little affect on arterial blood pressure while transection more caudally at the level of the obex produces levels of blood pressure similar to that observed after spinal cord transection. This indicates that the medulla is essential in maintaining sympathetic activity and blood pressure. Attempts to specifically localize the site that is critical in maintaining blood pressure have proved elusive. Lesions encompassing large areas of the medulla often fail to affect sympathetic activity or blood pressure. Recent studies, however, indicate that a restricted area in the rostral ventrolateral medulla is necessary for the maintenance of blood pressure. These studies are summarized below and reviewed in more detail elsewhere [1, 13, 14].

Attention was initially focused on the rostral ventrolateral medulla by the observation that direct application of pentobarbital or glycine to the ventrolateral surface of the medulla reduced blood pressure to levels seen after spinal cord transection [15, 16]. Electrical stimulation of the rostral ventrolateral area elicits pressor responses which are accompanied by increases in sympathetic activity [17–19]. Microinjection of excitatory amino acids in restricted regions of the rostral ventrolateral medulla mimics the pressor effects of electrical stimulation. Thus, neuronal cell bodies rather than fiber tracts mediate the increases in blood pressure and sympathetic activity [18–23]. Discrete bilateral lesions in the rostral ventrolateral area of the medulla reduce blood pressure to the level observed following spinal cord transection [17, 24, 25]. Neuroanatomical findings [18, 20, 26–29] indicate that a population of neurons located in the area described by the microinjection and lesion experiments project to the IML of the spinal cord (i.e., site of SPN). Electrophysiological studies have identified neurons within this region which exhibit discharges that are correlated to sympathetic nerve activity, receive baroreceptor inputs and terminate in the IML [30–36]. These neurons are located in a discrete area extending from the caudal pole of the facial nucleus caudally to the rostral third of the inferior olive. The region is limited dorsally by the nucleus ambiguus and medially by the inferior olive. Taken together, these observations indicate that this discrete area of the rostral ventrolateral medulla is essential for the maintenance of tonic sympathetic activity and therefore blood pressure.

3.2 Neurotransmitters contained within the descending pathway from the rostral ventrolateral medulla to sympathetic preganglionic neurons

3.2.1 Epinephrine

The preceding discussion summarizes a large body of evidence which indicates that a population of neurons in the rostral ventrolateral medulla provides a tonic excitatory drive to SPN. There is a striking similarity between the location of the tonic vasomotor region of the rostral ventrolateral medulla and the distribution of the C1 epinephrine cell group [13, 18, 19, 25]. Anatomical evidence indicates that C1 epinephrine neurons project to the IML [29, 37]. The magnitude of the pressor response elicited by chemical or electrical stimulation of the rostral ventrolateral medulla correlates with the placement of the electrode in the C1 area. Moreover, stimulation of medullary sites which contain the descending C1 axons also produce pressor responses [19]. Localized bilateral lesions coinciding precisely with the C1 area cause a decrease in blood pressure comparable to levels observed after spinal cord transection. These findings led to the hypothesis that descending epinephrine neurons from the C1 area of the rostral ventrolateral medulla provide a tonic excitatory input to SPN which is responsible for the maintenance of arterial blood pressure.

In opposition to this hypothesis, however, is the finding that microiontophoretic application of epinephrine consistently inhibits the firing of SPN [38–40]. The epinephrine-induced inhibition of SPN firing is blocked by the α_2-antagonists yohimbine and piperoxane [39, 40]. This indicates that the inhibition is mediated via an α_2-adrenergic receptor. This observation is supported by autoradiographic studies which show that α_2-adrenergic receptors are highly concentrated over clusters of SPN [41]. Ross et al. [29] suggest that the results of microiontophoretic studies must be viewed cautiously since epinephrine released from nerve terminals may act on receptors distant from those acted on by iontophoretic epinephrine. Thus, epinephrine may produce inhibition by acting on receptors located on SPN soma and excitation via receptors located on distal dendrites or antecedent interneurons. Indeed, excitatory and inhibitory effects of norepinephrine have been observed, occasionally on the same SPN [42]. However, microiontophoresis of α_2-antagonists fail to block the excitation of SPN elicited by stimulation of pressor sites in the C1 area of the rostral ventrolateral medulla (Morrison, personal communication). In addition, Sangdee and Franz [43] found that two selective inhibitors of central epinephrine synthesis (LY 134046 and SKF 64139) enhanced descending intraspinal transmission to SPN. These data suggest that bulbospinal epinephrine neurons depress rather than enhance the excitability of SPN.

The most convincing argument that epinephrine-containing C1 neurons mediate a sympathoexcitatory function is the remarkable correlation between the distribution of the medullospinal epinephrine neurons and the location of rostral ventrolateral pressor sites. However, this is only correlative data and should be viewed with caution in light of recent anatomical data. Helke et al. [44] identified a group of substance P immunoreactive neurons in the immediate vicinity of C1 epinephrine neurons that project to the IML. In fact, recent studies indicate that only about fifty percent of the neurons in the rostral ventrolateral medulla which project to the IML contain epinephrine [1, 37]. Electrophysiological investigations in the rat support this contention. The rostral ventrolateral medulla contains neurons which project to the IML and are baroreceptor sensitive. These neurons can be subdivided into two groups [45]. The first group is characterized by axonal conduction velocities of 3.5–8.0 m/s, a maximum discharge rate of 15–35 spikes/s, exhibit pacemaker potentials under certain experimental conditions and are not affected by the α_2-agonist clonidine [1, 46, 47]. The second group exhibits axonal conduction velocities of 0.4–0.8 m/s, a slow discharge rate and an extreme sensitivity to the inhibitory effects of clonidine. Individual neurons project to both the thoracic spinal cord and the central tegmental tract in the pons [45]. The characteristics of the latter group are consistent with the idea that these cells are the C1 epinephrine-containing neurons [1, 45, 48]. Interestingly, two distinct groups of medullospinal neurons arising in the rostral ventrolateral medulla have not been identified in the cat [30, 34, 35]. The C1 area also contains neurons whose cell bodies are immunoreactive to serotonin [49], acetylcholine [50], met- and leu-enkephalin [51], somatostatin [52], neurotensin [53], neuropeptide Y [54], cholecystokinin [55], vasoactive intestinal peptide [56] and thyrotropin releasing hormone [57]. Some of these neurons have been shown to project to the IML. Thus, the C1 area contains several types of neurons which project to IML. This fact, coupled with the observation that microiontophoretic epinephrine consistently inhibits SPN, leads to the conclusion that C1 epinephrine neurons have a sympathoinhibitory rather than a sympathoexcitatory function. However, this conclusion must be made tentatively since C1 epinephrine neurons contain other peptides such as NPY [58] and substance P [59]. The latter has been shown to excite SPN (see below). It is possible that these neurons are functionally heterogeneous having different effects on SPN depending on the type and number of receptors found on specific SPN. Finally, Sved reports [60] that epinephrine levels in the IML are less than 0.2 pg/mg tissue. Thus epinephrine levels are less than 0.05 % of the norepinephrine content in the IML. He concludes that nerve terminals that contain the PNMT enzyme are not necessarily adrenergic. Further experiments are required to demonstrate the functional importance of epinephrine neurons.

3.2.2 Substance P

Substance P is concentrated in the IML of the spinal cord [44]. Autoradiographic binding studies indicate that the IML contains a dense distribution of high affinity substance P binding sites that are located on SPN [61]. The origin of the substance P input to the IML is controversial. A major source of input appears to arise from substance P neurons located in the rostral ventrolateral medulla. This was first demonstrated by experiments in which unilateral electrolytic lesions of the rostral ventrolateral medulla bilaterally reduced the content of substance P by 40 % in IML [44]. Subsequent anatomical studies confirm that substance P neurons located in the rostral ventrolateral medulla project to the IML. At least some of these neurons also contain epinephrine [59]. In addition, substance P is colocalized with serotonin (5-HT)-containing neurons in medullary raphe nuclei. Although these neurons project to the spinal cord, they have not been shown to innervate the IML [62]. Finally, a portion of the substance P input to the IML arises from spinal interneurons [63, 64]. Microiontophoretic application of substance P excites SPN [65, 66]. The onset of the excitatory response is delayed (i.e., 30 s) and continues long after the application of substance P is terminated (i.e., 30–320 s). Similarly, intrathecal administration of substance P or stable agonist analogs of the peptide increase blood pressure, renal sympathetic nerve activity, plasma catecholamines and total peripheral resistance [67–70]. The pressor response to intrathecal substance P is not secondary to the perception of noxious stimuli [71].

The studies cited above have led to the widely accepted conclusion that substance P acts to excite SPN. However, the importance of the rostral ventrolateral medullary substance P neurons in the maintenance of tonic sympathetic activity is much more controversial. Intrathecal administration of substance P antagonists (e.g. [D-Pro^2, D-Trp^{7,9}]-substance P or [D-Pro^4, D-Trp^{7,9}]-substance P) causes dose-dependent decreases in arterial blood pressure. Higher doses of these antagonists reduce blood pressure to levels observed following spinal cord transection [72]. Microinjection of excitatory amino acids into the rostral ventrolateral medulla elicits a pressor response which is associated with an increase in the release of immunoreactive substance P from the spinal cord. Both the pressor response and the release of substance P can be prevented by intrathecal administration of a substance P antagonist [73]. Microinjection of the GABA antagonist bicuculline into the rostral ventrolateral medulla elicits a sympathetic-mediated pressor response which can be blocked by intrathecal administration of a substance P antagonist [74]. These and similar studies [for a review see ref. 75] led to the hypothesis that substance P neurons arising in the ventral lateral medulla function as a major tonic excitatory pathway to SPN. This hypothesis hinges on the specificity of the substance P antagonists at the level of the SPN. This issue is critical since the D-amino acid antagonists of substance P lack specificity for substance P in the isolated spinal cord [76] and have neurotoxic [77] and local anesthetic properties [78]. Yusof and Coote [68] have attempted to determine the specificity of these antagonists. They studied the effects of intrathecal administration of a substance P antagonist on the increase in renal nerve sympathetic activity elicited by intrathecally administered substance P, 5-HT and glutamate. The antagonist blocked the sympathoexcitatory effect of all these agents indicating that the antagonist was non-specific. In contrast, Yashpal et al. [70] report that intrathecal administration of a substance P antagonist blocks the pressor effect of intrathecal substance P but not that of intrathecal angiotensin II. However, the mechanism of the pressor action of angiotensin II was not determined (e.g. sympathoexcitation versus decreased spinal blood flow). Intrathecal administration of three different substance P antagonists to conscious rats led to a variety of sensory and motor dysfunctions. These included loss of motor function, responsiveness to mechanical and thermal stimuli and bladder function. These effects were associated with a dose-dependent necrosis of spinal cord tissue [79]. These effects likely result from decreases in spinal cord blood flow which is associated with intrathecal administration of substance P antagonists [80]. The spinal cord vasoconstriction produced by intrathecal substance P antagonists is not mediated by a substance P receptor [80]. In summary, available evidence supports a sympathoexcitatory role of substance P in the spinal cord. The importance of this system in the tonic maintenance of sympathetic activity is questionable.

3.2.3 Serotonin

The most lateral 5-HT cells of the B_1 and B_3 cell groups lie just medial to the pressor area of the rostral ventrolateral medulla. These laterally positioned 5-HT neurons have been postulated to play an important role in maintaining tonic vasomotor activity [81]. 5-HT neurons lying in the lateral medulla have been shown to project to the IML [82]. Chemical or electrical stimulation of this area results in a pressor response and this is attenuated by pretreatment with the 5-HT neurotoxin 5,7- dihydroxytryptamine (5,7-DHT[81]). More recent studies, however, did not confirm these observations [74, 83]. A recent study by Minson et al. [84] helps to explain these inconsistent findings. They found that glutamate microinjections into the area of the epinephrine-containing neurons evoked a pressor response which was not altered by 5,7-DHT. After ablation of the area of the epinephrine-containing neurons by electrolytic lesion, more medial microinjections of glutamate elicited a pressor response which was presumably mediated by 5-HT. These data indicate that the lateral wings of the B_1 and B_3 5-HT cell groups have a sympathoexcitatory function which is independent of the more laterally positioned sympathoexcitatory neurons in the rostral ventrolateral medulla. The role of 5-HT in the regulation of sympathetic activity is described in more detail below.

3.2.4 Glutamate

Intrathecal administration of the excitatory amino acid antagonist kynurenic acid eliminates spontaneous sympathetic activity and blocks the sympathoexcitatory effect of electrical stimulation of the rostral ventrolateral medulla. The effects of kynurenic acid are reversed by intrathecal administration of the excitatory amino acid kainic acid. This observation suggests that the depressor effects of kynurenic acid are produced by a specific action on a spinal excitatory amino acid receptor [85]. Similarly, intrathecal injection of a second excitatory amino acid antagonist, HA-966, also eliminates sympathetic nerve discharge and reduces arterial blood pressure to levels observed following spinal cord transection [3]. Microiontophoresis of excitatory amino acids excite SPN [86]. Glutamate depolarization of presumed SPN is typically associated with a decrease in membrane resistance and is unaffected by tetrodotoxin in a slice preparation. Focal electrical stimulation elicits a fast EPSP that is identical to the ionic mechanisms involved in the glutamate depolarization and is blocked by glutamate antagonists of the kainate/quisqualate subtype [87]. These data are consistent with the hypothesis that medullospinal neurons arising in the rostral ventrolateral medulla utilize an excitatory amino acid as their neurotransmitter. This is supported by the observation that microiontophoresis of kynurenic acid directly onto SPN blocked the excitation produced by stimulation of the rostral ventrolateral medulla (Morrison, personal communication).

4 Caudal ventrolateral medulla

4.1 Function in cardiovascular regulation

The caudal ventrolateral medulla corresponds with the external formation of the nucleus ambiguus at the level of the obex and overlaps at its most rostral extent with the sympathoexcitatory neurons of the rostral ventrolateral medulla. This area contains the A1 noradrenergic cell group [1]. Electrical or chemical stimulation of this area produces cardiovascular effects opposite those observed in more rostral portions of the ventrolateral medulla. Microinjection of glutamate into the caudal ventrolateral medulla results in decreases in arterial blood pressure and sympathetic nerve discharge [1, 13, 14, 88, 89]. Conversely, microinjection of hyperpolarizing agents such as GABA or the GABA agonist muscimol results in increases in blood pressure, heart rate and sympathetic activity [90, 91]. This occurs even when blood pressure is maintained below the threshold for arterial baroreceptor activation [92]. Electrolytic or kainic acid lesions of the caudal ventrolateral medulla result in increases in blood pressure, heart rate, sympathetic nerve discharge and plasma vasopressin levels [1, 13, 14, 93, 94]. Collectively, these data indicate that the caudal ventrolateral medulla exerts a tonic inhibition of sympathetic activity and arginine vasopressin release from the posterior pituitary.

4.2 Regulation of rostral ventrolateral medulla

There is increasing evidence to indicate that neurons in the caudal ventrolateral medulla tonically inhibit medullospinal sympathoexcitatory neurons in the rostral ventrolateral medulla. Anatomical studies indicate that neurons arising in the caudal ventrolateral medulla project to the rostral ventrolateral medulla [95]. Microinjections of tetrodotoxin into the rostral ventrolateral medulla prevent the increases in arterial blood pressure and sympathetic nerve activity observed following lesions of the caudal ventrolateral medulla [93]. Microinjections of muscimol into the rostral ventrolateral medulla block the vasopressor effects of muscimol microinjections into the caudal ventrolateral medulla [96]. Evidence exists to suggest that A1 noradrenergic neurons tonically inhibit sympathoexcitatory neurons in the rostral ventrolateral medulla. Microinjection of norepinephrine or clonidine into the rostral ventrolateral medulla results in an α_2-receptor-mediated decrease in arterial blood pressure [97, 98]. Microinjection of tyramine, a catecholamine releasing agent, into the rostral ventrolateral medulla produces hypotension [93]. The effect of tyramine can be blocked by pretreatment with the catecholamine uptake inhibitor impiramine. Destruction of noradrenergic terminals in the rostral ventrolateral medulla by 6-hydroxydopamine (6-OHDA) abolishes the effect of tyramine and prevents the increase in arterial blood pressure normally seen after lesion of the caudal ventrolateral medulla [93]. These data suggest that A1 noradrenergic neurons tonically inhibit sympathoexcitatory neurons in the rostral ventrolateral medulla. This conclusion however must be made tentatively. The results obtained with adrenergic agonists and tyramine might be explained by a direct action on rostral ventrolateral epinephrine neurons. In this regard, monoamine neurons possess autoreceptors and regulate their discharge via axon collaterals which project back to their soma [99]. In addition, microinjection of 6-OHDA produces local damage which includes non-catecholamine-containing neurons. Furthermore, Day et al. [89] have shown that depressor responses are elicited from sites immediately medial to the A1 area but not from the area containing the noradrenergic neurons. This observation has recently been confirmed [100]. Finally, non-adrenergic neurons in the caudal ventrolateral medulla project to the spinal cord [18, 101]. Thus it is possible that not all vasodepressor effects elicited from the caudal ventrolateral medulla are mediated via the rostral ventrolateral medulla. The role of the A1 noradrenergic neurons in mediating the depressor response remains to be determined. Evidence exists to indicate that cholinergic mechanisms in the rostral ventrolateral medulla play a role in mediating responses elicited from the caudal ventrolateral medulla. Microinjections of the δ-receptor agonist D-Ala^2-D-Leu^5-enkephalin (DADLE) into the caudal ventrolateral medulla produce a hypertensive response which is reversed by blockade of cholinergic receptors in the rostral ventrolateral medulla [102]. Microinjection of the cholinergic agonist carbachol or the acetylcholine esterase inhibitor physostigmine into the rostral ventrolateral medulla elicits a marked increase in arterial blood pressure which is mediated by an increase in sympathetic nerve activity [103]. Bilateral microinjections of 3,4-diaminopyridine which releases acetylcholine from cholinergic nerve terminals evokes a pressor response [104]. Intravenous administration of physostigmine produces a similar increase in arterial blood pressure. The pressor response is prevented by microinjections of tetrodotoxin, lidocaine or a muscarinic antagonist into the rostral ventrolateral medulla [105, 106]. Pressor responses elicited by the cholinergic agonists appear to be mediated by muscarinic M_2 receptors [107]. Recent anatomical data indicate that cholinergic interneurons in the rostral ventrolateral medulla make synaptic contact with descending sympathoexcitatory neurons [106, 108]. These data suggest that cholinergic interneurons in the rostral ventrolateral medulla provide an excitatory input to descending sympathoexcitatory neurons. These interneurons may mediate the pressor effects of intravenous physostigmine and microinjections of DADLE into the caudal ventrolateral medulla.

4.3 Regulation of vasopressin release

Anatomical studies indicate that neurons in the caudal ventrolateral medulla project to the regions of the supraoptic and paraventricular nuclei of the hypothalamus. These nuclei contain the arginine-vasopressin (AVP) neurosecretory cells. Double-labeling experiments show that over 80 % of these ascending neurons are A1 norepinephrine-containing neurons [109, 110]. Early studies showed that electrical stimulation of the A1 area produced an increase in the release of

AVP [111]. In addition, topical application of GABA antagonists on the ventrolateral surface of the medulla resulted in an increase in AVP secretion which was not attributable to changes in arterial blood pressure [112]. These data suggest that neurons in the caudal ventrolateral medulla are involved in the ascending control of AVP release. More recent studies indicate that microinjection of muscimol into the caudal ventrolateral medulla inhibits the release of AVP in response to hemorrhage and constriction of the inferior vena cava [113]. Conversely, microinjections of L-glutamate or the GABA antagonist bicuculline cause dose-dependent increases in plasma AVP [114]. Finally, microinjection of norepinephrine into the paraventricular nucleus enhances AVP release [115]. These studies suggest that A1 norepinephrine neurons stimulate the release of AVP.

Electrophysiology studies provide more support for this hypothesis. Electrical stimulation of the A1 region elicits an increase in firing of AVP-containing neurons in the paraventricular nucleus and in the supraoptic nucleus. The excitatory effect of A1 stimulation is abolished by local administration of 6-OHDA [116, 117]. Interestingly, pretreatment with 6-OHDA fails to alter basal levels of plasma AVP [118]. In addition, the excitatory effect of A1 stimulation on AVP secretory neurons in the paraventricular nucleus is blocked by microiontophoretic application of the α-adrenergic antagonist phentolamine [119]. Norepinephrine enhances the excitability and promotes bursting activity in supraoptic nucleus neurosecretory neurons via an α_1-receptor mechanism in vitro [120]. Similarly, stimulation of norepinephrine neurons in the locus coeruleus also stimulates AVP release [121]. Taken together, these data support the concept that ascending noradrenergic neurons facilitate the release of AVP. Since 6-OHDA pretreatment fails to alter basal levels of AVP, norepinephrine may act as a modulator to maintain the sensitivity of neurosecretory neurons to stimuli that normally elicit AVP release [118].

The role of the caudal ventrolateral medulla in regulating AVP release is likely more complicated than is indicated by the above discussion. For example, the data presented above suggest that lesions of the caudal ventrolateral medulla would result in a decreased release of AVP. In fact, however, lesion of the caudal ventrolateral medulla results in large increases in plasma AVP [122, 123]. These results may be due to injury discharge in noradrenergic pathways which enhance AVP secretion [53]. Moreover, microiontophoretic application of norepinephrine onto AVP secretory neurons inhibits rather than excites the discharges of these neurons. More recently, it has been demonstrated that the majority of PVN neurons antidromically activated by stimulation of the posterior pituitary are excited by microiontophoretic norepinephrine. In contrast, PVN neurons antidromically activated from the caudal ventrolateral medulla are consistently inhibited by norepinephrine [124]. Superfusion of norepinephrine in a hypothalamic slice preparation excites AVP secretory neurons in the supraoptic and paraventricular nuclei. The excitatory response is mediated via an α_1-receptor since the response can be blocked by prazosin. Following prazosin, norepinephrine inhibits rather than excites AVP neurons [125]. These data suggest that norepinephrine has a dual action on AVP secretory neurons. Similarly, norepinephrine produces both excitation and inhibition of the firing of SPN (see below). The differences between the effects of iontophoretically applied norepinephrine in vivo and superfused norepinephrine in vitro may be explained if the different techniques expose different populations of receptors to norepinephrine. The story is further complicated by the fact that norepinephrine neurons innervate both AVP-positive and closely adjacent AVP-negative neurons [126]. Electrical stimulation of the A1 area inhibits non-AVP neurons in the paraventricular nucleus via a β-receptor [119]. Thus, the actual role of the A1 norepinephrine neurons in regulating AVP release requires additional studies.

5 A5-Noradrenergic cell group

The major noradrenergic input to the SPN in the IML of the spinal cord arises from the A5 cell group of the ventrolateral pons [127, 128]. Over 90 % of the neurons in the A5 area project to the thoracic spinal cord. At least 90 % of the spinally projecting neurons in the A5 area contain norepinephrine [128]. The A5 noradrenergic cell group also sends projections to the central nucleus of the amygdala, perifornical area of the hypothalamus, midbrain periaqueductal gray, parabrachial nucleus and the nucleus tractus solitarius. The A5 area receives inputs from the paraventricular nucleus, the Kolliker-Fuse nucleus, the parabrachial nucleus, the intermediate and caudal portion of the nucleus of the solitary tract and the A1 area [129].

Thus, virtually all areas which project to, or receive inputs from, the A5 area are involved in cardiovascular regulation.
Electrical and glutamate stimulation of the A5 area results in increases and decreases in blood pressure, respectively [130, 131]. Intraventricular administration of 6-OHDA abolishes the depressor response to chemical stimulation but the pressor response to electrical stimulation remains intact [131]. These data suggest that the pressor response following electrical stimulation is mediated by non-catecholamine fibers that pass through the A5 area while the depressor response is mediated by A5 noradrenergic neurons. The decrease in arterial blood pressure following chemical stimulation of the A5 area is associated with a decrease in cardiac output, decreases in vascular resistance in skeletal muscle and increases in mesenteric and skin vascular resistance. The hemodynamic effects associated with glutamate stimulation of the A5 area are prevented by pretreatment with 6-OHDA [132]. Loewy et al. [131] found that individual A5 catecholamine neurons project to both the IML and the nucleus tractus solitarius. Microinjection of 6-OHDA into either of these areas blunt the A5 depressor response. This suggests that both areas contribute to the decrease in blood pressure following chemical stimulation of A5 norepinephrine neurons. Microinjection of 6-OHDA directly into the A5 area produces extensive destruction of norepinephrine neurons but fails to alter arterial blood pressure in conscious animals [133].

5.1 Norepinephrine regulation of SPN firing

The data described above suggest that A5 norepinephrine neurons inhibit SPN in the IML of the spinal cord. However, the exact nature of the interaction between A5 noradrenergic neurons and SPN is likely obscured by the diffuse innervation of A5 neurons to nuclei involved in cardiovascular regulation. More direct information regarding the role of norepinephrine in the regulation of SPN firing comes from microiontophoretic studies. Microiontophoretic application of norepinephrine consistently inhibits the firing rate of SPN [38–40, 134, 135]. The inhibitory effects of microiontophoretically applied norepinephrine are blocked by α_2-adrenergic antagonists (i.e., yohimbine and piperoxane) but not by the α_1-receptor antagonist prazosin or by β-receptor antagonists [38, 40]. This indicates that the inhibitory effects of norepinephrine are mediated by α_2-adrenergic receptors. Consistent with this view is the observations that microiontophoretic application of the α_2-agonist clonidine inhibits the firing rate of SPN while the α_1-agonist phenylephrine has no effect [38]. In addition, α_2-adrenergic receptors are highly concentrated over clusters of SPN in the IML [41].

Although the above data suggest that noradrenergic neurons inhibit the firing of SPN, there is evidence suggesting the opposite. Administration of α_1-adrenergic receptor antagonists (i.e., prazosin, WB-4101 or ketanserin) inhibit spontaneous sympathetic nerve discharge via an action in the spinal cord [136–138]. Since α_1-receptors are thought to mediate excitatory effects of norepinephrine in the central nervous system [139], these data indirectly support a sympathoexcitatory role of norepinephrine in the spinal cord. Intrathecal administration of norepinephrine results in an inhibition followed by an excitation of renal sympathetic activity. The inhibitory response is blocked by α_2-receptor antagonists while the excitatory response is blocked by α_1-receptor antagonists [140]. Administration of the catecholamine precursor L-dopa increases excitability in spinal sympathetic pathways [141]. More direct evidence for an excitatory function of norepinephrine at the level of SPN comes from work done in spinal cord slice preparations. Superfusion of a spinal slice with norepinephrine causes a membrane depolarization in antidromically identified SPN and results in repetitive cell discharges [87, 142–144]. Pretreating the slices with α_1-receptor antagonists but not α_2- or β-receptor antagonists prevent the depolarizing effect of norepinephrine [87].

Thus data exist to support either an excitatory or an inhibitory role of norepinephrine in regulating SPN discharge. This controversy may be resolved by recent studies in which the spinal slice preparation was employed. Nishi et al. [87, 145] found that SPN in a slice preparation exhibit a fast excitatory postsynaptic potential (EPSP) (or rarely a fast inhibitory postsynaptic potential [IPSP]) in response to single focal stimulation. Trains of repetitive stimuli produced a slow EPSP which was occasionally accompanied by a slow IPSP. The slow EPSP was always associated with an increased input resistance, disappeared at levels of anodal hyperpolarization exceeding -80 mV and was specifically blocked by the α_1-receptor antagonist prazosin. Norepinephrine superfusion produced a depolarization that shared the same characteristics of the slow EPSP. The slow IPSP was not normally observed until the slow EPSP was eliminated by prazosin. Following the use of prazosin, norepinephrine produced a hyperpolarization. Both the slow IPSP and the norepinephrine-induced hyperpolarization were accompanied by a decreased input resistance, reversed at -90 mV and abolished by the α_2-adrenergic receptor antagonist yohimbine. This data supports the hypothesis that noradrenergic neurons can both excite and inhibit SPN. The inhibitory interaction is mediated by α_2-receptors located on or near the soma. The excitatory interaction occurs through α_1-receptors located on distal dendrites. Since the recording electrode of large multibarreled pipettes used in microiontophoretic experiments must be near the soma to record action potentials, iontophoresis of norepinephrine appears to have only an inhibitory effect. In contrast, norepinephrine reaches the excitatory receptors located on dendritic trees in a superfused slice preparation. Although the above hypothesis is attractive and fits well with present data, it will be difficult to prove. Alternatively, the effects of noradrenergic neurons are mediated via one subtype of α-receptor, while epinephrine-containing neurons selectively interact with the other subclass of α-receptor. In this case, microiontophoretic application of a catecholamine might not give an accurate idea of the physiologic function of the neurotransmitter. In either case, the fact that lesions of the A5 area fail to alter arterial blood pressure [133] indicates that this area is not essential in maintaining tonic sympathetic nerve activity and likely functions as a modulatory system (see below).

6 Medullary raphe nuclei

Midline medullary raphe nuclei are often referred to as the medullary “depressor area” since electrical stimulation of this area typically produces marked decreases in blood pressure and sympathetic nerve discharge [146, 147]. Microinjection of L-glutamate at these depressor sites result in a decrease in arterial blood pressure and inhibition of spontaneous sympathetic activity [35]. Electrolytic lesion of the midline from the obex to the level of the genu of the facial nerve results in an increase in sympathetic nerve discharge [148]. These data indicate that neurons in the medullary raphe nuclei provide a tonic inhibition of sympathetic nerve activity. In support of this conclusion, electrophysiology studies have identified a group of midline raphe neurons whose tonic discharges are temporally related to sympathetic activity and are increased during baroreceptor reflex activation [149–151]. Thus, these neurons likely subserve a sympathoinhibitory function. These sympathoinhibitory neurons have been shown to project to the IML [150, 151]. Recent studies indicate that the midline medulla is not simply a sympathoinhibitory area, but rather is heterogenous with respect to autonomic function [see refs 3,152,153 for review]. Electrical stimulation of the midline medulla results in increases as well as decreases in arterial blood pressure and sympathetic nerve activity [154, 155]. The pressor effect of electrical stimulation is mimicked by microinjection of L-glutamate indicating that neuronal cell bodies rather than fibers of passage mediate the sympathoexcitatory response [152, 153]. Electrophysiology studies have identified sympathoexcitatory neurons in the midline medulla (i.e., exhibit discharges which are temporally related to sympathetic activity and are inhibited during baroreceptor activation [149, 151]). These neurons project to the spinal cord but have not been shown to terminate in the IML [150].

6.1 GABA mediation of midline depressor response

Available evidence indicates that GABA mediates the depressor response elicited by stimulation of the midline medulla. Picrotoxin, an agent that blocks the Cl^- channel associated with GABA receptor activation, abolishes the decreases in arterial blood pressure and sympathetic nerve activity observed following midline stimulation [146]. Similar effects were seen with bicuculline, a direct GABA receptor antagonist. The blockade of the sympathoinhibitory effect of midline stimulation was not a result of blood pressure changes produced by the antagonists and was specific for inhibition elicited from the midline. These data suggest that the midline medullary depressor response is mediated via a GABAergic inhibition of central sympathetic neurons. This hypothesis is supported by the observation that the benzodiazepine diazepam potentiates the sympathoinhibitory response to midline stimulation [146]. In this regard, diazepam potentiates the effects of GABA in the central nervous system.
A recent study suggests that the sympathoinhibition elicited by stimulation of midline medullary depressor sites results from a GABA-

mediated inhibition of medullospinal sympathoexcitatory neurons located in the rostral ventrolateral medulla [35]. Electrical stimulation of midline depressor sites inhibited the firing of electrophysiologically identified medullospinal sympathoexcitatory neurons in the rostral ventrolateral medulla. Microiontophoresis of bicuculline blocked the inhibition elicited from the midline and increased the firing rate of sympathoexcitatory neurons. In contrast, microiontophoresis of glutamate increased the firing rate of sympathoexcitatory neurons but failed to affect the midline-evoked inhibition of neuronal firing. Stimulation of midline depressor sites excited a second group of spontaneously active neurons in the rostral ventrolateral medulla. These neurons were often located in the same recording field as the sympathoexcitatory neurons. The neurons were excited by midline stimulation with an onset latency of 21 ms while sympathoexcitatory neurons were inhibited with an onset of 23 ms. These data suggest that neuronal elements in medullary raphe nuclei inhibit sympathoexcitatory medullospinal neurons in the rostral ventrolateral medulla by activating closely adjacent GABAergic interneurons. This interpretation fits with the observation that the region of the rostral ventrolateral medulla containing sympathoexcitatory neurons has a high concentration of GABAergic cell bodies [156, 157]. Furthermore, this inhibition is tonically active and of non-baroreceptor origin. This is indicated by the observations that 1) midline lesions increase sympathetic activity but have no effect on baroreceptor reflexes [148], 2) the putative GABA interneurons fire spontaneously and are not affected by baroreceptor activation [35] and 3) microiontophoretically applied bicuculline increases the discharge rate of sympathoexcitatory neurons [35]. A possible role of GABA in mediating baroreceptor inhibition in the ventrolateral medulla is discussed below.

6.2 Serotonin mediation of midline pressor response

The midline area of the lower brain stem contains the B1, B2 and B3 5-HT cell groups which have been demonstrated to project to the IML of the spinal cord [82, 158]. Indeed, the IML contains high concentrations of 5-HT and a study combining receptor autoradiography with retrograde labeling of SPN indicates that 5-HT receptors are highly concentrated over SPN [159]. Medullospinal 5-HT neurons often cross in the spinal cord several segments rostral to their termination. Both crossed and uncrossed 5-HT axons send out large numbers of collaterals [160]. The area of the midline medulla that contains 5-HT neurons projecting to the IML corresponds to the medullary depressor region [146, 147, 161]. The close association between 5-HT descending neurons and midline sites that elicit vasodepressor responses when stimulated has led to the conclusion that the descending 5-HT pathway inhibits SPN [162, 163]. The finding that stimulation of presumed 5-HT containing axons in the dorsolateral funiculus of the spinal cord inhibits sympathetic activity supports this hypothesis [164]. In addition, administration of 5-HT precursors produces a dose-dependent depression of spinal sympathetic reflexes [141]. Finally, a large body of pharmacological evidence involving administration of 5-HT, 5-HT precursors or 5-HT synthesis inhibitors suggests that 5-HT neurons inhibit sympathetic activity. These data have been summarized in an excellent review [165]. The above data are all indirect assessments of the role of 5-HT in regulating the firing of SPN and may in fact be misleading. For example, the depressor response elicited by stimulation of the midline medulla is mediated by GABA rather than by 5-HT (see above). In addition, the 5-HT precursor tryptophan has been used in many studies to suggest that 5-HT neurons function to inhibit sympathetic activity [165]. Recent studies indicate, however, that the hypotensive effect of tryptophan is mediated via non-serotonergic mechanisms [166].

Microiontophoretic application of 5-HT excites SPN [86, 134, 135]. The excitatory effect of 5-HT on SPN is blocked by microiontophoretic application of the 5-HT antagonists methysergide and metergoline [86]. More importantly, iontophoretic 5-HT antagonists decrease the spontaneous discharge rate of SPN in intact, but not in spinal transected animals [86]. Similarly, intravenous administration of methysergide and metergoline inhibits spontaneous sympathetic activity in intact animals but not in animals depleted of 5-HT [167]. These observations indicate that medullospinal 5-HT neurons provide a tonic excitatory input to SPN. Support for a direct excitatory action of 5-HT on SPN also comes from slice studies in which superfusion of 5-HT depolarizes SPN. The depolarization is associated with an increase in membrane resistance [87, 168]. The depolarizing action of 5-HT persists in the presence of tetrodotoxin or in the absence of external Ca^{2+}. This indicates that 5-HT acts directly on SPN [168].

Recent studies indicate that the pressor response elicited by stimulation of the midline medulla is mediated by medullospinal 5-HT neurons. The pressor response produced by midline stimulation is associated with an increase in sympathetic nerve activity. Intravenous or intrathecal administration of methysergide or metergoline blocks the sympathoexcitatory response elicited from midline medullary raphe nuclei. In contrast, the antagonists fail to block the sympathoexcitatory response to electrical stimulation of pressor sites in the rostral ventrolateral medulla. Sympathoinhibitory responses elicited from the midline medulla are either unaffected or potentiated by 5-HT antagonists [155]. The 5-HT uptake inhibitor chlorimipramine potentiates the sympathoexcitatory response elicited from the raphe but not from the rostral ventrolateral medulla [155]. Stimulation of the B3 cell group causes an increase in blood pressure. The pressor response is associated with an increase in 5-HT release in the thoracic spinal cord [169]. These data indicate that medullospinal 5-HT neurons provide a tonic excitatory input to SPN [see refs 152, 153 for reviews]. Intravenous or intrathecal administration of the 5-HT_1 and 5-HT_2 receptor antagonist methysergide blocks the pressor response to electrical stimulation of 5-HT cell bodies [155, 170]. In contrast, intrathecal administration of 5-HT_2 and 5-HT_3 receptor antagonists fail to block the 5-HT elicited pressor response [170]. Iontophoretic application of the 5-HT_{1A} agonist 8-OH DPAT has no effect on SPN firing but blocks the excitatory effect of 5-HT [172]. These data suggest that receptor mediating 5-HT-induced excitation of SPN is related to the 5-HT_1 receptor subtype.

Medullospinal 5-HT neurons have recently been electrophysiologically characterized [151]. Medullospinal 5-HT neurons share several important physiological and pharmacological characteristics with identified 5-HT neurons in the dorsal raphe nucleus [173] including: 1) an extremely regular discharge rate of approximately 1 spike/s, 2) spike durations of greater than 2 ms, 3) axonal conduction velocities which are appropriate for transmission through unmyelinated axons, 4) inhibition of neuronal firing produced by microiontophoretic 5-HT and 5) sensitivity to the inhibitory effect of the 5-HT_{1A} agonist 8-OH DPAT. Medullospinal 5-HT neurons were antidromically activated by electrical stimulation of the IML. Interestingly, the discharges of medullary 5-HT neurons are not temporally related to sympathetic nerve activity and are not affected by baroreceptor re-

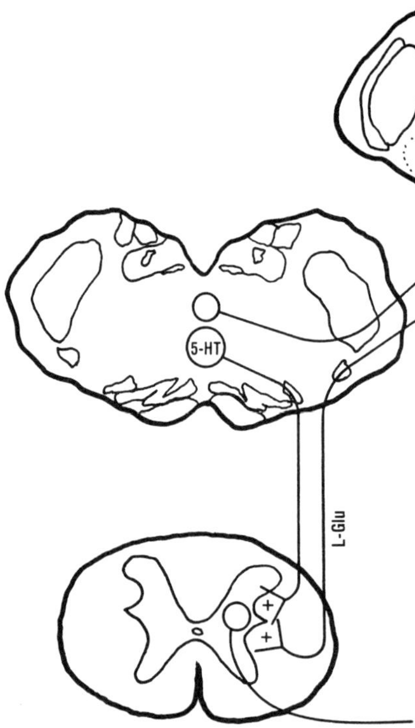

Figure 2
Summary of the role of the midline medulla in cardiovascular regulation. 5-HT neurons project to sympathetic preganglionic neurons in the IML and provide a tonic excitatory input to these neurons. Other midline neurons project to GABAergic interneurons in the rostral ventrolateral medulla. GABAergic interneurons provide a tonic non-baroreceptor inhibition of medullospinal sympathoexcitatory neurons (SE). Reprinted from *Cardiovascular Pharmacology,* ed. M. J. Antonaccio, Raven Press, New York.

flex activation. Thus, these neurons receive little or no afferent inputs from central baroreceptor or sympathetic pathways [151]. Since these neurons provide a tonic excitatory input to SPN (see above), this observation suggests that 5-HT acts as a modulator to increase the excitability of SPN (see below). Midline pathways involved in cardiovascular regulation are summarized in Figure 2.

Substance P and thyrotropin hormone (TRH) have been demonstrated to coexist with medullary 5-HT neurons which project to the IML [174, 175]. In addition, substance P, TRH and 5-HT coexist in nerve terminals which appear to make contact with SPN [176]. The functional significance of colocalization of neurotransmitters within a single neuron is not known and co-release from defined nerve endlings has so far not been demonstrated in the central nervous system. Nevertheless, it is worth noting that substance P is thought to excite SPN (see above). Microiontophoretically applied TRH also weakly excites SPN suggesting that this neuropeptide may function in sympathoexcitatory processes [177]. Studies also suggest that medullospinal 5-HT neurons also contain enkephalin [174] and GABA [178, 179] although termination in the IML has not been demonstrated. Intrathecal administration of dynorphin decreases blood pressure, heart rate and sympathetic activity [180]. Enkephalin depresses sympathetic activity in both an intraspinal excitatory pathway and a spinal reflex pathway [181]. Naloxone blocks a portion of the post excitatory depression following SPN discharge [182]. These data suggest that opioids may inhibit activity at the level of the SPN. Finally, GABA-like terminals make synaptic contacts with SPN soma and dendrites [3]. GABAergic neurons likely provide a tonic inhibitory input to SPN since iontophoretic GABA inhibits SPN firing while iontophoretic bicuculline increases spontaneous SPN discharge [183, 184]. Interestingly, GABA has been shown to co-exist with 5-HT in neurons which project to the spinal cord [185].

7 Dorsal raphe nucleus

Several studies suggest that 5-HT neurons in the dorsal raphe nucleus play a role in the central regulation of blood pressure. For example, electrical stimulation of the dorsal raphe nucleus elicits increases in arterial blood pressure and sympathetic nerve activity [186, 187]. Depletion of brain 5-HT attenuates the pressor response to raphe stimulation while 5-HT uptake inhibition potentiates the response [186]. Furthermore, the pressor response is blocked by administration of the 5-HT antagonist metergoline [187]. Microinjection of metergoline into the anterior hypothalamus/preoptic area (AH/PO) attenuates the pressor response to dorsal raphe stimulation [188]. Microinjection of 5-HT into the AH/PO elicits a pressor response [188]. Microinjection of the neurotoxin 5,7-DHT into the dorsal raphe blocks the effects of stimulation of the raphe and results in enhanced pressor responses to microinjections of 5-HT into the AH/PO [189]. These studies suggest that 5-HT neurons in the dorsal raphe can facilitate sympathetic blood pressure possibly through an action in the AH/PO. A hypothalamic cholinergic mechanism has been implicated in the dorsal raphe pressor response [188]. Additional studies suggest that 5-HT neurons in the dorsal raphe nucleus may play a facilitatory role in the regulation of vasopressin and renin secretion [190, 191].

8 Paraventricular nucleus

The peripheral pressor and antidiuretic actions of AVP are well known [for review see ref. 192]. In recent years, anatomical studies have shown that in addition to the classic neurosecretory pathways of the posterior pituitary, paraventricular neurons project to nuclei that are involved in central autonomic regulation. These neurons are located in the parvocellular subdivision of the paraventricular nucleus and send descending projections to the medulla and spinal cord [193–195]. Approximately 15 % of these neurons send axon collaterals to both the dorsal medulla (i.e. nucleus tractus solitarius [NTS] and dorsal motor nucleus of the vagus) and to the spinal cord [196]. The remaining neurons appear to project solely to either the dorsal medulla or the spinal cord. Immunohistochemistry has been used to demonstrate that a portion of the descending neurons contain AVP or oxytocin and neurophysin [193, 196, 197]. However, the majority of the fibers has not been characterized as to neurotransmitter content [198]. Electrophysiology studies confirm the existence of PVN pathways to the NTS, the ventrolateral medulla and the spinal cord [199–201]. Descending PVN neurons have been shown to receive baroreceptor inputs [202]. Ascending inputs from brain stem catecholamine cell groups have already been discussed. Thus, the PVN receives afferent input from medullary cardiovascular sites and exerts regulatory effects on the cardiovascular system through release of vasopressin into the peripheral circulation and through efferent signals to the medulla and to the spinal cord. Immunohistochemical studies show that spinally projecting neurons from the PVN contain oxytocin and vasopressin. These two putative neurotransmitters are contained in the IML [194, 196, 203, 204]. An early study reports that microiontophoresis of AVP and oxytocin, as well as electrical stimulation of the PVN, inhibits the firing of SPN [205]. More recent studies fail to support these observations. Electrical stimulation of the parvocellular subdivision of the PVN elicits a pressor response which is accompanied by an increase in sympathetic activity and vasoconstriction in the mesenteric, renal and skeletal muscle vascular beds [198, 206]. Backman and Henry [207] found that microiontophoretic application of AVP increased the firing of SPN. Superfusion of AVP depolarizes putative SPN in a slice preparation [208]. $Vasopressin_1$ antagonists block the depolarizing effect of AVP while a $vasopressin_2$ agonist has little effect. AVP-induced depolarizations are partially reduced, but never eliminated, by a low Ca/high Mg solution or by tetrodotoxin. This suggests that AVP excites SPN by a direct depolarization and by an indirect effect via the release of an excitatory transmitter [208]. The role of AVP in mediating the pressor response to PVN stimulation has been investigated. Intrathecal administration of as little as 1 pmole of AVP elicits a pressor response which is accompanied by vasoconstriction in renal, mesenteric and hindquarter vascular beds [198, 209]. The effect is due to an action of vasopressin within the spinal cord since blockade of peripheral vasopressin receptors had no effect on the response. Intrathecal administration of a vasopressin antagonist at doses which had no non-specific depressant effects blocked the pressor response to intrathecal AVP. The same dose of the AVP antagonist failed to block the pressor response to stimulation of the PVN. Assuming the antagonist had access to the same group of receptors affected by neurally released AVP, these data suggest that the cardiovascular effects produced by stimulation of the PVN do not depend on the release of vasopressin from spinal cord nerve terminals [198]. Support for this conclusion is found in studies performed using Brattleboro rats which lack hypothalamic and spinal AVP [198]. Stimulation of the PVN in Brattleboro rats elicits a pressor response which is identical to that observed in Long Evans control rats [198]. Thus, although AVP appears to depolarize SPN, the role that this transmitter plays in regulating the firing of SPN remains unclear. In addition, the identity of other spinal transmitters arising in the PVN is unknown. The role of AVP in modulating baroreceptor reflexes at the level of the NTS is described below.

9 Nucleus of the solitary tract

Stretch receptors in the major arteries and cardiac atria and ventricles detect changes in arterial pressure, central venous pressure and left ventricular pressure, respectively. The cell bodies of these viscerosensory afferent neurons are located in the petrosal and nodose ganglia. Baroreceptor afferent fibers enter the central nervous system in the IX and X cranial nerves and terminate primarily in the dorsomedial subdivision of the NTS [10]. The basic function of the baroreceptor reflex in buffering against acute changes in arterial blood pressure is reviewed elsewhere [11, 210]. The following sections deal with 1) identification of neurotransmitters that convey sensory information to the NTS, 2) neurotransmitters in the NTS that modulate baroreceptor afferent information and 3) neurotransmitters that mediate baroreceptor inhibition of central sympathetic neurons.

9.1 Putative neurotransmitters in baroreceptor afferent nerves

Two neurotransmitters have been considered as likely candidates for the baroreceptor afferent nerves: L-glutamic acid and substance P (Figure 3). Microinjection of L-glutamate into the NTS mimics baroreceptor reflex activation by eliciting a dose-dependent hypotension, bradycardia and apnea [211]. Microinjection of the glutamate antagonist glutamic acid diethylester (GDEE) into the NTS results in increases in arterial blood pressure and heart rate and blocks the baroreceptor-mediated bradycardia produced by intravenous injection of phenylephrine [212]. NTS microinjections of GDEE or a second glutamate antagonist, HA-966, but not saline block baroreceptor reflexes as judged by 1) inhibition of sympathetic activity elicited by stimulation of baroreceptor afferent nerves, 2) the baroreceptor-mediated locking of sympathetic slow waves to the cardiac cycle and 3) the inhibition of sympathetic activity associated with an intravenous pressor dose of phenylephrine [213]. The sympathoinhibitory response to afferent stimulation was restored after NTS microinjections of GDEE by increasing the intensity of afferent stimulation or by directly stimulating the NTS. These data suggest that the impairment of baroreceptor reflexes produced by GDEE was independent of mechanical damage or local anesthetic effects [213]. Microinjections of a fourth glutamate antagonist, kynurenic acid, also blocks the baroreceptor-mediated bradycardia associated with phenylephrine pressor responses [92]. Biochemical studies indicate that the intermediate area of the NTS contains a high-affinity uptake system for inactivation of L-glutamate and a high concentration of this excitatory amino acid. These data are consistent with this area being richly innervated by glutaminergic neurons [214]. Unilateral removal of the nodose ganglion results in a 50 % reduction in the uptake of L-glutamate bilaterally into homogenates of NTS and a reduction in the content of glutamate in the NTS [214]. However, a recent study was unable to confirm these observations [215]. Electrical stimulation of baroreceptor afferent nerves elicits hypotension and bradycardia which is associated with an increase in release of L-glutamate from the NTS [216]. Microinjection of the excitatory amino acid antagonist kynurenic acid abolishes baroreceptor reflexes elicited by electrical stimulation of the aortic depressor nerve. In addition, kynurenic acid blocks the hypotension and bradycardia produced by NTS microinjections of N-methyl-D-aspartate, kainate and α-amino-3-hydroxy-5-methylisoxazole-4-propionic acid but does not block glutamate effects [217]. Similarly, antagonists of the N-methyl-D-aspartate, kainate and quisqualate receptors block the effects of solitary tract afferent stimulation in a slice preparation [218]. Collectively, these data suggest that an excitatory amino acid may function as a neurotransmitter released by baroreceptor afferents in the NTS. The exact identity of the excitatory amino acid has not been determined.

A large body of evidence also suggests that substance P may be released by primary baroreceptor afferents in the NTS. The area of the NTS which is innervated by primary baroreceptor afferents contains significantly more substance P than do other areas of the nucleus [219]. Both the petrosal and nodose ganglia contain substance P [219]. Unilateral nodose ganglionectomy results in a decline of substance P in the ipsilateral intermediate and caudal areas of the NTS [220]. Similarly, nodose ganglionectomy has been reported to significantly reduce substance P- and neurokinin A-like immunoreactivity in the NTS [221]. Discrete varicose nerve fibers containing substance P have been identified in the peripheral regions known to contain baro- and chemo-receptor afferent nerves [220, 222]. Substance P is released in a calcium-dependent manner from NTS tissue slices in vitro during potassium depolarization and in the presence of capsaicin [223]. The latter observation is significant because capsaicin has been demonstrated to evoke the release of substance P from regions receiving sensory afferents but not from other nonsensory regions of the central nervous system [224]. More recently, substance P has been shown to be released in the NTS following aortic nerve stimulation using the technique of in vivo microdialysis [225]. Microinjection of substance P into the NTS has been reported to mimic baroreceptor reflex activation by producing a decrease in arterial blood pressure, heart rate and sympathetic nerve activity [226]. However, other studies have not been able to confirm these observations [227]. A recent study showing that small microinjected doses of substance P (0.1–10 ng) resulted in hypotension, bradycardia and apnea while higher doses (100 ng) had no effect may help to explain these discrepant studies [228]. Microiontophoretic application of substance P results in an increase in firing of NTS neurons [229]. Finally, microinjection of capsaicin (which is presumed to cause the release of substance P from nerve terminals) into the NTS produces a cardiovascular response which is identical to baroreceptor reflex activation [226]. Thus data exists to support the hypothesis that both L-glutamic acid and substance P are contained within primary baroreceptor afferents which terminate in the NTS.

9.2 Putative neurotransmitters in NTS which modulate baroreflexes

The NTS is richly innervated by monoamine, amino acid and peptidergic containing neurons. A large body of work has attempted to identify putative neurotransmitters which modulate baroreceptor reflexes within the NTS. Typically, these studies employ the technique of microinjection of agonists and sometimes antagonists into the NTS. Many inconsistencies regarding transmitter effects have been observed. This is due most likely to differences in injection sites within the NTS and to differences in doses and volumes employed. Although most studies have identified pharmacological effects in the NTS, few have attempted to define the physiological relevance of neurotransmitters contained within the NTS. The following is a very brief review of this literature.

The NTS is heavily innervated by catecholamine containing neurons. Microinjection of either norepinephrine or epinephrine into the NTS consistently produces a decrease in arterial blood pressure and heart rate [163]. Microiontophoretic application of norepinephrine or epinephrine inhibits the firing of NTS neurons regardless of the response of these neurons to baroreceptor reflex activation [230]. The firing of NTS neurons is also inhibited by microiontophoretic application of the α_1-receptor agonist methoxamine and the α_2-receptor agonist clonidine. Furthermore the inhibitory effects on iontophoretic catecholamines are reduced by both α_1- and α_2-receptor antagonists [230, 231]. These data suggest that the inhibitory effect of catecholamines in the NTS is mediated by both α_1- and α_2-receptors. Destruction of the catecholamine innervation of the NTS by microinjection of 6-OHDA results in a permanent lability of arterial blood pressure, despite preservation of reflex bradycardia [232]. These data suggest that catecholamines modulate baroreceptor reflexes.

The NTS is heavily innervated by 5-HT-containing neurons. However, microinjection of 5-HT into the NTS produces either pressor [233, 234] or depressor [235, 236] responses. Both the pressor and depressor effects of microinjected 5-HT can be blocked by the 5-HT antagonist metergoline [233, 235, 236]. In addition, the magnitude of the 5-HT-induced pressor response is potentiated by microinjection of the 5-HT uptake inhibitor fluoxetine [233]. The hypotension and bradycardia produced by 5-HT is blocked by NTS microinjection of the 5-HT_2 receptor antagonist ketanserin. Microinjection of 5-HT_1 agonists such as 8-OH DPAT and RU-24969 have no cardiovascular effects [236]. Bilateral NTS microinjections of ketanserin produce an increase in the level and the variability of arterial blood pressure but do not block the baroreceptor reflex arc [236]. Similarly, metergoline does not affect baroreceptor reflexes [235]. These data suggest that although serotonergic mechanisms in the NTS may be involved in the modulation of cardiovascular activity, serotonin is not integral to the baroreceptor reflex arc. Furthermore, the variability in the response to microinjections suggests that 5-HT may have multiple actions in the NTS.

Acetylcholine is also found in the NTS. Microinjections of acetylcholine into the NTS elicit a baroreceptor reflex-like response which includes a decrease in arterial blood pressure and a bradycardia. However, microinjections of cholinergic antagonists into the NTS fail to block the baroreceptor reflex suggesting that cholinergic processes are not an integral part of baroreceptor pathways in the NTS [214].

Recent studies indicate that NTS neurons involved in cardiovascular regulation are tonically inhibited by GABA. Microinjection of GABA or the GABA agonist muscimol into the NTS elicits an increase in arterial blood pressure [237, 238]. The increase in arterial blood pressure results from an increase in sympathetic nerve activity and an elevation of plasma AVP levels [237]. Similar results are observed following microinjection of the GABA uptake inhibitor nipecotic acid into the NTS [237, 238]. These effects may be mediated by both GABAA and GABAB receptors [237–239]. In contrast, microinjection of the GABA antagonist bicuculline results in a decrease in arterial blood pressure [237, 238]. Microiontophoretically applied bicuculline increases the firing rate of NTS neurons which receive afferent input from the carotid sinus nerve [240]. Finally, microinjections of muscimol or nipecotic acid into the NTS reduce the depressor response to aortic nerve stimulation [241]. These data are consistent with the hypothesis that NTS neurons involved in baroreceptor pathways are under tonic GABAergic inhibition. Interestingly, a recent study indicates that electrical stimulation of the hypothalamic defense area elicits a GABAergic inhibition of NTS neurons [242].

The NTS contains many neuropeptides which may act to modulate baroreceptor afferent processing. Microinjections of AVP into the NTS consistently increase arterial blood pressure and spontaneous sympathetic nerve discharge [243, 244]. Pretreatment with microinjected AVP antagonists block the pressor effect of AVP [243]. Microinjection of an AVP antagonist into the NTS reduces the increase in arterial blood pressure elicited by electrical stimulation of the PVN [245]. Microinjections of AVP antagonists alone have no cardiovascular effects [244], suggesting that the AVP innervation of the NTS is not tonically active.

Microinjection of angiotensin II (Ang) into the NTS has been reported to produce both pressor and depressor effects [246–248]. The action of Ang may be a function of the microinjected dose since at least three investigators now report that low doses of microinjected Ang produce depressor responses while higher doses elicit pressor responses [246, 247, 249]. Microinjection of low doses of the amino terminal angiotensin heptapeptide (Ang 1–7) similarly reduced arterial blood pressure [249]. This is significant since Ang 1–7 lacks a direct vascular action. These data suggest that the Ang receptors in the NTS are distinct from those found in the vasculature. Microinjection of the Ang antagonist saralasin has been reported to produce a small increase in arterial blood pressure [248] but also has been shown to enhance baroreceptor reflexes [250]. Thus the role of Ang in the NTS remains unclear. The effects of blood-borne AVP and Ang on central baroreceptor and sympathetic pathways are discussed below.

Microinjection of Met- or Leu-enkephalin into the NTS produces an increase in blood pressure and heart rate. The opioid antagonist naloxone blocks these actions [251]. Microinjection of β-endorphin into the NTS has been reported to produce a depressor response [252]. Microinjection of either TRH or somatostatin into the NTS results in a decrease in arterial blood pressure and a bradycardia [234, 253]. Low doses of calcitonin gene-related peptide decreased blood pressure and heart rate when microinjected into the NTS and potentiated the depressor response to microinjection of norepinephrine [254]. Finally, microinjection of NPY into the NTS results in a dose-related decrease in arterial blood pressure and heart rate [255].

9.3 Neurotransmitters mediating baroreceptor inhibition of sympathetic neurons

The importance of the ventrolateral medulla in the regulation of the cardiovascular system has already been described. A large body of evidence indicates that this area is critical in mediating the baroreceptor inhibition of sympathetic nerve discharge. The ventrolateral medulla is heavily innervated by neurons arising in the NTS [20, 95]. The NTS areas projecting to the ventrolateral medulla correspond to the zones in which baroreceptor afferents contained within the IXth and Xth cranial nerves terminate [10, 20]. Indeed, electrophysiology studies indicate that medullospinal sympathoexcitatory neurons in the rostral ventrolateral medulla receive a baroreceptor input [30–35]. Electrolytic or kainic acid lesions of the rostral ventrolateral medulla in animals previously receiving a contralateral NTS lesion abolished reflex responses to carotid stretch or vagal stimulation [25]. These data suggest that the integrity of the rostral ventrolateral medulla is required to maintain baroreceptor reflexes.

Evidence also suggests that the caudal ventrolateral medulla is necessary to maintain baroreceptor reflexes. Microinjection of the GABA agonist muscimol into the caudal ventrolateral medulla hyperpolarizes the area (i.e., silences neurons without interfering with axonal conduction), increases arterial blood pressure and blocks baroreceptor reflexes [91, 256]. Similarly, microinjection of the glutamate antagonist kynurenic acid into the caudal ventrolateral medulla blocks baroreceptor reflexes [92, 256]. Lesions of the NTS do not alter the GABA content of the caudal ventrolateral medulla [257]. These data are compatible with the hypothesis that second-order baroreceptor neurons arising in the NTS project to the caudal ventrolateral medulla. These neurons utilize a glutamate-like transmitter to excite inhibitory neurons which project to the rostral ventrolateral medulla (Figure 3). The nature of the neurotransmitter utilized in this pathway is unknown.

Present evidence suggests that GABA mediates baroreceptor-induced inhibition of sympathoexcitatory neurons in the rostral ventrolateral medulla (Figure 3). It is not known if GABAergic neurons are intrinsic to the rostral ventrolateral medulla or if they project to this area from the caudal ventrolateral medulla. The distribution of GABA neurons in the ventrolateral medulla has been described [157]. Microinjection of the GABA antagonist bicuculline into the rostral ventrolateral medulla increases arterial blood pressure and produces a 48 % reduction in the baroreceptor-mediated depressor response to aortic depressor nerve stimulation [258]. Bicuculline applied to the ventral surface of the medulla oblongata in the area of the rostral medulla reduces the inhibition of renal sympathetic nerve activity produced by baroreceptor reflex activation [259]. However, neither of these studies controlled for the large increase in sympathetic activity produced by bicuculline. Thus it is possible that the reduction in baroreceptor inhibition is a simple consequence of the dramatic increase in sympathetic activity which might lead to a decrease in the ability to inhibit sympathetic neurons. Recently, Sun and Guyenet found that microiontophoresis of bicuculline blocked the baroreceptor-induced inhibition of medullospinal sympathoexcitatory neurons in the rostral ventrolateral medulla [36]. Microiontophoretically applied glutamate produced an increase in neuronal firing comparable to that observed with bicuculline, but did not significantly alter the baroreceptor-mediated inhibition of the sympathoexcitatory neurons. These data suggest that GABA mediates the baroreceptor inhibition of rostral ventrolateral sympathoexcitatory neurons (Figure 3). Whether the GABA cell bodies are located in the caudal ventrolateral medulla or are contained within the rostral ventrolateral medulla remains to be determined. In contrast to the above studies, we found that GABA antagonists given intravenously or

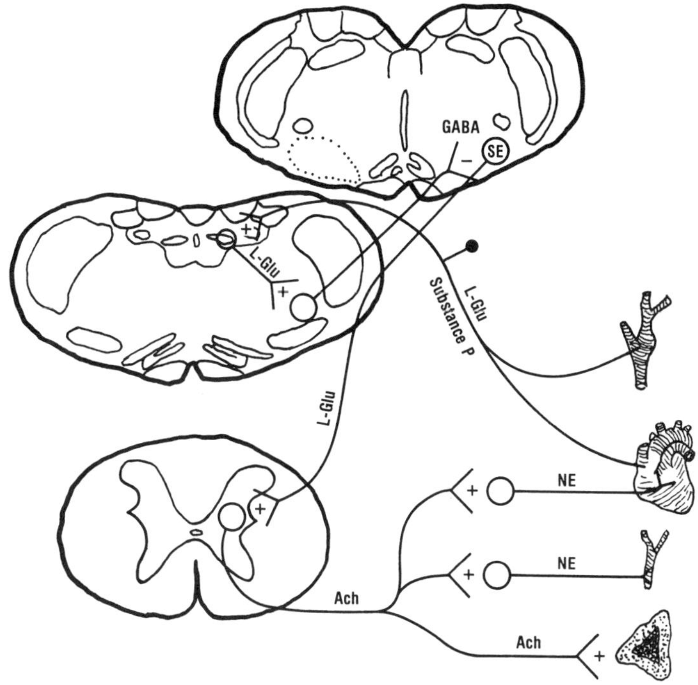

Figure 3
Model of the neural circuitry involved in the baroreceptor reflex. Baroreceptor afferents arising in the aortic arch and carotid sinus terminate in the NTS. Evidence suggests that L-glutamate and/or substance P acts as the transmitter in these afferent pathways. A proposed glutaminergic pathway arising in the NTS activates interneurons in the caudal ventrolateral medulla which project to the rostral ventrolateral medulla. This pathway activates GABAergic interneurons which inhibit glutaminergic medullospinal sympathoexcitatory neurons. The exact location of the GABA interneuron is unknown. The transmitter description is made tentatively and awaits confirmation by further studies. Reprinted from *Cardiovascular Pharmacology*, ed. M. J. Antonaccio, Raven Press, New York.

microinjected into the rostral ventrolateral medulla failed to block the baroreceptor-mediated inhibition of inferior cardiac sympathetic nerve discharge [34]. GABA antagonists also did not block baroreceptor-mediated temporal locking of sympathetic slow waves to the cardiac cycle [260]. Intravenous picrotoxin failed to block the baroreceptor-mediated inhibition of medullospinal sympathoexcitatory neurons recorded in the rostral ventrolateral medulla. In contrast, picrotoxin antagonized the GABA-mediated inhibition of sympathoexcitatory neurons produced by stimulation of the midline medulla and the inhibitory effect of microiontophoretically applied GABA

[34]. Thus, picrotoxin was present in sufficient concentration to antagonize the inhibition produced by GABA. The discrepancy between this and previous studies has not been resolved. An interesting ancillary observation made in this study was that midcollicular transection of the brain stem did not alter sympathetic activity but prevented the increase in sympathetic activity produced by intravenous picrotoxin. These data suggest that forebrain GABAergic neurons tonically inhibit excitatory pathways projecting to sympathetic neurons in the brain stem [34].

Activation of cardiopulmonary receptors by electrical stimulation of the afferent vagus nerve results in an inhibition of sympathetic nerve discharge. Intravenous administration of picrotoxin blocks vagal-induced inhibition of sympathetic activity suggesting that the inhibition is mediated by GABA [260]. More recently, it was demonstrated that the sympathoinhibitory effect of vagal afferent stimulation resulted from inhibition of rostral ventrolateral medullary sympathoexciatory neurons [261]. Furthermore, microiontophoresis of bicuculline blocks the vagal-induced inhibition of these neurons [261]. These data indicate that stimulation of cardiopulmonary afferents results in a GABA-mediated inhibition of sympathoexcitatory neurons in the rostral ventrolateral medulla.

Neurosecretory neurons in the supraoptic and paraventricular nuclei are inhibited during baroreceptor reflex activation. Studies suggest that GABA mediates this inhibition. For example, the baroreceptor-mediated inhibition of firing of AVP-containing neurons in the supraoptic nucleus observed during increases in arterial blood pressure are prevented by microiontophoresis of bicuculline [262]. Similarly, microiontophoresis of bicuculline blocks the baroreceptor-induced inhibition of AVP secretory neurons in the paraventricular nucleus [263]. The data suggest that the baroreceptor inhibition of AVP-containing secretory neurons is mediated by GABA.

10 Neurohumoral control

Hormones circulating in the blood can influence the activity of central nervous system pathways involved in cardiovascular regulation via an action on blood brain barrier deficient circumventricular organs. The role of blood-borne Ang in influencing central autonomic regulation has recently been reviewed [264] and is described only briefly here. Infusion of Ang into the vertebral artery produces a rapid rise in arterial blood pressure. Venous infusion of the same dose of Ang has no effect. The pressor effect of intravenous Ang is blunted by lesion of the area postrema (i.e., medullary circumventricular organ). Chronic infusions of Ang produce an elevation in arterial blood pressure which is reduced in area postrema lesioned animals. Furthermore, area postrema lesions blunt the development of two-kidney, one-clip renal hypertension as well as the increase in arterial blood pressure seen following acute unilateral renal artery stenosis [255, 265]. Finally, both the organum vasculosum of the lamina terminalis and the subfornical organ have been shown to play a critical role in body fluid homeostasis by integrating information derived from blood-borne Ang with input arising from vascular pressure and volume receptors [266]. Thus Ang gains access to the central nervous system via circumventricular organs where it plays a role in cardiovascular regulation.

Circulating AVP has also been shown to influence cardiovascular regulation by acting within the central nervous system to modify the reflex control of sympathetic nerve discharge [see ref. 267 for review]. For example, intravertebral infusion of AVP causes a greater baroreceptor-mediated reduction in heart rate than equivalent doses administered intravenously. In addition, the augmentation of the AVP pressor response following baroreceptor denervation far exceeds that observed with intravenously infused phenylephrine. These data indicate that circulating AVP acts to sensitize baroreceptor reflexes. More recently, studies indicate that at equal intravenous pressor doses, AVP produces a much greater reflex inhibition of renal sympathetic nerve activity than does phenylephrine. The enhanced sympathoinhibitory effect of AVP is not observed in area postrema lesioned animals [267]. The effects of AVP in the area postrema are mediated by the V1 AVP receptor [268]. Osmotically stimulated AVP also augments baroreceptor-mediated inhibition through an action in the area postrema [268]. Thus, circulating AVP enters the brain via the area postrema and functions to potentiate baroreceptor-mediated inhibition of sympathetic nerve activity.

11 Centrally-acting antihypertensive agents

11.1 α-Receptor agonists and antagonists

The therapeutically useful agents clonidine and α-methyl dopa act in the central nervous system to inhibit spontaneous sympathetic nerve discharge and therefore lower arterial blood pressure [see ref. 269 for review]. At the molecular level it is thought that clonidine acts by virtue of its α_2-adrenergic receptor agonist properties. α-Methyl dopa is metabolized to form α-methyl norepinephrine or α-methyl epinephrine within noradrenergic or adrenergic synaptic terminals, respectively. Like clonidine, the affinity of these false transmitters for α_2-receptors is greater than that of norepinephrine or epinephrine. Thus, potential sites of action of these compounds include any neuron involved in cardiovascular regulation which contains α_2-receptors.

Clonidine is thought to act in the brain stem or spinal cord since midcollicular transection does not affect the drug's hypotensive and sympatholytic actions [269]. One possible site of action is at the level of catecholamine neurons. It is thought that α_2-receptors located on the somatodendritic portion of central catecholamine neurons mediate a catecholamine inhibition of neuronal firing [139]. Clonidine has been shown to inhibit the firing rate of catecholamine neurons located in the locus coeruleus, A5, A2, C1 and C2 cell groups [1, 139]. Thus, the consequence of clinidine activation of α_2-receptors on catecholamine neurons is a widespread reduction in catecholamine release in the central nervous system. The net effect of decreased catecholamine release is dependent on the overall effects of catecholamines on central cardiovascular neurons. Massive depletion of brain catecholamines does not markedly affect the central antihypertensive properties of clonidine. This indicates that an action of clonidine to inhibit central catecholamine neurons is not sufficient to explain the drug's action. Furthermore, the observation suggests that catecholamine neurons both excite and inhibit central sympathetic neurons (see below).

The fact that clonidine is effective in catecholamine depleted animals suggests that the drug acts on postsynaptic α_2-receptors to produce its sympatholytic effects [270]. Studies indicate that the rostral ventrolateral medulla and SPN located in the IML of the spinal cord are likely sites of action of clonidine. Microinjection of clonidine into the rostral ventrolateral medulla produces a marked hypotensive response which is associated with an inhibition of spontaneous sympathetic nerve discharge [271–273]. Granata et al. [272] suggested that the active site of clonidine in the rostral ventrolateral medulla was in the immediate vicinity of the C1 epinephrine cell group. Indeed, microiontophoresis of clonidine inhibits the firing of presumed epinephrine neurons in the rostral ventrolateral medulla [48]. However, since epinephrine neurons may inhibit rather than excite SPN (see above), this observation does not necessarily account for the hypotensive activity of clonidine. Recently, Gatti et al. [273] found that microinjection of small amounts of clonidine into the nucleus reticularis lateralis of the ventrolateral medulla but not into the nucleus reticularis rostroventrolateralis (i.e., the C1 area) produced a marked hypotension. Microinjection of the α_2-receptor antagonist idazoxan into the nucleus reticularis lateralis reduced the hypotensive effects of systemically administered clonidine by 65 % [273]. These data suggest that a site in the nucleus reticularis lateralis is responsible for a large portion of the hypotensive action of clonidine. Evidence also supports a spinal site of action of clonidine. For example, clonidine has been shown to reduce activity in spinal sympathetic pathways [274]. In addition, microiontophoresis of clonidine inhibits the firing of SPN via an α_2-adrenergic receptor [38].

The effects of adrenergic agents on sympathetic activity has shed light on the role of catecholamine neurons in the central regulation of blood pressure. In addition to the sympatholytic actions of α_2-receptor agonists, studies have demonstrated that α_1-receptor antagonists inhibit sympathetic activity [136–138] while α_2-receptor antagonists increase sympathetic discharge [275]. The sympatholytic properties of selective α_1-receptor antagonists explain why these agents fail to produce an expected tachycardia [137]. The fact that α_1-receptor antagonists reduce sympathetic activity suggests that noradrenergic neurons excite sympathetic neurons in the central nervous system since it appears that α_1-receptors mediate the excitatory effects of norepinephrine [139]. Thus, a portion of clonidine's action may be explained by inhibiting the firing of norepinephrine neurons which normally act to facilitate central sympathetic neurons (possibly at the level of SPN [3, 42]). In this regard, clonidine reduces sympathetic activity in catecholamine depleted animals; however, the dose required for inhibition is at least three times greater than in control animals [270]. The α_2-adrenergic antagonists (e.g. piperoxane, yohimbine and rauwolscine) likely increase sympathetic activity by acting postsynaptically to block the direct inhibitory effects of catecholamines and by acting at α_2-receptors on noradrenergic soma to produce an increase in firing of these cells [275].

11.2 Serotonin$_{1A}$ agonists

Medullary 5-HT neurons provide a tonic excitatory input to SPN (see above). Thus, agents which antagonize the effects of 5-HT might be expected to act at the level of the SPN to inhibit sympathetic nerve discharge. In the cat, the 5-HT antagonists methysergide and metergoline inhibit spontaneous nerve activity recorded either from SPN [86] or a peripheral sympathetic nerve [167]. However, these agents fail to lower arterial blood pressure in rodents or man perhaps because of their partial agonist properties [3]. Recent studies indicate that the 5-HT$_{1A}$ agonist 8-OH DPAT acts in the central nervous system to inhibit spontaneous sympathetic activity and lower arterial blood pressure [276–278]. The sympatholytic effects of 8-OH DPAT are likely mediated by a 5-HT$_{1A}$ receptor since 1) other structurally dissimilar 5-HT$_{1A}$ agonists produce similar effects [276, 278] and 2) administration of the 5-HT$_{1A}$ antagonist spiperone prevents or reverses the sympatholytic effects of 8-OH DPAT [276]. In an early clinical trial, the 5-HT$_{1A}$ agonist flesinoxan has been shown to lower arterial blood pressure in hypertensive patients. This suggests a usefulness for these compounds in the treatment of hypertension (Wouters, personal communication).

The mechanism of the sympatholytic action of 8-OH DPAT has begun to be explored. Like catecholaminergic neurons, the soma and dendrites of serotonergic neurons contain 5-HT receptors which mediate a feedback 5-HT inhibition of the firing of 5-HT neurons (i.e., autoreceptor inhibition [173]). Recent studies indicate that the 5-HT "autoreceptor" may be identical to the 5-HT$_{1A}$ receptor in the dorsal raphe nucleus [279] and in medullary raphe nuclei [280]. Microiontophoresis of 8-OH DPAT has been shown to inhibit the firing of electrophysiologically identified 5-HT neurons in the dorsal raphe [281] and in raphe pallidus and obscurus [151]. In the latter study these 5-HT neurons were demonstrated to project to the IML of the spinal cord [151]. This observation led to the hypothesis that 8-OH DPAT produces its central sympatholytic effects by inhibiting the firing of

5-HT medullospinal sympathoexcitatory neurons. This would result in disfacilitation of sympathetic nerve discharge [151, 276]. In support of this hypothesis, mid-collicular transection does not affect the inhibition of sympathetic activity observed following administration of 8-OH DPAT [172]. However, recent work supports a postsynaptic action of 8-OH DPAT. Large electrolytic lesions encompassing the midline area of the lower brain stem from the level of the obex through the level of the facial nucleus fails to alter the sympatholytic effect of 8-OH DPAT. Since this lesion destroyed the great majority of 5-HT cell bodies which project to the spinal cord, these data suggest that 8-OH DPAT must act at a postsynaptic site to inhibit sympathetic nerve discharge [282]. This hypothesis was critically addressed by determining the inhibitory effects of 8-OH DPAT on activity recorded simultaneously from the inferior cardiac sympathetic nerve and from medullary 5-HT neurons which project to the IML. Medullary 5-HT neuronal activity is more sensitive to the inhibitory effect of 8-OH DPAT than is sympathetic nerve activity. Indeed, low doses of 8-OH DPAT completely suppress the firing of medullary 5-HT neurons but have no effect on sympathetic nerve discharge [282]. Collectively, these data provide strong evidence that direct inhibition of medullospinal 5-HT neuronal firing is not sufficient to explain the central sympatholytic effects of 8-OH DPAT. A postsynaptic action of 8-OH DPAT has been demonstrated on SPN [172]. As previously noted, microiontophoresis of 5-HT excites SPN [86]. Microiontophoresis of 8-OH DPAT has little effect on SPN firing but blocks the excitatory effect of iontophoretic 5-HT. Thus, at the level of the SPN, 8-OH DPAT acts as a 5-HT antagonist [172]. Microinjection of 8-OH DPAT into the ventrolateral area of the medulla produces a marked hypotensive response [283, 284]. Intravenous administration of 8-OH DPAT inhibits the firing of sympathoexcitatory neurons in the rostral ventrolateral medulla. Moreover, the inhibition of unit firing produced by 8-OH DPAT exactly parallels the shutoff of inferior cardiac nerve activity [172]. In contrast, microiontophoretic application of 8-OH DPAT onto sympathoexcitatory neurons in the rostral ventrolateral medulla fails to affect the firing rate of these neurons. These data suggest that 8-OH DPAT may act on central sympathetic neurons which lie antecedent to the ventrolateral sympathoexcitatory neurons. Alternatively, 8-OH DPAT may act on distal dendrites of the rostral ventrolateral sympathoexcitatory neurons [172]. Thus, in a manner analogous to α_2-adrenergic agonists, 5-HT_{1A} agonists inhibit the firing of 5-HT neurons, but a postsynaptic site of action appears to explain the sympatholytic effects of this class of centrally-acting antihypertensive agents.

12 Summary and conclusions

The last decade has seen tremendous progress in determining the nature of the neurotransmitters which regulate central nervous system pathways involved in the regulation of blood pressure. Investigations are now pursuing the identity and functional importance of neurotransmitters contained within pathways shown to be important in cardiovascular regulation. In addition, several key components of the brain stem networks involved in the control of sympathetic activity have been identified. For example, numerous studies indicate the importance of neurons located in the rostral ventrolateral medulla in the regulation of SPN. Indeed, this area contains medullospinal sympathoexcitatory neurons which represent the final site of integration of many brain stem and reflex pathways involved in the regulation of sympathetic nerve activity [1]. The neurotransmitter which is utilized by this medullospinal pathway remains unknown. Epinephrine, substance P and glutamate have all been hypothesized as primary chemical mediators in the descending pathway from the brain stem to SPN. Interestingly, lesions of, or antagonists to, epinephrine, substance P, glutamate and 5-HT neurons all abolish sympathetic activity and reduce blood pressure to a level similar to that in a spinal animal. Clearly, not all these transmitters are primary mediators of sympathetic information carried from the brain stem to the spinal cord. It is likely that monoamines and neuropeptides act in the IML, as in other area of the central nervous system, as neuromodulators to set the level of excitability of SPN rather than relaying sympathetic information over a functionally specific medullospinal pathway. This conclusion is supported by the observation that midline medullary 5-HT neurons provide a tonic excitatory input to SPN, but receive no afferent inputs from other central sympathetic or baroreceptor pathways. However, the firing of 5-HT neurons appears to relate to the state of vigilance of the animal [285]. This suggests that 5-HT neurons may lower the threshold of SPN to sympathetic inputs during states of wakefulness. In addition, the time course of the norepinephrine-mediated slow EPSPs and IPSPs in SPN [42] is consistent with a gain-setting function. By analogy, epinephrine is likely to act as a neuromodulator in the IML rather than to serve as the primary mediator of sympathetic information descending from the rostral ventrolateral medulla. Similarly, it is difficult to imagine that an agent with such a long duration of excitatory action as substance P [65, 66] could serve as the primary descending transmitter in a system where moment to moment changes in activity are essential. It is more likely that substance P functions to set the excitability of SPN. Pharmacologic antagonism of any of the excitatory neuromodulators might act to decrease, at least temporarily, the excitability of SPN to the point where primary sympathetic activity from the brain stem could not excite SPN. This accounts for the wide variety of pharmacologic agents that act to eliminate sympathetic activity and reduce arterial blood pressure.

On the basis of the above arguments, the most logical candidate for a transmitter mediating primary excitatory sympathetic information from the brain stem to SPN would be an excitatory amino acid. Fast EPSPs in SPN appear to be mediated by glutamate [42]. In addition, excitatory amino acid antagonists markedly inhibit sympathetic activity [3, 85]. The rapid time course of glutamate effects is consistent with a system in which activity changes from moment to moment. In more general terms, evidence is beginning to build to suggest that excitatory amino acids are important in mediating activity in sympathetic and baroreceptor pathways throughout the brain stem. In addition, it appears that GABA provides a tonic inhibition to most if not all pathways involved in the central regulation of blood pressure. GABA inhibition appears to be of both baroreceptor and non-baroreceptor origin. The role of neuropeptides in regulating central sympathetic pathways is beginning to be recognized. The significance of the co-localization of peptides and monoamines is unknown. Future research will determine the overall importance of these putative transmitters in cardiovascular regulation. In addition, the nature of neurotransmitters mediating cardiovascular responses from many areas of the central nervous system has not been investigated. These are but a few of the many unresolved issues in the field of the central regulation of blood pressure.

References

1 P. G. Guyenet: Role of the ventral medulla oblongata in blood pressure regulation. In: A. D. Loewy and K. M. Spyer, eds. *Autonomic nervous system: central regulation of autonomic function,* New York: Oxford University Press, in press.

2 G. L. Gebber: Brainstem systems involved in cardiovascular regulation. In: W. C. Randall, ed. *Nervous control of cardiovascular function,* Oxford: Oxford University Press, 1984; 346–368.

3 R. B. McCall: Effects of putative neurotransmitters on sympathetic preganglionic neurons. In: R. M. Berne, ed. *Annual Reviews of Physiology,* vol. 50. Palo Alto: Annual Reviews Inc., 1988; 553–564.

4 A. D. Loewy and J. J. Neil: The role of descending monoaminergic systems in central control of blood pressure. *Fed. Proc.* 1981; 40: 2778–2785.

5 P. G. M. Luiten, G. J. Ter Horst and A. B. Steffens: The hypothalamus, intrinsic connections and outflow pathways to the endocrine system in relation to the control of feeding and metabolism. *Prog. in Neurobiol.* 1987; 28: 1–54.

6 C. E. Fulwiler and C. B. Saper: Subnuclear organization of the efferent connections of the parabrachial nucleus in the rat. *Brain Research Rev.* 1984; 7: 229–259.

7 C. B. Saper and D. Levisohn: Afferent connections of the median preoptic nucleus in the rat: anatomical evidence for a cardiovascular integrative mechanism in the anteroventral third ventricular (AV3V) region. *Brain Research* 1983; 288: 21–32.

8 J. A. Andrezik, V. Chan-Palay and S. L. Palay: The nucleus paragigantocellularis lateralis in the rat: demonstration of afferents by the retrograde transport of HRP. *Anat Embryiol.* 1981; 161: 373–390.

9 C. B. Saper: Reciprocal parabrachial-cortical connections in the rat. *Brain Research* 1982; 242: 33–40.

10 J. S. Schwaber: Neuroanatomical substrates of cardiovascular and emotional-autonomic regulation. In: A. Magro, W. Osswald, D. Reis and P. Vanhoutte, eds. *Central and peripheral mechanisms of cardiovascular regulation,* New York: Plenum Press, 1986; 353–384.

11 M. J. Brody: Central nervous system mechanisms of arterial pressure regulation. *Federation Proc.* 1986; 45: 2700–2706.

12 M. Shioya and J. Tanaka: Inputs from the nucleus of the solitary tract ot subfornical organ neurons projecting to the paraventricular nucleus in the rat. Brain Research 1989; 483: 192–195.

13 D. J. Reis, C. Ross, A. R. Granata and D. A. Ruggiero: Role of C1 area of rostroventrolateral medulla in cardiovascular control. In: J. P. Buckley and C. M. Ferrario, eds. *Brain peptides and catecholamines in cardiovascular regulation,* New York: Raven Press, 1987; 1–14.

14 F. R. Calaresu and C. P. Yardley: Medullary basal sympathetic tone. In: R. M. Berne, ed. *Annual reviews of physiology,* vol. 50. Palo Alto: Annual Reviews Inc., 1988; 511–524.

15 W. Feldberg and P. G. Guertzenstein: A vasodepressor effect of pentobarbitone sodium. *J. Physiol.* 1972; 224: 83–103.

16 P. G. Guertzenstein and A. Silver: Fall in blood pressure from discrete regions of the ventral surface of the medulla by glycine and lesions. *J. Physiol.* 1974; 242: 489–503.

17 R. A. L. Dampney and E. A. Moon: Role of ventrolateral medulla in vasomotor response to cerebral ischemia. *Am. J. Physiol.* 1980; 239: H349–358.

18 C. A. Ross, D. A. Ruggerio, T. H. E. Joh, D. H. Park and D. H. Reis: Adrenaline synthesizing neurons in the rostral ventrolateral medulla: a

possible role in tonic vasomotor control. *Brain Research* 1983; 273: 356–361.
19 C. A. Ross, D. A. Ruggerio and D. H. Park, et al.: Tonic vasomotor control by the rostral ventrolateral medulla: effect of electrical or chemical stimulation of the area containing C1 adrenaline neurons on arterial pressure, heart rate and plasma catecholamines and vasopressin. *J. Neurosci* 1984; 4: 474–494.
20 R. A. L. Dampney, A. K. Goodchild, L. G. Robertson and W. Montgomery: Role of ventrolateral medulla in vasomotor regulation: a correlative anatomical and physiological study. *Brain Research* 1982; 249: 223–235.
21 P. J. Gatti, W. P. Norman, A. M. T. Da Silva and R. A. Gillis: Cardiorespiratory effects produced by microinjecting L-glutamic acid into medullary nuclei associated with the ventral surface of the feline medulla. *Brain Research* 1986; 381: 281–288.
22 R. N. Willette, P. P. Barcas, A. J. Krieger and H. N. Sapru: Vasopressor and depressor areas in the rat medulla: identification by microinjection of L-glutamate. *Neuropharmacol.* 1983; 22: 1071–1079.
23 R. N. Willette, S. Punnen-Grandy, A. J. Krieger and H. N. Sapru: Differential regulation of regional vascular resistance by the rostral and caudal ventrolateral medulla in the rat. *J. Auton. Nerv. Syst.* 1987; 18: 143–151.
24 E. E. Benarroch, A. R. Granata, D. Ruggiero, D. H. Park and D. J. Reis: Neurons of the C1 area mediate cardiovascular responses initiated form the ventral medullary surface. *Am. J. Physiol.* 1986; 250: R932–R945.
25 A. R. Granata, D. A. Ruggiero, D. H. Park, T. H. E. Joh and D. H. Reis: Brain stem area with C1 epinephrine neurons mediates baroreflex vasodepressor responses. *Am. J. Physiol.* 1985; 248: H547–H567.
26 K. Amendt, J. Czachurski, K. Dembowsky and H. Seller: Neurons within the "chemosensitive area" on the ventral surface of the brainstem which project to the intermediolateral column. *Pflugers Arch.* 1978; 375: 289–292.
27 A. D. Loewy, J. H. Wallach and S. McKellar: Efferent connections of the ventral medulla oblongata in the rat. *Brain Research Rev.* 1981; 3: 63–80.
28 W. W. Blessing, A. K. Goodchild, R. A. L. Dampney and J. P. Chalmers: Cell groups in the lower brainstem of the rabbit projecting to the spinal cord, with special reference to catecholamine-containing neurons. *Brain Research* 1981; 221: 35–55.
29 C. A. Ross, D. A. Ruggiero, T. H. E. Joh, D. H. Park and D. J. Reis: Rostral ventrolateral medulla: selective projections to the thoracic autonomic cell column from the region containing C1 adrenaline neurons. *J. Comp. Neurol.* 1984; 228: 168–185.
30 S. M. Barman and G. L. Gebber: Axonal projection patterns of ventrolateral medullospinal sympathoexcitatory neurons. *J. Neurophysiol.* 1985; 53: 1551–1566.
31 D. L. Brown and P. G. Guyenet: Cardiovascular neurons of brainstem with projections to spinal cord. *Am. J. Physiol.* 1984; 247: R1009–R1016.
32 P. G. Guyenet and D. L. Brown: Nucleus paragigantocellularis lateralis and lumbar sympathetic discharge in the rat. *Am. J. Physiol.* 1986; 250: R1081–R1094.
33 R. M. McAllen: Identification and properties of subretrofacial bulbospinal neurones: a descending cardiovascular pathway in the cat. *J. Auton. Nerv. Syst.* 1986; 17: 151–164.
34 R. B. McCall: Lack of involvement of GABA in baroreceptor-mediated sympathoinhibition. *Am. J. Physiol.* 1986; 253: R1065–R1073.
35 R. B. McCall: GABA-mediated inhibition of sympathoexcitatory neurons by midline medullary stimulation. *Am. J. Physiol.* 1988; 255: R605–R615.
36 M. K. Sun and P. G. Guyenet: GABA-mediated baroreceptor inhibition of reticulospinal neurons. *Am. J. Physiol.* 1985; 249: R672–R680.

37 D. C. Tucker, C. B. Saper, D. A. Ruggiero and D. J. Reis: Organization of central adrenergic pathways: I. Relationships of ventrolateral medullary projections to the hypothalamus and spinal cord. *J. Comp. Neurol.* 1987; 259: 591–603.
38 P. G. Guyenet and J. B. Cabot: Inhibition of sympathetic preganglionic neurons by catecholamines and clonidine: mediation by an adrenergic receptor. *J. Neurosci.* 1981; 1: 908–917.
39 P. G. Guyenet and R. L. Stornetta: Inhibition of sympathetic preganglionic discharges by epinephrine and α-methylepinephrine. *Brain Research* 1982; 235: 271–283.
40 K. Kadzielawa: Inhibition of the activity of sympathetic preganglionic neurones and neurones activated by visceral afferents, by alpha-methylnoradrenaline and endogenous catecholamines. *Neuropharmacol.* 1983; 22: 3–17.
41 A. V. Seybold and R. P. Elde: Receptor autoradiography in thoracic spinal cord: correlation of neurotransmitter binding sites with symphatoadrenal neurons. *J. Neurosci.* 1984; 4: 2533–2542.
42 S. Nishi, M. Yoshimura and C. Polosa: Synaptic potentials and putative transmitter actions in sympathetic preganglionic neurons. In: J. Ciriello, F. R. Calaresu, L. P. Renaud, C. Polosa, eds.: *Organization of the autonomic nervous system central and peripheral mechanisms,* New York: Alan R. Liss, 1987; 15–26.
43 C. Sangdee and D. N. Franz: Evidence for inhibition of symphathetic preganglionic neurons by bulbospinal epinephrine pathways. *Neurosci. Lett.* 1983; 37: 167–173.
44 C. J. Helke, J. J. Neil, V. J. Massari and A. D. Loewy: Substance P neurons project from the ventral medulla to the IML in the rat. *Brain Research* 1982; 243: 147–152.
45 J. R. Haselton and P. G. Guyenet: Electrophysiological characeterization of putative C1 adrenergic neurons in the rat. Neuroscience 1989; 30: 199–214.
46 M. K. Sun, J. T. Hackett and P. G. Guyenet: Symphatoexcitatory neurons of rostral ventrolateral medulla exhibit pacemaker proterties in the presence of a glutamate-receptor antagonist. Brain Research 1988; 438: 23–40.
47 M. K. Sun, B. S. Young, J. T. Hackett and P. G. Guyenet: Rostral ventrolateral medullary neurons with intrinsic pacemaker properties are not catecholaminergic. Brain Research 1988; 451: 345–349.
48 M. K. Sun and P. G. Guyenet: Effect of clonidine and GABA on the discharges of medullospinal sympathoexcitatory neurons in the rat. *Brain Research* 1986; 368: 1–19.
49 B. L. Jacobs, P. J. Gannon and E. C. Azmitia: Atlas of serotonergic cell bodies in the cat brainstem: an immunocytochemical analysis. *Brain Research Bull.* 1984; 13: 1–31.
50 I. M. Willenberg, R. Dermietzel, A. G. Leibstein and M. Effenberger: Mapping of cholinoceptive (nicotinoceptive) neurons in the lower brainstem: with special reference to the ventral surface of the medulla. *J. Auton. Nerv. Syst.* 1985; 14: 287–298.
51 J. C. Finley, J. L. Maderdrut and P. Petrusz: The immunocytochemical localization of enkephalin in the central nervous system of the rat. *J. Comp. Neurol.* 1981; 198: 541–565.
52 J. C. Finley, J. L. Maderdrut, L. J. Roger and P. Petrusz: The immunocytochemical localization of somatostatin-containing neurons in the rat central nervous system. *Neuroscience* 1981; 6: 2173–2192.
53 J. Ciriello, M. M. Caverson and C. Polosa: Function of the ventrolateral medulla in the control of the circulation. *Brain Reserarch Rev.* 1986; 111: 359–391.

54 B. M. Chronwall, D. A. Di Maggio and V. J. Massari et al.: The anatomy of neuropeptide-Y-containing neurons in rat brain. *J. Neurosci.* 1985; 15: 1159–1181.
55 P. W. Mantyh and S. P. Hunt: Evidence for cholecystokinin-like immunoreactive neurons in the rat medulla oblongata which project to the spinal cord. *Brain Research* 1984; 291: 49–54.
56 A. G. Leibstein, R. Dermietzel, I. M. Willenberg and R. Pauschert: Mapping of different neuropeptides in the lower brainstem of the rat: with special reference to the ventral surface. *J. Auton. Nerv. Syst.* 1985; 14: 299–313.
57 R. L. Eskay, R. T. Long and M. Palkovits: Localization of immunoreactive thyrotropin releasing hormone in the lower brainstem of the rat. *Brain Research* 1983; 277: 159–162.
58 T. Hokfelt, J. M. Lundberg and K. Tatemoto, et al.: Neuropeptide Y (NPY) and FMRFamide neuropeptide-like immunoreactivities in catecholamine neurons of the rat medulla oblongata. *Acta. Physiol. Scand.* 1983; 117: 315–318.
59 R. G. Lorenz, C. B. Saper, D. L. Wong, R. D. Ciaranello and A. D. Loewy: Co-localization of substance P and phenylethanolamine N-methyltransferase-like immunoreactivity in neurons of the ventrolateral medulla that project to the spinal cord: potential role in control of vasomotor tone. *Neurosci. Lett.* 1985; 55: 255–260.
60 A. F. Sved: PNMT-containing catecholaminergic neurons are not necessarily adrenergic. Brain Research 1989; 481: 113–118.
61 C. J. Helke, C. G. Charlton and R. G. Wiley: Studies on the cellular localization of spinal cord substance P receptors. *Neuroscience* 1986; 19: 523–533.
62 O. Johansson, T. Hokfelt and B. Pernow, et al.: Immunohistochemical support for three putative transmitters in one neuron: co-existence of 5-hydroxytryptamine, substance P and thyrotropin releasing hormone-like immunoreactivity in medullary neurons projecting to the spinal cord. *J. Neurosci.* 1981; 6: 1857–1881.
63 B. M. Davis, J. E. Krause, J. F. McKelry and J. B. Cabot: Effects of spinal lesions on substance P levels in the rat sympathetic preganglionic cell column: evidence for local spinal regulation. *Neurosci.* 1984; 13: 1311–1326.
64 B. M. Davis, J. E. Krause, N. Bogan and J. B. Cabot: Intraspinal substance P-containing projections to the sympathetic preganglionic neuropil in pigeon, Columbia livia: high-performance liquid chromatography, radioimmunoassay and electron microscopic evidence. Neuroscience 1988; 26: 655–668.
65 S. B. Backman and J. L. Henry. Effects of substance P and thyrotropin-releasing hormone on sympathetic preganglionic neurones in the thoracic intermediolateral nucleus of the cat. *Can. J. Physiol. Pharmacol.* 1983; 62: 248–251.
66 M. P. Gilbey, K. E. McKenna and L. P. Schramm: Effects of substance P on sympathetic preganglionic neurones. *Neurosci. Lett.* 1983; 41: 157–159.
67 C. J. Helke, E. T. Phillip and J. T. O'Neil: Regional peripheral and CNS hemodynamic effects of intrathecal administration of a substance P receptor agonist. *J. Auton. Nerv. Syst.* 1987; 21: 1–7.
68 A. P. M. Yusof and J. H. Coote: The action of a substance P antagonist on sympathetic nerve activity in the rat. *Neurosci. Lett.* 75: 329–333.
69 J. R. Keeler, C. G. Charlton and C. J. Helke: Cardiovascular effects of spinal cord substance P: Studies with a stable receptor agonist. *J. Pharmacol. exp. Therap.* 1985; 233: 755–760.
70 K. Yasphal, S. Gauthier and J. L. Henry: Substance P given intrathecally at the spinal T_9 level increases arterial pressure and heart rate in the rat. *J. Auton. Nerv. Syst.* 1987; 18: 93–103.

71 H. Hassesian and R. Couture: Cardiovascular responses induced by intrathecal substance P in the conscious freely moving rat. J. Cardiovasc. Pharmacol. 1989; 13: 594–602.

72 A. D. Loewy and W. B. Sawyer: Substance P antagonists inhibit vasomotor responses elicited from ventral medulla in rat. *Brain Research* 1982; 245: 379–383.

73 Y. Takano, J. E. Martin, S. E. Leeman and A. D. Loewy: Substance P immunoreactivity released from spinal cord after kainic acid excitation of the ventral medulla oblongata: a correlation with increases in blood pressure. *Brain Research* 1984; 291: 168–172.

74 J. R. Keeler and C. J. Helke: Spinal cord substance P mediates bicuculline-induced activation of cardiovascular responses from the ventral medulla. *J. Auton. Nerv. Syst.* 1985; 13: 19–33.

75 A. D. Loewy: Substance P neurons of the ventral medulla: their role in the control of vasomotor tone. In: R. Hainsworth, R. J. Linden and P. N. McWilliam, Mary DASG, eds. *Cardiogenic Reflexes,* Oxford: Oxford University Press, 1987; 269–285.

76 T. E. Salt, G. J. De Vries, R. E. Rodriguez, P. M. B. Cahusac, R. Morris and R. G. Hill: Evaluation of (D-Pro^2,D-Trp^{7,9})-substance P as an antagonist of substance P responses in the rat central nervous system. *Neurosci. Lett.* 1982; 30: 291–295.

77 T. Hokfelt, S. Vincent and L. Hellsten, et al.: Immunohistochemical evidence for a "neurotoxic" action of (D-Pro^2,D-Trp^{7,9}) substance P, an analogue with substance P antagonistic activity. *Acta. Physiol. Scand.* 9181; 113: 571–573.

78 C. Post, J. A. Karlsson, F. G. Butterworth, C. G. A. Persson and G. R. Strichartz: Local anaesthetic effects of substance P (SP) analogues in vitro. In C. C. Jordan, P. Oehme, eds. *Substance P: metabolism and biological activity,* London, Taylor and Francis, 1985; 227–241.

79 B. F. Cox, R. L. Schelper, F. M. Faraci and M. J. Brody: Autonomic, sensory, and motor dysfunction following intrathecal administration of three substance P antagonists. Exp. Brain Research 1988; 70: 61–72.

80 C. J. Helke and E. T. Phillips: Substance P antagonist-induce spinal cord vasoconstriction: effects of thyrotropin-releasing hormone and substance P agonists. Peptides 1988; 9: 1307–1315.

81 Howe P. R. C., D. M. Kuhn, J. B. Minson, B. H. Stead and J. P. Chalmers: Evidence for a bulbospinal serotonergic pressor pathway in the rat brain. *Brain Research* 1983; 270: 29–36.

82 A. D. Loewy and S. McKellar: Serotonergic projections form the ventral medulla to the intermediolateral cell column in the rat. *Brain Research* 1981; 211: 146–152.

83 G. A. Head and P. R. C. Howe: Effects of 6-hydroxydopamine and the PNMT inhibitor LY134046 on pressor responses to stimulation of the subretrofacial nucleus in anaesthetized stroke-prone spontaneously hypertensive rats. *J. Auton. Nerv. Syst.* 1987; 18: 213–224.

84 J. B. Minson, J. P. Chalmers, A. C. Caon and B. Renaud: Separate areas of rat medulla oblongata with populations of serotonin-and adrenaline-containing neurons alter blood pressure after L-glutamate stimulation. *J. Auton. Nerv. Syst.* 1987; 19: 39–50.

85 P. G. Guyenet, M. K. Sun and D. L. Brown: Role of GABA and excitatory aminoacids in medullary baroreflex pathway. In: J. Ciriello, F. R. Calaresu, L. P. Renaud, C. Polosa, eds. *Organization of the autonomic nervous system: central and peripheral mechanisms.* New York: Alan Liss Inc. 1987; 215–225.

86 R. B. McCall: Serotonergic excitation of sympathetic preganglionic neurons: a microiontophoretic study. *Brain Research* 1983; 289: 121–127.

87 S. Nishi, M. Yoshimura and C. Polosa: Synaptic potentials and putative transmitter actions in sympathetic preganglionic neurons. In: J. Ciriello, F. R. Calaresu, L. P. Renaud and C. Polosa, eds. *Organization of the autonomic nervous system: central and peripheral mechanisms*. New York: Alan Liss Inc. 1987; 15–26.

88 W. W. Blessing and D. J. Reis: Inhibitory cardiovascular function of neurons in the caudal ventrolateral medulla of the rabbit: relationship to the area containing A1 noradrenergic cells. *Brain Research* 1982; 253: 161–171.

89 T. A. Day, A. Ro and L. P. Renaud: Depressor area within caudal ventrolateral medulla of the rat does not correspond to the A1 catecholamine cell group. *Brain Research* 1983; 279: 299–302.

90 W. W. Blessing and D. J. Reis: Evidence that GABA- and glycine-like inputs inhibit vasodepressor neurons in the caudal ventrolateral medulla of the rabbit. *Neurosci. Lett.* 1983; 37: 57–62.

91 R. N. Willette, A. J. Krieger, P. P. Barcas and H. N. Sapru: Medullary-aminobutyric acid (GABA) receptors and the regulation of blood pressure in the rat. *J. Pharmacol. Exp. Therap.* 1983; 226: 893–899.

92 P. G. Guyenet, T. M. Filtz and S. R. Donaldson: Role of excitatory aminoacids in rat vagal and sympathetic baroreflexes. *Brain Research* 1987; 407: 272–284.

93 A. R. Granata, M. Kumada and D. J. Reis: Sympathoinhibition by A1-noradrenergic neutrons is mediated by neurons in the C1 area of the rostral medulla. *J. Auton. Nerv. Syst.* 1985; 14: 387–395.

94 J. M. Elliott, V. Kapoor, M. Cain, M. J. West and J. P. Chalmers: The mechanism of hypertension and bradycardia following lesions of the caudal ventrolateral medulla in the rabbit: the role of sympathetic nerves, circulating adrenaline, vasopressin and renin. *Clin. Exp. Hyperten.* 1985; A7: 1059–1082.

95 C. A. Ross, D. A. Ruggiero and D. J. Reis: Projections form the nucleus tractus solitarii to the rostral ventrolateral medulla. *J. Comp. Neurol.* 1985; 242: 511–534.

96 R. N. Willette, S. Punnen, A. J. Krieger and H. N. Sapru: Interdependence of rostral and caudal ventrolateral medullary areas in the control of blood pressure. *Brain Research* 1984; 321: 169–174.

97 P. Bousquet, J. Feldman, R. Boch and J. Schwartz: Central cardiovascular effects of alpha-adrenergic drugs: differences between catecholamines and imidazolines. *J. Pharmacol. exp. Therap.* 1985; 230: 232–236.

98 J. N. Sinha, S. Gurtu, K. Sharma and K. P. Bhargava: An analysis of the α-adrenoceptor modulation of vasomotor tone at the level of lateral medullary pressor area (LMPA). *Naunyn-Schmied Arch. Pharmacol.* 1985; 330: 163–168.

99 G. K. Aghajanian: The modulatory role of serotonin at multiple receptors in brain. In: B. L. Jacobs, A. Gelperin, eds. *Serotonin Neurotransmission and Behavior*, Cambridge: The MIT Press, 1981; 156–185.

100 A. C. Bonham and I. Jeske: Cardiorespiratory effects of DL-homocysteic acid in caudal ventrolateral medulla. Am. I. Physiol. 1989; 256: H688–H696.

101 S. McKellar and A. D. Loewy: Efferent projections of the A1 catecholamine cell group in the rat: an autoradiographic study. *Brain Research:* 1982; 241: 11–29.

102 S. Punnen and H. N. Sapru: Blockade of cholinergic receptors in the C1 area abolishes hypertensive response to opiates in the A1 area of the ventrolateral medulla. *Brain Research* 1985; 336: 180–186.

103 R. N. Willette, S. Punnen, A. J. Krieger and H. N. Sapru: Cardiovascular control by cholinergic mechanisms in the rostral ventrolateral medulla. *J. Pharmacol. exp. Therap.* 1984; 231: 457–463.

104 K. Sundaram and H. Sapru: Cholinergic nerve terminals in the ventrolateral medullary pressor area: pharmacoligical evidence. J. Auton. Nerv. Syst. 1988; 22: 221–228.
105 S. Punnen, R. N. Willette, A. J. Krieger and H. N. Sapru: Medullary pressor area: site of action of intravenous physostigmine. *Brain Research* 1986; 382: 178–184.
106 R. Giuliano, D. A. Ruggiero, S. Morrison, P. Ernsberger and D. J. Reis: Cholinergic regulation of arterial pressure by the C1 area of the rostral ventrolateral medulla. *J. Neurosci.* 1989; 9: 923–942.
107 K. Sundaram, A. J. Krieger and H. Sapru: M2 muscarinic receptors mediate pressor responses to cholinergic agonists in the ventrolateral medullary pressor area. Brain Research 1988; 449: 141–149.
108 D. J. Reis, S. Morrison and D. A. Ruggiero: The C1 area of the brainstem in tonic and reflex control of blood pressure. State of the art lecture. *Hypertension* 1988; 11: 8–13.
109 W. W. Blessing, C. B. Jaeger, D. A. Ruggiero and D. J. Reis: Hypothalamic projections of medullary catecholamine neurons in the rabbit: a combined catechlamine fluorescence and HRP transport study. *Brain Research Bull.* 1982; 9: 279–286.
110 P. E. Sawchenko and L. W. Swanson: The organization of noradrenergic pathways from the brainstem to the paraventricular and supraoptic nuclei in the rat. *Brain Research Rev.* 1982; 4: 275–325.
111 E. Mills and S. C. Wang: Liberation of antidiuretic hormone: location of ascending pathways. *Amer. J. Physiol.* 1964; 207: 1399–1404.
112 W. Feldberg and S. M. Rocha: Vasopressin release produced in anaesthetized cats by antagonists of GABA and glycine. *Brit. J. Pharmacol.* 1978; 62: 99–106.
113 W. W. Blessing and J. O. Willoughby: Inhibiting the rabbit caudal ventrolateral medulla prevents baroreceptor-initiated secretion of vasopressin. *J. Physiol. (Lond.)* 1985; 367: 253–265.
114 W. W. Blessing and J. O. Willoughby: Excitation of neuronal function in rabbit caudal ventrolateral medulla elevates plasma vasopressin. *Neurosci. Lett.* 1985; 58: 189–194.
115 A. Benetos, I. Gavras and H. Gavras: Noripinephrine applied in the paraventricular hypothalamic nucleus stimulates vasopressin release. *Brain Research* 1986; 381: 322–326.
116 T. A. Day, A. V. Ferguson and L. P. Renaud: Facilitatory influence of noradrenergic afferents on the excitability of rat paraventricular nucleus neurosecretory cells. *J. Physiol. (Lond.)* 1984; 355: 237–249.
117 T. A. Day and L. P. Renaud: Electrophysiological evidence that noradrenergic afferents selectively facilitate the activity of supraoptic vasopraoptic vasopressin neurons. *Brain Research* 1984; 303: 233–240.
118 B. J. Davis, M. L. Blair, J. R. Sladek and C. D. Sladek: Effects of lesions of hypothalamic catecholamines on blood pressure, fluid balance, vasopressin and renin in the rat. *Brain Research* 1987; 405: 1–15.
119 J. Tanaka, H. Kaba, H. Saito and K. Seto: Inputs from the A1 noradrenergic region to hypothalamic paraventricular neurons in the rat. *Brain Research* 1985; 355: 368–371.
120 J. C. R. Randle, C. W. Bourque and L. P. Renaud: α-Adrenergic activation of rat hypothalamic supraoptic neurons maintained in vitro. *Brain Research* 1984; 307: 374–378.
121 A. F. Sved: Pontine pressor sites which release vasopressin. *Brain Research* 1986; 369: 143–150.
122 T. Imizumi, A. R. Granata, E. E. Benarroch, A. F. Sved and D. J. Reis: Contributions of arginine vasopressin and the sympathetic nervous system to fulminating hypertension after destruction of neurons of caudal ventrolateral medulla in the rat. *J. Hyperten.* 1985; 3: 491–501.

123 J. Minson, J. Chalmers, V. Kappor, M. Cain and A. Caon: Relative importance of sympathetic nerves and of circulating adrenaline and vasopressin in mediating hypertension after lesions of the caudal ventrolateral medulla in the rat. *J. Hyperten.* 1986; 4: 273–281.
124 Y. I. Kim, C. A. Dudley and R. L. Moss: Inhibitory effect of norepinephrine on the single-unit activity of caudally projecting paraventricular neurons. Synapse 1989; 3: 213–224.
125 H. Yamashita, R. E. J. Dyball, K. Inenaga and H. Kannan: The effects of noradrenaline on supraoptic and paraventricular cells of mice in vitro. In: J. Ciriello, F. R. Calaresu, L. P. Renaud and C. Polosa, eds. *Organization of the autonomic nervous system: central and peripheral mechanisms.* New York: Alan Liss Inc. 1987; 417–423.
126 A. J. Silverman, Yu A. Hou and B. J. Oldfield: Ultrastructural identification of noradrenergic nerve terminals and vasopressin-containing neurons of the paraventricular nucleus in the same thin section. *J. Histochem. Cytochem.* 1983; 31: 1151–1156.
127 A. D. Loewy, S. McKellar and C. B. Saper: Direct projections from the A5 catecholamine cell group to the intermediolateral cell column. *Brain Research* 1979; 174: 309–314.
128 C. E. Byrum, R. Stornetta and P. G. Guyenet: Electrophysiological properties of spinally-projecting A5 noradrenergic neurons. *Brain Research* 1984; 303: 15–29.
129 C. E. Byrum and P. G. Guyenet: Afferent and efferent connections of the A5 noradrenergic cell group in the rat. *J. Comp. Neurol.* 1987; 261: 529–542.
130 J. J. Neil and A. D. Loewy: Decreases in blood pressure in response to L-glutamate microinjections in the A5 catecholamine cell group. *Brain Research* 1982; 241: 271–278.
131 A. D. Loewy, L. Marson, D. Parkinson, M. A. Perry and W. B. Sawyer: Descending noradrenergic pathways involved in the A5 depressor response. *Brain Research* 1986; 386: 313–324.
132 K. A. Stanek, J. J. Neil, W. B. Sawyer and A. D. Loewy: Changes in regional blood flow and cardiac output after L-glutamate stimulation of A5 cell group. *Amer. J. Physiol.* 1984; 246: H44–H51.
133 R. L. Stornetta, P. G. Guyenet and R. McCarty: Modulation of autonomic outflow by pontine A5 noradrenergic neurons. In: K. Nakamura, ed. *Brain and blood pressure control.* Elsevier Science Publishers, 1986; 23–28.
134 J. H. Coote, V. H. McLeod, S. Fleetwood-Walker and M. P. Gilbey: The response of individual sympathetic preganglionic neurones to microelectrophoretically applied endogenous monoamines. *Brain Research* 1981; 215: 1135–1145.
135 W. C. DeGroat and R. W. Ryall: An excitatory action of 5-hydroxytryptamine on sympathetic preganglionic neurons. *Exp. Brain Research* 1967; 3: 299–305.
136 R. B. McCall and L. T. Harris: Characterization of the central sympathoinhibitory action of ketanserin. *J. Pharmacol. exp. Therap.* 1987; 241: 736–740.
137 R. B. McCall and S. J. Humphrey: Evidence for a central depressor action of postsynaptic α_1-adrenergic receptor antagonists. *J. Auton. Nerv. Syst.* 1981; 3: 9–23.
138 R. B. McCall and M. R. Schuette: Evidence for an alpha-1 receptor-mediated central sympathoinhibitory action of ketanserin. *J. Pharmacol. exp. Therap.* 1984; 228: 704–710.
139 G. K. Aghajanian and M. A. Rogawski: The physiological role of α-adrenoceptors in the CNS: new concepts from single-cell studies. *Trends in Pharmacol. Sciences* 1983; 4: 315–317.

140 H. Shi, D. I. Lewis and J. H. Coote: Effects of activating spinal alpha-adrenoreceptors on sympathetic nerve activity in the rat. *J. Auton. Nerv. Syst.* 1988; 23: 69–78.

141 B. D. Hare, R. J. Neumayr and D. N. Franz: Opposite effects of L-dopa and 5-HTP on spinal sympathetic-reflexes. *Nature* 1972; 239: 336–337.

142 R. C. Ma and N. J. Dun: Norepinephrine depolarizes lateral horn cells of neonatal rat spinal cord in vitro. *Neurosci. Lett.* 1985; 60: 163–168.

143 M. Yoshimura, C. Polosa and S. Nishi: Noradrenaline modifies sympathetic preganglionic neuron spike and afterpotential. *Brain Research* 1986; 362: 370–374.

144 M. Yoshimura, C. Polosa and S. Nishi: Noradrenaline induces rhythmic bursting in sympathetic preganglionic neurons. *Brain Research* 1987; 420: 147–151.

145 M. Yoshimura, C. Polosa and S. Nishi: Slow IPSP and the noradrenaline-induced inhibition of the cat sympathetic preganglionic neuron in vitro. *Brain Research* 1987; 419: 383–386.

146 R. B. McCall and S. J. Humphrey: Evidence for GABA mediation of sympathetic inhibition evoked from midline medullary depressor sites. *Brain Research* 1985; 339: 356–361.

147 S. C. Wang and S. W. Ranson: Autonomic responses to electrical stimulation of the lower brain stem. *J. Comp. Neurol.* 1939; 71: 437–455.

148 R. B. McCall and L. T. Harris: Sympathetic alterations after midline medullary raphe lesions. *Amer. J. Physiol.* 1987; 253: R91–R100.

149 S. F. Morrison and G. L. Gebber: Classification of raphe neurons with cardiac-related activity. *Amer. J. Physiol.* 1982; 243: R49–R59.

150 S. F. Morrison and G. L. Gebber: Raphe neurons with sympathetic-related activity: baroreceptor responses and spinal connections. *Amer. J. Physiol.* 1984; 246: R338–R348.

151 R. B. McCall and M. E. Clement: Identification of serotonergic and sympathetic neurons in medullary raphe nuclei. *Brain Research* 1988; 477: 172–182.

152 R. B. McCall and L. T. Harris: Role of serotonin and serotonin receptor subtypes in the central regulation of blood pressure. In R. H. Rech and G. A. Gudelsky, eds. *5-HT agonists as psychoactive drugs,* Ann. Arbor: NPP Books, 1988; 1433–1462.

153 R. B. McCall: Role of serotonin in the regulation of sympathetic nerve discharge. In: P. Saxena and P. Bevan, eds. *Cardiovascular Pharmacology of 5-hydroxytryptamine: prospective therapeutic applications,* in press.

154 J. R. Adair, B. L. Hamilton, K. A. Scappaticci, C. J. Helke and R. A. Gillis: Cardiovascular responses to electrical stimulation of the medullary raphe area of the cat. *Brain Research* 1977; 128: 141–145.

155 R. B. McCall: Evidence for a serotonergically mediated sympathoexcitatory response to stimulation of medullary raphe nuclei. *Brain Research* 1984; 311: 131–139.

156 D. A. Ruggiero, M. P. Meeley, M. Anwarand and D. J. Reis: Newly identified GABAergic neurons in regions of the ventrolateral medulla which regulate blood pressure. *Brain Research* 1985; 339: 171–177.

157 M. Kihara and T. Kubo: Immunocytochemical localization of GABA containing neurons in the ventrolateral medulla oblongata of the rat. Histochemistry 1989; 91: 309–314.

158 A. D. Loewy: Raphe pallidus and raphe obscurus projections to the intermediolateral cell column in the rat. *Brain Research* 1981; 222: 129–132.

159 A. V. Seybold and R. P. Elde: Receptor autoradiography in toracic spinal cord: correlation of neurotransmitter binding sites with sympathoadrenal neurons. *J. Neurosci.* 1984; 4: 2533–2542.

160 B. W. Newton and R. W. Hamill: Immunohistochemical distribution of serotionin in spinal autonomic nuclei: I. Fiber patterns in the adult rat. J. Comp. Neurol. 1989; 279: 68–81.

161 B. L. Jacobs, P. J. Gannon and E. C. Azmitia: Atlas of serotonergic cell bodies in the cat brainstem: an immunocytochemical analysis. *Brain Research Bull.* 1984; 13: 1–31.

162 M. P. Gilbey, J. H. Coote, V. H. McLeod and D. F. Peterson: Inhibition of sympathetic activity by stimulating in the raphe nuclei and the role of 5-hydroxytryptamine in this effect. *Brain Research* 1981; 226: 131–142.

163 P. R. C. Howe: Blood pressure control by neurotransmitters in the medulla oblongata and spinal cord. *J. Auton. Nerv. Syst.* 1985; 12: 95–115.

164 J. H. Coote and V. H. McLeod: The spinal route of sympatho-inhibitory pathways descending from the medulla oblongata. *Pfluger Arch.* 1975; 359: 335–347.

165 D. M. Kuhn, W. A. Wolf and W. Lovenberg: Review of the central serotonergic neutral system in blood pressure regulation. *Hypertension* 1980; 2: 243–255.

166 W. A. Wolf and D. M. Kuhn: Antihypertensive effects of L-tryptophan are not mediated by brain serotonin. *Brain Research* 1984; 295: 356–359.

167 R. B. McCall and S. J. Humphrey: Involvement of serotonin in the central regulation of blood pressure: evidence for a facilitating effect on sympathetic nerve activity. *J. Pharmacol. exp. Therap.* 1982; 222: 94–102.

168 R. C. Ma and N. J. Dun: Excitation of lateral horn neurons of the neonatal rat spinal cord by 5-hydroxytryptamine. *Develop. Brain Research* 1986; 24: 89–98.

169 P. M. Pilowsky, V. Kapoor, J. B. Minson, M. J. West and J. P. Chalmers: Spinal cord serotonin release and raised blood pressure after brainstem kainic acid injection. *Brain Research* 1986; 366: 354–357.

170 E. H. Mills, J. B. Minson and J. P. Chalmers: The effect of intrathecal serotonergic antagonists on the pressor response to stimulation of the brainstem in the rat. Clin. exp. Hypertens. 1989; 11: 265–276.

171 M. E. Clement and R. B. McCall: Studies on the site and mechanism of the sympathoexcitatory action of 5-HT_2 agonists. Brain Research 1989; submitted.

172 M. E. Clement and R. B. McCall: Studies on the site and mechanism of the sympatholytic action of 8-OH DPAT. Brain Research 1989; submitted.

173 G. K. Aghajanian and R. Y. Wang: Physiology and pharmacology of central serotonergic neurons. In M. A. Lipton, A. Di Mascio and K. F. Killam, eds. *Psychopharmacology: a generation of progress,* New York: Raven Press, 1978; 171–183.

174 R. M. Bowker, K. N. Westlund, M. C. Sullivan, J. F. Wilber and J. D. Coulter: Descending serotonergic, peptidergic and cholinergic pathways from the raphe nuclei: a multiple transmitter complex. *Brain Research* 1983; 288: 33–48.

175 C. J. Helke, S. C. Sayson, J. R. Keeler and C. G. Charlton: Thyrotropin releasing hormone-immunoreactive neurons project from the ventral medulla to the intermediolateral cell column: partial coexistence with serotonin. *Brain Research* 1986; 381: 1–7.

176 N. M. Appel, M. W. Wessendorf and R. P. Elde: Thyrotropin-releasing hormone in spinal cord: coexistence with serotonin and with substance P in fibers and terminals apposing identified preganglionic sympathetic neurons. *Brain Research* 1987; 415: 137–143.

177 S. B. Backman and J. L. Henry: Effect of substance P and thyrotropin-releasing hormone on sympathetic preganglionic neurones in the upper thoracic intermediolateral nucleus of the cat. *Can. J. Physiol. Pharmacol.* 1984; 62: 248–251.

178 D. E. Milhorn, T. Hokfelt, K. Seroogy, W. Oertel, A. A. J. Verhofstad and J. Y. Wu: Immunohistochemical evidence for colocalization of γ-aminobutyric acid and serotonin in neurons of the ventral medulla oblongata projecting to the spinal cord. *Brain Research* 1987; 410: 179–185.
179 D. E. Milhorn, T. Hokfelt, A. A. Verhofstad and L. Terenius: Individual cells in the raphe nuclei of the medulla oblongata in rat that contain immunoreactivities for both serotonin and enkephalin project to the spinal cord. *Exp. Brain Research* 1989; 75: 536–542.
180 C. W. Xie, J. Tang and J. S. Han: Clonidine stimulated the release of dynorphin in the spinal cord of the rat: a possible mechanism for its depressor effects. *Neurosci. Lett,* 1986; 65: 224–228.
181 D. N. Franz, B. D. Hare and K. L. McCloskey: Spinal sympathetic neurons: possible sites of opiate-withdrawal suppression by clonidine. *Science* 1982; 215: 1643–1645.
182 K. E. McKenna and L. P. Schramm: Mechanisms mediating the silent period. Studies in the isolated spinal cord of the neonatal rat. *Brain Research* 1985; 329: 233–240.
183 S. B. Backman and J. L. Henry: Effects of GABA and glycine on sympathetic preganglionic neurons in the upper thoracic intermediolateral nucleus of the cat. *Brain Research* 1983; 277: 365–369.
184 F. J. Gordon: Spinal GABA receptors and central cardiovascular control. *Brain Research* 1985; 328: 165–169.
185 D. E. Millhorn, T. Hokfelt, K. Seroogy and A. A. Verhofstad: Extent for colocalization of serotonin and GABA in neurons of the ventral medulla oblongata in rat. Brain Research 1988; 461: 169–174.
186 D. M Kuhn, W. A. Wolf and W. Lovenberg: Pressor effects of electrical stimulation of the dorsal and median raphe nuclei in anesthetized rats. *J. Pharmacol. exp. Therap.* 1980; 214: 403–409.
187 W. A. Wolf, D. M. Kuhn and W. Lovenberg: Pressor effects of dorsal raphe stimulation and intrahypothalamic application of serotonin in the spontaneously hypertensive rat. *Brain Research* 1981; 208: 192–197.
188 S. E. Robinson: Serotonergic-cholinergic interactions in blood pressure control in the rat. *Fed. Proc.* 1984; 43: 21–24.
189 S. E. Robinson, M. J. Austin and D. M. Gibbens: The role of serotonergic neurons in dorsal raphe, median raphe and anterior hypothalamic pressor mechanisms. *Neuropharmacol.* 1985; 24: 51–58.
190 M. Iovino and L. Steardo: Effect of substances influencing brain serotonergic transmission on plasma vasopressin levels in the rat. *Europ. J. Pharmacol.* 1985; 113: 99–103.
191 E. Gotoh, K. Murakami, T. D. Bahnson and W. F. Ganong: Role of brain serotonergic pathways and hypothalamus in regulation of renin secretion. *Amer. J. Physiol.* 1987; 253: R179–R185.
192 G. Valiquette: Posterior pituitary hormones and neurophysins. In: Motta M. ed. *The endocrine functions of the brain,* New York: Raven Press, 1980; 385–417.
193 L. W. Swanson: Immunohistochemical evidence for a neurophysin-containing autonomic pathway arising in the paraventricular nucleus of hypothalamus. *Brain Research* 1977; 128: 346–353.
194 L. W. Swanson and S. McKellar: The distribution of oxytocin- and neurophysin-stained fibers in the spinal cord of the rat and monkey. *J. Comp. Neurol.* 1979; 188: 87–106.
195 W. E. Armstrong, S. Warach, G. I. Hatton and T. H. McNeil: Subnuclei in the rat hypothalamic paraventricular nucleus: a cytoarchitectural, horseradish peroxidase and immunocytochemical analysis. *Neuroscience* 1980; 5: 1931–1958.

196 L. W. Swanson and H. G. J. M. Kuypers: The paraventricular nucleus of the hypothalamus: cytoarchitectonic subdivisions and organization of projections to the pituitary, dorsal vagal complex, and spinal cord as demonstrated by retrograde fluorescence double-labeling methods. *J. Comp. Neurol.* 1980; 194: 555–570.
197 M. V. Sofroniew: Vasopressin and oxytocin in the mammalian brain and spinal cord. *Trends in Neurosci.* 1983; 5: 467–472.
198 M. J. Brody, T. P. O'Neill and J. P. Porter: Role of paraventricular and arcuate nuclei in cardiovascular regulation. In: A. Magro, W. Osswald, D. Reis and P. Vanhoutte, eds. *Central and peripheral mechanisms of cardiovascular regulation,* New York: Plenum Press, 1986; 443–464.
199 H. Kannan and H. Yamashita: Connections of neurons in the region of the nucleus tractus solitarius with the hypothalamic paraventricular nucleus: their possible involvement in neural control of the cardiovascular system. *Brain Research* 1985; 329: 205–212.
200 M. M. Caverson, J. Ciriello and F. R. Calaresu: Paraventricular nucleus of the hypothalamus: an electrophysiological investigation of neurons projecting directly to intermediolateral nucleus in the cat. *Brain Research* 1984; 305: 380–383.
201 H. Yamashita, K. Inenaga and K. Koizumi: Possible projections from regions of paraventricular and supraoptic nuclei to the spinal cord: electrophysiological studies: *Brain Research* 1984; 296: 373–378.
202 F. R. Calaresu and J. Ciriello: Projections to the hypothalamus from buffer nerves and nucleus tractus solitarius in the cat. *Amer. J. Physiol.* 1980; 239: R130–R136.
203 V. Holets and R. Elde: The differential distribution and relationship of serotoninergic and peptidergic fibers to sympathoadrenal neurons in the intermediolateral cell column of the rat: a combined retrograde axonal transport and immunoflorescence study. *Neuroscience* 1982; 7: 1155–1174.
204 T. L. Krukoff, J. Ciriello and F. R. Calaresu: Segmental distribution of peptide- an D5-HT-like immunoreactivity in nerve terminals and fibers of the thoracolumbar sympathetic nuclei of the cat. *J. Comp. Neurol.* 1985; 240: 103–116.
205 M. P. Gilbey, J. H. Coote, S. Fleetwood-Walker and D. F. Peterson: The influence of the paraventriculo-spinal pathway, and oxytocin and vasopressin on sympathetic preganglionic neurons. *Brain Research* 1982; 251: 283–290.
206 J. Ciriello and F. R. Calaresu: Role of paraventricular and supraoptic nuclei in central cardiovascular regulation in the cat. *Amer. J. Physiol.* 1980; 239: R137–R142.
207 S. B. Backman and J. L. Henry: Effects of oxytocin and vasopressin on thoracic sympathetic preganglionic neurons in the cat. *Brain Research Bull.* 1984; 13: 679–684.
208 R. C. Ma and N. J. Dun: Vasopressin depolarizes lateral horn cells of the neonatal rat spinal cord in vitro. *Brain Research* 1985; 348: 36–43.
209 J. P. Porter and M. J. Brody: The paraventricular nucleus and cardiovascular regulation: role of spinal vasopressinergic mechanisms. *J. Hyperten.* 1986; 4 (suppl. 3): S181–S184.
210 P. Sleight: *Arterial baroreceptors and hypertension,* Oxford: Oxford Univ. Press., 1980.
211 W. T. Talman, M. H. Perrone and D. J. Reis: Evidence for L-glutamate as the neurotransmitter of baroreceptor afferent nerve fibers. *Science* 1980; 209: 813–815.
212 W. T. Talman, M. H. Perrone, P. Scher, S. Kwo and D. J. Reis: Antagonism of the baroreceptor reflex by glutamate diethyl ester, an antagonist to L-glutamate. *Brain Research* 1981; 217: 186–191.

213 S. J. Humphrey and R. B. McCall: Evidence that L-glutamic acid mediates baroreceptor function in the cat. *Clin. exp. Hyperten.* 1984; 6: 1311–1329.
214 W. T. Talman, A. R. Granata and D. J. Reis: Glutamatergic mechanisms in the nucleus tractus solitarius in blood pressure control. *Fed. Proc.* 1984; 43: 39–44.
215 J. R. Simon, S. K. Di Micco, J. A. Di Micco and M. H. Aprison: Choline acetyltransferase and glutamate uptake in the nucleus tractus solitarius and dorsal motor nucleus of the vagus: effect of nodose ganglionectomy. *Brain Research* 1985; 344: 405–408.
216 A. R. Granata and D. J. Reis: Release of [^{3}H]L-glutamine acid (L-Glu) an d[^{3}H]D-aspartic acid (D-Asp) in the area of nucleus tractus solitarius in vivo produced by stimulation of the vagus nerve. *Brain Research* 1983; 259: 77–93.
217 C. Leone and F. J. Gordon: Is L-glutamate a neurotransmitter of baroreceptor information in the nucleus of the tractus solitarius? J. Pharmacol. exp. Therap. 1989; 250: 953–962.
218 B. D. Miller and R. B. Felder: Excitatory amino acid receptors intrinsic to synaptic transmission in nucleus tractus solitarii. Brain Research 1988; 456: 333–343.
219 R. A. Gillis, C. J. Helke, B. L. Hamilton, W. P. Norman and D. W. Jacobowitz: Evidence that substance P is a neurotransmitter of baro- and chemoreceptor afferents in nucleus tractus solitarius. *Brain Research* 1980; 181: 476–481.
220 C. J. Helke, T. L. O'Donohue and D. M. Jacobowitz: Substance P as a baro-and chemoreceptor afferent neurotransmitter: immunocytochemical and neurochemical evidence in the rat. *Peptides* 1980; 1: 1–9.
221 A. Nagashima, Y. Takano, K. Tateishi, Y. Matsuooka, T. Hamaoka and H. Kaniya: Cardiovascular roles of tachykinin peptides in the nucleus tractus solitarii of rats. Brain Research 1989; 487: 392–396.
222 D. M. Jacobowitz and C. J. Helke: Localization of substance P immunoreactive nerves in the carotid body. *Brain Research Bull.* 1980; 5: 195–197.
223 C. J. Helke, D. M. Jacobowitz and N. B. Thoa: Capsaicin and potassium evoked substance P release from the nucleus tractus solitarius and spinal trigeminal nucleus in vitro. *Life Sci.* 1981; 29: 1779–1785.
224 C. J. Helke: Neuroanatomical localization of substance P: implications for central cardiovascular control. *Peptides* 1982; 3: 479–483.
225 D. A. Morilak, M. Morris and J. Chalmers: Release of substance P in the nucleus tractus solitarius measured by in vivo microdialysis in rabbit. Neurosci Lett. 1988; 94: 131–137.
226 G. Haeusler and R. Osterwalder: Evidence suggesting a transmitter or neuromodulatory role for substance P at the first synapse of the baroreceptor reflex. *Naunyn-Schmied Arch. Pharmacol.* 1980; 314: 111–121.
227 W. T. Talman and D. J. Reis: Baroreflex actions of substance P microinjected into the nucleus tractus solitarii in rat: a consequence of local distortion. *Brain Research* 1981; 220: 402–407.
228 T. Kubo and M. Kihara: Blood pressure modulation by substance P in the rat nucleus tractus solitarius. *Brain Research* 1987; 413: 379–383.
229 M. P. Morin-Surun, D. Jordan, J. Champagnat, K. M. Spyer and M. Denavit-Saudie: Excitatory effects of iontophoretically applied substance P on neurons in the nucleus tractus solitarius of the cat: lack of interaction with opiates and opioids. *Brain Research* 1984; 307: 3988–392.
230 P. D. Feldman and H. C. Moises: Adrenergic responses of baroreceptive cells in the nucleus tractus solitarii of the rat: a microiontophoretic study. *Brain Research* 1987; 420: 351–361.
231 P. D. Feldman and H. C. Moises: Electrophysiological evidence for alpha 1- and alpha 2-adrenoceptors in solitary tract nucleus. *Amer. J. Physiol.* 1988; 254: H756–H762.

232 D. J. Reis, T. H. Joh, M. A. Nathan, B. Renaud, D. W. Snyder and W. T. Talman: Nucleus tractus solitarii: catecholaminergic innervation in normal and abnormal control of arterial pressure. In: P. Myer, H. Schmitt, eds. *Nervous system and hypertension,* Toronto: Wiley-Flammarion, 1979; 147–164.
233 W. A. Wolf, D. M. Kuhn and W. Lovenberg: Blood pressure responses to local application of serotonergic agents in the nucleus tractus solitarii. *Europ. J. Pharmacol.* 1981; 69: 291–299.
234 D. A. Carter and S. L. Lightman: Cardio-respiratory actions of substance P, TRH and 5-HT in the nucleus tractus solitarius of rats: evidence for functional interactions of neuropeptides and amine neurotransmitters. *Neuropeptides* 1985; 6: 425–436.
235 R. Laguzzi, D. J. Reis and W. T. Talman: Modulation of cardiovascular and electrocortical activity through serotonergic mechanisms in the nucleus tractus solitarius of the rat. *Brain Research* 1984; 304: 321–328.
236 A. Shvaloff and R. Laguzzi: Serotonin receptors in the rat nucleus tractus solitarii and cardiovascular regulation. *Europ. J. Pharmacol.* 1986; 132: 283–288.
237 J. M. Catelli, W. J. Giakas and A. F. Sved: GABAergic mechanisms in nucleus tractus solitarius alter blood pressure and vasopressin release. *Brain Research* 1987; 403: 279–289.
238 T. Kubo and M. Kihara: Evidence for gamma-aminobutyric acid receptor-mediated modulation of the aortic baroreceptor reflex in the nucleus tractus solitarii of the rat. *Neurosci Lett.* 1988; 89: 156–160.
239 J. C. Sved and A. F. Sved: Cardiovascular responses elicited by gamma-amionbutyric acid in the nucleus tractus solitarius: evidence for action at the GABAB receptor. Neuropharmacology 1989; 28: 515–520.
240 P. N. McWilliam and S. L. Shepheard: A GABA-mediated inhibition of neurones in the nucleus tractus solitarius of the cat that respond to electrical stimulation of the carotid sinus nerve. Neurosci. Lett. 1988; 94: 321–326.
241 T. Kubo and M. Kihara: Evidence for the presence of GABAergic and glycine-like systems responsible for cardiovascular control in the nucleus tractus solitarii of the rat. *Neurosci. Lett.* 1987; 74: 331–336.
242 D. Jordan, S. W. Mifflin and K. M. Spyer: Hypothalamic inhibition of neuronses in teh nucleus tractus solitarius of the cat is GABA mediated. J. Physiol. (Lond.) 1988; 399: 389–404.
243 H. Matsuguchi, F. M. Sharabi, F. J. Gordon, A. K. Johnson and P. G. Schmid: Blood pressure and heart rate responses to microinjection of vasopressin into the nucleus tractus solitarius region of the rat. *Neuropharmacol.* 1982; 21: 687–693.
244 K. A. King and C. C. Pang: Cardiovascular effects of injections of vasopressin into the nucleus tractus solitarius in conscious rats. *Brit. J. Pharmacol.* 1987; 90: 531–536.
245 Q. J. Pittmann and L. G. Franklin: Vasopressin antagonist in nucleus tractus solitarius/vagal area reduces pressor and tachycardia responses to paraventricular nucleus stimulation in rats. *Neurosci. Lett.* 1985; 56: 155–160.
246 D. B. Averill, D. I. Diz, K. L. Barnes and C. M. Ferrario: Pressor responses of angiotensin II microinjected into the dorsomedial medulla of the dog. *Brain Research* 1987; 414: 294–300.
247 R. Casto and M. I. Phillips: Neuropeptide action in nucleus tractus solitarius: angiotensin specificity and hypertensive rats. *Amer. J. Physiol.* 1985; 249: R341–R347.
248 R. Rettig, D. P. Healy and M. P. Printz: Cardiovascular effects of microinjections of angiotensin II into the nucleus tractus solitarii. *Brain Research* 1986; 364: 233–240.

249 M. J. Campagnole-Santos, D. I. Diz, R. A. S. Santos, M. C. Khosla, K. B. Brosnihan and C. M. Ferrario: Cardiovascular effects of angiotensin (1–7) injected into the dorsal medulla of rats. Am. J. Physiol. 1989; 257: H 324–H329.
250 M. J. Campagnole-Santos, D. I. Diz and C. M. Ferrario: Baroreceptor reflex modulation by angiotensin II at the nucleus tractus solitarii. *Hypertension* 1988; 11: 167–171.
251 M. A. Petty and W. De Jong: Enkephalins induce a centrally mediated rise in blood pressure in rats. *Brain Research* 1983; 260: 322–325.
252 M. A. Petty and W. De Jong: Cardiovascular effects of β-endorphin after microinjection into the nucleus tractus solitarii of the anesthetized rat. *Europ. J. Pharmacol.* 1982; 81: 449–457.
253 L. Y. Koda, N. Ling, R. Benoit, S. G. Madamba and C. Bakhit: Blood pressure following microinjection of somatostatin related peptides into the rat nucleus tractus solitarii. *Europ. J. Pharmacol.* 1985; 113: 425–430.
254 M. Vallejo, S. Lightman and I. Marshall: Central cardiovascular effects of calcitonin generelated peptide: interaction with noradrenaline in the nucleus tractus solitarious of rats. Exp. Brain Research 1988; 70: 221–224.
255 C. J. Tseng, R. Mosqueda-Garcia, M. Appalsamy and D. Robertson: Cardiovascular effects of neuropeptide Y in rat brainstem nuclei. Circulation Research 1989; 64: 55–61.
256 R. W. Urbanski and H. N. Sapru: Putative neurontransmitters involved in medullary cardiovascular regulation. J. Auton. Nerv. Syst. 1988; 25: 181–193.
257 M. Kihara and T. Kubo: Cardiovascular effects of GABA system activating drugs injected into the caudal ventrolateral medulla of the rat. Arch. Int. Pharmacodyn Ther. 1988; 295: 67–79.
258 R. N. Willette, P. P. Barcas, A. J. Krieger and H. N. Sapru: Endogenous GABAergic mechanisms in the medulla and the regulation of blood pressure. *J. Pharmacol. exp. Therap.* 1984; 230: 34–39.
259 K. A. Yamada, R. M. McAllen and A. D. Loewy: GABA antagonists applied to the ventral surface of the medulla oblongata block the baroreceptor reflex. *Brain Research* 1984; 297: 175–180.
260 S. J. Humphrey and R. B. McCall: Evidence for γ-aminobutyric acid mediation of the sympathetic nerve inhibitory response to vagal afferent stimulation. *J. Pharmacol. exp. Therap.* 1985; 234: 288–297.
261 M.-K. Sun and P. G. Guyenet: Arterial baroreceptor and vagal inputs to sympathoexcitatory neurons in rat medulla. *Amer. J. Physiol.* 1987; 252: R699–R709.
262 J. H. Jhamandas and L. P. Renaud: Bicuculline blocks an inhibitory baroreflex input to supraoptic vasopressin neurons. *Amer. J. Physiol.* 1987; 252: R947–R952.
263 M. Kasai, T. Osaka, K. Inenaga, H. Kannan and H. Yamashita: Gamma-aminobutyric acid antagonist blocks baroreceptor-activated inhibition of neurosecretory cells in the hypothalamic paraventricular nucleus of rats. *Neurosci. Lett.* 1987; 81: 319–324.
264 K. L. Barnes and C. M. Ferrario: Differential effects of angiotensin II mediated by the area postrema and the anteroventral third ventricle. In: J. P. Buckley and C. M. Ferrario, eds. *Brain peptides and catecholamines in cardiovascular regulation,* New York: Raven Press, 1987; 289–300.
265 J. E. Faber and M. J. Brody: Central nervous system action of angiotensin during onset of renal hypertension in awake rats. *Amer. J. Physiol.* 1984; 247: H349–H360.
266 A. K. Johnson: The periventricular anteroventral third ventricle (AV3V): its relationship with the subfornical organ and neuronal systems involved in maintaining body fluid homeostasis. *Brain Research Bull.* 1985; 15: 595–601.

267 V. S. Bishop, E. M. Hasser and K. P. Undesser: Vasopressin and sympathetic nerve activity: involvement of the area postrema. In: J. P. Buckley and C. M. Ferrario, eds. *Brain peptides and catecholamines in cardiovascular regulation,* New York: Raven Press, 1987; 373–382.
268 E. M. Hasser, S. E. Di Carlo, R. J. Applegate and V. S. Bishop: Osmotically released vasopressin augments cardiopulmonary reflex inhibition of the circulation. *Amer. J. Physiol.* 1988; 254: R815–R820.
269 P. I. Korner and J. A. Angus: Central nervous control of blood pressure in relation to antihypertensive drug treatment. In: D. E. Austin, ed. *Anti-hypertensive Drugs,* New York: Permagon Press, 1982; 61–96.
270 G. Haeusler: Clonidine-induced inhibition of sympathetic nerve activity: no indication of a central presynaptic or an indirect sympathomimetic mode of action. *Naunyn-Schmied Arch. Pharmacol.* 1974; 286: 97–111.
271 P. Bousquet, J. Feldman and J. Schwartz: The medullary cardiovascular effects of imidazolines and some GABA analogues: a review *J. Auton. Nerv. Syst.* 1985; 14: 263–270.
272 A. R. Granata, Y. Numao, M. Kumada and D. J. Reis: A1 noradrenergic neurons tonically inhibit sympathoexcitatory neurons of C1 area in rat brain stem. *Brain Research* 1986; 377: 127–146.
273 P. J. Gatti, K. J. Hill, A. M. T. Da Silva, W. P. Norman and R. A. Gillis: Central nervous system site of action for the hypotensive effect of clonidine in the cat. *J. Pharmacol. exp. Therap.* 1988; 245: 373–380.
274 D. N. Franz, P. W. Madsen, R. G. Peterson and C. Sangdee: Functional roles of monoaminergic pathways to sympathetic preganglionic neurons. *Clin. exp. Hyperten.* 1982; A4: 543–562.
275 R. B. McCall, M. R. Schuette, S. J. Humphrey, R. A. Lahti and C. Barsuhn: Evidence for a central sympathoexcitatory action of alpha-2 adrenergic antagonists. *J. Pharmacol. exp. Therap.* 1983; 224: 501–507.
276 R. B. McCall, B. N. Patel and L. T. Harris: Effects of serotonin$_1$ and serotonin$_2$ receptor agonists and antagonists on blood pressure, heart rate and sympathetic nerve activity. *J. Pharmacol. exp. Therap.* 1987; 242: 1152–1159.
277 J. R. Fozard, A. K. Mir and D. N. Middlemiss: Cardiovascular response to 8-hydroxy-2-(di-N-propylamino) tetralin (8-OH DPAT) in the rat: site of action and pharmacological analysis. *J. Card. Pharmacol.* 1987; 9: 328–347.
278 A. G. Ramage and J. R. Fozard: Evidence that the putative 5-HT$_{1A}$ receptor agonists 8-OH DPAT and ipsapirone have a central hypotensive action that differs from that of clonidine in anesthetized cats. *Europ. J. Pharmacol.* 1987; 138: 179–191.
279 D. Verge, G. Daval, A. Patey, H. Gozlan, S. El Mestikawy and M. Hamon: Presynaptic 5-HT autoreceptors on serotonergic cell bodies and/or dendrites but not terminals are of the 5-HT$_{1A}$ subtype. *Europ. J. Pharmacol.* 1985; 113: 463–464.
280 M. R. Dashwood, M. P. Gilbey, D. Jordan and A. G. Ramage: Autoradiographic localization of 5-HT$_{1A}$ binding sites in the brainstem of the cat. *Brit. J. Pharmacol.* 1988; 94: 386P.
281 R. B. McCall, M. E. Clement and L. T. Harris: Studies on the mechanism of the sympatholytic effect of 8-OH DPAT: lack of correlation between inhibition of serotonin neuronal firing and sympathetic activity. *Brain Research* 1989; in press.
282 R. A. Gillis, K. Hill and J. S. Kirby, et al.: Effect of activation of CNS serotonin 1A receptors on cardiorespiratory function. *J. Pharmacol exp. Therap.* 1989; 248: 851–857.
283 M. Laubie, M. Drouillat, H. Dabire, C. Cherqui and H. Schmitt: Ventrolateral medullary pressor area: site of hypotensive and sympatho-inhibitory

effects of (+/−)8-OH-DPAT in anaesthetized dogs. Europ. J. Pharmacol. 1989; 160: 385–394.

284 J. S. Sprouse and G. K. Aghajanian: Electrophysiological responses of serotoninergic dorsal raphe neurons to 5-HT_{1A} and 5-HT_{1B} agonists. *Synapse* 1987; 1: 3–9.

285 C. A. Fornal and B. L. Jacobs: Physiological and behavioral correlates of serotonergic single-unit activity. In: N. N. Osborne and M. Hamon, eds. *Neuronal serotonin,* WK Chichester: John Wiley and Sons Ltd., 1988; in press.

Drugs affecting serotonin neurons

By Ray W. Fuller
Lilly Research Laboratories, Eli Lilly and Company, Lilly Corporate Center, Indianapolis, IN 46285, USA

1	Introduction	86
2	Sites of drug action	86
3	Serotonin uptake inhibitors	87
4	Serotonin releasers	88
5	Direct-acting serotonin receptor agonists	90
5.1	Indoles	90
5.2	Piperazines	90
5.3	Serotonin receptor heterogeneity	91
5.4	5HT-1A-Receptor-selective agonists	92
5.5	Other selective agonists	92
6	Serotonin receptor antagonists	93
7	Serotonin-depleting drugs	94
8	Drugs that are neurotoxic to serotonin neurons	95
9	Therapeutic uses of drugs affecting serotonin neurons	98
10	Clinical nontherapeutic uses of drugs affecting serotonin neurons	99
11	Summary and conclusions	100
	References	100

1 Introduction

Most of the currently identified neurons that make and release serotonin as their neurotransmitter are in the central nervous system (CNS), and those will be the focus in this chapter. Serotonin in the gastrointestinal tract is mainly in enterochromaffin cells of the mucosal epithelium, but there are serotonin neurons in the enteric nervous system [1]. Serotonin neurons in the CNS have their cell bodies mainly clustered along the midline (raphe areas) of the midbrain and brain stem, and they send widespread projections to many regions of the forebrain as well as downward to the spinal cord. Their projection to many parts of the CNS involved in many different physiologic functions [2, 3] allows serotonin neurons to influence many brain functions, and they may play some part in the pathologic state in diverse diseases. Also, drugs that modify the function of serotonin neurons have various therapeutic uses or potential therapeutic uses as well as being valuable pharmacologic tools in helping to understand physiologic roles of serotonin neurons.

2 Sites of drug action

Serotonin neurons synthesize serotonin from the amino acid tryptophan via the intermediate 5-hydroxytryptophan. The two enzymes involved are tryptophan 5-hydroxylase, a highly specific enzyme thought to be present only in serotonin-forming cells, and L-aromatic amino acid decarboxylase, a relatively nonspecific and ubiquitous enzyme. Serotonin is stored in granules or vesicles, from which it is released at nerve impulse into the synaptic cleft. After acting on the postsynaptic receptor to complete the process of neurotransmission across the serotonergic synaptic cleft, the serotonin is inactivated by being removed from the synaptic cleft via specific membrane uptake carriers, which transport it back into serotonergic nerve terminals. There it may be re-used in storage granules or degraded enzymatically by monoamine oxidase, the major product being 5-hydroxyindoleacetic acid. Serotonin receptors occur not only postsynaptically but also presynaptically on cell bodies or axon terminals of serotonin neurons. These receptors on serotonin neurons are called autoreceptors; the terminal autoreceptors have the physiologic role of sensing the concentration of serotonin in the synaptic cleft and modulating the further release and synthesis of serotonin. Drugs can intervene at several sites to affect serotonergic function. For instance, inhibitors of tryptophan 5-hydroxylase or **L**-aromatic amino acid decarboxylase reduce serotonin synthesis and decrease neuronal serotonin content. Inhibitors of monoamine oxidase inhibit serotonin metabolism and increase serotonin content, as well as the content of other monoamines. Drugs can act directly at postsynaptic or presynaptic serotonin receptors to mimic or to antagonize the action of serotonin. Drugs can deplete serotonin stores by releasing serotonin from storage granules. Inhibitors of the serotonin uptake carrier increase serotonin concentration in the synaptic cleft, thereby enhancing serotonergic function, as do serotonin-releasing drugs. Drugs can destroy serotonin neurons or terminals, producing long-lasting deficits in serotonergic function. These various classes of drugs have been highly useful in elucidating functional roles of serotonin neurons, and some have therapeutic applications.

3 Serotonin uptake inhibitors

During the past 15 years, many compounds have been identified and described that are selective inhibitors of the serotonin uptake carrier. These include alaproclate [4], CGP 6085A [5], citalopram [6], cyanoimipramine [7], femoxetine [8], fluoxetine [9], fluvoxamine [10], indalpine [11], Org 6582 [12], panuramine [13], paroxetine [14], pirandamine [15], RU 25591 [16], sertraline [17], SL 81.0385 [18], zimelidine [19] and others, and have been the subject of previous review articles [20, 21]. Drugs that inhibit the serotonin uptake carrier enhance serotonin function by causing serotonin molecules to remain in the synaptic cleft longer and to activate the postsynaptic receptors to a greater extent. Inhibition of the serotonin uptake carrier in brain slices or brain synaptosomes can be demonstrated *in vitro* through the use of radioactive serotonin. Selectivity can be demonstrated by showing that the compounds do not inhibit the uptake of other brain monoamines such as norepinephrine or dopamine and do not interfere with binding of radioligands to neurotransmitter receptors such as muscarinic cholinergic, histaminergic and alpha adrenergic receptors *in vitro* [22, 23]. Inhibition of the serotonin uptake carrier *in vivo* has been demonstrated by *ex vivo* techniques of showing inhibited serotonin uptake *in vitro* by synaptosomal or slice preparations from

animals treated with the uptake inhibitor [9, 24], or by showing that the carrier-dependent depletion of brain serotonin by agents such as p-chloromethamphetamine [25], p-chloroamphetamine [26], fenfluramine [27, 28] and H75/12 [29] is antagonized or prevented.
Increased extraneuronal concentrations of brain serotonin produced by uptake inhibitors have been demonstrated by cytofluorometric [30], *in vitro* voltammetric [31] and push-pull cannulae [32] techniques. Several functional consequences of uptake inhibitors apparently result from increased serotonergic activation of synaptic receptors. These consequences include a decrease in serotonin synthesis and in the firing of serotonin neurons [26, 33, 34], probably mainly as a result of increased activation of autoreceptors but perhaps due in part to trans-synaptic feedback loops.
Despite the decreased activity of serotonin neurons, serotonergic neurotransmission is enhanced after uptake inhibition, leading to behavioral and other changes, especially when uptake inhibitors are combined with serotonin precursors (tryptophan or 5-hydroxytryptophan). For example, serotonin uptake inhibitors increase serum corticosterone concentration in rats [35] by increasing corticotropin-releasing hormone secretion from the hypothalamus and adrenocorticotrophin release from the anterior pituitary [36]. Serotonin uptake inhibitors inhibit muricidal behavior in rats [37] and potentiate behavioral effects of 5-hydroxytryptophan such as head twitch in mice [38] and the discriminative cue stimulus property of 5-hydroxytryptophan in rats [39]. Serotonin uptake inhibitors have analgesic effects in some experimental paradigms and potentiate analgesic effects of opioid drugs [40, 41]. Serotonin uptake inhibitors decrease total food intake in rats [42], selectively suppress carbohydrate consumption [43] and suppress stress-induced [44] or insulin-induced [45] hyperphagia in rats.

4 Serotonin releasers

Drugs such as reserpine, tetrabenazine and Ro 4–1284 release serotonin from granular stores in nerve terminals, but they also release other brain monoamines, e. g., dopamine, norepinephrine and epinephrine. They are not useful as tools for specifically manipulating serotonergic function. p-Chloroamphetamine and fenfluramine are halogenated analogs of amphetamine that release serotonin selectively. Most of their acute effects result from release of serotonin into the synaptic cleft and activation of postsynaptic serotonin receptors [46–49].

p-Chloroamphetamine was developed as an anorectic drug [50] and was also tested as an antidepressant drug [51] but was never marketed. Fenfluramine has been marketed for many years in the racemic form as an anorectic drug, and recently d-fenfluramine has been introduced as a more specifically acting serotonergic anorectic drug.

Other substituted amphetamines have similar serotonin-releasing actions, e. g., H75/12 (4-methyl-α-ethyl-m-tyramine) and 3,4-methylenedioxymethamphetamine (MDMA) [52]. As mentioned above, the serotonin release these agents induce is carrier-dependent and can lead to depletion of brain serotonin. Antagonism of that depletion is a useful marker of inhibition of the serotonin uptake carrier. The acute functional effects of the drugs are also antagonized by pretreatment with uptake inhibitors, because the uptake inhibitor prevents the acute release of serotonin. As examples, pretreatment with fluoxetine blocks the discriminative stimulus properties of fenfluramine [53], hyperthermia caused by fenfluramine in rats kept at warm ambient temperature [54], elevation of serum corticosterone [55] and serum prolactin concentration [56] by fenfluramine, the initial elevation of plasma renin activity by fenfluramine [57], the increase in twitch frequency of the suprahyoideal muscles by fenfluramine in anesthetized rats [58] and the fenfluramine-induced increase in striatal acetylcholine concentrations in rats [59].

Serotonin-releasing drugs and serotonin uptake-inhibiting drugs have many pharmacologic similarities, because both increase synaptic concentrations of serotonin and enhance activation of postsynaptic serotonin receptors. Both classes of drugs decrease food intake and increase serum corticosterone concentrations, for example. There are subtle differences between the two classes of drugs, however. For instance, enhancement of serotonergic function by uptake inhibitors requires that the serotonin neurons be firing and releasing serotonin. The degree of enhancement of serotonergic function by uptake inhibitors may be limited by compensatory mechanisms that quickly come into play to dampen serotonin release as synaptic concentrations of serotonin build up. Thus the diminished firing of serotonin neurons limits the degree to which synaptic concentrations of serotonin are increased after uptake inhibition. In contrast, serotonin-releasing drugs increase synaptic concentrations of serotonin independently of firing of the serotonin neuron, the degree of increase in synaptic concentrations of serotonin being limited only by the availability of presynaptic stores of serotonin. As a result of these differences, serotonin-releasing drugs produce some effects that uptake inhibitors do not produce unless 5-hydroxytryptophan (or tryptophan) is combined with the uptake inhibitor to increase the influx of serotonin into the synaptic cleft. Examples of such effects are increases in serum prolactin concentration in male rats [34, 56, 60–62] and reduction of blood pressure in spontaneously hypertensive rats [63, 64].

5 Direct-acting serotonin receptor agonists

5.1 Indoles

Several indoleethylamines structurally related to serotonin can interact with serotonin receptors. These include N,N-dimethylserotonin [65], N,N-dipropylserotonin [66], tryptamine and its N-alkyl derivatives [67], bufotenin [68], 5-carboxamidotryptamine and its N,N-dipropyl derivative [69], α-methyl and 2-methylserotonin [70], and 5-methoxytryptamine and its N-alkylated derivatives [68]. Some of the indolealkylamines are naturally occurring compounds, whereas RU24969 [70] and related compounds and also indorenate (TR3369) [71] are indole-containing compounds created by medicinal chemists. Some of these indolealkylamines vary in their selective affinity for the various serotonin receptor subtypes (see below). N,N-Dimethyl-5-methoxytryptamine has been a useful indole substitute for serotonin in whole-animal studies because it crosses the blood brain barrier and is somewhat protected against metabolism.

5.2 Piperazines

Numerous 1-phenylpiperazines and related compounds are centrally acting serotonin receptor agonists [see 72]. Quipazine [73, 74] and MK-212 [75] were the first 1-arylpiperazines described to be serotonin agonists. 1-(m-Trifluoromethylphenyl) piperazine (TFMPP) [76] and 1-(m-chlorophenyl-) piperazine (mCPP) [77] were the most-studied members of a larger series of substituted 1-arylpiperazines that were described subsequently. mCPP had been patented as an anorectic drug before its mechanism of action was known [78]. Urapidil is a substituted o-methoxyphenylpiperazine whose antihypertensive effects may be mediated by activation of brain serotonin receptors [79]. Buspirone [80], ipsapirone (TVXQ 7821) [81] and gepirone [82] are di-N-substituted piperazines with serotonin agonist activity.

5.3 Serotonin receptor heterogeneity

As early as the 1950's, Gaddum and Picarelli [83] presented pharmacologic evidence for serotonin receptor heterogeneity, defining "M" and "D" receptors in peripheral tissues that responded to serotonin. In 1979, Peroutka and Snyder [84] used radioligand-binding methods to define two distinct types of serotonin receptors in brain, the 5HT-1 and 5HT-2 receptors. Soon thereafter, it was recognized that the 5HT-1 binding site was not homogeneous [85], and now 5HT-1A, 5HT-1B, 5HT-1C and 5HT-1D receptors have been described and differentiated in brain. In addition, the 5HT-3 receptor has been found to be similar or identical to the "M" receptor of Gaddum, and the 5HT-2 receptor now seems to be pharmacologically identical to the "D" receptor of Gaddum [86].

There is no reason to believe that our understanding of serotonin receptor subtypes is complete at this time, and already evidence for additional subtypes of serotonin receptors is building [see 87]. The 5HT-1A [88], 5HT-2 [89] and 5HT-1C [90, 91] receptors have been cloned, and eventually the most complete and informative classification of serotonin receptors may be on the basis of protein structure [87, 92]. These different receptor subtypes are distributed differently among anatomic regions in brain, and different receptor subtypes presumably receive serotonergic signals in various neuronal pathways in brain. The multiple receptors offer a tantalizing opportunity for drug development, and the currently intense search for agonists and antagonists that act selectively on these receptor subtypes will provide important pharmacologic tools for elucidating the physiologic functions of these receptors [93].

Uptake inhibitors and releasers, by increasing synaptic concentrations of serotonin, probably increase activation of all of these receptor subtypes located synaptically. Direct-acting agonists, in contrast, may have selectivity and thus activate only one or certain ones of the receptor subtypes. Consequently, different direct agonists may share some specific actions but not all actions of indirect agonists (uptake inhibitors and releasers).

5.4 5HT-1A-Receptor-selective agonists

Although many of the previously known serotonin receptor agonists have relatively nonselective affinity for multiple serotonin receptor subtypes, there are some highly selective agonists of 5HT-1A receptors available currently. Perhaps the most selective of these is 8-hydroxy-2-(di-n-propylamino) tetralin [68, 94], which has stereoselective affinity for 5HT-1A receptors [95] and has been widely used as a prototypic 5HT-1A receptor agonist in exploring functions of 5HT-1A receptors [93]. Buspirone is an anxiolytic drug whose mechanism of action was unknown for many years and was at one time thought to involve dopamine [96]; in the past few years, buspirone was found to have high affinity for the 5HT-1A receptor [97, 98], and now it is generally believed that 5HT-1A receptor activation accounts for the anxiolytic effects of buspirone [99]. Several structural analogs of buspirone, including gepirone [82], ipsapirone [100] and SM-3997 [101] also are selective 5HT-1A receptor agonists and have anxiolytic effects in humans and/or animals [102]. Some other direct-acting serotonin agonists also have higher affinity for 5HT-1A receptors than for other serotonin receptor subtypes, including LY165163 [103, 104] and MDL 72832 [105].

5.5 Other selective agonists

Highly selective agonists for receptor subtypes other than 5HT-1A receptors are not yet available, and these are needed as pharmacologic tools. Although m-chlorophenylpiperazine and m-trifluoromethylphenylpiperazine have sometimes been referred to as selective 5HT-1B receptor agonists [106, 107], insufficient selectivity between 5HT-1B and other serotonin receptors such as 5HT-1A [108, 109], 5HT-1C [110] and 5HT-2 [68] receptors prevent these agents from being very useful in discriminating among serotonin receptor subtypes [111]. Perhaps the most selective 5HT-1B receptor agonist available to date is CGS 12066B, 7-trifluoromethyl-4(4-methyl-1-piperazinyl-)-

pyrrolo [1, 2a] quinoxaline, although the compound does not have high potency [109]. RU 24969 is sometimes referred to as a selective 5HT-1B agonist [112], but radioligand-binding studies show it has similar affinity for 5HT-1A and 5HT-1B receptors [68]. Agonists with some selectivity toward 5HT-2 receptors include 4-bromo-2,5-dimethoxyamphetamine [113, 114] and its iodo analog [115]. Indolealkylamines that are hallucinogenic have particular affinity for the 5HT-2 receptor [65]. 2-Methylserotonin is used as a relatively selective agonist at 5HT-3 receptors [69]. GR43175, 3-(2-dimethylamino)ethyl-N-methyl-1H-indole-5-methane sulfonamide or sumatriptan [116], and 5-carboxamidotryptamine [117] are used as selective agonists for 5HT-1 receptor subtypes.

6 Serotonin receptor antagonists

Antagonists of serotonin receptor-mediated functions have been known almost as long as serotonin itself. Antagonists led to the first recognition of serotonin receptor heterogeneity in the mid-1950's, when Gaddum and his colleagues defined "M" and "D" receptors in peripheral tissues based on differential antagonism of indoleamine-mediated effects [83]. In recent years the focus has been on identifying antagonists with high selective affinity toward specific serotonin receptor subtypes.

Serotonin antagonists that have been considered to be selective for the 5HT2 receptor include ketanserin [118], ritanserin [119, 120], altanserin [119], LY53857 [121], LY281067 [122], ICI 169, 369 [123], ICI 170, 809 [124], setoperone [119, 120], MDL 11, 939 [125] and irindalone [126]. However, because the 5HT-1C receptor has high homology and pharmacologic similarity to the 5HT-2 receptor [87, 92], many of these antagonists also are highly potent blockers of 5HT-1C receptors [127], and it is now important to find antagonists that discriminate better between these two receptor subtypes.

Various other serotonin antagonists that block 5HT2 receptors but have less selectivity among serotonin receptor subtypes include metergoline [128], methysergide [68, 129], cyproheptadine [68, 129], pizotifen [130], methiothepin [131], danitracen [132], mianserin [133], cinanserin [134] and benzoctamine [135]. Trazodone is an interesting example of a drug that is a potent 5HT-2 receptor antagonist [136], but is metabolized [137] to m-chlorophenylpiperazine, a serotonin receptor agonist [77]. Thus a mixture of pharmacologic effects, including serotonergic influences in opposite directions, has been demonstrated over a range of trazodone doses in animals [138]. The molecular basis for the therapeutic actions of trazodone in depression [139] remains unknown.

Selective antagonists of 5HT-3 receptors have been developed in recent years, including MDL 72222 [140], ICS 205-930 [69], GR 38032F [141], BRL 24924 and BRL 43694 [142, 143], zacopride [144] and LY278584 [145, 146]. The use of some of these compounds as radioligands has made it possible to demonstrate the presence of 5HT-3 receptors in brain [146, 147].

Several α and β adrenoreceptor-blocking drugs have high affinity for 5HT-1A and 5HT-1B receptors [148–150], and some such as pindolol and penbutolol have been useful experimentally as antagonists of these receptor subtypes [151–153]. Some of these compounds are partial agonists at 5HT-1A receptors and may show properties of agonists or antagonists in different experimental paradigms [154, 155]. There have been attempts to synthesize analogs of these drugs as more selective 5HT-1A receptor ligands [156–158].

7 Serotonin-depleting drugs

The short-term depletion of brain serotonin can be produced by drugs such as reserpine that impair granular storage or by inhibitors of serotonin synthesis. Since reserpine, tetrabenazine and related compounds deplete other brain monoamines in addition to serotonin, specific inhibitors of serotonin synthesis are generally used for short-term depletion of serotonin. Tryptophan 5-hydroxylase, the initial enzyme in serotonin biosynthesis localized only in serotonin-forming cells, is the preferred target of synthesis inhibitors. The most widely used inhibitor of tryptophan 5-hydroxylase is p-chlorophenylalanine [159]. Other inhibitors of tryptophan 5-hydroxylase include 6-fluorotryptophan [160].

p-Chlorophenylalanine has been used for more than two decades to produce long-lasting but reversible depletion of brain serotonin as a means of investigating a role of serotonin in physiologic functions [161], in pathologic changes [162] or in drug actions [163]. Often serotonin depletion by p-chlorophenylalanine produces similar functional effects as serotonin depletion by other classes of agents such as reserpine, fenfluramine and 5,7-dihydroxytryptamine [e. g., 164], but sometimes differences are noted among classes of serotonin-depleting drugs [e. g., 165, 166]. Among several possible explanations for these differences, including different degrees of serotonin depletion in various brain and spinal cord serotonergic pathways, one is that lesions in serotonin neurons are expected to reduce cotransmitters as well as serotonin, whereas p-chlorophenylalanine would reduce serotonin concentration by inhibiting serotonin synthesis but would not reduce intraneuronal concentrations of cotransmitters in serotonin neurons [167].

Recently evidence has been presented that some serotonin axons in rat brain are not sensitive to p-chlorophenylalanine [168]. The insensitive axons were especially prominent in the limbic system, in cranial motor and parasympathetic nuclei, and some had common morphological features such as large varicosities and location adjacent to neural somata suggestive of axosomatic synapses.

8 Drugs that are neurotoxic to serotonin neurons

5,6- [169] and 5,7-dihydroxytryptamine [170] are neurotoxic to serotonin neurons in an analogous way to the neurotoxic effect of 6-hydroxydopamine on catecholamine neurons [171]. These hydroxylated analogs of the natural transmitters have affinity for the membrane uptake carriers on serotonin and catecholamine neurons, respectively. They are easily autoxidized molecules that give rise to hydrogen peroxide, free radicals and other reactive species that lead to cytotoxicity [172]. 5,6- and 5,7-Dihydroxytryptamines are accumulated via the membrane uptake carrier into serotonin neurons and have destructive effects on those neurons, just as 6-hydroxydopamine is accumulated via the membrane uptake carrier into catecholamine neurons and destroys those neurons.

For nearly 20 years, 5,6- and more recently 5,7-dihydroxytryptamine have been used to lesion serotonin neurons in brain. They have been valuable tools in elucidating functional roles of serotonin neurons, especially particular neuronal tracts. Prior lesioning of serotonin neurons with these neurotoxins attenuates or abolishes many functional effects of indirect-acting serotonin agonists (uptake inhibitors or releasers) while often enhancing effects of direct-acting agonists by causing supersensitivity of postsynaptic receptors [see 173].

p-Chloroamphetamine is not only a serotonin-releasing drug whose acute effects are mediated by enhanced activation of postsynaptic serotonin receptors, but p-chloroamphetamine at higher doses can actually be neurotoxic. The first clue to this came in 1970 when Frey [174] reported that brain serotonin in rats was depleted rapidly and remained depleted for as long as 4 weeks after only 5 days of treatment with p-chloroamphetamine. Sanders-Bush et al. [175] extended these findings and focused on the long duration of p-chloroamphetamine's actions. Now it is documented that many immunofluorescent serotonin-containing fibers are eliminated by p-chloroamphetamine and that all parameters specifically associated with serotonin neurons that have been measured are reduced for weeks or months after a few doses or even a single dose of p-chloroamphetamine. Parameters shown to remain decreased for long periods after administration of p-chloroamphetamine or its structural congeners include, in addition to serotonin content, the content of 5-hydroxyindoleacetic acid (the metabolite of serotonin), serotonin turnover, tryptophan hydroxylase, serotonin uptake capacity and serotonin uptake carriers labeled with radioligands [175–178]. Thus it seems clear that p-chloroamphetamine can be neurotoxic to brain serotonergic nerve axon terminals, especially to the fine projections with minute varicosities arising from dorsal raphe serotonin neurons that can be distinguished from coarse-beaded fibers arising from medial raphe nuclei [179].

In contrast to 5,6- and 5,7-dihydroxytryptamine, which do not penetrate the blood-brain barrier and must be applied intraventricularly or directly into brain tissue, p-chloroamphetamine can be given systemically to deplete serotonin in the CNS [180]. The neuroanatomic specificity of p-chloroamphetamine effects differs from that of the dihydroxytryptamines [180, 181].

In addition to p-chloroamphetamine, some other substituted amphetamines that release serotonin acutely produce long-lasting effects on brain serotonin neurons similar to those of p-chloroamphetamine, apparently reflecting neurotoxicity. These analogs include fenfluramine [182, 183] and 3,4-methylenedioxymethamphetamine (MDMA) [178, 184]. Because fenfluramine is used therapeutically in humans and MDMA is used recreationally in humans, the possible hazards of these compounds have received attention recently [185–187], but no direct evidence of any sort exists to support the idea that neurotoxic effects occur in humans as they do in rodents [185–187] and in non-human primates [188].

The type of adverse consequences that might be expected if brain serotonergic neurons or terminals were destroyed is not certain. The degree to which serotonergic function is impaired by p-chloroamphetamine and related drugs when they deplete brain serotonin in laboratory animals has not been studied extensively. Rats treated with p-chloroamphetamine a week or more earlier such that brain serotonin is depleted by half or more are relatively normal in appearance. There have been some demonstrations of impaired or altered cerebral function in these animals, however. For instance, Vorhees et al. [189] reported that rats treated with a 5-mg/kg i. p. dose of p-chloroamphetamine showed hypoactivity and increased defecation in open-field testing for as long as 30 days. They showed facilitated acquisition in a shock avoidance Y-maze task for up to 15 days. Brain serotonin was still reduced by 41 % after 38 days.

Grabowska and Michaluk [190] reported that p-chloroamphetamine injected at 2 mg/kg daily for 5 days did not affect locomotion measured three days after the last dose in rats but it intensified the stimulation caused by apomorphine. Those findings and other data in the paper were taken by the authors as confirming "our hypothesis about the possible inhibitory role of serotonin in the apomorphine-induced locomotor stimulation in rats".

Ogren et al. [191] reported perhaps the most convincing evidence to date of impaired serotonergic function after p-chloroamphetamine-induced depletion of brain serotonin. They gave two doses of p-chloroamphetamine hydrochloride (10 mg/kg), which resulted in 63 % depletion of serotonin concentration in whole brain. Pretreatment with zimelidine completely prevented the depletion of serotonin. Eight days after p-chloroamphetamine treatment, rats failed in the acquisition of a two-way conditioned avoidance response. Rats that had been pretreated with zimelidine and then given p-chloroamphetamine acquired the avoidance response at least as well as control rats; zimelidine completely prevented the depletion of brain serotonin by p-chloroamphetamine.

Yamaguchi et al. [192] reported potentiation of apomorphine-, methamphetamine- and phencyclidine-induced dopaminergic behaviors (sniffing, licking, gnawing, biting) in rats after p-chloroamphetamine (20 mg/kg i. p.) given 72 and 48 hours before the drug challenge.

Kollner et al. [193] gave one i. p. injection of p-chloroamphetamine or of fenfluramine to rats on day 8 of postnatal life, which they had earlier found to produce a depletion of brain serotonin for several months, and observed that the rats were unable to learn a self-shaped avoidance behavior in a peripheral field avoidance test. The investigators concluded that serotoninergic neuron groups participate in the inhibition of incorrect responses and differentiation learning.

Nabeshima et al. [194] gave one dose of p-chloroamphetamine (10 mg/kg i. p.) 13 days before drug challenge and observed that the intensity of head-weaving, turning, forepaw treading, hind-limb abduction and Straub tail induced by 5-methoxy-dimethyltryptamine and the intensity of head-twitch, turning and backpedalling induced by phencyclidine were markedly increased. By contrast, serotonin-mediated behaviors induced by an acute dose of p-chloroamphetamine were attenuated. Tritiated serotonin and ketanserin binding sites were increased.

9 Therapeutic uses of drugs affecting serotonin neurons

Because serotonin neurons project to many parts of the brain and spinal cord and can influence these regions involved in CNS regulation of many bodily functions, drugs that alter serotonergic function have current or potential uses in many diseases.

Uptake inhibitors are antidepressant drugs and are useful in treating other psychiatric diseases such as obsessive-compulsive disorder [195, 196], panic disorder [197, 198], body-dysmorphic disorder [199] and alcoholism [200, 201] and in treating obesity [202, 203] and bulimia [204]. Releasers are useful in treating some of these same disorders, especially obesity (fenfluramine). p-Chloroamphetamine has also been reported to be effective in treating depression [51].

5HT-1A receptor selective agonists are useful in treating anxiety [99, 205] and apparently also depression [206] but not obsessive-compulsive disorder [207].

5HT-2 receptor antagonists may be useful in treating cardiovascular diseases based on their block of vascular and perhaps platelet 5HT2 receptors [118, 208–211] and may be useful in psychiatric disorders such as schizophrenia [212], anxiety [213, 214] and depression [215] based on their central effects.

5HT-3 receptor antagonists have shown efficacy in clinical trials as anti-emetic agents in patients undergoing cancer chemotherapy [216, 217] and may also be useful in treating anxiety [218] and schizophrenia [219].

Different classes of serotonergic drugs may be useful in treating migraine [220]. Serotonin uptake inhibitors have been evaluated, based on the idea that their depletion of platelet serotonin may prevent triggering of migraine attack by serotonin coming from blood platelets [221]. Several antimigraine drugs, e. g., dihydroergotamine, methysergide and pizotifen, are 5HT-2 receptor antagonists, and that may be involved in their mechanism of therapeutic efficacy, although since they are partial agonists their agonist activity has also been suggested as relevant to their antimigraine use [222, 223]. Many 5HT-2 receptor antagonists are also antagonists at 5HT-1C receptors, and that may account for their effectiveness in migraine. Recently 5HT-3 antagonists, thought to act by inhibiting sensory nerve transmission of pain messages [224], have been reported to have antimigraine efficacy [225]. Two new indole agonists at 5HT-1 receptors, GR-43175 and AH-25086, are under development as antimigraine drugs acting directly on cerebral blood vessels [226–228].

10 Clinical nontherapeutic uses of drugs affecting serotonin neurons

Some serotonergic drugs are being used in clinical studies, not for treatment of a disease, but as pharmacologic probes. The drug is given to elicit a measurable response, the magnitude of which is taken as a marker of central serotonergic function. Agents being used in this way include direct agonists – m-chlorophenylpiperazine [229] and MK-212 [230], a serotonin releaser – fenfluramine [231], and the serotonin precursors – tryptophan [232] and 5-hydroxytryptophan [233]. Responses being measured include increases in serum hormone concentrations, notably cortisol, growth hormone and prolactin, and body temperature or behavioral changes [229]. Studies have been done to probe central serotonergic function in various disease states, e. g., depression [233], agoraphobia and panic disorder [234] and obsessive-compulsive disorder [235]; after substance abuse [236, 237]; and after acute [238] or chronic [239] treatment with drugs.

11 Summary and conclusions

Advances in serotonin pharmacology, the development of drugs that intervene at specific sites to modify serotonergic function, have accompanied advances in the understanding of physiologic roles of serotonin present in neurons and elsewhere and of serotonin receptors that are widely distributed in brain and many peripheral tissues. The pharmacologic advances have sometimes been stimulated by developments in serotonin physiology, such as the recognition of multiple serotonin receptor subtypes, and in other cases have been a major factor in providing new insights into physiologic roles of serotonin. Drugs that modify serotonin function have a variety of therapeutic applications currently and many more potential therapeutic uses to be explored in the future. Having drugs that act with high specificity or selectivity on particular enzymes in serotonin biosynthesis, on particular serotonin receptors, or at other sites such as uptake carriers for serotonin not only offers the hope of improved clinical therapy in diseases caused by abnormal serotonergic function or in which alteration of serotonergic function can alleviate symptoms, but also provides valuable pharmacologic tools for learning more about serotonin physiology and probing the functional status of serotonergic systems. The next few years promise to yield important new serotonergic drugs.

References

1 M. D. Gershon: In Biology of Serotonergic Transmission, N. N. Osborne, ed. 1982, John Wiley & Sons Ltd, Chichester, pp. 363–399.
2 I. Tork. In The Rat Nervous System, Vol. 2. Hindbrain and Spinal Cord. G. Paxinos, ed. Academic Press, Sydney, pp. 43–78 (1985).
3 E. C. Azmitia and P. J. Gannon. In Advances in Neurology, Vol. 43: Myoclonus, pp. 407–468, Raven Press, New York (1986).
4 S.-O. Ogren, A.-C. Holm, H. Hall and U. H. Lindberg: J. Neural Trans. *59,* 265 (1984).
5 P. C. Waldmeier, P. A. Baumann and L. Maitre, J. Pharmacol. Exp. Ther. *211,* 42 (1979).
6 J. Hyttel: Psychopharmacology *51,* 225 (1977).
7 S. Gasic, A. Korn and L. M. O. Omer: Eur. J. Clin. Pharmacol. *21,* 357–361 (1982).
8 J. Buus Lassen, R. F. Squires, J. A. Christensen and L. Molander: Psychopharmacologia *42,* 21 (1975).
9 D. T. Wong, F. P. Bymaster, J. S. Horng and B. B. Molloy: J. Pharmacol. Exp. Ther. *193,* 804 (1975).
10 V. Claassen, J. E. Davies, G. Hertting and P. Placheta: Brit. J. Pharmacol. *60,* 505 (1977).

11 G. Le Fur, M. Kabourche and A. Uzan: Life Sci. *23,* 1959 (1978).
12 M. F. Sugrue, I. Goodlet an S. E. Mireylees: Eur. J. Pharmacol. *40,* 121 (1976).
13 P. A. Blurton, A. M. Broadhurst, J. A. Cross, C. Ennis, M. D. Wood and M. G. Wylie: Neuropharmacology *23,* 1049 (1984).
14 J. Buus Lassen: Psychopharmacology *57,* 151 (1978).
15 W. Lippmann and T. A. Pugsley: Pharmacol. Res. Commun. *8,* 387 (1976).
16 C. Dumont, J. Laurent, A. Grandadam and J. R. Boissier: Life Sci. *28,* 1939 (981).
17 B. K. Koe, A. Weissmann, W. M. Welch and R. G. Browne: J. Pharmacol. Exp. Ther. *226,* 686 (1983).
18 B. Scatton, Y. Claustre, D. Graham, T. Dennis, A. Serrano, S. Arbilla, C. Pimoule, H. Schoemaker, D. Bigg and S. Z. Langer: Drug Develop. Res. *12,* 29 (1988).
19 S. B. Ross, H. Hall, A. L. Renyi and D. Westerlund: Psychopharmacology *72,* 219 (1981).
20 L. Lemberger, R. W. Fuller and R. L. Zerbe: Clin. Neuropharmacol. *8,* 299 (1985).
21 R. W. Fuller and D. T. Wong: Ann. N. Y. Acad. Sci., in press.
22 D. T. Wong, F. P. Bymaster, L. R. Reid and P. G. Threlkeld: Biochem. Pharmacol. *32,* 1287 (1983).
23 E. Richelson and A. Nelson: J. Pharmacol. Exp. Ther. *230,* 94 (1984).
24 S. B. Ross and A. L. Renyi: Neuropharmacology *16,* 57 (1977).
25 J. L. Meek, K. Fuxe and A. Carlsson: Biochem. Pharmacol. *20,* 707 (1971).
26 R. W. Fuller, K. W. Perry, H. D. Snoddy and B. B. Molloy: Eur. J. Pharmacol. *28,* 233 (1974).
27 D. Ghezzi, R. Samanin, S. Bernasconi, G. Tognoni, M. Gerna and S. Garattini: Eur. J. Pharmacol. *24,* 205 (1973).
28 T. A. Pugsley and W. Lippman: Arch. int. Pharmacodyn. *228,* 322 (1977).
29 A. Carlsson, H. Corrodi, K. Fuxe and T. Hokfelt: Eur. J. Pharmacol. *5,* 357 (1969).
30 M. A. Geyer, W. J. Dawsey and A. J. Mandell: J. Pharmacol. Exp. Ther. *207,* 650 (1978).
31 C. A. Marsden, J. Conti, E. Strope, G. Curzon and R. N. Adams: Brain Res. *171,* 85 (1979).
32 X.-M. Guan and W. J. McBride. Brain Res. Bull. *21,* 43 (1988).
33 R. W. Fuller and D. T. Wong: Fed. Proc. *36,* 2154 (1977).
34 J. A. Clemens, B. D. Sawyer and B. Cerimele: Endocrinology *100,* 692 (1977).
35 R. W. Fuller: Neuroendocrinology *47,* 389 (1981).
36 D. M. Gibbs and W. Vale: Brain *280,* 176 (1983).
37 V. A. Molina, S. Gobaille and P. Mandel: Aggress. Behav. *12,* 201 (1986).
38 R. Ortmann, P. C. Waldmeier, E. Radeke, A. Felner and A. Delini-Stula: Naunyn-Schmiedeberg's Arch. Pharmacol. *311,* 185 (1980).
39 R. J. Barrett, M. A. Blackshear and E. Sanders-Bush: Psychopharmacology *76,* 29 (1982).
40 R. B. Messing, L. Phebus, L. A. Fisher and L. D. Lytle: Psychopharmacol. Commun. *1,* 511 (1975).
41 A. A. Larson and A. E. Takemori: Life Sci. *21,* 1807 (1977).
42 D. T. Wong and R. W. Fuller: Int. J. Obesity *11,* (Suppl. 3), 125 (1987).
43 J. J. Wurtman and R. J. Wurtman: Science *198,* 1178 (1977).
44 S. M. Antelman, N. Rowland and D. Kocan: In Anorectic agents: mechanisms of action and tolerance (S. Garattini and R. Samanin, ed.), New York, Raven Press (1981), pp. 45–62.

45 M. O. Carruba, S. Ricciardi, P. Spano and P. Mantegazza: Life Sci. *36,* 1739 (1985).
46 M. E. Trulson and B. L. Jacobs: Eur. J. Pharmacol. *36,* 149 (1976).
47 J. H. Growdon: Neurology *27,* 1074 (1977).
48 E. C. Hwang and M. H. Van Woert: Biochem. Pharmacol. *29,* 3163 (1980).
49 P. H. Hutson and G. Curzon: Brit. J. Pharmacol. *96,* 801 (1989).
50 C. Kaergaard Nielsen, M. P. Magnussen, E. Kampmann and H.-H. Frey: Arch. int. Pharmacodyn. *170,* 428 (1967).
51 H. M. Van Praag, T. Schut, E. Bosma and R. Van den Bergh: Psychopharmacologia *20,* 66 (1971).
52 J. F. Nash, Jr., H. Y. Meltzer and G. A. Gudelsky: J. Pharmacol. Exp. Ther. *245,* 873 (1988).
53 J. F. McElroy and R. S. Feldman: Psychopharmacology *83,* 172 (1984).
54 M. F. Sugrue: Brit. J. Pharmacol. *81,* 651 (1984).
55 J. F. McElroy, J. M. Miller and J. S. Meyer: J. Pharmacol. Exp. Ther. *228,* 593 (1984).
56 L. D. Van de Kar, J. H. Urban, K. D. Richardson and C. L. Bethea: Neuroendocrinology *41,* 283 (1985a).
57 L. D. Van de Kar, K. D. Richardson and J. H. Urban: Neuropharmacology *24,* 487 (1985b).
58 B. V. Clineschmidt and J. C. McGuffin: Eur. J. Pharmacol. *50,* 369 (1978).
59 S. Consolo, H. Ladinsky, A. S. Tirelli, V. Crunelli, R. Samanin and S. Garattini: Life Sci. *25,* 1975 (1979).
60 L. Krulich: Life Sci. *17,* 1141 (1975).
61 A. Quattrone, G. Di Renzo, G. Schettini, G. Tedeschi and F. Scopacasa: Eur. J. Pharmacol. *49,* 163 (1978).
62 R. W. Fuller, H. D. Snoddy and D. W. Robertson: Pharmacol. Biochem. Behav. *30,* 715 (1988).
63 R. W. Fuller, D. R. Holland, T. T. Yen, K. G. Bemis and N. B. Stamm: Life Sci. *25,* 1237 (1979).
64 R. W. Fuller, T. T. Yen and N. B. Stamm: Clin. Exp. Hypertension 3, 497 (1981).
65 R. A. Lyon, M. Titeler, M. R. Seggel and R. A. Glennon: Eur. J. Pharmacol. *145,* 291 (1988).
66 R. A. Glennon, M. Titeler, R. A. Lyon and R. M. Slusher: J. Med. Chem. *31,* 867 (1988).
67 G. Engel, M. Gothert, D. Hoyer, E. Schlicker and K. Hillenbrand: Naunyn-Schmiedeberg's Arch. Pharmacol. *332,* 1 (1986).
68 D. Hoyer: J. Receptor Res. *8,* 59 (1988).
69 B. P. Richardson, G. Engel, P. Donatsch and P. A. Stadler: Nature *316,* 126 (1985).
70 C. Euvrard and J. R. Boissier: Eur. J. Pharmacol. *63,* 65 (1980).
71 E. Hong: In Molecular Basis of Drug Action (T. P. Singer and R. N. Ondarza, eds.), pp. 247–252, Elsevier-North Holland, New York (1981).
72 R. W. Fuller, in 5-HT Agonists as Psychoactive Drugs (R. H. Rech and G. A. Gudelsky, eds.), 1988, NPP Books, Ann Arbor, MI. pp. 35–59.
73 E. Hong and E. G. Pardo: J. Pharmacol. Exp. Ther. *153,* 259 (1966).
74 E. Hong. L. F. Sancilio, R. Vargas and E. G. Pardo: Eur. J. Pharmacol *6,* 274 (1969).
75 B. V. Clineschmidt, J. C. McGuffin and A. B. Pflueger: Eur. J. Pharmacol. *44,* 65 (1977).
76 R. W. Fuller, H. D. Snoddy, N. R. Mason and B. B. Molloy: Eur. J. Pharmacol. *52,* 11 (1978).
77 R. Samanin, T. Mennini, A. Ferraris, C. Bendotti, F. Borsini and S. Garattini: Naunyn-Schmiedeberg's Arch. Pharmacol. *308,* 159 (1979).

78 L. R. Moser, J. A. Kalser and R. A. Hardy, Jr. U. S. Patent 3, 253, 989 (1966).
79 R. A. Gillis, K. J. Kellar, J. A. Quest, I. J. Namath, A. Martino-Barrows, K. Hill, P. J. Gatti and K. Dretchen: Drugs *35* (Suppl. 6), 20 (1988).
80 A. S. Eison and D. L. Temple, Jr.: Am. J. Med. *80* (Suppl. 3B), 1 (1986).
81 J. Traber, M. A. Davies, W. U. Dompert, T. Glaser, T. Schuurman and P.-R. Seidel: Brain Res. Bull. *12,* 741 (1984).
82 A. S. Eison, M. S. Eison, M. Stanley and L. A. Riblet: Pharmacol. Biochem. Behav. *24,* 701 (1986).
83 J. H. Gaddum and Z. P. Picarelli: Brit. J. Pharmacol. *12,* 323 (1957).
84 S. J. Peroutka and S. H. Synder: Mol. Pharmacol. *16,* 687 (1979).
85 N. W. Pedigo, H. I. Yamamura and D. L. Nelson: J. Neurochem. *36,* 220 (1981).
86 S. S. Wang and S. J. Peroutka: In The Serotonin Receptors. E. Sanders-Bush, ed. Humana Press, Clifton, NJ, pp. 3–20 (1989).
87 A. W. Schmidt and S. J. Peroutka: FASEB J. *3,* 2242 (1989).
88 A. Fargin, J. R. Raymond, M. J. Lohse, B. K. Kobilka, M. G. Caron and R. J. Lefkowitz: Nature 335: 358 (1988).
89 D. B. Pritchett, A. W. J. Bach, M. Wozny, O. Taleb, R. Dal Toso, J. C. Shih and P. H. Seeburg: The EMBO Journal *7,* 4135 (1988).
90 H. Lubbert, B. J. Hoffman, T. P. Snutch, T. Van Dyke, A. J. Levine, P. R. Hartig, H. A. Lester and N. Davidson: Proc. Nat. Acad. Sci. *84,* 4332 (1987).
91 D. Julius, A. B. MacDermott, R. Axel and T. M. Jessell: Science 241: 558 (1988).
92 P. R. Hartig: Trends in Pharmacol. Sci. *10,* 64 (1989).
93 C. A. Marsden, A. J. Sleight, K. C. F. Fone, J. V. Jonson, F. Crespi, K. F. Martin, J. C. Garrett and G. W. Bennett: Comp. Biochem. Physiol. *93A,* 107 (1989).
94 S. Hjorth, A. Carlsson, P. Lindberg, D. Sanchez, H. Wikstrom, L.-E. Arvidsson, U. Hacksell and J. L. G. Nilsson. J. Neural Trans. *55,* 169 (1982).
95 L. Bjork, B. B. Hook, D. L. Nelson, N.-E. Anden and U. Hacksell: J. Med. Chem. *32,* 779 (1989).
96 D. P. Taylor, L. A. Riblet, H. C. Stanton, A. S. Eison, M. S. Eison and D. L. Temple, Jr: Pharmacol. Biochem. Behav. *17* (Suppl. 1), 25 (1982).
97 T. Glaser and J. Traber: Eur. J. Pharmacol. *88,* 137 (1983).
98 S. J. Peroutka: Biol. Psychiat. *20,* 971 (1985).
99 J. Traber and T. Glaser: Trends in Pharmacol. Sci. *8,* 432 (1987).
100 T. Glaser and J. Traber: Naunyn-Schmiedeberg's Arch. Pharmacol. *329,* 211 (1985).
101 H. Shimizu, N. Karai, A. Hirose, T. Tatsuno, H. Tanaka, Y. Kumasaka and M. Nakamura: Japan. J. Pharmacol. *46,* 311 (1988).
102 B. A. McMillen, E. A. DaVanzo, S. M. Scott and A. H. Song: Drug Develop. Res. *12,* 53 (1988).
103 K. B. Asarch, R. W. Ransom and J. C. Shih: Life Sci. *36,* 1265 (1985).
104 R. W. Fuller, H. D. Snoddy and B. B. Molloy: J. Pharmacol. Exp. Ther. *239,* 454 (1986).
105 A. K. Mir, M. Hibert, M. D. Tricklebank, D. Middlemiss, E. J. Kidd and J. R. Fozard: Eur. J. Pharmacol. *149,* 107 (1988).
106 M. A. Sills, B. B. Wolfe and A. Frazer: J. Pharmacol. Exp. Ther. *231,* 480 (1984).
107 G. A. Kennett, C. T. Dourish and G. Curzon: Eur. J. Pharmacol. *141,* 429 (1987).
108 M. Hamon, J.-M. Cossery, U. Spampinato and H. Gozlan: Trends in Pharmacol. Sci. *7,* 336 (1986).

109 R. F. Neale, S. L. Fallon, W. C. Boyar, J. W. F. Wasley, L. L. Martin, G. A. Stone, B. S. Glaeser, C. M. Sinton and M. Williams: Eur. J. Pharmacol. *136,* 1 (1987).
110 G. A. Kennett and G. Curzon: Brit. J. Pharmacol. *94,* 137 (1988).
111 P. Schoeffter and D. Hoyer: Naunyn-Schmiedeberg's Arch. Pharmacol. *339,* 675 (1989).
112 G. A. Kennett and G. Curzon: Psychopharmacology *96,* 103 (1988b).
113 M. Titeler, K. Herrick, R. A. Lyon, J. D. McKenney and R. A. Glennon: Eur. J. Pharmacol. *117,* 145 (1985).
114 W. A. Heller and J. M. Baraban: Eur. J. Pharmacol. *138,* 115 (1987).
115 R. A. Glennon, M. R. Seggel, W. H. Soine, K. Herrick-Davis, R. A. Lyon and M. Titeler: J. Med. Chem. *31,* 5 (1988).
116 P. P. A. Humphrey, W. Feniuk, M. J. Perren, H. E. Connor, A. W. Oxford, I. H. Coates and D. Butina: Brit. J. Pharmacol. *94,* 1123 (1988).
117 H. E. Connor, W. Feniuk, P. P. A. Humphrey and M. J. Perren: Brit. J. Pharmacol. *87,* 417 (1986).
118 F. Awouters: Drug Develop. Res. *6,* 263 (1985).
119 P. A. J. Janssen: J. Cardiovasc. Pharmacol. *7* (Suppl. 7), S2 (1985).
120 L. E. J. Kennis, J. Vandenberk, J. M. Boey, J. C. Mertens, A. H. M. Van Heertum, M. Janssen and F. Awouters: Drug Develop. Res. *8,* 133 (1986).
121 M. L. Cohen, K. D. Kurz, N. R. Mason, R. W. Fuller, G. P. Marzoni and W. L. Garbrecht: J. Pharmacol. Exp. Ther. *235,* 319 (1985).
122 M. L. Cohen, R. W. Fuller, K. D. Kurz, C. J. Parli, N. R. Mason, D. B. Meyers, J. K. Smallwood and R. E. Toomey: J. Pharmacol. Exp. Ther. *244,* 106 (1988).
123 T.P. Blackburn, C. W. Thornber, R. J. Pearce and B. Cox: FASEB J. *2,* A1404 (1988).
124 T. P. Blackburn, C. W. Thornber, R. J. Pearce and B. Cox: Eur. J. Pharmacol. *150,* 247 (1988).
125 M. W. Dudley, N. L. Wiech, F. P. Miller, A. A. Carr, H. C. Cheng, L. E. Roebel, N. S. Doherty, H. I. Yamamura, R. C. Ursillo and M. G. Palfreyman: Drug Develop. Res. *13,* 29 (1988).
126 J. Arnt, K. P. Bogeso, V. Boeck, A. V. Christensen, N. Dragsted, J. Hyttel and T. Skarsfeldt: Drug Develop. Res. *16,* 59 (1989).
127 E. Sanders-Bush and M. Breeding: J. Pharmacol. Exp. Ther. *247,* 169 (1988).
128 C. Beretta, R. Ferrini and A. H. Glasser: Nature *207,* 421 (1965).
129 J. H. Jacoby, J. J. Poulakos and G. F. Bryce: Neuropharmacology *17,* 299 (1978).
130 E. Przegalinaki, L. Baran, W. Palider and J. Siwanowicz: Psychopharmacology *62,* 295 (1979).
131 M.-A. Monachon, W. P. Burkard, M. Jalfre and W. Haefely: Naunyn-Schmiedeberg's Arch. Pharmacol. *274,* 192 (1972).
132 J. Maj, L. Baran, J. Michaluk, W. Palider and A. Rawlow: Pol. J. Pharmacol. Pharm. *30,* 407 (1978).
133 B. B. Vargaftig, J. L. Coignet, C. J. de Vos, H. Grijsen and I. L. Bonta: Eur. J. Pharmacol. *16,* 336 (1971).
134 A. R. Furgiuele, J. P. High and Z. P. Horowitz: Arch. int. Pharmacodyn. *155,* 225 (1965).
135 E. Przegalinski, L. Baran, W. Palider and K. Bigajska: Pol. J. Pharmacol. Pharm. *30,* 781 (1978).
136 R. W. Fuller, H. D. Snoddy and M. L. Cohen: Neuropharmacology *23,* 539 (1984).
137 S. Caccia, M. H. Fong, S. Garattini and M. G. Zanini: J. Pharm. Pharmacol. *34,* 605 (1982).

138 J. Maj, W. Palider and A. Rawlow: J. Neural Trans *44,* 237 (1979).
139 R. N. Brogden, R. C. Heel, T. M. Speight and G. S. Avery: Drugs *21,* 401 (1981).
140 J. R. Fozard: Naunyn-Schmiedeberg's Arch. Pharmacol. *326,* 36 (1984).
141 A. Butler, J. M. Hill, S. J. Ireland, C. C. Jordan and M. B. Tyers: Brit. J. Pharmacol. *94,* 397 (1988).
142 P. L. R. Andrews and J. Hawthorn: Neuropharmacology *27,* 1367 (1987).
143 E. A. Boyle, W. D. Miner and G. J. Sanger: Brit. J. Cancer *56,* 227 (1987).
144 M. Cohen, W. Bloomquist, J. S. Gidda and W. Lacefield: J. Pharmacol. Exp. Ther. *248,* 197 (1989).
145 P. Fludzinski, D. A. Evrard, W. E. Bloomquist, W. B. Lacefield, W. Pfeifer, N. D. Jones, J. B. Deeter and M. L. Cohen: J. Med. Chem. *30,* 1535 (1987).
146 D. T. Wong, D. W. Robertson and L. R. Reid: Eur. J. Pharmacol. *166,* 107 (1989).
147 G. J. Kilpatrick, B. J. Jones and M. B. Tyers: Nature *330,* 746 (1987).
148 A. B. Norman, G. Battaglia and I. Creese: Mol. Pharmacol. *28,* 487 (1985).
149 A.Pazos, G. Engel and J. M. Palacios: Brain Res. *343,* 403 (1985).
150 E. Edwards and P. M. Whitaker-Azmitia: Neuropharmacology *26,* 93 (1987).
151 E. Schlicker, M. Gothert and K. Hillenbrand: Naunyn-Schmiedeberg's Arch. Pharmacol. *331,* 398 (1985).
152 M. D. Tricklebank, C. Forler and J. R. Fozard: Eur. J. Pharmacol. *106,* 271 (1985).
153 D. N. Middlemiss: Eur. J. Pharmacol. *120,* 51 (1986).
154 S. Hjorth and A. Carlsson: Eur. J. Pharmacol. *129,* 131 (1986).
155 G. Maura, M. Ulivi and M. Raiteri: Neuropharmacology *26,* 713 (1987).
156 M. F. Hibert, M. W. Gittos, D. N. Middlemiss, A. K. Mir and J. R. Fozard: J. Med. Chem. *31,* 1087 (1988).
157 P. C. Waldmeier, M. Williams, P. A. Baumann, S. Bischoff, M. A. Sills and R. F. Neale: Naunyn-Schmiedeberg's Arch. Pharmacol. *337,* 609 (1988).
158 M. E. Pierson, R. A. Lyon, M. Titeler, P. Kowalski and R. A. Glennon: J. Med. Chem. *32,* 859 (1989).
159 B. K. Koe and A. Weissman: J. Pharmacol. Exp. Ther. *154,* 499 (1966).
160 S. Miwa, M. Fujiwara, K. Lee and M. Fujiwara: J. Neurochem. *48,* 1577 (1987).
161 G. A. Marks and H. P. Roffwarg: Pharmacol. Biochem. Behav. *31,* 509 (1988).
162 H. M. Pappius, R. Dadoun and M. McHugh: J. Cereb. Blood Flow Metab. *8,* 324 (1988).
163 S. Hunskaar, J. H. Rosland and K. Hole: Eur. J. Pharmacol. *160,* 83 (1989).
164 J. M. Barnes, N. M. Barnes, B. Costall, R. J. Naylor and F. D. Tattersall: Neuropharmacology *27,* 783 (1988).
165 P. K. Eide,K. Hole, O. G. Berge and O. J. Broch: Brain Res. *440,* 42 (1988b).
166 P. K. Eide, K. Hole and O. J. Broch: Acta physiol. Scand. *132,* 419 (1988a).
167 N. A. Sharif, A. C. Towle, D. R. Burt, R. A. Mueller and G. R. Breese: Brain Res. *480,* 365 (1989).
168 I. Tohyama, M. Kameyama and H. Kimura: Neuroscience *26,* 971 (1988).
169 H. G. Baumgarten, A. Bjorklund and L. Lachenmayer: Acta Physiol. Scand. *373,* 1 (1971).

170 H. G. Baumgarten and L. Lachenmayer: Z. Zellforsch. *135,* 399 (1972).
171 G. Jonsson: Ann. Rev. Neurosci. *3,* 169 (1980).
172 A. K. Sinhababu and R. T. Borchardt: Neurochem. Int. *12,* 273 (1988).
173 J. H. Jacoby and L. D. Lytle (eds.): Ann. N. Y. Acad. Sci. *305,* 1 (1978).
174 H.-H. Frey, In Amphetamines and Related Compounds (E. Costa and S. Garattini, eds.), pp. 343–347 (1970).
175 E. Sanders-Bush, J. A. Bushing and F. Sulser: Eur. J. Pharmacol. *20,* 385 (1972).
176 E. Sanders-Bush, J. A. Bushing and F. Sulser: J. Pharmacol. Exp. Ther. *192,* 33 (1975).
177 R. W. Fuller and B. B. Molloy: Adv. Biochem. Psychopharmacol. *10,* 195 (1974).
178 G. Battaglia, S. Y. Yeh, E. O'Hearn, M. E. Molliver, M. J. Kuhar and E. B. DeSouza: J. Pharmacol. Exp. Ther. *242,* 911 (1987).
179 L. A. Mamounas and M. E. Molliver: Exp. Neurol. *102,* 23 (1988).
180 R. W. Fuller: Ann. N. Y. Acad. Sci. *305,* 178 (1978).
181 R. Alesci and P. Bagnoli: Brain Res. *450,* 259 (1988).
182 J. A. Harvey and S. E. McMaster: Psychopharmacol. Commun. *1,* 217 (1975).
183 N. M. Appel, J. F. Contrera and E. B. DeSouza: J. Pharmacol. Exp. Ther. *249,* 928 (1989).
184 C. J. Schmidt, L. Wu and W. Lovenberg: Eur. J. Pharmacol. *124,* 175 (1986).
185 C. R. Schuster, M. Lewis and L. S. Seiden: Psychopharmacol. Bull. *22,* 148 (1986).
186 G. A. Ricaurte, L. E. DeLanney, I. Irwin and J. W. Langston: Brain Res. *446,* 165 (1988).
187 D. M. Barnes: Science *243,* 29 (1989).
188 G. A. Ricaurte, L. S. Forno, M. A. Wilson, L. E. DeLanney, I. Irwin, M. E. Molliver and J. W. Langston: J. Am. Med. Assoc. *260,* 51 (1988).
189 C. V. Vorhees, G. J. Schaefer and R. J. Barrett: Pharmacol. Biochem. Behav. *3,* 279 (1975).
190 M. Grabowska and J. Michaluk: Pharmacol. Biochem. Behav. *2,* 263 (1974).
191 S. O. Ogren, C. Kohler, S. B. Ross and B. Screbro: Neurosci. Lett. *3,* 341 (1976).
192 K. Yamaguchi, T. Nabeshima and T. Kameyama: J. Pharmacobio-Dyn. *9,* 179 (1986).
193 O. Kollner, U. Kollner, R. Gob and F. Klingberg: Biomed. Biochim. Acta *47,* 997 (1988).
194 T. Nabeshima, K. Yamaguchi, K. Ishikawa, H. Furukawa and T. Kameyama: Res. Commun. Subst. Abuse *10,* 37 (1989).
195 R. Levine, J. S. Hoffman, E. D. Knepple and M. Kenin: J. Clin. Psychopharmacol. *9,* 281 (1989).
196 C. Benkelfat, D. L. Murphy, J. Zohar, J. L. Hill, G. Grover and T. R. Insel: Arch. Gen. Psychiat. *46,* 23 (1989).
197 J. M. Gorman, M. R. Liebowitz, A. J. Fyer, D. Goetz, R. B. Campeas, M. R. Fyer, S. O. Davies and D. F. Klein: J. Clin. Psychopharmacol. *7,* 329 (1987).
198 J. A. den Boer, H. G. Westenberg, W. D. Kamerbeek, W. M. Verhoeven and R. S. Kahn: Int. Clin. Psychopharmacol. *2,* 21 (1987).
199 E. Hollander, M. R. Liebowitz, R. Winchel, A. Klumker and D. F. Klein: Am. J. Psychiat. *146,* 768 (1989).
200 C. A. Naranjo, E. M. Sellers and M. O. Lawrin: J. Clin. Psychiat. *47* (Suppl.) 16 (1986).

201 C. A. Naranjo, E. M. Sellers, J. T. Sullivan, D. V. Woodley, K. Kadlec and K. Sykora: Clin. Pharmacol. Ther. *41,* 266 (1987).
202 J. P. Smedegaard, P. Christiansen and B. Skrumsager: Int. J. Obesity *5,* 377 (1981).
203 L. R. Levine, S. Rosenblatt and J. Bosomworth: Int. J. Obesity *11* (Suppl. 3), 185 (1987).
204 C. P. Freeman and M. Hampson: Int. J. Obesity *11* (Suppl. 3), 171 (1987).
205 J. P. Feighner and W. F. Boyer: Psychopathology *22* (Suppl. 1), 21 (1989).
206 D. S. Robinson, D. R. Alms, R. C. Shrotriya, M. Messina and P. Wickramaratne: Psychopathology *22* (Suppl. 1), 27 (1989).
207 M. A. Jenike and L. Baer: Am. J. Psychiat. *145,* 1285 (1988).
208 D. S. Houston and P. M. Vanhoutte: Drugs *31,* 149 (1986).
209 D. D. Heistad, D. G. Harrison, M. L. Armstrong, D. Piegors, K. Breese and M. Meegan: Int. J. Cardiol. *14,* 205 (1987).
210 N. Hollenberg: Am J. Hypertens. *1,* 312S (1988).
211 P. Golino, J. H. Ashton, J. McNatt, P. Glas-Greenwalt, S. K. Yao, R. A. O'Brien, L. M. Buja and J. T. Willerson: Circulation *79,* 911 (1989).
212 D. L. S. Ceulemans, Y. Gelders, M. L. Hoppenbrouwers, A. Reyntjens and P. Janssen, Psychopharmacology *85,* 319 (1985).
213 D. L. S. Ceulemans, M.-L. J. A. Hoppenbrouwers, Y. G. Gelders and A. J. M. Reyntjens: Pharmacopsychiat. *18,* 303 (1985).
214 G. M. Bressa, S. Marini and S. Gregori: Int. J. Clin. Pharm. Res. *7,* 111 (1987).
215 E. Klieser and W. H. Strauss: Pharmacopsychiat. *21,* 391 (1988).
216 J. Cassidy, V. Raina, C. Lewis, L. Adams, M. Soukop, W. G. Rapeport, B. D. Zussman, E. M. Rankin and S. B. Kaye: Brit. J. Cancer *58,* 651 (1988).
217 M. G. Kris, R. J. Gralla, R. A. Clark and L. B. Tyson: J. Nat. Cancer Inst. *81,* 42 (1989).
218 B. J. Jones, B. Costall, A. M. Domeney, M. E. Kelly, R. J. Naylor, N. R. Oakley and M. B. Tyers: Brit. J. Pharmacol. *93,* 985 (1988).
219 B. Costall, A. M. Domeney, R. J. Naylor and M. B. Tyers: Brit. J. Pharmacol. *92,* 881 (1987).
220 S. J. Peroutka: Ann. Neurol. *23,* 500 (1988).
221 E. Syvalahti, P. Kangasniemi and S. B. Ross: Curr. Ther. Res. *25,* 299 (1979).
222 M. Fanciullachi, G. Granchi and F. Sicuteri: Headache *16,* 226 (1976).
223 J.-C. Lamar and J.-E. Hardebo: Eur. J. Pharmacol. *60,* 263 (1979).
224 B. P. Richardson, P. Donatsch, G. Engel and R. Gier: J. Pharmacol. *17,* 99 (1986).
225 C. Loisy, S. Beorchia, V. Centonze, J. R. Fozard, P. J. Schechter and G. P. Tell: Cephalalgia *5,* 79 (1985).
226 A. Doenicke, J. Brand and V. L. Perrin: Lancet *1,* 1309 (1988).
227 A. A. Parsons, E. T. Whalley, W. Feniuk, H. E. Connor and P. P. A. Humphrey: Brit. J. Pharmacol. *96,* 434 (1989).
228 P. R. Saxena and M. D. Ferrari: Trends in Pharmacol. Sci. *10,* 200 (1989).
229 E. A. Mueller, D. L. Murphy and T. Sunderland: J. Clin. Endocrinol. Metab. *61,* 1179 (1985).
230 M. T. Lowy and H. Y. Meltzer: Biol. Psychiat. *23,* 818 (1988).
231 L. J. Siever, D. L. Murphy, S. Slater, E. de la Vega and S. Lipper: Life Sci. *34,* 1029 (1984).
232 P. J. Cowen: Am. J. Psychiat. *145,* 740 (1988).
233 T. Koyama, M. T. Lowy and H. Y. Meltzer: Am. J. Psychiat. *144,* 344 (1987).
234 D. S. Charney, S. W. Woods, W. K. Goodman and G. R. Heninger: Psychopharmacology *92,* 14 (1987).

235 D. S. Charney, W. K. Goodman, L. H. Price, S. W. Woods, S. A. Rasmussen and G. R. Heninger: Arch. Gen. Psychiat. *45,* 177 (1988).
236 D. H. Fishbein, D. Lozovsky and J. J. Jaffe: Biol. Psychiat. *25,* 1049 (1989).
237 L. H. Prive, G. A. Ricaurte, J. H. Krystal and G. R. Heninger: Arch. Gen. Psychiat. *46,* 20 (1989).
238 C. Foresta, G. Scanelli, M. Indino, G. Federspil and C. Scandellari: Clin. Endocrinol. *22,* 539 (1985).
239 B. Shapira, A. Reiss, N. Kaiser, S. Kindler and B. Lerer: J. Affective Disord. *16,* 1 (1989).

The role of cytokines in macrophage activation

By Richard M. Schultz
Lilly Research Laboratories, Indianapolis, IN 46285, USA

1	Introduction	110
1.1	On the definition of the activated state	111
2	Cytokines as inducers of macrophage activation	111
2.1	Priming activity	113
2.2	Agents that function as "second signals"	116
2.3	Cytokines other than interferon that activate macrophages	117
3	Molecular mechanisms of macrophage activation	119
3.1	Signal transduction and interferon-γ	119
3.2	Signal transduction and LPS	122
4	Factors regulating macrophage activation	123
5	Monokines as effector molecules of tumor cytotoxicity	125
6	Therapeutic considerations	128
6.1	Macrophage tumor content	128
6.2	Macrophage functional state within tumor	129
7	Concluding remarks	131
	References	132

1 Introduction

It is well established that activated macrophages play an important role as effector cells in the body's defense against tumor cells and microorganisms [for reviews, see references 1–3). These cells can recognize and kill neoplastic cells, under conditions that normal cells are spared [4, 5]. Although the exact mechanism(s) by which macrophages recognize tumor cells are still unclear, macrophages appear to respond to tumor cell properties whose expression is intimately linked to the neoplastic state. Correlations exist in various animal models between the ability of synthetic and biologic agents to enhance host resistance to cancer and their capacity to activate macrophages to the tumoricidal state [6–8]. Macrophages appear to represent a major effector of tumor surveillance against metastatic tumor spread [9–11]. These studies have suggested new strategies for therapy of metastatic disease using *in situ* activated macrophages [12]. Over the last decade much has been learned about the inducers and complex biochemical events that underlie the effector functions of mononuclear phagocytes as they encounter environmental stimuli. These studies have been greatly aided by recent advances in molecular biology and monoclonal antibody technique. The pharmacologic control of macrophage activation offers possibilities for stimulating the host defensive role of mononuclear phagocytes in resistance to neoplasia and infectious microorganisms or in ameliorating the deleterious host effects of these cells in disorders such as rheumatoid arthritis and other inflammatory tissue diseases.

In this review, six major items will be discussed; namely, cytokines as inducers and regulators of macrophage activation (Section 2), signal transduction pathways involved in macrophage activation (Section 3), negative regulation of macrophage activation (Section 4), monokines as effector molecules of macrophage activation (Section 5), therapeutic considerations (Section 6), and concluding remarks (Section 7). It is hoped that this article will reflect the basic information available as well as show the great complexity involved in altering the functional activity of the mononuclear phagocyte system.

1.1 On the definition of the activated state

The term "activated macrophage" was first introduced by Mackaness in the 1960s to explain the enhanced nonspecific resistance to intracellular growing bacteria observed in animals recovering from infection with pathogens such as *Listeria monocytogenes* [13]. This concept could be traced back to Metchnikoff [14], who observed in the early 1900s that mononuclear phagocytes from animals resistant to certain bacteria were "in a condition of superactivity" for ingesting and killing these microbes. This type of immunity has been shown to be nonspecific in its expression, since macrophages from animals immunized against a particular bacterium display increased microbicidal activity against a variety of unrelated bacteria. The acquisition of this microbicidal activity depends on the generation of a specific T cell-mediated immune response [15, 16], through secretion of lymphokines [17, 18]. This concept was further extended in the early 1970s by the findings of Evans and Alexander [19] and Hibbs and coworkers [20] that these angry "activated" macrophages were also efficient at killing neoplastic cells. Over more recent times, this term has been defined operationally by investigators to reflect different, possibly unrelated, phenotypic alterations in macrophage function or morphology induced by a plethora of biologic and synthetic agents. It has become increasingly clear that exogenous agents can induce functional heterogeneity, in that activated states may differ markedly with regard to the effect macrophages have on potential pathogens or tumor cells. To avoid ambiguity, the use of the term "activation" in this review will be restricted to the induction of cytotoxic activity in macrophages for tumor cells.

2 Cytokines as inducers of macrophage activation

Observations from several different laboratories suggest that development of macrophage tumoridical function occurs after a series of reactions [21–25]. Hibbs and colleagues demonstrated that specific environmental signals acted sequentially to activate marcrophages [21]. They showed that tumor killing by activated macrophages is not a highly determined biologic event, but a relative capability influenced by the local environment: The macrophage cytotoxic effector system is modulated by serum and other environmental factors that can either enhance or suppress tumor killing, and tumor cytotoxicity occurs only when a regulating threshold drops to a critically low level.

Meltzer and coworkers suggested that the reaction sequence can be conceptually divided into three phases: precursor differentiation, priming, and trigger reactions [23]. In the first phase, immature, blood-derived mononuclear phagocytes are recruited and differentiated into competent, lymphokine-responsive cells by various factors at the inflammatory site. Ruco and Meltzer [26] observed that immature, peroxidase-positive mononuclear phagocytes that accumulate during nonspecific inflammatory reactions are 10- to 20-fold more responsive to lymphokine activation than equal numbers of more differentiated resident tissue macrophages. The expression of nonspecific effector function is completed after two additional phases, in which the macrophages first enter into a receptive or primed state by certain lymphokine signals and are then triggered by other signals to develop full functional activity. These signals form the basis of a regulatory system that sets the threshold and determines the onset of macrophage effector activity.

The interaction of lymphokine and gram-negative bacterial endotoxin (LPS) has been used as a model system that illustrates the two-stage reaction sequence for induction of macrophage effector function [22]. In this model, macrophages are first exposed to the lymphokine, macrophage-activating factor (MAF) and enter into a primed state in which they are not yet cytotoxic, but can be triggered by low levels of LPS to develop full functional activity. This interaction was shown to be synergistic and critically dependent on treatment sequence and interval [27]. LPS was active only if given simultaneously with or after lymphokine treatment; LPS exposure before lymphokine treatment was ineffective. The capacity of lymphokine-treated macrophages to express cytotoxic activity after LPS exposure decayed with time in culture and was lost by 24 hr. Several investigators have noted that interferon-γ is the principal species of MAF in lymphokine supernatants [28–30].

Many studies have demonstrated the ability of a single agent to render macrophages tumoricidal [31–35]. The requirement for one or multiple signals for activating macrophages is influenced by a number of variables, including the type of cytotoxicity assay employed, the quality of animal care and lack of intercurrent infections, the level of endotoxin contamination in cell culture reagents and glassware, and the macrophage source (strain differences in susceptibility to activation stimuli and differences between inflammatory versus resident populations of cells etc.). Russell and associates noted the need to standardize the way in which macrophage-mediated, nonspecific cytotoxicity is measured for comparison of test results between laboratories [36]. Pace and coworkers noted that the capacity of murine interferon-γ to activate macrophages without the need for a second triggering stimulus is related to assay conditions and, most especially, the strain of mouse used to provide the macrophages [37].

2.1 Priming activity

In a variety of assay systems, two signals are required to activate macrophages for tumor cell killing [22–25]. The first signal is a T cell-derived lymphokine, denoted macrophage-activating factor (MAF), that primes the macrophage and makes it more receptive to a diverse group of triggering agents that can induce full cytocidal activity. Macrophages that are primed or triggered can be characterized on the basis of distinct phenotypic characteristics [38, 39]. Ruco and Meltzer [22] and Sone and Fidler [24] demonstrated that a defined sequence of events had to exist with combinations of MAF and lipopolysaccharide (LPS) or muramyl dipeptide (MDP) for induction of macrophage cytotoxicity. In these studies, macrophages initially exposed to MAF enter into a receptive state in which they are much more sensitive to the effects of the second activating stimulus rather than inducing full activation. Since cell culture reagents are frequently contaminated with LPS [40, 41], these data cast doubts on earlier studies which demonstrated that MAF activity was due to a single pharmacologic entity.

Several laboratories, using a variety of experimental techniques, have shown that interferon-γ, the interferon produced by T lymphocytes [42], functions as the principal MAF for tumor cytotoxicity [28–30, 43, 44]. In 1982, two studies showed that MAF produced by antigen- and mitogen-stimulated splenocyte cultures was indistinguishable from interferon-γ by several criteria including chromatography on a variety of separatory media, kinetics of temperature and pH inactivation, and immunochemical reactivity [45, 46]. Several investigators were unable to distinguish between MAF and interferon-γ on a biosynthetic basis using several normal T-cell lines and T-cell hybridomas [47–49]. The final identification of interferon-γ as the principal MAF came in 1983 as a result of the cloning of the interferon-γ gene by Gray and Goeddel [50]. When tested for biologic activity, this first recombinant interferon-γ was shown to display MAF activity [28, 29]. The MAF activity was acid labile and was neutralized by rabbit anti-interferon-γ. Moreover, the amount of MAF activity in the recombinant preparation was comparable to that of conditioned media from murine splenocytes [28] and from the murine T-cell hybridoma, 24/G 1 [51]. Schreiber and coworkers showed that hamster monoclonal IgG antibodies to recombinant murine interferon-γ inhibited 100% of the MAF, microbicidal, and Ia-inducing activities from lymphokine preparations produced by mitogenic stimulation of normal splenic cells [43]. These studies do not rule out the possibility that other cytokines unrelated to interferon-γ can act as MAF and will be discussed later.

Torres and Johnson investigated the apparent one-signal activation of bone marrow-derived macrophages for P815 mastocytoma killing by treatment with LPS or by the synthetic double-stranded polyribonucleotide polyinosinic acid-polycytidylic acid (polyl: C) [52]. They found that "direct" activation by either LPS or polyl: C was still a two-signal process. Based on antibody neutralizations, the first signal was probably mediated by LPS or polyl: C induced α/β interferon in the macrophage cultures, and the second signal was that of a direct effect of LPS or polyl: C on the cell. Similarly, Schultz and Chirigos had earlier observed that antibodies against type I interferons neutralized that ability of L-cell interferon, *Brucella abortus* ether extract, *Salmonella typhimurium* LPS, and polyanions to activate macrophages [53]. Pace and associates showed that all three interferons were able to prime macrophages for tumor killing, but that interferon-γ is approximately 500 to 1000-fold more active than either interferon-α or -β [54]. Moreover, macrophages stimulated with combinations of interferon-β and interferon-γ were more sensitive to the trigger signal provided by LPS than were cells primed with either interferon alone [55]. Two classes of interferon-γ receptors have been demonstrated, and interferon-α and interferon-β selectively and competitively blockade the high-affinity interferon-γ receptor [56]. Celada and colleagues noted that interferon-γ internalization and receptor recycling are essential in the induction of nonspecific tumoricidal activity by macrophages [57, 57].

Bancroft and associates noted a T-cell-independent mechanism of macrophage activation by interferon-γ in mice with the severe combined immunodeficiency (scid) mutation [59, 60]. They utilized neutralizing monoclonal antibodies for murine interferon-γ to show that interferon-γ can be produced by scid mice in the absence of lymphocyte-mediated immunity, and that this interferon is important for macrophage activation during infection with *Listeria* [59]. Moreover, they showed that interferon-γ secretion *in vitro* was preceded by the rapid production of tumor necrosis factor (TNF) and was inhibited by addition of neutralizing monoclonal antibody to TNF [60].

Lemaire and co-workers demonstrated that macrophages can be activated by microbes or molecules arising from microbial degradation in the absence of lymphocytic products [61]. The ability to respond to such signals has a decisive evolutionary advantage. Trehalose dimycolate, a glycolipid extracted from *Mycobacterium tuberculosis,* has also been shown to prime macrophages so that they respond *in vitro* to low doses of LPS for antitumor activity [62]. Macrophages elicited by an injection of killed group C Streptococci, but not by thioglycollate broth, were also in a primed state. Primed macrophages were shown to express several biochemical markers previously described as specific for activated macrophages [63, 64] (e.g., low levels of β-galactosidase and alkaline phosphodiesterase) and, although they were not tumoricidal, they had the capacity to release large amounts of H_2O_2 [62].

Lambert and Paulnock utilized macrophage cell lines as a model system to dissect the biochemical and functional mosaic of the macrophage activation process [65]. They showed that a given homogeneous macrophage population expresses a distinct subset of functional activities in response to single, defined activating signals such as interferon-γ and LPS. Furthermore, these interferon-treated macrophage lines display a variety of macrophage surface antigens, enzymes, and secreted products, although a particular pattern of such activation-associated markers cannot reproducibly be used to predict the ability of an activated cell to perform a particular function. The authors suggested that macrophage cell lines expressing differential response patterns following interferon-γ stimulation provide a valuable system for dissection of the molecular and cell biology of macrophage activation. Lambert and Paulnock also showed that RAW 265.7, a cell line exhibiting a strict requirement for both interferon-γ

and LPS in the development of tumor cytolytic activity, acquires the primed stage after γ-irradiation [66]. The effects of γ-irradiation on the RAW 264.7. cell line are strikingly similar to those resulting from incubation with interferon-γ, with the exception that the irradiation-induced primed cell intermediate is stable and responsive to LPS triggering for at least 24 hr.

2.2 Agents that function as "second signals"

Hibbs and associates proposed that MAF provided a "priming" signal to the macrophage, which alone did not render the macrophage tumoricidal, but lowered the tumoricidal threshold in response to bacterial LPS [2]. A variety of "second signals" other than LPS trigger MAF- or interferon-γ-primed macrophages to develop full functional activity: These include muramyl dipeptide [24, 67, 68], a bacterial cell wall skeleton of bacillus Calmette-Guerin [68], heat-killed *Listeria monocytogenes* [36], a synthetic acyltripeptide (FK-565) [69], and poly(I) · poly(C) [70]. It should be pointed out that many of these agents are active in the absence of lymphokine for inducing tumor cytostatic properties of macrophages [reviewed in Ref. 3, 35]. Antibody neutralization studies have suggested that macrophage-derived interferons function as endogenous second signals in these systems [52, 53]. Studies have suggested that type 1 (nonimmune) interferons can synergize with interferon-γ for inducing cytotoxic properties of macrophages [55, 71]. Similarly, type I and type II murine interferons have been demonstrated to potentiate each other with respect to in vitro antiviral activity [72].

Nagao and co-workers examined macrophage activation by recombinant murine interferon-γ and a synthetic MDP derivative [73]. They observed that interferon-γ-primed macrophages display augmented binding of the MDP derivative to macrophages resulting in synergistic macrophage activation. This increase in binding resulted from enhanced endocytosis rather than the appearance of specific binding sites for the MDP derivative. Akagawa and Tokunaga reported that the number of LPS binding sites on murine lung macrophages was increased by interferon-γ, resulting in recovery of macrophage responsiveness to activation by LPS [74]. It is presently unclear whether the major role of the priming stimulus is to increase cellular availability to the triggering agent. In contrast, Marino and Adams

presented evidence that the augmented capacity for binding tumor cells is induced by lymphokines and that a major part of induction of priming for cytolysis by MAF is induction of such binding [75]. Encapsulation of various macrophage activators, including MDP [76–78], FK-506 [79], and poly(I) · poly(C) [80] in liposomes markedly increases their potency *in vitro*. Control studies have suggested that participation of macrophage cell surface receptors is not required, and that activation results from interaction of these agents with intracellular targets [78]. Moreover, their efficacy at activating macrophages could be further enhanced by addition of suboptimal levels of MAF within the same liposomes [77, 80].
Meltzer and colleagues suggested that an LPS-like signal was present in lymphokine supernatants [27]. They demonstrated that macrophages exposed to low concentrations of lymphokine enter into a receptive state in which they can be triggered by high concentrations of lymphokine. The latter activity was not due to contaminating endotoxin in the lymphokine preparation. Krammer and associates [81–83] described a new lymphokine activity, macrophage cytotoxicity inducing factor 2 (MCIF2), in the T cell mitogen-induced supernatant of a murine T cell clone in long-term culture. The factor eluted from a Sephadex G-100 column in three m. w. forms (10, 34, and 100 KD), it was acid labile (ph 2 to 4) and heat sensitive (80 min at 56 C), and was not constitutively secreted. This factor coexisted with interferon-γ in the same supernatant and synergized with interferon-γ for induction of tumoricidal and schistosomulicidal activities in resident murine peritoneal macrophages.

2.3 Cytokines other than interferon that activate macrophages

Several investigators have identified a lymphokine with MAF activity distinct from interferon-γ produced by human T-cell leukemia virus-positive cell lines [84–86]. This factor was not neutralized by monoclonal antibodies to interferon-γ and was resistant to heat and acid treatments which abolished antiviral activity. The factor was produced constitutively and activated macrophages in the absence of detectable LPS.
Meltzer and associates have characterized MAF released by phorbol myristate acetate (PMA)-treated EL-4, a murine T cell line [87, 88]. They demonstrated two separate factors that activated macrophage

for cytotoxicity: one was indistinguishable from interferon-γ by a variety of criteria and the other was clearly distinct (no antiviral activity, 23,000 m. w., pH 2 stable and unaffected by anti-interferon-γ-serum) [87]. Sutton and co-workers demonstrated that this latter MAF activity was due to B cell stimulatory factor-1 (interleukin 4) in the EL-4 conditioned media [89]. Crawford and associates showed that thioglycollate-elicited macrophages express about 2300 high affinity (Ka $\sim 2 \times 10^{10} M^{-1}$) IL-4 receptors/cell [90]. Macrophage tumoricidal activity induced by IL-4 but not by interferon-γ was inhibited > 90 % with monoclonal anti-IL-4 antibody. Somers and Erickson demonstrated that IL-4 can activate macrophages, but the state of activation differs from that of interferon-γ [91]. They showed that IL-4 appears to selectively prime macrophages for TNF production without initiating other tumoricidal mechanisms.

Several studies have suggested that granulocyte/macrophage colony stimulating factor (GM-CSF) can activate macrophages [92–95]. Grabstein and co-workers showed that recombinant human GM-CSF activated human peripheral blood monocytes to become tumoricidal against the human melanoma cell line A375 [92]. In contrast to interferon-γ, no exogenous signal like LPS was required for the activation of monocytes by GM-CSF. GM-CSF also inhibited replication of *Trypanosoma cruzi* in cultured peritoneal macrophages [93, 94], but was much less potent than interferon-γ [94]. Wing and colleagues demonstrated that recombinant human GM-CSF enhanced monocyte cytotoxicity and tumor necrosis factor-α (TNF-α) secretion in cancer patients [95]. Other investigators noted that GM-CSF alone was incapable of stimulating TNF-α release, but primed macrophages for elevated TNF-α production in response to interferon-γ [96, 97]. Cannistra and co-workers suggested that human GM-CSF and interleukin 3 stimulate macrophage killing indirectly by enhancing expression of TNF mRNA, thereby leading to augmented TNF protein secretion in response to a second activation signal [98].

Although Talmadge and colleagues demonstrated that TNF alone activated macrophages for cytotoxicity [99], most reports show that TNF acts synergistically with interferon-γ to induce macrophage killing of tumor cells [100–105]. The ability of macrophages to release TNF upon stimulation with bacterial endotoxins or viral infection suggests that TNF could act in an autocrine manner [100, 101, 106]. Other studies by Chen and co-workers suggest that interferon-γ synergizes with TNF and with IL-1 and requires the presence of both monokines to induce antitumor cytotoxic activity in macrophages [103].

3 Molecular mechanisms of macrophage activation

As discussed earlier, macrophage activation for tumor cytotoxicity is acquired in a complex, multistep fashion by exposure of macrophages to a priming signal such as interferon-γ, followed by a second signal such as LPS [22–25, 27]. This induction of competence to destroy tumor cells is stringently controlled by dose, sequence, and timing requirements with which these signals must be applied. Adams and associates have characterized macrophage activation as a selective and stringently controlled set of alterations in various physiologic capacities and properties, which culminates in the development of enhanced potential for executing a complex function [107, 108]. The biochemical mechanisms by which macrophages become activated to the tumoricidal state are poorly understood.

3.1 Signal transduction and interferon-γ

Since macrophages require 2–4 hr of exposure to interferon-γ for priming [39], investigators have focused on this period of time in attempts to identify changes generated by receptor occupancy. Adams and co-workers have identified two biochemical changes in macrophages which fit this criterion and are initiated by active concentrations of ligand [reviewed in Ref. 107–109]. First, interferon-γ was demonstrated to induce a rise in intracellular levels of Ca^{++}, manifested by enhanced efflux of radiolabeled calcium from preloaded cells [110]. This efflux of Ca^{++} initiated by physiologic concentrations of interferon is first seen 5–10 min. after initiation of treatment and is complete by 10–20 min. Second, interferon-γ enhances the potential activity of protein kinase C (PK-C) [111]. This change was not induced by interferon-α, interferon-β, CSF-1, LPS or a variety of other macrophage stimulatory signals. It is seen as early as 1 hr after interferon-γ treatment, peaks at 3–4 hr and declines by 6–12 hr. The change does not apparently result from an increased number of enzyme molecules, but rather from an increase in the catalytic efficiency of existing enzymes. Enzyme from interferon-γ-treated macrophages phosphorylates substrate protein 2–4 times more rapidly than enzyme from control cells after stimulation with diacylglycerol [112].

To investigate the role of calcium in this process, several investigators have evaluated the effect of calcium channel blockers and calmodulin antagonists on the acquisition of macrophage tumoricidal properties [113–115]. Wright and co-workers showed that activation of thioglycollate-stimulated C57Bl/6 mouse peritoneal macrophages by combinations of macrophage activating factor or interferon-γ plus LPS was inhibited in a dose-dependent fashion by the calcium channel blockers nifedipine and verapamil [113]. These agents also blocked the influx of ^{45}Ca into macrophages activated by MAF plus LPS. Macrophage activation was also inhibited by chlorpromazine, W-7, and calmidazolium at concentrations known to perturb calmodulin function. Gorecka-Tisera and associates determined the concentration of cytosolic-free calcium with aequorin in RAW-264 macrophages activated *in vitro* with lymphokine and LPS [115]. Treatments of these cells with optimal doses of stimulants, which evoked the development of cytolytic activity, also induced a rise in Ca^{++}. Cytolytic activity in RAW-264 macrophages was inhibited by two intracellular calcium antagonists, TMP-8 and ruthenium red. This inhibition could be reversed by high concentrations of extracellular calcium. Mecham *et al.* demonstrated that binding of calmodulin to the microfilament correlates with induction of tumoricidal responses in mouse peritoneal macrophages [116].

Over the past few years, much attention has been given to a signal transduction pathway that involves a calcium ion (Ca^{++}) and phospholipid-dependent protein kinase known as protein kinase C. In this pathway, the exposure of cells to certain receptor-mediated stimuli initiates a series of molecular events that ultimately lead to the phospholipase-dependent cleavage of phosphotidylinositol 4,5-biphosphate into inositol-1,4,5-triphosphate and 1,2-diacylglycerol (DAG) [117]. IP_3 increases cytolosic Ca^{++} levels, and in the presence of both Ca^{++} and phospholipid, DAG directly activates protein kinase C. Several investigators have utilized agents that mimic the action of certain second messengers to explore the biochemical transductional events initiated by interferon-γ. Somers and co-workers observed that phorbol myristate acetate (PMA) and the ionophore A23187 cooperatively reproduced the ability of interferon-γ to prime macrophages for tumoricidal function [110]. Neither agent alone was able to prime macrophages. Control experiments showed that these agents acted on macrophages, and target susceptibility to destruction by macrophages was not altered by PMA and A23187. Only active phorbol esters, which are known to stimulate protein kinase C, were able to cooperate with A23187 to induce priming. They also observed that Quin-2/AM, an intracellular chelator of Ca^{++}, blocked priming by interferon-γ. In summary, their data further supports the concept that priming of macrophage for cytotoxic function involves, at least in part, alterations in protein kinase C and in levels of intracellular Ca^{++}. Celada and Schreiber provided additional experimental evidence for the hypothesis that changes in intracellular Ca^{++} and activation of the calcium ion- and phospholipid-dependent protein kinase C were required for macrophage activation by interferon-γ [118]. They showed that phenothiazines and R24571, known antagonists of calcium-binding proteins and therefore nonspecific inhibitors of protein kinase C, block in a dose-dependent manner the induction of macrophage cytocidal activity. They reproduced the findings of Somers and colleagues [110] that macrophages exposed to both active phorbol esters and Ca^{++} ionophores were elevated into the primed activation state such that in the presence of LPS or heat killed *Listeria monocytogenes* the cells were triggered to express full levels of tumoricidal activity. Johnson and Torres showed that A23187 mimics interferon-γ in the process of evoking cytolytic responses in mouse bone marrow macrophages [119] Similarly, Drysdale and associates showed that A23187 activated macrophages for tumor cell killing, and the activated cells produce a soluble cytotoxic factor that is similar, if not identical, to tumor necrosis factor [120].

Radzioch and Varesio observed that the protein kinase C inhibitors (H-7 and retinal) block the activation of macrophages by interferon-β but not by interferon-γ [121]. The effectiveness of both drugs in inhibiting PK-C in intact macrophages was evaluated by measuring the inhibition of induction of c-fos mRNA by L-α-1-oleoyl-2-acetoyl-sn-3-glycerol, a process that has been shown to be dependent on PK-C activation. Their results suggest that different pathways of activation are triggered by interferon-γ and interferon-β, the former being independent from and the latter dependent on intact PK-C activity. The possibility that two different pathways are involved in the activation of macrophages is further supported by Blasi *et al.* [122], who noted the requirement for protein synthesis for induction of macrophage tumoricidal activity by interferon-α and interferon-β, but not by interferon-γ.

Among the late events that follow the signal transduction are changes in ribosomal RNA metabolism, a common feature of macrophages activated *in vitro* and *in vivo* [123–126]. Upon treatment with interferon-γ, cytotoxic macrophage display a decrease in total RNA synthesis [124, 125] and an altered processing of the ribosomal RNA, leading to an increase in the steady state levels of ribosomal RNA precursors and imbalanced maturation of 28 S and 18 S ribosomal RNA [123, 126]. These results indicate that posttranscriptional modification of RNA metabolism occurs in activated macrophage, which may be important for the expression of cytotoxicity. Varesio proposed that dsRNA-dependent enzymes, that are induced by interferons, are involved in the expression of cytotoxic activity by macrophages [123]. They suggested by computer analysis that rRNA precursors can fold in secondary structures containing very stable dsRNA regions. Thus, Varesio speculated that these rRNA precursors could be a source of dsRNA for the activation of dsRNA-dependent enzymes, and the accumulation of rRNA precursors in activated macrophages would provide adequate amounts of dsRNA.

3.2 Signal transduction and LPS or other second signals

Uhing and Adams [108] presented evidence that LPS utilizes a variety of signal events in mediating its diverse effects on macrophages. It induces production of inositol trisphosphate and diacylglycerol and mobilization of intracellular calcium in macrophages [127]. Drysdale and colleagues suggested that macrophages could be activated by LPS without influx of extracellular calcium or redistribution of intracellular calcium [120]. LPS also causes an initial cytosolic acidification in interferon-γ-treated macrophages, followed by alkalinization [128]. In addition, LPS initiates a characteristic pattern of heightened phosphorylation of 5 distinct proteins (i. e., pp28, pp33, pp38, pp67, and pp103) in murine peritoneal macrophages [129]. Their data strongly indicate that LPS induces phosphorylation mediated by protein kinase C. When macrophages are pulse-labeled with ^{35}S-methionine, at least 7 newly synthesized proteins (i. e., 85, 80, 75, 65, 57, 42, and 38 kD) are detected on one-dimensional gels;

many of which are detected on two-dimensional gels [130, 131]. The proteins are seen within 30 min after the macrophages are pulsed with LPS, and their synthesis stops within 2–6 hr after LPS is applied even in the continuous presence of stimulus. Due to their early appearance, brief synthesis, and short half-lives, Adams and Hamilton [107] suggested that these proteins provide regulatory function, rather than mediate effector functions in macrophages. Aderem demonstrated that LPS promotes the covalent attachment of myristic acid to a set of macrophage proteins [132]. One of the myristoylated proteins is a 68K protein kinase C substrate which associates with membranes upon myristoylation. LPS was also demonstrated to activate expression of other early genes in macrophages by "Northern" analyses [107, 133], causing 5- to 10-fold rises in the mRNA for several "competence" genes (i. e., c-fos, c-myc, JE, and KC). Okai showed that RNA synthesis of macrophages was considerably increased early after LPS stimulation (0–3 hr), but drastically decreased by 20 hr of treatment [125]. He suggested that such modulation of RNA synthesis was regulated by some factors affecting RNA polymerase reactions.

4 Factors regulating macrophage activation

Studies using subcutaneous inflammatory lesions [134] or peritoneal exudates [26] have indicated that macrophages are much more susceptible to activation signals during the acute phase of inflammation. The onset of resistance to activation in these studies coincided with a marked decline in the number of peroxidase-positive macrophages. The accumulation of peroxidase-positive macrophages in inflammatory lesions results from immigration of young, peroxidase positive blood monocytes from the circulation [26, 135]. These studies suggest that maintenance of tumoricidal activity requires continuous influx of new lymphokine-responsive blood-derived macrophages into the inflammatory site. Poste and Kirsh showed that changes in macrophage function *in vivo* result from events that operate only at a local level and do not affect the ability of macrophages to be recruited and activated in inflammatory lesions elsewhere in the same animal [134]. The expression of tumoricidal activity by activated macrophages is similarly short-lived and irreversible, since previously activated cells cannot be reactivated by lymphokines [134]. Numerous studies have indicated that the cytotoxic activity of macrophages is rapidly lost *in vitro,* regardless of the source of macrophages or the methods used to activate the cells [136, 137]. The mechanisms responsible for the decay of tumoricidal activity and subsequent resistance to further lymphokine stimulation are not well understood. Schultz and colleagues demonstrated that E-type prostaglandins (PGE) inhibit macrophage tumoricidal activity and suggested that endogenously-produced PGE act in negative feedback mechanisms to intrinsically modify macrophage functional activity [138–140]. In support for this concept, Taffet and Russell showed that treatment of LPS-stimulated macrophages with 10^{-6}M indomethacin (or other cyclooxygenase inhibitors) prevented both PGE synthesis and the shut-off of macrophage cytolytic activity [136]. Chronic administration of indomethacin in mice has also been demonstrated to prolong macrophage activation *in vivo* induced by the interferon inducer, maleic anhydride-divinyl ether copolymer [141].

Several reports show that interferons inhibit prostaglandin synthesis in macrophages [142–144]. Dore-Duffy and co-workers demonstrated that both leukocyte- and fibroblast-derived interferon were potent inhibitors of monocyte prostaglandin synthesis at low physiological concentrations [142]. Boraschi *et al.* presented evidence suggesting that interferon-β decreased prostaglandin production in macrophages by blocking phospholipase activation [143]. Russell and Pace reported that gamma interferon changes the sensitivity of activated macrophages to the negative regulatory effects of PGE, thereby maintaining activation, as manifested by the continued expression of tumor cell killing by these cells [145, 146]. Thus, interferons are important to host defense not only because they participate in the induction of macrophage activation for tumoricidal activity but also because they help to maintain the activated state once it has been induced.

Aside from PGE, the cytotoxic action of activated macrophages appears to be regulated by a number of other inhibitory molecules in their local environment. Naturally occurring agents, such as a serum component of high molecular weight that is isolated with a low-density lipoprotein [147], glucocorticoid hormones [148, 149, 150], estrogen metabolites [151], proteinases [21], and agents leading to intracellular cyclic AMP accumulation [152], have been shown to suppress macrophage cytotoxicity against tumor cells. Similarly, macrophage cholesterol content serves as a negative regulator in macrophage tumor cytotoxicity [153, 154]. Wright and co-workers showed that cyclosporin-A can directly modulate the acquisition and expression of macrophage tumoricidal properties [155]. The identification of these varied inhibitors raises the possibility that host resistance against cancer may be enhanced by controlling their levels in the microenvironment of the tumor. Pharmacological doses of human interferon-γ has been shown to markedly enhance plasma cortisol levels in patients with advanced metastatic cancer, further demonstrating the complexities involved in enhancing macrophage functional activity *in vivo* [156].

Tumor promoters and products of neoplastic cells have also been demonstrated to interfere with macrophage function. Keller and associates reported that phorbol esters and other tumor promoters of the polyfunctional diterpene ester type covering tigliane, ingenane, and daphnane derivatives, were effective in preventing the lymphokine-induced enhancement of cytolytic activity in resting macrophages (157–159). These findings support the concept that tumor promoters, apart from directly stimulating the outgrowth of transformed cells, may function via interference with natural antitumor effector systems. Schultz demonstrated that macrophages treated with phorbol 12-myristate 13-acetate, a potent trigger for PGE_2 release, were refractory to activation by LPS unless PGE_2 production was blocked by indomethacin (70).

5 Monokines as effector molecules of tumor cytotoxicity

The mechanism of macrophage cytotoxic activity depends to a certain extent on the assay system and particular target cell used. For example, monokines like TNF or IL-1 cause lysis or growth inhibition of tumor target cells over 24–72 h [160–162], whereas macrophage-derived H_2O_2 lyses sensitive targets rapidly over a period of hours [163, 164]. A better understanding of the nature of cytotoxic reactions at a molecular and biochemical level helps to explain discrepancies observed between one system and another.

Several investigators demonstrated that tumor necrosis factor (TNF) functions as an effector molecule in macrophage-mediated cytotoxicity [165–169]. Urban and co-workers showed that tumor cells susceptible to activated macrophages were killed by recombinant TNF-α, whereas variant tumor cells resistant to activated macrophages after selection *in vitro* or *in vivo* were resistant to killing by TNF-α [165]. These macrophage-resistant variants were also resistant to the effects of another cytotoxic protein, B-cell lymphotoxin, which is structurally related to TNF-α. Similar results were obtained in their studies regardless of whether they measured short-term or long-term cytotoxic effects of activated macrophages. Similarly, Feinman and associates observed a good correlation between the sensitivity of A673 human rhabdomyosarcoma clones to the cytotoxicity of TNF and their susceptibility to killing by monocytes [166]. They showed that a TNF-specific neutralizing monoclonal antibody reduced monocyte killing of parental A673 cells and of a TNF-sensitive clone of A673 cells. Decker and colleagues investigated different macrophage populations for their ability to secrete TNF and to lyse TNF-susceptible tumor cells [167]. They demonstrated that TNF secretion, although a feature of all activated macrophage populations, was not an absolute requirement for the killing of the TNF-sensitive Wehi 164 target. TNF-dependent lysis of tumor cells could be performed by activated macrophages that had been fixed with paraformaldehyde before the addition of target cells. Macrophage cytotoxicity against this cell but not against the TNF-resistant P815 mastocytoma, was completely inhibitable by a specific anti-TNF serum.

The mechanism of the cytotoxic action of TNF has been the focus of intense investigation. Cells that are sensitive to TNF bear receptors for TNF, although the number of receptors do not correlate with sensitivity [170]. Interferon-γ has been demonstrated to enhance TNF cytotoxicity by increasing the number of receptors on target cells [171, 172]. Kull and Cuatrecasas demonstrated a possible requirement for internalization in the mechanism of in vitro cytotoxicity of TNF by showing that inhibitors of lysosomal activity and cytoskeletal-disrupting agents all depressed sensitivity [173]. Haranaka and co-workers suggested that TNF exerts its effect by enhancing endogenous tumor lysosomal activity and showed that the increment in cellular respiration paralleled the susceptibility to TNF [174]. Similarly, Watanabe and co-workers showed that the cytotoxicity of TNF on L-M cells, a line of tumorigenic mouse fibroblasts, was clearly reduced by lysosomotropic agents and hydroxyl radical scavengers [175]. Aggarwal and associates reported that two inhibitors of ADP-ribosylation, namely 3-aminobenzamide and nicotinamide, prevented TNF-mediated cytotoxicity of L929 murine fibrolasts [176, 177]. The possibility that protein synthesis might be involved in target resistance is suggested by observations that target cells treated with inhibitors of protein or mRNA synthesis are rendered more susceptible to the cytotoxic action of TNF [160, 178].

Activated macrophages are also good producers of IL-1 [179, 180]. IL-1 has been shown to be cytostatic [181–183] and cytolytic [184–186] for several tumor cell lines *in vitro*, and it has thereby been suggested that IL-1 is involved in macrophage-mediated cytotoxicity. Onozaki and co-workers were the first to report that highly purified, monocyte-derived IL-1β was directly cytocidal for A375 melanoma cells [184]. Lachman *et al.* confirmed these findings by showing that homogeneous recombinant human IL-1, but not recombinant TNF, was cytotoxic for A375 cells [185]. Endo and colleagues found that only A375 clones that possess specific receptors for IL-1 were sensitive to the cytotoxic action of IL-1 [181]. Inamura *et al.* presented evidence that membrane-associated IL-1α, expressed on human activated monocytes, could explain at least one of the mechanisms of monocyte-mediated cytotoxicity [180].

The mechanism(s) by which IL-1 mediates cytotoxicity is not well understood. Endo and coworkers showed that recombinant human IL-1 inhibited the proliferation of A375 human melanoma cells at concentrations that exhibited inhibitory effects on ornithine decarboxylase (ODC) activity [181]. Moreover, they found that putrescine, a nontoxic product of the ODC pathway, could prevent the cytostatic effect of IL-1 on these tumor target cells. This observation indicated that inhibition of the ODC pathway is causally related to the antiproliferative action of IL-1 on A375 cells. Using a simple and sensitive bioassay for IL-1β-mediated A375.6 melanoma cytolysis, Schultz *et al.* sought to delineate intracellular biochemical events involved in IL-1 action [186]. Their data was consistent with other findings that phospholipase A_2 activation [179, 187, 188] and enhancement of protein kinase C activity [189, 190] may explain the cytolytic activity of IL-1 and TNF as well as many of the pleiotropic biologic effects of these monokines.

Other secretory products of activated macrophages, including reactive oxygen intermediates [191], neutral proteases [192, 193], thymidine [194], arginase [195], $C3_a$ [196], and interferon-α/β [197, 198] may adversely affect the growth and viability of cells *in vitro*. The target cell spectrum of most of these molecules, including TNF and IL-1, is much more restricted than the large array of neoplastic cells susceptible to macrophage-mediated cytotoxicity. Evidence for the involvement of these agents as effectors of macrophage cytotoxicity has been reviewed elsewhere [3, 35, 117] and will not be covered here.

6 Therapeutic considerations

In this review, I have attempted to show that macrophage activation is a complex cascade of cellular alterations that is controlled by an elaborate web of interacting microenvironmental signals. Several factors influence both the macrophage concentration as well as the functional state of macrophages in and around the tumor site. As discussed earlier, macrophage functional activity is largely dependent on the balance between opposing chemical signals which determine the extent of macrophage activation in the microenvironment of the tumor. In addition, the macrophage functional state is also dependent on the degree of maturity of resident macrophages in the tumor [26, 134] and genetic factors which determine the responsiveness to activation stimuli [199].

6.1 Macrophage tumor content

A variety of factors also influence the macrophage concentration in and around the tumor site. Pike and Snyderman demonstrated that certain tumors produce soluble factors that interfere with macrophage chemotaxis [200]. The macrophage content of tumors appears to be closely associated with therapeutic efficacy of agents that enhance nonspecific host resistance to neoplasia (e. g., glucan [201], BCG [202], *Brucella abortus* ether extract [203], and pyran copolymer [204]. Evans observed that necrotic tumors do not show a higher level of macrophages than healthy tumors, suggesting that the ingestion of dead cells itself is not an adequate explanation for the macrophage infiltration [205]. Eccles and Alexander showed that the macrophage tumor content is inversely related to the rate of spontaneous metastasis [206]. Similarly, Wood and Gillespie increased the metastatic potential of murine chemically-induced fibrosarcomas by depleting macrophages from the tumor cell suspensions [207]. By contrast, control animals that received tumor cell suspensions containing their

normal complement of macrophages invariably developed progressive localized tumors. Furthermore, a variety of antimacrophage agents including silica, carrageenan, and 2-chloroadenosine promote tumor growth and metastasis formation *in vivo* [69, 208, 209]. Collectively, these studies strongly suggest that tumor-associated macrophages modulate tumor growth and metastasis.
Since intimate cell contact appears to be an absolute requirement for the destruction of target cells by activated macrophages [210], successful therapy is dependent on achieving adequate cell density at the tumor site. Eccles and Alexander suggested that macrophages enter the tumor as part of a specific reaction of the host directed against tumor-associated antigens [206]. They showed that the macrophage content of tumors is directly related to immunogenicity using a variety of sarcomas. Evidence for the effectiveness of activated macrophages in controlling tumor metastasis *in vivo* has been obtained in adoptive transfer experiments [211–213]. In these experiments, the intravenous injection of *in vitro*-activated macrophages into syngeneic mice bearing lymph node and lung metastases produced a significant reduction in the metastatic tumor burden. Piessens demonstrated that macrophages activated *in vitro* by lymphocyte mediators inhibited local tumor growth *in vivo* [214].

6.2 Marcrophage functional state within tumor

The mere presence of macrophages within tumors is not sufficient to bring about tumor regression. Russell and associates showed that macrophages isolated from regressing Moloney sarcomas are more cytotoxic than those recovered from progressing sarcomas [215, 216]. These nontumoricidal macrophages isolated from progressively growing Moloney sarcomas became tumoricidal in the presence of picogram or nanogram quantities of endotoxin [217], suggesting modulation of an activated macrophage tumoricidal threshold *in vivo* that may determine whether or not tumor rejection occurs. Several other investigators have shown that macrophages harvested directly from tumor tissue are cytotoxic for tumor cells *in vitro* [217–219].
The presence of cytotoxic activated macrophages in tissue tends to be a local phenomenon, confined to the anatomical compartment that contains high levels of inducing agent or antigen [210]. Macrophage activation is a short-lived phenomenon and requires the con-

tinuous influx of new peroxidase-positive macrophages from the circulation for optimal therapeutic effect. Murine peritoneal macrophages exposed to lymphokine preparations demonstrate tumoricidal activity after as little as four hours, but lose this capacity by 24 hours, although they remain fully viable. Macrophage activation appears to be self-limiting due to inhibitory molecules, such as E-type prostaglandins produced by the activated cells [136, 138–140]. The ability to control local production of agents that negate macrophage chemotaxis and functional activity may allow for improved therapy of cancer utilizing macrophage activating agents.

Fidler and co-workers have shown that systemic administration of lymphokines [220] or muramyl dipeptide [221] encapsulated in liposomes leads to the activation of tumoricidal properties in macrophages and is associated with destruction of established pulmonary and lymph node metastases originating from a s. c. melanoma. These liposome preparations have been shown to localize mainly in reticuloendothelial cells *in vivo* [222]. Dose-response measurements indicated that liposome-encapsulated preparations of these activators induce maximum levels of macrophage activation at significantly lower doses than that needed for equivalent activation by nonencapsulated (free) preparations of these materials [reviewed in Ref. 222]. Fidler and colleagues noted that negatively charged multilamellar (MLV) liposomes (phosphatidylcholine: phosphatidylserine, 3:7 mol ratio) localize efficiently in the lung and that macrophage-activating agents encapsulated within such liposomes can successfully activate lung macrophages *in situ*. Therapeutic agents encapsulated within liposomes can thus be targeted, albeit passively, to macrophages, allowing a reduction in the amount of drug needed to activate macrophages, and thereby reducing the likelihood of toxicity problems arising from action on cells other than macrophages. The evidence that macrophages were involved in the antimetastatic effect of these liposomes included observations that 1) antimacrophage agents inhibit the therapeutic effect and 2) macrophages activated *in vitro* by liposome-encapsulated MDP and then injected i. v. into mice bearing experimental lung metastases also significantly inhibit lung metastases [223]. It is interesting to note that lymphokines encapsulated within liposomes can induce activation of intratumoral macrophages that are completely refractory to activation by free lymphokines [12]. Moreover, studies on the mechanism of activation by liposome-encapsulated activators indicate that participation of macrophage cell surface receptors is not required, suggesting that activation results from interaction of interferon-γ or MDP with intracellular receptors [78, 224].

7 Concluding remarks

The activated macrophage appears to be ideally suited as a primary antineoplastic surveillance cell for a variety of reasons [reviewed in Ref. 2]. Although the activated macrophage tumoricidal reaction is nonspecific, an elemental sort of discrimination exists in that normal cells are not harmed. The activated macrophage tumoricidal reaction may have a primitive phylogenetic origin, since macrophages are activated by bacterial agents and cellular products in the complete absence of immunoglobulins and lymphocytes. Furthermore, it is clear that macrophages, themselves, can intrinsically modify their functional activity through the elaboration of various secretory products, including α/β interferons and E-type prostaglandins. Macrophages are widely distributed in tissues, and the macrophage plasma membrane is a highly sophisticated sensory organelle capable of detecting subtle changes in its environment.

Over the past decade, much has been learned about the inducers and complex biochemical events that underlie the effector functions of mononuclear phagocytes. Interferon-γ has emerged as a major MAF in enhancing macrophage antitumor and antimicrobial effector function. The potential application of macrophage activation in cancer therapy has been further supported by studies demonstrating antitumor effects against established pulmonary and lymph node metastases utilizing liposome-encapsulated macrophage activators [220, 221].

Several studies have demonstrated potential advantages to using the tumoricidal properties of activated macrophages in clinical cancer therapy. Macrophage-mediated cytolysis of tumorigenic cells, at least *in vitro,* occurs independently of such tumor cell characteristics as antigenicity, invasiveness, drug sensitivity, and metastatic potential [12, 225–229]. Thus, activated macrophages are able to kill cells that are resistant to killing by other components of the host defense system and various anticancer drugs. The major limitation of this strategy for cancer therapy, as with any other modality, is the size of the tumor burden. It has been stressed that the potential therapeutic potential of macrophage activation will not be the destruction of massive tumor burdens but rather the destruction of micrometastases and the residual tumor cell burden that remain after elimination of the majority of tumor cells by other means such as surgery, chemotherapy, and radiation [1, 12, 230]. It is with that caveat that we await the exciting clinical results utilizing novel approaches to macrophage activation.

References

1 S. D. Deodhar and B. P. Barna: Cleve. Clin. Q. *53,* 223 (1986).
2 J. B. Hibbs, H. A. Chapman, Jr. and J. B. Weinberg: J. Reticuloendothel. Soc. *24,* 549 (1978).
3 R. M. Schultz and M. A. Chirigos: Advances in Pharmacology and Chemotherapy, vol. 17, p. 157. Academic Press, Inc., New York 1980.
4 J. B. Hibbs: J. Natl. Cancer Inst. *53,* 1487 (1974).
5 I. J. Fidler: Isr. J. Med. Sci. *14,* 177 (1978).
6 D. Juy and L. Chedid: Proc. Natl. Acad. Sci. USA, *72,* 4105 (1975).
7 M. Bruley-Rosset, I. Florentin and G. Mathe: Agents Action *6,* 251 (1976).
8 R. M. Schultz, J. D. Papamatheakis, J. Luetzeler, and M. A. Chirigos: Cancer Res. *37,* 3338 (1977).
9 I. J. Fidler: Cancer Res. *34,* 1074 (1974).
10 L. A. Liotta, G. Gattozzi, J. Kleinerman and G. Saidel: Br. J. Cancer *36,* 369 (1977).
11 R. M. Schultz and M. G. Altom: J. Immunopharmacol. *5,* 277 (1983).
12 I. J. Fidler and G. Poste: Springer Semin. Immunopathol. *5,* 161 (1982).
13 G. B. Mackaness: J. Exp. Med. *120,* 105 (1964).
14 E. Metchnikoff: Immunity in Infectious Diseases, p. 324. Johnson Reprint Corporation, New York 1968.
15 G. B. Mackaness: J. Exp. Med. *129,* 973 (1969).
16 F. C. Lane and E. R. Unanue: J. Exp. Med. *135,* 1104 (1972).
17 M. L. Lohmann-Matthes, F. G. Ziegler and M. Fischer: Eur. J. Immunol. *3,* 56 (1973).
18 R. E. Fowley, I. M. Fajardo, J. L. Leibowitch, and J. R. David: J. Exp. Med. *138,* 952 (1973).
19 R. Evans and P. Alexander: Nature New Biol. *236,* 168 (1972).
20 J. B. Hibbs, L. H. Lambert and J. S. Remington: Nature New Biol. *235,* 48 (1972).
21 J. B. Hibbs, R. Taintor, H. A. Chapman Jr., and J. Weinberg: Science *197,* 279 (1977).
22 L. P. Ruco and M. S. Meltzer: J. Immunol. *121,* 2035 (1978).
23 M. S. Meltzer, Occhionero, M. and L. P. Ruco: Fedn. Proc. *41,* 2198 (1982).
24 S. Sone and I. J. Fidler: J. Immunol. *125,* 2454 (1980).
25 J L. Pace and S. W. Russell: J. Immunol. *126,* 1863 (1981).
26 L. P. Ruco and M. S. Meltzer: J. Immunol. *120,* 1054 (1978).
27 M. S. Meltzer, L. P. Ruco, J. K. Lazdins, and E. J. Leonard: Mediation of Cellular Immunity in Cancer by Immune Modifiers, p. 65. Raven Press, New York 1981.

28 R. M. Schultz, and W. J. Kleinschmidt: Nature *305,* 239 (1983).
29 J. L. Pace, S. W. Russell, B. A. Torres, H. M. Johnson, and P. W. Gray: J. Immunol. *130,* 2011 (1983).
30 R. D. Schreiber: Contemp. Top. Immunbiol. *13,* 171 (1984).
31 P. Alexander, and R. Evans: Nature (London) New Biol. *232,* 76 (1971).
32 R. M. Schultz, J. D. Papamatheakis, and M. A. Chirigos: Cell. Immunol. *29,* 403 (1977).
33 R. M. Schultz, J. D. Papamatheakis, and M. A. Chirigos: Science *197,* 674 (1977).
34 R. M. Schultz, M. A. Chirigos, and U. I. Heine: Cell. Immunol. *35*: 84 (1978).
35 R. M. Schultz: Lymphokine Reports *1,* 63 (1980).
36 S. W. Russell, J. L. Pace, L. Varesio, E. Blasi, A. Celada, R. D. Schreiber, R. M. Schultz, A. P. Stevenson, C. C. Stewart, and S. J. Stewart: J. Leukocyte Biol. *40,* 801 (1986).
37 J. L. Pace, L. Varesio, S. W. Russell, and E. Blasi: J. Leukocyte Biol. *37,* 475 (1985).
38 W. J. Johnson, P. A. Marino, R. D. Schreiber, and D. O. Adams: J. Immunol. *131,* 1038 (1983).
39 D. O. Adams, and T. A. Hamilton: Annu. Rev. Immunol. *2,* 283 (1984).
40 S. M. Shigi, and R. Mishell: J. Immunol. *115,* 741 (1975).
41 D. Taramelli, H. T. Holden, and L. Varesio: J. Immunol. Meth. *37,* 225 (1980).
42 J. S. Youngner, and S. B. Salvin: J. Immunol. *111,* 1914 (1973).
43 R. D. Schreiber, L. J. Hicks, A. Celada, N. A. Buchmeier, and P. W. Gray: J. Immunol. *134,* 1609 (1985).
44 C. F. Nathan, H. W. Murray, M. E. Wiebe, and B. Y. Rubin: J. Exp. Med. *158,* 670 (1983).
45 W. J. Kleinschmidt, and R. M. Schultz: J. Interferon Res. *2,* 291 (1982).
46 W. K. Roberts, and A. Vasil: J. Interferon Res. *2,* 519 (1982).
47 A. Kelso, A. L. Glasebrook, O. Kanagawa, and K. T. Brunner: J. Immunol. *129,* 550 (1982).
48 K. W. Talmadge, H. Gallati, F. Sinigaglia, A. Waltz, and G. Garotta: Eur. J. Immunol. *16,* 1471 (1986).
49 M. B. Prystowsky, J. M. Ely, D. I. Beller, L. Eisenberg, J. Goldman, M. Goldman, E. Goldwasser, J. Ihle, J. Quintans, H. Remold, S. N. Vogel, and F. W. Fitch: J. Immunol. *129,* 2337 (1982).
50 P. W. Gray, and D. V. Goeddel: Proc. Natl. Acad. Sci. USA *80,* 5842 (1983).
51 J. L. Pace, S. W. Russell, R. D. Schreiber, A. Altman, and D. H. Katz: Proc. Natl. Acad. Sci. USA *80,* 3782 (1983).
52 B. A. Torres, and H. M. Johnson: Biochem. biophys. Research Comm. *131,* 395 (1985).
53 R. M. Schultz, and M. A. Chirigos: Cell. Immunol. *48,* 52 (1979).
54 J. L. Pace, S. W. Russell, P. A. Leblanc, and D. M. Murasko: J. Immunol. *134,* 977 (1985).
55 J. L. Pace: J. Leukocyte Biol. *44,* 514 (1988).
56 R. Yoshida, H. W. Murray, and C. F. Nathan: J. Exp. Med. *167,* 1171 (1988).
57 A. Celada, and R. D. Schreiber: J. Immunol. *139,* 147 (1987).
58 A. Celada, P. W. Gray, E. Rinderknecht, and R. D. Schreiber: J. Exp. Med. *160,* 55 (1984).
59 G. J. Bancroft, R. D. Schreiber, G. C. Bosma, M. J. Bosma, and E. R. Unanue: J. Immunol. *139,* 1104 (1987).
60 G. J. Bancroft, K. C. F. Sheehan, R. D. Schreiber, and E. R. Unanue: J. Immunol. *143,* 127 (1989).

61 G. Lemaire, G. Barrat, J. C. Drapier, T. Grand-Perret, M. Lepoivre, J. P. Tenu, and J. F. Petit: Ann. Inst. PasteUr/Immunol. *137C,* 218 (1986).
62 T. Grand-Perret, M. Lepoivre, J.-F. Petit, and G. Lemaire: Eur. J. Immunol. *16,* 332 (1986).
63 P. G. Morahan, P. J. Edelson, and K. Gass: J. Immunol. *125,* 1312 (1980).
64 G. Lemaire, J. C. Drapier, J. P. Tenu, A. Bouchahda, M. Lepoivre, and J. F. Petit: J. Reticuloendothel. Soc. *32,* 87 (1982).
65 L. E. Lambert, and D. M. Paulnock: Cell. Immunol. *120,* 401 (1989).
66 L. E. Lambert, and D. M. Paulnock: J. Immunol. *139,* 2834 (1987).
67 I. Saiki, and I. J. Fidler: J. Immunol. *135,* 684 (1985).
68 S.-I. Nagao, K. Sato, and Y. Osada: Cancer Immunol. Immunother. *23,* 41 (1986).
69 R. M. Schultz and M. G. Altom: J. Immunopharmacol. *8* 515 (1986).
70 R. M. Schultz: Adv. Immunopharmacol. *2,* 161 (1983).
71 R. M. Schultz: J. Interferon Res. *2,* 459 (1982).
72 A. Zerial, A. G. Hovanessian, S. Stefanos, K. Huygen, G. H. Werner, and E. Falcoff: Antiviral Res. *2,* 227 (1982).
73 S. Nagao, K. Sato, and Y. Osada: Jpn. J. Cancer Res. (Gann) *78,* 80 (1987).
74 K. S. Akagawa, and T. Tokunaga: J. Exp. Med. *162,* 1444 (1985).
75 P. A. Marino, and D. O. Adams: J. Immunol. *128,* 2816 (1982).
76 S. Sone, and I. J. Fidler: Cell. Immunol. *57,* 42 (1981).
77 S. Sone, and I. J. Fidler: J. Immunol. *125,* 2454 (1980).
78 I. J. Fidler: Cancer Res. *45,* 4714 (1985).
79 S. Sone, S. Mutsuura, M. Ogawara, T. Utsugi, and E. Tsubura: Cancer Immunol. Immunother. *18,* 169 (1984).
80 C. Pidgeon, R. D. Schreiber, and R. M. Schultz: J. Immunol. *131,* 311 (1983).
81 P. H. Krammer, C. F. Kubelka, W. Falk, and A. Ruppel: J. Immunol. *135,* 3258 (1985).
82 U. Hamann, and P. H. Krammer: Eur. J. Immunol. *15,* 18 (1985).
83 C. F. Kubelka, A. Ruppel, D. Gemsa, and P. H. Krammer: Immunobiol. *171,* 311 (1986).
84 E. S. Kleinermann, R. Zicht, P. S. Sarin, R. C. Gallo, and I. J. Fidler: Cancer Res. *44,* 4470 (1984).
85 J. C. Lee, L. Rebar, P. Young, F. W. Ruscetti, N. Hanna, and G. Poste: J. Immunol. *136,* 1322 (1986).
86 M. P. A. Jones, D. E. Gunapala, E. Matutes, D. Catovsky, and A. R. M. Coates: Cell. Immunol. *113,* 361 (1988).
87 M. S. Meltzer, W. R. Benjamin, and J. J. Farrar: J. Immunol. *129,* 2802 (1982).
88 M. S. Meltzer, M. J. Gilbreath, R. M. Crawford, R. D. Schreiber, and C. A. Nacy: Cell. Immunol. *107,* 340 (1987).
89 C. Sutton, P. Depledge, L. Bawden, A. Carne, M. Meltzer, V. Newton, and L. Vodinelich: J. Biol. Stand. *17,* 65 (1989).
90 R. M. Crawford, D. S. Finbloom, J. Ohara, W. E. Paul, and M. S. Meltzer: J. Immunol. *139,* 135 (1987).
91 S. D. Somers, and K. L. Erickson: Proc. Am. Assoc. Cancer Res. *29,* 368 (1988).
92 K. H. Grabstein, D. L. Urdal, R. J. Tushinski, D. Y. Mochizuki, V. L. Price, M. A. Cantrell, S. Gillis, and P. J. Conlon: Science *232,* 506 (1986).
93 P. J. Morrissey, K. H. Grabstein, S. G. Reed, and P. J. Conlon: Int. Arch. Allergy Appl. Immunol. *88,* 40 (1989).
94 S. G. Reed, C. F. Nathan, D. L. Pihl, P. Rodricks, K. Shanebeck, P. J. Conlon, and K. H. Grabstein: J. Exp. Med. *166,* 1734 (1987).

95 E. J. Wing, D. M. Magee, T. L. Whiteside, S. S. Kaplan, and R. K. Shadduck: Blood *73,* 643 (1989).
96 P. H. Hart, G. A. Whitty, D. S. Piccoli, and J. A. Hamilton: J. Immunol. *141,* 1516 (1988).
97 S. Heidenreich, J.-H. Gong, A. Schmidt, M. Nain, and D. Gemsa: J. Immunol. *143,* 1198 (1989).
98 S. A. Canistra, E. Vellenga, P. Grosvek, R. Alessandro, and J. D. Griffin: Blood *71,* 672 (1988).
99 J. E. Talmadge, H. Phillips, M. Schneider, T. Rowe, R. Pennington, O. Bowersox, and B. Lenz: Cancer Res. *48,* 544 (1988).
100 K. Hori, M. J. Ehrke, K. Mace, and E. Mihich: Cancer Res. *47,* 5868 (1987).
101 I. Esparza, D. Mannel, A. Ruppel, W. Falk, and P. H. Krammer: J. Exp. Med. *166,* 589 (1987).
102 K. Hori, M. J. Ehrke, K. Mace, D. Maccubbin, M. J. Doyle, Y. Otsuka, and E. Mihich: Cancer Res. *47,* 2793 (1987).
103 L. Chen, Y. Suzuki, and E. F. Wheelock: J. Immunol. *139,* 4096 (1987).
104 S. Heidenreich, M. Weyers, J.-H. Gong, H. Sprenger, M. Nain, and D. Gemsa: J. Immunol. *140,* 1511 (1988).
105 K. Hori, E. Mihich, and M. J. Ehrke: Cancer Res. *49,* 2606 (1989).
106 R. Philip, and L. B. Epstein: Nature (Lond.) *323,* 86 (1986).
107 D. O. Adams, and T. A. Hamilton: Immunol. Rev. *97,* 5 (1987).
108 R. J. Uhing, and D. O. Adams: Agents Actions *26,* 9 (1989).
109 T. A. Hamilton, and D. O. Adams: Immunology Today *8,* 151 (1987).
110 S. D. Somers, J. E. Weiel, T. A. Hamilton, and D. O. Adams: J. Immunol. *136,* 4199 (1986).
111 T. A. Hamilton, D. Becton, S. D. Somers, P. A. Gray, and D. O. Adams: J. Biol. Chem. *260,* 1378 (1985).
112 D. L. Becton, D. O. Adams, and T. A. Hamilton: J. Cell. Physiol. *125,* 485 (1986).
113 B. Wright, I. Zeidman, R. Greig, and G. Poste: Cell. Immunol. *95,* 46 (1985).
114 H. K. Nguyen, C. A. Cain, and W. A. F. Tompkins: J. Biol. Response Modifiers *6,* 154 (1987).
115 A. M. Gorecka-Tisera, K. W. Snowdowne, and A. B. Borle: Cell. Immunol. *100,* 411 (1986).
116 J. O. Meecham, M. M. Soong, C. A. Cain, S. Koem, J. Goff, and W. A. F. Tompkins: J. Immunol. *134,* 3516 (1985).
117 M. J. Berridge, and R. F. Irvine: Nature *312,* 315 (1984).
118 A. Celada, and R. D. Schreiber: J. Immunol. *137,* 2373 (1986).
119 H. M. Johnson, and B. A. Torres: Proc. Natl. Acad. Sci. USA *82,* 5959 (1985).
120 B.-E. Drysdale, R. A. Yapundich, M. L. Shin, and H. S. Shin: J. Immunol. *138,* 951 (1987).
121 D. Radzioch, and L. Varesio: J. Immunol. *140,* 1259 (1988).
122 E. Blasi, R. B. Herberman, and L. Varesio: J. Immunol. *132,* 3226 (1984).
123 L. Varesio: Ann. Inst. Pasteur/Immunol. *137C,* 235 (1986).
124 L. Varesio: J. Immunol. *132,* 2683 (1984).
125 Y. Okai: Immunol. Lett. *8,* 261 (1984).
126 L. Varesio, M. Clayton, D. Radzioch, B. Bottazzi, and E. Blasi: J. Interferon Res. *7,* 696 (1987).
127 V. Prpic, J. E. Weiel, S. D. Somers, B. Herman, S. Gonias, S. V. Pizzo, T. A. Hamilton, and D. O. Adams: J. Immunol. *139,* 526 (1987).
128 V. Prpic, G. Gawdi, B. Herman, R. J. Uhing, and D. O. Adams: FASEB J. *2,* A916 (1988).

129 J. E. Weiel, T. A. Hamilton, and D. O. Adams: J. Immunol. *136,* 3012 (1986).
130 T. A. Hamilton, M. J. Jansen, S. D. Somers, and D. O. Adams: J. Cell. Physiol. *128,* 9 (1986).
131 P. A. Johnston, M. M. Jansen, S. D. Somers, D. O. Adams, and D. O. Adams: J. Immunol. *138,* 1551 (1987).
132 A. A. Aderem: J. Cell Sci. Suppl. *9,* 151 (1988).
133 M. Introna, T. A. Hamilton, R. E. Kaufman, D. O. Adams, and R. C. Bast: J. Immunol. *137,* 2711 (1986).
134 G. Poste, and R. Kinsh: Cancer Res. *39,* 2582 (1979).
135 R. Van Furth: Mononuclear Phagocytes in Immunity, infection, and Pathology, p. 161. Blackwell Scientific Publications, Oxford 1975.
136 S. M. Taffet, and S. W. Russell: J. Immunol. *126,* 424 (1981).
137 I. J. Fidler, J. H. Darnell, and M. B. Budmen: J. Immunol. *117,* 666 (1976).
138 R. M. Schultz, N. A. Pavlidis, W. A. Stylos, and M. A. Chirigos: Science *202,* 320 (1978).
139 R. M. Schultz, J. N. Stoychkov, N. Pavlidis, M. A. Chirigos, and Z. L. Olkowski: J. Reticuloendothel. Soc. *26,* 93 (1979).
140 R. M. Schultz: Med. Hypotheses *6,* 831 (1980).
141 R. M. Schultz, and M. G. Altom: J. Immunopharmacol. *5,* 277 (1983).
142 P. Dore-Duffy, W. Perry, and H.-H. Kuo: Cell. Immunol. *79,* 232 (1983).
143 D. Boraschi, S. Censini, M. Bartalini, G. Scapigliati, G. Barbarulli, E. Vicenzi, M. B. Donati, and A. Tagliabue: J. Immunol. *132,* 1987 (1984).
144 J. L. Browning, and A. Ribolini: J. Immunol. *138,* 2857 (1987).
145 S. W. Russell, and J. L. Pace: Molec. Immunol. *21,* 249 (1984).
146 S. W. Russell, and J. L. Pace: J. Leukocyte Biol. *35,* 291 (1984).
147 H. A. Chapman, and J. B. Hibbs: Science *197,* 282 (1977).
148 J. B. Hibbs: The Macrophage in Neoplasia, p. 83. Academic Press, New York 1976.
149 R. M. Schultz, M. A. Chirigos, J. N. Stoychkov, and N. A. Pavlidis: RES. J. Reticuloendothel. Soc. *26,* 83 (1979).
150 R. M. Schultz, S. K. W. Nanda, and M. G. Altom: J. Immunol. *135,* 2040 (1985).
151 R. W. Pfeifer, and R. M. Patterson: J. Immunopharmacol. *7,* 247 (1985).
152 R. M. Schultz, N. A. Pavlidis, J. N. Stoychkov, and M. A. Chirigos: Cell. Immunol. *42,* 71 (1979).
153 S. I. Slager, L. D. Madden, M. S. Meltzer, S. Bara, and M. J. Mamula: Cell. Immunol. *77,* 52 (1983).
154 S. I. Slager, and M. S. Meltzer: Cell. Immunol. *80,* 10 (1983).
155 B. Wright, R. Greig, and G. Poste: Biochem. biophys. Res. Commun. *123,* 710 (1984).
156 E. Spath-Schwalbe, F. Porzsolt, W. Digel, J. Born, B. Kloss, and H. L. Fehm: Immunopharmacol. *17,* 141 (1989).
157 R. Keller: Nature *282,* 729 (1979).
158 R. Keller: Natural Cell-Mediated Immunity, Vol. 2. Academic Press, New York 1982.
159 R. Keller, R. Keist, W. Adolf, H. J. Opferkuch, R. Schmidt, and E. Hecker: Exp. Cell. Biol. *50,* 121 (1982).
160 M. R. Ruff, and G. E. Gifford: Lymphokines, voll. II, p. 235. Academic Press, New York 1981.
161 K. Onozaki, K. Matsuchima, B. B. Aggarwal, and J. J. Oppenheim: J. Immunol. *135,* 3962 (1985).
162 L. B. Lachman, C. A. Dinarello, N. D. Llansa, and I. J. Fidler: J. Immunol. *136,* 3098 (1986).

163 C. F. Nathan, L. H. Brukner, S. C. Silverstein, and Z. A. Cohn: J. Exp. Med. *149,* 84 (1979).
164 C. F. Nathan, and S. Tsunawaki: Biochemistry of Macrophages, p. 211. Pittman, London (Ciba Foundation Symposium 118) 1986.
165 J. L. Urban, H. Michael-Shepard, J. L. Rothstein, and B. J. Sugarman: Proc. Natl. Acad. Sci. USA *83,* 5233 (1986).
166 R. Feinman, D. Henriksen-DeStefano, M. Tsujimoto, and J. Vilcek: J. Immunol. *138,* 635 (1987).
167 T. Decker, M.-L. Lohmann-Matthes, and G. E. Gifford: J. Immunol. *138,* 957 (1987).
168 J. Nissen-Meyer, E. Hofsli, T. Espevik, and R. Austgulen: Nat. Immun. Cell Growth Reg. *7,* 266 (1988).
169 H. W. L. Ziegler-Heitbrock, A. Moeller, R. P. Linke, J. G. Haas, E. P. Rieber, and G. Riethmueller: Cancer Res. *46,* 5947 (1986).
170 M. Tsujimoto, Y. K. Yip, and J. Vilcek: Proc. Natl. Acad. Sci. USA *82,* 7626 (1985).
171 N. Watanabe, Y. Niitsu, N. Yamauchi, H. Umeno, H. Sone, H. Neda, and I. Urushizaki: J. Biol. Resp. Modif. *7,* 24 (1988).
172 M. Tsujimoto, R. Feinman, and J. Vilcek: J. Immunol. *137,* 2272 (1986).
173 F. C. Kull, and P. Cuatrecasas: Cancer Res. *41,* 4885 (1981).
174 K. Haranaka, N. Satomi, A. Sakurai, and R. Haranaka: Tumor Necrosis Factor, and Related Cytotoxins, p. 140. Wiley Chichester (Ciba Foundation Symposium 131) 1987.
175 N. Watanabe, Y. Niitsu, H. Nedo, H. Sone, N. Yamauchi, M. Maeda, and I. Urushizaki: Immunopharmacol. Immunotoxical. *10,* 109 (1988).
176 S. Agarwal, B.-E. Drysdale, and H. S. Shin: J. Immunol. *140,* 4187 (1988).
177 B.-E. Drysdale, S. Agarwal, and H. S. Shin: Prog. Allergy *40,* 111 (1988).
178 M. Kirstein, W. Fiers, and C. Baglioni: J. Immunol. *137,* 2277 (1986).
179 C. A. Dinarello: Rev. infect. Dis. *6,* 51 (1984).
180 N. Inamura, S. Sone, A. Okubo, E. Kunishige, M. Nakanishi, and T. Ogura: Cancer Immunol. Immunother. *28,* 164 (1989).
181 Y. Endo, K. Matsushima, and J. J. Oppenheim: Immunobiol. *172,* 316 (1986).
182 D. Lovett, B. Kozan, M. Hadam, K. Resch, and D. Gemsa: J. Immunol. *136,* 340 (1986).
183 V. Ruggiero, and C. Baglioni: J. Immunol. *138,* 661 (1987).
184 K. Onozaki, K. Matsushima, B. B. Aggarwal, and J. J. Oppenheim: J. Immunol. *135,* 3962 (1985).
185 L. B. Lachman, C. A. Dinarello, N. D. Llansa, and I. J. Fidler: J. Immunol. *136,* 3098 (1986).
186 R. M. Schultz, J. J. Howbert, and R. A. Archer: Immunopharmacol. Immunotoxicol., in press (1989).
187 J. Chang, S. C. Gilman, and A. J. Lewis: J. Immunol. *136,* 1283 (1986).
188 R. M. Burch, J. R. Connor, and J. Axelrod: Proc. Natl. Acad. Sci. USA *85,* 6306 (1988).
189 Y. Zhang, J.-X. Lin, Y. K. Yip, and J. Vilcek: Proc. Natl. Acad. Sci. USA *85,* 6802 (1988).
190 P. M. Rosoff, N. Savage, and C. A. Dinarello: Cell. *54,* 73 (1988).
191 C. F. Nathan, S. C. Silverstein, L. H. Brukner, and Z. A. Cohn: J. Exp. Med. *149,* 100 (1979).
192 D. O. Adams: J. Immunol. *124,* 286 (1980).
193 D. O. Adams, K.-J. Kao, R. Farb, and S. V. Pizzo: J. Immunol. *124,* 293 (1980).
194 M. J. Stadecker, J. Calderon, M. L. Karnovsky, and E. R. Unanue: J. Immunol. *119,* 1738 (1977).
195 G. A. Currie: Nature (London) *273,* 758 (1978).

196 H. U. Schorlemmer, D. Bitter-Suermann, and A. C. Allison: Immunol. *32,* 929 (1977).
197 J. Taylor-Papadimitriou: Interferon, Vol. 2, p. 13 Academic Press, London 1980.
198 B. Y. Rubin, and S. L. Gupta: Proc. Natl. Acad. Sci. USA *77,* 5928 (1980).
199 D. Boraschi, and M. S. Meltzer: Cell. Immunol. *45,* 188 (1979).
200 M. C. Pike, and R. Snyderman: J. Immunol. *117,* 1243 (1976).
201 P. W. A. Mansell, H. Ichinose, R. J. Reed, E. T. Krementz, R. McNamee, and N. R. Diluzio: J. Natl. Cancer Inst. *54,* 571 (1975).
202 M. G. Hanna, M. J. Snodgrass, B. Zbar, and H. J. Rapp: Natl. Cancer Inst. Monogr. *39,* 71 (1973).
203 R. M. Schultz, N. A. Pavlidis, and M. A. Chirigos: Cancer Res. *38,* 3427 (1978).
204 R. M. Schultz, J. D. Papamatheakis, J. Luetzeler, P. Ruiz, and M. A. Chirigos: Cancer Res. *37,* 358 (1977).
205 R. Evans: Transplantation *14,* 468 (1972).
206 S. A. Eccles, and P. Alexander: Nature (London) *250,* 667 (1974).
207 G. W. Wood, and G. Y. Gillespie: Int. J. Cancer *16,* 1022 (1975).
208 R. Keller: J. Natl. Cancer Inst. *57,* 1355 (1976).
209 R. M. Schultz, J. C. Tang, D. C. DeLong, E. W. Ades, and M. G. Altom: Cancer Res. *46,* 5624 (1986).
210 J. B. Hibbs: The Macrophage in Neoplasia, p. 83. Academic Press, New York 1976.
211 I. J. Fidler: Cancer Res. *34,* 1074 (1974).
212 E. Den Otter, H. F. J. Dullens, H. Van Lovern, and E. Pells: The Macrophage, and Cancer, p. 119. Econoprint, Edinburgh 1977.
213 L. A. Liotta, G. Gattozzi, J. Kleinerman, and G. Saidel: Br. J. Cancer *36,* 369 (1977).
214 W. F. Piessens: Fedn. Proc. *26,* 440A (1978).
215 S. W. Russell, and A. T. McIntosh: Nature (London) *268,* 69 (1977).
216 S. W. Russell, W. F. Doe, and C. G. Cochrane: The Macrophage in Neoplasia, p. 199. Academic Press. Inc., New York 1976.
217 R. Evans: Br. J. Cancer *28* Suppl. 1, 19 (1973).
218 Mantovani, A.: Int. J. Cancer, *22,* 741 (1978).
219 C. C. Stewart, and K. L. Beetham: Int. J. Cancer *22,* 152 (1978).
220 I. J. Fidler: Science *208,* 1469 (1980).
221 I. J. Fidler, S. Sone, W. E. Fogler, and Z. Barnes: Proc. Natl. Acad. Sci. USA *78,* 1680 (1981).
222 I. J. Fidler, A. Raz, W. E. Fogler, R. Kirsh, P. Bugelski, and G. Poste: Cancer Res. *40,* 4460 (1980).
223 I. J. Fidler, Z. Barnes, W. E. Fogler, R. Kirsh, P. Bugelski, and G. Poste: Cancer Res. *42,* 496 (1982).
224 I. J. Fidler: Ann. Inst. Pasteuer/Immunol. *137C,* 212 (1986).
225 I. J. Fidler, and G. Poste: Manual of Macrophage Methodology, p. 509. Marcel Dekker, New York 1981.
226 I. J. Fidler, and A. Raz: Lymphokines, Vol. 3, p. 345. Academic Press, New York 1981.
227 J. E. Talmadge, M. Key, and I. J. Fidler: J. Immunol. *126,* 2245 (1981).
228 J. Rhodes: Cancer Immunol. Immunother. *7,* 211 (1980).
229 R. S. Kerbel: Nature *280,* 358 (1979).
230 R. M. Schultz: Lancet *8208,* 1362 (1980).

Benzodiazepine augmentation of the treatment of disruptive psychotic behavior

By David A. Solomon, Edison Miyawaki and Carl Salzman

Dr. Solomon is a Psychiatrist in Residence at Brown University, Providence, Rhode Island, USA; Dr. Miyawaki is a Clinical Fellow in Psychiatry at Harvard Medical School, Boston, Massachusetts, USA; Dr. Salzman is an Associate Professor of Psychiatry at Harvard Medical School and Director of Psychopharmacology, Massachusetts Mental Health Center, 74 Fenwood Road, Boston, Massachusetts 02115, USA.

Psychiatry has searched for a humane and effective treatment for the disruptive behavior and agitation that accompanies acute psychosis. Prior to the introduction of psychotropic drugs, baths, physical restraints, wet packs, and ECT [1] were used to manage agitated and assaultive conduct. Early pharmacological therapies included insulin coma, bromides, and barbiturates. Although useful for quieting some patients, these physical and pharmacologic approaches had serious complications and often lost effectiveness over time.

Current pharmacologic approaches to control aggressive, assaultive, and violent behavior include lithium [2, 3], betablockers [4–8], and carbamazepine [9]. The effectiveness of these treatments, however, is primarily for the long-term management of affective and behavioral instability, rather than for the immediate control of disruptive behavior. For acute management, high doses of neuroleptics, administered over a brief period of time, have been effective in controlling assaultive and aggressive behavior. Reschke et al. [11] found that control of severe disruptive behavior in schizophrenics (n = 50) could be achieved within two hours by giving intramuscular (IM) haloperidol, either 2 or 5 mg, at 30-minute intervals. Anderson et al. [12] administered higher doses of IM haloperidol (5 mg every hour or 10 mg every 30 minutes) to acutely psychotic schizophrenics and manics (n = 24) over a three-hour period. Marked improvement in nearly half of the patients, irrespective of treatment group, was noted. Both studies reported only mild to moderate side effects, which included hypotension, drowsiness, dry mouth, blurred vision, dizziness, and extrapyramidal symptoms.

Although therapeutically effective, serious side effects have been associated with the use of haloperidol. Drug-induced dysphoria as well as akathisia [13] commonly develop, and may lead to noncompliance. Akathisia sometimes exacerbates agitation and psychosis, resulting in suicide and homicide [14, 15]. Potentially life-threatening reactions, as well, have occurred following the use of high dose haloperidol [16]. There has been one report of a non-lethal cardiac arrest [17] and one of sudden death presumed to be the result of drug-induced laryngeal spasm, leading to cardiac arrest via vagal reflexes [18].

In the pursuit of less toxic methods for treating severe disruptive behavior associated with psychosis, psychiatrists have increasingly used benzodiazepines. These drugs provide sedation, and are associated with less respiratory depression than barbiturates. One of the first attempts to control agitation with benzodiazepines was a non-randomized study by McClish et al. [19] of 79 patients who underwent open-heart surgery. Severe anxiety, nervous tension, disorientation, confusion, depression, hostility, and hallucinations were significantly lower in the postoperative period in patients who received intravenous diazepam (2.5 to 5.0 mg every four hours). The length of treatment was unspecified. Abel and Reis [20] gave intravenous diazepam (0.1 mg/kg) to 15 cardiac surgery patients as needed for anxiety from 16–72 hours after surgery. They achieved calming effects without adverse hemodynamic effects. After injecting diazepam over a 30-second period, there was no significant change in heart rate, right atrial pressure, systemic vascular resistance, or cardiac index for the following 30 minutes. It should be noted that in both the McClish and Abel and Reis studies, benzodiazepines were used to control delirium rather than agitation associated with a primary psychotic disorder.

There has been a recent accumulation of reports on the use of benzodiazepines to control severe disruptive behavior associated with psychosis. Modell et al. [21] studied patients with manic agitation who were being treated with lithium in an open trial of lorazepam. Four bipolar patients were admitted to an inpatient unit, and were started on lithium 1200 mg per day; doses were adjusted to achieve therapeutic levels. In addition, they were given lorazepam 2–4 mg orally, intramuscularly, or intravenously every two hours as needed to control agitation. Initially, one patient required 15–30 mg of lorazepam per day for five days, another required 8–20 mg per day for four days, and the other patients required lower doses to control agitation. There were no episodes of hypotension, respiratory depression, syncope, increased aggressive behavior, or reports of benzodiazepine withdrawal. All patients were discharged on lithium alone. The authors suggest that lorazepam may be superior to antipsychotics for augmenting lithium treatment of manic agitation. Modell [22] subsequently described use of lorazepam to treat agitation in 100 patients with heterogeneous diagnoses (mania, schizophrenia, major depression, delirium, and situational anxiety) in various settings (general and state hospital psychiatric units, general hospital medical and surgical units, and the emergency room). Lorazepam was given orally, intramuscularly, intravenously, or in combination, in daily doses which ranged from 2 to 120 mg. Despite the anecdotal nature of the report, there are clinically useful findings. Given orally, lorazepam achieves its peak effect in two hours, intramuscularly in one hour, and intravenously in 10 minutes. Nonpsychotic agitation was relieved with 2–4 mg, by any route, given 2 to 3 times daily. Psychotic agitation required two to three times that dosage. Patients who were both psychotic and had a recent history of alcohol or other sedative abuse required five to ten times the dosage given to nonpsychotic, nontolerant patients. There was no association between age and dosage requirements. The most frequent side effects were ataxia (50 %), nausea or vomiting (25 %), and confusion or disorientation (10 %). It was not stated whether the side effects were dose related. There were no episodes of respiratory or cardiovascular depression which required medical intervention, nor were there episodes of withdrawal.

Chouinard et al. [45, 46] have found clonazepam to be a useful single agent in the management of manic agitation. In data originally reported in 1983, 12 acutely manic patients were treated with clonazepam or lithium in a double-blind, crossover design: half of the patients were randomly assigned to ten days of treatment with clonazepam followed immediately by ten days of lithium; others received the treatments in reverse order. High doses of clonazepam (up to 16 mg per day) were used. The crossover occurred without washout between trials, so it is not clear whether study design really clarified differences in efficacy between the drugs. Clonazepam patients received a lower total dose of PRN haloperidol, and had lower ratings for extrapyramidal symptoms. Clinical status was evaluated with the Clinical Global Impression Scale (mania) and an Inpatient Multidimensional Psychiatric Scale modified for mania. No scale was used specifically for an assessment of psychosis.

Benzodiazepines have also been used to augment the antipsychotic effect of neuroleptic drugs, with mixed success [30–33]. Clinical ratings in these studies suggest a therapeutic effect of benzodiazepine and neuroleptic compared to neuroleptic alone especially within first 1–2 weeks of treatment. Despite suggestions that benzodiazepines may play a role in the acute management of psychosis and agitation, other authors [34–38] have noted that benzodiazepines may result in paradoxical increases in excitation. Such instances of behavioral disinhibition have required discontinuation of benzodiazepines in some cases [34, 35].

Bick and Hannah [23] conducted a retrospective chart review of ten violent patients whose agitated behavior required chemical restraint on two separate occasions, once with haloperidol 10 mg IM and once with lorazepam 2 mg IM. They concluded that the benzodiazepine was at least as effective as the antipsychotic based on the number of hours each patient spent in a seclusion room following each incident and injection: an average of 6.1 hours when given haloperidol and 1.9 hours when given lorazepam. The authors acknowledge the limitations of their study: patients with different and unspecified diagnoses were included and the drugs were administered in a non-blind manner. In one case report, Ward et al. [24] repeatedly treated acute psychotic agitation in a chronic schizophrenic patient. The average time to calm the patient was significantly less for the eight times lorazepam was given when compared to the three times antipsychotic medication was administered. In another case report [25], lorazepam (1.5 mg per day) aborted a manic episode after three doses in a patient with organic mania secondary to closed head injury. When the medication was stopped, the patient relapsed, but the mania remitted within 24 hours when lorazepam was again administered. In a report of three cases of psychotic agitation, Mendoza et al. [26] describe rapid behavioral control with IM midazolam. The onset of sedation and sleep occurred within ten minutes for two patients given 2.5 mg and 15 minutes for the third patient given 3 mg. Salzman et al. [27] reviewed the treatment records of 46 inpatients given IM neuroleptic for behavior considered dangerous to themselves or others during a six-month period in 1982. The group was compared to an age- and diagnosis-matched cohort of 53 inpatients treated with either IM neuroleptic and lorazepam or IM neuroleptic alone during a six-month period in 1984. The 1982 cohort received a mean dose of 408 mg of chlorpromazine equivalents for each chemical restraint, while the 1984 cohort received a mean dose of 217 mg of chlorpromazin equivalents. In addition to the statistically significant difference in mean doses, comparison revealed that the cumulative dose of IM neuroleptic per patient for the six-month period was significantly lower for the 1984 group. The authors suggested that the addition of lorazepam allowed for a reduction in the amount of neuroleptic necessary to control severe disruptive behavior. Their conclusions, however, are limited by the retrospective and open nature of the study, the absence of rating scales to evaluate effects, and lack of control over diagnosis, type of disruptive behavior, length of treatment, and follow-up.

In an article discussing various methods used to treat psychotic agitation, including ward milieu, physical restraints, seclusion, neuroleptics, and benzodiazepines, Campbell and Simpson [28] mention preliminary results of a prospective study of lorazepam versus haloperidol. Approximately 50 patients receiving daily doses of haloperidol were given either lorazepam 2 mg IM or haloperidol 5 mg IM for acute psychotic agitation. No difference between the agents has been reported after some 200 administrations. Information is not available about methodology or rating instruments. Trevino et al. [10] have described combined use of IM haloperidol and lorazepam in the control of agitation in 20 patients. Among successfully treated patients, doses of haloperidol ranged from 5 to 65 mg (median 5 mg); doses of lorazepam ranged from 4 to 24 mg (median 4 mg). The authors were specifically interested in the time required to achieve pharmacologic control over agitation. Time to satisfactory response varied between 30 to 210 minutes (median 60 minutes). No episodes of acute dystonia were noted. The authors concluded that the combination of haloperidol and lorazepam was effective and safe in most patients with severe agitation.

Results of our own controlled, randomized, and double-blind study [29] suggest that haloperidol and lorazepam are equally effective in behavioral control. The setting was the Massachusetts Mental Health Center, a state-owned and operated community mental health facility that serves as a major psychiatric teaching hospital of Harvard Medical School. Fifty subjects, all inpatients on a locked unit, were enrolled. All suffered from schizophrenia, schizoaffective disorder, or bipolar disorder. Patients with serious medical illnesses, organic brain syndromes, and substance abuse disorders were excluded. Each patient received a daily dose of a neuroleptic, and were randomly assigned to receive lorazepam 2 mg IM or haloperidol 5 mg IM if they required chemical restraint for control of acute psychotic agitation after being taken to a seclusion room. Each subject was assessed with the Brief Psychiatric Rating Scale (BPRS) and Overt Aggression Scale (OAS) at baseline and on the first, second, seventh, and fourteenth days after the chemical restraint. The results suggest that 2 mg of IM lorazepam are equal to 5 mg of IM haloperidol in controlling assaultive, violent, agitated disruptive behavior in psychotic patients [Tables 1–4].

The mechanism of behavioral control that is obtained by combining benzodiazepines with neuroleptics is not clear. Lorazepam is sedating, and some of the behavioral control may be a non-specific sedative effect of both the benzodiazepine and neuroleptic. Lorazepam

Table 1
Outcome: Total BPRS

	Lorazepam	Haldol
x̄ Baseline	55.76 (n = 29)	60.61 (n = 28)
Day 1	48.68 (n = 28)	47.89 (n = 28)
Day 14	42.19 (n = 21)	45.29 (n = 21)

Table 2
Outcome: BPRS Hostility scale

	Lorazepam	Haldol
x̄ Baseline	15.17 (n = 30)	16.96 (n = 28)
Day 1	12.05 (n = 22)	11.00 (n = 18)
Day 14	9.85 (n = 20)	10.19 (n = 21)

Table 3
Outcome: Total OAS

	Lorazepam	Haldol
x̄ Baseline	9.04 (n = 27)	6.76 (n = 25)
1 hour	2.15 (n = 26)	1.32 (n = 25)
2 hours	1.33 (n = 24)	1.00 (n = 24)
24 hours	2.12 (n = 17)	1.40 (n = 15)

Table 4
Outcome: Side effects (24 hours after injection)

	Lorazepam	Haldol
Number of patients with extrapyramidal symptoms	1	5
$\bar{x}$ Degree of sedation	1.84	1.32
Miscellaneous	0	1 (syncope)

has been often used in clinical studies for several reasons. It is available in parenteral forms and is well absorbed after intramuscular administration. It is highly potent and has a short serum half-life (10–20 hours) without active metabolites. It is not dependent on hepatic metabolism for elimination. Although lorazepam has specific utility in the clinical setting, it appears that benzodiazepines as a class may exert a modulatory effect on dopamine, a neurotransmitter that may play a role in agitated and assaultive behavior. Lesion studies [39, 40] have found that dopamine-rich areas of the frontal cortex may regulate neurotransmitter systems subcortically. In basal ganglia, decreased levels of 5-hydroxytryptamine (5-HT) and gamma-aminobutyric acid (GABA) as well as increased dopamine (DA) turnover in the striatum and limbic system were observed in rats who underwent stereotactic lesioning of the medial prefrontal cortex. GABA and prefrontal catecholamine systems were therefore implicated in regulation of limbic and nigrostriatal DA systems. Another group [41, 42] examined the effect of neuroleptics alone and in combination with benzodiazepines on behavior and subcortical catecholamine turnover in rats. Diazepam was found to potentiate neuroleptic effects both on behavior and on DA and norepinephrine (NE) turnover. The authors concluded that diazepam potentiation might be related to GABA inhibition of striatal DA and other catecholamine systems. Other animal studies [43, 44] have used triazolobenzodiazepines and clonazepam–benzodiazepines which, in addition to having effects on GABA-ergic transmission, have been cited as having relatively specific effects on NE and 5-HT, respectively. Benzodiazepines added to neuroleptic treatment of psychosis in humans have been associated with significant reductions in plasma homovanillic acid (HVA), a major metabolite of DA, during the early weeks of treatment, but not in subsequent weeks. Early reduction in core psychotic symptoms was associated with decreases in plasma HVA [31]. These data suggest that DA may be implicated in a mechanism of psychotic agitation, and that modulation of DA by enhanced GABA activity may result in a diminution of such behavior.

Benzodiazepines are an important part of the psychopharmacologic armamentarium used to modulate acute psychotic agitation. Based on uncontrolled studies, case reports, and double-blind comparisons, benzodiazepines have been shown to be effective adjuncts to ongoing neuroleptic treatment for control of severely agitated and disruptive behavior associated with psychosis. The use of additional doses of neuroleptics to manage such behavior improves the patient's psychosis, but also increases the likelihood of various side effects. Treatment with benzodiazepines is relatively safe, and is associated with fewer side effects. Use of benzodiazepines in acute management does not appear to be associated with tolerance or dependency in the long term. Further studies of benzodiazepines as a modality in the treatment of psychotic agitation are clearly indicated.

References

1 C. Salzman: The use of ECT in the treatment of schizophrenia. Am. J. Psychiatry 137: 1032–1040, 1980.
2 M. H. Sheard, J. L. Marini and C. I. Bridges et al.: The effect of lithium in impulsive aggressive behavior in man. Am. J. Psychiatry 133: 1409–1413, 1976.
3 R. I. Shader, A. H. Jackson and L. M. Dodes: The antiaggressive effects of lithium in man. Psychopharmacologia 40: 17–24, 1974.
4 D. T. Williams, R. Mehl and S. Yudofsky et al.: The effect of propranolol on uncontrolled rage outbursts in children and adolescents with organic brain dysfunction. J. Am. Acad. Child Psychiatry 21: 129–135, 1982.
5 J. J. Ratey, R. Morrill and G. Oxenkrug: Use of propranolol for provoked and unprovoked episodes of rage. Am. J. Psychiatry 140: 1356–1357, 1983.
6 S. C. Yudofsky, L. Stevens and J. Silver et al.: Propranolol in the treatment of rage and violent behavior associated with Korsakoff's psychosis. Am. J. Psychiatry 141: 114–115, 1984.
7 R. M. Greendyke, D. R. Kantor and D. B. Schuster et al.: Propranolol treatment of assaultive patients with organic brain disease. J. Nerv. Ment. Dis. 174: 290–294, 1986.
8 P. J. Sorgi, J. J. Ratey and S. Polakoff: Beta-adrenergic blockers for the control of aggressive behaviors in patients with chronic schizophrenia. Am. J. Psychiatry 143: 775–776, 1986.
9 S. E. Hyman and G. W. Arana: Carbamazepine and other anticonvulsants, in: Handbook of Psychiatric Drug Therapy. Boston/Toronto, Little, Brown and Company, 1987, p. 104.

10 E. S. Garza-Trevino and L. Hollister: Haloperidol combined with lorazepam for agitation, in: Continuing Medical Education Syllabus and Scientific Proceedings. D. C. Washington, A. P. A. Press, 1988, p. 75.
11 R. W. Reschke: Parenteral haloperidol for rapid control of severe, disruptive symptoms of acute schizophrenia. Dis. Nerv. Syst. 35: 112–115, 1974.
12 W. H. Anderson, J. C. Kuehnle and D. M. Catanzano: Rapid treatment of acute psychosis. Am. J. Psychiatry 133: 1076–1078, 1976.
13 R. Van Putten, P. R. A. May and S. R. Marder: Response to antipsychotic medication: the doctor's and the consumer's view. Am. J. Psychiatry 141: 16–19, 1984.
14 T. Van Putten and S. R. Marder: Behavioral toxicity of antipsychotic drugs. J. Clin. Psychiatry 48: 9 (Suppl.) 13–19, 1987.
15 J. J. Ratey and C. Salzman: Recognizing and managing akathisia. Hosp. Community Psychiatry 35: 975–977, 1984.
16 C. Salzman and S. A. Hoffman: Rapid Tranquilization. Hosp. Community Psychiatry 33: 346, 1082.
17 F. Huyse and R. S. Van Schijndel: Haloperidol and cardiac arrest (letter). Lancet II: 568–569, 1988.
18 J. Modestin, R. Krapf and W. Boker: A fatality during haloperidol treatment: mechanism of sudden death. Am. J. Psychiatry 138: 1616–1617, 1981.
19 A. McClish, D. Andrew and L. Tetreault: Intravenous diazepam for psychiatric reactions following open heart surgery. Can. Anaesth. Soc. J. 15: 63–79, 1968.
20 R. M. Abel and R. L. Reis: Intravenous diazepam for sedation following cardiac operations: clinical and hemodynamic assessments. Anesth. Analg. 50: 244–248, 1971.
21 J. G. Modell, R. H. Lenox and S. Weiner: Inpatient clinical trial of lorazepam for the management of manic agitation. J. Clin. Psychopharmacol 5: 109–113, 1985.
22 J. G. Modell: Further experience and observations with lorazepam in the management of behavioral agitation (letter). J. Clin. Psychopharmacol 6: 385–387, 1986.
23 P. A. Bick and A. L. Hannah: Intramuscular lorazepam to restrain violent patients (letter). Lancet I: 206, 1986.
24 M. E. Ward, S. R. Saklad and L. Ereshefsky: Lorazepam for the treatment of psychotic agitation (letter). Am. J. Psychiatry 143: 1195–1196, 1986.
25 A. Dever and E. Schweizer: Rapid remission of organic mania after treatment with lorazepam (letter). J. Clin. Psychopharmacol 8: 227–228, 1988.
26 R. Mendoza and A. H. Djenderedjian, J. Adams et al.: Midazolam in acute psychotic patients with hyperarousal. J. Clin. Psychiatry 48: 291–292, 1987.
27 C. Salzman, A. I. Green and F. Rodriguez-Villa et al.: Benzodiazepines combined with neuroleptics for management of severe disruptive behavior. Psychosomatics 27 (Suppl.): 17–21, 1986.
28 R. Campbell and G. M. Simpson: Alternative approaches in the treatment of psychotic agitation. Psychosomatics 27 (Suppl.): 23–26, 1986.
29 C. Salzman, D. A. Solomon and E. Miyawaki et al.: Use of intramuscular lorazepam to control severe disruptive behavior: a double-blind trial. Presented at ACNP, San Juan, December, 1988.
30 O. M. Wolkowitz, D. Pickar, A. R. Doran, A. Breier, J. Tarell and S. M. Paul: Combination alprazolam-neuroleptic treatment of the positive and negative symptoms of schizophrenia. Am. J. Psychiatry 143: 85–87, 1986.
31 O. M. Wolkowitz, A. Breier, A. Doran, J. Kelsoe, P. Lucas, S. M. Paul and D. Pickar: Alprazolam augmentation of the antipsychotic effects of fluphenazine in schizophrenic patients. Arch. Gen. Psych. 45: 664–671, 1988.

32 J. G. Csernansky, L. Lombrozo, G. D. Gulevich and L. E. Hollister: Treatment of negative schizophrenic symptoms with alprazolam: a preliminary open-label study. J. Clin. Psychopharmacol 4: 349–351, 1984.
33 J. G. Csernansky, S. J. Riney, L. Lombrozo, J. E. Overall and L. E. Hollister: Double-blind comparison of alprazolam, diazepam, and placebo for the treatment of negative schizophrenic symptoms. Arch. Gen. Psych. 45: 655–659, 1988.
34 R. L. Binder: Three case reports of behavioral disinhibition with clonazepam. Gen. Hosp. Psychiatry 9: 151–153, 1987.
35 C. N. Karson, D. R. Weinberger, L. Bigelow and R. J. Wyatt: Clonazepam treatment of chronic schizophrenia: negative results in a double-blind, placebo-controlled trial. Am. J. Psychiatry 139: 1627–1628, 1982.
36 W. J. Burke: Benzodiazepine-induced hypomania? (letter). J. Clin. Psychopharmacol 7: 356, 1987.
37 L. S. Cohen and J. F. Rosenbaum: Clonazepam: new uses and potential problems. J. Clin. Psychiatry 48: 10 (Suppl.) 50–55, 1987.
38 J. C. Pecknold and D. Fleury: Alprazolam-induced manic episode in two patients with panic disorder. Am. J. Psychiatry 143: 652–653, 1986.
39 C. J. Pycock, R. W. Kerwin and C. J. Carter: Effect of lesion of cortical dopamine terminals on subcortical dopamine receptors in rats. Nature 286: 74–77, 1980.
40 C. J. Pycock, C. J. Carter and R. W. Kerwin: Effect of 6-hydroxydopamine lesions of the medial prefrontal cortex on neurotransmitter systems in subcortical sites in the rat. J. Neurochemistry 34: 91–99, 1980.
41 R. B. Rastogi, Y. D. Lapierre, R. L. Singhal and N. P. V. Nair: Evidence for involvement of dopamine in the action of diazepam: potentiation of haloperidol and chlorpromazine action in the brain, in F. Cattabeni et. al. (eds.): Long-Term Effects of Neuroleptics (Adv. Biochem. Psychopharmacol, vol. 24), New York, Raven Press, 1980, pp. 283–289.
42 R. L. Singhal, R. B. Rastogi and Y. D. Lapierre: Diazepam potentiates the effect of neuroleptics on behavioural activity as well as dopamine and norepinephrine turnover: do benzodiazepines have antipsychotic potency? J. Neural Transmission 56: 127–138, 1983.
43 P. Jenner, D. Chadwick, E. H. Reynolds and C. D. Marsden: Clonazepam-induced changes in 5-hydroxytryptamine (5HT) metabolism in animals and man. J. Pharm. Pharmacol 27 (Suppl. 2): 1: 28, 1975.
44 W. Kostowski, L. Valzelli and G. Baiguerra: Effect of chronic administration of alprazolam and adinozolam on clonidine- or apomorphine-induced aggression in laboratory rodents. Neuropharmacology 25: 757–761, 1986.
45 G. Chouinard, S. N. Young and L. Annable: Antimanic effect of clonazepam. Biol. Psychiatry 18: 451–466, 1983.
46 G. Chouinard: Clonazepam in acute and maintenance treatment of bipolar affective disorder. J. Clin. Psychiatry 48: 10 (Suppl.) 29–36, 1987.

The oral antiarrhythmic drugs

By Philip J. Podrid, Lisa Mendes, Scott L. Beau and John S. Wilson

Department of Medicine and Section of Cardiology, Boston University School of Medicine, Boston, MA 02118, USA

1	Normal action potential	152
2	Mechanisms of arrhythmia generation	154
3	Antiarrhythmic drugs	156
3.1	Classification of antiarrhythmic agents	157
3.2	Class I, membrane-stabilizing or local anesthetic agents	157
4	Aggravation of arrhythmia	160
4.1	Exacerbation of congentive heart failure	162
5	Review of individual agents	163
5.1	Flecainide	163
5.2	Encainide	168
5.3	Propafenone	173
5.4	Indecainide	178
5.5	Mexiletine	180
5.6	Tocainide	185
5.7	Moricizine (Ethmozine)	188
5.8	Recainam	191
5.9	Sotalol	194
5.10	Amiodarone	198
5.11	Pirmenol	203
5.12	Quinidine	207
5.13	Disopyramide	213
5.14	Procainamide	219
	References	224

Over the past few years, many new antiarrhythmic drugs have been developed, expanding the choices available for treating arrhythmia. However, at the same time, there has been confusion about how to choose an appropriate drug for the individual patient. Unfortunately, there are no reliable guidelines available for drug selection and there are no drugs which are specific for certain arrhythmias or groups of patients. Drug therapy is empiric and the beneficial or harmful effects of one drug do not provide any information about the response to any other agent. It is therefore necessary to know about the actions and side effects of each drug.
This chapter reviews the basic mechanisms for arrhythmogenesis, the electrophysiologic effects of the antiarrhythmic drugs, their pharmacologic properties and the clinical results of therapy with the specific agents available for the treatment of arrhythmia.

1 Normal action potential

The insertion of microelectrodes into the myocardial cell has defined 5 phases of the action potential generated by the myocardium and resulting in contraction. The action potential is a result of movement of different ions into or out of the cell causing changes in the membrane potential [1, 2]. While the membrane is in a resting state, an electrode placed within the cell records a transmembrane potential of –90 mV (with reference to the extracellular fluid), which is the resting membrane potential or phase 4. This potential is defined by the relationship between intra and extracellular potassium. In the atrial and ventricular myocardium and His-Purkinje tissue, which generate a fast action potential, there is no spontaneous change in this resting membrane potential over time and thus phase 4 is flat. The transmembrane potential is maintained at –90 mV by an active, energy-dependent, ATPase sodium-potassium pump [3]. Since the extracellular sodium concentration is much higher than that present intracellularly, and the intracellular potassium concentration is much greater than the extracellular value, there is a tendency for an influx of sodium ions and an efflux of potassium ions which would cause the potential to spontaneously become less negative or depolarized. The ATPase sodium potassium pump exchanges sodium for potassium, maintaining the resting potential. When the membrane is activated by an electrical stimulus, the sodium channel begins to open, permitting movement of sodium ions into the cell and the membrane potential becomes less negative. When the voltage reaches –60 mV, or threshold potential, the channel is widely open and there is a rapid influx of sodium ions, causing the membrane potential to become markedly positive. This rapid influx of sodium produces phase 0 of the action potential and represents membrane depolarization. The upstroke velocity of phase 0 or V_{max} determines the rate of impulse conduction through the myocardium, that is the rate of membrane depolarization. As V_{max} becomes faster, depolarization is more rapid. V_{max} of phase 0 is affected by many factors, especially antiarrhythmic drugs which block the sodium channel and interfere with inward sodium ion movement, thereby reducing V_{max} and slowing the rate of impulse conduction. Also important for determining the rate of depolarization is the resting membrane potential. When it is less negative, or closer to the threshold potential, the sodium gradient is reduced and the influx rate of sodium into the cell is slower, decreasing the rate of impulse conduction. In contrast, if the membrane is hyperpolarized and the resting potential more negative, the sodium gradient and the rate of its influx are increased, increasing V_{max} of phase 0.

The maximum amplitude of the action potential or phase 1 is approximately +20 mV and this value rapidly falls to a potential of 0 mV which defines the beginning of phase 2, the plateau phase. This represents the beginning of membrane repolarization and during this phase there is inactivation or closure of the sodium channel and sodium no longer fluxes into the cell. However, at the same time, there is activation or opening of the calcium channel and calcium ions begin to move slowly into the cell, maintaining the membrane potential at approximately 0. Activation of calcium ion influx begins when the membrane potential is approximately –50 mV, but since the movement of this ion is much slower, it is responsible for the phase 2 plateau.

After a brief period, there is activation of potassium currents and this ion begins to flow out of the cell. At the same time, sodium and calcium ions no longer flow in and with the extrusion of positive potassium ions, the membrane potential becomes more negative again or repolarizes. This represents phase 3 of the action potential or final repolarization of the membrane with return to its resting potential.

The pacemaker tissue of the heart, the sinoatrial and atrioventricular nodes, and possibly other tissue which is abnormal such as ischemic or damaged myocardium, generate a slow action potential [4]. In this tissue, phase 4 is not flat, but it spontaneously becomes less negative as a result of gradual inactivation of the sodium-potassium pump. As sodium ions gradually "leak" into the cell, the membrane potential becomes less negative and at –50 mV, there is activation and opening of the calcium ion channel and this ion begins to move into the cell. This inward movement is relatively slow and therefore, the upstroke velocity of phase 0 is slower than that observed with sodium ion movements which are rapid. As a result, impulse conduction is slow. The antiarrhythmic drugs alter the action potential by interfering with the movement of these ions into the cell and thus, effecting membrane depolarization, repolarization and automaticity. In this way, they affect the electrophysiologic mechanisms responsible for arrhythmogenesis.

2 Mechanisms of arrhythmia generation

There are 3 proposed mechanisms to explain the occurrence of arrhythmia [5]. These mechanisms are theoretic and it is not certain which, if any, are responsible for the clinical arrhythmia in a given patient. Establishing a mechanism for any specific arrhythmia is difficult, but of no clinical importance for it is not helpful in the identification of an effective agent. The 3 mechanisms are enhanced automaticity, triggered automaticity and reentry.

Enhanced automaticity. As previously described, the resting membrane potential, or phase 4, is maintained by an active ATPase sodium potassium pump which is energy-dependent. If this pump is inactivated, such as occurs with ischemia, or if the influx of positively charged ions increases, such as occurs with sympathetic stimulation, the resting membrane potential becomes less negative and approaches the threshold potential more rapidly. As a result of this enhanced automaticity, action potential generation is faster. When this occurs in the normal pacemaker tissue, there is a physiologic increase in heart rate, or sinus tachycardia. However, enhanced automaticity may also occur in damaged tissue within the atrial or ventricular myocardium [6]. If the rate of impulse formation in this tissue is increased such that it generates an action potential more rapidly than the normal pacemaker tissue, an automatic or ectopic arrhythmia may occur.

Triggered automaticity. As previously indicated, microelectrodes placed within the cell have defined 5 normal phases of the action potential. However, under certain conditions, these techniques have also documented low amplitude oscillations triggered by the action potential and occurring at the end of membrane repolarization early during phase 4 [7]. These low amplitude oscillations or delayed after-potentials result from calcium ion influxes occurring after the membrane has returned to its resting or baseline potential. If these calcium fluxes are augmented, such as occurs with digitalis drugs or sympathetic stimulation, the amplitude of these oscillations increases and if sufficiently high, may reach threshold potential, generating a spontaneous action potential [8]. If this is repeated, a repetitive, sustained arrhythmia may result.

Re-entry. It has been proposed that the mechanism most often causing clinical arrhythmia is re-entry which results from an impulse continuously circulating around a circuit formed within the myocardium as a result of damaged and normal tissue coexisting in the same area and linked proximally and distally [9]. Under normal circumstances, the conduction velocity and refractory periods within the myocardium are uniform. However, if some of the tissue is damaged, its conductivity is reduced and refractory period prolonged. A premature impulse, conducted normally in the normal tissue, may be blocked in the abnormal tissue which is still refractory. If these 2 regions are anatomically linked proximally and distally, forming a circuit, the impulse reaches the distal end of the circuit via the normal tissue and if the abnormal tissue has been repolarized, it is now capable of being stimulated and will conduct the impulse in a retrograde direction. Since retrograde conduction is slow, the impulse may reach the proximal part of this circuit after the normal tissue has become repolarized and is no longer refractory and it will re-enter the normal pathway. If this repeats itself, a sustained repetitive arrhythmia is generated.

Antiarrhythmic drugs may be effective regardless of the mechanism since they affect membrane automaticity, conductivity and refractoriness.

3 Antiarrhythmic drugs

Therapy with antiarrhythmic agents continues to be the most frequently used approach for the suppression and prevention of arrhythmia. While the antiarrhythmic agents available have some electrophysiologic properties in common, they represent a diverse group of drugs and each one is unique. The beneficial or adverse effects of one drug do not predict the effect of any other agent. It is therefore important to review the pharmacologic properties of each one of these drugs.

Table 1
Classification of antiarrhythmic drugs

Class	Action	ECG changes	Agents
1-Membrane stabilizers	Sodium ion inhibition		
1A	Slight reduction in upstroke velocity Prolongation of repolarization	Slight QRS widening Marked QT lengthening	Quinidine Disopyramide Procainamide (Amiodarone) (Recainam)
1B	No change in upstroke velocity No change in shortening repolarization	None	Lidocaine Mexiletine Tocainide Ethmozine
1C	Marked slowing of upstroke velocity No change in repolarization	QRS widening	Encainide Flecainide Propafenone Recainam Indecainide Ethmozine
2-Beta blockers	Blocks catecholamine action	PR prolongation Slow heart rate QT shortening	Propranolol Acebutolol Metoprolol Nadolal Timolol Pindolol Atenolol (Propafenone) (Amiodarone)
3-	Inhibits potassium fluxes Prolongation of repolarization	QT prolongation	Amiodarone Sotalol N-acetyl Procainamide
4-Calcium channel blockers	Calcium channel blockade	PR prolongation	Verapamil Diltiazem (Amiodarone)

3.1 Classification of antiarrhythmic agents (Table 1)

There are several ways in which antiarrhythmic drugs may work to alter the electrophysiologic properties of the myocardial membrane, interfering with its ability to generate or sustain an arrhythmia. On the basis of these different mechanisms, Vaughn-Williams proposed a classification of drugs and those sharing similar effects were placed into the same class [10]. This classification, however, did not imply that drugs in the same class were identical or interchangeable, but was meant to present different mechanisms by which these agents worked.

3.2 Class I, membrane-stabilizing or local anesthetic agents

Each of the drugs in this class poison or block the sodium channel and thereby inhibit the rapid inward flux of sodium ions [1, 11]. Since sodium ion movement is responsible for generating the rapid upstroke velocity (V_{max}) of phase 0 of the action potential or membrane depolarization, the result is a reduction of V_{max} of phase 0 and a decrease in the velocity of impulse conduction through the myocardium. The net effect is a prolongation of the time for membrane depolarization. Additionally, each class 1 drug prolongs the effective refractory period of the myocardium, thereby making the tissue less excitable. Lastly, these agents reduce the slope of phase 4 of the action potential which defines membrane automaticity. It is during this phase that abnormal tissue may generate an action potential and the membrane active drugs suppress this automaticity. Therefore, the effects common to each of the class 1 drugs are a reduction in membrane conductivity, excitability and automaticity. Despite these similarities, there are a number of properties which differ among the various agents and this forms the basis for a subclassification into 1A, 1B, 1C [12]. The major differences among these subclasses relate to the intensity of depression of V_{max} of phase 0 and the effect of the drug on the action potential duration or repolarization time. Additionally, class 1 drugs display the property of use-dependent depression of V_{max}, that is, the reduction in the upstroke velocity, or inhibition of inward sodium currents, is progressively more intense as the heart rate increases [13]. There are differences among drugs of the 3 subclasses in regard to the rate of onset of this use-dependent effect.

Drugs of the 1A subclass, including quinidine, disopyramide, procainamide and pirmenol slow V_{max} only modestly, correlating with an intermediate onset and offset of rate or use-dependent block. As a result of its electrophysiologic properties, the conduction velocity and rate of membrane depolarization are only moderately slowed and the QRS width only slightly prolonged. This is dose-related. However, these agents markedly prolong the action potential duration and time for repolarization, resulting in marked QT or JT prolongation. The 1B drugs, lidocaine, mexiletine and tocainide, have little effect on V_{max} and the onset and offset of the depression are rapid.
Therefore, there is little slowing of impulse conduction velocity and the QRS is not altered. As the action potential duration is not changed or may be slightly reduced, repolarization is not substantially changed and the JT not altered or may be slightly shortened. The class 1C drugs, including encainide, flecainide, propafenone and indecainide have a marked effect on depressing V_{max}, markedly slowing impulse conduction and depolarization time. The onset and offset of the use-dependent block are slow. As a result, there is substantial prolongation of the QRS duration. The action potential duration or time for repolarization is not changed and the JT is unaffected.
The class 2 drugs are the beta-adrenergic blocking agents which include: propranolol, timolol, metroprolol, nadolol, pindolol, atenolol and acebutolol. The only antiarrhythmic effects produced by this class of drugs result from prevention of the catecholamine actions on the heart [14]. These agents do not produce any direct electrophysiologic effects on the myocardium at the doses used clinically, although at high doses some of these agents do exert local anesthetic or membrane-stabilizing effects, similar to the class 1 agents. Catecholamines produce significant effects on the ventricular myocardium including an increase in V_{max} and the velocity of impulse conduction, shortening of the refractory period and an increase in membrane excitability, and an increase in automaticity. These changes are opposite to those resulting from the membrane active agents. Beta blockers, by blunting these catecholamine-mediated changes, will potentiate the actions of the membrane-stabilizing drugs and are best used as adjunctive therapy [15]. The beta blocking drugs have only a limited role as primary antiarrhythmic agents and are not first

line drugs for any arrhythmia except when catecholamines are implicated in its genesis and maintenance [16].

While each of the beta blockers have the same effects, there are differences among them in regard to their pharmacologic properties, which determine drug metabolism, clearance and side effects [17]. Some of the agents are highly lipid soluble and as a result, they are rapidly metabolized by the liver and have short half-lives. Other agents are not lipid soluble, are completely cleared by the kidney as unchanged drug and have a long half-life. Additionally, those agents which are lipid soluble penetrate tissue well, especially the central nervous system, and can cause side effects as a result. Lastly, there are other properties of beta blockers which differ among these agents, specifically beta receptor specificity ($beta_1$ or $beta_2$) and intrinsic sympathomimetic activity. Such properties are also important for determining side effects as well as the hemodynamic profile of the agent.

The class 3 agents work primarily by prolonging the action potential duration, repolarization duration and membrane refractory periods as a result of blockade of potassium ion fluxes into the cell [18]. The action potential duration is markedly increased and the refractory period prolonged which results in a reduction of membrane excitability. This is termed the „antifibrillatory“ action of this class of drug which includes amiodarone, bretylium and sotalol [19]. As a result of their effect on the action potential duration and repolarization, the JT interval is prolonged on the surface ECG.

The class 4 agents are the calcium channel blockers including verapamil and diltiazem. These agents inhibit the slow influx of calcium ions into the cell and therefore have their major action on that tissue which generates the slow action potential as a result of calcium ion movement [20]. This is primarily the pacemaker tissue of the heart, i. e. the sinus and atrioventricular nodes. By interfering with calcium fluxes, these agents reduce impulse generation as well as the refractory periods of pacemaker tissue [21]. The class 4 drugs are primarily effective for those arrhythmias which involve the sinus and AV nodes, i. e. supraventricular arrhythmias [22, 23]. The manifestations of these effects on the surface ECG include sinus bradycardia and prolongation of the PR interval.

4 Aggravation of arrhythmia

Although most of the side effects caused by the antiarrhythmic drugs are dose- and blood-level-related and thus predictable, there are some toxic reactions which are not dose-related, but are idiosyncratic. Among the most frequent and serious is the potential for these agents to aggravate arrhythmia [24]. Aggravation of arrhythmia is of 3 forms [25]:

1. Exacerbation of the underlying arrhythmia. This may be a statistical and perhaps benign increase in the number of ventricular premature beats, couplets or runs of nonsustained ventricular tachycardia (VT). Previously occurring episodes of a sustained ventricular or supraventricular tachyarrhythmia may become more frequent, more rapid, of longer duration, more difficult to terminate or incessant and unable to be terminated. Nonsustained VT may be converted to a sustained ventricular tachycardia. Such events are more serious.
2. Provocation of a new arrhythmia for the patient. This includes torsade des pointes, which is a polymorphic VT associated with QT prolongation, or polymorphic VT without QT prolongation. The first occurrence of sustained VT, ventricular fibrillation (VF) or supraventricular tachycardia also represents types of arrhythmia aggravation. This form of arrhythmia aggravation is serious and potentially life-threatening.
3. Development of a bradyarrhythmia due to impairment of sinus or AV nodal function or resulting from other conduction abnormalities.

One of the difficulties in establishing arrhythmia aggravation relates to the known random variability of arrhythmia. When a new arrhythmia occurs, there is no question about the relationship to the drug. However, when there is a statistical increase in the frequency of arrhythmia or some other change, implicating the antiarrhythmic drug may be difficult unless the patient is re-challenged. However, if arrhythmia aggravation occurs within a few days of initiating therapy or changing the dose, the drug is a likely cause.

When the above criteria are applied, aggravation of arrhythmia occurs in 9 % of studies in which noninvasive ambulatory monitoring and exercise testing are used and in 18 % of electrophysiologic stud-

Table 2
Aggravation of arrhythmia incidence

	Incidence (%)	
Drug	Noninvasive	Invasive
Amiodarone	4	–
Disopyramide	6	5
Encainide	15	27
Ethmozine	11	14
Flecainide	12	–
Indecainide	19	–
Mexiletine	7	20
Procainamide	9	21
Propafenone	8	15
Quinidine	15	20
Tocainide	8	5

ies (Table 2) [26, 27]. The most common form of arrhythmia aggravation is the occurrence of a new arrhythmia.

Although arrhythmia aggravation is not predictable based on any clinical parameters, including ECG changes, drug dose or blood levels, it has been reported to be more common in patients who have a previous history of sustained VT or VF and in patients with a history of congestive heart failure and a reduced left ventricular ejection fraction [28–30]. While the antiarrhythmic drugs have many electrophysiologic properties in common, aggravation of arrhythmia with one drug does not predict the occurrence of this toxic reaction with any other drug, even if of the same subclass [28]. Exercise testing is an important method for exposing this potentially serious complication. In approximately a third of cases, aggravation occurs during exercise testing at a time when monitoring documents arrhythmia suppression [28]. Exercise testing produces a number of important physiologic changes which may be responsible for the provocation of this complication [31]. During exercise, there is activation of the sympathetic nervous system and an increase in circulating catecholamines. Additionally, there are changes in pH, serum potassium and the development of myocardial ischemia which may interact with the action of antiarrhythmic drugs or exacerbate the underlying inhomogeneity of the myocardium resulting in the potential for enhancing arrhythmia. The idiosyncratic nature of this complication urges cautious use of these agents with continuous monitoring of their effects.

4.1 Exacerbation of congestive heart failure

Each of the antiarrhythmic agents is negatively inotropic and has the potential to depress myocardial function and provoke congestive heart failure (Table 3) [32]. In addition to their direct myocardial effects, many of these agents have peripheral vascular effects which may offset or exacerbate their negative inotropic actions [33, 34].

Table 3
Hemodynamic effects of antiarrhythmic drugs

Drug	PAP/PCWP	SVR	Contractility	SV/CF	LVEF
Quinidine	0	↓	0 or sl ↓	0	0
Procainamide	0	↓	0 or sl↓	0	0
Disopyramide	↑	↑	↓	↓	↓
Pirmenol	↑	↑	↓	↓	↓
Mexiletine	0	sl↓	0 or sl↓	0	0
Tocainide	0	sl↓	0 or sl↓	0	0
Ethmozine	sl↑	0	0 or sl↓	0	0
Encainide	0	↑	sl↓	sl↓	0
Flecainide	↑	↑	↓	↓	↓
Propafenone	↑	↑	↓	↓	sl↓
Indecainide	↑	↑	↓	↓	↓
Recainam	sl↑	0	↓	sl↓	sl↓
Amiodarone	0	↓	0 or sl↓	0	0
Sotalol	0	0	0 or sl↓		0 or sl↓

In general, each drug does cause some reduction in left ventricular contractility. If the agent is also a peripheral vasodilator, which reduces peripheral vascular resistance or afterload, stroke volume and cardiac output may be maintained. If there is peripheral vasoconstriction, either by direct effect or indirectly because of the depression of left ventricular contractility, there is an increase in afterload and stroke volume and cardiac output may be further reduced.
The incidence of congestive heart failure provocation varies from one drug to another [32]. However, regardless of the agent, patients with a prior history of failure, a markedly reduced left ventricular ejection fraction or an underlying cardiomyopathy are at greatest risk for this complication [32]. It is unusual for patients with intact left ventricular function to experience this complication.

5 Review of individual agents

5.1 Flecainide [35] (Table 4)

Electrophysiology

The major electrophysiologic action of flecainide is sodium channel blockade which accounts for its membrane-stabilizing properties [36, 37]. In isolated Purkinje fibers or ventricular muscle, this inhibition of the inward sodium current results in slowing of the upstroke velocity (phase 0) of the action potential and a decrease in action potential amplitude. Similar to other class 1 agents, sodium channel blockade with flecainide is use-dependent, related to the frequency of stimulation, that is, an increase in stimulus frequency results in a more marked depressant effect on upstroke velocity [12]. Like other class 1C drugs, but unlike the 1A and 1B agents, flecainide binds to and dissociates from the sodium channel slowly. This prolonged recovery time may account for the slowing of His-Purkinje and ventricular conduction even at physiologic heart rates. Flecainide also increases action potential duration and effective refractory periods in the ventricular and atrial myocardium, but decreases these parameters in Purkinje fibers [36, 37]. This disparity in the electrophysiologic effects between Purkinje fibers and ventricular tissue may be important for the development of reentrant circuits and therefore, may be a mechanism for arrhythmia aggravation.

Table 4
Flecainide

Dose	100–200 mg BID oral
Protein binding	40 %
Clearance	75 % hepatic metabolism – inactive metabolites 25 % renal unchanged
Half-life	Up to 27 hours
Blood levels	< 1000 μg/ml
Drug interactions	cimetidine, digoxin
Indications	ventricular and supraventricular arrhythmia
Side effects	CNS, GI, cardiac

Consistent with the in vitro electrophysiologic effects of flecainide, conduction is slowed in all portions of the intact heart, especially the His-Purkinje system and ventricular myocardium. Electrophysiologic studies have shown that flecainide markedly slows atrial, AV nodal, His-Purkinje and intraventricular conduction resulting in lengthening of the AH and HV interval [36]. There is also a small, but significant, increase in the effective refractory period of atrial and ventricular muscle while there is a marked increase in retrograde refractoriness of the fast AV nodal pathway in addition to slowing of conduction. The drug also slows impulse conduction and prolongs the refractory period of accessory bypass tract tissue. The implication of these electrophysiologic changes is that flecainide may be useful for the treatment and prevention of supraventricular tachyarrhythmias as well as ventricular arrhythmias [38]. Flecainide does not have a depressant effect on sinus node function when normal, but it may exacerbate pre-existing disease by prolonging sinus node recovery time.

Changes in the surface electrocardiogram (ECG) reflect the depressant effect of flecainide which is dose-dependent. The QRS and PR intervals are prolonged while the JT interval, a measurement of repolarization, is unchanged.

Pharmacology

Flecainide is currently available only for oral use with the recommended dosage being 100 to 200 mg twice per day. The drug is rapidly and completely absorbed from the GI tract with peak plasma levels being achieved 3 to 4 hours following administration. Neither food nor antacids affect the absorption of this drug [39]. Only 40 % of flecainide is protein-bound, therefore plasma levels do not fluctuate with changes in plasma protein levels. Approximately 75 % of flecainide undergoes hepatic biotransformation into metabolites which are devoid of antiarrhythmic activity. The remaining 25 % is excreted unchanged by the kidney and this is pH-dependent. An alkaline urine (pH > 7) is associated with decreased renal clearance of flecainide and higher plasma levels of the drug.

The average half-life of flecainide is approximately 13 to 20 hours, but may be as long as 27 hours, allowing for BID dosing. As a result of its long half-life, at least 96 hours are required for a steady-state level to be achieved and dose changes should not be more frequent than every four days. Patients with congestive heart failure and renal disease have decreased plasma clearance of flecainide and the initial dose should be decreased by 25 to 50 % with the dosage increased no more frequently than every four days. Dialysis does not effectively clear flecainide from the body.

Loading doses of the drug are not indicated and can potentially expose the patient to unnecessary toxicity. The therapeutic plasma level of flecainide is reportedly 200 to 1000 ng/ml. While the plasma level is unrelated to the efficacy of the drug for an individual patient, the incidence of side effects is higher when the level is > 1000 ng/ml. Little data are available about the effects of flecainide administered in combination with other commonly used medications; however, a few important interactions have been described [35]. Cimetadine has been found to decrease the nonrenal clearance of flecainide by 13 % to 27 % in normal healthy patients and its elimination half-life is increased. Digoxin levels have been reported to increase minimally with flecainide therapy, but this is clinically unimportant. When flecainide and beta blockers are administered together, the plasma concentration of both agents increase which may potentiate the hypotensive and negative inotropic effects of these two agents in patients with pre-existing left ventricular dysfunction. Caution should be exercised when flecainide is administered with other cardioactive or hepatically metabolized agents, especially in patients with underlying cardiac disease.

Hemodynamics

Flecainide exerts a direct depressant effect on left ventricular function and this is related to dose and plasma level [40]. Following an intravenous dose of flecainide, left ventricular ejection fraction is reduced by 10 % to 15 % in patients with and without pre-existing left ventricular dysfunction. There is also a concomitant decrease in stroke volume, cardiac output and an increase in pulmonary capillary wedge pressure. Heart rate and blood pressure remain unaffected. Long-term oral therapy is also associated with a reduction in left ventricular function.

The incidence of new onset congestive heart failure in patients with normal ventricular function is low, approximately 2 %. However, up to 15 % of patients with compromised left ventricular function and preexisting congestive heart failure will have aggravation of heart failure with oral therapy [40–42]. Therefore, flecainide must be used with caution in patients known to have depressed left ventricular function, a reduced ejection fraction and congestive heart failure.

Efficacy

Flecainide has been reported to be very effective for the treatment of ventricular premature beats (VPBs) and nonsustained VT. Multiple placebo controlled trials have demonstrated that flecainide suppresses over 85 % of VPBs in approximately 80 % of patients with a wide range of cardiac disease [43, 44]. In addition, nonsustained VT is completely abolished in over 95 % of patients treated with flecainide. A dose of 100 to 200 mg twice daily is the usual one necessary to achieve this degree of arrhythmia suppression.

In comparative trials, flecainide has been shown to be more effective than either quinidine or disopyramide [45, 46]. While 85 % of patients have 80 % VPB reduction with flecainide, quinidine produces a similar reduction in only 57 %. Flecainide is also more effective than quinidine in suppressing couplets and nonsustained VT. Similar results are obtained when disopyramide is compared to flecainide. Additionally, the incidence of side effects with short- and long-term flecainide therapy is less than that resulting from quinidine and disopyramide.

Although flecainide is very effective for suppressing asymptomatic VPBs and nonsustained VT, patients with a recent myocardial infarction treated with flecainide are at an increased risk for an arrhythmic death as reported by the Cardiac Arrhythmia Suppression Trial [47]. In this study, there was a significant increase in the number of deaths among patients receiving flecainide (and encainide) (7,7 %) compared to placebo (3,0 %). Based on this study, flecainide is now restricted for use only in patients with refractory-sustained ventricular tachyarrhythmias (VT or VF). However, in such patients, the efficacy of flecainide is less well established and only 12–20 % are rendered noninducible when electrophysiologic techniques are employed while 29 % experienced a worsening of arrhythmia [48, 49]. Other investigators have reported a greater rate of noninducibility. All studies using electrophysiologic techniques to assess drug efficacy have reported that flecainide significantly decreases the rate of induced VT and this usually, but not always, results in better hemodynamic tolerance of the arrhythmia. However, hemodynamic collapse may occur even if the VT rate is slowed due to the marked depression of impulse conduction and the potential for heterogeneous depolarization and activation of the myocardium resulting in disorganized contraction.

Although not yet approved for supraventricular arrhythmias, flecainide has been reported to be effective for a wide spectrum of supraventricular arrhythmias including re-entrant supraventricular tachycardia, atrial flutter and atrial fibrillation even when these arrhythmias are associated with the Wolff-Parkinson-White (WPW) syndrome [38, 50, 51].

Side effects

The most frequent side effects associated with flecainide are related to the central nervous system and are usually dose-related [52]. They include dizziness, visual disturbances and tremor. These adverse effects have been reported in up to 40 to 60 % of patients and are usually easily managed with a reduction in dose. Only 10 % of patients will require discontinuation of flecainide because of these side effects. Gastrointestinal side effects are also common and include nausea, abdominal discomfort and vomiting.

Since flecainide markedly decreases AV and infranodal conduction, there is a prolongation of the PR and QRS intervals. In general, these changes in conduction are asymptomatic, but bundle branch block and complete heart block have been reported. Sinus node dysfunction may occur with flecainide use, especially in those with preexisting disease. Flecainide is a potent negatively inotropic agent which may precipitate or exacerbate congestive heart failure [40, 41]. This is an important concern in patients with underlying myocardial dysfunction and a history of heart failure.

Arrhythmia aggravation is the most serious side effect associated with flecainide use and has been reported in up to 20 % of patients assessed noninvasively [48, 49]. Like other class 1C agents, the drug has the potential to cause incessant VT or VF which is refractory to all modes of therapy including defibrillation. Initial studies reported that this complication was most common in patients with a reduced ejection fraction, CHF and a history of sustained VT or VF [29, 30]. However, results from CAST indicate that arrhythmia aggravation is a major complication in post infarction patients being treated with flecainide for asymptomatic ventricular arrhythmias [47]. In light of these results, flecainide is no longer indicated for the treatment of asymptomatic or symptomatic nonsustained arrhythmias and should be used with caution in patients with life-threatening arrhythmias.

The risk of arrhythmia aggravation in other patients, such as those with supraventricular arrhythmias or chronic stable CAD, is still unknown.

Table 5
Encainide

Dose	25–50 mg TID or QID oral
Clearance	hepatic to active metabolites
Half-life	3-hour parent 8–10 hours metabolites
Blood levels	variable
Drug interactions	cimetidine, beta blockers
Indications	ventricular and supraventricular arrhythmia
Side effects	CNS, cardiac

5.2 Encainide [53, 54] (Table 5)

Electrophysiology

As a class I membrane-stabilizing drug, encainide exerts sodium channel blocking effects and exhibits electrophysiologic properties which are typical of 1C antiarrhythmic agents [55]. Both encainide and its metabolites, which are active, have potent electrophysiologic activity on normal and ischemic atrial and ventricular myocardium which accounts for its clinical efficacy for supraventricular and ventricular arrhythmias.

Studies in the isolated nonischemic Purkinje fiber and myocardial cell report that the drug slows the upstroke velocity (phase 0) of the action potential by blocking fast sodium channels, thereby exerting a direct membrane-stabilizing action. In addition, the amplitude of the action potential is reduced, but the effective refractory period of this tissue remains unchanged. Similar electrophysiologic changes are seen with O-desmethyl encainide (ODE) and 3-methoxy-O-desmethyl encainide (MODE), the two major metabolites [56]. However, ODE is reported to slow upstroke velocity and decrease action potential amplitude to a greater degree than either the parent compound or MODE [57]. All three compounds have a more pronounced depressant effect on ischemic myocardial cells compared to normal tissue.

The major electrophysiologic effect of encainide on the intact heart is prolongation of the HV interval. There is a dose-dependent decrease in His-Purkinje and ventricular conduction velocity, but no significant change in ventricular repolarization or refractoriness [55]. As with other agents, the depression of conduction is use-dependent, being more profound at faster heart rates. In patients with an accessory pathway, the refractory period of the bypass tract tissue is also lengthened. Heart rate, sinus node recovery time, AV nodal conduction and atrial and AV nodal refractoriness are not affected by a short-term intravenous encainide infusion. More significant electrophysiologic changes are seen with chronic oral therapy, and these are probably secondary to the active metabolites of encainide, especially ODE, which have been shown to markedly prolong AV nodal conduction and ventricular refractoriness in the intact heart model [57]. Changes on the surface ECG after encainide therapy reflect its depressant effects on the impulse conduction velocity through all tissue. There is significant lengthening of the PR and QRS intervals. There is no change in the JT segment suggesting that repolarization is not affected with encainide therapy.

Pharmacology

Encainide is available only for oral use and the recommended dose is 25 to 50 mg TID or QID up to a maximum dose of 200 mg per day. The drug is rapidly absorbed from the GI tract and transported to the liver where further metabolism occurs [58]. The rate of metabolism is genetically linked [59]. In 90 % of patients, encainide is rapidly metabolized in the liver into two clinically active compounds, ODE and MODE. ODE is the metabolite that accumulates during long-term therapy and is the primary compound responsible for the antiarrhythmic effects of the drug [57]. While the half-life of encainide is only 3 hours, the elimination half-life of ODE is 8 to 10 hours, allowing for TID dosing. In the remaining 10 % of the population, a genetic defect results in the absence of the hepatic enzyme which rapidly degrades encainide. Although encainide, the parent compound, has weaker antiarrhythmic effects than ODE, it accumulates in high levels in these patients and is responsible for the electrophysiologic activity and arrhythmia suppression. The half-life of encainide in these slow metabolizers is 6 to 14 hours. Therefore, despite variability in encainide metabolism, the maintenance dose and dosing intervals are approximately the same in both groups.

In both rapid and slow metabolizers, encainide and its metabolites are partially cleared by the kidney. In rapid metabolizers, 21 % to

50 % of the encainide dose is eliminated in the urine in the form of ODE. Poor metabolizers excrete over 50 % of the drug in the form of encainide [57].

Therapy with encainide should be initiated with a low dose and increased no more frequently than every three to four days. This allows time for steady-state levels of ODE and MODE to be achieved. As can be expected from the variability in metabolism, plasma levels of encainide and its metabolites are of little value in guiding therapy or establishing an effective dose.

In patients with advanced liver disease who are rapid metabolizers, the conversion of encainide to its active metabolites is decreased [60]. However, the subsequent degradation of ODE and MODE is also impaired such that the electrophysiologic and clinical effects of encainide in this group of patients are not significantly different from those in patients with normal hepatic function and no adjustment in encainide dosage is necessary. No information is yet available on the pharmacokinetics of encainide in slow metabolizers with significant hepatic dysfunction.

In contrast, rapid metabolizers who have renal disease have impaired excretion of both encainide and its metabolites [61, 62]. Studies have shown that despite a reduction in encainide dose, patients with renal impairment have significantly higher levels of ODE and MODE than patients with normal renal function. Therefore, in the presence of renal impairment, therapy should be initiated at the lowest possible dose with long dosing intervals. Dosage increase should be slow with careful attention to toxic side effects.

Few interactions have been reported when encainide is administered along with other commonly prescribed medications [61]. Cimetadine has been shown to increase plasma levels of encainide and its metabolites by 30 to 40 %. The concurrent use of beta blockers may worsen the neurological side effects associated with encainide therapy as well as its negative inotropic actions. There are no clinically significant interaction between encainide and other cardioactive agents, such as digoxin or verapamil. There is very little experience with the concomitant use of encainide and other antiarrhythmic agents. Because of the possible additive effects on cardiac conduction and myocardial contractility, caution should be exercised when combination therapy with encainide is used.

Hemodynamics

Compared to other type 1C agents, encainide appears to exert less negative inotropic effects [63]. Patients with normal and moderately impaired left ventricular function, but no failure, receiving long-term encainide therapy do not have a significant reduction in ejection fraction at rest or with exercise at doses necessary to achieve arrhythmia suppression. There is also no significant change in heart rate, blood pressure, pulmonary capillary wedge pressure or stroke volume. However, there have been reports of new onset or worsened congestive heart failure with encainide therapy, occurring in up 6 % of patients with left ventricular dysfunction and a history of congestive heart failure [32]. In one study, a single large dose of drug caused significant changes in all invasive hemodynamic parameters and an exacerbation of failure in many patients with significant left ventricular function and a history of heart failure [64]. Therefore, encainide should be used with caution in patients with depressed left ventricular function and a history of heart failure.

Efficacy

Similar to other class 1C agents, encainide is highly effective for suppressing spontaneously occurring VPBs and nonsustained VT. Multiple placebo controlled trials have demonstrated that encainide effectively suppresses over 80 % of VPBs, 99 % of couplets and all runs of nonsustained VT when compared to placebo [58, 65–67]. Comparative trials have reported that both quinidine and disopyramide are less effective than encainide for suppressing these arrhythmias [68]. The efficacy of encainide in the treatment of malignant, sustained tachyarrhythmias is less well established. When evaluated noninvasively with ambulatory monitoring and exercise testing, 48–52 % of patients with a history of refractory-sustained VT or VF have suppression of spontaneous arrhythmia with encainide therapy [69]. However, only 10–20 % of patients with sustained ventricular tachyarrhythmias are rendered noninducible when electrophysiologic techniques are used [70, 71]. As with flecainide, the rate of induced VT is slower as compared to baseline. There appears to be no correlation between encainide efficacy and dose or blood level of drug or electrocardiographic changes. It has been reported that encainide is more effective in patients without a history of congestive heart failure and with a left ventricular ejection fraction ›35 % than in those with impaired left ventricular function.

Encainide has been reported to be effective in the treatment of supraventricular tachyarrhythmias, especially those associated with an accessory pathway [72–75]. In patients with such arrhythmias, including those with the WPW syndrome complicated by atrial fibrillation or reentrant SVT, 40 % are noninducible by electrophysiologic studies during oral encainide therapy and during long-term follow-up 65 % of these patients remain asymptomatic and free of recurrence. In patients with AV nodal re-entrant tachycardia, 68 % will be noninducible on encainide therapy. In the remaining patients, the induced tachycardia will often be with a slower rate, asymptomatic and self-terminating.

Side effects

Encainide is generally well tolerated although approximately 20 % of patients will experience minor side effects such as nausea, dizziness, tremor, parasthesias, and fatigue [69, 76]. These symptoms are usually mild, dose-related and do not necessitate discontinuation of the drug.

The most serious complications of encainide use are cardiac, especially arrhythmia aggravation which has been reported in up to 25 % of patients [26, 27, 29, 30, 77]. As with other antiarrhythmic agents this adverse effect is more frequent among patients with a history of sustained VT or VF, depressed left ventricular function and a history of congestive heart failure. Arrhythmia aggravation also occurs in patients with a history of VPBs and nonsustained VT, although initial trials involving patients with these arrhythmias reported an incidence of only 4 %. However, results from CAST indicate that patients post infarction with VPBs being treated with encainide had an increased incidence of sudden death, a result of arrhythmia aggravation. There have been few reports of arrhythmia aggravation in patients receiving encainide for symptomatic SVT and the incidence is about 1 %. There does not appear to be a relationship between QRS prolongation and arrhythmia aggravation. In addition, dosage of encainide and serum levels of the drug and its metabolites do not correlate with this cardiac complication.

Other cardiac side effects of encainide therapy include AV nodal and infranodal conduction abnormalities and the precipitation or worsening of congestive heart failure.

5.3 Propafenone [78, 79] (Table 6)

Table 6
Propafenone

Dose	150–300 mg TID oral
Clearance	hepatic – first pass effect-metabolites have slight activity
Half-life	6–10 hours
Blood levels	variable
Drug interactions	digoxin, cimetidine, coumadin
Indications	ventricular and supraventricular arrhythmia
Side effects	cardiac, CNS, GI

Electrophysiology
Propafenone is a new 1C antiarrhythmic agent that is effective for the treatment of ventricular and supraventricular arrhythmias. Propafenone is a membrane-stabilizing agent with unusual electrophysiologic properties. Similar to other class 1C agents, its dominant action is blockade of fast sodium channels resulting in a reduction in upstroke velocity of phase 0 of the action potential [80]. The amplitude of the action potential is also decreased, but its duration is not significantly prolonged. Thus, the drug does not alter membrane repolarization. As with flecainide and encainide, the effects of propafenone on the sodium channel are use-dependent being more apparent at faster heart rates.
Unlike the other 1C drugs, propafenone possesses both weak beta blocking and calcium channel blocking activity [81]. In vitro and in vivo studies have documented that propafenone inhibits the chronotropic response to isoproterenol, a potent beta agonist. Based on these studies, it has been estimated that propafenone is 1/40 as potent as propranolol. However, the plasma concentration of propafenone achieved is more than 40 times greater than what is achieved with propranolol [82]. In addition, propafenone has been reported to reduce exercise-induced tachycardia in normal volunteers, further suggesting this agent processes clinically important beta blocking properties.
Propafenone also exerts an inhibitory action on slow inward calcium channels [83, 84]. In vitro studies have demonstrated that propafenone decreases the amplitude of delayed after potentials which are calcium-dependent. Phase 4 depolarization and action potential am-

plitude in the sinus and AV nodes are also depressed with propafenone, secondary in part to the calcium channel blocking activity of this agent.

Clinically, propafenone prolongs the conduction times and refractory periods in the atria, AV node, His-Purkinje system, ventricle and accessory tract tissue [85, 86]. This results in lengthening of the PR and QRS intervals on the surface ECG. As with flecainide and encainide, repolarization is not significantly affected by propafenone and therefore, the JT interval is unchanged.

Pharmacology

Propafenone is administered by the oral route with the recommended dosage being 150 to 300 mg TID. An intravenous form is also available for investigational use. Following near complete absorption of propafenone from the GI tract, 99 % of the drug undergoes first pass hepatic metabolism, yielding two major metabolites, 5-hydroxy propafenone and N-dibutyl propafenone [87, 88]. The former exhibits some antiarrhythmic activity although its potency is not well established.

Propafenone exhibits nonlinear pharmacokinetics when standard doses are administered. As the dose increases, hepatic metabolic sites become saturated resulting in decreased metabolism of the drug, a prolongation of its half-life, and an increase in blood levels. Larger doses of the drug given less frequently may therefore be more beneficial than the same dose administered more often in smaller amounts.

Similar to encainide, the rate of propafenone metabolism is variable and genetically predetermined [89]. Approximately 90 % of the population will rapidly metabolize propafenone into its two major metabolites. The elimination half-life in this group is approximately six hours. In the remaining 10 % of patients who slowly metabolize this drug, the half-life of propafenone is 10 to 16 hours. As a result of the variability in propafenone metabolism and its nonlinear kinetics, therapy should be initiated with a low dose and increased only every three to four days to allow steady state levels to be achieved. Monitoring blood levels of propafenone is of no clinical value for determining efficacy or toxicity.

There are a number of drug interactions reported with propafenone, although their clinical significance is uncertain [90]. There have been a few studies reporting a digoxin-propafenone interaction with an increase of digoxin levels by 37–63 % [91]. However, data are not conclusive and the clinical relevance not clear. Cimetidine causes an increase in propafenone levels by 24 %, but this is of little clinical importance. Propafenone will increase coumadin levels and the prothrombin time and coumadin dose should be reduced and prothombin times carefully monitored. As with other antiarrhythmic drugs, there are studies reporting a beneficial antiarrhythmic effect when propafenone is combined with other agents including beta blockers, quinidine, procainamide, amiodarone and sotalol [92].

Hemodynamics
Propafenone is a negative inotropic agent which has been shown to have a depressant effect on left ventricular function [93]. Following an intravenous infusion of propafenone, there is a significant reduction in myocardial contractile force and blood pressure and a concomitant increase in pulmonary capillary wedge pressure, and pulmonary and systemic vascular resistance. With chronic oral propafenone therapy, patients with normal baseline left ventricular function do not have a significant reduction in function. However, those patients with an ejection fraction <50 %, may have a 10 to 20 % reduction in ejection fraction. Despite this reduction, the incidence of new or worsening congestive heart failure is very low, reported to be 5–9 % [32, 94]. However, propafenone should be administered with caution to patients with a history of congestive failure and poor left ventricular function.

Efficacy
Similar to other 1C agents, propafenone is effective for the suppression of all forms of spontaneous ventricular ectopy including VPBs, couplets, and nonsustained VT [91, 95, 96]. Multiple placebo-controlled trials have reported that oral propafenone reduces VPBs by over 80 % in approximately 80 % of patients, and abolishes nearly all couplets and runs of nonsustained VT. When compared with disopyramide, propafenone suppressed VPBs by greater than 80 % in 59 % of patients, while only 43 % responded to disopyramide [97]. Propafenone was also shown to completely abolish nonsustained VT in 89 % of patients whereas only 55 % had elimination of these forms with disopyramide. Quinidine and propafenone are reported to be

equally effective in the suppression of VPBs and runs of nonsustained VT [98]. However, quinidine was associated with more frequent extracardiac side effects requiring discontinuation of therapy.
Propafenone efficacy in patients with reccurent sustained VT or VF is less well established. When patients with a history of a sustained hemodynamically significant ventricular tachyarrhythmia who manifest a high density of spontaneous arrhythmias are assessed by exercise testing and ambulatory monitoring, propafenone is effective in approximately 50 % of patients when drug efficacy is defined as total elimination of runs of nonsustained VT, 90 % reduction in the frequency of couplets, and 50 % reduction in VPBs [99, 100]. Like other 1C agents, propafenone appears to be less effective when assessed by electrophysiologic testing and only 15 to 20 % of patients are rendered noninducible [85, 101, 102]. However, in those patients with sustained ventricular tachycardia which is still inducible, the cycle length of the induced arrhythmia is prolonged and the tachycardia usually better tolerated hemodynamically. In addition, several studies have reported a low recurrence rate of symptomatic arrhythmia, and a low mortality among patients who remain inducible on propafenone therapy when rate slowing is achieved. Therefore, other factors in addition to the presence or absence of inducibility may be important in determining which patients will benefit from long-term propafenone therapy.
Propafenone slows conduction through the atria, AV node and accessory pathway tissue and has been found to be useful in the treatment of supraventricular arrhythmias including re-entrant supraventricular tachycardia, atrial fibrillation and arrhythmias associated with WPW syndrome [103, 104]. In patients with AV re-entrant tachycardia associated with WPW, intravenous propafenone (2 mg/kg) terminates and prevents reinduction of the arrhythmia in approximately 65 % of patients assessed by electrophysiologic techniques [105]. During long-term therapy with propafenone, 40 % of these patients remain asymptomatic and without recurrence of arrhythmia. In an additional 40 %, the tachycardia occurs less frequently and was usually slow and self-terminating [106]. Propafenone has also been useful in the treatment of paroxysmal atrial fibrillation and atrial flutter, although the studies involve only small numbers of patients and are uncontrolled [103, 104]. In approximately 55 % to 60 % of patients with this arrhythmia, propafenone appears to be effective for

decreasing the frequency of symptomatic arrhythmia. When the atrial arrhythmia does recur, it is often better tolerated because slowing of AV nodal conduction results in a decrease in the ventricular rate.

Side effects

Up to 30 % of patients receiving propafenone experience mild side effects including dizziness, nausea, constipation and abnormal metallic taste resulting from secretion of the drug in saliva by the salivary glands [107]. As a result of the drug's depressive effect on conduction, potentially serious conduction abnormalities can occur, such as sinus node arrest or complete AV nodal or intraventricular block. Because propafenone is a negative inotropic agent, left ventricular function can be impaired resulting in new or worsening congestive heart failure [32, 93, 94]. However, this complication is unusual in patients with normal ventricular function and no clinical history of congestive heart failure.

The most serious complication with propafenone therapy is arrhythmia aggravation, which has been observed in patients with benign as well as life-threatening ventricular arrhythmias. Approximately 10 % of patients with a history of sustained VT or VF, especially in association with heart failure and a reduced ejection fraction, will have aggravation of their underlying arrhythmia [27, 28]. In patients without such features, aggravation occurs in about 4 % [107]. EKG changes and drug levels do not predict the patient at greatest risk for this complication. Propafenone was not investigated by CAST and its efficacy and toxicity in post-infarction patients are unknown. It differs electrophysiologically from flecainide and encainide, and appears to cause less arrhythmia aggravation in patients with sustained VT or VF or those with heart failure. While it is possible that propafenone would not have the same adverse effects in post-infarction patients as encainide and flecainide, this has not been established.

5.4 Indecainide (Table 7)

Table 7
Indecainide

Dose	50–100 mg TID oral
Clearance	hepatic – metabolites inactive
Half-life	10–12 hours
Blood levels	–
Drug interactions	–
Indications	ventricular arrhythmia – supraventricular
Side effects	GI, CNS, cardiac

Electrophysiology
Indecainide is an investigational 1C drug, but there are little published data about its efficacy and safety. Similar to other class 1C agents, the drug's major electrophysiologic effect is slowing of impulse conduction through atrial and ventricular myocardial tissue by blocking sodium channels [108, 109]. The action potential duration and thus membrane repolarization and the effective refractory period of the myocardium are not significantly altered. As a result of the depression changes in membrane conductivity, the PR and QRS intervals are prolonged, while the JT segment remains unchanged. Although indecainide exerts a marked depressant effect on atrial, AV nodal and ventricular conduction, sinus node function and automaticity are unaffected.

Pharmacology
Indecainide is available only for oral use with doses of 50 to 100 mg TID reported effective [110, 111]. The drug is metabolized by the liver to inactive metabolites and its half-life is 10 to 12 hours [112]. The role of blood levels, and the recommended frequency of dose adjustment are not established. Drug interactions associated with indecainide use have not yet been reported.

Hemodynamics
Similar to other 1C agents, indecainide is negatively inotropic and has been shown to depress left ventricular function in patients with normal as well as those with a reduced left ventricular ejection fraction. In one small study, patients with an ejection fraction > 50 % had

a 20 % reduction in ventricular function as assessed by radionuclide ventriculogram [111]. A similar reduction was seen in patients with a depressed ejection fraction <50 %. The clinical importance of this change is unknown as there was no long-term follow-up and the reported experience with indecainide is limited.

Efficacy

There have been only a few studies of indecainide use, but it appears to be effective in suppressing all forms of ventricular ectopy. Approximately 85 % of patients on indecainide therapy will have a 90 % reduction in VPBs [112, 113]. A similar percentage will have complete abolition of runs of nonsustained VT. When compared to disopyramide, indecainide suppressed VPBs by 93 %, while disopyramide was associated with an 80 % reduction [114]. Both drugs were found to be equally efficacious for the treatment of nonsustained VT.

Indecainide appears less effective in patients with a history of sustained ventricular arrhythmias. When evaluated by ambulatory monitoring and exercise testing, only 25 % of patients had a significant reduction in spontaneous arrhythmia [111]. When electrophysiologic techniques are used, the response rate is even lower, approximately 10 %, similar to other class 1C drugs [110]. However, as expected, the cycle length of induced VT is significantly prolonged.

Side effects

The noncardiac side effects associated with indecainide include nausea, vomiting, abnormal taste, dizziness and blurred vision. Further experience with this agent is necessary before the side effect profile can be completely described. As with other class 1C drugs, conduction abnormalities, congestive heart failure, and arrhythmia aggravation have been reported with indecainide use [110, 111]. However, the incidence of these side effects and patient population most at risk are yet to be determined.

5.5 Mexiletine [116] (Table 8)

Table 8
Mexiletine

Dose	150–400 mg po TID
Clearance	hepatic 85 % – inactive metabolites renal 15 %
Half-life	10–12 hours
Blood levels	0.5–2 μg/ml
Drug interactions	phenytoin, barbiturates, rifampin, lidocaine
Indications	ventricular arrhythmia
Side effects	GI, CNS

Electrophysiology
Mexiletine is a membrane-stabilizing drug which, like other drugs of the 1B class, produces blockade of the rapid sodium channel. As a result of this inhibition in the movement of sodium ions into the cell, there is a slowing of the upstroke velocity of phase 0 resulting in a reduction in impulse conduction through the myocardium [115, 116A]. However, the effect of this drug is relatively weak and therefore, the reduction in upstroke velocity only slight. Mexiletine is a lidocaine congener and similar to lidocaine it does not effect the action potential duration [117]. It is therefore classified as a 1B agent along with lidocaine and tocainide. Similar to other membrane active agents, mexiletine demonstrates use dependence and like other agents of the 1B subclass, it rapidly binds to the receptor and the onset of sodium channel blockade is rapid [117]. Its dissociation is also rapid. In view of its rapid onset and offset of action, it has been suggested that this agent is more effective for rapid ventricular tachyarrhythmias [118]. Mexiletine exerts its electrophysiologic effects primarily on the His-Purkinje system and ventricular muscle [119, 120]. The drug has no action on atrial tissue and has no effects on sinus or AV nodal function. Mexiletine does, however, have some effect on the accessory pathway and may slow conduction through this tissue. Based on these above properties, the role of mexiletine is specifically for the treatment of ventricular arrhythmia and it has no beneficial effect against atrial tachyarrhythmias except perhaps in some cases of those associated with the WPW syndrome.
Given the relatively weak effect of this drug on inhibiting sodium ion influx and slowing of conduction and the lack of effect on action

potential duration or repolarization, it can be predicted that the drug has no significant effect on the surface electrocardiogram and at the usual dosage used, there are no changes in the PR, QRS or QT intervals.

Pharmacology

Mexiletine is only available for oral use and the recommended dose is 150 to 400 mg PO TID. The most frequently used doses, however, are 200 to 300 mg TID [121]. The intravenous form of the drug is investigational and is not yet in clinical use. Mexiletine is rapidly and completely absorbed from the gastrointestinal tract with peak levels being achieved in 1½ to 2 hours [121–124]. The rate of absorption of mexiletine is decreased by narcotics, antacids and atropine-like drugs [125, 126]. It has also been reported that the coadministration of the drug with food will delay absorption and the time to reach peak levels, but not impair the amount of drug absorbed or the maximal level achieved.

The elimination half-life of mexiletine ranges from 10 to 16 hours and its clearance is approximately 6 ml/minute/kg. Approximately 85 % of the drug is metabolized by the liver to inactive metabolites while 15 % of the drug is excreted unchanged by the kidney [121–123]. Renal excretion is pH-dependent and in the presence of urinary alkalosis the renal excretion of the drug is impaired and the drug blood levels increase [127]. The presence of severe hepatic or renal dysfunction will cause an increase in blood level of mexiletine and therefore, the dose of the drug should be reduced in such conditions. The presence of congestive heart failure will reduce both hepatic and renal clearance of the drug, prolonging the half-life [128]. It has also been reported that drug half-life is increased in the presence of an acute myocardial infarction [129]. Although the therapeutic half-life of mexiletine is reported to be 0.5 to 2 μg/ml, blood levels do not correlate with either efficacy or toxicity and are not clinically useful [130].

There have been a number of drug interactions reported with mexiletine. Drugs that induce hepatic enzymes and increase hepatic metabolic activity such as phenytoin, phenobarbital, rifampin and primidone will increase the rate of drug metabolism, therefore reducing its half-life [131, 132]. Mexiletine has not been reported to alter serum digoxin levels [133]. Likewise, there does not appear to be any interaction between mexiletine and coumadin.

There have been a number of reports about the beneficial effects of combining mexiletine with other class 1 antiarrhythmic agents, particularly disopyramide and quinidine [134]. Such combinations enhance the antiarrhythmic efficacy of these agents and permit a reduction in dose, decreasing the frequency and severity of side effects. Also reported is a beneficial effect of combining mexiletine with beta blocking drugs [135].

Hemodynamics

In several animal studies, therapeutic doses of mexiletine cause no significant depressive effects on left ventricular function, although at higher doses, mexiletine does exert a negative inotropic effect [136]. However, as observed with several other drugs, such as quinidine and procainamide, mexiletine decreases peripheral vascular resistance and afterload, offsetting its negative inotropic effect and maintaining cardiac output.

There have been a few clinical studies in which the hemodynamic effects of oral mexiletine administered to patients with ventricular arrhythmia have been reported. When invasive hemodynamic measurements are obtained, the drug does not significantly alter any hemodynamic parameters [137]. However, many of these studies involved patients with normal left ventricular function. Studies in patients with an impaired left ventricular ejection fraction have also reported that mexiletine does not significantly reduce left ventricular ejection fraction [138]. However, in patients with significant left ventricular dysfunction and a history of congestive heart failure, the drug may exacerbate heart failure due to its mild negative inotropic effect [32]. The incidence of aggravation of congestive heart failure is low, occurring in approximately 2 to 3 % of patients.

Efficacy

Many of the initial reports in which mexiletine was administered involved the intravenous use of the drug. Mexiletine was reported to be effective for suppressing all forms of ventricular arrhythmia when given by the intravenous route.

Unfortunately, in these earlier studies, efficacy was evaluated by ECG rhythm strips or with observation of a bedside monitor and ambulatory monitoring was generally not used. It was reported that the intravenous drug was effective in 70 % to 80 % of patients [139, 140].

Studies in which oral mexiletine was administered confirm that the drug is effective for suppressing all forms of spontaneous ventricular arrhythmia. Approximately 50 % to 60 % of patients will have >75 % reduction in VPBs [141, 142]. Many of these studies involved patients with frequent VPBs, but no history of a serious, sustained ventricular tachyarrhythmia. There have been a few studies in such patients and approximately 60 % of those with a history of sustained VT or VF who demonstrate high density of spontaneous ventricular arrhythmia on ambulatory monitoring will respond to mexiletine [130]. When the drug is evaluated invasively with electrophysiologic testing however, it has a low efficacy rate like other agents and approximately 10 to 15 % of patients will be rendered noninducible during mexiletine therapy [143, 144]. In addition to the beneficial effect of this agent on ventricular arrhythmia during short-term use, studies involving long-term therapy with mexiletine have reported continued suppression of arrhythmia. In a large study involving 107 patients discharged on mexiletine, which was an effective agent based on noninvasive methods, the annual sudden death rate was 3.6 % [145].

There have been several studies in which mexiletine has been administered to patients post myocardial infarction in an attempt to prevent serious ventricular tachyarrhythmias and sudden death in the post infarction period [146]. In none of these studies did mexiletine reduce mortality when compared to placebo. Indeed, IMPACT reported a higher mortality among patients receiving drug compared to those on placebo, although this was not a statistically significant difference [147].

Mexiletine has no role for the treatment of supraventricular tachyarrhythmias. An exception however may be WPW associated with atrial fibrillation. Since mexiletine may slow conduction through the accessory pathway and prolong its refractory period, the drug may reduce the ventricular response rate during atrial fibrillation and may be useful for the termination of other supraventricular tachyarrhythmias associated with this condition. However, there have been no systematic studies of the drug for this indication.

Mexiletine has been compared to other antiarrhythmic drugs and has been reported as effective as quinidine and disopyramide [148, 149]. Although the drug is a lidocaine congener, the effect of lidocaine on arrhythmia does not predict the effect of mexiletine. Likewise, mexiletine and tocainide have similar structures and share the same electrophysiologic properties, but the response to one agent does not predict the effect of the other. Therefore, these two agents are not interchangeable.

Side effects

Similar to lidocaine, mexiletine causes frequent toxicity involving the gastrointestinal and central nervous systems [130, 145]. The most frequent side effects are neurologic reported to occur in up to 50 % of patients. Most of these side effects are dose-related and reduction in dose is usually effective for eliminating these side effects. Among the most common are dizziness, lightheadedness, speech disturbances, ataxia, tremors, and rarely, seizures. Since many of these side effects are similar to those produced by lidocaine, the administration of intravenous lidocaine to patients receiving oral mexiletine should be done cautiously using low dose so as to avoid the precipitation of neurologic side effects.

Other neurologic side effects due to mexiletine are not dose-related, but are a result of the drug itself. These include, memory impairment, altered personality, behavioral changes, sleep disturbance and other psychologic problems.

The next most common side effects are gastrointestinal occurring in up to 30 % of patients [130, 145]. Gastrointestinal toxicity is generally due to a direct effect of the drug on the stomach and can be prevented or reduced in severity by the coadministration of the drug with food or antacids. The most common gastrointestinal side effects are nausea, vomiting, anorexia, abdominal discomfort and diarrhea.

The most serious side effects from mexiletine involve the cardiovascular system. As with other drugs, aggravation of arrhythmia is produced by this agent in approximately 8 % of patients [24]. Congestive heart failure is very uncommon occurring in only 2 to 3 % of patients and primarily in those who have a history of congestive heart failure, a reduced ejection fraction and cardiomegaly [32]. Bradycardia, hypotension and AV nodal blockade are relatively infrequent, each occurring in less than 1 % of patients. Other side effects caused by mexiletine are uncommon and include urinary retention, rash, musculoskeletal complaints, and asymptomatic laboratory abnormalities including neutropenia, a positive ANA and elevated liver-function tests.

5.6 Tocainide (Table 9)

Table 9
Tocainide

Dose	200–800 mg TID oral
Clearance	65 % hepatic – inactive metabolites 35 % renal
Half-life	10–14 hours
Blood levels	4–10 µg/ml
Drug interactions	lidocaine
Indications	ventricular arrhythmia
Side effects	GI, CNS, hematologic, pulmonary

Electrophysiology
Similar to mexiletine and lidocaine, the other class 1B drugs, tocainide exerts a relatively weak effect on the sodium channel and the inhibition of sodium ion influx is only mild [150]. As a result, there is only a slight reduction in the phase 0 upstroke velocity of the action potential. The drug does not significantly alter the action potential duration. Similar to the other class 1 drugs, tocainide has use dependency with rapid onset and offset of block. The depression of conduction is more pronounced at faster heart rates and indeed this agent may be more effective for suppressing and preventing those ventricular tachyarrhythmias that are more rapid. Tocainide exerts its electrophysiologic effects primarily on the His-Purkinje and ventricular muscle tissue and has no effects on atrial tissue [151]. Therefore, tocainide is primarily effective for the treatment of ventricular arrhythmia and it has no role in the treatment of supraventricular tachyarrhythmias. Like lidocaine, the drug may exert some effect on accessory pathway tissue resulting in a slowing of conduction along this pathway and prolonging its refractory period. It may therefore, have a role for terminating and preventing certain arrhythmias associated with WPW syndrome. When ventricular activation in atrial fibrillation is conducted aberrantly through the accessory pathway, tocainide may slow the ventricular response rate or in fact completely block conduction down this pathway, normalizing conduction.
As expected from its electrophysiologic effects, tocainide causes no change in PR, QRS, or QT intervals on the surface electrocardiogram.

Pharmacology

Tocainide is only available for oral use, although an intravenous form of the drug has undergone clinical evaluation [152]. The oral dose of tocainide reported to be effective for suppressing ventricular arrhythmia is 200 to 800 mg TID, with the usual dose being 400 to 600 mg TID. The drug is rapidly and nearly completely absorbed from the gastrointestinal tract and peak levels are achieved within 1 to 2 hours after drug administration [153, 154]. The half-life of tocainide varies from 10½ to 14 hours and the therapeutic plasma level is reported to be 4 to 10 μg/ml. However, as with other drugs, blood levels are of no importance for guiding an effective drug dose or for predicting toxic reactions.

Unlike lidocaine, tocainide does not have a first pass effect in the liver and it is slowly metabolized by the liver to inactive metabolites. Approximately 35 % of the drug is excreted unchanged by the kidney. Therefore, in the presence of renal or hepatic insufficiency, the dose of drug should be reduced to avoid toxicity. There are no clinically important drug-drug interactions reported with tocainide.

Hemodynamics

Data about the hemodynamic effects of tocainide are limited. In animal studies, tocainide exerts negative inotropic effects and does reduce left ventricular contractility [155]. However, the drug also causes a small reduction in peripheral vascular resistance and usually cardiac output is not altered. Clinical studies in which the intravenous form was administered have reported that the drug does not alter hemodynamic parameters [156, 157]. There is a decrease in left ventricular contractility when the drug is infused into patients with an acute myocardial infarction. Data about the oral drug are limited however. Although the drug is only a very weak negatively inotropic agent, there have been cases of congestive heart failure precipitated by tocainide, although the incidence is low, approximately 2 % to 4 % [32]. Congestive heart failure exacerbation occurs predominately in patients with a previous history of clinical congestive heart failure.

Clinical experience

Tocainide is indicated for the treatment of ventricular arrhythmia and it has no role for supraventricular tachyarrhythmias. The majority of studies reported in literature have involved patients with benign, but frequent, ventricular premature beats. In this population of patients, tocainide significantly reduces the frequency of ventricular premature beats in 50 % to 70 % of patients responding [52, 158]. Several reports involving patients with a recent myocardial infarction have also reported it to be an effective agent for suppressing all forms of ventricular arrhythmia in this setting. There have been a few reports of tocainide for the treatment of refractory ventricular tachyarrhythmias and the drug is effective in approximately 45 % of patients when noninvasive methods (ambulatory monitoring and exercise testing) have been used to document drug efficacy [159]. As with other antiarrhythmic drugs, tocainide is less effective when evaluated by electrophysiologic techniques and only 10 to 15 % of patients are rendered noninducible [160].

Studies involving tocainide in patients with a recent myocardial infarction have failed to document any effect of this drug on survival [161]. However, these studies were short-term, involved small numbers of patients and were not arrhythmia suppression trials since arrhythmia before and during drug therapy was not adequately evaluated or established and drug was continued even if arrhythmia persisted.

Although tocainide and lidocaine are structurally similar, the response to intravenous lidocaine does not necessarily predict the effect of oral tocainide [162, 163]. While failure to respond to intravenous lidocaine does predict ineffectiveness of tocainide, a positive response to intravenous lidocaine does not correlate well with the same response to oral tocainide. As with mexiletine, intravenous lidocaine should be administered with extreme caution, using small doses, to patients receiving oral tocainide so as to avoid the precipitation of neurologic toxicity.

In one study, tocainide and mexiletine were randomly administered to 79 patients with a history of serious ventricular arrhythmia [164]. Each patient thereafter received the other agent. Drug effect was evaluated with either noninvasive or invasive methods. The response to one drug did not predict the response to the other agent. Therefore it may be concluded that mexiletine and tocainide are not interchangeable despite a similarity in structure and electrophysiologic effects.

Side effects

Similar to lidocaine and mexiletine, the most frequent side effects due to tocainide involve the central nervous system. Neurologic toxicity is reported in up to 50 % of patients [162, 163, 165]. The majority of neurologic side effects are dose-related and include lethargy, tremor, nystagmus, gait disturbance, dizziness, lightheadedness, ataxia, paresthesias and slurred speech. Other neurologic side effects including hallucination, change in personality, memory impairment, emotional lability and insomnia are not necessarily dose-related but are due to the drug itself.

The next most common side effects are gastrointestinal occurring in up to 40 % of patients [160, 161, 163]. These include nausea, vomiting, abdominal discomfort, diarrhea and anorexia. These gastrointestinal side effects can often be prevented or made less severe by coadministration of the drug with food or antacids. These interventions do not inhibit the absorption of the drug.

Cardiovascular side effects include aggravation of arrhythmia [24], exacerbation of congestive heart failure which is infrequent [32] and rarely, conduction abnormalities. Other side effects which are rare, include rash, LFT abnormalities, and positive ANA. More recently reported are agranulocytosis and pulmonary infiltrates. These two side effects have been associated with severe complications and for this reason the use of tocainide has been restricted for life-threatening arrhythmias.

5.7 Moricizine (Ethmozine) [166] (Table 10)

Electrophysiology

Moricizine is a phenothiazine derivative which exerts membrane effects typical of the class 1 membrane active or local anesthetic agents [167, 168]. However, it possesses characteristics which resemble both class 1B and 1C agents. The drug inhibits the inward sodium currents during phase 0 and it markedly slows the upstroke velocity, slowing impulse conduction through all cardiac tissue including the atrium, His-Purkinje fibers and ventricular myocardium. As with other drugs moricizine demonstrates use dependency with slow onset and offset kinetics and its depressant effects are more pronounced at faster heart rates. These effects are very similar to class 1C agents.

Table 10
Moricizine (Ethmozine)

Dose	200–400 mg TID oral
Protein binding	95 %
Clearance	hepatic – first pass effect-metabolite activity
Half-life	3–4 hours normals up to 13 hours in patients
Blood levels	–
Drug interactions	cimetidine
Indications	ventricular and supraventricular arrhythmia
Side effects	GI, CNS, anticholinergic, cardiac

However, the drug does not alter the action potential duration or may actually shorten it, similar to the class 1B drugs. The effective refractory periods of the His-Purkinje system and ventricular myocardium are prolonged by the drug [169, 170]. Moricizine also slows conduction through and prolongs the refractory period of accessory pathway tissue and it may be effective for the supraventricular arrhythmias associated with the WPW syndrome. As expected, from its electrophysiologic effects, moricizine may also be an effective agent against supraventricular tachyarrhythmias not due to WPW.
Correlating with its electrophysiologic effects, moricizine causes a prolongation of the PR and QRS intervals without altering the QT interval.

Pharmacology
Moricizine is available only for oral administration and the doses currently undergoing investigation are 200 to 400 mg TID. Absorption of the drug is rapid and peak levels are achieved in 1½ to 2 hours after the administration of the oral dose [171, 172]. Although moricizine is almost completely absorbed from the gastrointestinal tract, only 30 to 40 % of the dose is bioavailable because of extensive presystemic metabolism or a first pass effect in the liver. The drug is extensively metabolized by the liver to multiple metabolites which have not been well characterized [172]. It has been reported that at least one of these metabolites may possess some antiarrhythmic activity. Since the drug is extensively metabolized by the liver, the dose should be reduced in the presence of significant hepatic dysfunction.

In healthy subjects, the half-life of the drug is 3 to 4 hours, but is increased up to 13 hours in patients with underlying heart disease. Moricizine is extensively (approximately 95 %) bound by plasma proteins and minor changes in protein binding can therefore result in significant alterations in the free fraction of this agent.
There are little data available about possible drug interactions with moricizine, but a few have been described [173]. The drug does not interfere with digoxin or coumadin. Cimetidine does decrease the clearance of moricizine, resulting in an increased half-life. Therefore, when coadministered with this agent, the dose of moricizine should be reduced.

Hemodynamics
A number of animal studies have reported that moricizine is a mild negatively inotropic drug. However, several clinical studies have failed to document any significant change in left ventricular ejection fraction or invasive hemodynamic parameters [174]. Exacerbation of congestive heart failure is uncommon, occurring in approximately 1 % to 2 % of patients and it is generally seen only in those with a previous history of congestive heart failure. Congestive heart failure is rarely precipitated in patients with a normal left ventricular ejection fraction. In patients with more serious underlying heart disease and congestive heart failure who have experience life-threatening ventricular tachyarrhythmias, the incidence of congestive heart failure exacerbation is higher, approximately 4 to 6 % [32]. The precipitation of congestive heart failure does not appear to be dose-related.

Efficacy
Ethmozine has been reported to be highly effective for the suppression of all forms of ventricular arrhythmia [175, 176]. In comparative trials with quinidine and disopyramide, moricizine has been reported to be equally effective for the suppression of VPBs, couplets and runs of nonsustained VT and up to 80 % of patients will have greater than a 75 % reduction in total VPBs [176, 177]. The drug is somewhat less effective for suppressing ambient arrhythmia in patients with more serious, underlying heart disease, congestive heart failure and a history of serious life-threatening ventricular tachyarrhythmias. Only about 40 % of such patients will respond to moricizine [178].

Similar to other antiarrhythmic drugs, moricizine is less effective when evaluated invasively in the electrophysiologic laboratory. It is reported that only about 15 to 20 % of patients evaluated by electrophysiologic testing will be render noninducible on moricizine [170]. However, even when the TV remains inducible, the drug often causes a significant reduction in the VT rate similar to drugs of the 1C subclass. As with other drugs, it has been reported that moricizine is more effective in patients with better left ventricular function, while its efficacy in patients with a markedly reduced ejection fraction is somewhat less. The outcome of long-term therapy with moricizine has not been reported. However it does appear that if the drug continues to be well tolerated during long-term therapy there is a reduction in recurrence rate of significant ventricular arrhythmia [178].

Side effects
Toxicity resulting from moricizine has been reported in approximately 25 % of patients [178, 179]. The most frequent side effects are gastrointestinal including nausea, vomiting, abdominal discomfort and diarrhea. These appear to be a direct effect of the drug. Neurologic side effects include tremor, ataxia, lightheadedness and dizziness and these are often dose-related. Phenothiazine-like side effects including dry mouth and urinary retention are infrequently observed and appear to be dose-related. Serious side effects involve the cardiovascular system and include congestive heart failure and conduction abnormalities which are infrequent. As with all other drugs, aggravation of arrhythmia is a complication reported in about 8 % of patients [27].

5.8 Recainam (Table 11)

Electrophysiology
Recainam is a new class 1 antiarrhythmic drug which has a unique electrophysiologic profile. Like other local anesthetic or membrane active agents, the drug inhibits the rapid influx of sodium ions into the myocardium, decreasing the upstroke velocity of phase 0 [180, 181]. The depression of the upstroke velocity is more pronounced at faster rates, the property of use dependence. The rate at which block of the inward movement of sodium ion develops is similar to that observed with the 1A agents and is faster than that observed with the

Table 11
Recainam

Dose	200–600 mg TID oral
Protein binding	poor
Clearance	70 % renal unchanged 30 % hepatic no active metabolites
Half-life	up to 10 hours
Blood levels	2–5 μg/ml
Drug interactions	–
Indications	ventricular arrhythmia – supraventricular
Side effects	CNS, GI, cardiac

1C drugs [182]. The drug slows conduction in all cardiac tissue. However, it does not change the action potential duration of ventricular or atrial myocardium, similar to the 1B drugs. Thus, the drug has kinetics of use-dependent block which are similar to 1A drugs, but exerts effects on repolarization which are similar to 1B and 1C drugs, making it difficult to classify this agent based on the Vaughn-Williams classification.
Recainam increases the effective refractory period of atrial tissue, but does not alter the refractory period of the ventricular myocardium. Based on these electrophysiologic effects, recainam prolongs the PR and QRS intervals, but does not alter the QT.

Hemodynamics
Recainam is a mild negative inotropic agent. Although data are as yet limited, it appears that the intravenous drug does not significantly alter systemic blood pressure, but it does cause a slight increase in pulmonary artery pressures, left ventricular and diastolic pressure and a reduction in stroke volume, stroke work index and cardiac output [183]. Data about the oral drug are as yet limited, but a slight reduction in left ventricular ejection fraction has been noted although the precipitation of congestive heart failure is infrequent, observed in less than 1 % of patients studied to date. However, most of the patients entered into these studies did not have left ventricular dysfunction.

Pharmacology
As part of investigational studies, recainam has been administered intravenously and orally. The oral doses used are 200 to 600 mg TID,

while the doses administered by the intravenous route are 4 to 4.5 mg/kg for loading and 0.5 to 1.0 mg/kg/h as a maintenance infusion [183, 184]. When given orally, the drug is well absorbed with a 67 % bioavailability. The elimination half-life of the drug is 3 to 6 hours when administered intravenously and 10.2 hours (range 4–20) when given by the oral route [185]. The effective concentration reported is 2 to 5 μg/ml and steady state levels are achieved by 72 hours. There is a linear relationship between dose and blood levels. The drug is poorly protein bound. Approximately 70 % of the drug is excreted unchanged by the kidney while 16 % is excreted unchanged by the liver. The remainder of the drug is metabolized by the liver to at least 3 metabolites which do not possess antiarrhythmic activity.
There are no drug interactions reported as yet.

Efficacy
Data about the efficacy of recainam are as yet limited as only relatively few patients have been studied. When an oral dose of 600 mg TID is administered, the drug produces a 86 % suppression of nonsustained VT and a 75 % decrease in VPBs [184–186]. As with other drugs, it has been observed that the complex forms are abolished at a lower dose of drug.
When evaluated by electrophysiologic testing, the intravenous drug prevents the induction of VT in only 12 % of patients, while in an additional 15 %, there is a partial response, defined as a slowing of the VT cycle length by >100 m m/sec [184]. When the oral drug is administered, noninducibility is achieved in 5 % while a partial effect (i. e. rate slowing) occurs in an additional 20 % of patients.
As expected from its electrophysiologic properties, recainam does have a beneficial effect for supraventricular arrhythmia although its role remains to be defined. When evaluated by electrophysiologic testing, the oral drug prevents the induction of AV nodal re-entrant tachycardia in 60–70 % of patients while 45 % of patients with supraventricular tachycardia associated with an accessory pathway are rendered noninducible. Its role for prevention of atrial fibrillation is as yet undefined.

Side effects
Data about adverse side effects due to recainam are as yet limited, but they appear to be infrequent. Like other agents, the most com-

mon complaints are neurologic and gastrointestinal including dizziness, nausea and headaches [184–186]. Less frequently reported are rash, blurred vision, paresthesias, abdominal pain, diarrhea, weakness and fatigue. Cardiac side effects are the most frequent and most serious. Conduction abnormalities, primarily first or second degree AV block are observed in 10 % of patients, hypotension in 1 %, CHF in 1 % and aggravation of arrhythmia in approximately 12 % of patients.

5.9 Sotalol (Table 12)

Table 12
Sotalol

Dose	80–240 mg BID oral
Clearance	renal – unchanged
Half-life	6–18 hours
Blood levels	–
Drug interactions	–
Indications	ventricular and supraventricular arrhythmia
Side effects	CNS, cardiovascular

Electrophysiology
Sotalol is a unique antiarrhythmic drug possessing both "class 2" and "class 3" properties since it is a racemic mixture of the d and l isomer. The drug has an interesting electrophysiologic profile and is useful for treating both supraventricular and ventricular tachyarrhythmia.
Sotalol possesses class 2 antiarrhythmic properties characteristic of other beta blockers and therefore, it blocks the effects of catecholamines on the myocardium [187, 188]. In addition, sotalol has "class 3" properties, resulting in an inhibition of potassium ion fluxes. These are characterized by lengthening of the action potential duration in all cardiac tissue and a prolongation of the effective and absolute refractory period. This depressant effect of sotalol on electrophysiologic properties of the myocardium is concentration-dependent, but levels necessary for these changes are higher than those needed for beta blockade [189]. Interestingly, the class 3 actions are independent of the beta blocking effects as the d-isomer, which does not have beta blocking capabilities, has class 3 electrophysiologic properties which are identical to the racemic drug [190, 191].

When its electrophysiologic effects haven been evaluated clinically, sotalol has been observed to lengthen the monophasic action potential of ventricular and atrial tissue, as well as increase the effective refractory period of atrial, AV nodal, His-Purkinje, ventricular and accessory pathway tissue [192]. These effects are manifest in the surface ECG as lengthening of the ST interval. As a result of the beta blocking effect, the PR interval is prolonged and sinus rate decreased.

Pharmacokinetics

Sotalol is a competitive non-selective beta receptor antagonist which also has independent, direct membrane effects [187, 188]. The drug is generally available as a racemic mixture of its stereoisomers, with both isomers having similar electrophysiologic effects. However, the l-isomer is 50 times more potent with regards to its beta blocking properties while it is the d-isomer that causes the class 3 action [187, 193]. Interestingly, sotalol has no intrinsic sympathomimetic properties.

Sotalol is undergoing investigation in both oral and intravenous forms. The oral dose is 80–240 mg BID. The daily dose of drug should be slowly increased by 80–160 mg every 3 or 4 days up to a maximum of 480 mg/day. Careful monitoring of the hemodynamic and electrophysiologic effects are necessary at higher doses. The intravenous form is generally administered as a rapid infusion to terminate supraventricular or ventricular arrhythmias and for evaluation of the drug at electrophysiologic testing. An infusion of 0.2 mg/kg–2.0 mg/kg given slowly over 5–15 minutes is generally safe and effective.

The drug is well absorbed when administered orally, although more slowly than other beta blockers, with plasma concentration peaking 2–3 hours after ingestion [194]. Its bioavailability is nearly 100 % and once in the blood, the drug is cleared unchanged primarily by the kidney. In the presence of renal impairment, the dose of drug must be reduced [195]. Sotalol has the lowest lipid solubility of the beta blockers and has no known active metabolites [196, 197]. The serum half-life of the drug is 6–17.7 hours. No drug-drug interactions have been reported although it has been used in combination with other antiarrhythmic agents.

Hemodynamic effects

The hemodynamic effects of intravenous and oral sotalol differ from that of other beta blockers, although it has beta blocking effects and can reduce left ventricular contractility [198]. There is less depression of left ventricular function with sotalol because of its class 3 electrophysiologic properties. As a result of prolonging phase 3 there is an increased influx of calcium ions resulting in an increase in myocardial contractility, opposing the beta blocking effects. However, the drug does decrease the heart rate, and as a result reduces cardiac output without altering stroke volume or systemic vascular resistance [199, 200]. The d-isomer may prove to have even less depressant effect since it is devoid of beta blocking properties.

Efficacy

As a result of its unique and broad electrophysiologic properties, sotalol has been reported to be effective for terminating and preventing many supraventricular tachyarrhythmias including atrial flutter, atrial fibrillation, paroxysmal supraventricular tachycardia and supraventricular tachyarrhythmias secondary to atrioventricular accessory pathways [201–205]. In addition, its beta blocking effects result in a slowing of sinus rate and decrease in the ventricular response rate during atrial fibrillation and flutter.

An intravenous bolus of sotalol has been reported to restore normal sinus rhythm in the majority of patients with supraventricular tachycardia, but less frequent in those with atrial flutter and fibrillation [201]. With a continuous infusion over 2 hours, the number of patients who respond is increased. The drug has been effective for preventing the induction of supraventricular tachycardia, including those associated with WPW, in patients with a history of these arrhythmias. Chronic oral therapy has been reported to have the same efficacy. Sotalol may be particularly useful for arrhythmias associated with WPW since the drug increases the refractory period of the atria and ventricles, depresses conduction through the AV node as well as depressing both retrograde and antegrade conduction in the accessory paths [206, 207].

Recently, d-sotalol, which is devoid of beta blocking properties, has been reported to be effective and safe for therapy of supraventricular tachycardia, atrial flutter and atrial fibrillation [208].

Sotalol has shown activity against ventricular arrhythmia and is a promising agent for these arrhythmias. VPBs are reduced from 50–86 % in approximately two thirds of patients receiving chronic therapy [209–211]. Runs of nonsustained ventricular tachycardia are reduced by 90–95 %.

The effect of sotalol in patients with life-threatening and sustained ventricular tachyarrhythmias is variable. When evaluated with electrophysiologic studies, sotalol renders 20–62 % of patients noninducible although, in most studies, efficacy is observed in 20–25 % [212–215]. Similar results have been reported with d-sotalol [216]. In addition, sotalol may be useful in patients refractory to other antiarrhythmic agents.

The clinical outcome during long-term therapy seems to correlate with results of electrophysiologic testing, although some patients who are noninducible on sotalol will have recurrence of arrhythmia [213, 214]. Response to intravenous infusion of sotalol has been reported to be predictive of response to the oral form, but data are limited. Long-term controlled trials are however lacking at this point.

Side effects

Adverse effects from sotalol are primarily the result of its beta blocking effects and have been reported in over 50 % of patients [188, 193]. Similar to other beta blockers, approximately 20 % discontinues the drug because of side effects. These include mental depression, fatigue, impotency, headache and dry mouth. Of note, central nervous system side effects are uncommon since sotalol is hydrophilic and has no lipid solubility. More serious cardiovascular side effects including hypotension, bradycardia, high degree AV block and exacerbation of heart failure may occur, but are less frequently observed than what is reported with other beta blockers. Since the drug has class 3 activity and prolongs the JT interval, torsade des pointes has been reported [217, 218]. This adverse reaction is seen more frequently in the setting of hypokalemia, toxic levels or with concomitant use of other drugs which prolong the JT interval. Other forms of arrhythmia aggravation have also been reported. As with some other beta blockers, sotalol has been reported to be associated with retroperitoneal fibrosis [219].

5.10 Amiodarone [220] (Table 13)

Table 13
Amiodarone

Dose	600–1800 mg oral loading, 200–400 mg daily maintenance
Clearance	hepatic – active metabolite deiodination
Half-life	up to 120 days
Blood levels	1–2 μg/ml
Drug interactions	digoxin, coumadin, antiarrhythmic drugs, beta blockers, calcium channel blockers, anesthetic agents
Indications	ventricular and supraventricular arrhythmias
Side effects	cardiac, neurologic, GI, hematologic, dermatologic, pulmonary, thyroid, ophthalmologic

Electrophysiology

Amiodarone is a unique and complex agent which is most often used for the treatment of life-threatening ventricular tachyarrhythmias. However, the drug has several important actions that make it effective against supraventricular arrhythmias as well.

Amiodarone exerts a number of electrophysiologic effects, the most important being a prolongation of repolarization and refractory periods in all cardiac tissue, a result of inhibition of potassium ion fluxes [221–223]. It is therefore a class 3 agent. The increase in repolarization and refractory periods result in a reduction in membrane excitability, the antifibrillatory property of the drug [224].

Amiodarone is also a potent sodium channel blocker and will slow the upstroke velocity of phase 1 of the action potential when the sodium channels are activated. Consequently, the drug has a marked effect on slowing depolarization as well as prolonging repolarization. In addition, the drug directly depresses the automaticity of the sinus node and AV nodes [222]. Amiodarone has other actions which also are important including noncompetitive beta blocking activity and calcium channel blocking effects [223, 225]. The clinical importance of these actions are as yet unknown. Lastly, the drug contains substantial amounts of iodine accounting for 38 % of its weight. It is possible that some of its antiarrhythmic actions are the result of antithyroid effects.

As predicted from its electrophysiologic effects, the drug causes significant changes on the surface ECG. The most impressive are a sinus bradycardia, often resistant to atropine or catecholamines, prolongation of the PR and QT intervals and a pronounced U wave, not associated with hypokalemia.

Pharmacokinetics

Amiodarone hydrochloride is a benzofuran derivative which contains iodine (75 mg per 200 mg tablet) and is structurally similar to thyroxin. The drug is available for oral use, but there is also an intravenous form which is investigational.

The gastrointestinal absorption of amiodarone is slow and incomplete, with a peak serum level reached four to seven hours after oral dosing [226–228]. First pass intestinal mucosal and liver metabolism are responsible for the low bioavailability which is approximately 50 % of the administered dose. The drug is highly protein and lipid bound and is widely distributed throughout all adipose tissue [229, 230] with an estimated volume of distribution of 50 liters. As a result of this large body store of drug, an extended period of administration with a loading dose, approximately 15 grams, is necessary to saturate the adipose storage sites [230]. The drug is deiodinated as part of its clearance, but the major route of elimination is via hepatic metabolism and bile excretion with negligible renal clearance [228, 230, 231]. The major metabolite in humans is desethylamiodarone which has activity equivalent to the parent compound [229]. It has been reported that a serum concentration of at least 1–2 μg/ml is necessary for drug efficacy; however there is considerable overlap between the concentrations associated with efficacy and toxicity [232]. The elimination half-life of amiodarone is highly variable, ranging from 26 days to as long as 6 months, with a mean of 52 days owing to its extensive storage and avid binding to poorly perfused tissue [227]. This property has many clinical implications, particularly with regard to toxicity.

The onset of action of amiodarone is delayed because of its poor absorption, low bioavailability and large volume of distribution into adipose tissue. Therefore, when given orally, a loading dose of drug is recommended [230]. In general, 1200–1800 per day for 1–2 weeks is administered followed by 800 mg for up to 4 weeks and then a maintenance dose of 200–400 mg/day [226, 233].

Intravenous amiodarone has been safely administered as a rapid infusion of 5 mg/kg over 15–30 minutes for life-threatening ventricular

arrhythmias or atrial arrhythmias associated with the WPW syndrome [234]. After the loading dose, a continuous infusion of up to 1 g per day is used and when possible, therapy with the oral drug begun.
Amiodarone is associated with a number of important drug interactions [235]. Commonly observed is an interaction with coumadin and the prothrombin times may be prolonged and erratic, especially during the loading period. As a rule, the coumadin dose must be reduced to prevent an excessively prolonged prothrombin time. Amiodarone will raise serum digoxin levels and can result in clinically significant digoxin toxicity. The drug may also increase the serum levels of other antiarrhythmic drugs including quinidine, procainamide, mexiletine and propafenone. Additionally, there may be added toxic effects on the ECG from these combinations, especially prolongation of the QT interval. Potentiation of the effects of anesthetic agents has been observed, primarily more pronounced hypotension and bradycardia. Combined use with beta blockers and calcium channel blockers may result in potentiation of sinus and AV nodal depression as well as enhanced negative inotropic action.

Hemodynamics
As a result of its diverse pharmacologic properties, amiodarone exerts important hemodynamic effects [223, 236, 237]. By direct action on smooth muscle, calcium and alpha receptor blockade, the drug dilates the coronary arteries and increases coronary artery blood flow [223]. Indeed, amiodarone was initially used as an antianginal agent. The drug also directly produces peripheral artery vasodilatation and reduces afterload and blood pressure. Amiodarone is negatively inotropic and reduces the force of left ventricular myocardial contraction. However, as a result of the reduction in afterload, stroke volume and cardiac output are generally maintained [223, 237]. With chronic therapy, there is usually no change in left ventricular ejection fraction and drug-induced congestive heart failure is unusual, with a reported incidence of 2–3 %.
Amiodarone is very effective against all forms of ventricular arrhythmia. It reduces the frequency of ventricular premature beats by over 80 % and decreases the number of runs of ventricular tachycardia by over 90 % [238–240]. Unfortunately, there are no trials comparing amiodarone with other antiarrhythmic drugs and the majority of

available data involve patients with sustained ventricular tachycardia or fibrillation refractory to other antiarrhythmic agents [241–243]. The reported efficacy of amiodarone for preventing recurrence is 60–80 %. Unfortunately, the studies available differ in regard to previous therapy, dose of amiodarone used, methods of evaluation, duration of therapy before evaluation, the use of adjunctive antiarrhythmic drugs and the duration of follow-up. In most of the studies amiodarone is the last agent tested, being reserved for the most refractory patients. It must be emphasized that most of the available studies have followed patients for only an average of one year. While studies have reported that 80 % of patients are alive and free of side effects after 1–2 years of therapy, studies in which follow-up is longer have reported a significant recurrence rate at the end of 3 years. Nonetheless, for patients who can tolerate the drug, amiodarone is a useful agent when the ventricular tachyarrhythmia is resistant to other standard therapy or when other drugs are not well tolerated. One group in whom the drug may be of particular benefit are patients with hypertrophic cardiomyopathy where several investigators have reported improved survival in such patients, even when treated prophylactically [244]. However, this is still controversial as data are conflicting.

There remains uncertainty over the utility of electrophysiologic testing for predicting a successful outcome. Several studies have reported that continued inducibility during drug therapy does not necessarily predict clinical failure [245, 246, 247], while noninducibility is highly predictive of freedom from recurrence [248, 249]. It has been observed that the electrophysiologic test is of use when other parameters are considered. Most important, are a slowing of the ventricular tachycardia rate during therapy and prolongation of the refractory period [250, 251]. The suppression of runs of nonsustained ventricular tachycardia on ambulatory monitoring also predicts a good outcome although this is still controversial [240]. Important is the timing for evaluation of efficacy which is variable given the long and unpredictable time before the onset of drug action. Guidelines include ECG manifestations of drug activity (PR and QT prolongation, sinus bradycardia and U waves) and the administration of at least 15–20 grams of drug [252].

While amiodarone has been most often evaluated for life-threatening ventricular tachyarrhythmias, it is effective for preventing most supraventricular arrhythmias including atrial fibrillation, flutter and supraventricular tachycardia in patients with or without an accessory pathway [253–255]. Interestingly, doses lower than those needed for ventricular arrhythmias are effective for the prevention of supraventricular tachyarrhythmia. Up to 90 % of patients will respond to the drug. The drug may be particularly beneficial in patients with supraventricular arrhythmias associated with an accessory pathway. By lengthening the refractory period of the accessory pathway, atrial and ventricular myocardium, AV node, and His-Purkinje fibers. Amiodarone can prevent supraventricular arrhythmias or control the ventricular response rate of atrial fibrillation and flutter in virtually all patients with WPW [256].

Toxicity

After more than 20 years of clinical use with amiodarone, the side effect profile of this drug is well established and the overall incidence of side effects as high as 80 % [225, 257–259]. During long-term follow-up, as many as 18 % of patients require discontinuation of the drug because of toxicity [221, 257, 258]. The mechanism of toxicity is multifactorial and may be a result of direct accumulation of amiodarone or iodine, cellular phospholipidosis secondary to inhibition of phospholipases, free radical formation, altered cellular function or immunologic injury. As a result of its pharmacologic properties, side effects are not observed early, but occur after weeks or months of drug therapy. Importantly, the incidence of side effects increases over time as more drug accumulates [257].

Cardiovascular toxicity includes accentuation of the normal electrophysiologic effects of the drug, including bradycardia and conduction defects, worsening of congestive heart failure and aggravation of arrhythmia [259, 260]. The latter has been reported in 3–5 % of patients and can be difficult to distinguish from the natural history of the underlying cardiac disease given the long time course of drug action [225, 260, 261]. Since the drug prolongs the QT interval, amiodarone can also cause torsades des pointes, although this is rare and usually associated with hypokalemia or concomitant therapy with quinidine [262]. The drug may increase the defibrillation and cardioversion threshold, especially important when an implantable defibrillator is present.

Pulmonary toxicity, manifested by pneumonitis, infiltrates or fibrosis, is the most serious noncardiac side effect and occurs in up to 15 % of patients [225, 263–265]. The pathogenesis is probably multifactorial and appears to be dose dependent. While pulmonary toxicity can be fatal, it usually responds to discontinuation of the drug. Occasionally, steroids are necessary to increase the rate of healing. The pulmonary toxicity may be insidious in onset, associated with nonspecific complaints of cough, fatigue, low grade fever or shortness of breath. It has been confused with congestive heart failure or pneumonia. Respiratory failure and death have been reported when the diagnosis is missed or the toxicity advanced. Unfortunately, pulmonary function tests, often abnormal as a result of the drug, are of no use for predicting this toxic side effect and the chest X-ray is the best screening technique.

Another serious and potentially fatal complication is amiodarone induced hepatitis. While mild to moderate elevations of transaminases occur in 15–20 % of patients, severe hepatitis is rare [257, 258]. Other gastrointestinal side effects are constipation and nausea which occur frequently when higher doses are used. Thyroid function test abnormalities are common, occurring in up to 30 % of patients, as a result of the iodine. Amiodarone does cause clinically important hypo- or hyperthyroidism each of which has been reported in 5–10 % of patients [266].

Other side effects are generally less serious. Neurologic toxicity includes tremor, ataxia, peripheral neuropathy, fatigue and weakness. Corneal microdeposits are universal, but rarely serious [267]. Amiodarone has also been associated with photosensitivity and blue-grey skin discoloration [225, 232, 239, 241].

While toxic effects are common with the use of amiodarone, these effects must be weighed against the potential benefit of this very effective antiarrhythmic agent, particularly in patients with refractory life-threatening ventricular arrhythmia.

5.11 Pirmenol [268] (Table 14)

Electrophysiology

Pirmenol hydrochloride is a new antiarrhythmic agent synthesized in the late 1970's. Pirmenol exhibits effects on the fast action potential similar to other class 1 membrane active antiarrhythmic agents [269,

Table 14
Pirmenol

Dose	50–300 mg BID or TID
Protein binding	80–90 %
Clearance	30–50 % renal unchanged remainder hepatic
Half-life	6–12 hours
Blood levels	–
Drug interactions	rifampin
Indications	ventricular and supraventricular arrhythmia
Side effects	GI, anticholinergic, cardiac

270]. The drug depresses the maximum rate of depolarization (V_{max} of phase 0) and reduces action potential amplitude, membrane responsiveness and automaticity. Similar to the the class 1A agents, it prolongs the action potential duration and effective refractory period. Pirmenol's effect on phase 0 upstroke velocity is frequency- and voltage-dependent, having relatively slow onset and offset kinetics, also similar to the other 1A agents [271]. Although pirmenol suppresses the automaticity of tissue generating a slow action potential the clinical importance of this is unclear [272]. Atrioventricular (AV) conduction time is prolonged, but the delay of impulse conduction occurs in the atrial and His-Purkinje tissue, while conduction in the AV node itself is not directly slowed.

In clinical studies, IV pirmenol has little direct effect on sinus node function, although there may be a minimal increase in heart rate, related to a combination of an anticholinergic effect and a reflex increase in sympathetic tone resulting from modest negative inotropic activity [273, 274]. Since the drug has no effect on AV nodal conduction time, the AH and PR intervals are not significantly prolonged. IV pirmenol does prolong infranodal and intraventricular conduction and there is prolongation in the HV interval which is statistically significant, but clinically irrelevant [273]. QRS duration may be prolonged by approximately 10 %, while the QTc interval is more substantially prolonged as a result of the prolongation of the action potential.

The only consistently observed effect of the oral drug is QTc lengthening, and the PR and QRS intervals remain essentially unchanged.

Pharmacology

Pirmenol has been administered by the intravenous and oral routes. In several studies, intravenous pirmenol was given as a single infusion of 50–150 mg over 30 minutes [274]. Several clinical trials reported that oral pirmenol, 100–600 mg/day in 2 divided doses, is effective for suppressing VPBs while higher oral doses 200–250 mg TID may be necessary in patients with sustained VT [275–277].

Absorption of pirmenol after an oral dose is moderately rapid, with peak plasma concentrations occurring 1–2 hours after drug administration [278, 279]. The bioavailability of oral drug is approximately 85 %, suggesting the absence of a significant first-pass effect.

The half-life of pirmenol is similar for both the oral and intravenous route of administration and follows linear or first-order pharmacokinetics. The mean half-life ranges from 6.5–11.3 hours, but there is considerable variability [280, 281]. Despite this there has been no correlation between half-life and measurements of renal function.

The metabolism of pirmenol has not been fully characterized, but the drug appears to be partly excreted unchanged by the kidney and approximately 30–50 % of the dose appears in the urine [282]. Renal clearance of free drug far exceeds creatinine clearance, suggesting that renal tubular secretion of drug occurs. The nonrenal route of excretion involves a hepatic mechanism, although the metabolic pathway is not established. There is evidence for enterohepatic recycling of pirmenol and a "second peak", indicating reintroduction of drug into central compartment, has been observed in man. To date, there has been no data regarding pirmenol metabolites.

Pirmenol is 80–90 % protein bound and the binding ratio correlates well with the acute phase reactant, alpha-acid glycoprotein. Despite this extensive protein binding, the drug has a large volume of distribution, indicating significant tissue penetration of pirmenol. This finding is in agreement with studies which reported that tissue concentration in the liver, kidney and lung exceed plasma concentrations by 10–100-fold.

The only drug interaction reported to date with pirmenol is with rifampin which, by inducing hepatic enzymes, increases the plasma clearance of the drug, decreasing its half-life [283].

Hemodynamics

In canine studies, intravenous pirmenol depresses myocardial contractility and results in a reflex increase in total peripheral vascular resistance [284]. However, with oral administration, hemodynamic effects are negligible.

There have been only a few hemodynamic studies with pirmenol in man. Following the intravenous administration of pirmenol there is a significant increase in heart rate, diastolic blood pressure and peripheral vascular resistance [284]. There is a significant decrease in LVEF measured by radionuclide angiography reflecting a mild negative inotropic effect of intravenous pirmenol even in patients without a history of CHF and those who have well-preserved left ventricular function at baseline [285]. Intravenous pirmenol has recently been reported to cause hypotension and CHF, but these events occurred in patients with pre-existing LV dysfunction and a clinical history of heart failure.

A prolonged IV infusion of drug has been associated with an insignificant increase in heart rate, mean atrial pressure and a slight reduction in ejection fraction. Similar hemodynamic effects occur with chronic oral therapy although data are limited.

Clinical efficacy

Pirmenol is effective for ventricular and supraventricular arrhythmia. It appears to have an antifibrillatory effect on the atria, and has been reported to revert AF and restore normal sinus rhythm in 60 % of patients 2–16 minutes following IV bolus administration [287]. The drug has also been reported to suppress symptomatic APBs by >80 %, but the number of patients studied is small [288]. The IV drug was reported effective for restoring sinus rhythm in a small number of patients with AV nodal re-entrant SVT [289]. Pirmenol slows conduction in and prolongs the refractory period of the accessory pathway. However, in patients with WPW, pirmenol only infrequently prevents arrhythmia induction, but it does significantly increase the cycle length of the tachycardia [290]. As a result of its effect on the AV node, the drug slows the ventricular response during atrial fibrillation.

When administered acutely to patients with VPBs, pirmenol is effective in 75–90 % when efficacy is defined as >90 % reduction in VPBs [276, 291]. Approximately 50–60 % of initial responders continue to have significant VPB reduction after 1–2 years follow-up [275, 277, 292, 293].

In patients with sustained ventricular tachycardia or ventricular fibrillation, oral pirmenol has been effective in 20–30 % of patients when evaluated by programmed electrophysiologic testing [273, 294]. In the patients who remain inducible, pirmenol significantly increases the VT cycle length similar to other drugs. However, the predictive accuracy of EP testing for oral pirmenol is uncertain, as the number of patients studied to date has been small.

Side effects
Pirmenol has been relatively well tolerated in clinical studies, although the number of patients studied is small. Side effects have been mild and generally do not require discontinuation of drug. The most commonly reported adverse effect is a bitter or metallic taste occurring in up to 15 % of patients [274, 275, 272]. It appears to be related to secretion of the drug by the salivary glands [295]. Other mild side effects, which are dose-related, result from the anticholinergic properties of the drug, and include dry mouth, urinary hesitancy, constipation and blurred vision [296]. The incidence of these adverse effects is greater at higher doses of pirmenol. Reports of aggravation of arrhythmia are anecdotal, and its actual incidence has yet to be determined. The results of multicenter studies in patients with VPBs suggests a 2–3 % incidence, but the incidence in patients with more serious arrhythmias is unknown. Reports of hepatic toxicity and elevated LFTs are also anecdotal.

5.12 Quinidine (Table 15)

Table 15
Quinidine

Dose	sulfate 200–400 mg TID or QID oral gluconate 324 mg TID oral
Protein binding	70–90 %
Clearance	70–90 % hepatic – metabolites have minimal activity 10–30 % renal
Half-life	5–8 hours
Blood levels	2–5 μg/ml
Drug interactions	digoxin, barbiturates, phenytoin, rifampin, cimetidine, vasodilators, verapamil
Indications	ventricular and supraventricular arrhythmias
Side effects	GI, hematologic, CNS, cardiac

Electrophysiology
Quinidine is a cinchona alkaloid and was the first antiarrhythmic agent used clinically. Although originally prescribed for the treatment of atrial fibrillation, it is also effective for other supraventricular as well as ventricular arrhythmias.

Quinidine exerts both direct effects on the myocardium and indirect effects mediated by autonomic changes resulting from its vagolyic (atropine-like) activity and alpha-adrenergic blocking effect [297–299]. In healthy atrial, His-Purkinje, and ventricular myocardial tissue, quinidine decreases the upstroke velocity of phase 0 of the action potential reducing impulse conduction velocity through the membrane [300, 301]. The effective refractory period (ERP) of all tissue is lengthened and membrane excitability therefore reduced. The drug also exerts a direct depressant effect on the sinoatrial (SA) node as well as the conduction through and refractory period of the atrioventricular (AV) node [302, 303]. The firing rate or automaticity of abnormal atrial tissue, Purkinje fibers and ventricular myocardium is significantly reduced due to a decrease in the slope of spontaneous phase-4 depolarization [304]. Quinidine significantly lengthens the action potential duration of atrial, His-Purkinje and ventricular myocardial tissue and accessory pathway [301, 305]. These effects are more pronounced in diseased or ischemic tissue [306]. As a result of these electrophysiologic effects, quinidine causes a slight, dose-related increase in PR and QRS intervals, but a marked lengthening of the QT duration [307].

Pharmacology

Quinidine is primarily administered by the oral route. Intravenous (IV) or intramuscular (IM) preparations are available although infrequently used or required [308]. The dose of IV quinidine is 0.5 mg/kg/min up to a total dose of 10 mg/kg. There are 3 quinidine salts available for oral use: the sulfate, gluconate and polygalacturonate. The quinidine content varies among these preparations and additionally, there are differences in absorption, blood levels achieved and half-life. The dosage of quinidine sulfate usually administered is 200–400 mg every 6 hours, although daily dosage may range from 800–2400 mg [309]. The therapeutic blood levels range from 2–5 μg/ml. If an effective concentration is rapidly required, a loading dose of 600 mg of quinidine sulfate can be given orally. When administered by the oral route, bioavailability is 70 % and peak plasma levels are achieved within 1–2 hours after a dose of quinidine sulfate and 3–4 hours after quinidine gluconate. The major route of elimination of drug is via hepatic metabolism, while 10–30 % of a dose of quinidine is excreted unchanged in the urine [309, 310]. Although data are sparse, it appears that the major metabolites of quinidine, mono- and dihydroxyquinidine, have far less antiarrhythmic effects compared to that of the parent drug [311, 312].

The half-life for quinidine after oral administration is 5–8 hours, but may range from 3–19 hours. The induction of hepatic metabolic enzyme activity will shorten the half-life. The half-life is not altered in patients with renal failure, probably owing to an equivalent reduction in clearance and volume of distribution which is 2–4 liters/g in normals [313]. However, clearance is minimally reduced and half-life increased in patients with CHF [314]. Dosage does not have to be adjusted in patients with mild to moderate renal failure, but patients on hemodialysis require smaller doses once steady state conditions are achieved. In the presence of hepatic dysfunction, protein binding of quinidine is reduced, free quinidine levels are increased and efficacy and toxicity occur with lower doses [315]. Additionally, quinidine metabolism is decreased causing an increase in blood levels and half-life. In contrast, when hepatic metabolism is augmented, a higher dose of quinidine may be required.

About 70–90 % of quinidine is bound to plasma proteins including alpha-1 acid glycoprotein (AAG) and albumin [315]. While 80 % of absorbed quinidine is bound to albumin, clinically significant changes in quinidine binding are more often due to changes in the serum level of AAG, which is frequently elevated in a number of common diseases, including acute myocardial infarction. An increase in protein binding may also be seen in renal failure, while a decrease may occur in hepatic failure and other disease states which cause a reduction in protein concentration and a concomitant increase in quinidine levels. Serum alkalosis also increases the binding of free quinidine and lactate or bicarbinate have been used to acutely treat cardiac toxic effects [316].

A number of drug interactions with quinidine have been reported. The administration of quinidine to patients receiving maintenance digoxin therapy results in elevated plasma levels of digoxin [317]. When a loading dose of quinidine is administered, a substantial increase in serum digoxin concentration may be achieved within 24 hours, while in the absence of such a loading dose, a steady elevation of the plasma digoxin level occurs over 1–2 weeks [318]. The magnitude of the increase in digoxin concentration depends upon the dose of quinidine administered, but at a therapeutic plasma quinidine concentration, mean plasma digoxin levels are increased 2–2.5-fold. The clinical effect of such increased digoxin levels remains controversial, although evidence suggests that there is an increased cardiac effect and the occurrence of digoxin toxicity [319]. The mechanisms of this interaction involve a decrease in renal and extra-renal clearance of digoxin, altered absorption of digoxin from the GI tract and possibly altered distribution and binding of digoxin.

Anticonvulsant drugs, primarily phenobarbital and phenytoin, and rifampin induce hepatic enzymes and will increase the metabolism and clearance of quinidine, while cimetidine impairs its elimination [320]. Quinidine, because of its alpha-adrenergic blocking effect, can cause hypotension and this may be exacerbated by the administration of vasodilating agents. Quinidine interacts with coumadin and may cause an increase in the prothombin time [321]. A recently noted interaction between quinidine and verapamil has been described in which a clinically significant increase in quinidine half-life, peak level and steady state level was observed during verapamil abministration.

Hemodynamics

Studies in dogs indicate that IV quinidine in therapeutic doses is a ganglionic blocker and selectively inhibits vasoconstriction mediated by alpha-adrenergic stimuli [297, 298]. While the drug reduces afterload at doses greater than 20 mg/kg, as a result of vasodilitation, quinidine appears to directly depress myocardial contractility [322]. However, when given at therapeutic doses, quinidine does not cause a reduction in stroke volume or cardiac output as a result of the decrease in afterload.

Clinical observations in man have confirmed the animal data. Intravenous quinidine causes peripheral vasodilatation in man due to its alpha-adrenergic blocking effect, but at therapeutic doses does not exert any significant direct negative inotropic effects on the heart [323, 324]. In patients with a congestive cardiomyopathy, a slow IV infusion of quinidine does not reduce left ventricular contractility [325]. Effects after oral administration, although qualitatively similar, appear to be less pronounced than those seen with IV dosing of quinidine. Congestive heart failure is an unusual complication of quinidine therapy.

Efficacy

Quinidine's initial therapeutic use was for the conversion of atrial fibrillation and the maintenance of sinus rhythm. Results from a number of studies have suggested that the efficacy of quinidine ranges from 30–88 % [326, 327]. Factors associated with a lower success rate of conversion include duration of atrial fibrillation (> 6 months), underlying rheumatic heart disease and presence of cardiomegaly on chest X-ray. Maintenance of sinus rhythm from quinidine has also varied considerably with early relapse occurring more commonly in patients with a longer duration of atrial fibrillation prior to conversion or when quinidine is discontinued for any reason such as intolerable side effects. These data are, however, difficult to interpret because of the heterogeneous patient populations studied, varied dosing regimens of quinidine employed and the lack of appropriately randomized, controlled studies. More recently, several randomized, controlled studies have reported a beneficial effect of quinidine (particularly with long-acting preparations) compared to placebo in the maintenance of sinus rhythm after elective cardioversion [327]. The use of quinidine in atrial flutter parallels its use in atrial fibrillation.
Quinidine, perhaps because it decreases the frequency of atrial premature beats as well as a result of its effect on atrial and AV nodal conduction and refractoriness, can prevent paroxysmal attacks of AV nodal re-entrant tachycardia [328]. In the Wolff-Parkinson-White (WPF) syndrome, quinidine slows anterograde and retrograde conduction and increases the refractoriness of the accessory pathway [329]. Therefore, it is effective for SVT due to WPW and can prolong the shortest R-R interval between two pre-excited beats in atrial fibrillation.
In contrast to the paucity of placebo-controlled or comparative studies of quinidine for supraventricular tachyarrhythmias, there have been many comparative trials of quinidine against other antiarrhythmic drugs, a result of FDA requirements for the introduction of new antiarrhythmic agents [148, 330–332]. When assessed by 24-hour ambulatory monitoring, oral quinidine, at doses of 200–400 mg q6h, is effective for suppressing the frequency of VPBs, ventricular couplets and nonsustained ventricular tachycardia. Quinidine is generally as effective as the comparative agent, usually an investigational drug and 32–77 % of patients respond with a reduction in VPBs defined as a > 70 % or > 85 % decrease. Quinidine is also equivalent in its effect

to other drugs for the suppression of couplets and nonsustained ventricular tachycardia and 41–80 % of patients respond. Combinations of quinidine and other antiarrhythmic drugs, especially the 1B drugs, have been reported to suppress chronic ventricular arrhythmia in a greater percentage of patients than when monotherapy is used [334]. Data about the use of quinidine for the treatment of malignant ventricular tachyarrhythmias when evaluated by EPS are limited, but it has been reported to prevent induction of VT in 15–25 % of patients [308, 335, 336]. In one study, despite the presence of clinical heart failure in 68 % of patients, IV quinidine prevented arrhythmia induction in 26 % of patients, while VT cycle length increased in the remaining 41 patients.

Adverse effects

As with other drugs, the most serious side effects from quinidine are cardiac. Quinidine syncope is the result of a ventricular tachyarrhythmia, called torsade des pointes and can occur as a result of quinidine overdosage or more commonly as an idiosyncratic reaction [337–339]. The incidence of this form of arrhythmia aggravation is between 1–4 %. The rhythm has a distinct morphology with the QRS axis always changing or twisting. A prolonged QTc is a prerequisite for this arrhythmia and may either be idiosyncratic or dose related. The occurrence of torsade appears to be favored by hypokalemia, and in vitro experiments have suggested an etiologic role for early afterdepolarizations. The majority of episodes of this arrhythmia occur within three days of initiating treatment, although it has also been observed after long-term therapy.

Other "arrhythmogenic" effects of quinidine include an increase in episodes of ventricular tachycardia, sinus arrest and sino-atrial block in patients with sick sinus syndrome and an increase in the ventricular rate during atrial flutter, a result of slowing of the atrial flutter rate and the occurrence of 1:1 AV nodal conduction.

Noncardiac side effects are very common. Cinchonism, including headache, tinnitus, blurry vision and GI upset can be mild or more severe involving diplopia, photophobia, altered color perception, hot flushed skin, nausea, vomiting, diarrhea, abdominal pain, confusion, delirium and psychosis. These toxic side effects are dose-related [340]. The most common side effects from quinidine are GI. In the Boston Collaborative Drug Surveillance Program, 14 % of 652 pa-

tients had side effects, and half of these adverse reactions were gastrointestinal, predominantly diarrhea [341]. Studies have reported the incidence of diarrhea to be over 40 %.
Several hypersensitivity mediated syndromes have been reported with quinidine. Fever usually occurs within the first three weeks of therapy, but may be seen as late as one year after the initiation of the drug [342]. Hepatitis is uncommon and liver biopsy reveals granulomatous changes.
Cutaneous reactions include urticaria, scarlatiform or morbilliform eruptions, eczematous reactions, erythrodermic exacerbation of psoriasis, and exfoliative dermatitis. Systemic lupus erythematous has been reported and occurs a few days to a few years after the start of quinidine therapy. In general, symptoms disappear completely a few weeks after discontinuing the drug. Hematologic reactions also occur with quinidine and include thrombocytopenic purpura, due to antiplatelet antibodies [343] and a hemolytic anemia which is also immunologically mediated.

5.13 Disopyramide (Table 16)

Table 16
Disopyramide

Dose	100–200 mg TID or QID
Protein binding	variable
Clearance	70 % renal unchanged 15 % GI unchanged 15 % hepatic – inactive metabolites
Half-life	5–8 hours
Blood levels	2–5 μg/ml
Drug interactions	rifampin, phenytoin
Indications	ventricular and supraventricular arrhythmia
Side effects	cardiac, GI, urologic, anticholinergic

Electrophysiology
Disopyramide is a class 1A agent and like quinidine, it exerts both direct and indirect (anticholinergic) effects on the conducting tissue of the heart [344]. The drug's net effect is to some extent, dependent upon the degree of autonomic tone present during disopyramide administration.

Disopyramide has been reported to possess electrophysiologic properties similar to those of quinidine and procainamide. In isolated canine Purkinje fibers, disopyramide causes an increase in action potential duration and effective refractory period while there is a decrease in maximal diastolic potential, action potential amplitude, maximal upstroke velocity of phase 0 (V_{max}), conduction velocity and the rate of spontaneous phase-4 depolarization [345, 346]. As with other class 1 agents, these effects are enhanced by an increase of extracellular potassium concentrations, while the magnitude of these effects is diminished in the presence of lowered extracellular potassium.

In the AV node and isolated rabbit atria, disopyramide slows the spontaneous firing rate of pacemaker cells in part by a reduction in the slope of phase 4 slow diastolic depolarization, but also as a result of prolongation of the action potential duration [348]. The drug reduces the frequency of the sinus node discharge in a dose-dependent fashion.

Although canine studies have demonstrated a dose-related increase in AV nodal refractory periods and conduction time, these effects have not been observed in human studies. In fact, changes in AV nodal refractory periods in man are variable, while AV nodal conduction time is unchanged [349, 350]. It has been postulated that there are 2 mechanisms to explain this apparent discrepancy between animal and human AV nodal response to disopyramide. Disopyramide exerts anticholinergic effects which may offset the direct depressive activity of the drug. Alternatively, since none of the canine studies measured His bundle electrograms, it is possible that the apparent AV nodal effects in the animal studies were actually the result of drug effects on the His Purkinje system. It is most likely that the vagolytic effects of the drug have some influence on AV nodal properties.

In humans, disopyramide administration causes no significant change in sinus node recovery time [351, 352]. The sinus rate may increase, but when patients are pretreated with atropine, there is a consistent decrease in sinus node automaticity, suggesting that there is a vagolytic effect on the SA node. The refractory periods of the atrial and ventricular myocardium are increased, while the refractoriness of the His Purkinje system is either not altered or increased [352, 353]. Most studies have reported that disopyramide has little effect on conduction time within the His Purkinje fibers. Conduction time through accessory pathway tissue is however prolonged [354, 355]. In animals, the drug has been reported to increase PR and QRS intervals in a concentration-dependent fashion, whereas in studies, the most consistent findings have been little if any change in the PR and QRS, but a significant prolongation of the QT interval.

Pharmacology
Disopyramide is primarily administered orally and the dose is 100–200 mg TID or QID. This dose is decreased for patients in renal failure and is 50–100 mg po q12 hours [356, 357]. There is a long acting, controlled release, preparation available and when used, the same daily dose is administered in two divided doses [358]. The therapeutic blood level of disopyramide is reported to be 2–5 μg/ml, but the usefulness of monitoring the plasma concentration is debatable, since the measured blood level is total concentration of disopyramide while the pharmacologic activity depends on the free fraction of drug [359]. Also, the pharmacokinetics of disopyramide are nonlinear so that a change in total drug concentration results in a greater, but unpredictable change in the pharmacologically active free concentration [360].

The absorption of disopyramide when given orally is complete and rapid and peak plasma concentrations are achieved at 30 minutes to 3 hours. However, because a first pass effect which removes approximately 15 % of the oral dose, systemic bioavailability is 80–85 % of that which would be expected after intravenous administration [361]. Some studies have reported intersubject variability in disopyramide absorption when the controlled release preparation is used, but this has not been a consistent finding [361]. Peak plasma concentrations occur later with the controlled release preparation (3.4 to 4.9 hours), but bioavailability is not different.

Disopyramide is concentrated in fat, liver and spleen where the tissue concentration is 7–15 times higher than in plasma. In canine studies, the drug concentration in myocardial tissue is 4 times greater than in plasma.

The major route of elimination of disopyramide is via the kidneys, with up to 67 % of the oral dose being recovered unchanged in the urine [356]. In studies using radiolabeled drug, 80 % of disopyramide is recovered in the urine and 15 % in the feces as unchanged drug.

The major metabolic pathway is hepatic dealkylation to form the mono-N-dealkylated metabolite which appears to have less than 50 % of the antiarrhythmic activity of the parent compound [357, 362]. Administration of rifampin or phenytoin increases the metabolism of disopyramide with a subsequent decrease in the percentage of unchanged drug excreted in the urine [363, 364].

Clearance of disopyramide is similar in healthy subjects and patients with CHF, but is decreased in patients with renal insufficiency [365]. Elimination half-life in normal subjects is 4.4–7.8 hours, but is more variable in those with heart disease and ranges from 6–15 hours. The elimination half-life is longer in elderly subjects and is markedly prolonged in patients with renal dysfunction and elimination is indirectly associated with the creatinine clearance [365, 366]. Interindividual variations in the volume of distribution and total elimination requires that dosing in renal failure be individualized. During hemodialysis, elimination kinetics are not apparently different from those of patients with severe renal failure not undergoing dialysis.

The extent of protein binding is concentration dependent and there is considerable interindividual variability [367, 368]. Albumin binds 5–10 % of disopyramide and its mono-N-dealkylated metabolite, with a major percentage of these compounds being bound by alpha-1-acid glycoprotein. In cirrhotic patients or those with severe hepatic dysfunction the free fraction of disopyramide is greater at any given total drug concentration, compared to that found in healthy subjects [369, 370].

There is no interaction between disopyramide and digoxin, but disopyramide does potentiate the effects of warfarin. Both phenytoin and rifampin may increase the metabolism of disopyramide. There are no other drug-drug concentrations reported [363, 364].

Hemodynamics

Disopyramide is a potent negative inotropic agent [372, 373]. Animal and human studies have reported that the drug significantly depresses left ventricular contractility and results in a decrease in left ventricular stroke volume, stroke work and cardiac output while left ventricular end diastolic pressures increase [374, 375]. In contrast to quinidine and procainamide, disopyramide does not have ganglionic blocking effects and does not produce peripheral vasodilation or a decrease in afterload [375]. Indeed, as a result of a decrease in stroke volume and cardiac output, there is a reflux increase in systemic vascular resistance and afterload which can further reduce cardiac output. As a result of the negative inotropic effects of the drug, congestive heart failure can occur, especially in patients with a previous history of left ventricular dysfunction in whom the incidence is as high as 55 % [376].

Disopyramide may also provoke congestive heart failure in patients with intact left ventricular function, although this is not frequently seen. Studies have reported that the drug can significantly impair the contractility of abnormal myocardium while normal segments are not significantly effected [377, 378].

Efficacy

Disopyramide is effective for atrial arrhythmias and probably is as effective as quinidine although studies are limited [379]. When used intravenously, disopyramide has been reported to convert atrial fibrillation to sinus rhythm in 20–71 % of patients, but the number of patients involved is small [380]. The oral drug is effective for preventing recurrent atrial fibrillation during long-term maintenance [381, 382]. During electrophysiologic testing, disopyramide may prevent an initiation of atrial fibrillation in some patients while in others, atrial fibrillation may actually be easier to initiate because of the effect of the drug on the conduction – refractoriness relationship of atrial tissue [383].

Studies of disopyramide for atrial flutter are scarce, but it has been observed that the intravenous drug will slow the atrial rate and may decrease ventricular response rate [384]. Similar to the experience with other drugs, disopyramide is less effective for reverting atrial flutter, but it may facilitate conversion to sinus rhythm when overdrive pacing is used [384]. Additionally, the drug used for maintenance may prevent recurrences.

Disopyramide has been reported to be effective for supraventricular tachycardia and in approximately 40 % of patients, the intravenous drug will terminate the arrhythmia [385, 386]. The oral drug prevents recurrent supraventricular tachycardia clinically as well as preventing its induction in the electrophysiology laboratory in almost 60 % of patients. Its major effect is depression of conduction in the retrograde limb of the re-entrant circuit.

Since disopyramide also alters the electrophysiologic properties of the accessory pathway, it can be expected that the drug will prevent supraventricular tachycardia associated with a pre-excitation syndrome (Lown-Ganong-Levine or Wolff-Parkinson-White). The intravenous drug has been shown to terminate this arrhythmia, while the oral drug is effective for preventing reinduction of the arrhythmia [354, 387, 388] largely as a result of its effect on retrograde conduction in the accessory pathway. Since the drug also prolongs the antegrade refractory period of the accessory pathway, conduction through it is impaired and the ventricular response rate during atrial fibrillation is slowed.

Disopyramide has been reported to be effective for suppressing all forms of ventricular arrhythmia including VPBs, couplets and runs of VT. In uncontrolled trials, both intravenous and oral disopyramide reduce VPB frequency by >70 % in 65–85 % of patients [389, 390]. In placebo-controlled trials the oral drug is effective for all forms of arrhythmia [391]. Oral disopyramide has been used in a number of trials in which it has been compared to an investigational drug [391–393]. While it is equivalent in efficacy to some of these agents, in most of the trials disopyramide it is less effective than the comparative drug. However, reduction in runs of nonsustained VT is usually equivalent with both agents.

The role of disopyramide for preventing sustained VT or VF is less well defined as there are only a few studies involving a small number of patients [394–396]. When given intravenously, the drug will terminate VT in 50–75 % of patients. However, some of these studies also involved patients with nonsustained VT. In the electrophysiology laboratory, 20–35 % of patients are rendered noninducible. In the nonresponders, there is a decrease in the rate of the induced VT. During long-term therapy, the drug continues to be effective in the majority of patients who respond initially.

Side effects

The most serious side effects from disopyramide are cardiac. Like the other class 1A antiarrhythmic agents, it can cause prolongation of the QT interval with associated torsade des pointes [397, 398]. Arrhythmia aggravation occurs early in the course of treatment.

Other forms of arrhythmia aggravation have been reported and are unpredictable, as they occur in the absence of other side effects and with blood levels of disopyramide that are in the therapeutic range [399].

Conduction abnormalities are uncommon, although AV nodal block and bundle branch block have been reported [389, 390]. In patients with severe renal disease, standard doses of disopyramide can cause excessively high blood levels and electromechanical dissociation and cardiovascular collapse have been reported [400]. Since disopyramide is a potent negative inotropic agent, CHF has been provoked, especially in patients with a history of CHF [375, 376]. In such patients, disopyramide should not be used unless absolutely necessary.

The most frequent side effects due to disopyramide result from its anticholinergic action and include dry mouth, blurring of vision, constipation, close angle glaucoma and urinary hesitancy [389–402]. These side effects tend to be mild, well tolerated and transient. Infrequently, they result in discontinuation of therapy. Symptomatic management or reduction in dosage is often sufficient. It has been observed that these anticholenergic side effects may be less frequent with the long-acting preparation, perhaps because of lower peak and steady state blood levels (358). Several investigators have reported that cholinergic agents, particularly mestinone, will reverse or prevent these side effects, resulting in better patient tolerance [403].

Disopyramide should be administered with caution to patients who have a pre-existing history of prostate problems, as urinary retention may be more common in these patients [402]. Less common side effects include nausea, vomiting, diarrhea, abdominal pain, fatigue and headache, with some reports of fasting hypoglycemia, acute psychosis, cholestatic jaundice and precipitation of uterine contractions during pregnancy also being noted.

5.14 Procainamide (Table 17)

Electrophysiology

Procainamide has electrophysiologic effects which are similar to those of quinidine and disopyramide and it is therefore a class 1A agent. It decreases the rate of upstroke velocity of phase 0 of atrial and ventricular muscle, Purkinje fibers and tissue of the accessory pathway, therefore decreasing the rate of membrane depolarization [404]. As a result, there is a decrease in membrane conduction velocity. As with other agents, these effects are use-dependent [405]. The

Table 17
Procainamide

Dose	oral – 500–1000 mg q4–6 hours (short acting) 750–1500 mg TID or QID (long acting)
Protein binding	15–25 %
Clearance	hepatic – acetylation to active metabolite – NAPA
Half-life	1½–3 hours – procainamide 6–7 hours – NAPA
Blood levels	4–10 μg/ml NAPA – variable
Drug interactions	none known
Indications	ventricular and supraventricular arrhythmias
Side effects	CNS, GI, cardiac, lupus-like syndrome, hematologic

drug also delays repolarization and prolongs the effective refractory period (ERP) of all cardiac tissue, thereby reducing membrane excitability [406]. There is prolongation of the action potential duration, although this is less marked than what is observed with quinidine. This electrophysiologic effect is primarily the result of an active metabolite, N-acetyl procainamide (NAPA) which, unlike procainamide, exerts primarily class 3 activity [407, 408]. The drug also reduces automaticity by decreasing the slope of phase 4 spontaneous depolarization. In general, procainamide does not affect normal sinus node function, but in the presence of underlying sinus node dysfunction the drug further depresses its function and can prolong sinus node recovery time [409].

Procainamide directly depresses the conduction through and the refractory period of the AV node, especially in the retrograde direction [410, 411]. The drug also has a mild vagolytic effect which results in a shortening of ERP and an increase in conduction velocity which may partially or completely reverse the direct depressant action [412]. As a result of these electrophysiologic effects, there is a variable effect on the AH interval and a prolongation of the HV interval as observed during electrophysiologic studies. On the surface ECG, procainamide minimally prolongs the PR and QRS intervals, usually in a dose-related fashion. Prolongation of the QT is more significant however, it is less marked in comparison to the prolongation observed with quinidine.

Pharmacology

Procainamide is available for both oral and intravenous use. The intravenous dose is generally 100 mg by infusion every 5 minutes up to 1–1.5 grams guided by QRS width, blood pressure and for those patients in VT, a slowing of the rate of the tachycardia.

Thereafter, a maintenance dose of 1–4 mg/min can be continued if necessary [412]. The oral dose is 500–1000 mg every 4–6 hours, although on occasion, a daily dose of up to 12 grams has been used. Unfortunately, absorption is rapid (30–60 minutes) and procainamide has a short half-life of 1½–3 hours therefore necessitating more frequent dosing. There is a long-acting preparation now available which, because of a special coating, prevents rapid absorption of the drug [414, 415]. As a result of the slow absorption, blood levels are maintained for a longer period of time, although the half-life is unchanged. Procainamide is 75–95 % bioavailable and is 15–25 % protein bound. The blood levels reported to be therapeutic range from 4–10 μg/ml.

Procainamide is rapidly metabolized in the liver by acetylation to one major metabolite (NAPA) which possesses significant antiarrhythmic activity. However, its electrophysiologic effects are different from procainamide [407, 408]. Similar to the situation with hydralazine and isoniazid (an antituberculosis agent) the metabolism by acetylation is generally determined. Procainamide is metabolized quickly in some patients (rapid acetylators) in whom NAPA levels are higher than procainamide blood levels when measured simultaneously 2 hours after dose [418]. In the slow acetylators NAPA blood levels are lower than procainamide levels. It is however, not useful to add the blood levels together to determine the actual drug level since these 2 agents have different electrophysiologic effects on the heart.

The half-life of NAPA is longer, approximately 6–7 hours [419]. NAPA is cleared by the kidney and therefore it may accumulate in patients with renal impairment [420]. Procainamide levels can also increase in patients with renal impairment and the half-life is prolonged.

There are no known drug interactions with procainamide.

Hemodynamics

Procainamide is negatively inotropic and by direct action it can depress left ventricular contractility resulting in an increase in pulmonary capillary wedge pressure and left ventricular end diastolic pressure and a reduction in stroke volume and stroke work [421, 422]. However, the drug does also have ganglionic blocking action and causes peripheral vasodilatation. This reduction in afterload main-

tains cardiac output. In general, procainamide does not cause congestive heart failure although failure may occur in patients with very low left ventricular ejection fractions [423, 424].
As a result of the drug's ganglionic blocking activity, hypotension may occur and this has been most often observed when the drug is administered intravenously.

Efficacy
Procainamide is widely used for the treatment of ventricular as well as supraventricular arrhythmias. When administered intravenously, procainamide is effective for reverting atrial fibrillation to sinus rhythm in 40–50 % of patients [425–427], although it is of greatest benefit when the arrhythmia is of recent onset, such as after cardiac surgery. Therapy with oral procainamide is, however, less effective for reverting atrial fibrillation or maintaining sinus rhythm. This may be because blood levels achieved with intravenous drug are much higher than those achieved with oral drug. Also, the short half-life results in more fluctuations of blood levels. The drug is not very effective for reverting atrial flutter, although it does slow the rate of the atrial flutter and may enhance the efficacy of rapid atrial overdrive pacing for reverting atrial flutter [428].
When given intravenously to patients with supraventricular tachycardia due to AV nodal re-entry, procainamide reverts the arrhythmia in 70–80 % of patients as a result of depression of retrograde conduction within the AV node [409, 429]. The majority of these patients remain noninducible during a continuous intravenous infusion of procainamide. There are no well-controlled studies utilizing the oral drug for long-term therapy, but it appears to be of benefit for some patients. Procainamide is of use in patients with supraventricular arrhythmias due to an accessory pathway as a result of its effect on the AV node as well as the accessory pathway tissue [430, 431]. As a result of prolongation of the refractory period of the accessory pathway, it may block this pathway, normalizing conduction. This is primarily seen in patients with a longer refractory period of the accessory pathway tissue [432].
The major role for intravenous and oral procainamide is in the therapy of a wide range of ventricular arrhythmias and there are many studies reporting the drug efficacy [433, 434]. A number of investigators have reported that 50–60 % of patients with frequent VPBs have

a significant reduction of this arrhythmia when ambulatory monitoring and exercise testing are utilized [435]. The drug appears to be as effective as quinidine for arrhythmia suppression. Unfortunately, many of these studies were carried out before there were strict criteria for defining drug efficacy. Recent comparative trials between procainamide and other antiarrhythmic agents which are usually investigational have reported that oral procainamide is only effective in 20–40 % of patients when efficacy is defined as ≥ 75 % reduction of VPBs [436–438]. Efficacy of procainamide can be enhanced when it is combined with another agent such as quinidine, mexiletine, propafenone or amiodarone.

The role of procainamide for sustained VT or VF is not clear. While the intravenous drug is effective for terminating sustained VT, the drug prevents the induction of sustained VT in only 15–25 % of patients [439–441]. Unfortunately, there are only a few studies in such patients and they are not well controlled nor are comparative data available. It was initially reported that the response to procainamide predicted the response to other class 1A agents [442], but this has not been confirmed by others [443]. Indeed, the response to IV procainamide may not predict the effect of the oral drug, probably as a result of the active metabolite NAPA [444].

Toxicity

The most frequent side effects resulting from procainamide, reported in 10–50 % of patients, are gastrointestinal and include abdominal pain, anorexia, nausea and vomiting, Also frequent are central nervous system side effects including insomnia, headache, confusion, dizziness, depression, psychosis and hallucinations [433, 434, 445]. Allergic reactions, especially fever and skin rash, are less common, and unpredictable but occur 2–3 weeks after initiating drug. Also uncommon, but potentially life-threatening, is agranulocytosis or pancytopenia which is idiosyncratic [446] and may be allergic or immunologically mediated. It may occur days to weeks after drug initiation. Cardiac side effects are primarily conduction abnormalities and arrhythmia aggravation, including QT prolongation and torsade des pointes [447].

The most frequent side effect is a lupus-like syndrome which has been reported in 15–25 % of patients although 50–80 % will develop a positive ANA titer after 1 year of therapy [448]. It has been observed that patients who are slow acetylators and thus have low NAPA levels and higher procainamide levels are more likely to develop this side effect while rapid acetylators, who have lower procainamide and higher NAPA levels, are less likely to develop this complication [418]. Simultaneous blood levels of procainamide and NAPA are useful for establishing acetylator phenotype. Also useful is the ANA titer which becomes positive earlier and rises more quickly in the slow acetylators and correlates with an increased risk of developing lupus [448]. The clinical manifestations of this lupus-like syndrome include arthralgias and arthritis, pneumonitis and pericarditis. Renal and central nervous system involvement is rare. To distinguish between systemic lupus and the drug-induced form, it has been helpful to measure anti DNA antibodies which are common in systemic lupus and absent in drug-induced lupus as well as complement levels which are normal in drug-induced lupus, but low in systemic lupus.

References

1 C. M. Chen, L. S. Gettes, B. G. Katzung: Effect of lidocaine and quinidine on steady state characteristics and recovery kinetics of dv/dt_{max} in guinea pig ventricular myocardium. Circ Res 37:20, 1975.
2 M. R. Rosen, A. L. Wit: Electropharmacology of antiarrhythmic drugs. Am Heart J 106:829, 1983.
3 B. F. Hoffman, M. R. Rosen: Cellular mechanism for cardiac arrhythmias. Circ Res 49:1, 1981.
4 A. Norma, H. Irisawa, S. Kokubun, H. Kotake, M. Nishimara, Y. Watanabe: Slow current systems in the AV node of the rabbit heart. Nature 285:228, 1980.
5 A. L. Wit, M. R. Rosen: Pathophysiologic mechanisms of cardiac arrhythmias. Am Heart J 106:798, 1983.
6 R. F. Gilmour, J. J. Heger, E. N. Prystowsky, D. P. Zipes: Cellular electrophysiological abnormalities of diseased human ventricular myocardium. Am J Cardiol 51:137, 1983.
7 P. F. Cranefield: Action potentials, afterpotentials and arrhythmias. Circ Res 41:415, 1977.
8 A. L. Wit, P. F. Cranefield: Triggered activity in cardiac muscle fibers of the simian mitral valve. Circ Res 38:85, 1976.
9 A. L. Wit, P. F. Cranefield: Reentrant excitation as a cause of cardiac arrhythmias. Am J Physiol 235:H1, 1978.
10 E. M. Vaughan Williams: Classification of antiarrhythmic drugs. In: E. Sandoe, E. Flensted-Jensen, E. Olsen, eds. Symposium on cardiac arrhythmias. Sodertalje, Sweden: AB Astra, pp. 449–472, 1970.
11 B. F. Hoffman, M. R. Rosen, A. L. Wit: Electrophysiology and pharmacology of cardiac arrhythmias: 7. cardiac effects of quinidine and procainamide. Am Heart J 90:117, 1975.
12 D. C. Harrison: Antiarrhythmic drug classification: new science and practical applications. Am J Cardiol 56:185, 1985.

13 L. M. Hondeghem, B. C. Katzang: Time and voltage dependent interactions of antiarrhythmic drugs with cardiac sodium channels. Biochem Biophys Acta 472:383, 1977.
14 A. L. Wit, B. F. Hoffman, M. R. Rosen: Electrophysiology and pharmacology of cardiac arrhythmias: 9. Cardiac electrophysiologic effects of beta-adrenergic receptor stimulation and blockade: Part C. Am Heart J 90:797, 1975.
15 G. S. Hirsowitz, P. Podrid, S. Lampert, J. Stein, B. Lown: The use of beta blocking agents as adjunct therapy in the treatment of malignant ventricular arrhythmia. J Am Coll Cardiol 3:619, 1984.
16 P. J. Podrid, B. Lown: Use of beta-adrenergic blocking agents for cardiac arrhythmias. In: J. P. Kostis, E. A. DeFelice, eds. Beta blockers in the treatment of cardiovascular disease. New York: Raven Press, pp. 121–154, 1984.
17 B. N. C. Prichard, C. W. I. Owens: Clinical pharmacology of beta-adrenoceptor blocking drugs. In: J. P. Kostis, E. A. DeFelice, eds. Beta blockers in the treatment of cardiovascular disease. New York: Raven Press, pp. 1–56, 1984.
18 R. P. Swenson: Inactivation of potassium current in quid axon by a variety of quaternary ammonium ions. J Gen Physiol 77:255, 1981.
19 M. B. Bacaner: Bretylium totylate for suppression of induced ventricular fibrillation. Am J Cardiol 17:528, 1966.
20 A. L. Wit, P. Cranefield: Effects of verapamil in the sinoatrial and atrioventricular nodes of the rabbit and the mechanism by which it arrests reentrant atrioventricular nodal tachycardia. Circ Res 35:413, 1974.
21 M. R. Rosen, A. L. Wit, B. F. Hoffmann: Electrophysiology and pharmacology of cardiac arrhythmias: 6. Cardiac effects of verapamil. Am Heart J 89:665, 1975.
22 B. N. Singh, H. S. Hecht, K. Nademanee: Electrophysiologic and hemodynamic actions of slow channel blocking compounds. Prog Cardiovasc Dis 25:103, 1982.
23 R. L. Sung, B. Elser, R. G. McAllister: Intravenous verapamil for termination of reentrant supraventricular tachycardias. Ann Intern Med 93:682, 1980.
24 V. Velebit, P. J. Podrid, B. Lown, B. Cohen, T. B. Graboys: Aggravation and provocation of ventricular arrhythmias by antiarrhythmic drugs. Circ 65:886, 1982.
25 D. P. Zipes: Proarrhythmic effects of antiarrhythmic drugs. Am J Cardiol 59:26E, 1987.
26 R. F. Poser, P. J. Podrid, F. Lombardi, B. Lown: Aggravation of arrhythmias induced by antiarrhythmic drugs during electrophysiologic testing. Am Heart J 110:4, 1985.
27 P. J. Podrid, S. Lampert, T. B. Graboys, C. M. Blatt, B. Lown: Aggravation of arrhythmia by antiarrhythmic drugs – Incidence and predictors. Am J Cardial 59:31E, 1987.
28 W. Slater, S. Lambert, P. J. Podrid, B. Lown: Clinical predictors of arrhythmia worsening by antiarrhythmic drugs. Am J Cardiol 61 :349, 1988.
29 J. Morganroth: Risk factors for the development of proarrhythmic events. Am J Cardiol 59:32E, 1987.
30 J. Morganroth, L. N. Norowitz: Flecainide – its proarrhythmic effect and expected changes on the surface electrocardiogram. Am J Cardiol 53:893, 1984.
31 P. J. Podrid, F. J. Vendetti, P. A. Levine, M. D. Klein: The role of exercise testing in evaluation of arrhythmias. Am J Cardiol 62:24H, 1988.
32 J. Ravid, P. J. Podrid, S. Lampert, B. Lown: Congestive heart failure induced by six of the new antiarrhythmic drugs. J Am Coll Cardiol 14:1326, 1989.

33 L. D. Nelson, P. G. Schmid, D. Holmsten, A. L. Mark, D. D. Heistad, F. M. Abboud. Effects of quinidine on venous responses to adrenergic and nonadrenergic constrictor stimuli: indirect evidence of two sites of action in vascular smooth muscle. Proc Soc Exp Biol Med 146:409, 1974.
34 A. J. Leach, J. E. Brown, P. W. Armstrong: Cardiac depression by intravenous disopyramide in patients with left ventricular dysfunction. Am J Med 68:839, 1980.
35 Bigger (Ed): A symposium: Flecainide Acetate. Am J Cardiol 53: (suppl B), 1984.
36 A. B. Hodess, W. P. Follansbee, J. F. Spear, E. N. Moore: Electrophysiological effects of a new antiarrhythmic agent, flecainide, on the intact canine heart. J Cardiovasc Pharmacol 1:427, 1979.
37 B. N. Singh, K. Nademanee, M. A. Josephson, N. Ikeda, N. Venkatish, L. Kannan: The electrophysiology and pharmacology of verapamil, flecainide and amiodarone: Correlations with clinical effects and antiarrhythmic actions. In Garfein OB. Clinical Pharmacology of Cardiac Antiarrhythmic Agents. Classical and recent concepts reevaluated. New York: NY Academy of Sciences, Vol 432, pp. 210–235, 1984.
38 P. I. Hoff, A. Tronstadt, B. Oil, O. J. Ohm: Electrophysiology and clinical effects of flecainide for recurrent paroxysmal supraventricular tachycardias. Am J Cardiol 62:585, 1988.
39 G. J. Conrad, R. E. Ober: Metabolism of flecainide. Am J Cardiol 53:41B, 1984.
40 M. A. Josephson, N. Ikeda, B. N. Singh: Effects of flecainide on ventricular function: clinical and experimental correlations. Am J Cardiol 53:95B, 1984.
41 M. A. Josephson, S. Kaul, J. Hopkins, D. Kvam, B. Singh: Hemodynamic effects of intravenous flecainide relative to the level of ventricular function in patients with coronary artery disease. Am Heart J 109:41, 1985.
42 P. H. J. M. Dunselman, J. H. Ingma, L. M. van Wijk, A. Disk, J. Wesseling, K. I. Lie: Hemodynamic and antiarrhythmic effects of intravenous flecainide acetate in chronic congestive heart failure. Drugs (suppl 4):58, 1985.
43 J. L. Anderson, J. R. Stewart, B. A. Perry, D. D. van Hamersveld, T. A. Johnson, G. J. Conard: Oral flecainide acetate for the treatment of ventricular arrhythmias. N Engl J Med 305:473, 1981.
44 M. Hodges, J. M. Haugland, G. Granrud, G. J. Conard, R. W. Asinger, F. L. Mikell: Suppression of ventricular ectopic depolarizations by flecainide acetate, a new antiarrhythmic agent. Circulation 65:879, 1982.
45 J. Kjekshus, J. Bathen, O. M. Orning, L. Storstein: A double blind crossover comparison of flecainide acetate and disopyramide phosphate in the treatment of ventricular premature beats. Am J Cardiol 53:72B, 1984.
46 The Flecainide-Quinidine Research Group. Flecainide versus quinidine for treatment of chronic ventricular arrhythmias. A multicenter clinical trial. Circ 67:1117, 1983.
47 The Cardiac Arrhythmia Suppression Trial (CAST) Investigators. Preliminary report: effect of encainide and flecainide on mortality in a randomized trial of arrhythmia suppression after myocardial infarction. N Engl J Med 321:406, 1989.
48 P. R. Reid, L. S. C. Griffith, E. V. Platia, E. S. Ord: Evaluation of flecainide acetate in management of patients at high risk of sudden cardiac death. Am J Cardiol 53:108B, 1988.
49 Flecainide Ventricular Tachycardia Study Group. Treatment of resistant ventricular tachycardia with flecainide acetate. Am J Cardiol 57:1299, 1986.

50 S. S. Kim, R. Lal, R. Ruffy: Treatment of paroxysmal reentrant supraventricular tachycardia with flecainide acetate. Am J Cardiol 58:80, 1986.
51 J. L. Anderson, E. L. L. Pritchett (Ed.): International Symposium on Supraventricular Arrhythmias: Focus on flecainide. 62: Suppl D, 1988.
52 G. D. Gentzkow, J. Y. Sullivan: Extracardiac adverse effects of flecainide. Am J Cardiol 53:101B, 1984.
53 J. Morganroth, D. C. Harrison (Ed.): A Symposium – Encainide. Am J Cardiol 58: Suppl C, 1986.
54 G. V. Naccarelli, H. J. J. Wellens (Ed.): A Symposium: The use of encainide in supraventricular tachycardias. Am J Cardiol 62: Suppl C, 1988.
55 M. Sami, J. W. Mason, G. Oh, D. C. Harrison: Canine electrophysiology of encainide, a new antiarrhythmic drug. Am J Cardiol 43:1149, 1979.
56 R. A. Winkle, F. A. Peters, R. E. Kates, D. C. Harrison: Possible contribution of encainide metabolites to the antiarrhythmic efficacy of encainide. Am J Cardiol 51:1182, 1983.
57 J. T. Barbey, K. A. Thompason, D. S. Echt, R. L. Woosley, D. M. Roden: Antiarrhythmic activity, electrocardiographic effects and pharmacokinetics of the encainide metabolites, 0-desmethyl encainide and 3 methoxy-0-desmethyl encainide in man. Circ 77:380, 1988.
58 R. A. Winkle, F. Peters, R. E. Kates, C. Tucker, D. C. Harrison: Clinical pharmacology and antiarrhythmic efficacy of encainide in patients with chronic ventricular arrhythmias. Circ 64:290, 1981.
59 T. Wang, D. M. Roden, H. T. Wolfendin, R. L. Woosley, A. J. J. Wood, G. R. Wilkenson: Influence of genetic polymorphism on the metabolites and disposition of encainide in man. J Pharmacol Exp Ther 228:605, 1984.
60 R. H. Bergstrand, T. Wang, D. M. Roden, G. R. Avant, W. W. Sutton, L. A. Siddoway, H. Wolfenden, R. L. Woosley, G. R. Wilkenson, A. J. J. Wood: Encainide disposition in patients with chronic cirrhosis. Clin Pharmacol Ther 40:148, 1986.
61 A. H. Bergstrand, T. Wang, D. M. Roden, W. S. Stone, H. T. Wolfinden, R. L. Woosley, G. R. Wilkinson, A. J. J. Wood: Encainide disposition in patients with renal failure. Clin Pharmacol Ther 40:64, 1986.
62 B. D. Quart, D. G. Gallo, M. H. Sami, A. J. J. Wood: Drug interaction studies and encainide use in renal and hepatic impairment. Am J Cardiol 58:104C, 1986.
63 M. H. Sami, V. A. Derbekyan, R. Lisbona: Hemodynamic effects of encainide in patients with ventricular arrhythmia and poor ventricular function. Am J Cardiol 52:507, 1983.
64 S. S. Gottlieb, M. L. Kukin, M. Yushak, N. Medina, M. Packer: Adverse hemodynamic and clinical effects of encainide in severe chronic heart failure. Ann Int Med 110:505, 1989.
65 J. Morganroth, P. E. Pool, R. Miller, P. H. Hsu, I. K. Lee, D. M. Clark: Dose responsive range of encainide for benign and potentially lethal ventricular arrhythmias. Am J Cardiol 57:769, 1986.
66 D. M. Roden, B. R. Stots, S. B. Higgins, R. F. Mayol, R. E. Gammans, J. A. Oates, R. L. Woosley: Total suppression of ventricular arrhythmias by encainide. N Engl J Med 302:877, 1980.
67 R. DiBianco, R. D. Fletcher, A. I. Cohen, J. S. Gottdiener, S. N. Singh, J. H. Katz: Treatment of frequent ventricular arrhythmia with encainide: assessment using serial ambulatory electrocardiograms, intracardiac electrophysiologic studies, treadmill exercise tests and radionuclide angiographic studies. Circ 65:1134, 1982.
68 J. Morganroth, J. C. Somberg, P. E. Pool, P. H. Hsu, I. K. Lee, J. Durkee: Comparative study of encainide and quinidine in the treatment of ventricular arrhythmia. J Am Coll Cardiol 7:9, 1986.

69 T. Tordjman, P. J. Podrid, E. Raeder, B. Lown: Safety and efficacy of encainide for malignant ventricular arrhythmias. Am J Cardiol 58:87C, 1986.
70 L. N. Horowitz: Encainide in lethal ventricular arrhythmias evaluated by electrophysiologic testing and decrease in symptoms. Am J Cardiol 58:83C, 1986.
71 J. W. Mason, F. Peters: Antiarrhythmic efficacy of encainide in patients with refractory recurrent ventricular tachycardia. Circ 63:670, 1981.
72 W M. Miles, D. P. Zipes, R. L. Rinkenberger, M. L. Markel, E. N. Prystowsky, A. H. Dougherty, J. J. Heger, G. V. Naccarelli: Encainide for treatment of atrioventricular reciprocating tachycardia in the Wolff Parkinson White Syndrome. Am J Cardiol 62:20C, 1988.
73 G. V. Naccarelli, W. M. Jackman, M. Akhtar, R. L. Rinkenberger, K. J. Friday, A. H. Dougherty, P. Tchou, J. A. Yiung-Lei-Wak: Efficacy and electrophysiologic effects of encainide for atrioventricular nodal reentrant tachycardia. Am J Cardiol 62:312, 1988.
74 K. P. Kunzl, K. H. Kuck, M. Schluter: The effect of encainide and flecainide on chronic ectopic atrial tachycardia. J Am Coll Cardiol 7:1121, 1986.
75 R. L. Rinkenberger, G. V. Naccarelli, W. M. Miles, M. L. Markel, A. H. Dougherty, E. N. Prystowsky, J. J. Heger, D. P. Zipes: Encainide for atrial fibrillation associated with Wolff Parkinson White Syndrome. Am J Cardiol 62:264, 1988.
76 L. F. Soyka: Safety of encainide for treatment of ventricular arrhythmias. Am J Cardiol 58:96C, 1986.
77 R. A. Winkle, J. W. Mason, J. C. Griffin, D. Ross: Malignant ventricular tachyarrhythmias associated with the use of encainide. Am Heart J 102:857, 1981.
78 P. J. Podrid (Ed.): Symposium on Propafenone. J Electrophy 1:501, 1987.
79 Zipes (Ed.): Symposium on Propafenone. Am J Cardiol 54: Suppl D, 1984.
80 M. Kohlhardt: Block of sodium currents by antiarrhythmic agents. Analysis of the electrophysiologic effects of propafenone. Am J Cardiol 54:, 1984.
81 H. Muller-Piltzer, G. Greger, G. Neugebauer, M. Hollman: Beta blocking and electrophysiologic effects of propafenone in volunteers. Eur Heart J 6:123, 1985.
82 A. A. McLeod, G. L. Stiles, D. G. Shand: Demonstration of beta adrenoceptor blockade by propafenone hydrochloride: Clinical pharmacologic, radioligand binding and adenylate cyclase activation studies. J Pharmacol Exp Ther 228:461, 1984.
83 F. Ledda, L. Mantelli, S. Manzini, S. Amerini, A. Mugelli: Electrophysiological and antiarrhythmic properties of propafenone in isolated cardiac preparations. J Cardiovasc Pharmacol 3:1162, 1981.
84 I. D. Dukes, E. M. Vaughan Williams: The multiple modes of action of propafenone. Eur Heart J 5:115, 1984.
85 S. J. Connolly, R. E. Kates, C. S. Lebsack, D. S. Echt, J. W. Mason, R. A. Winkle: Clinical effifacy and electrophysiology of oral propafenone for ventricular tachycardia. Am J Cardiol 52:1208, 1983.
86 L. Seipel, Breithardt. Propafenone – a new antiarrhythmic drug. Eur Heart J 1:309, 1980.
87 N. A. Smith, R. E. Kates, D. C. Harrison: The clinical pharmacology of propafenone. J Electrophysiol 1:517, 1987.
88 S. J. Connolly, R. E. Kates, C. S. Lebsack: Clinical pharmacology of propafenone. Circ 68:589, 1983.

89 C. A. Siddoway, D. M. Roden, R. L. Woosley: Clinical pharmacology of propafenone: Pharmacokinetics, metabolism and concentration – response relations. Am J Cardiol 54:9D, 1984.
90 J. L. Stohler, P. R. Kowey, R. A. Marinchak, T. D. Friehling: Drug interactions with propafenone. J Electrophysiol 1:568, 1987.
91 D. M. Salerno, G. Granrud, P. Sharkey, R. Asinger, M. A. Hodges: A controlled trial of propafenone for the treatment of frequent and repetitive ventricular premature contractions. Am J Cardiol 53:77, 1984.
92 R. L. Klein, F. I. Marcus: Efficacy of propafenone when used in combination antiarrhythmic therapy. J Electrophysiol 1:575, 1987.
93 B. J. Baker, H. Dinh, D. Kroskey, N. D. B. de Soza, M. C. Murphy, J. A. Franciosa: Effect of propafenone on left ventricular ejection fraction. Am J Cardiol 54:20D, 1984.
94 M. A. Brodsky, B. J. Allen, D. Abate, W. L. Henry: Propafenone therapy for ventricular tachycardia in the setting of congestive heart failure. Am Heart J 110:794, 1985.
95 N. deSoyza, C. Terry, M. L. Murphy: Effects of propafenone in patients with stable ventricular arrhythmias. Am Heart J 108:285, 1984.
96 M. Hodges, D. Salerno, G. Granrud: Double blind placebo-controlled evaluations of propafenone in suppressing ventricular ectopic activity. Am J Cardiol 54:45D, 1984.
97 F. Naccarelli, D. Bracchetti, M. Palmieri: A Comparison of propafenone and disopyramide for threatment of chronic ventricular arrhythmias: palacebo controlled, double-blind, randomized crossover study. Am Heart J 109:833, 1985.
98 H. A. Dinh, M. L. Murphy, B. J. Baker: Efficacy of propafenone compared with quinidine in chronic ventricular arrhythmias. Am J Cardiol 55:1520, 1985.
99 P. J. Podrid, B. Lown: Propafenone: A new agent for ventricular arrhythmia. J Am Coll Cardiol 4:117, 1984.
100 L. Cueni, P. J. Podrid: Propafenone therapy in patients with serious ventricular arrhythmia. Noninvasive evaluation of efficacy. J Electrophysiol 1:548–560, 1987.
101 D. A. Chilson, J. J. Heger, D. P. Zipes, E. N. Prystowsky: Electrophysiologic effects and clinical efficacy of oral propafenone therapy in patients with ventricular tachycardia. J Am Coll Cardiol 5:1407, 1985.
102 L. Fananpazir, E. N. Prystowsky: Propafenone – clinical electrophysiology and efficacy in patients with documented or potentially life threatening arrhythmias. – J Electrophysiol 1:543–547, 1987.
103 S. C. Hammil: Use of propafenone in patients with supraventricular tachycardia. J Electrophysiol 1:561, 1987.
104 P. Coumel, S. F. Leclercq, P. Assayag: European experience with the antiarrhythmic effifacy of propafenone for supraventricular and ventricular arrhythmias. Am J Cardiol 54:60D, 1984.
105 E. N. Shen, E. Keung, E. Huycke: Intravenous propafenone for termination of reentrant supraventricular tachycardia: A placebo controlled randomized, double-blind, crossover study. Ann Int Med 105:655, 1986.
106 G. Breithardt, M. Borggrefe, E. Wiebringhaus, L. Seipel: Effect of propafenone in the Wolff-Parkinson-White Syndrome: Electrophysiologic findings and long-term follow-up. Am J Cardiol 54:29D, 1984.
107 S. Ravid, P. J. Podrid, B. Novrit: Safety of long-term propafenone therapy for cardiac arrhythmia: experience with 774 patients. J Electrophysiol 1:580, 1987.
108 P. D. Dennis, E. M. Vaughan-Williams: Effects of rabbit cardiac potentials of aprindine and indecainide, a new antiarrhythmic agent in normoxia and hypoxia. Br J Pharmacol 85:11, 1985.

109 P. D. Dennis, E. M. Vaughan-Williams: The effects of indecainide , a new antiarrhythmic drug on nodal tissues in the isolated rabbit heart. J Cardiovasc Pharmacol 8:1, 1986.

110 L. N. Horowitz, S. R. Spielman, L. R. Webb, J. Morganroth, A. M. Greenspan: The clinical electrophysiology of intravenous indecainide. Am Heart J 110:784, 1985.

111 P. J. Podrid, S. Hohnloser, B. Lown: Noninvasive evaluation of indecainide for serious ventricular tachyarrhythmias. Am J Cardiol 61:764, 1988.

112 J. Krejci, G. Granrud, D. Salerno, D. Gresbach, M. Hodges: Indecainide, a dose-ranging efficacy trial for treatment of ventricular ectopic depolarization (abstr.). Circ 68:III–311, 1986.

113 L. A. Siddoway, R. H. Bergstrand, R. K. Drumm, D. M. Roden, R. L. Woosley: Antiarrhythmic efficacy, safety and pharmacokinetics of indecainide (abstr.). Circ 68:III, 1986.

114 J. Morganroth, for Indecainide-Disopyramide Research Group: Comparative efficacy and safety of indecainid versus disopyramide on ventricular arrhythmias (abstr.). Circ 74:II–104, 1986.

115 I. Yamaguchi, B. N. Singh, W. J. Mandel: Electrophysiologic actions of mexiletine on isolated rabbit atra and canine ventricular muscle and Purkinje fibers. Cardiovasc Res 13:288, 1979.

116 P. J. Podrid, J. Ruskin (Ed.): The role of oral mexiletine in the management of ventricular arrhythmias. Am Heart J 107:1053, 1984.

116A F. M. Weld, J. T. Bigger, D. Swistel: Elextrophysiological effects of mexiletine (KO 1173) on ovine cardiac Purkinje fibers. J Pharmacol Exp Ther 210:222, 1979.

117 S. Hereng, A. Bodewci, A. Wollenberger: Sodium current in freshly isolated and in cultured single rate myocardial cells. Frequency and voltage dependent block by mexiletine. J Mol Cell Cardiol 15:431, 1983.

118 J. P. Allen, J. G. Kelly, R. G. G. James, R. G. Shanks, S. A. Zaidi: Comparison of the effects of lignocaine and mexiletine on experimental ventricular arrhythmias. Postgrad Med J 53:35, 1977.

119 M. McComish, D. Kitsin, C. Robinson, D. E. Jewitt: Clinical electrophysiologic effects of mexiletine. Postgrad Med J 53 (suppl 1):85, 1977.

120 J. C. Roos, D. U. A. Paalman, A. J. Dunning: Electrophysiologic effects of mexiletine in man. Postgrad Med J 53 (suppl 1):92, 1977.

121 N. P. S. Campbell, J. G. Kelly, A. A. J. Adgey, R. G. Shanks: The clinical pharmacology of mexiletine. Br J Clin Pharmacol 6:103, 1978.

122 A. Pottage: Oral dosage schedules for mexiletine. Postgrad Med J 53 (suppl 1):155, 1977.

123 V. Haselbarth, J. E. Doevendans, M. Wolf: Kinetics and bioavailability of mexiletine in healthy subjetcs. Clin Pharmacol Ther 29:729, 1981.

124 A. H. Beckett, E. C. Chidomere: The distribution, metabolism and excretion of mexiletine in man. Postgrad Med J 53 (suppl 1):60, 1977.

125 P. Herzog, W. Holtermuller, T. Kasper, T. Meinertz, D. Trenk, E. Jahnchen: Absorption of mexiletine after treatment with gastric antacids. Br J Clin Pharmacol 14:746, 1982.

126 L. M. V. Weng, P. J. Meffin, J. J. Grygiel, J. Smith, D. J. Birkett: Effect of metoclopramide and atropine on the absorption of orally administered mexiletine. Br J Clin Pharmacol 9:505, 1980.

127 M. A. Kiddie, C. M. Kaye, P. Turner: The influence of urinary pH in the elimination of mexiletine. Br J Clin Pharmacol 1:229, 1974.

128 E. B. Leahy, E. G. U. Giardina, J. T. Bigger: Effect of ventricular failure in steady state kinetics of mexiletine. Clin Res (abstr.) 26:239, 1980.

129 P. J. Pentikainen, M. O. Halenin, M. J. Helin: Pharmacokinetics of oral mexiletine in patients with acute myocardial infarction. Eur J Clin Pharmacol 25:773, 1983.

130 P. J. Podrid, B. Lown: Mexiletine for ventricular arrhythmias. Am J Cardiol 47:895, 1981.
131 E. J. Begg, P. M. Chinwah, C. Webb, R. O. Day, D. N. Wade: Enhanced metabolism of mexiletine after phenytoin administration. Br J Clin Pharmacol 14:219, 1982.
132 P. J. Pentikainen, I. H. Koivula, H. A. Hiltunen: Effect of rifampicin treatment on the kinetics of mexiletine. Eur J Clin Pharmacol 23:261, 1982.
133 E. B. Leahey, J. A. Reiffel, E. G. V. Giardina, et al.: The effect of quinidine and other oral antiarrhythmic drugs on serum digoxin: A prospective study. Ann Intern Med 92:605, 1980.
134 H. J. Duff, D. Roden, R. K. Primm, J. A. Oates, R. L. Woosley: Mexiletine in the treatment of resistant ventricular arrhythmias: Enhancement of efficacy and dose reduction of dose related side effects by combination with quinidine. Circ 67:1124, 1983.
135 E. B. Leahey, R. H. Heissenbuttel, E. G. V. Giardina, T. J. Bigger: Combined mexiletine and propranolol treatment of refractory ventricular tachycardia. Br Med J 281:357, 1980.
136 J. Carlier: Hemodynamic, electrocardiographic and toxic effects of the intravenous administration of increasing doses of mexiletine in the dog. Comparison with similar effects produced by other antiarrhythmics. Acta Cardiol 25 (suppl):81, 1980.
137 K. I. Saunamaki: Hemodynamic effects of a new anti-arrhythmic agent, mexiletine (Ko 1173) in ischemic heart disease. Cardiovasc Res 9:788, 1975.
138 J. Stein, P. J. Podrid, B. Lown: Effects of oral mexiletine on left and right ventricular function. Am J Cardiol 54:575, 1984.
139 A. Waleffe, H. E. Kulbertus: The efficacy of intravenous mexiletine on ventricular ectopic activity. Acta Cardiol 32:269, 1977.
140 R. G. Talbot, J. Nimmo, D. G. Jalian, R. A. Clark, J. M. M. Neilson, L. F. Prescott: Treatment of ventricular arrhythmias with mexiletine (Ko 1173). Lancet 2:399, 1973.
141 R. G. Talbot, D. G. Julian, L. F. Prescot: Long-term treatment of ventricular arrhythmia with oral mexiletine. Am Heart J 91:58, 1976.
142 N. P. S. Campbell, J. A. Kelly, A. A. Adgey, et al.: The clinical pharmacology of mexiletine. Br J Clin Pharmacol 6:103, 1978.
143 M. H. Schoenfeld, I. Whitford, B. McGovern, H. Garan, J. N Ruskin: Oral mexiletine in the treatment of refractory ventricular arrhythmias: The role of electrophysiology techniques. Am Heart J 107:1071, 1984.
144 L. E. Waspe, H. L. Waxman, A. E. Buxton, M. E. Josephson: Mexiletine for control of drug-resistant ventricular tachycardia: Clinical and electrophysiologic results in 44 patients. Am J Cardiol 51:1175, 1983.
145 J. Stein, P. J. Podrid, S. Lampert, G. Hirsowitz, B. Lown: Long-term mexiletine for ventricular arrhythmia. Am Heart J 107:1091, 1984.
146 H. G. Smyllie, J. W. H. Doar, C. D. Head, et al.: A trial of intravenous and oral mexiletine in acute myocardial infarction. Eur J Clin Pharmacol 26:537, 1984.
147 IMPACT Research Group: International mexiletine and placebo antiarrhythmic coronary trial: I. Report on arrhythmia and other findings. J Am Coll Cardiol 4:1148, 1984.
148 J. B. Singh, A. M. Rasul, A. Skah, et al.: Efficacy of mexiletine in chronic ventricular arrhythmia compared with quinidine: A single-blind, randomized trial. Am J Cardiol 53:84, 1984.
149 G. Breithard, L. Seipel, J. Lersmacher: Comparative study of the antiarrhythmic efficacy of mexiletine and disopyramide in patients with chronic ventricular arrhythmias. J Cardiovasc Pharmacol 4:276, 1982.

150 E. N. Moore, J. F. Spear, L. N. Horowitz, H. S. Feldman, R. A. Moller: Electrophysiologic properties of a new antiarrhythmic drug – tocainide. Am J Cardiol 49:703, 1978.
151 J. L. Anderson, J. W. Mason, R. A. Winkle, P. J. Meffin, K. F. Fowles, F. Peters, D. C. Harrison: Clinical electrophysiologic effects of tocainide. Circ 57:685, 1978.
152 D. G. McDevitt, R. S. Nies, G. R. Wilkinson, R. F. Smith, R. F. Woosley, J. A. Oates: Antiarrhythmic effects of lidocaine congener, tocainide, 2 amino-2', 6' propionoxylidide, in man. Clin Pharmacol Ther 19:396, 1976.
153 R. A. Winkle, P. J. Meffin, J. W. Fitzgerald, D. C. Harrison: Clinical efficacy and pharmacokinetics of a new orally effective antiarrhythmic, tocainide. Circ 54:884, 1976.
154 C. Graffner, T. B. Conradson, S. Hofvendahl, L. Ryden: Pharmacokinetics of tocainide following intravenous and oral administration in healthy subjetcs and in patients with acute myocardial infarction. Clin Pharmacol Ther 27:64, 1980.
155 D. J. Coltart, T. B. Berndt, R. Ternoff, D. C. Harrison: Antiarrhythmic circulatory effects of Astra W3609J: A new lidocaine like agent. Am J Cardiol 34:35, 1974.
156 K. Swedberg, J. Pehrson, L. Ryden: Electrocardiographic and hemodynamic effects of tocainide in man. Eur J Pharmacol 19:100, 1979.
157 R. A. Winkle, J. L. Anderson, F. Peter, P. J. Meffin, R. E. Fowler, D. C. Harrison: The hemodynamic effects of intravenous tocainide in patients with heart disease. Circ 57:787, 1978.
158 R. L. Woosley, D. G. McDevitt, A. S. Nies, R. F. Smith, G. R. Wilkenson, J. A. Oates: Suppression of ventricular ectopic depolarizations by tocainide. Circ 56:980, 1977.
159 P. J. Podrid, B. Lown: Tocainide for refractory symptomatic ventricular arrhythmias. Am J Cardiol 49:1279, 1982.
160 L. N. Horowitz, M. R. Josephson, A. Farshidi: Human electrophysiology of tocainide, a lidocaine congener. Am J Cardiol 42:276, 1980.
161 L. Ryden, K. Arnman, T. B. Conradson, S. Hofuendahl, O. Mortenson, P. Smedgard: Prophylaxis of ventricular tachyarrhythmias with intravenous and oral tocainide in patients with and recovering from acute myocardial infarction. Am Heart J 100:1006, 1980.
162 R. A. Winkle, J. W. Mason, D. C. Harrison: Tocainide for drug resistant ventricular arrhythmias. Efficacy, side effects and lidocaine responsiveness for predicting tocainide success. Am Heart J 100:1031, 1980.
163 S. H. Hohnloser, H. W. Lang, E. A. Raeder, P. J. Podrid, B. Lown: Short and long term tocainide therapy for malignant ventricular tachyarrhythmias. Circ 73:143, 1986.
164 M. Hession, R. Blum, P. J. Podrid, B. Lown: Mexiletine and tocainide: Does response to one predict response to the other? J Am Coll Cardiol 7:338, 1986.
165 H. R. Horn, Z. Hadidian, J. L. Johnson, H. G. Vassalo, J. H. Williams, M. D. Young: Safety evaluation of tocainide in the American Emergency Use Program. Am Heart J 100:1037, 1980.
166 Lown (ed.) A Symposium: Ethmozine (moricizine HCl) – A new antiarrhythmic agent. Am J Cardiol 60:suppl F, 1987.
167 Z. P. Senova, M. V. L'Vov: Mechanism of the antiarrhythmic effect of ethmozine. Farmakol Toksikol 36:703, 1973.
168 L. V. Rosenshtraukh, E. P. Anyukhovsky, V. V. Nesterenko, A. L. Undrovinas, K. K. Shugushev, V. F. Portnoy, N. A. Burnshev: Electrophysiologic aspects of moricizine HCL. Am J Cardio 60:27F, 1987.

169 A. S. Smitnev, K. K. Shugushi, L. V. Rosenshtraukh: Clinical electrophysiologic and antiarrhythmic efficacy of moricizine ECL. Am J Cardiol 60:40F, 1987.
170 C. R. C. Wyndham, C. M. Pratt, D. E. Mann, V. F. Runova: Electrophysiology of ethmozine (moricizine HCL) for ventricular tachycardia. Am J Cardiol 60:67F, 1987.
171 N. V. Kaverina, Z. P. Senova, V. S. Mitrophanrv, G. McMahon, J. O. Humphries, D. T. Mason, R. L. Williams: Pharmacology of ethmozin: a new antiarrhythmic drug Farmacol Toksikol 22:182, 1972.
172 R. L. Woosley, J. Morganroth, R. N. Fogoros, et al.: Pharmacokinetics of moricizine HCL. Am J Cardiol 60:35F, 1987.
173 H. L. Kennedy, A. J. J. Wood, R. T. MacFarland: Drug interactions with ethmozine (moricizine HCL). Am J Cardiol 60:79F, 1987.
174 C. M. Pratt, P. J. Podrid, A. A. Seals, J. B. Young, M. Hession, S. Lampert, M. Quiones, B. Lown: Effect of ethmozine (moricizine HCL) on ventricular function using echocardiographic, hemodynamic and radionuclide assessments. Am J Cardiol 60:73F, 1987.
175 P. J. Podrid, A. Lyakishev, B. Lown, N. Mazur: Ethmozine, a new antiarrhythmic drug for suppressing ventricular premature complexes. Circ 61:450, 1980.
176 S. N. Singh, R. DiBianco, J. S. Gottdiener: Effects of moricizine hydrochloride in inducing high frequency ventricular arrhythmia: Results of a prospective, controlled trial. Am J Cardiol 53:745, 1984.
177 C. M. Pratt, J. B. Young, M. J. Francis, A. A. Taylor, H. J. Norton, L. English, D. E. Mann, H. Kopelen, M. A. Quiones, R. Roberts: Comparative efficacy of disopryramide and ethmozine in suppressing complex ventricular arrhythmias by use of a double-blind, placebo controlled longitudinal crossover design. Circ 69:288, 1984.
178 M. J. Hession, S. Lampert, P. J. Podrid, B. Lown: Ethmozine (moricizine HCL) therapy for complex ventricular arrhythmias. Am J Cardiol 60:59F, 1987.
179 J. Morganroth, C. M. Pratt, H. L. Kennedy: Efficacy and tolerance of ethmozine (moricizine HCL) in placebo controlled trials. Am J Cardiol 60:48F, 1987.
180 J. L. Bergey, T. Salkowsk, D. R. Mirch: Antiarrhythmic, hemodynamic and cardiac electrophysiologic evaluation of N-(2,6-dimethyphenyl)-N'[3(1-methylithylamino)-propyl] urea (wy-42, 362) Arznlim Forschung 33:1258, 1983.
181 J. J. Colatsky, C. B. Bird, J. A. Knowles: Cardiac electrophysiology of the antiarrhythmic agent, recainam (Wy 42, 362) in anesthetized dogs: Relation to plasma and myocardial concentrations. J Cardiovasc Pharmacol 11:308, 1988.
182 R. Takikawa, K. Kamiya, R. Kato, B. N. Singh: Electrophysiologic effects of a new antiarrhythmic agent, recainam, in isolated canine and rabbit myocardial fibers. J Am Coll Cardiol 11:875, 1988.
183 R. M. Luceri, A. Castellanos, L. Zaman: Efficacy and safety of intravenous Wy-42, 362, a new antiarrhythmic agent for ventricular arrhythmias (abstr.). Clin Res 32:244A, 1984.
184 M. I. Anastasiou-Nana, J. L. Anderson, E. M. Hampton, J. N. Nanas, B. M. Heath: Recainam, a potent new antiarrhythmic agent: effects in complex ventricular arrhythmias. J Am Coll Cardiol 8:427, 1986.
185 R. M. Luceri, T. Selby, S. F. Sesenwine, A. Castellanos, L. Zamon: Pharmacokinetics and safety of a new type 1 antiarrhythmic agent (abstr.). Clin Res 33:285A, 1985.
186 J. L. Anderson, M. I. Anastasiou-Nana, B. M. Heath, R. L. MenCove, J. N. Nanas, J. Freedman: Efficacy of recainam, a new antiarrhythmic drug for control of ventricular arrhythmia. Am J Cardiol 60:281, 1987.

187 B. N. Singh, P. Deedwania, K. Nademanee, A. Ward, E. M. Sorkin: Sotalol: A review of its pharmacodynamic and pharmacokinetic properties, and therapeutic use. Drugs 34:311, 1987.
188 B. N. Singh, E. M. Vaughan-Williams: A third class of antiarrhythmic action: effects on atrial and ventricular intracellular potentials and other pharmacological actions on cardiac muscle of MJI 999 and AM 3676. Br J Pharmacol 39:675, 1970.
189 T. Wang, R. M. Bergstrand, K. A. Thompson, L. A. Siddoway, H. J. Duff, R. L. Woosley, D. M. Roden: Concentration-dependent pharmacologic properties of sotalol. Am J Cardiol 57:1160, 1986.
190 R. Kato, S. Yabeic, N. Ikeda, R. Kannan, B. N. Singh: Electrophysiologic effects of dextro and levo isomers of sotalol in isolated cardiac muscle and their in vivo pharmacokinetics. J Am Coll Cardiol 7:116, 1980.
191 J. M. McComb, B. McGovern, J. B. McGowan, J. N. Ruskin, H. Garan: Electrophysiologic effects of d-sotalol in humans. J Am Coll Cardiol 10:211, 1987.
192 P. Touboul, G. Atallah, G. Kirkorian, M. Lamaud, P. Moleur: Clinical electrophysiology of intravenous sotalol, a beta blocking drug with class III antiarrhythmic properties. Am Heart J 107:888, 1984.
193 B. N. Singh, K. Nademanee: Sotalol: A beta blocker with unique antiarrhythmic properties. Am Heart J 114:121, 1987.
194 M. Anttila, M. Arstila, M. Pfeffer, R. Tikkanen, V. Vallinkoski: Human pharmacokinetics of sotalol. Acta Pharmacologica et toxicologica 39:118, 1976.
195 H. K. Sudquist, M. Anttila, J. Forrstrom, A. Kasenan: Serum levels and half-life of sotalol in chronic renal failure. Ann Clin Res 7:442, 1975.
196 J. M. Cruickshank: The clinical importance of cardioselectivity and lipophilicity in beta blockers. Am Heart J 100:160, 1980.
197 K. Schnelle, G. Klein, A. Schinz: Studies on the pharmacokinetics and pharmacodynamics of beta-adrenergic blocking agent sotalol in normal man. J Clin Pharmacol 19:516, 1979.
198 H. Brooks, J. Banas, S. Meister, M. Szues, J. Dalen, L. Dexter, Sotalol – induced beta blockade in cardiac patients. Circ 42:99, 1970.
199 I. Hutton, A. R. Lorimar, W. S. Millis, D. McCall, J. M. Reid, T. V. D. Lawrie: Hemodynamics and myocardial function after sotalol. Br Heart J 34:787, 1972.
200 J. J. Mahmarian, M. S. Verani, T. Hohmann, R. Hill, B. C. Thornton, R. Bolli, J. B. Young, R. Roberts, C. M. Pratt: The hemodynamic effects of sotalol and quinidine: analysis by use of rest and exercise gated radionuclide angiography. Circ 76:324, 1987.
201 K. K. Teo, M. Harlem, J. H. Horgan: Sotalol infusion in the treatment of supraventricular tachyarrhythmias. Chest 87:113, 1985.
202 I. Rizos, J. Senges, R. Javernig, W. Lengfelder, E. Czygan, J. Brachmann, W. Kubler: Differential effects of sotalol and metoprolol on induction of paroxysmal supraventricular tachycardia. Am J Cardiol 53:1022, 1984.
203 T. J. Campbell, T. P. Gavaghan, J. J. Morgan: Intravenous sotalol for the treatment of atrial fibrillation and flutter after cardiopulmonary bypass: comparison with disopyramide and digoxin in a randomized trial. Br Heart J 54:86, 1985.
204 J. H. Janssen: Prevention and treatment of supraventricular tachycardia early following coronary artery bypass surgery. Angiology 36:575, 1985.
205 P. Tuboul, G. Atallah, G. Krkorian, P. Lavaud, M. P. Mathieu, A. Dellinger: Effects of intravenous sotalol in patients with atrioventricular accessory pathways. Am Heart J 114:545, 1987.
206 K. P. Kunze, M. Schluter, K. H. Kuck: Sotalol in patients with Wolff-Parkinson-White Syndrome. Circ 75:1050, 1987.

207 L. B. Mitchell, D. G. Wyse, H. J. Duff: Electropharmacology of sotalol in patients with Wolff-Parkinson-White Syndrome. Circ 76:810, 1987.
208 D. I. Sahar, J. A. Reiffel, J. T. Bigger, A. Squatrito, G. A. Kidwell: Efficacy, safety, and tolerance of d-sotalol in patients with refractory supraventricular tachyarrhythmias. Am Heart J 117:562, 1989.
209 D. P. Myburgh, A. P. Goldman, J. Cartoon, J. M. Schamroth: The efficacy of sotalol in suppressing ventricular ectopic beats. S Afr Med J 56:295, 1979.
210 D. Burckhardt, M. Pfisterer, A. Hoffmann, F. Burkart, H. Emmeregger, M. Jost, P. Bolli, F. R. Buehler: Effects of the beta-adrenoceptor blocking agent sotalol on ventricular arrhythmias in patients with chronic ischemic heart disease. Cardiol 70:114, 1983.
211 J. L. Anderson, J. C. Askins, E. M. Gilbert, R. H. Miller, D. L. Keefe, J. C. Somberg, R. A. Freedman, L. R. Haft, J. W. Mason, J. N. Lessem: Multicenter trial of sotalol for suppression of frequent, complex ventricular arrhythmias: a double-blind, randomized, placebo-controlled evaluation of two doses. J Am Coll Cardiol 8:752, 1986.
212 L. J. Jordaens, D. L. Palmer Am Clement: Low-dose oral sotalol for monomorphic ventricular tachycardia: effects during programmed electrical stimulation and follow-up. Eur Heart J 10:218, 1989.
213 M. A. Ruder, T. Ellis, C. Lebsack, R. M. Mead, N. A. Smith, R. A. Winkle: Clinical experience with sotalol in patients with drug-refractory ventricular arrhythmias. J Am Coll Cardiol 13:145, 1989.
214 R. Gonzales, M. M. Scheinman, J. M. Herre, J. C. Griffin, M. J. Sauve, H. Sharkey: Usefulness of sotalol for drug-refractory malignant ventricular arrhythmias. J Am Coll Cardiol 12:1568, 1988.
215 M. A. Kopelman, L. Woosley, J. T. Lee, D. M. Roden, D. S. Echt: Electrophysiologic effects of intravenous and oral sotalol for sustained ventricular tachycardia secondary to coronary artery disease. Am J Cardiol 61:1006, 1988.
216 J. Schwartz, K. Crocker, J. Wynn, J. C. Somberg: The antiarrhythmic effects of d-sotalol. Am Heart J 114:539, 1987.
217 H. D. Gossinger, P. Siostrzonek, R. Schmoliner, G. Grimm, U. Jager, H. Mosslacher: Sotalol-induced torsades des pointes in a patient with pre-existent normal response to programmed stimulation. Eur Heart J 8:1351, 1987.
218 K. H. Kuck, K. P. Kunze, W. Bleifeld: Sotalol-induced torsade de pointes. Am Heart J 107:179, 1984.
219 M. Laakso, I. Arvala, S. Tervonen, M. Sotaravta: Retroperitoneal fibrosis associated with sotalol. Br Med J 285:1085, 1982.
220 B. N. Singh, D. P. Zipes (Ed.): Proceedings of the symposium – Amiodarone: Basic concepts and clinical applications. Am Heart J 106:787, 1983.
221 D. P. Zipes, E. N. Prystowsky, J. J. Heger: Amiodarone: electrophysiologic actions, pharmacokinetics and clinical effects. J Am Coll Cardiol 3:1059, 1984.
222 N. Groupil, J. Lenfant: The effects of amiodarone on the sinus node activity of the rabbit heart. Eur J Pharmacol 39:23, 1976.
223 W. F. Lubbe, M. L. McFadyen, C. A. Muller, M. Worthington, L. M. Opie: Protective action of amiodarone against ventricular fibrillation in the isolated perfused rat heart. Am J Cardiol 43:533, 1979.
224 B. N. Singh: Amiodarone: historical development and pharmacologic profile. Am Heart J 106:788, 1983.
225 J. W. Mason: Drug therapy: amiodarone. N Eng J Med 316:455, 1987.
226 D. W. Holt, G. T. Tucker, P. R. Jackson, G. C. A. Storey: Amiodarone: pharmacokinetics. Am Heart J 106:840, 1983.

227 P. C. Adams, D. W. Holt, C. A. Storey, A. R. Morley, J. Callaghan, W. F. Campbell: Amiodarone and its desethyl metabolite: tissue distribution and morphologic changes during long-term therapy. Circ 72:1064, 1985.
228 D. W. Holt, G. T. Tucker, A. C. Jackson, G. C. A. Storey: Amiodarone pharmacokinetics. Am Heart J 106:840, 1983.
229 A. Berdeauz, A. Roche, T. Labaille, B. Giroux, A. Edovard, J. F. Giudicelli. Tissue extraction of amiodarone and N-desethylamiodarone in man after a single oral dose. Br J Clin Pharmacol 18:759, 1984.
230 L. A. Siddoway, C. B. McAllister, G. R. Wilkinson, D. M. Roden, R. L. Woosley: Amiodarone dosing: a proposal based upon its pharmacokinetics. Am Heart J 106:951, 1983.
231 F. Andreasen, H. Agerback, P. Bjerregaard, H. Gotzsche: Pharmacokinetics of amiodarone after intravenous and oral administration. Eur J Clin Pharmacol 19:293, 1981.
232 J. J. Heger, E. N. Prystowsky, D. P. Zipes: Relationships between amiodarone dosage, drug concentrations and adverse side effects. Am Heart J 106:931, 1983.
233 M. M. Rotmensch, B. Belhassen, B. N. Swanson, D. Shoshani, S. R. Spielman, A. J. Greenspan, A. M. Greenspan, P. H. Vlasses, L. N. Horowitz: Steady-state serum amiodarone concentrations: relationships with antiarrhythmic efficacy and toxicity. Ann Int Med 101:462, 1984.
234 I. Helmy, J. M. Serre, G. Gee, H. Sharkey, P. Malone, M. J. Sauve, J. C. Griffin, M. M. Scheinman: Use of intravenous amiodarone for emergency treatment of life-threatening ventricular arrhythmias. J Am Coll Cardiol 12:1015, 1988.
235 F. I. Marcus: Drug interactions with amiodarone. Am Heart J 106:924, 1983.
236 E. J. Kozinski, J. B. Albin, E. Young, S. M. Lewis, S. LeLand Jr.: Hemodynamic effects of intravenous amiodarone. J Am Coll Cardiol 4:565, 1984.
237 C. T. Haffajie, J. C. Love, J. S. Alpert, G. K. Asdourian, K. C. Sloan: Efficacy and safety of long-term amiodarone in treatment of cardiac arrhythmias: Dosage experience. Am Heart J 106:935, 1983.
238 M. B. Rosenbaum, P. A. Chiale, M. S. Halpern, G. J. Nau, J. Przybylski, R. J. Levi, J. O Lazzari, M. V. Elizari: Clinical efficacy of amiodarone as an antiarrhythmic agent. Am J Cardiol 38:934, 1976.
239 P. J. Podrid, B. Lown: Amiodarone therapy in symptomatic, sustained refractory atrial and ventricular tachycarrhythmias. Am Heart J 101:374, 1981.
240 E. P. Veltri, P. R. Reid, E. V. Platia, L. S. C. Griffith: Amiodarone in the treatment of life-threatening ventricular tachycardia: The role of Holter monitoring in predicting long-term clinical efficacy. Am Coll Cardiol 6:806, 1985.
241 R. N. Fogoros, K. P. Anderson, R. A. Winkle, C. D. Swerdlow, J. W. Mason: Amiodarone: clinical efficacy and toxicity in 96 patients with reccurrent, drug-refractory arrhythmias. Circ 68:88, 1983.
242 T. Peter, A. Hamer, W. J. Mandel, D. Weiss: Evaluation of amiodarone therapy in the treatment of drug resistant cardiac arrhythmias: long-term follow-up. Am Heart J 106:943, 1983.
243 C. I. Haffajee, J. C. Love, A. T. Canada, L. J. Lesko, G. Asdourian, J. S. Alpert: Clinical pharmacokinetics and efficacy of amiodarone for refractory tachyarrhythmias. Circ 67:1347, 1983.
244 W. J. McKenna, C. M. Oakley, D. M. Krikler, J. F. Goodwin: Improved survival with amiodarone in patients with hyperthrophic cardiomyopathy and ventricular tachycardia. Br Heart J 53:412, 1985.

245 J. J. Heger, E. N. Prystowsky, W. M. Jackman, G. V. Naccarelli, K. A. Warfel, R. L. Rinkenberger, D. P. Zipes: Amiodarone: clinical efficacy and electrophysiology during long-term therapy for reccurent ventricular tachycardia or ventricular fibrillation. N Engl J Med 305:539, 1981.
246 H. L. Waxman, W. C. Groh, F. E. Marchlinski, A. E. Buxton, L. M. Sadowski, L. N. Horowitz, M. E. Josephson, J. A. Kaster: Amiodarone for control of sustained ventricular tachyarrhythmia: clinical and electrophysiologic effect in 51 patients. Am J Cardiol 50:1066, 1982.
247 G. C. Nademanee, J. A. Hendrickson, D. S. Cannon, B. N. Goldreyer, B. N. Singh: Control of refractory life-threatening ventricular arrhythmias by amiodarone. Am Heart J 101:759, 1981.
248 B. McGovern, G. Garan, R. F. Malacoff: Long-term clinical outcome of ventricular tachycardia or fibrillation treated with amiodarone. Am J Cardiol 53:1558, 1984.
249 G. V. Naccarelli, N. S. Fineberg, D. P. Zipes, J. J. Heger, G. Duncan, E. N. Prystowsky. Amiodarone: risk factors for recurrence of symptomatic ventricular tachycardia identified at electrophysiologic study. J Am Coll Cardiol 6:814, 1985.
250 A. Kadish, F. E. Marchlinski, M. E. Josephson, A. E. Buxton. Amiodarone: Correlation of early and late electrophysiologic studies with outcome. Am Heart J 112:1134, 1986.
251 L. N. Horowitz, A. M. Greenspan, S. R. Spielman, C. R. Webb, J. Morganroth, H. Rotmensch, N. M. Sokoloff, A. P. Rae, B. L. Segal, H. R. Kay: Usefulness of electrophysiologic testing in evaluation of amiodarone therapy for sustained ventricular tachyarrhythmias associated with coronary artery disease. Am J Cardiol 55:367, 1985.
252 P. J. Podrid: Practical follow-up guidelines for patients treated with amiodarone: Clin Cardiol 10 (suppl 1):107, 1987.
253 E. Rowland, D. M. Krikler: Electrophysiological assessment of amiodarone in the treatment in patients with resistant paroxysmal tachycardias. Br Heart J 44:82, 1980.
254 D. E. Ward, A. J. Camm, R. A. J. Spurrell: Clinical antiarrhythmic effects of amiodarone in patients with resistant paroxysmal tachycardias. Br Heart J 44:91, 1980.
255 T. B. Graboys, P. J. Podrid., B. Lown: The efficacy of amiodarone for refractory supraventricular tachyarrhythmias. Am Heart J 106:870, 1983.
256 H. J. J. Wellens, P. Brugada, H. Abdollah: Effect of amiodarone in paroxysmal supraventricular tachycardia with or without Wolff-Parkinson-White Syndrome. Am Heart J 106:876, 1983.
257 E. A. Raeder, P. J. Podrid, B. Lown: Side effects and complications of amiodarone therapy. Am Heart J 109:975, 1985.
258 L. Harris, W. J. McKenna, E. Roland, D. M. Krikler: Side effects and possible contraindications of amiodarone use. Am Heart J 106:916, 1983.
259 F. Morady, M. J. Sauve, P. Malone, E. N. Shen, A. B. Schwartz, A. Bhandari, E. Keung, R. J. Sung, M. M. Scheinman: Long-term efficacy and toxicity of high-dose amiodarone therapy for ventricular tachycardia or ventricular fibrillation. Am J Cardiol 58:975, 1983.
260 S. Sclarovsky, R. F. Lewin, O. Kracoff, B. Strasberg, A. Arditti, J. Agman: Amiodarone-induced polymorphous ventricular tachycardia. Am Heart J 105:6, 1983.
261 B. McGovern, H. Garan, E. Kelly, J. N. Ruskin: Adverse reactions during treatment with amiodarone. Br Med J 287:175, 1983.
262 C. Maro, T. Romero, M. A. Lorris-Peiretti: Amiodarone and hypokalemia: a dangerous combination. Int J Cardiol 13:356, 1986.
263 W. J. Marin, E. D. Rosenow: Amiodarone pulmonary toxicity: recognition and pathogenesis (part 1 & 2). Chest 93:1067 & 1242, 1988.

264 L. Rakita, S. M. Sobol, N. Mostow, T. Vrobel: Amiodarone pulmonary toxicity. Am Heart J 106:906, 1983.
265 F. E. Marchlinski, T. S. Gansler, H. L. Waxman, M. E. Josephson: Amiodarone pulmonary toxicity. Ann Int Med 97:839, 1982.
266 B. N. Singh, K. Nademanee: Amiodarone and thyroid function: clinical implications during antiarrhythmic therapy. Am Heart J 106:857, 1983.
267 D. V. Ingram: Ocular effects in long-term amiodarone therapy. Am Heart J 106:902, 1983.
268 H. Just (Ed.), A symposium: Pirminol – a unique, new antiarrhythmic agent. Am J Cardiol 59 (suppl H), 1987.
269 R. F. Reder, P. Danil, M. R. Rosen: Effects of pirmenol HCL on electrophysiologic properties of cardiac Purkinje fibers. Eur J Pharmacol 61:321, 1980.
270 I. D. Dukes, E. M. Vaughan Williams, P. D. Dennis: Electrophysiological and cardiovascular effects of pirmenol, a new class 1 antiarrhythmic drug. J Cardiovasc Pharmacol 8:227, 1986.
271 H. Nakaya, N. Tohse, M. Kanno: Frequency and voltage dependent depression of maximum upstroke velocity of action potentials by pirmenol in guinea pig ventricular muscles. Jap J Pharmacol 48:423, 1988.
272 H. Kotaki, S. Hiai, T. Kinugaua, I. Ito, J. Yamasaki, J. Hasigawa, H. Masheba: Effects of pirmenol hydrochloride in the spontaneous action potentials and membrane current systems of rabbit sinoatrial node cells. J Electrocardiol 21:355, 1988.
273 L. B. Liem, D. A. Clay, M. C. Franz, C. D. Swendlow: Electrophysiology and antiarrhythmic efficacy of intravenous pirmenol in patients with sustained ventricular tachyarrhythmias. Am Heart J 113:1390, 1987.
274 M. J. Reiter, S. C. Hammill, D. G. Shand, C. Verghese, E. McCarthy, E. L. Pritchett. Efficacy, safety and pharmacokinetics of a concentration maintaining regimen of intravenous pirmenol. Am J Cardiol 52:83, 1983.
275 M. deBuitleir, B. J. Grevey, T. Johnson, W. H. Kou, S. D. Nelson, S. Schmaltz, F. Morady: Long-term efficacy of oral pirmenol in suppressing ventricular premature depolarizations. Am Heart J 116:379, 1988.
276 J. L. Anderson, J. R. Lutz, J. M. Nappi: Pirmenol for control of ventricular arrhythmias, oral dose ranging and short-term maintenance study. Am J Cardiol 53:522, 1984.
277 J. Baguena Candela: for the Pirmenol Multicenter Study Group. Effectiveness of pirmenol hydrochloride in suppressing high frequency ventricular ectopy: a placebo controlled, double blind multicenter study. Am J Cardiol 59:48H, 1987.
278 T. G. Lee, A. D. Goldberg, T. Chang, M. T. Jerkland, G. J. Yakatan, E. L. Johnson, J. G. Toole, S. Goldstien: Pharmacokinetics and efficacy of pirmenol hydrochloride in the treatment of ventricular dysrhythmia. J Cardiovasc Pharmacol 5:632, 1983.
279 S. C. Hammill, D. G. Shand, F. E. Harrell, J. M. Zimmerman, M. J. Reiter, C. Verghese, E. L. Pritchett: Pirmenol Kinetics and effective oral dose. Clin Pharmacol Ther 32:686, 1982.
280 K. A. Ellenbogin, S. F. Roark, A. L. Sintetos, M. S. Smith, E. A. McCarthy, W. M. Smith, R. E. Kates, E. L. Pritchett: Pharmacodynamics of oral pirmenol Clin Pharm Ther 42:405, 1987.
281 D. C. Garg, N. S. Jallad, S. Singh, K. F. Ng, D. J. Weidler: Efficacy and pharmacokinetics of oral pirmenol, a new antiarrhythmic drug. J Clin Pharm 28:812, 1988.
282 T. Chang: Preclinical and clinical pharmacokinetics of pirmenol. Am J Cardiol 59:15H, 1987.
283 K. A. Stringer, A. B. Cetnarowski, A. Goldfarb, M. E. Lebsack, T. Chang, A. J. Sidman: Enhanced pirmenol elimination by refampin. J Clin Pharm 28:1094, 1988.

284 K. G. Morris, H. R. Philipp: Hemodynamic effects of intravenous pirmenol. Angiol 39:321, 1988.
285 L. K. Toivonen, M. S. Nieminen: Hemodynamic effects of pirmenol, a new antiarrhyhmic agent, assessed noninvasively. Int J Cardiol 10:111, 1986.
286 L. K. Toivonen, M. S. Nieminen, V. Manninen, H. Frick: Conversion of paroxysmal atrial fibrillation to sinus rhythm by intravenous pirmenol. Am J Cardiol 59:39H, 1987.
287 L. K. Toivonen: Effect of pirmenol on premature atrial contractures. Ann Clin Res 18:106, 1986.
288 L. K. Toivonen: Pirmenol in termination of paroxysmal supraventricular tachycardia. Clin Cardiol 9:369, 1986.
289 J. E. Brown, R. Yee, G. J. Klein: Acute electrophysiologic effects of pirmenol in normal subjects and in patients with Wolff Parkinson White syndrome. Am J Cardiol 57:775, 1986.
290 A. Buitleir, M. Crevey B., T. Johnson, J. Baerman, F. Morady: Efficacy of oral pirmenol in suppressing ventricular premature depolarization. Ir Med J 80:60, 1987.
291 D. J. Farnham: A multicenter dose response study of pirmenol hydrochloride in patients with ventricular premature contractions. Am J Cardiol 59:43H, 1987.
292 E. M. Hampton, J. L. Anderson, T. R. Lutz, J. M. Nappi: Initial and long-term outpatient experience with pirmenol for control of ventricular arrhythmias. Eur J Clin Pharmacol 31:15, 1986.
293 A. R. Easley, D. E. Mann, M. J. Reiter, M. J. Sakun, S. M. Sullivan, S. A. Magro, J. C. Luck, C. R. Wgndham. Electrophysiologic evaluation of pirmenol for sustained ventricular tachycardia secondary to coronary artery disease. Am J Cardiol 58:86, 1986.
294 B. F. Johnson, V. Shekar, T. Woodman, A. T. Canada: Salivary concentrations of pirmenol as a possible cause of unpleasant taste. Br J Clin Pharmacol 22:613, 1986.
295 S. H. Hamill, D. G. Shand. P. A. Routledge, M. C. Hendman, J. T. Baker, E. L. Pritchett: Pirmenol, a new antiarrhythmic agent: initial study of efficacy, safety and pharmacokinetics. 65:369, 1982.
296 M. C. Hendman, J. T. Baker, E. L. Pritchett: Pirmenol, a new antiarrhythmic agent: initial study of efficacy, safety and pharmacokinetics. Circ 65:369, 1982.
297 P. Alboni, R. Cappato, N. Paparella, R. Pirani, G. C. Candini, A. M. Tomcesi: Electrophysiologic effects and mechanisms of action of oral quinidine in patiens with sinus bradycardia and first degree AV block. Eur Heart J 8:1080, 1987.
298 P. G. Schmid, L. D. Nelson, A. C. Mark, D. D. Heistad, F. M. Abboud: Inhibition of adrenergic vasoconstriction by quinidine. J Pharmacol Exp Ther 188:124, 1974.
299 J. W. Mason, R. A. Winkle, A. K. Reder, E. B. Stenson, D. C. Harrison: The electrophysiologic effects of quinidine in the transplanted human heart. J Clin Invest 59:481, 1977.
300 F. M. Weld, J. Coronilas, J. N. Rottmann, J. T. Bigger: Mechanism of quinidine induced depression of maximum upstroke velocity in ovine cardiac Purkinje fibers. Circ Res 50:369, 1982.
301 B. F. Hoffman, M. R. Rosen, A. L. Wit: Electrophysiology and pharmacology of cardiac arrhythmias. VII Cardiac effects of quinidine and procainamide. Am Heart F 89:804, 1975.
302 M. E. Josephson, S. F. Seides, W. P. Botsford, G. M. Weisfogel, M. Akhtar, A. R. Caracta, S. H. Lau, A. N. Damato: The electrophysiologic effects of intramuscular quinidine in the atrioventricular conducting system in man. Am Heart J 87:55, 1974.

303 Y. Watanabe, L. S. Dreifres: Interactions of quinidine and potassium in atrioventricular transmission. Circ Res 20:434, 1967.
304 L. M. Hondegham, A. D. Grant, R. A. Jensen: Antiarrhythmic drug action. Selective depression of Hypoxic Cardiac cells. Am Heart J 87:602, 1974.
305 T. Colatsky: Mechanisms of action of lidocaine and quinidine on action potential duration in rabbit cardiac Purkinje fibers. Circ Res 50:17, 1982.
306 C. M. Hodeghem: Effects of lidocaine, phenytoin and quinidine in the ischemic canine myocardium. J Electrocardiol 9:203, 1976.
307 R. H. Heisenbuttel, J. T. Bigger: The effect of oral quinidine in intraventricular conduction in man. Correlation of plasma quinidine with changes in QRS duration. Am Heart J 80:453, 1970.
308 C. D. Swerdlow, J. O. Yu, E. Jacobson, S. Mann, R. A. Winkle, J. C. Griffen, D. L. Ross, J. W. Mason: Safety and efficacy of intravenous quinidine. Am J Med 75:36, 1983.
309 D. J. Greenblatt, H. J. Pfeifer, H. R. Ochs, K. Franke, P. S. MacLaughlin, T. W. Smith, J. Koch-Weser: Pharmacokinetics of quinidine in humans after intravenous, intramuscular and oral administration. J Pharm Exp Ther 202:365, 1977.
310 K. A. Conrad, B. L. Moch, L. A. Chidsey: Pharmacokinetic studies of patients with arrhythmias. Circ 55:1, 1977.
311 K. H. Palmer, B. Marten, B. Baggett, M. E. Wall: The metabolic fate of orally administered quinidine gluconate in humans. Brochen Pharmacol 18:1845, 1969.
312 B. B. Brodie, J. E. Baer, L. C. Craig: Metabolic products in anchona alkaloids in human urine. J Biol Chem 188:567, 1951.
313 D. E. Drayer, D. T. Lowenthal, K. M. Restivo, A. Schwartz, C. E. Look, M. M. Reidenberg: Steady state levels of quinidine active metabolites in cardiac patients with varying degress of renal function. Clin Pharm Ther 24:31, 1978.
314 K. M. Kessler, D. T. Lowenthal, H. Warner, T. Gibson, W. Briggs, M. M. Reidenberg: Quinidine elimination in patients with congestive heart failure or poor renal function. N Engl J Med 290:70, 1974.
315 H. L. Lonn, R. J. Luchi: Some quantitative aspects of the binding of quinidine and related quinoline compounds by human serum albumin. J Clin Invest 40:509, 1961.
316 J. Bellet, G. Hamden, A. Somlyo, R. Lara: The reversal of cardiotoxic effects of quinidine by molar sodium lactate. An experimental study. Am J Med Sci 237:165, 1959.
317 E. B. Leahey, J. A. Reiffel, R. H. Heissinbuttel, R. E. Drusin, W. P. Lovejoy, T. B. Bigger: Enhanced cardiac effect of digoxin during quinidine treatment. Arch Int Med 139:519, 1979.
318 G. G. Belz, P. E. Aust, W. Doering, M. Heinz, B. Schneider: Pharmacodynamics of a single dose of quinidine during chronic digoxin treatment. Eur J Clin Pharmacol 22:117, 1982.
319 N. J. warner, E. B. Leahey, T. J. Hougan, J. T. Bigger, T. W. Smith: Tissue digoxin concentrations during the quinidine digoxin interaction. Am J Cardiol 51:1717, 1983.
320 J. L. Data, G. R. Wilkinson, A. S. Nies: Interaction of quinidine with anticonvulsant drugs. N Engl J Med 244:699, 1976.
321 J. Koch-Weser: Quinidine induced hypoprothombanemic hemorrhage in patients in chronic warfarin therapy. Ann Int Med 68:511, 1968.
322 E. T. Angelakos, E. P. Hastings: The influence of quinidine and procainamide on myocardial contractility in vivo. Am J Cardiol 6:761, 1960.
323 W. R. Markiowicz, R. Winkle, G. Binitti, R. Kernoff, D. C. Harrison: Normal myocardial contracticity state in the presence of quinidine. Circ 53:101, 1976.

324 R. A. O'Rourke, L. D. Horowitz: Effects of chronic oral quinidine on left ventricular performance. Am Heart J 101:769, 1981.
325 D. White, M. Crawford, R. O'Rourke, J. Karliner, J. Gorwit: Effects of quinidine in left ventricular performance in normals and patients with congestive cardiomyopathy. Circ 56 (Suppl III):180, 1977.
326 T. Sodermark, B. Johnson, A. Olsson, L. Oro, H. Wallin, V. Edhag, A. Sjorgren, M. Danielsson, G. Rosenhamer: Effect of quinidine on maintaining sinus rhythm after conversion of atrial fibrillation in flutter. Bri Heart J 37:486, 1975.
327 K. Rasmussen, H. Wang, D. Fausa: Comparative efficacy of quinidine and verapamil in the maintenance of sinus rhythm after DC cardioversion of atrial fibrillation. A controlled clinical trial. Acta Med Scand 645 (suppl):23, 1981.
328 D. Wu, J. S. Hung, C. T. Kuo, K. S. Hsu, W. B. Shieh: Effects of quinidine on atrioventricular nodal reentrant paroxysmal tachycardia. Circ 64:823, 1981.
329 T. D. Sillers, R. W. F. Campbell, T. M. Bashore, J. J. Gallagher: Effects of procainamide and quinidine sulfate in the Wolff Parkinson White syndrome. Circ 55:15, 1977.
330 J. Morganroth, for the Mexiletine-Quinidine Research Group: Comparative efficacy and safety of oral mexiletine and quinidine in benign or potentially lethal ventricular arrhythmias. Am J Cardiol 60:1276, 1987.
331 J. Morganroth, J. C. Somberg, P. E. Pool, P. H. Hsu, I. K. Lee, J. Purkee: Comparative study of encainide and quinidine in the treatment of ventricular arrhythmias. J Am Coll Cardiol 7:9, 1986.
332 M. Hodges: Role of propafenone in the treatment of ventricular premature beats. J Electrophysiol 1:536, 1987.
333 J. Morganroth, H. Heinter: Comparative efficacy and safety of short acting and sustained release quinidine in the treatment of patients with ventricular arrhythmia. Am Heart J 110:1176, 1985.
334 H. J. Duff, D. M. Roden, R. K. Primm, J. A. Oates, R. E. Woosley: Mexiletine in treatment of resistant ventricular arrhythmia enhancement of efficacy and dose related side effects by combination with quinidine. Circ 67:1124, 1983.
335 J. P. DiMarco, H. Garan, J. N. Ruskin: Quinidine for ventricular arrhythmia. Value of electrophysiologic testing. Am J Cardiol 51:9, 1985.
336 V. Torres, D. Flowers, D. Miara, J. Sombeog: Intravenous quinidine by intermittent bolus for electrophysiologic studies in patients with ventricular tachycardia. Am Heart J 108:1437, 1984.
337 A. H. Selzer, H. W. Wray: Quinidine syncope: Paroxysmal ventricular fibrillation occurring during treatment of chronic atrial arrhythmias. Circ 30:17, 1964.
338 R. W. Koster, H. J. J. Wellens: Quinidine induced ventricular flutter and fibrillation without digitalis therapy. Am J Cardiol 38:519, 1976.
339 D. M. Roden, R. L. Woosley, K. Primm: Incidence of clinical features of the quinidine associated long QT syndrome. Implications for patient care. Am Heart J 111:1088, 1986.
340 M. Sokolow, A. C. Edgar: Blood quinidine concentrations as a guide in the treatment of cardiac arrhythmias. Circ 1:576, 1950.
341 I. S. Cohen, J. Hershel, S. I. Cohen: Adverse reactions to quinidine in hospitalized patients. Findings based on data from the Boston Collaborative Drug Surveillance Program. Prog Cardiovasc Dis 20:151, 1977.
342 W. H. Stinson, V. A. Mikusick: Felsrile reactions to quinidine. Am J Med Sci 221:440, 1951.
343 R. F. Larson: The mechanism of quinidine purpura. Blood 8:16, 1953.

344 J. S. Birkhead, E. M. Vaughan Williams: Dual effect of disopyramide on atrial and atrioventricular conduction and refractory periods. Brit Heart J 39:657, 1977.
345 T. Kus, B. I. Sasyniuk: Electrophysiologic actions of disopyramide phosphate in canine ventricular muscle and Purkinje fibers. Circ Res 37:844, 1975.
346 A. Sekeya, E. M. Vaughan Williams: A comparison of the antifibrillatory actions and effects in intracellular cardiac potentials of pronethalol, disopyramide and quinidine. Brit J Pharmacol 21:473, 1963.
347 B. I. Sasyniuk, T. Kus: Cellular electrophysiologic changes induced by disopyramide phosphate in normal and infarcted hearts. J Int Med Res 4 (Suppl 1):20, 1976.
348 P. Danilo, M. R. Rosen: Cardiac effects of disopyramide. Am Heart J 62:53, 1976.
349 T. A. English, A. J. Camm: The direct electrophysiologic effects of disopyramide phosphate in the transplanted human heart. Circ 67:38, 1983.
350 P. R. Wilkinson, J. Desai, J. Hollister, R. Gonzalez, J. A. Abbott, M. M. Scheinman: Electrophysiologic effects of disopyramide in patients with atrioventricular nodal dysfunction. Circ 66:1211, 1982.
351 A. LaBarre, H. C. Strauss, M. M. Scheinman, A. G. Wallace: Electrophysiologic effects of disopyramide phosphate in sinus node function in patients with sinus node dysfunction. Circ 59:226, 1979.
352 M. E. Josephson, A. R. Caracta, S. H. Law, J. J. Gallagher, A. N. Damato: Electrophysiologic evaluation of disopyramide in man. Am Heart J 86:771, 1973.
353 J. M. Desai, M. M. Sheinman, R. W. Peters, J. O'Young: Electrophysiologic effects of disopyramide in patients with bundle branch block. Circ 59:215, 1979.
354 C. R. Kerr, E. N. Prystowsky, W. M. Smith, C. Cook, J. J. Gallagher: Electrophysiologic effects of disopyramide phosphate in patients with Wolff Parkinson White syndrome. Circ 65:869, 1982.
355 R. A. J. Spurrell, C. W. Thornburn, J. Camm, E. Sowton, D. C. Deuchar: Effects of disopyramide in electrophysiologic properties of specialized conduction system in man and in accessory atrioventricular pathway in Wolff Parkinson White syndrome. Br Heart J 37:861, 1975.
356 A. Karim: The pharmakinetics of Norpace. Angiol 26:85, 1975.
357 P. H. Hinderling, E. R. Garrett: Pharmacokinetics of the antiarrhythmic disopyramide in healthy humans. J Pharmacokinet Biopharm 4:199, 1976.
358 L. A. Siddoway, J. T. Barbey, D. M. Roden, R. L. Woosley: Pharmacologic evaluation of standard and controlled release of disopyramide. Angiol 38:184, 1987.
359 C. M. Shaw, J. H. Doherty, H. L. Waxman, M. E. Josephson: The pharmacokinetic and pharmacodynamic effects of varying the free fraction of disopyramide. Angiol 38:192, 1987.
360 D. B. Haughey, J. J. Lima: Influence of concentration dependent protein binding in serum concentrations and urinary excretion of disopyramide and its metabolite following oral administration. Biopharm Drug Dispos 4:103, 1983.
361 J. M. Bryson, B. Whiting, J. R. Lawrence. Disopyramide serum and pharmacologic effect kinetics applied to the assessment of bioavailability. Br J Clin Pharmacol 6:409, 1978.
362 J. Bonde, N. A. Graudal, L. E. Pedersan, S. Balsiv, H. R. Angelo, T. L. Svendsen, J. P. Kampmann: Kinetics of disopyramide in decreased hepatic function. Eur J Clin Pharmacol 31:73, 1986.

363 R. P. Kapil, J. E. Axilson, E. L. Mansfield, D. J. Edwards, B. McErlane, M. A. Mason, D. Lalka, C. R. Kerr: Disopyramide pharmacokinetics and metabolism: effect of inducers. Brit J Clin Pharmacol 24:781, 1987.
364 M. L. Aitio, L. Mansury, E. Tala, Haataja, O. Aitio: The effect of enzyme induction on the metabolism of dysopyramide in man. Brit J Clin Pharmacol 11:279, 1981.
365 M. Burk, U. Peters: Dispyramide kinetics in renal impairment: determinants of interindividual variability. Clin Pharmacol Ther 34:331, 1983.
366 M. J. Sevka, J. J. Matthews, C. H. Nightingale, M. W. Izard, A. Fieldman, M. S. Chow: Disopyramide hemodialysis and kinetics in patients requiring long term hemodialysis. Clin Pharmacol Ther 29:322, 1981.
367 J. Braum, F. Sorgel, W. P. Grath, S. Oie: Bioavailability of disopyramide in normal volunteers using unbound concentrations. Eur J Clin Pharmacol 32:625, 1987.
368 S. M. Bryson, J. R. Lawrence, W. H. Steele, B. C. Campbell, H. L. Elliott, D. J. Sumner: The influence of protein binding in disopyramide clearance. Eur J Clin Pharmacol 23:453, 1982.
369 D. B. Haughey, L. J. Kraft, G. R. Matzke, W. F. Keane, C. F. Halstenson: Protein binding of disopyramide and elevated alpha-1-acid glycoprotein concentration in serum obtained from dialysis patients and renal transplant recipients. Am J Nephrol 5:35, 1985.
370 H. Echizen, S. Sarma, W. Umeda, T. Ishizaki: Protein binding of disopyramide in liver cirrhosis and in nephrotic syndrome. Clin Pharmacol Ther 40:274, 1986.
371 E. Haworth, A. K. Burroughs: Disopyramide and warfarin interaction. Br Med J 2:866, 1977.
372 A. J. Leach, J. E. Brown, P. W. Armstrong: Cardiac depression by intravenous disopyramide in patients with left ventricular dysfunction. Am J Med 68:839, 1980.
373 R. DiBianco, J. S. Gottdiener, S. N. Singh, R. D. Fletcher: A review of the effects of disopyramide phosphate on left ventricular function and the peripheral circulation. Angiol 38:174, 1987.
374 P. P. Mather: Cardiovascular effects of new antiarrhythmic agent, disopyramide phosphate. Am Heart J 84:764, 1972.
375 G. B. Jensen, B. Sigard, Uhrenholt: Circulatory effects of intravenous disopyramide in heart failure. J Int Med Res 4 (Suppl 1) 42, 1976.
376 P. J. Podrid, A. Schoenberger, B. Lown: Congestive hart failure caused by oral disopyramide. N Engl J Med 302:694, 1980.
377 P. R. Kowey, P. L. Friedman, P. J. Podrid, J. Zeilonka, B. Lown, J. Wynne, B. L. Holman: Use of radionuclide ventriculography for assessment of changes in myocardial performance induced by disopyramide phosphate. Am Heart J 104:769, 1982.
378 J. S. Gottdiener, R. DiBianco, R. Bates, B. J. Sauerbrum, R. D. Fletcher: Effects of disopyramide on left ventricular function: assessment by radionuclide (ineangiography). Am J Cardiol 51:1554, 1983.
379 M. Arif, J. C. Ladlaw, C. Oshrain, P. W. Willis, C. H. Nissin, D. J. McDermott, W. S. Smith, A. Karim, R. R. Wilson: A randomized, double blind, parallel group comparison of disopyramide phosphate and quinidine on patients with cardiac arrhythmias. Angiol 34:39, 1983.
380 J. Camm, D. Ward, R. A. Spurrell: The effects of intravenous disopyramide phosphate in reentrant paroxysmal tachycardia. Br J Clin Pharmacol 8:441, 1979.
381 G. Hartel, A. Louhija, A. Konttimen: Disopyramide in the prevention of reccurrent atrial fibrillation after electroconversion. Clin Pharmacol Ther 15:551, 1974.

382 B. W. Karlson, I. Tortensson, C. Alyorn, S. O. Jansson, L. E. Peterson: Disopyramide in the maintenance of sinus rhythm after electroconversion of atrial fibrillation: A placebo controlled one year follow-up study. Euro Heart J 9:284, 1988.

383 M. Ito, S. Onodera, J. Hashimoto, H. Noshiro, S. Shinoda, M. Nagashima, H. Suzaki: Effect of disopyramide on initiation of atrial fibrillation and relation to effective refractory period. Am J Cardiol 63:561, 1989.

384 P. Della-Bella, C. Tondo, G. Marenzi, C. M. Cipolla, F. Doni, S. Graji, A. Rimondini, A. Salvioni, M. D. Griazzi: Facilitating influence of disopyramide on atrial flutter termination by overdrive pacing. Am J Cardiol 61:1046, 1988.

385 P. Brugada, H. J. Wellens: Effects of intravenous and oral disopyramide on paroxysmal atrioventricular nodal tachycardia. Am J Cardiol 53:88, 1984.

386 S. Swiryn, R. A. Bauernfeind, C. R. Wyndham, R. C. Dhengra, I. Palileo, B. Strasberg, K. M. Rosen: Effects of oral disopyramide phosphate on induction of paroxysmal supraventricular tachycardia. Circ 64:169, 1981.

387 D. H. Bennett: Disopyramide in patients with the Wolff Parkinson White syndrome and atrial fibrillation. Chest 74:624, 1978.

388 R. A. Spurrell, C. W. Thorburn, J. Camm, E. Sowton, D. C. Deuchar: Effects of disopyramide in electrophysiologic properties of specialized conduction systems in man and on accessory atrioventricular pathway in Wolff Parkinson White syndrome. Br Heart J 37:861, 1975.

389 P. W. Willis: The clinical scope of disopyramide seven years after introduction – an overview. Angiol 38:165, 1987.

390 F. Morady, M. M. Scheinman, J. Desai: Disopyramide. Ann Inter Med 96:337, 1982.

391 F. Narcarella, D. Bracchetti, M. Palmieri, I. Cantelli, P. Bertaccini, E. Ambrosioni: Comparison of propafenone and disopyramide for treatment of chronic ventricular arrhythmias: placebo controlled, double-blind, randomized crossover study. Am Heart J 109:833, 1985.

392 C. M. Pratt, S. M. Batman, J. B. Young, M. Knoll, L. D. English: Antiarrhythmic efficacy of Ethmozine (moricizine HCI) compared with disopyramide and propranolol. Am J Cardiol 60:52F, 1987.

393 J. Kjekshus, J. Bathen, O. M. Orning, L. Storstein: A double blind, crossover comparison of flecainide acitate and disopyramide phosphate in the treatment of ventricular premature complexes. Am J Cardiol 53:72B, 1984.

394 B. B. Lerman, H. L. Waxman, A. E. Buxton, M. E. Josephson: Disopyramide: evaluation of electrophysiologic effects and clinical efficacy in patients with sustained ventricular tachycardia or ventricular fibrillation. Am J Cardiol 51:759, 1983.

395 L. A. Vismara, Z. Vera, R. R. Miller, D. K. Glason: Efficacy of disopyramide phosphate in the treatment of refractory ventricular tachycardia. Am J Cardiol 39:1028, 1977.

396 D. G. Bendett, E. L. Pritchett, A. G. Wallace, J. J. Gallagher: Recurrent ventricular tachycardia in man: evaluation of disopyramide therapy by intracardiac electrical stimulation. Eur J Cardiol 9:255, 1979.

397 N. Riccioni, M. Castiglioni, C. Bartolonili: Disopyramide induced QT prolongation and ventricular tachyarrhythmias. Am Heart J 105:870, 1983.

398 D. Tzivoni, D. Keren, S. Stern, S. Gottleib: Disopyramide induced Torsade de Pointes. Arch Inter Med 141:946, 1981.

399 R. S. Meltzer, E. W. Robert, M. McMorrow, R. P. Martin: Atypical ventricular tachycardia as a manifestation of disopyramide toxicity. Am J Cardiol 42:1049, 1978.

400 J. M. Desai, M. M. Scheinman, D. Hirschfeld, R. Gonzales, R. W. Peters: Cardiovascular collapse associated with disopyramide therapy. Chest 79:545, 1981.
401 J. L. Bauman, J. Gallastegui, B. Strasberg, S. Swiryn, J. Hoff, W. J.Welch, R. A. Bauernfeind: Long-term therapy with disopyramide phosphate: side effects and effectiveness. Am Heart J 111:654, 1986.
402 L. H. Danziger, J. R. Horn: Disopyramide induced urinary retention: Report of nine cases and review of the literature. Arch Inter Med 143:1683, 1983.
403 S. L. Teichmann, J. D. Fisher, J. A. Matos, S. G. Kim: Disopyramide – pyridostigmine: report of a beneficial drug interaction. J Cardiovasc Pharmacol 7:108, 1985.
404 B. F. Hoffman: The action of quinidine and procainamide in single fiber of drug ventricle and specialized conducting system. Ann Acad Bras Clin 29:365, 1958.
405 F. Morady, L. A. DiCarlo, J. M. Baerman, R. B. Krol: Rate-dependent effects of intravenous lidocaine, procainamide and amiodarone on intraventricular conduction. J Am Coll Cardiol 6:179, 1985.
406 H. I. Worske, F. I. Belford, F. N. Fastier, C. M. L. Brooks: The effect of procainamide on excitability, refractoriness and conduction in the mammelian heart. J Pharmacol Exp. Ther 107:134, 1953.
407 D. M. Roden, S. B. Reele, S. B. Higgins, G. R. Wilkinson, R. F. Smith, J. A. Oates, R. L. Woosley: Antiarrhythmic efficacy, pharmacokinetics and safety of N-acetyl procainamide in human subjetcs: comparison with procainamide. Am J Cardiol 460:463, 1980.
408 K. H. Dongman, B. F. Hoffman: Effects of N-acetyl procainamide on cardiac Purkinje fibers. Pharmacol 20:150, 1978.
409 D. Goldberg, J. A. Reiffel, J. C. Davis, E. Gang, F. Livelli, J. T. Bigger: Electrophysiologic effects of procainamide on sinus node function in patients with and without sinus node disease. Am Heart J 103:75, 1982.
410 M. Shenasa, C. J. Gilbert, D. H. Schmidt, M. Akhtar: Procainamide and retrograde atrioventricular nodal conduction in man. Circ 65:355, 1982.
411 M. E. Josephson, A. R. Caracta, M. H. Riccuiti, S. H. Lau, A. N. Damato: Electrophysiologic properties of procainamide in man. Am J Cardiol 33:596, 1974.
412 W. D. M. Paton, J. W. Thompson: The ganglion blocking action of procainamide. Br J Pharmacol 22:143, 1964.
413 E. G. V. Giardina, R. H. Heissenbutel, J. T. Bigger: Intermittent intravenous procainamide to treat ventricular arrhythmias: Correlation of plasma concentration with effect on arrhythmia, electrocardiogram and blood pressure. Ann Intern Med 78:183, 1973.
414 E. G. V. Giardina, P. E. Fenster, J. T. Bigger, M. Mayersohn, D. Perrier, F. I. Marcus: Efficacy, plasma concentrations and adverse effects of a new sustained release procainamide preparation. Am J Cardiol 46:855, 1980.
415 C. Gaffner, G. Johnson, J. Sjogren, Pharmacokinetics of procainamide intravenously, orally as conventional and slow release tablets. Clin Pharmacol Ther 17:414, 1975.
416 J. Koch Weser, S. M. Klein: Procainamide dosage schedules, plasma concentrations and clinical effects. J Am Med Assoc 215:1454, 1971.
417 E. G. V. Giardina, J. Dreifuss, J. T. Bigger, J. M. Shaw, E. C. Schreiber: Metabolism of procainamide in normal and cardiac subjects. Clin Pharmacol Ther 19:339, 1976.
418 R. L. Woosley, D. E. Drayer, M. M. Reidenberg, A. S. Nies, K. Carr, J. E. Oates: Effect of acetylator phenotype on the rate at which procainamide induces antinuclear antibodies and the lupus syndrome. N Engl J Med 298:1157, 1978.

419 J. M. Strong, J. S. Dutcher, W. K. Lie, A. J. Atkinson: Pharmacokinetics in man of the N-acetylated metabolite of procainamide. J Pharmacokinet Biopharm 3:223, 1975.
420 D. E. Drayer, D. J. Lowenthal, R. L. Woosley, A. J. Nies, A. Schwartz, M. M. Reidenberg: Cummulation of N-acetylprocainamide, an active metabolite of procainamide in patients with impaired renal function. Clin Pharmacol Ther 22:63, 1977.
421 J. J. C. Lertora, D. Block, D. P. Stec: Effects of N-acetylprocainamide and procainamide on myocardial contractile force, heart rate ad blood pressure. Proc Soc Exp Biol Med 161:332, 1979.
422 E. Angelakos, E. Hastings: The influence of quinidine and procainamide on myocardial contractility in vivo. Am J Cardiol 5:791, 1960.
423 J. R. Burton, M. T. Matthew, P. W. Armstrong: Comparative effects of lidocaine and procainamide on acute impaired hemodynamics. Am J Med 61:215, 1962.
424 R. L. McClendon, W. R. Hansen, J. M. Kinsman: Hemodynamic changes following procanamide administration. Am J Med Sci 222:375, 1951.
425 P. E. Fenster, K. A. Comess, R. Marsh, C. Katzenberg, W. D. Hager: Conversion of atrial fibrillation to sinus rhythm by acute intravenous procainamide infusion. Am Heart J 106:501, 1983.
426 G. Miller, J. L. Weinberg, A. Pick: The effect of procainamide in clinical ventricular fibrillation and flutter. Circ 6:41, 1952.
427 S. W. Halpern, G. Ellrodt, B. H. Singh, W. J. Mandel: Efficacy of intravenous procainamide infusion in converting atrial fibrillation to sinus rhythm: Relation to left atrial size. Br Heart J 44:589, 1980.
428 B. Olshansky, K. Okumura, P. G. Hess, R. W. Henthorn, A. L. Waldo: Use of procainamide with rapid atrial pacing for successful conversion of atrial flutter to sinus rhythm. J Am Coll Cardio 11:359, 1988.
429 P. Wu, P. Dines, R. Bauernfeind, R. Kehar, F. Amat-y-Leon, K. M. Rosen: Effects of procainamide on atrioventricular nodal reentrant paroxysmal tachycardia. Circ 57:1171, 1978.
430 W. I. Mandel, M. M. Laks, K. Obayashi, H. Hayakawa, W. Daley: The Wolff Parkinson White Syndrome: Pharmacologic effects of procainamide. Am Heart J 90:744, 1975.
431 T. D. Sellers, R. W. Campbell, T. M. Bashore, J. J. Gallagher: Effects of procainamide and quinidine sulfate in the Wolff Parkinson White syndrome. Circ 35:15, 1977.
432 H. J. J. Wellens, S. Braat, P. Brugada, A. P. M. Gorgels, F. W. Bar: Use of parocainamide in patients with the Wolff Parkinson White syndrome to disclose a short refractory period of the accessory pathway. Am J Cardiol 50:1087, 1982.
433 H. J. Kayden, B. B. Brodel, J. M. Skele: Procainamide: a review. Circ 15:118, 1957.
434 T. J. Bigger, R. A. Heissenbuttel: The use of procainamide and lidocaine in the treatment of cardiac arrhythmias. Prog Cardiovasc Dis 11:515, 1969.
435 R. A. Winkle, A. H. Gradman, J. W. Fitzgerald, P. A. Bill: Antiarrhythmic drug effects assessed from ventricular arrhythmia reduction in the ambulatory electrocardiogram and treadmill test: comparison of propanolol, procainamide and quinidine. Am J Cardiol 42:473, 1978.
436 K. Arakawa, Y. Poi, K. Hashiba, T. Mitsuoka, T. Yanaga, Y. Hata, N. Ogawa: Well controlled comparative study of the clinical efectiveness of intravenous mexiletine and procainamide in ventricular premature contractions. Am Heart J 25:357, 1984.
437 C. Lidell, N. Rehnquist, A. Sjoren, R. J. Yli-Uotila, P. K. Ronnevik: Comparative efficacy of oral sotalol and procainamide in patients with

chronic ventricular arrhythmias: A multicenter study. Am Heart J 109:970, 1985.
438 K. Nademanee, G. Feld, J. Hendrickson, V. Intarachot, C. Yale, M. K. Heng, B. H. Singh: Mexiletine: Double blind comparison with procainamide in PVC suppression and open label sequential comparison with amiodarone in life-threatening ventricular arrhythmias. Am Heart J 110:923, 1985.
439 A. M. Greenspan, L. N. Horowitz, S. R. Spielman, M. E. Josephson: Large dose procainamide for ventricular tachyarrhythmia. Am J Caridol 46:453, 1980.
440 T. R. Engel, A. del Gonzalez, S. G. Meister, W. S. Frankle: Effect of procainamide on induced ventricular tachycardia. Clin Pharmacol Ther 24:274, 1978.
441 H. J. J. Willens, F. W. Bar, K. I. Lie, D. R. Duren, H. J. Dohmen: Effect of procainamide, propranolol and verapamil on mechanism of tachycardia in patients with chronic recurrent ventricular tachycardia. Am J Cardiol 40:579, 1977.
442 H. C. Waxman, A. E. Buxton, L. M. Sadowski, M. E. Josephson: The response to procainamide during electrophysiologic study for sustained ventricular tachyarrhythmia predicts the response to other medications. Circ 67:30, 1983.
443 D. G. Wyse, B. Mitchell, H. J. Duff: Procainamide, disopyramide and quinidine. Discordant antiarrhythmic effects during crossover comparison in patients with inducible ventricular tachycardia. J Am Coll Cardiol 9:882, 1987.
444 D. S. Oseran, E. S. Gang, M. E. Rosenthal, W. J. Mandel, T. Peter: Electropharmacologic testing in sustained ventricular tachycardia associated with coronary heart disease: Value of the response to intravenous procainamide in predicting the response to oral procainamide and oral quinidine treatment. Am J Cardiol 56:883, 1985.
445 P. H. Lawson, H. Jick: Adverse reactions to procainamide. Br J Clin Pharmacol 4:507, 1977.
446 B. Strasberg, S. Sclarovsky, A. Erdberg, C. E. Duffy, W. Lam, S. Swiryn, J. Agmon, K. M. Rosen: Procainamide induced polymorphous ventricular tachycardia. Am J Cardiol 47:1309, 1981.
447 K. Rothman, E. L. Amorosi: Procainamide induced agranulocytosis and thrombocytopenia. Arch Intern Med 139:246, 1979.
448 B. D. Kosowsky, J. Taylor, B. Lown, R. F. Ritchie: Long-term use of procainamide following acute myocardial infarction. Circ 42:1204, 1973.

Dioxopiperazines: Chemistry and biology

By Donald T. Witiak and Yong Wei
Division of Medicinal Chemistry, College of Pharmacy, and The Comprehensive Cancer Center, The Ohio State University, Columbus, Ohio 43210, USA

1 Introduction . . . 251
2 2,3-Dioxopiperazines (**1**) . . . 252
2.1 β-Lactam antibiotics with 2,3-dioxopiperazine side chains . . . 253
2.1.1 Synthesis and antibiotic properties . . . 253
2.1.2 Vitamin K-related activities . . . 256
2.1.3 Disulfiram-related activities . . . 257
2.1.4 Lipophilicity and biological activity . . . 257
2.1.5 Penicillin-binding proteins; interactions . . . 258
2.1.6 Radiolabeled 2,3-dioxopiperazines; metabolism . . . 259
2.1.7 Gastrointestinal absorption . . . 259
2.1.8 Antibiotic stability studies . . . 260
2.1.9 Non-antibiotic-related biological properties of β-lactams . . . 261
2.1.10 Synergism with antitumor drugs . . . 261
2.2 Non-β-lactam synthetic 2,3-dioxopiperazines . . . 262
2.2.1 Syntheses and biological properties . . . 262
2.2.2 The quinoxaline-2,3[1*H*,4*H*]diones . . . 266
2.2.3 Chemistry of potential medicinal utility . . . 267
3 2,5-dioxopiperazines (**2**) . . . 269
3.1 Cyclodipeptide sources . . . 270
3.2 Importance to agriculture . . . 272
3.3 Pharmacological properties . . . 274
3.3.1 Cholinesterase inhibition . . . 274
3.3.2 Ergopeptide-related compounds . . . 275
3.3.3 Platelet-activating factor (PAF) inhibitors . . . 276
3.3.4 Thyrotropin-releasing hormone metabolite . . . 277
3.3.5 2,5-Dioxopiperazine-containing analgetics . . . 278
3.3.6 Cyclodipeptide metabolites . . . 278
3.3.7 Bacteriostatic N-chloramine analogues . . . 278
3.4 Simple cyclodipeptide syntheses . . . 279
3.5 Cyclodipeptides in amino acid asymmetric synthesis . . . 281
3.6 Naturally and related unnaturally occurring 2,5-dioxopiperazines: chemistry and biology . . . 282
3.6.1 Mycelianamide syntheses and related chemistry . . . 282
3.6.2 Picrorocillin chemistry . . . 284
3.6.3 Echinulin, brevianamide, and related chemistry . . . 284
3.6.4 Piperazines obtained from 2,5-dioxopiperazine precursors . . . 286
3.6.5 Bicyclomycin syntheses and biology; related chemistry . . . 289
3.6.6 Epidithio- and polythio-2,5-dioxopiperazines; biosynthesis and synthetic chemistry of natural and unnatural compounds . . . 294
3.6.7 *Cyclo*(Pro)$_2$ chemistry . . . 301
4 2,6-Dioxopiperazines (**3**) . . . 302
4.1 Early antitumor studies . . . 302

4.2 Conformationally constrained analogues 304
4.3 N,N'-Bis(morpholinomethyl) and related derivatives 307
4.4 Cytotoxic and cytokinetic effects of razoxane, its (+) enantiomorph ICRF-187, and the bis(morpholinomethyl) derivative bimolane 312
4.5 Mutagenic effects 313
4.6 Immunosuppressive properties 314
4.7 Treatment of psoriasis 314
4.8 Embryogenic effects 315
4.9 Synergism with antitumor drugs 315
4.10 Synergism with radiation on tumor growth 317
4.11 Antimetastatic properties 319
4.12 Amelioration of anthracyclin-induced toxicity in animals 324
4.13 Amelioration of acetominophen-induced toxicities in Syrian golden hamsters 330
4.14 Protection against alloxan diabetes in animals 330
4.15 Chelation of biometals and mechanism of action 331
4.16 Uptake and pharmacokinetics 336
4.17 Toxicity in phase 1 clinical studies 338
4.18 Combination with radiation therapy in humans 340
4.19 Protection against doxorubicin-induced cardiotoxicity in humans . 340
4.20 Clinical studies with morpholinomethyl derivatives 341
References 342

1 Introduction

Dioxopiperazines have three regioisomeric forms (**1–3**) which are derived from fundamentally different bond-disconnection precursors. The 2,3-dioxopiperazines (**1**) may be viewed as resulting from reaction of a disubstituted ethylenediamine and an α-dicarbonyl fragment such as oxalate. 2,5-Dioxopiperazines (**2**) are cyclodipeptides obtained via condensation of two α-amino acids. 2,6-Dioxopiperazines (**3**) may be viewed as cyclized imide derivatives of the corresponding di(carboxymethyl)amines. Hopefully, this review shall serve investigators not only as a summary of important progress, but as a catalyst for new biomedical research initiatives involving one or more classes of these interesting heterocycles.

All three regioisomeric forms are found in biologically active compounds, but in markedly different biological categories. Currently, the 2,3-dioxo series finds significance mainly as modifiers of penicillin- and cephalosphorin-related antibiotics. Some relatively simple 2,3-dioxopiperazines not bonded to β-lactam structures have antibacterial, antiviral and antitumor activities. Related quinoxaline-2,3[1*H*,4*H*]diones possess important central nervous system activities, serve as coccidiostats, and are catabolic intermediates in the enzymatic degradation of riboflavin. The 2,5-dioxo series is found in a variety of natural and unnatural compounds of varying structural complexity. Biological properties of these cyclodipeptides are as varied as the structures in which the 2,5-dioxohetero ring is found. Because of their sometimes structural complexity and the importance of their biological activities, this series has received considerable attention from synthetic organic chemists. The 2,6-dioxo regioisomers mainly generate biological significance as bis(2,6-dioxopiperazine) analogues derived from open chain chelating agents related to ethylenediaminetetraacetic acid (EDTA). These compounds are of interest because of their synergistic antitumor effects when used in combination with other antineoplastic drugs. Additionally, these dioxopiperazines exhibit stereoselective anti- and pro-metastatic properties, and most importantly ameliorate the toxicity of potent antitumor drugs such as the anthracycline antibiotics.

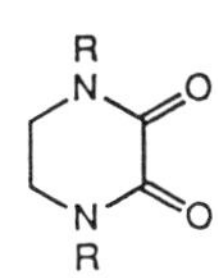

1

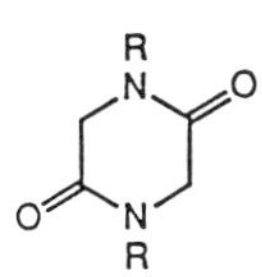

2

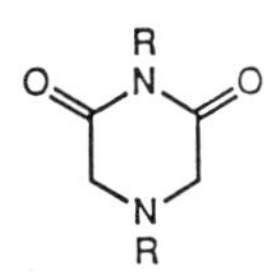

3

2 2,3-Dioxopiperazes (1)

Physical organic chemistry has long played a major role in medicinal chemical research, and the importance of this science to our understanding of 2,3-dioxopiperazine chemistry and biology is no exception. These compounds belong to a class of α-dicarbonyl compounds [1], investigated by photoelectron [2–8], ultra-violet/emission [9–10], and nuclear magnetic resonance (NMR) [11] spectroscopy. Photoelectron spectra of five-, six-, and seven-membered cyclic α-dicarbonyl compounds (**6–8**) wherein the oxamide unit is forced to adopt an *s-cis* (**5**) or twisted *s-cis* conformation are available. When compared to acylic *s-trans* oxamides (**4a,b**), the influence of the *s-trans-s-cis* conformational change on the ionization is masked by the effects of the alkyl parts of the ring systems [2]. Observed differences in the spectra of **4b** and **7**(R = H) cannot be interpreted solely on the bases of conformational changes. Whereas ionization energies in photoelectron spectra of cyclic *s-cisoid*-oxamides can be accounted for by molecular orbital energies derived from *ab initio* calculations, empirical corrections are necessary. The spectrum of 7(R = H) is well resolved; calculated ionization energies agree with experimental values for the three lowest ionization events. N-methylation shifts all ionizations towards lower energies; π_+ and π_- bands are more sensitive than n_+ and n_- bands to N-alkylation [2,5]. Thus, a qualitative difference in the degree of twist of the oxamide unit in different cyclic compounds may be obtained, and such information should be useful in drug design. Differential effects observed in photoelectronic spectra resulting from compounds having different lengths of cyclic alkyl chains prohibits a quantitative analysis [2].

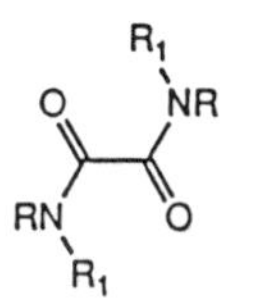

s-trans
4a: $R_1 = R = H$
4b; $R_1 = H, R_2 = Me$

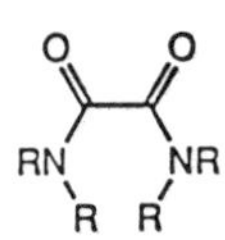

s-cis
5

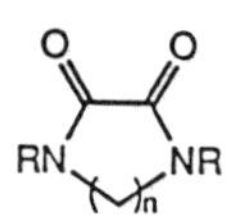

6; n = 1
7; n = 2
8; n = 3

2.1 β-Lactam antibiotics with 2,3-dioxopiperazine side chains

Important antibiotics such as piperacillin (**9a**) [12–15] and cefoperazone (**10**) [16–19] contain the 2,3-dioxopiperazine moiety. Numerous articles describe the antimicrobial, metabolic, pharmacokinetic, and clinical properties of the penicillin and cephalosporin containing dioxopiperazines, most often in comparison with antibiotics of completely different structure, or with penicillins and cephalosporins having other side chains. Review of this literature is beyond the scope of this work, but a relatively more recent noninclusive list of lead articles is available for the interested reader [20–146]. Additionally, many gas-liquid chromatographic (GLC) assays for these and structurally related antibiotics are accessible [147–165]. In this section, emphasis is placed mainly upon how 2,3-dioxopiperazines serve the goals of Medicinal Chemistry. Subsequent sections focus on the 2,5- and 2,6-dioxopiperazine regioisomers.

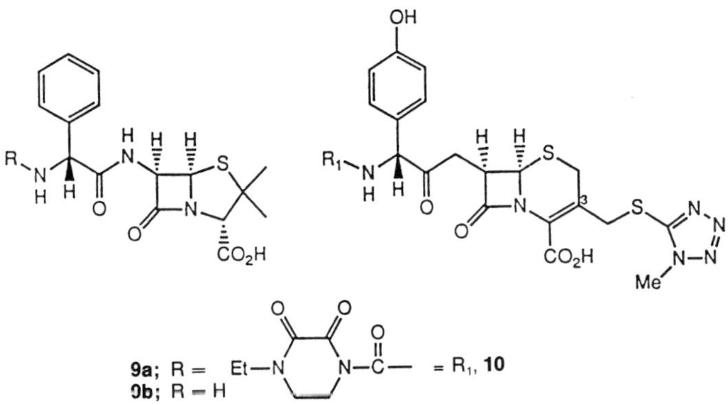

2.1.1 Synthesis and antibiotic properties

Impetus for the introduction of the 2,3-dioxopiperazinyl function into antibiotics was provided by the superior antibacterial activity observed against Gram-negative bacilli such as *Pseudomonas aeruginosa. P. aeruginosa* is an important pathogen, and is frequently found in neutropenic patients whose defense mechanisms may be severely compromised [45]. A mathematical solution derived to a pharmacodynamic model for the survival of this organism in medium containing 0.64 mM piperacillin is satisfied for up to 24 hours [166]. Structure-activity studies reveal the 2,3-dioxo system to be most efficacious of several oxopiperazines investigated [167]. Comparison of β-lactam antibiotics containing regioisomeric dioxopiperazine side chains reveal an order of potency 2,3- > 2,5- > 2,6-dioxo substitution against *Staphylococcus aureus, Escherichia coli, P. aeruginosa, Proteus vulgaris, Klebsiella pneumonia and Serratia marcescens* [168]. In other qualitative structure-activity studies the order of decreasing potency of oxopiperazines is 2,3- > 2- > unsubstituted > 3 > 2,5- > 2,6 [167]. Such compounds exhibit activity at μg/ml levels against both Gram-positive and Gram-negative bacteria [169–170].

Syntheses for 2,3-dioxopiperazine-substituted penicillins are straight forward. Conversion of the appropriate dioxopiperazine to its N-substituted carbonyl chloride and subsequent reaction with D-aminobenzylpenicillin **9b** affords the respective target [168]. Also, semisynthetic penicillins and cephalosporins are prepared by activation of 2,3-dioxopiperazine-substituted carboxylic acids with chlorothioformates to produce mixed thioanhydrides which upon reaction with β-lactam amines afford desired antibiotics in low to average yield [171]. Piperacillin analogue **12**, prepared in 21% yield using 4-ethyl-2,3-dioxopiperazin-l-ylcarbonyl chloride and α-amino penicillin **11**, is contaminated with N-ethyl-2,3-dioxopiperazine which is not easily removed [172]. However, acylation of zwitterion **13** with the carbonyl chloride affords **14** which serves as precursor to a separable mixture of diastereoisomeric penicillins **15** in high yield [172]. Hydrogenolysis of the

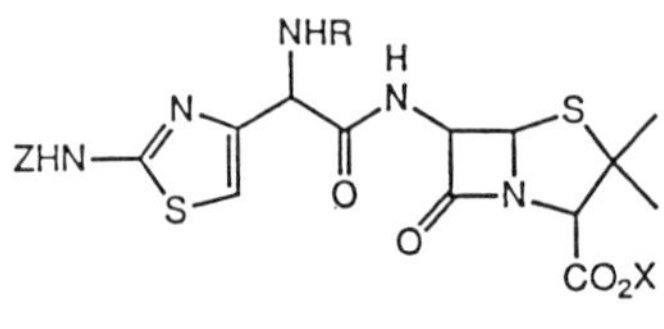

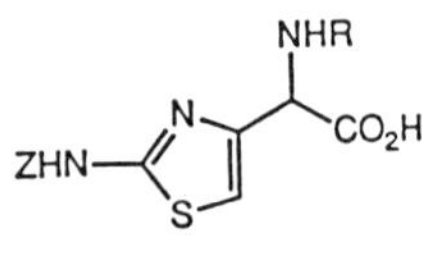

11; R = Z = X = H

12; R = 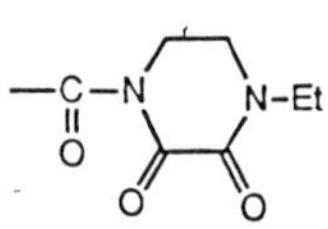

, Z = X = H

13; Z = t-butyloxycarbonyl, R = H

14; Z = t-butyloxycarbonyl, R = 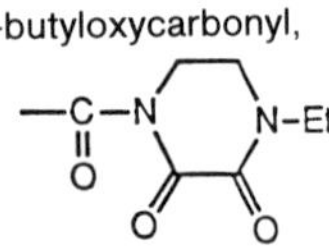

15; R = 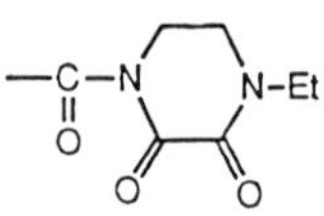

, Z = t-butyloxycarbonyl, X = CH_2Ph

separated isomers generates target **12**. The more active isomer of **12**, which exhibits antibiotic properties similar to piperacillin, is derived from the more polar protected intermediate diastereomer of **15**. Many 6α-substituted piperacillin analogues are prepared via the 6α-hydroxymethyl and 6α-formyl species derivatized with the classic 2,3-dioxopiperazine side chain [173]. Esters of 6α-carboxypenicillins having 2,3-dioxopiperazine side chains are available by oxidation of 6α-formyl hemiacetals [174]. Derivatization also affords 6α-carbamoyl penicillins. Synthesis and study of antibacterial activities of 7α-formamidocephalosporins such as **16** having various substituents at C(3) is of interest because of their broad spectrum of activity, their β-lactamase-stable properties, and their effectiveness against Gram-negative bacteria, including *P. aeruginosa* [175]. 7α-methoxy cefoperazone analogues also have a broad spectrum of activity against Gram-negative organisms [176], and are prepared directly from the demethoxy compound possessing the desired 2,3-dioxopiperazine side chain using LiOMe/t-BuOC1 [177]. 7α-Formamido cephalosporins such as **17** have potential clinical utility [178]. β-Lactam modification by 6α-methoxy functions improves activity of piperacilline-related antibiotics against β-lactamase-producing strains of *E. coli and Proteus mirabidis* [179]. However, use of a 6α-formamido function in place of a 6α-methoxy group increases β-lactamase stability without compromising intrinsic activity [180].

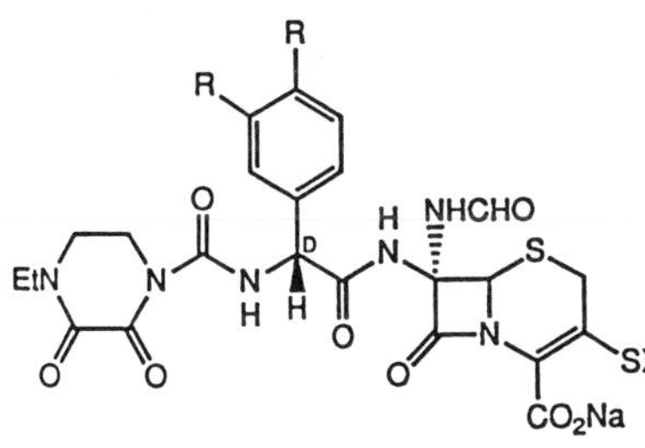

16; R = oxygen analogues, X = heteroaromatic rings

17; R = AcO, X = 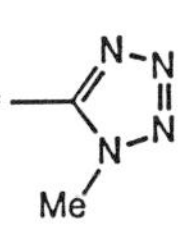

Antibacterial l-carbacephem analogues **19** containing 2,3-dioxopiperazine and other heterocyclic side chains are prepared from optically active azetidinone **18**, the chiral carbon of which is derived from penicillin [181]. 2,3-Dioxopiperazine and other heterocyclic ring-containing side chain (R) derivatives of 3-amino-2-azetidinone-1-sulfonate (**20**) are prepared by degradation of penicillins to the 3-amino-β-lactam followed by acetylation and sulfonation [182]. The antibiotic activity spectrum is markedly influenced by the nature of R. Also available via trimethylsilyl activated-dioxopiperazines are 2,3-dioxopiperazine analogues of β-amino-β-lactams (**21**), but optically active diastereomers of **21** have no antibacterial activity [183]. Electronegativity effects of CF_3 on antibacterial properties in 3-trifluoromethyl-3-cephem-4-carboxylic acids possessing various heterocyclic side chains also are of interest [184].

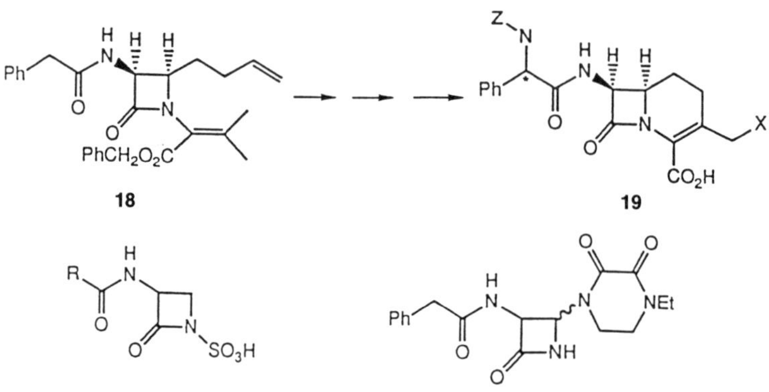

18 **19** **20** **21**

2.1.2 Vitamin K-related activities

Among several heterocyclic substitutions examined at C(3), the methyltetrazolylthio group in cefoperazone (**10**) generally provides for the most efficacious antibiotics. In one clinical study suppression of vitamin K-dependent clotting factor biosynthesis seemed to be independent of the presence of the tetrazolylthio side chain [185]. Nonetheless, antibiotics containing this side chain inhibit microsomal vitamin K epoxide reductase in rats [186]. Hypoprothrombinemia in vitamin K-deficient female rats is caused by β-lactam antibiotics having this and related thiol-heterocycles in the 3 position of the cephem nucleus [187]. These effects on coagulation seem not to be related to the 2,3-dioxopiperazine heterocycle. Cefoperazone (**10**), which has the methyltetrazolylthio side chain, as well as other antibiotics containing such side chains were shown to inhibit hepatic vitamin K epoxide reductase in patients thereby providing for a lower nutritional-vitamin K status predisposing individuals to hypoprothrombinemia [188].

2.1.3 Disulfiram-related activities

Disulfiram-like effects of β-lactam antibiotics on alcohol-metabolizing systems in rats also is likely due to the methyltetrazolylthio heterocycle released in vivo [189,190], rather than to any intrinsic activity related to the 2,3-dioxopiperazine side chain or other structural features of the antibiotic. Thus, cefoperazone contains the tetrazolylthio group, and this but not other 2,3-dioxopiperazine-containing antibiotics, display disulfiram-like inhibitory effects on alcohol metabolism [191].

2.1.4 Lipophilicity and biological activity

The antibacterial potency of pipercillin (**9a**)-related analogues, wherein the N(4) 2,3-dioxopiperazine nitrogen is substituted with chains of 1–7 carbons increases with increasing lipophilicity [192,193]. However, penetration of the outer membrane of the microorganism is independent of lipophilicity thereby leading to the suggestion that penicillins pass through the outer membrane via a non-porin pathway [194]. β-lactam antibiotics with exceptionally bulky side chains, such as piperacillin, exhibit much slower penetration rates through membranes of *E. coli* K-12 than expected from their hydrophobicity [195]. Most recently, studies [46] indicate that ionic charge, hydrophobicity, and molecular weight all influence β-lactam drug uptake into microorganisms, possibly via porin channels [195]. β-lactamase stability, blood level concentrations, protein binding, and chronic toxicity in mice correlate with increasing N(4) piperazino chain length and partition coefficient [193,194]. Also, encapsulation of piperacillin by liposomes prepared with phosphatidylcholine and cholesterol (1:1) protects the β-lactam antibiotic from hydrolysis by staphylococcal β-lactamase, thus increasing its activity against *Staphylococcus aureus* [196]. Presumably, the structurally related bis(norpiperacillin) exhibits improved activity compared to piperacillin against Gram-negative bacteria owing to increased stability to the β-lactamases produced by the organisms [197]. Against Gram-positive organisms, the bis(norpenicillins) are generally two to five-fold less active than the corresponding penicillins [198].

The lipophilicities of the injectable form of many antibiotics including piperacillin (log P = –3.30) and cefoperazone (log P = –3.39) are available [199]. For a limited series of these antibiotics, a linear relationship is observed between log P and log permeability data for human and rat brain thus enabling brain concentrations in man and rats to be predicted [200]. Additionally, placental transfer of a heterogeneous series of drugs including piperacillin can be described as a function of drug lipophilicity [200]. In one study utilizing a limited series of penicillins, piperacillin was least hydrophilic as determined using reverse phase thin layer chromatography [201]. Increases in hydrophilicity are related to higher serum concentrations, and to higher urinary and lower biliary excretion of these drugs. Cefoperazone, a drug which is mainly eliminated by the biliary tract in rat, inhibits biliary phospholipid and cholesterol secretion, a phenomenon which seems not be due to impairment of either mixed micelle formation within the canalicular lumin nor to interaction with lipid vesicles [202].

2.1.5 Penicillin binding proteins; interactions

Interaction of β-lactam antibiotics with penicillin-binding proteins (PBP) is of considerable interest. The activity spectrum of a given antibiotic may be due in large part to interaction with these membrane proteins known to be involved in peptidoglycan biosynthesis [203,204]. The streptomyces K15 DD-peptidase, a membrane-bound PBP which performs effective transpeptidation reactions on carbonyl-donor and amino-acceptor co-substrates, undergoes reaction with β-lactam carbonyl donors such as piperacillin (ID_{50} = 9 μM at 37 °C and in detergent-containing potassium phosphate buffer, pH 5) [205]. Cefoperazone binding to *E. Coli* PBP-3 and *P. aeruginosa* PBP-3, proteins which participate in cell division, is likely related to formation of filamentous cells by the antibiotic [204]. Although outermembrane permeability and β-lactamase stability are minimally affected by 2,3-dioxopiperazine N(4) chain length, affinity to PBPs, particularly PBP-3 increases with increases in length of the N(4) substitution [192]. Minimum inhibitory concentrations (MIC) and ID_{50} values correlate with increases in affinity for PBP in *E. coli* W3110 and *K. pneumonia* Y-50 [194]. The low affinity of *S. falcalis* PBP, relative to that of *S. au-*

reus PBP, towards most *β*-lactam antibiotics is likely responsible for the resistance of the former to these drugs [206]. Piperacillin-related monobactams of the general structure **22** exhibit good binding to PBP-3 of *E. coli* (0.1 μg/ml) and PBPs 1,2, and 3 of *S. aureus* (~1 μg/ml) [207].

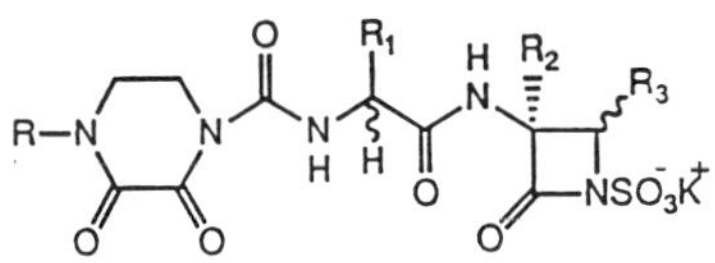

22

2.1.6 Radiolabeled 2,3-dioxopiperazines; metabolism

Labeled 2,3-dioxopiperazine rings are used in *β*-lactam antibiotic metabolic studies [208]. The 2,3-dioxopiperazine side chain of 7*α*-[^{14}C]methoxy-labeled cephalosporin antibiotics is known to undergo both metabolic hydrolysis and separation possibly as N-ethyl-2,3-dioxopiperazine from the parent structure in humans and rodents, but following intravenous administration 92–98% of the radioactivity in urine, bile and feces is attributable to the unmetabolized drug [209].

2.1.7 Gastrointestinal absorption

Diethylpyrocarbonate (DEP) is thought to modify the specific transport protein amino acid residue necessary for interaction with *α*-amino functions of *β*-lactam antibiotics [210]. With cefoperazone, a drug which does not have an *α*-amino group and has poor oral absorption, DEP-treatment of intestinal brushborder membranes has no inhibitory effect on the intestinal uptake of the antibiotic [210]. The extent of rectal absorption of the *β*-lactam antibiotics, including piperacillin and cefoperazone, coadministered with sodium caprate correlates with the permeability of the drugs to cellulose membranes [211]. Interestingly, absorption rectally can be promoted by concurrent administration of various carbocylic acids; sodium salts of these substances having log P values of 4.2–4.8 are most effective as promoters of *β*-lactam rectal absorption [211].

2.1.8 Antibiotic stability studies

[^{1}H] NMR analysis of the 2,3-dioxopiperazine ring in conjunction with a comprehensive study of *β*-lactam antibiotics and their degradation products is available [212]. Whereas many antibiotics remain stable in untreated biological fluids and plasma for one week at 4 °C, and for at least three weeks at −20 °C, cefoperazone is comparatively unstable and requires immediate processing [165]. Pipericillin is among those *β*-lactam antibiotics which undergo negligible degradation in aqueous solution at 37 °C for 7 days to form N-formylpenicillamine [213]. Piperacillin also is relatively thermal stable in the solid state when compared to eight other penicillins [214]. Stability seems to correlate positively with the lipophilic character of the molecule. Piperacillin (and cefotoxime), but not ampicillin, is stable at 29 °C in an L-amino acid solution prepared for total parenteral nutrition of newborn infants [215]. Alternatively, in pH 9.0 buffer solution, piperazinyl ring cleavage is facile compared to *β*-lactam hydrolysis which readily takes place in acid media [216]. Degradation studies over a wide pH range reveal that the 2,3-dioxopiperazine ring is hydrolyzed 20 times faster than the *β*-lactam function at alkaline pH [217]. BRL 36650 (**23**), a 6*α*-formamidopenicillin containing the 2,3-dioxopiperazine side chain [218], is stable as a dry solid and in buffered aqueous solutions kept at low temperature, but under mildly acidic aqueous conditions undergoes C(5)-C(6) bond cleavage as the major degradation pathway [219]. Base-catalyzed (pH 10) opening of the 2,3-dioxopiperazine ring is rapid [217,219].

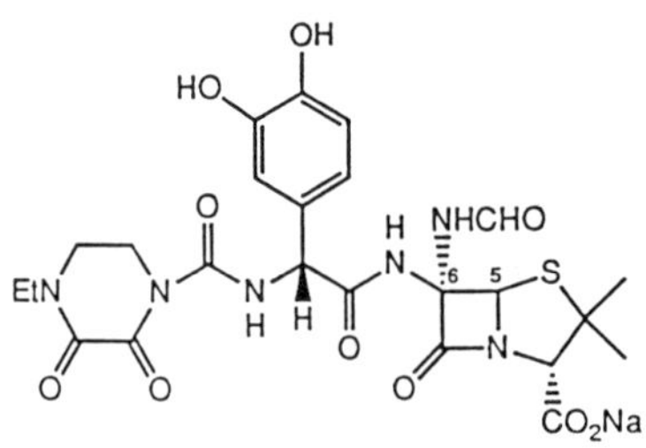

23

2.1.9 Non-antibiotic-related biological properties of β-lactams

These β-lactams have other interesting pharmacological properties not related to their antibacterial activities. Piperacillin (20 mmol/L) inhibits epinephrine-induced platelet aggregation by 80% [220], but prolongation of bleeding time seems not to be a problem following 7 days of drug administration [221]. Cefoperazone (6 gm/day) has no effect in response to several initiators of platelet aggregation, and also has no significant effect on template bleeding times [222]. In vitro, certain ureidopenicillins, including piperacillin, stimulate production of interleukin-1 by macrophages [223] and suppress lymphocyte transformation responses to concanavalin A [224], but the mechanism of these effects and their clinical significance remain to be determined.

2.1.10 Synergism with antitumor drugs

Synergism between antibiotics and anticancer drugs continues to be of interest. However, nearly all combinations of β-lactam antibiotics with doxorubicin display neither synergistic nor antagonistic actions [225]. Among the antibiotics tested, piperacillin is more frequently synergistic than are other antibiotics. This synergy is most pronounced in *P. vulgaris* compared to other microorganisms studied. Synergism between cefoperazone and antineoplastic drugs against *E. Coli* or *K. pneumoniae* cannot be demonstrated [225]. Among four antimicrobial agents recently investigated, piperacillin displays the greatest degree of synergism in combination with antineoplastic agents (mitomycin C, bleomycin, doxorubicin and 5-fluorouracil) against 108 clinical isolates of four species of Gram-negative bacilli [226]. Pipericillin is synergistic with 5-fluorouracil, and enhances 5-fluorouracil uptake into bacterial cells. Binding of piperacillin to PBP is not changed by the antineoplastic drug [227]. Unfortunately, piperacillin, as well as mezlocilline, rifampicin and doxycycline, four of twelve broad spectrum antibiotics studied, when administered prior to sarcoma tumor implantation in mice increase tumor growth rate and/or lung metastases [228].

2.2 Non-β-lactam synthetic 2,3-dioxopiperazines

2.2.1 Syntheses and biological properties

Outside the scope of semisynthetic antibiotics, 2,3-dioxopiperazines or their α-dithiocarbonyl analogues [1] have had limited success in the pharmaceutical arena. Interesting synthetic chemistry has outpaced efforts to research the biological significance of heterocyclic targets containing these rings. Thus, halogenated pyrazinones such as **24**, prepared from oxalyl chloride and α-methylaminoacetonitrile·HCl, undergo hydrolysis with KOH in aqueous dioxane to produce the 5-chloro-2,3-dioxopiperazine **25**. Methylation (CH_2N_2) affords O-and N-methylated products **26** and **27** [229]. Alternatively, reaction of **24** with KOEt generates the 3-ethoxy species **28** and reaction with ethanolic ammonia yields amino analogue **29**. Some older related materials have antimicrobial properties [230]. Related nucleoside derivatives of 2-oxo-5-methylpyrazine-4-oxide (i.e. **30**) inhibit the growth of leukemia L1210 cells in vitro by 50 % at 9×10^{-7}M and have marginal activity against *Streptococcus faecium* and *E. coli* [231]. But, in general little has been done of pharmaceutical interest to exploit such chemistry.

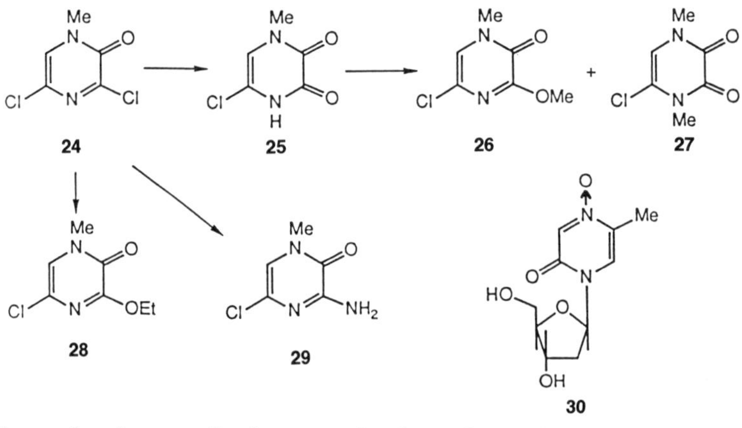

During investigations of the synthesis of 1,4-benzodiazepines [232,233], carbonylation of secondary amine **31** produces a small quantity of target **32** (9-30 %) and unexpected quinoxaline-2,3[1*H*,4*H*]dione **33** (36–42 %), the later likely arising from air oxidation of the intermediate quinoxalin-2-one [233]. Yields of desired benzodiazepine **32** can be increased by modifying reaction conditions or by altering the nucleophilicity of nitrogen in **31** by use of an acetamide in place of the benzylamine group.

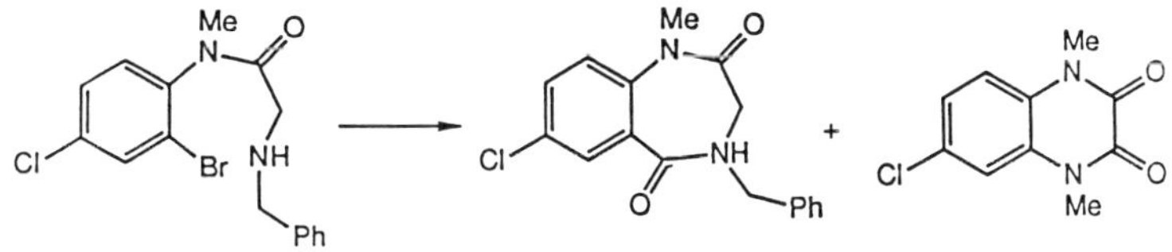

31 32 33

Several simple 2,3-dioxopiperazine analogues exhibit weak to moderate antieoplastic activity in small animals [234–236]. These include 1-(4-aminobenzyl)-2,3-dioxopiperazines (**34**) which have T/C values against L-1210 inoculated mice up to 177 % at 110 mg/kg [237]. Related derivatives, such as **35** exhibit MICs against HeLa cells of 0.1 and 0.39 μg/ml, respectively [236,238]. Potency of 1-(2-pyrimidinylamino)benzyl-substituted analogues is enhanced by insertion of an additional benzyl group on the N(4) dioxopiperazinyl nitrogen. 2,3-Dioxopiperazine analogues are generally more potent than related 2,5- or 2,6-dioxo compounds in this series, even though analogues such as alboursin (**36**) and 593A (**37**) have antitumor properties. The 2,6-dioxopiperazine, razoxane (section 4), also has weaker antitumor activity than **34**-related species in L1210-infected mice [239]. Analogue **38** has a simple structure, but possesses an MIC <0.05 μg/ml against HeLa cells in vitro and T/C = 236 % against L1210-infected mice at 20 mg/kg. Dioxopiperazines of the type **39** have in vitro inhibitory concentrations of 10–50 μg/ml with an LD_{50} of 3.0 g/kg in mice [240]. Alkenyl-substituted N-benzyl-2-3-dioxopiperazines of the type **40** possess antitumor activity in vitro and in vivo, but these compounds are not particularly efficacious [241]. The rationale for construction of such drugs is related to a desire to have the appropriate lipophilic-hydrophilic balance. Interestingly, the 2,3-dioxopiperazinyl group is relatively hydrophilic [193,242]. No correlation can be established relating cytotoxicity in vitro to antitumor activity in vivo. For **p**-diethylaminobenzyl-substituted-2,3-dioxopiperazines, successive N-dealkylation followed by acetylation of the free amine may account for the lower antitumor activity of the parent compound in vivo [241].

Compounds of the type **41** (T/C = 189 % at 200 mg/kg ip.) and **42** (T/C = 222 % at 50 mg/kg ip.) are among several heterocyclic derivatives [including the related epoxide of the bis(2,6-dioxopiperazine) razoxane (section 4) examined in P388 leukemic mice I which do not show significant activity [243,244]. The most promising compounds are de-

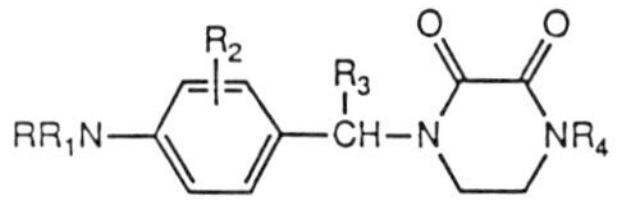

34; R = 2-pyrimidinyl, R_1 = $MeOCH_2$, $R_2 = R_3$ = H, R_4 = benzyl
35; R, R_1 = H, alkyl, cycloalkyl, aralkyl, acyl, thiocarbamate, alkylthioimidoyl, amidino, heterocyclic.
RR_1N = heterocyclic
R_2 = H, amino, alkyl, alkoxy,
R_3 = H, alkyl.
R_4 = alkyl, aryl, heterocyclic

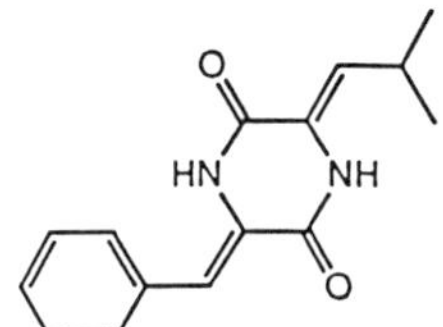

36

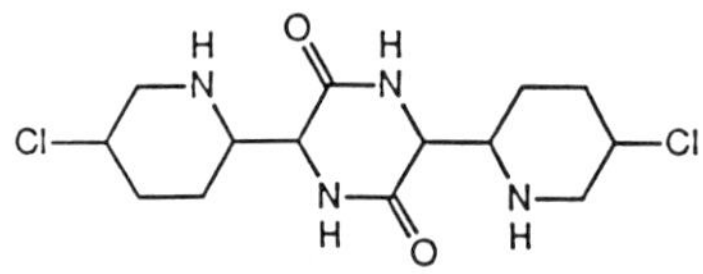

37

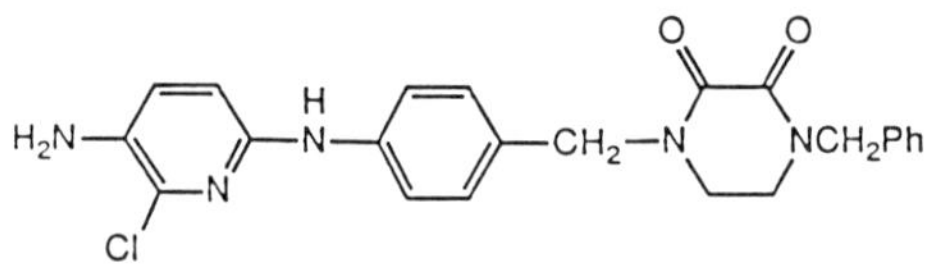

38

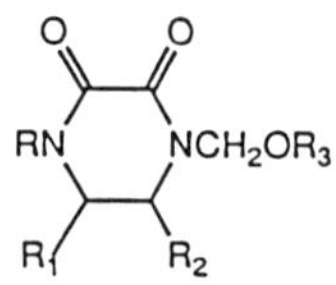

39; R = H, alkyl, alkenyl, etc.
$R_1 = R_2$ = H, alkyl or aryl,
R_1R_2 = alkylene
R_3 = H, alkyl or cycloalkyl

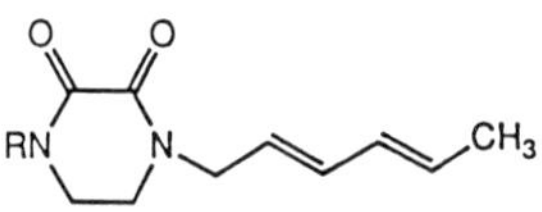

40; R = H, n-C_8H_{17}, $CH_3(CH=CH)_2(CH_2)_2$,

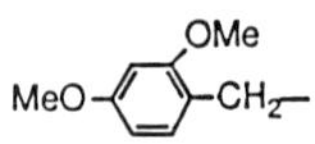

rivatives of heterocycles not belonging to the dioxopiperazine class [243,244]. However, related alkylating species such as 1-[2-chloroethyl-3-(4-substituted-2,3-dioxopiperazinyl]alkyl-1-nitrosourea (**43**) have antitumor properties. The stability of **43**-type analogues correlates with the length (n) of the methylene link between the piperazinyl group and the urea nitrogen. These compounds are related to the important antitumor nitrosoureas such as 1-(2-choroethyl)-3-cyclohexyl-1-nitrosourea (CCNU) [245]. Compounds of the type **44** and **45**, not having a nitroso function, display antitumor activity against solid Ehrlich carcinoma in vivo; after 5 days of treatment at 500 mg/kg the percent T/C for tumor weight was 58 for **44** and 56 for **45** [245]. Several 1,4-benzoquinone-substituted 2,3-dioxopiperazines (**46–48**) are available; analogues of the type **47** show better antineoplastic activity in

L1210 and S180 infected mice. Such compounds provide T/C values up to 168 % at doses of 2 mg/kg against L1210, and T/C values up to 248 % against S180 at 0.5 mg/kg [246].

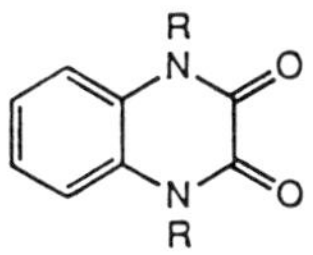

41

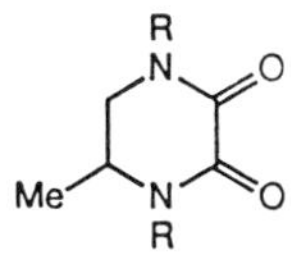

42

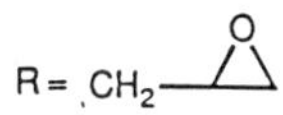

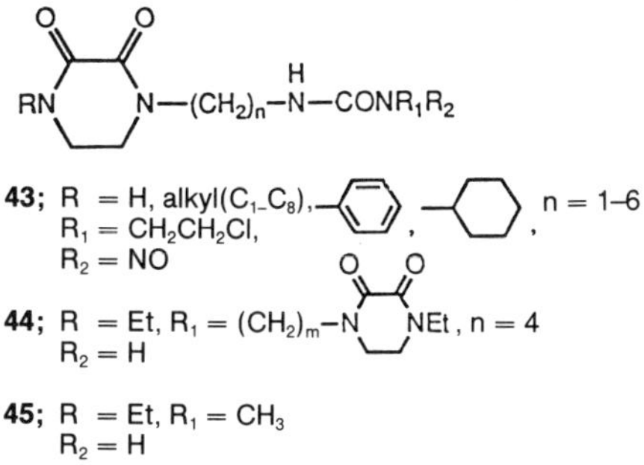

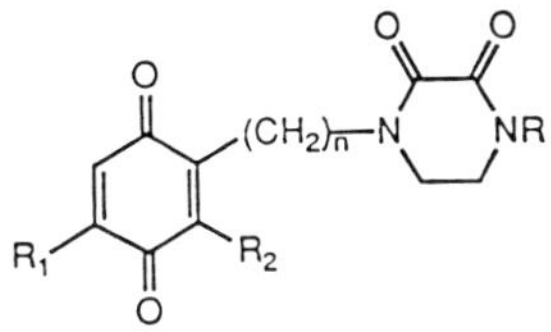

46

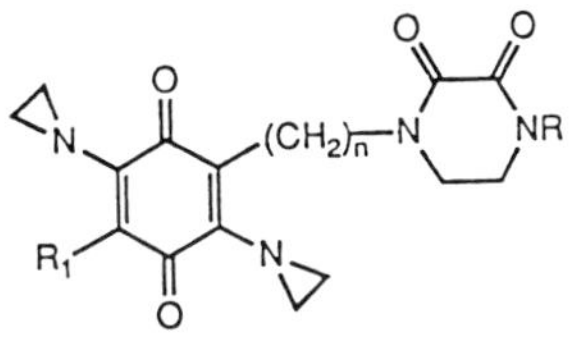

47

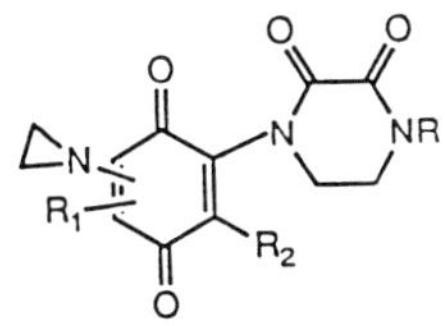

48

Several series of 2,3-dioxopiperazines of the type **49–51** have antibacterial and/or antiviral activity. Some are active against influenza, parainfluenza type 3 and respiratory syncytial virus [247,248]. The re-

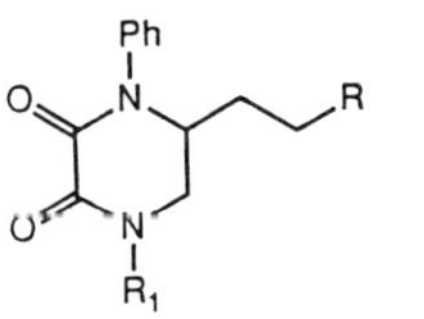

49; R = morpholino, Me_2N, R_1 = Me, Et, Me_2CH, cyclohexyl

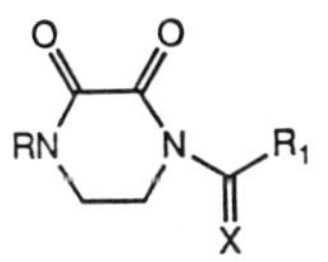

50; R = C_{1-8}alkyl, $ClCH_2CH_2$, Ph, etc., R_1 = Cl or $OCCl_3$, X = O or S

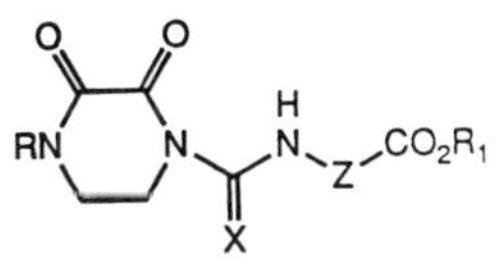

51; R = alkyl, R_1 = H or cation, R_2 = H or alkyl, X = O or S, Z = α-amino acid group

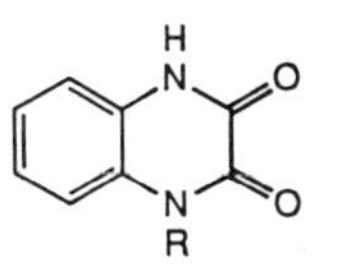

52a; R =

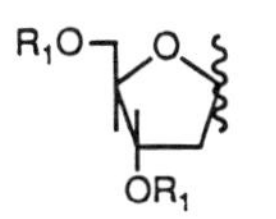

52b; R = $R_1O(CH_2)_2OCH_2$

lated quinoxaline-2,3[1*H*,4*H*]diones of general structures **52a,b** do not have antiviral activity when compared to 5-ethyl-2′-deoxyuridine [249].

2.2.2 The quinoxaline-2,3[1*H*,4*H*]diones

Synthetic substituted 2,3-dioxo- and related 2,3-dithiono[1*H*,4*H*] quinoxaline analogues (**53–56**) are known, but biological data seems not to be available [250]. 1,4-Bis(N-oxides) of 2,3-disubstituted quinoxalines, such as the antibacterial quinoxidine **57** and dioxidine **58**, undergo photochemical-induced rearrangement to produce 2,3-dioxo species such as **59** and other materials, but biological applications of such chemistry seem to be unresearched [251].

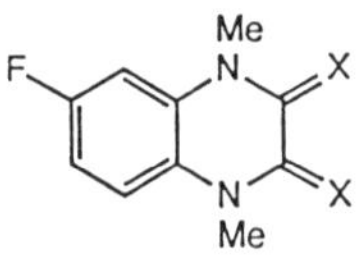

53; X = O
54; X = S

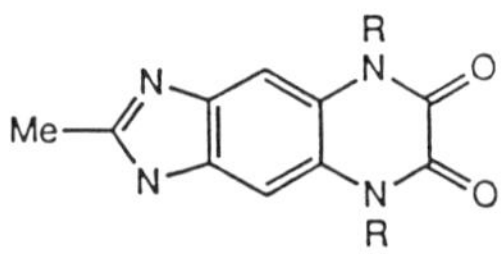

55; R = Me
56; R = H

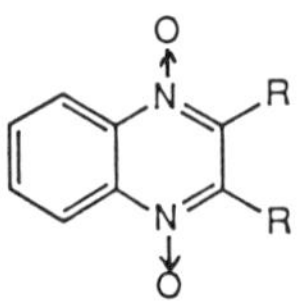

57; R = CH_2OCOMe
58; R = CH_2OH

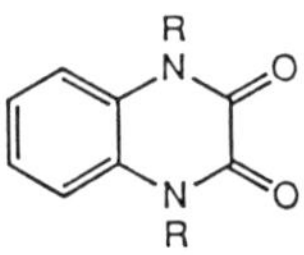

59; R = CH_2OCOMe or H

Quinoxaline-2,3(1*H*,4*H*)dione (**60**) exists mainly in the keto form in organic and neutral aqueous solvents [252]. Deprotonation (pka = 9.7 by spectrophotometric measurements) of the semienol form **61** generates anion **62** which at increasing pH seems to be stabilized via rearrangement to an intramolecularly hydrogen bonded form **63** [252]. How various tautomeric forms might relate to biological activity is unknown, but such information likely will become of increasing importance. Related substituted analogues are effective coccidiostats [253], have orally active sedative-hypnotic properties [254,255], serve as precursor for triazoloquinoxalines which possess antidepressant and antifatigue activities [256], and are catabolic intermediates formed during riboflavin degradation [257]. Model reactions involving alkaline ferricyanide oxidations of related dioxopiperazines mimic these enzymatic degradation steps [257].

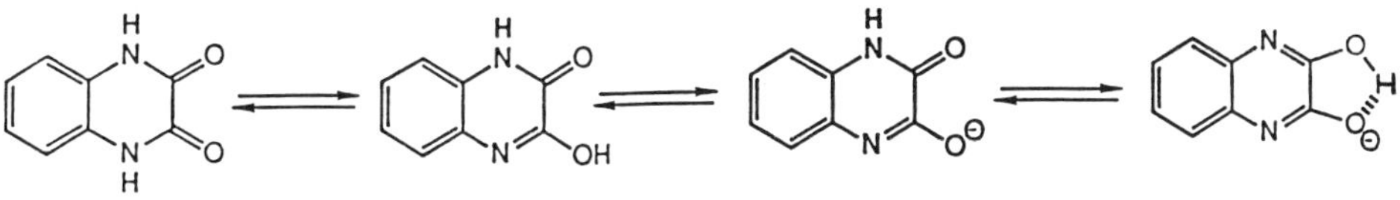

60 61 62 63

Appropriately substituted quinoxaline-2,3[1*H*,4*H*]diones also have significance in the excitatory amino acid (EAA) area [258,259]. Antagonists for central nervous system EAA such as glutamate are of contemporary interest [260–262]. At least three glutamate receptor subtypes are known. The N-methyl-D-aspartate (NMDA) receptor subtype has several classes of antagonists [263]. However, 6,7-dinitroquinoxaline-2,3(1*H*,4*H*)dione (DNQX,FG9041) and its 6-cyano-7-nitro analogue (CNQX,FG9065) are potent and competative antagonists at non-NMDA glutamate receptors [258,259]. These compounds inhibit [^{3}H]α-amino-3-hydroxy-5-methyl-4-isoxazole propionic acid binding to quisqualate receptor subtypes at 0.5 and 0.3 μM concentrations, and [^{3}H]kainate binding at 2.0 and 1.5 μM concentrations, respectively [259]. Such analogues also block the excitatory action of quisqualate and kainate, but not NMDA, on spinal neurons [259].

2.2.3 Chemistry of potential medicinal utility

Additional 2,3-dioxopiperazine syntheses of possible use to Medicinal Chemists follow: Photochemical rearrangement of the electron poor, vicinally dihalogenated 1,4-dihydropyrazine-2,3-dione **64** affords **65** (with thermal reversibility), and in the presence of H_2O generates **67** via tetrone **66** [264]. The lumiflavin **68** is converted into the spiro(imidazo-4,2-quinoxaline) **69** below pH 8.5, but between pH 8.5–10.5 hydrolysis generates quinoxaline-2,3-dione **70** accompanied by a small amount of spiro compound **69** [265]. In strong base formation of **70** is accompanied by a trace of tentatively characterized quinoxalinone **71**. The kinetic pK_{app} for the 4a-hydroxyl proton in **68** is 9.9 as determined by analysis of its pH-dependent hydrolysis.

N,N^1-Disubstituted 2,3-dioxopiperazine dioximes **72**, available from ethylenediamine derivatives and dichloroglyoxime, serve as precursors for N,N^1-disubstituted furazano- and furoxano[3,4-b]piperazines **73** and **74** [266]. Additionally, 3,4-diaminofurazan and glyoxal con-

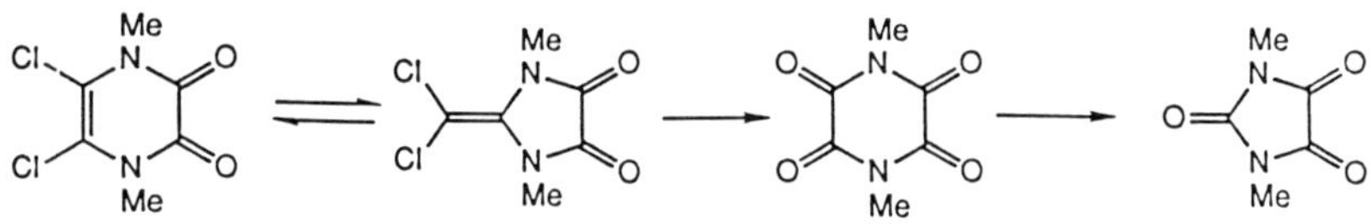

64 65 66 67

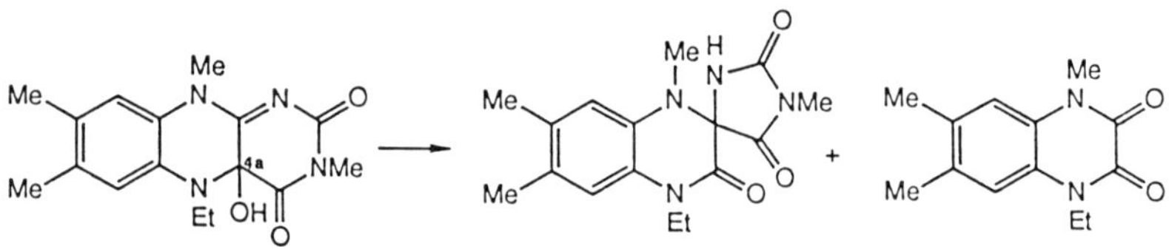

68 69 70

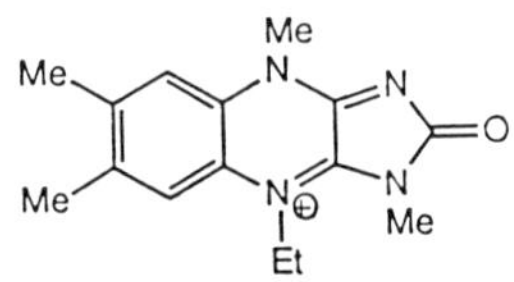

71

dense in acidic solution to produce tetracyclic analogue **75**. Reaction of α-D-glucopyranoside **76** with diethyl oxylate yields the 2,3-dioxo-piperazine derivative **77** which upon deblocking produces 2,3-diamino-2,3-dideoxy-2,3-N-oxalyl-D-glucose (**78**). Dioxo ring cleavage occurs in alkali and under acetylating conditions [267]. The medicinal potential of such chemistry seems not to have been explored in any detail.

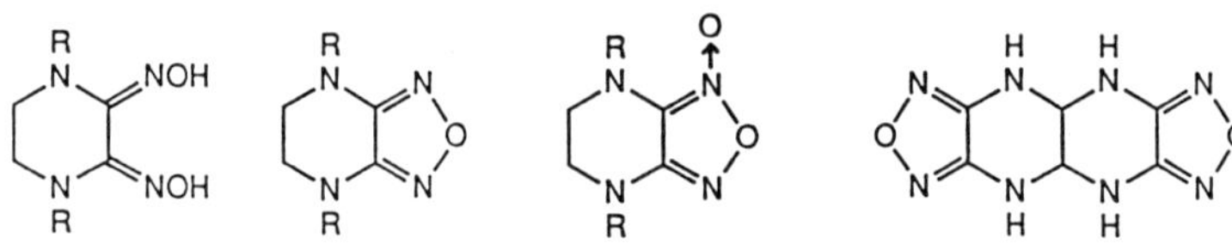

72 73 74 75

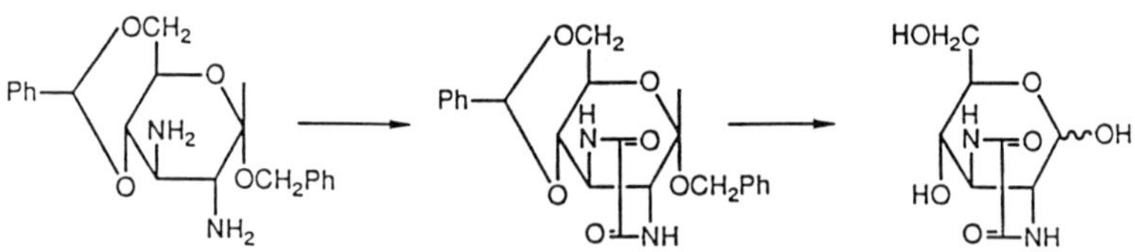

76 77 78

Piperazinetetrone **79** may be viewed as a composite of two α-dicarbonyl subunits; i.e. a cyclic diimide. Such materials have interesting photochemical properties [268], and rearrange to 4,5-dioxoimidazolidine-2-carboxylic acids upon heating in alcoholic solvents [269]. N,N^{1}-dimethylpiperazinetetrone (**80**) undergoes decarbonylation upon ultra-violet irradiation to afford N,N^{1}-dimethylimidazolinetrione (**81**) [264]. Irradiation, through a Pyrex filter with a high pressure mercury lamp, of N,N^{1}-dialkylpiperazinetetrones **82** in acetonitrile under argon produce cyclization products **83** and intramolecular redox products **84** [268].

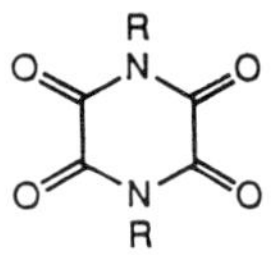

79

80 **81**

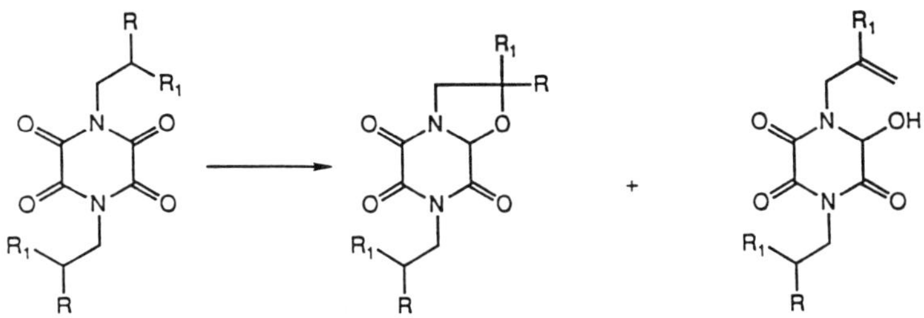

82a; R = R_1 = H
b; R = Me, R_1 = H
c; R = R_1 = Me

83a; R = R_1 = H
b; R = Me, R_1 = H
c; R = R_1 = Me

84a; R = Me, R_1 = H
b; R = R_1 = Me

3 2,5-Dioxopiperazines (2)

Numerous natural products contain the 2,5-dioxopiperazine (cyclodipeptide) ring. Their source, synthesis, biogenesis, and biological and physical-chemical properties have been reviewed into the early 1970's [270]. Theoretical [271–274] and spectroscopic [NMR, circular dichroism (CD)] [275] investigations of 2,5-dioxopiperazine ring conformational preferences continue to receive attention. These 6-membered heterocycles adopt a multitude of planar to highly puckered conformations in the solid state [273]. For N-methylated cyclic dipeptides, X-ray analysis [276] indicates that the optically active isomers *cyclo*-(N-methyl-L-Ala)$_2$ (or its racemate), *cyclo*(N-methyl-L-Val)$_2$, and *cyclo*(N-methyl-L-Phe)$_2$ adopt boat conformations with axial substituents on α carbons. The meso isomers, *cyclo*(N-methyl-L-Val-N-methyl-D-Val), and *cyclo*(N-methyl-L-Phe-N-methyl-D-Phe), assume chair conformations with axial substituents on α carbons [276]. 2,5-Dioxopiperazines, both in the solid state and in solution, serve as models for studies of polypeptides and proteins [276]. Theoretical explanations and helicity rules for chiroptical properties of 2,5-dioxopiperazines in context of other organic compounds are available [277]. CD provides the absolute conformation of the dioxopiperazine ring, but no information about the absolute configuration at the centers of chirality [277]. In the NMR, [^{13}C] carbonyl signals of cyclodipeptides show a 1.5–4.0 ppm upfield shift relative to the corresponding polypeptides; [^{15}N] NMR signals do not show such a relationship [278]. Substituent and neighboring group effects in the [^{15}N] spectra of cyclodipeptides differ from those of polypeptides [278]. Trifluoroacetyl derivatives of 2,5-dioxopiperazines are recommended for mass spectral analysis [279,280] since they are easier to prepare and are less susceptible to hydrolysis than trimethylsilyl derivatives [281]. Fragmentation pattern analyses are based on parent ion degradations; each compound within a series has a similar mass spectrum [279,280], but differences in observed spectra permit identification of individual dioxopiperazines.

3.1 Cyclodipeptide sources

Cyclodipeptides not only abound in nature, but often are unwanted side reaction products [282,283], or degradation products of oligo- and polypeptides. The optically active pyrolysis product **85** is formed by the degradation of Majusculamide B during isolation of the lipophilic constituents of the blue-green alga *Lyngbya majuscula* (Gomont) [284]. Alternatively, the structurally related natural product (2*S*,5*S*)-2,5-bis(benzyl)piperazine (**86**), isolated from the Mexican wild citrus relative *Zanthoxylum arborescens* [285] and available from optically pure 2,5-dioxopiperazine precursors via $LiAlH_4$ reduction [285], likely is biogenetically related to the L-phenylalanine-derived dioxopiperazine **87** by N-methylation and reduction [286]. (±)-*cis*-3,6-Dibenzyl analogue **88** is available via hydriodic acid or catalytic reduction of the corresponding dibenzylidine compound [287]. The trans isomer of **88**, several related analogues, and optically active (*S*, *S*)-**89** are known [287].

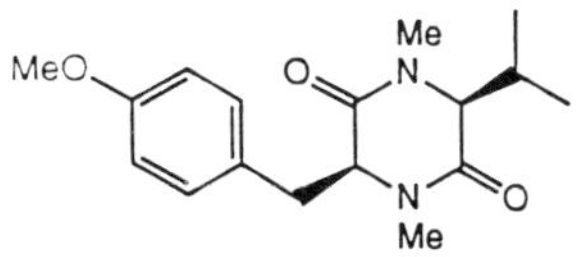

85

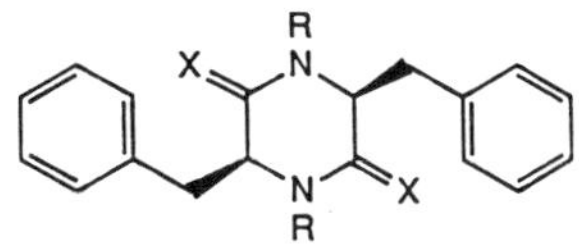

86; X = H_2, R = Me
87; X = O, R = H
88; X = O, R = Me
89; X = O, R = Ac

2,5-Dioxopiperazines generated by thermal degradation of several actinomycins can be identified by GLC and/or GLC-mass spectrometry. Pyrogram analysis provides an approach for sequencing the amino acid side chains of individual actinomcycin antibiotics [288]. Dioxopiperazine standards are prepared by cyclization of dipeptides or their esters. Under relatively mild conditions an orderly sequential degradation of the actinomycins occurs which precludes formation of overlapping dipeptide fragments.

The dioxopiperazine 5-benzyl-3,6-dioxo-2-piperazinylacetic acid (DOP) is one of the decomposition products of the dipeptide sweetner α-L-aspartyl-L-phenylalanine methyl ester·HCl (USAL, Aspartam hydrochloride). This pH- and temperature-dependent decomposition process involves demethoxylation and cyclization of the sweetner [289,290]. The optimum pH for sweetner stability is 2.5–4.5 [290]. GLC methodology is useful for analysis of resulting mixtures [291]; currently, the acceptable daily intake for DOP is 7.5 mg/kg body weight [292].

Propensity of tripeptides to undergo dimerization during attempts to generate 9-membered cyclotripeptide rings may be attributable to a favorable arrangement of two hydrogen-bonded peptide chains aligned in an antiparallel geometry so that the two pairs of chain ends are positioned for cyclodipeptide ring closure [293,294]. The cyclotripeptide, *cyclo*$(\text{L-Pro})_3$, may be synthesized in high yield since flexibility of the peptide backbone is reduced by consecutive proline residues [294–296]. Optically active *cyclo*$(\text{N-methyl-L-Ala})_3$ (25 %), the D, L, L-diastereomer (12 %), and the anticipated 2,5-dioxopiperazine (39 %) are generated from the linear tripeptide 2,4,5-trichlorophenyl ester with 4-dimethylaminopyridine serving as catalyst [294]. Tripeptides with secondary amide groups do not form cyclotripeptides **90** likely owing to equilibrium with unstable azacyclols **91**. Such azacyclols can be prepared from N-protected aminoacyl 2,5-dioxopiperazines **92** under well-controlled conditions [297]. Tripeptides having the Pro-X-Pro sequence **93** afford stable azacyclols (**94,95**) [294]. Cyclotripeptides **96** are formed via transannular isomerization. NMR analysis reveals these substances to be in equilibrium with azacyclols in polar solvents, but in addition cyclotripeptides may be isolated from nonpolar solvents. The all L *cyclo*(Pro-Ala-Pro)$_3$, *cyclo*(Pro-Val-Pro)$_3$, cyclo (Pro-Leu-Pro)$_3$ and cyclo (Pro-Phe-Pro)$_3$ are known, but the later crystallizes as the azacyclol [294].

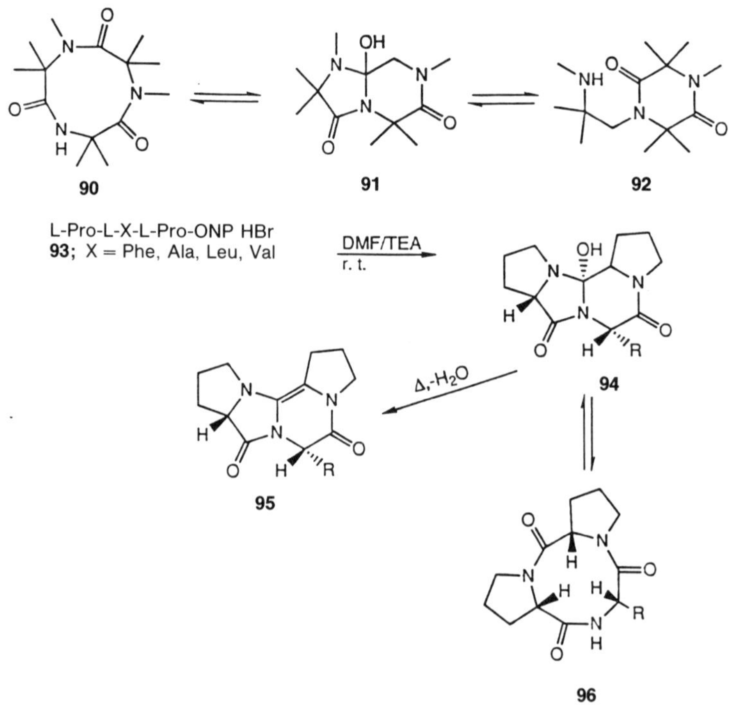

3.2 Importance to agriculture

Alkylating 2,5-dioxopiperazines are of interest because of their antitumor properties [298–302]. However, compounds containing the 2,5-dioxopiperazine structure have many other medicinally significant activities, and their toxicity can be important to agriculture. Maize meal cultures of *Aspergillus ustris* (Bainier) can cause acute toxicosus in day-old ducklings [303]. Six toxic dioxopiperazine analogues (**97–102**) are found in extracts of these cultures. 2*S*,9*S*,12*R*-12,13-Dihydro-

12-hydroxyaustamide (**102**), a relative of previously isolated austamides (**97–101**), is the most recently isolated toxin from maize meal infected with *A. ustris* [304]. Structural relationships to L-tryptophan and L-proline are evident from comparative analysis of CD spectra.

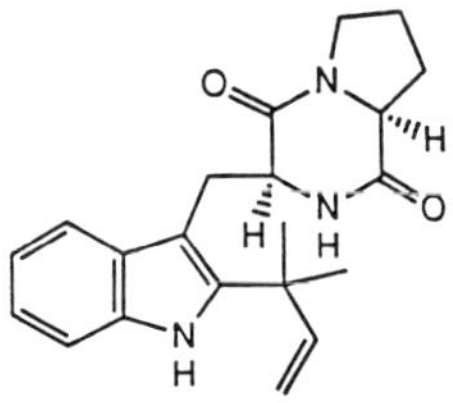

97

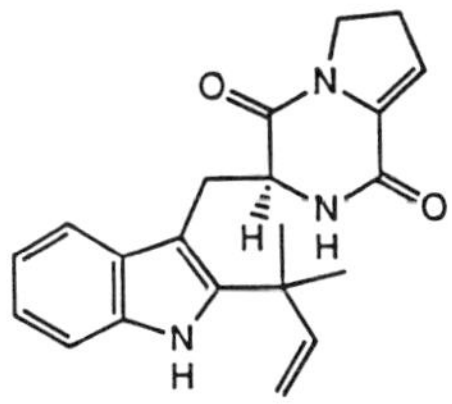

98

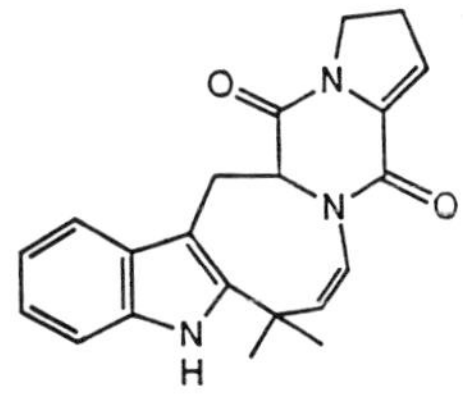

99

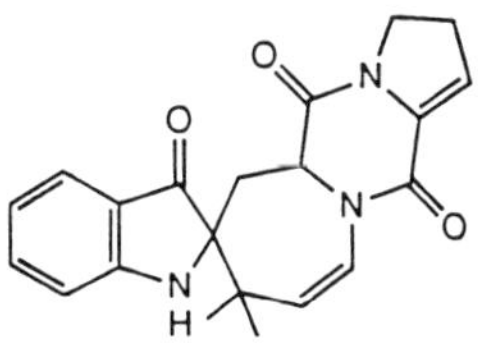

100 (austamide)

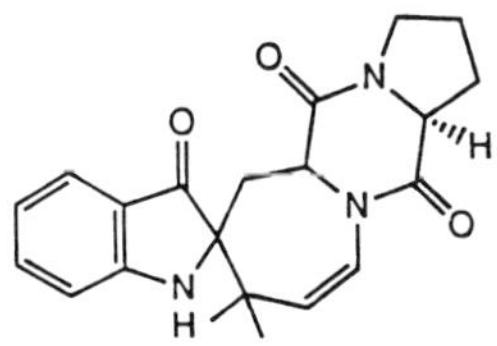

101 (12,13-dihydroaustamide)

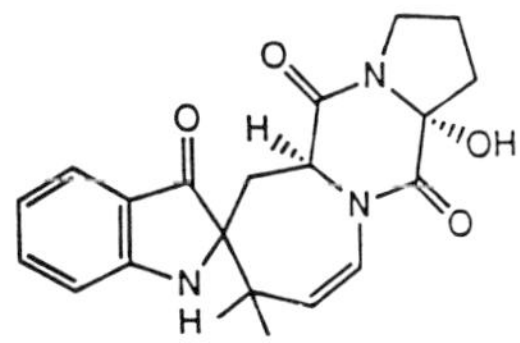

102

It is not known whether mycotoxins produced by *Cheatomium* species are significant in outbreaks of foodborn diseases of humans and animals, but because of their ubiquity in deteriorated plant materials (foods and feeds) they may represent a health risk [305]. Two of many mycotoxins, chaetoxin (**103**) and chetomin (**104**) are 2,5-dioxopiperazine containing compounds [305]. Screening for mycotoxin production by *Chaetomium* species and related fungi on rice culture has involved cytotoxicity tests using HeLa cells [305]; their production and cytotoxicity data have been reviewed [305,306].

Chetomin (**104**) is an epidithiodioxopiperazine produced by many *Chaetomium* species [305,307,308], for example by *C. cochliodes* isolated from soil samples collected in Nova Scotia [309]. Elucidation of the linkage between the quaternary *β*-indoline carbon and indole N was aided by [^{15}N]- and [^{13}C]-NMR spectroscopy [310]. The biological properties of this natural product are consistent with the hypothesis [311] that antibiotic production by soil fungi is a factor in the etiology of ovine ill-thrift [312]. Ill-thrift is characterized by a decline or cessation of growth by grazing lambs, a condition which coincides with gross changes in rumin metabolism. The chetomin disulfides are reduced to the tetrathiol compound **105** under the low redox conditions characteristic of rumin contents. This tetrathiol is bacteriostatic in vitro against five strains of functionally important rumin bacteria. The antibiotic stimulates acetate production from glucose by several organisms at the expense of the production of desirable soluble fermentation end products.

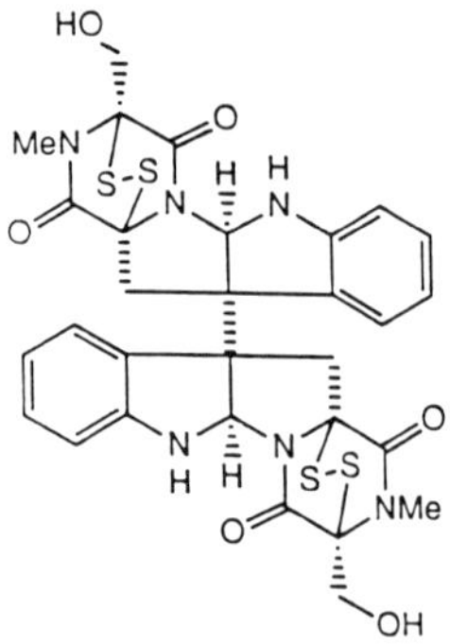

103

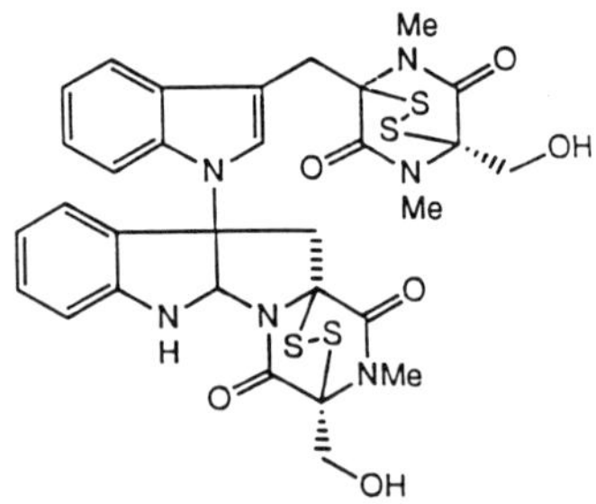

104

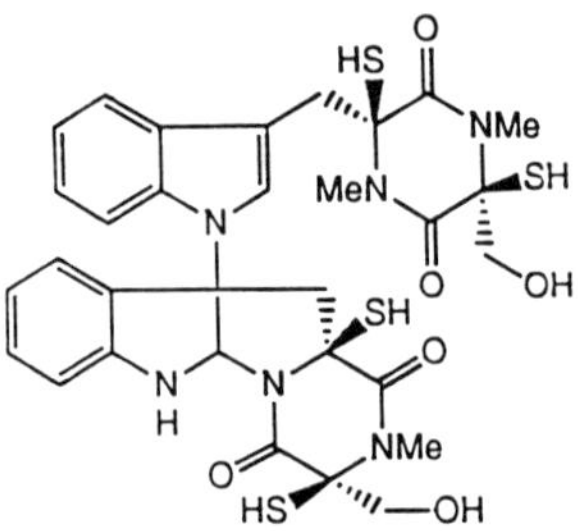

105

3.3 Pharmacological properties

3.3.1 Cholinesterase inhibition

The 2,5-dioxopiperazines represent a rich source of pharmacologically interesting compounds. For example, mycelianamide (**106**), an antibiotic isolated from *Penicillium* species No. C-81, competitively inhibits (ki = 4.0×10^{-6} M) butyrylcholinesterase isolated from *Pseudomonas polycolor* [313]. Mycelianamide also inhibits hog liver carboxylesterase (Ki = 3.3×10^{-5} M) and horse serum cholinesterase (Ki = 1.9×10^{-4} M), but not *Pseudomonas* or electric eel acetylcholinesterase.

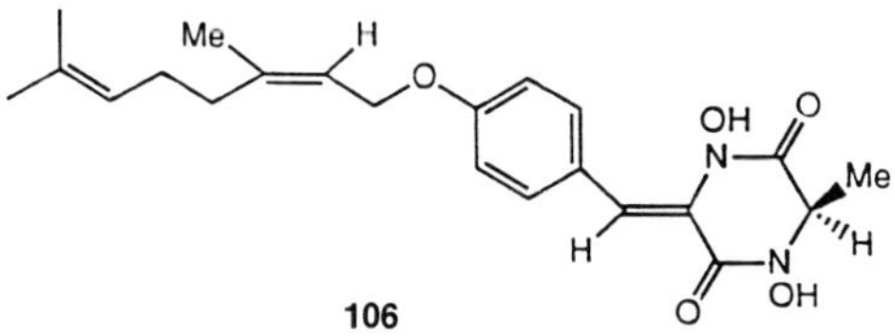

106

3.3.2 Ergopeptide-related compounds

Medicinal properties of ergopeptides and derivatives related to **107** are of interest; such compounds inhibit prolactin secretion [314,315]. Lysergic acid diethylamide (LSD) and many other analogues including lysergic acid (**108**)-related cyclic peptides ergocornine (**109**) and 2-bromo-α-ergocryptine (**110**) cause reductions in systemic levels of prolactin, and display antitumor activity in rats [315–320]. Dioxopiperazine-related hemiketal analogues **115** and **116** of ergocornine (**109**) are prepared from acyl chloride **111** and optically pure α-amino acid species **112** and **113** [315]. Hydrogenation of intermediate **114** affords targets **115** and **116**, but unfortunately these compounds are inactive in L1210 and P388 leukemic mice. Ergocornine also is only marginally active against leukemia P388.

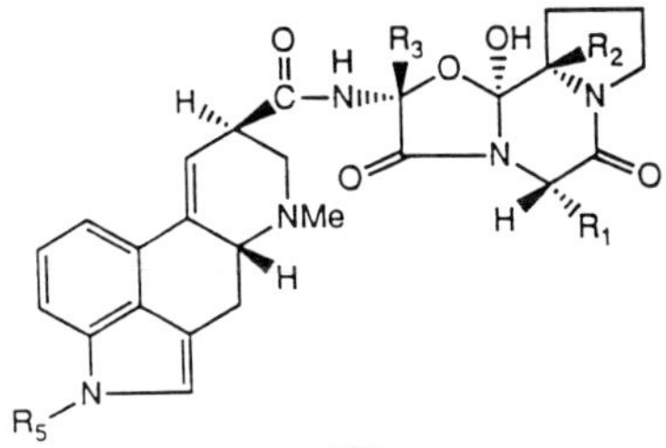

107

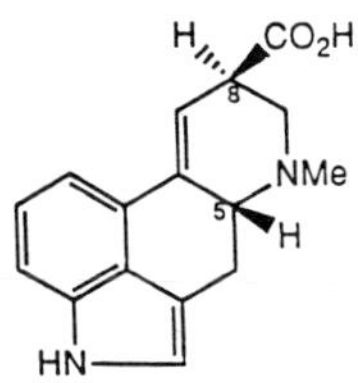

(+)-**108**
5(R), 8(S)

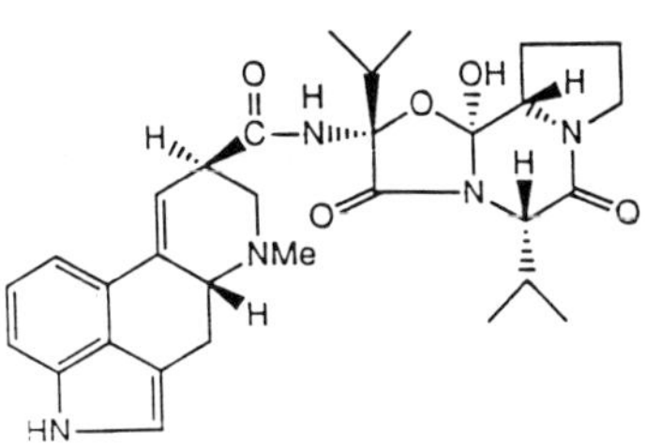

109

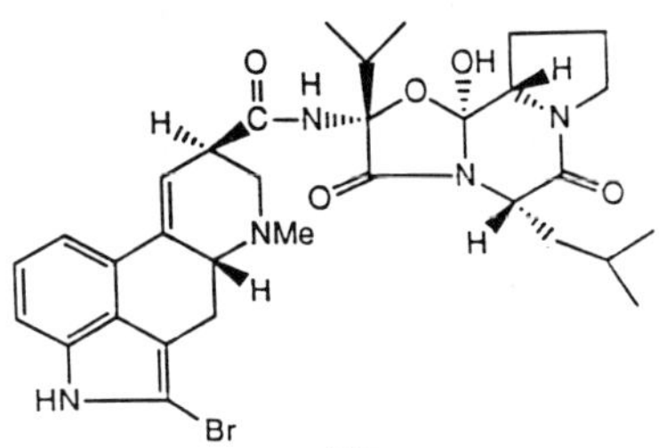

110

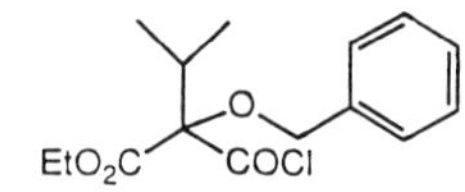

111

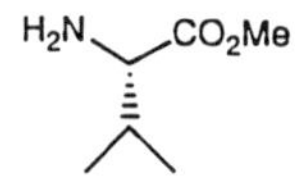

112

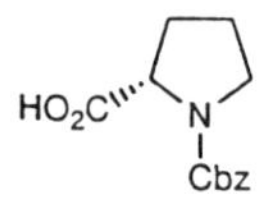

113

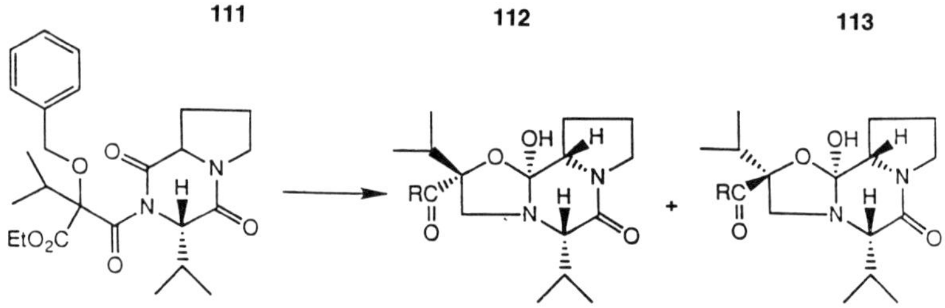

114 **115** **116**

3.3.3 Platelet-activating factor (PAF) inhibitors

2,5-Dioxopiperazine rings are found in a new class of platelet-activating factor (PAF) inhibitors [321,322] which may find use in allergic and inflammatory disorders. Three specific inhibitors of PAF, namely, FR 49175 (**117**), FR 106969 (**118**), and FR 900452 (**119**), which are isolated from the culture filtrates of *Penicillium* molds, contain 2,5-dioxopiperazines as a common structural element [321–325]. Studies in small animals suggest that two receptor types may be involved, platelet-mediated PAF-induced bronchoconstriction and vascular-mediated hypotension accompanied by increased vascular permeability [325,326]. Platelet-mediated bronchoconstriction is selectively inhibited by analogue **117** (FR 49175). The ED_{50} = 0.26 mg/kg following iv. administration to guinea pigs. The vinylogous dioxopiperazine **119** (FR 900452) significantly inhibits hypotension (57 %), PAF-induced vascular permeability (52 % at 1 mg/kg iv.) and bronchoconstriction in rodents [325]. At 100 mg/kg twice daily for 6 days, this analogue also significantly reduces urinary protein loss in puromycin-induced nephrotic rats [326].

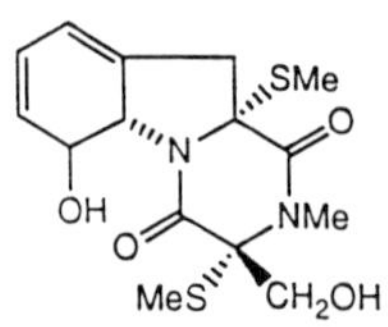

117 (FR 49175)

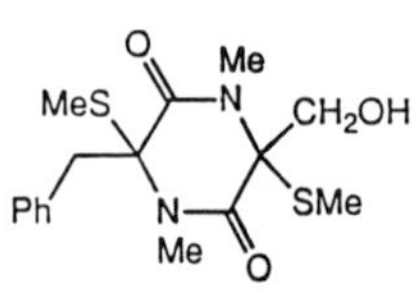

118 (FR 106969)

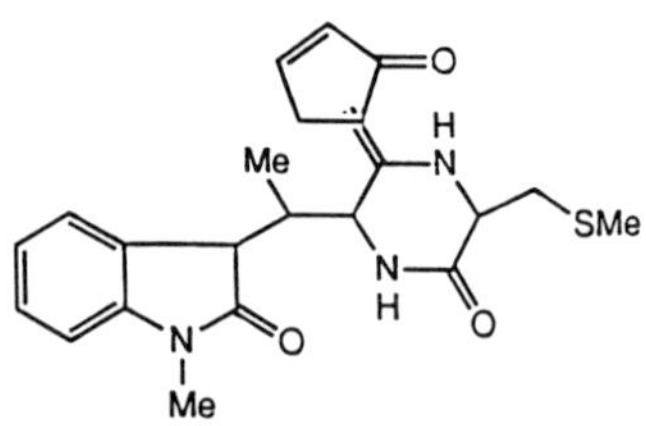

119 (900452)

Semisynthetic analogues **120–126** also possess PAF-inhibitory activities [327,328]. Of these, dioxopiperazine **124** (X = OMe, R = Ph; FR 65049) is a specific PAF inhibitor with an IC_{50} value of 0.69 μM. Naphthalene-substituted piperazine **120** [R = (*R*)-Me, X = $CH_2{=}CH$ with 3(*R*),6(*R*) configuration] is the most potent (IC_{50} = 0.18 μM) of the series against PAF-induced rabbit platelet aggregation, and does not block arachidonic acid- or DAP-induced aggregation. For N-methylindole structures of the type **120** (X = NMe), 3(*S*),6(*R*) configurations (*cis*) are more potent (IC_{50} = 7.9 μM) than those possessing the *S,S* (IC_{50} = 142 μM) or *R,R* (IC_{50} = 36 μM) configurations. The 3(*R*),6(*S*) enantiomer is virtually inactive (IC_{50} > 250 μM). Indole **126** may have commercial possibilities [328].

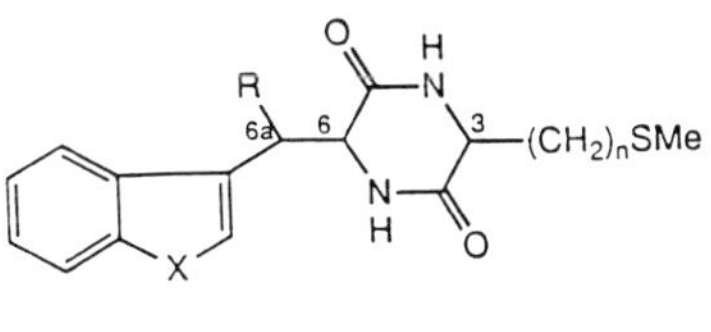

120

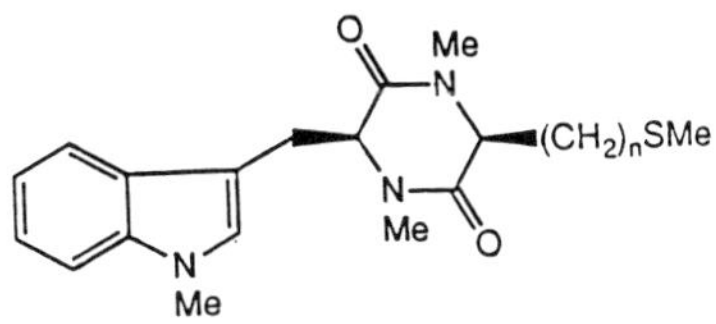

121

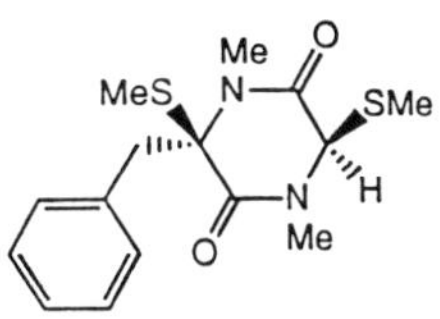

122

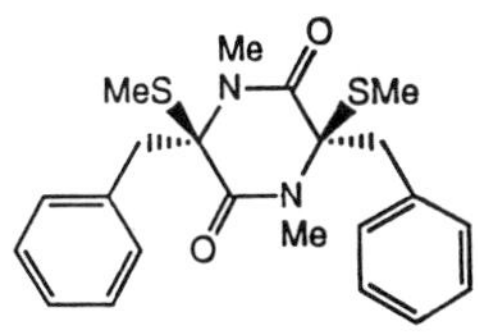

123

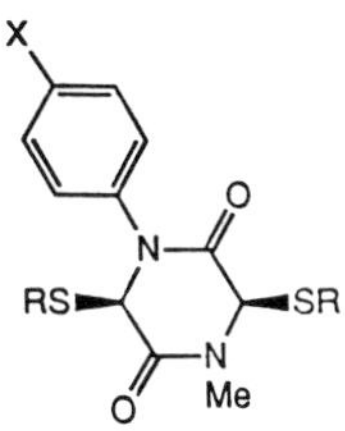

124

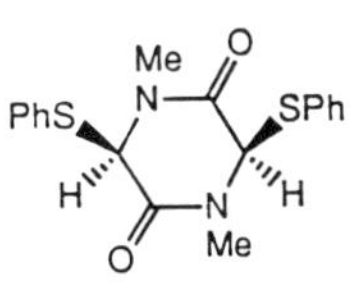

125

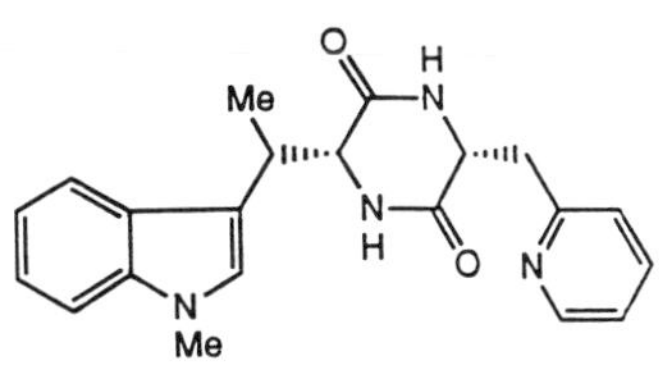

126

3.3.4 Thyrotropin-releasing hormone metabolite

A histidyl-proline dioxopiperazine metabolite of thyrotropin-releasing hormone is found in rat brain [329–331]. This cyclodipeptide inhibits ($Na^+ + K^+$)-ATPase [332] and protactin release [333,334]. In the central nervous system this unique compound elevates cGMP concentrations [335], produces hypothermia [336], and antagonizes ethanol-induced sedation [330]. Non-competitive inhibition of dopamine uptake in rat brain striatal synaptosomes is a function of $(Na^{+}+K^{+})$-ATPase inhibition by the cyclodipeptide [332]. This inhibition does not involve the ouabain binding site [332]. Rather, this compound inhibits at the vanadate (V^{+5}) binding region, a low affinity binding site for ATP which seems to be located on the inner surface of the plasma membrane [337].

3.3.5 2,5-Dioxopiperazine-containing analgetics

As analgetics, 2,5-dioxopiperazines are of less interest. Enkephalin-related 2,5-dioxopiperazine-containing peptides have reduced opiate-like activity or are completely inactive [338].

3.3.6 Cyclodipeptide metabolites

2,5-Dioxopiperazines also occur as metabolites of peptide drugs. Metabolites **128** and **129** are found in the blood and urine of humans and other animals [339] following oral administration of CV-3317 (**127**), a synthetic angiotensin converting enzyme inhibitor [340] having antihypertensive activity [341]. The major metabolite in serum is the hydrolyzed ester of **127** followed by dioxopiperazines **128** and **129**. In urine, hydroxylated species **129** is the major metabolite, followed by the hydrolyzed ester, the parent drug, and dioxopiperazine **128** [339].

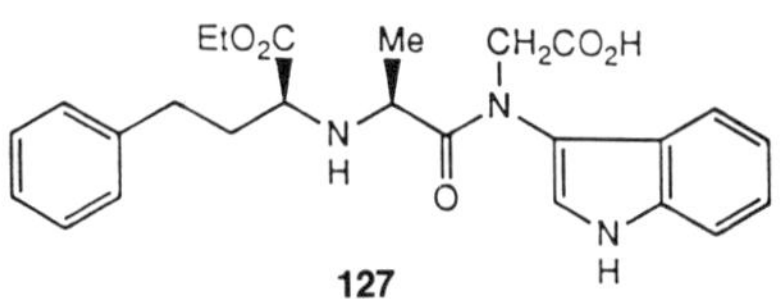

127

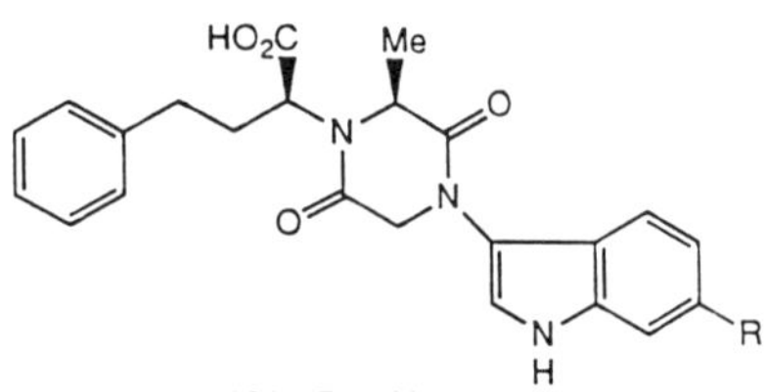

128; R = H
129; R = OH

3.3.7 Bacteriostatic N-chloramine analogues

In a large series of N-chloramines of low chlorine potential, comparative investigations reveal that bacteriostatic activity is decreased as the nitrogen-chlorine-bond polarity is decreased [342]. N-chloramines containing the more polar nitrogen-chlorine bond, as is the case for dioxopiperazine derivative **130**, are most susceptible to hydrolytic cleavage and provide for a pH-dependent antimicrobial activity against Gram-positive and Gram-negative organisms. These compounds have antimicrobial activity owing to Cl^+ donation to receptors [342]. The driving force for loss of Cl^+ is thought to be a result of anion stabilization by delocalization into the neighboring carbonyl group [343]. Ultraviolet photoelectron spectral analysis indicates that conformation **130**, wherein the chlorine is axial and the nitrogen lone pair is equatorial, is most stable. Non-bonded interactions between oxygen and chlorine are minimized in this form [343].

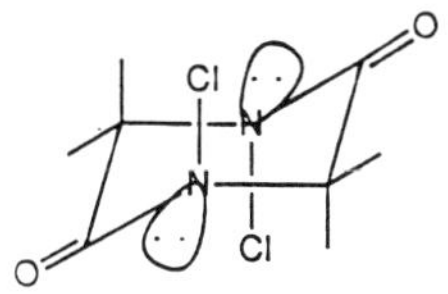

130

3.4 Simple cyclodipeptide syntheses

Syntheses for simple 2,5-dioxopiperazines from respective amino acids are available [344–348]. These materials are formed by thermal dehydration of α-amino acids [349], as byproducts of α-amino acid ester syntheses [349], rhodium chloride catalyzed cyclization of α-amino acid esters [350], or when heating dipeptide esters in toluene or free acids in phenol [348,351]. N,N-diphenyl- and dibenzyl-2,5-dioxopiperazines also are prepared in 64–88 % yield from α-halocarboxamides in the presence of phase transfer catalyst [CH_2Cl_2, 50 % aq.NaOH, Duolite A-109 (Cl-form) [352]. Fragmentation of tri- and tetrapeptides under alkaline conditions also yields cyclodipeptides [353,354]. During removal of benzyloxycarbonyl protecting groups from dipeptides containing α-aminoisobutyric acid (IBA) residues, only the corresponding dioxopiperazines are formed [355]. Possibly, this is related to geminal dialkyl affects of AIB moieties thereby forcing reacting functions to be in close proximity. Tripeptides containing AIB residues readily undergo cyclization during deprotection [356]. Also available by synthesis are the 7- and 8-thia-1,4-azabicyclo[4.3.0]nonane systems **131** and **132** [357]. Polycyclic dioxopiperazines **133–135** are synthesized from optically active thiazolidine-4-carboxylic acids by treatment with $MeSO_2Cl$ in dry pyridine [358]. Owing to facile methodologies, relatively simple, but interesting dioxopiperazines such as **136** are easily produced; at 0.001% concentrations, inhibition in vitro of *Botrytis cinerea* and *Fusarium culmorum* spore germination is observed [359].

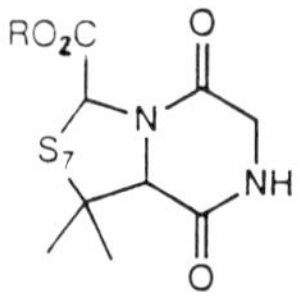

131

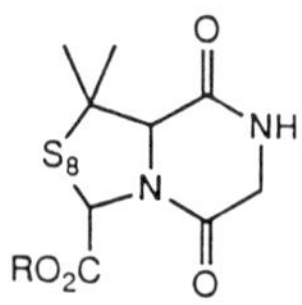

132

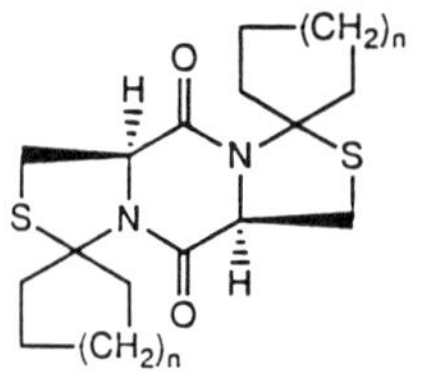

133

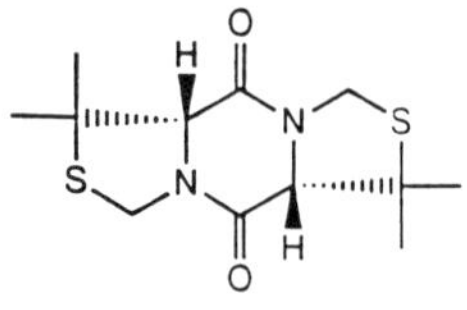

134

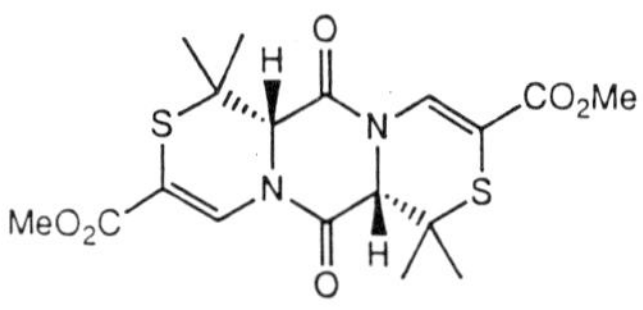

135

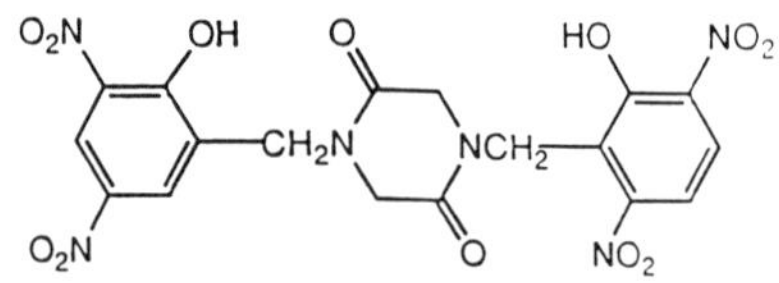

136

Hydroxycyclodipeptides **137–139** are readily synthesized from pyruvoyl amino acid amides [360]. Cyclization of an optically pure proline precursor affords kinetically controlled product **137** [3(*R*),8a(*S*)] with high asymmetric induction. Under aqueous acidic or basic conditions this isomer rearranges with high optical induction to the thermodynamically more stable species of opposite absolute configuration (*S*) at C(3). Under Lewis acid catalysis, thioacetic acid or ethyl mercaptan may be employed in place of hydroxy groups to afford optically active 3-acetylthio- or 3-ethylthiodioxopiperazines.

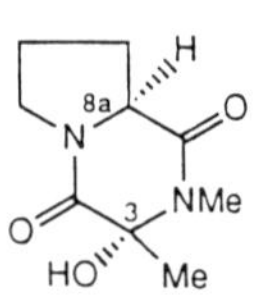

137

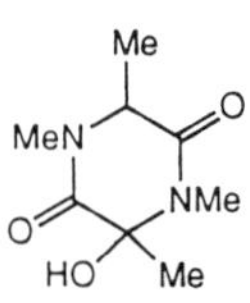

138

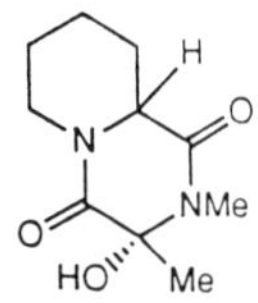

139

In other studies, 2,4-dioxopiperazine **140** serves as precursor for the spiro-oxolane **141**, which in turn is converted to trione **142** or hydroxydiones **143** and **144** [361]. Autoxidation of racemic *cis*- or *trans*-2,5-dioxopiperazines **145** afford racemic 2,3,5-trione **147**, but oxidation of a racemic cis-trans mixture of **146** with elemental sulfur produces racemic (*Z*)-2,5-dioxopiperazine **148** [362].

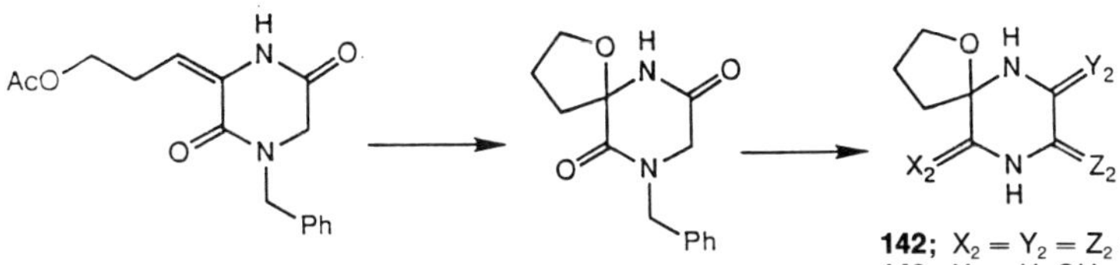

140 **141**

142; $X_2 = Y_2 = Z_2 = O$
143; $X_2 = H, OH$; $Y_2 = Z_2 = O$
144; $X_2 = Y_2 = O$; $Z = H$

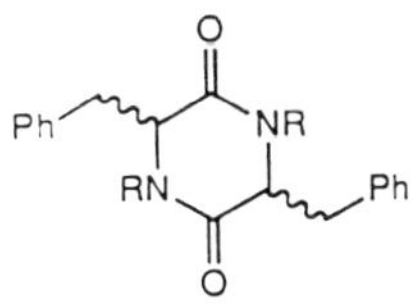

145; R = Me
146; R = Ac

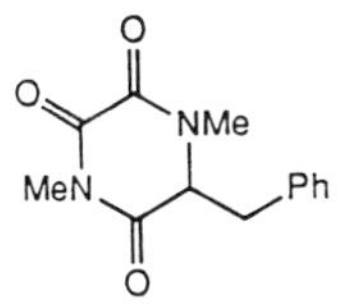

147

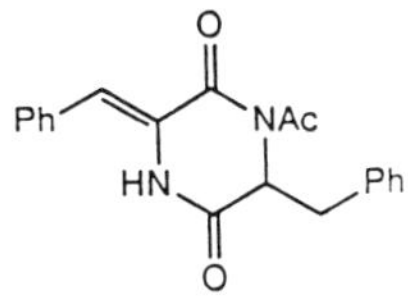

148

Finally, a multi-step method of producing 1,4-dihydroxy-2,5-dioxo-piperazines (**150**) and the Raney-Ni reduced parent structure **151** from 3,6-dialkyl-2,5-dichloropyrazines (**149**) is available [363]. Polar molecules having multiple hydrogen-bonding possibilities such as **150** may serve as useful adjuncts in drug development.

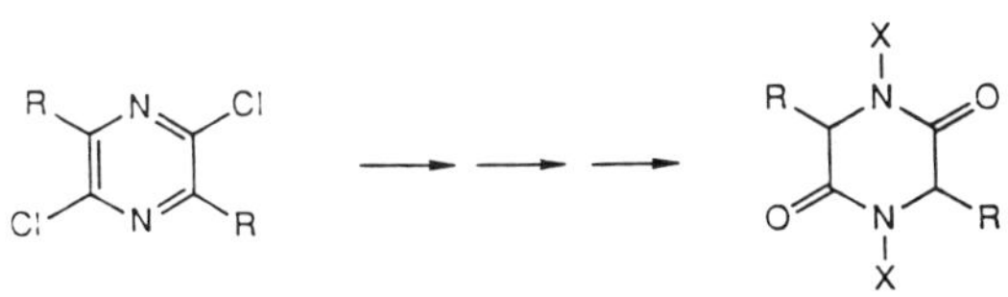

149

150; X = OH
151; X = H

3.5 Cyclodipeptides in amino acid asymmetric synthesis

Not only do amino acids serve as precursors for optically active and meso cyclodipeptides, but reverse is also possible. Efficient asymmetric syntheses of α-amino acids are available by hydrogenation of α,β-dehydroamino acids in cyclodipeptides [364,365]. Thus, hydrogenation (Pd black) of **152**, prepared from cyclodipeptides and aldehydes, followed by hydrolysis produces L-amino acids **153**. Chiral inductions in cyclodipeptides generated by catalytic hydrogenation of cyclic dehydrodipeptides are 84–98 % [365].

152 153

3.6 Naturally and related unnaturally occurring 2,5-dioxopiperazines: chemistry and biology

The 2,5-dioxopiperazine heterocyclic system is found in many unique natural products. For example, the N-methyltrichloroleucine cyclodipeptide derivative **154** is isolated from a variant of *Dysidea herbacea,* a dictyoceratid sponge, collected near Gladstone, Queensland/Australia [366], and belongs to a family of compounds derived from trichloromethyl amino acids [367,368]. The pentacyclic bipolaramide (**155**) is isolated from whole maize cultures of a toxigenic strain of *Bipolaris sorokiniana* (Shoem) [369], and has a structural resemblance to aranotins such as bisdethiobis(methylthio)acetylaranotin (**156**), a metabolite of *Aspergillus terrus* [370]. Bipolaramide likely is formed by epoxidation, rearomatization and cyclization of two phenylalanine residues [371].

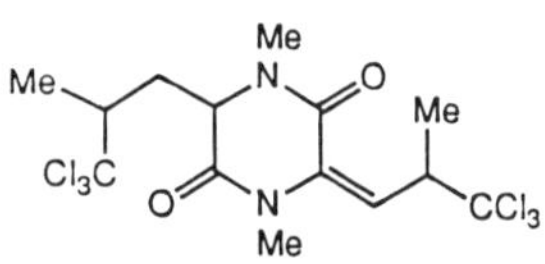

154

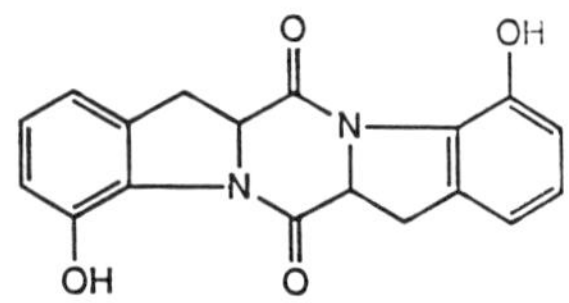

155

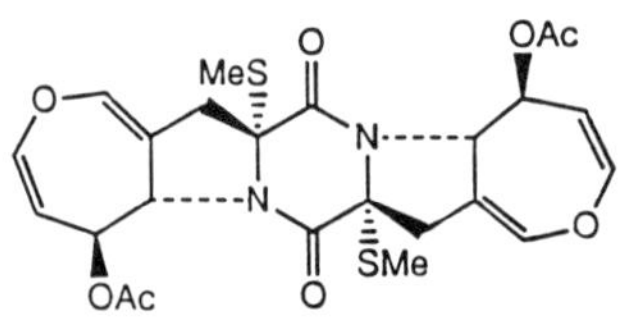

156

3.6.1 Mycelianamide syntheses and related chemistry

Syntheses for N-hydroxydioxopiperazines such as **158**, which is found in many natural products, may be accomplished in good yields from N-(benzyloxy)-α-amino acid amides **157** and pyruvoyl chloride [371].

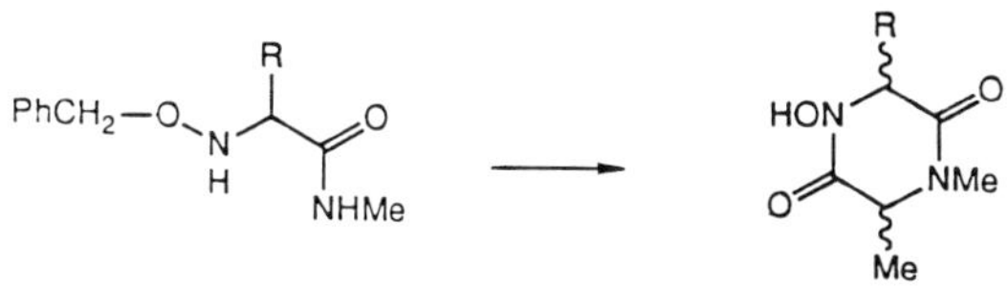

157 **158**

The antibiotic (±)-mycelianamide (**106**) is synthesized using **p**-geranyloxybenzaldehyde and ethyl nitroacetate to produce an intermediate isomeric mixture of α-nitrocinnamic esters (**159**) convertible to the N,N-dihydroxylated dioxopiperazine ring system [372]. The aromatic isomer **161** of degeranylmycelianamide **160** is of interest since 3-arylmethylene analogues of 2,5-dioxopiperazines exist in tautomeric equilibria with their aromatic counterparts [373]. Bis(N-oxides) of the type **161** are prepared by multi-step processes from 3-(4-hydroxybenzyl)-6-methyl-2,5-dioxopiperazines (**162**). Aromatic species **161** differs in NMR and polarity (thin layer chromatography) from **160**, and in neutral protic solvents, acid, or base does not yield the dioxopiperazine (**160**), possibly due in part to the instability of **161**. Acylation of either **160** or **161** gives the same major product (**163**) and an isomer, likely **164** [373].

159; R = geranyl

106; R = geranyl
160; R = H

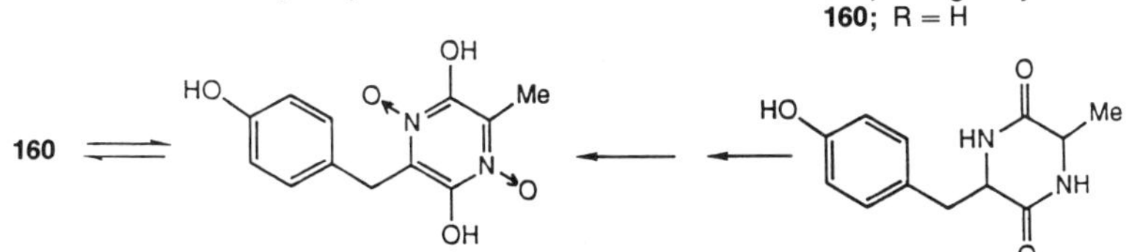

161 **162**

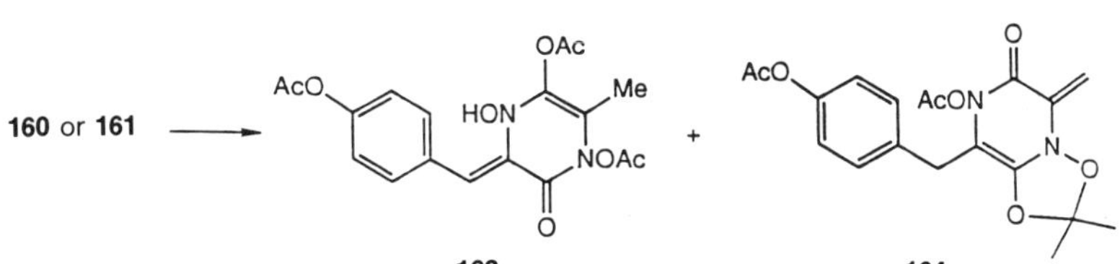

163 **164**

3.6.2 Picrorocillin chemistry

The synthesis of three possible isomers of methylanhydro-picrorocellin (**165–167**), and of the trans isomer of dimethylpicrorocellin (**168**) lead to the conclusion that the lichen (*Roccella fuciformis* D.C.) 2,5-dioxopiperazine, the once isolated but not reisolated picrorocellin, is **169**, rather than **170** [374,375]. Related studies of N,N[1]-substitution effects on $FeCl_3$-catalyzed photo-oxidation reactions of 2,5-dioxopiperazines also support **169** ($R = Me$, $R_1 = H$) as the proposed structure for picrorocellin [376]. References to solid and solution conformational preferences of such synthetic 2,5-dioxopiperazines are available [372].

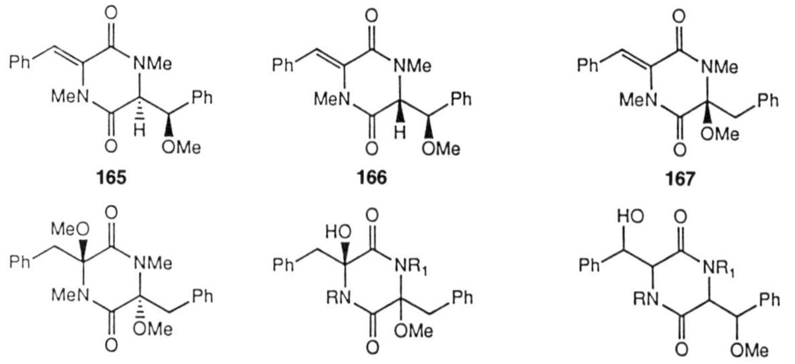

165 **166** **167**

168 **169**; R = H, R_1 = Me or R = Me, R_1 = H **170**; R = H, R_1 = Me or R = Me, R_1 = H

3.6.3 Echinulin, brevianamide, and related chemistry

A particularly interesting synthesis involves the construction of optically acitve echinulin (**174**) in eight steps from the bis(isopentenyl)aniline **171** and dioxopiperazine **173**; the overall yield of **174** from gramine intermediate **172** is 21% [377]. The trans indolyl epimer is obtained from **172** in 33% yield. (±)-Neoechinulin A (*Z* isomer, **177**) and its protected *E* isomer (**178**) are prepared by condensation of N-bis(methylthiomethyl) *cyclo*(Ala-Gly) (**176**) with aldehyde **175**. Deprotection of *Z* or *E* intermediates affords the racemic natural product **177** [378]. Neoechinulin A is also available by DDQ oxidation of a *cyclo*(Try-Ala) dioxopiperazine [379]. Neoechinulin (**179**) [380], and neoechinulins B (**180**) and C (**181**) [381] are also synthesized from the appropriate indol-3-aldehyde and di- or trioxopiperazine. Also, indol-3-pyruvate (**182**) may be converted to didehydrodioxopiperazine

183, an analogue of neoechinulin B (**184**) in nine steps [382]. (−)-Brevianamide E (**187**) and three of its stereoisomers are prepared by photo-oxygenations of (−)-deoxybrevianamide E (**186**) a C(9) epimer prepared from L-proline (**185**) [383].

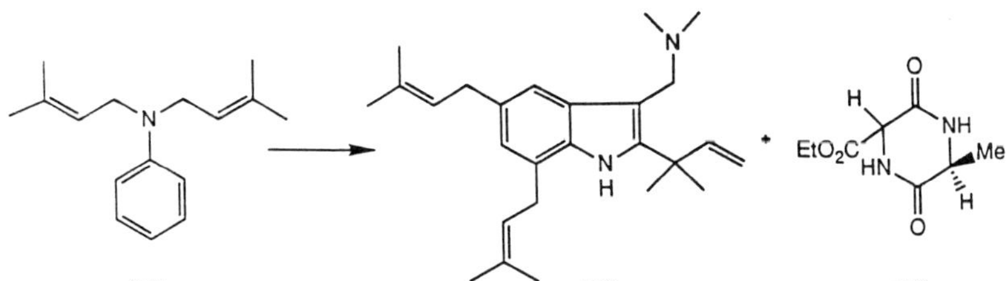

171 **172** **173**

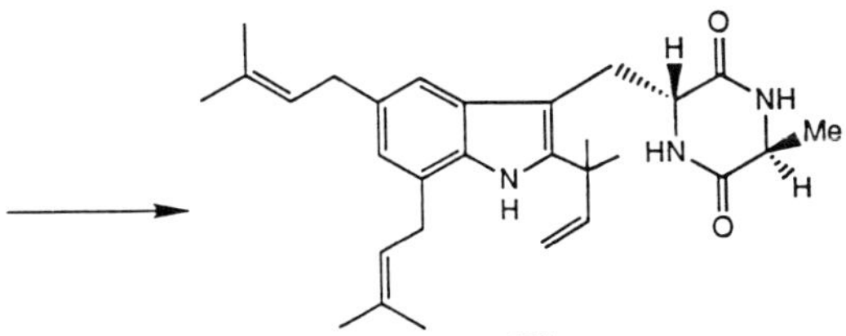

174

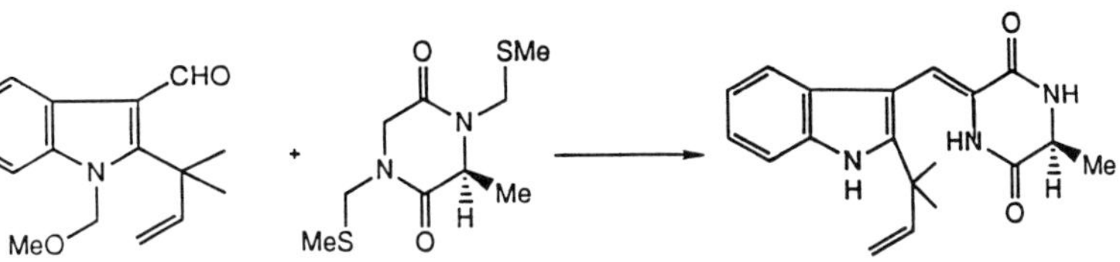

175 **176** **177**

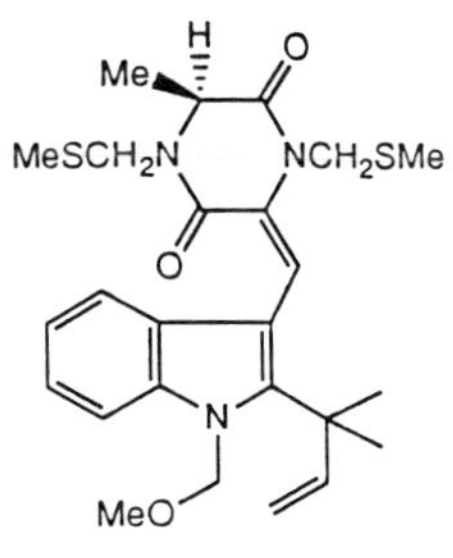

178

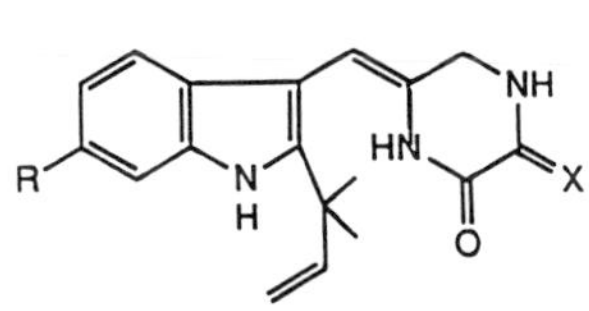

179; R = $CH_2CH = C(CH_3)_2$, X = O
180; R = H, X = CH_2
181; R = $CH_2CH = C(CH_3)_2$, X = CH_2

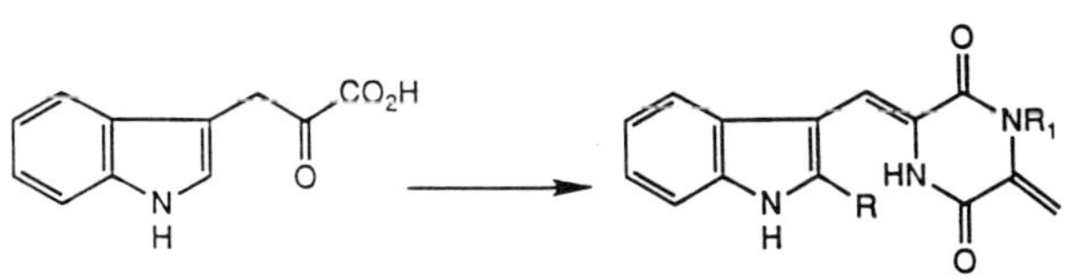

182

183; R = H, R_1 = Me
184; R = $C(CH_3)CH = CH_2$, R_1 = H

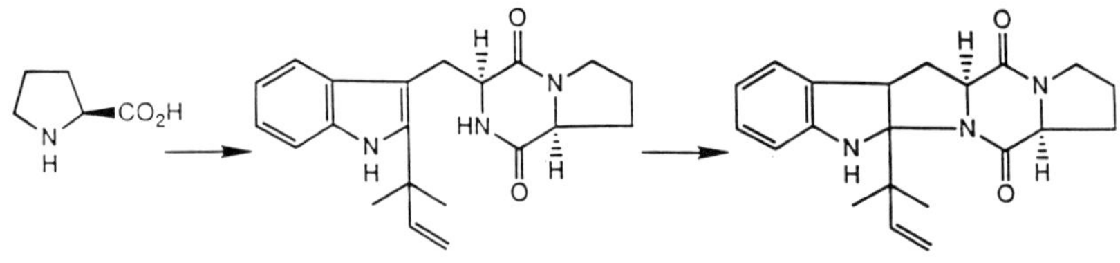

185 **186** **187**

3.6.4 Piperazines obtained from 2,5-dioxopiperazine precursors

Naturally and unnaturally occurring piperazines of biological importance may be obtained from 2,5-dioxopiperazine precursors. Thus, chemical [374,384,385] or electrochemical [386] reduction of dioxopiperazines is a straight forward way to prepare piperazine-containing compounds. Optically active dioxopiperazines, including several mixed *cyclo*$(Tyr)_2$ derivatives **188a–d**, serve as precursors to piperazines **189a–d** [387]. Reduction with borane in THF overcomes poor solubility problems of certain *cyclo*$(Tyr)_2$ compounds and is preferred to reduction with $LiAlH_4$. However, naturally occurring **189a**, isolated from *Zanthoxylium arborescens*, is prepared using either BH_3/THF [387] or $LiAlH_4$ reductions [374].

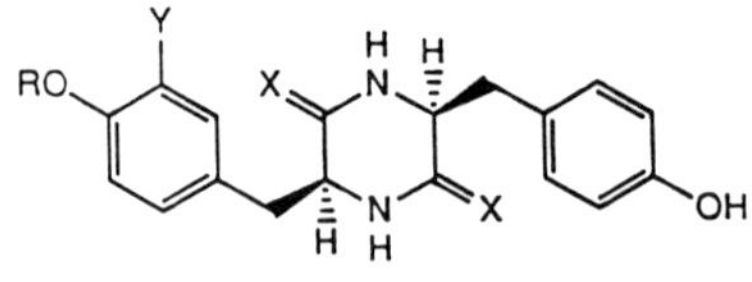

188a; R = Y = H, X = O
b; R = H, Y = Br, X = O
c; R = Me, Y = Br, X = O
d; R = Me, Y = I, X = O

189a; R = Y = H, X = H_2
b; R = H, Y = Br, X = H_2
c; R = Me, Y = Br, X = H_2
d; R = Me, Y = I, X = H_2

The 2,5-, and 2,6-dioxopiperazines **190** and **191** serve as synthetic intermediates for the preparation of diethylcarbamazine(**192**)-related antifilarial analogues **193** and **194** [388]. Such piperazines and several other analogues decrease blood microfilaremia levels in *Litomosoides carinii*-infected gerbils, but have no effect on the adult worms.

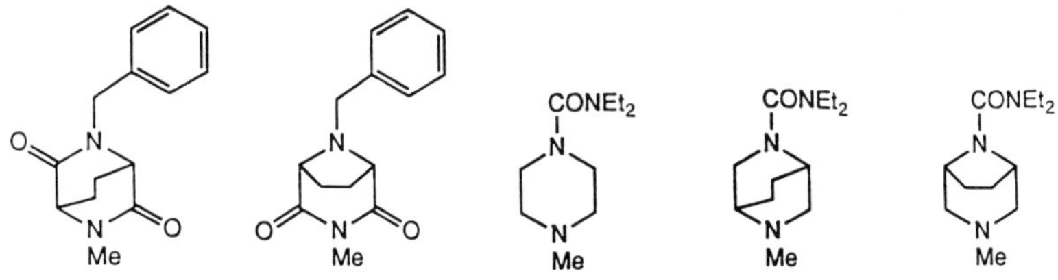

190 **191** **192** **193** **194**

(+)-Piperazinomycin (**202**), isolated as a minor metabolite of *Streptoverticillium olivoreticuli* subsp. *neoenacticus* [389] and characterized by X-ray crystallographic analysis [390], displays antimicrobial (yeast and fungi) properties [389] of considerable significance. Bromination of N-formyl-L-tyrosine methyl ester (**195**), conversion to intermediate dioxopiperazine **196**, and oxidation with thallium (III) nitrate produces an inseparable mixture which upon reduction (Zn/HOAc) generates three biphenyl ethers [**197**(19 %), **198**(5 %), and **199**(11 %)] [391]. Transformation of **197** into tris(benzyl)dioxopiperazine macrocyclic derivative **200** in five steps, reduction ($NaBH_4$-BF_3-OEt_2) affording tris(benzyl) piperazine derivative **201**, and catalytic hydrogenation yields target **202**.

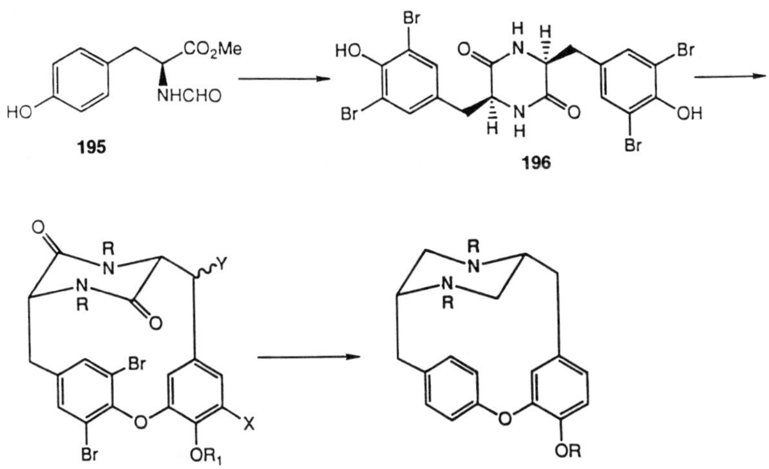

197; X = Br, Y = R = R_1 = H
198; X = OMe, Y = R = R_1 = H
199; X = Y = OMe, R = R_1 = H
200; X = Y = H, R = R_1 = CH_2Ph

201; R = CH_2Ph
202; R = H, $[\alpha]_D^{28}$ +24° (c 0.29, MeOH) (Synthetic); $[\alpha]_D^{23}$ +32° (c 0.74, MeOH) (Natural Product)

Bis(aryl)-substituted 2,5-dioxopiperazines of the type **203** are convertible to bicyclic species **205** [392]. Several structural variants of such multioxygenated bis(aryl) 2,5-dioxopiperazines (**203**) are available [393], and the chemistry is important for the synthesis of dimeric isoquinoline-quinone antibiotics including the saframycins [394], safracins [395], and renieramycins [396,397] which contain the 1,5-imino-3-benzazocine skeleton **206**. Regioselective benzylation of **204** occurs at the N(4) nitrogen [398]. Syntheses for saframycin β and related compounds using this chemistry are available [392,398]. Thus, tricyclic lactam **207** serves as precursor to amino acetal **208** which undergoes stereoselective intramolecular cyclization with trifluoroacetic acid to produce n-butyl ester **209**, precursor to ether ester **210** [393]. X-ray analysis shows C(9) of **210** epimeric with the saframycins. Epimerization of **210** and conversion to (±)-saframycin B (**209**) is accomplished in six steps.

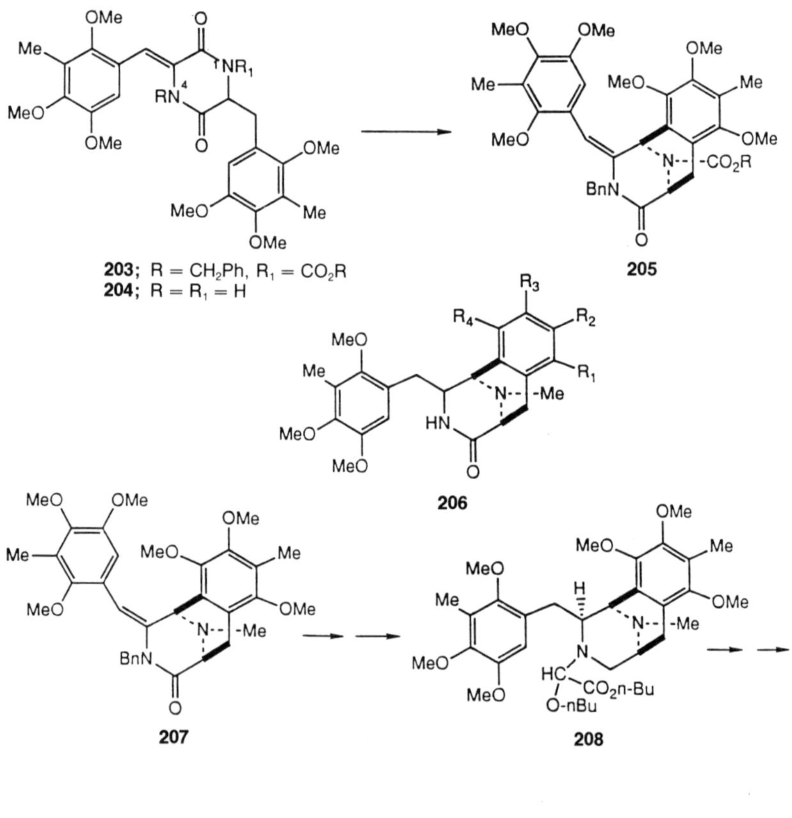

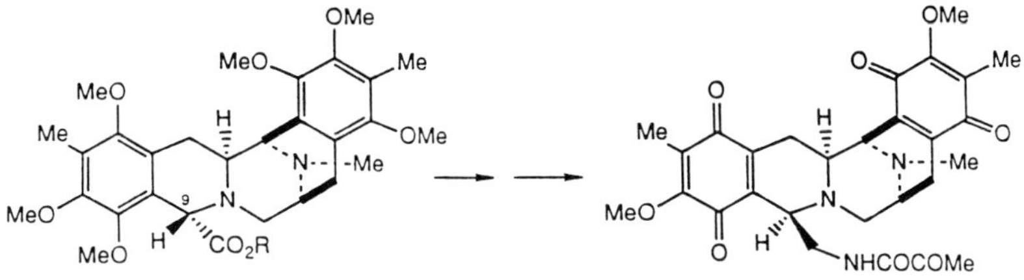

209; R = n-Bu
210; R = Et

211

3.6.5 Bicyclomycin syntheses and biology; related chemistry

Bicyclomycin (**212**), characterized by chemical [399] and X-ray [400,401] methods, is produced by *Streptomyces sapporonesis* [402] and *S. aizunensis* [403,404] and has a unique antibacterial spectrum [405]. This compound is biosynthetically derived by the oxidative cyclodimerization of leucine and isoleucine [406,407]. Bicyclomycin inhibits the growth of many Gram-negative bacteria [399,408–410], and in *E. coli* this antibiotic causes the formation of multinucleate and aseptate filaments, and osmotically fragile spheroplast-like formations [411]. Bicyclomycin does not inhibit protein and nucleic acid synthesis, but inhibits biosynthesis of peptidoglycan-linked lipoprotein bound to the outer membrane of the microorganisms [411]; this may not be a primary action [412]. [^{14}C]Bicyclomycin binds to several inner membrane proteins of *E. coli* [412], but not to outer membrane proteins. Benzylpenicillin does not compete with bicyclomycin for these bicyclomycin-binding proteins (BBP) and bicyclomycin similarly does not compete with benzylpenicillin for PBP binding. Bimolecular binding to all BBP is irreversible; antibiotic is not released from protein during three hours of incubation [412]. Since bicyclomycin blocks cell division without interacting with PBP-3, a more complex mechanism involving BBPs may be operative. The observation that bicyclomycin binds to *E. coli* and not to *B. subtilis* is in accord with its antibacterial spectrum [412].

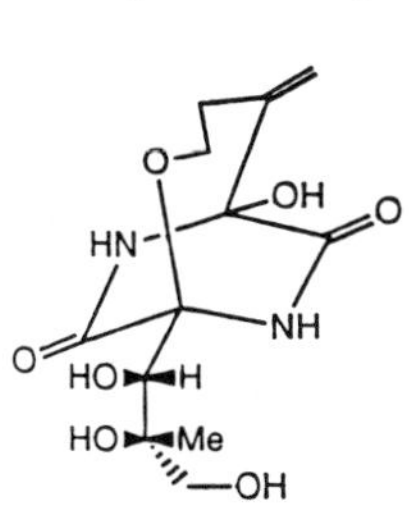

212

Syntheses for the (±)-N,N[1],O-trimethylbicyclomycin skeleton **215** are available via regiospecific cyclization of alcohol **213** to afford bicyclo [4.2.2] compound **214** followed by side chain introduction at the bridgehead position to produce the target (**215**) or related species [413]. Additionally, (±)-N,N[1],O-trimethylbicyclomycin **219** is available from major diastereomer **218a** generated as a 9:3:3:1 mixture of **218a–d** from (±)-**217** and (±)-**216** [414]. The total synthesis of (±)-bicyclomycin using related chemistry is achieved in 19 steps via intermediate benzyl-protected species **220** [415]. Details of this chemistry are available [416].

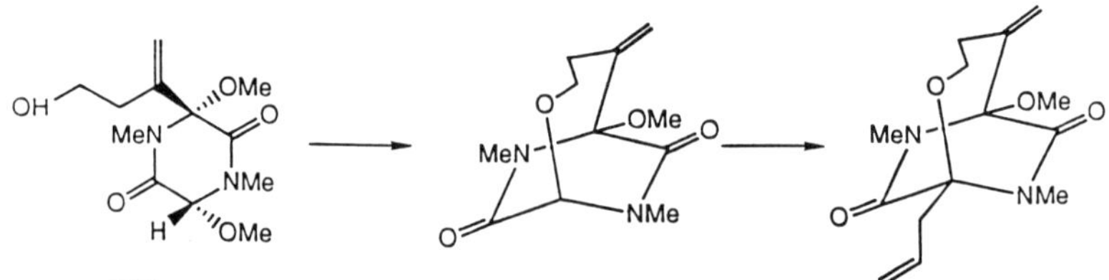

213 **214** **215**

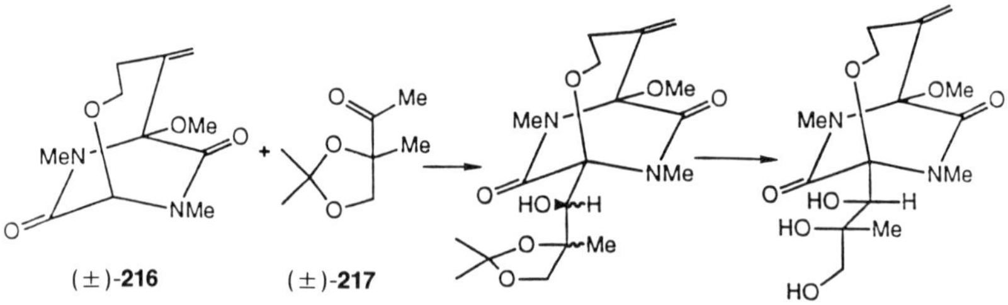

(±)-**216** (±)-**217** **218a-d** (diastereomers) (±)-**219**

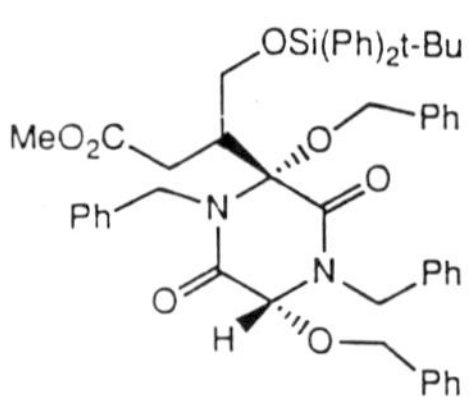

220

In other studies, bis(acetoxy) *α,β*-unsaturated acid **221** serves as precursor for dioxopiperazine **222**, which upon oxidative cyclization using phenylselenyl chloride produces bicyclic dioxopiperazine **223**, a substance possibly convertible to bicyclomycin-related compounds [417]. Rapid and efficient concomitant deprotection/cyclization of 3,6-disubstituted (silyloxypropyl) (pyridylthio) piperazinediones (**224**) in the presence of Hg(II) perchlorates produce the bridged bicyclic dioxopiperazines **225** [418]. Bicyclomycin model analogue **227** is prepared utilizing an efficient hydroxylation of simple bicyclic bridgehead carbanions [419]. Thus, treatment of **226** with t-BuLi in THF at –78 °C followed by addition of oxodiperoxymolybdenum (hexamethylphosphorictriamide) (pyridine) affords target **227** in 62% yield [419]. Similarly, generation of bridgehead carbanions of bicyclic dioxopiperazine **228** or the bicyclo[3.2.2] nor analogue using HMPA/LDA followed by quenching with various electrophiles significantly favors production of **228b** rather than **228c** or bifunctionalized systems **228d** [420]. Regio- and stereocontrolled total synthesis of bicyclomycin **212** in 12 chemical steps (4.6 % overall yield) is achieved via **230**, prepared from intermediate **229** in four steps [421]. Regioselective bridgehead hydroxylation involving anion generation under specific conditions and reaction with O_2, followed by aldol condensation with (±)-2,2,4-trimethyl-1,3-dioxolane-4-carboxaldehyde affords products ultimately convertible to bicyclomycin [421].

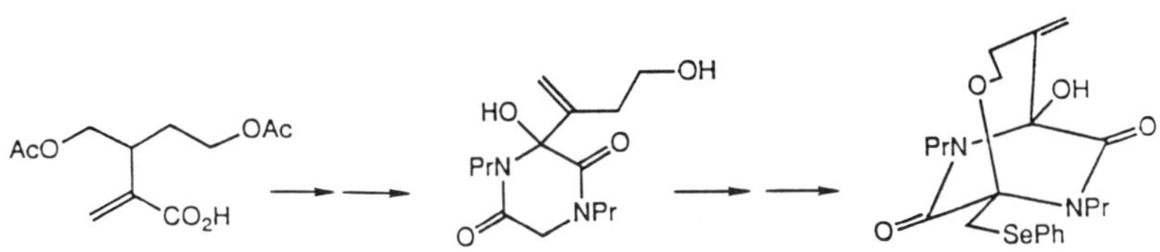

221 **222** **223**

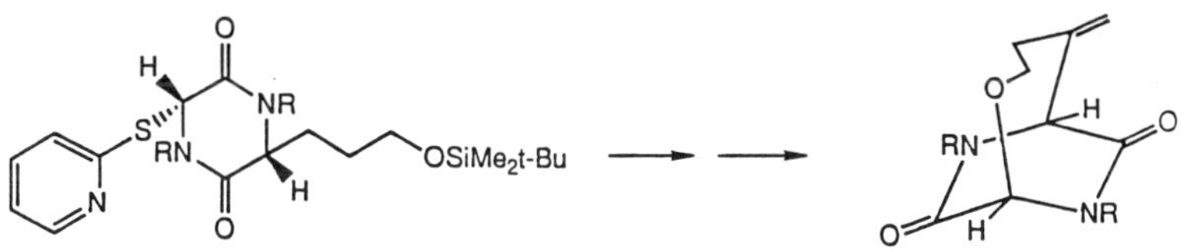

224 **225**

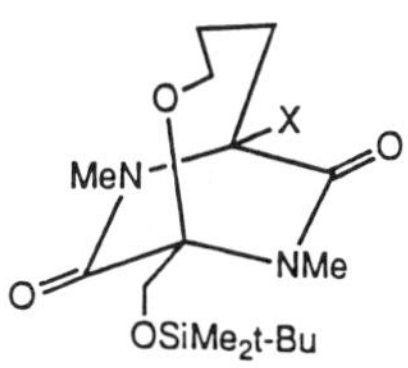

226; X = H
227; X = OH

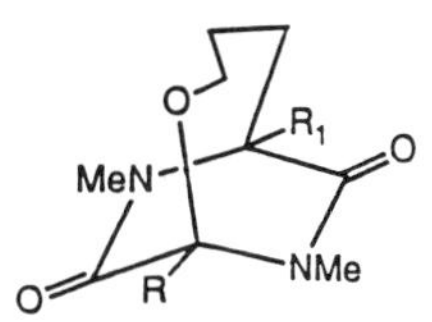

228a; R = R_1 = H
b; R = H, R_1 = substituent
c; R = substituent, R_1 = H
d; R = R_1 = substituent

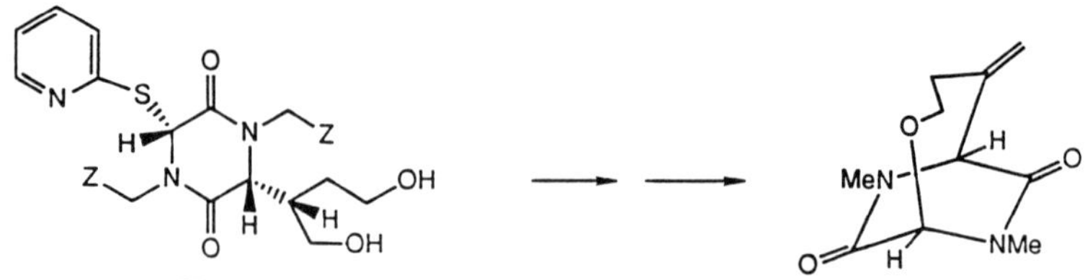

229: Z = Ph or MeOPh

230

Also, L-Cysteine thiazolidine methyl ester **231** serves as precursor to the exo methylene analogue **233**, suggested as an intermediate in a proposed bicyclomycin synthesis [422]. Retrograde Michael cleavage of the intermediate thiazole[3,4-a]pyrazine-5,8-dione **232** facilitates conversion to the ultimate target **233** [422]. d,l-Homoserine and N-p-methoxybenzylglycine ethyl ester serve as starting material for the elaboration of both bicyclo[2.2.2]- and bicyclo[3.2.2.]dioxopiperazines **234** and **235** [423], and N,N′-diacetyl-*cyclo*$(Gly)_2$ **234** is convertible to 6-hydroxy-7,9-bis(p-methoxybenzyl)-5-methylene-2-oxa-7,9-diazabicyclo[4.2.2]decane-8,10-dione (**237**) in 12 steps [424].

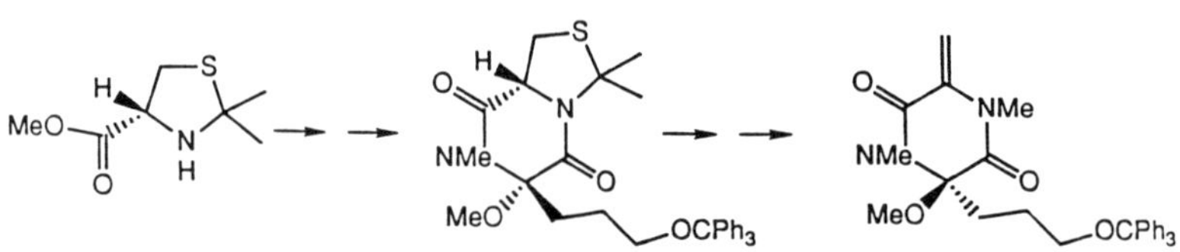

231 **232** **233**

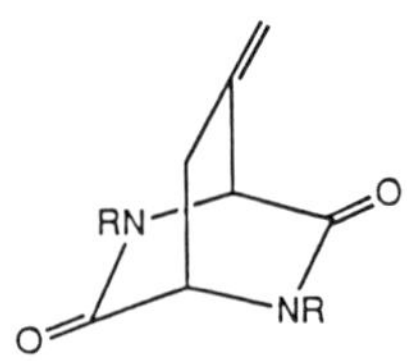

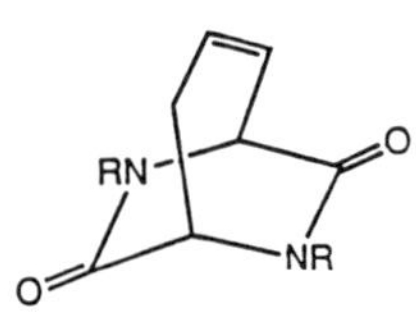

234 **235**

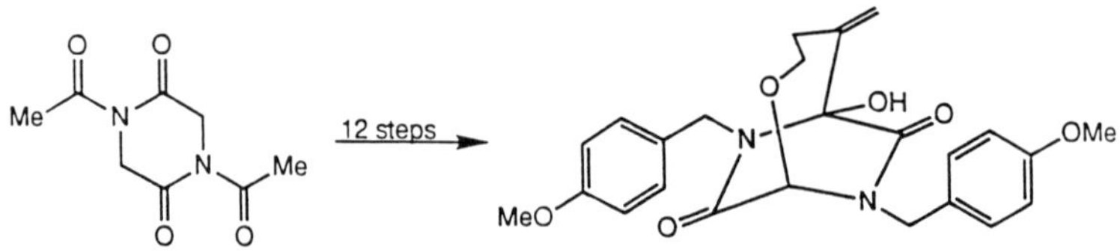

236 **237**

2-Oxa-7,9-diazabicyclo[4.2.2]decane-8,10-dione derivatives **239**, which are easily interconvertible with spiro[oxolane-2,2′-piperazine]-3′,6′-diones **241** via 2,5-dioxopiperazines **240**, are prepared from starting l-benzyl-3-(3-hydroxypropylidine)-2,5-dioxopiperazine **238** [425,426]. Both the desired 2-hydroxy-5-methoxy analogues of **239** ($R=PhCH_2, R_1=CH_3$) and unwanted 3-(3-t-butyloxypropyl)-5-methoxy-2,3,6-trioxypiperazine derivatives related to structure **240** are generated [425].

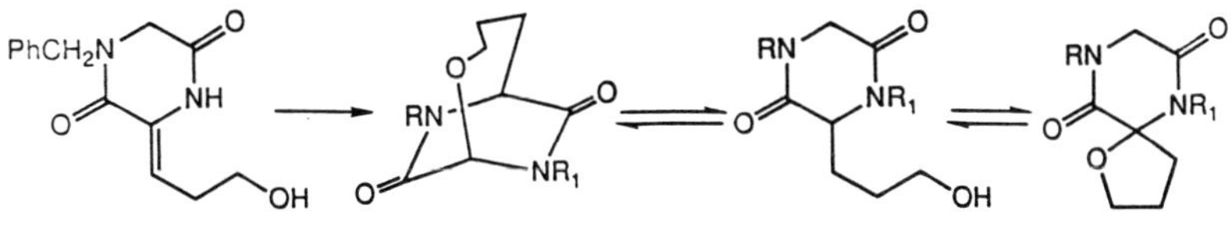

238 **239** **240** **241**

In other bicyclomycin-motivated syntheses, 2,5-dioxopiperazines **244** and **245** are prepared via their respective precursors **242** and **243**. Hydrolysis of THP analogue **245** affords the hydroxy compound **246** [427]. 2,5-Dioxopiperazine **242** is prepared via diazaspiro intermediate **248**; epoxide ring opening with magnesium isopropyl cyclohexamide gives **249** [428].

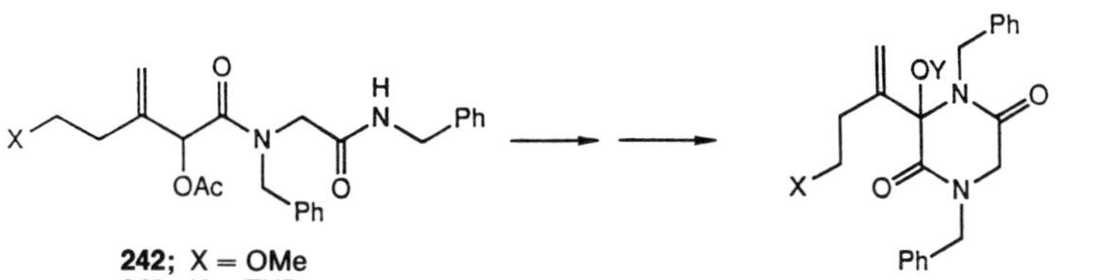

242; X = OMe
243; X = THP

244; X = OMe, Y = Ac
245; X = THP, Y = Ac
246; X = OH, Y = Ac
247; X = OMe, Y = H

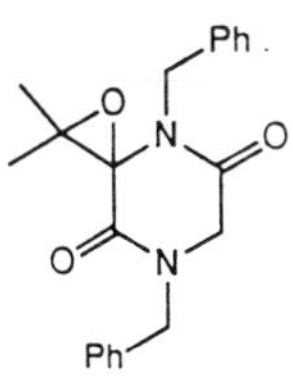

248

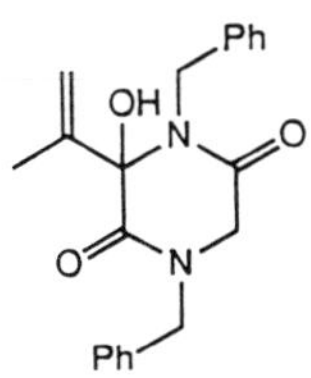

249

3.6.6 Epidithio- and polythio-2,5-dioxopiperazines; biosynthesis and synthetic chemistry of natural and unnatural compounds

Previously discussed are sulfur-containing dioxopiperazines of the chetomin type (**103–105**), which have toxicities important to agriculture [section 3.2], and thiomethyl-or thiophenyl-substituted dioxopiperazines (**117–125**), which are inhibitors of PAF [section 3.3.3]. Also, several other polycyclic sulfur-containing compounds (**131–135**) of potential pharmaceutical interest are known. In this section, the chemistry and biology of epidithio- and polythio-2,5-dioxopiperazines are summarized.

Reviews covering the literature into the mid 1960's [429] and mid 1970's [430] of the chemistry and biochemistry of sulfur-containing epipolythio-2,5-dioxopiperazines are available. The antimicrobial agent, hyalodendrin (**250**), is an example of an epidithio compound isolated from liquid cultures of a *Hyphomycete* species of *Hyalodendron* [431], and also from the fermentation broth of *Penicillium turbatum* [432]. This compound has a broad spectrum of activity against human bacteria and fungi pathogens [432], and against fungi associated with the deterioration and decay of forest products and trees. Hyalodendrin also inhibits growth in vitro of *Ceratocytis ulmi,* the causal agent of Dutch elm disease, a disease which has destroyed numerous beautiful trees in North America. Hyalodendrin has low phytotoxicity, and an LD_{50} in mice of 75 mg/kg [431]. The stereoselective synthesis of racemic gliovicin (**118**, FR 106969), an open chain analogue of **250**, from sarcosine anhydride (**251**), is achieved in 38 % overall yield via successive sulfenylation and benzylation of the respective enolates derived from protected alcohol **252** [433].

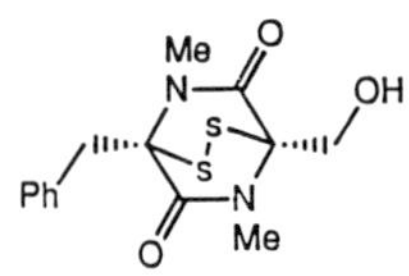

250

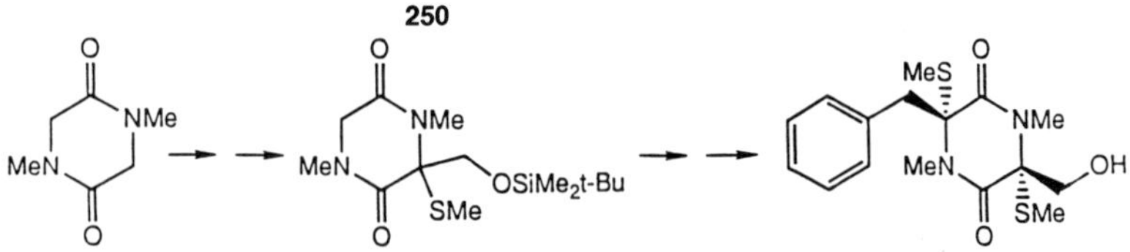

251 **252** **118**

Several *geminal* dithols **254** react with 3,6-dibromo-2,5-dioxopiperazine (**253**) to yield bicyclo[3.2.2] analogues **255**, and thioacetic acid adds to 3,6-dimethylene systems **256** to afford bis(acetylthio) compounds (**257**) [434]. However, attempts to convert bicyclo analogue **255** to epidithio analogues were unsuccessful; oxidation of **257** provides mixtures of epimono- and epidithio-2,5-dioxopiperazines [434].

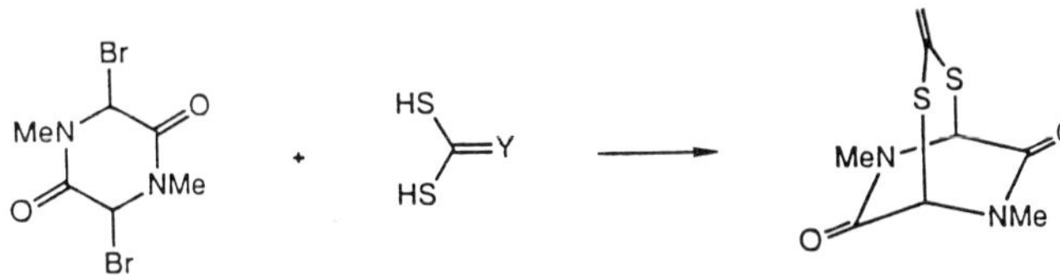

253 **254**; Y = NNHPh, NPh, NEt, CHCOPh, $(CH_2Ph)_2$, or Me and t-Bu **255**

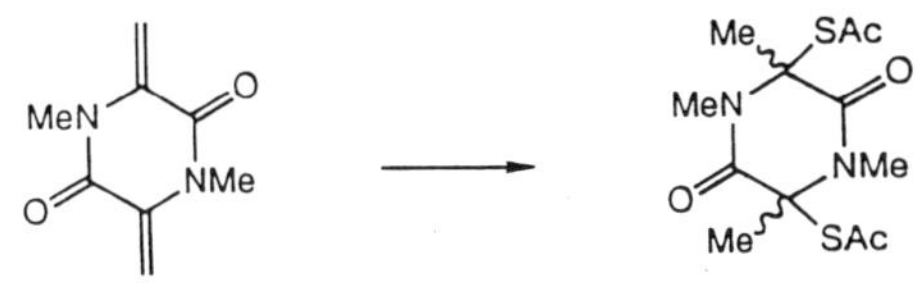

256 **257**

The epidithiopiperazinedione system is common to fungal metabolites such as the gliotoxins, sporidesmins, aranotins, verticillins, hyalodendrins chaetomins, orijzachlorins, and the melinacidins [270]. Such compounds possess potent antiviral, antifungal, antibacterial, antitumor and cytotoxic properties [270,430,432,433,435]. Interestingly, when carbon chiralities are *R*, the disulfide bridge is a left-handed helix; when these carbons are *S* the bridge is right-handed [432]. In epidithiodioxopiperazines, the disulfide bridge adopts a preferred conformation wherein each of the sulfur atoms is situated closer to the adjacent carbonyl carbon than to the nitrogen atom [432]. However, in epitetrathiodioxopiperazines such as **258** and **259** the tetrasulfide bridge is twisted such that the two central sulfur atoms are in close contact with nitrogen atoms [436,437]. Thus, the fungitoxic twist boat dioxopiperazine analogue, hyalodendrin (**250**) [438] exhibits solid state structural features [439] similar to N,N′-dimethyl-3,6-epitetrathio-2,5-dioxopiperazine and sporidesmin G etherate (**259**) [437]. The crystalline structure [440] for epidithio fungal metabolite epicorazine B (**260**) is available [441] and compared to epicorazine A (**261**) [442].

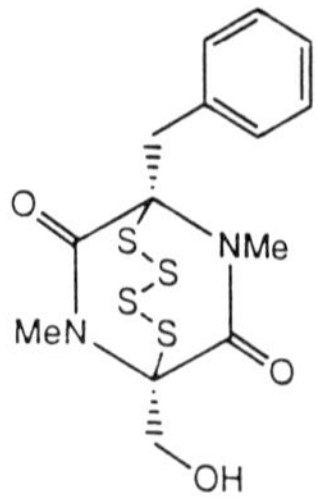

258

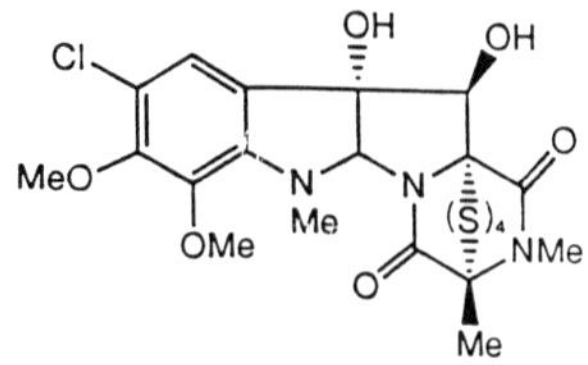

259

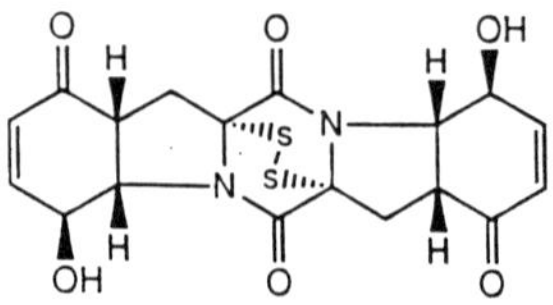

260

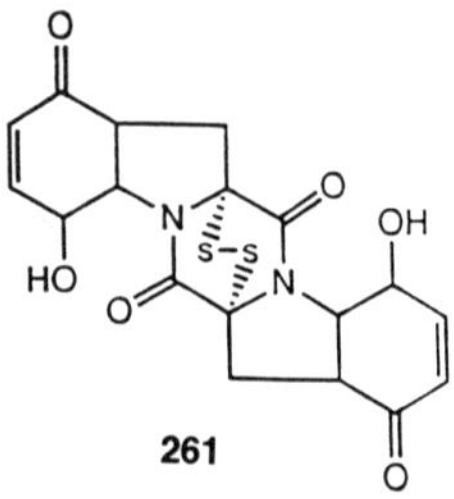

261

Recently isolated sulfur-bridged 2,5-dioxopiperazine natural products include gliovirin (**262**) and FA-2097 (**263**) from *Gliocladium virens* [443–446]. Such compounds are selectively active against members of the Oomycetes. Also, the interesting epitrithia analogue **264** has been isolated along with **250** and **118** from an unidentified fungus [447]. Epidithia and trithia analogues (**250** and **264**) are relatively active antifungal agents whose toxicity (mouse ip. LD_{50} range of 24–50 mg/kg) spectrum and potency (MIC ~ 1.6–25 mg/ml) are similar to that of gliotoxin (MIC ~ 3.1–25 mg/ml). These compounds, like gliotoxin, primarily have Gram-positive antibacterial activity. Alkylthio analogues such as **118** are inactive against fungi and bacteria, and are relatively non-toxic to mice [447].

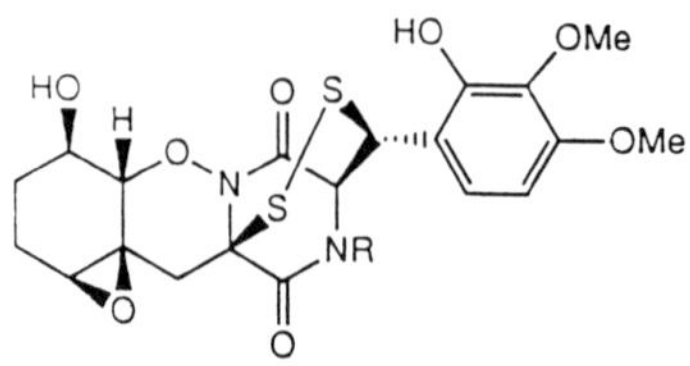

262; R = H
263; R = Me

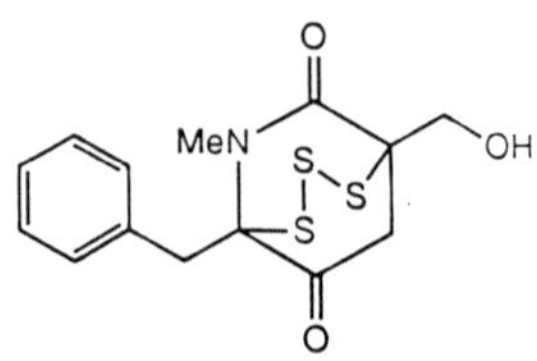

264

Synthetic methods developed for the construction of epidithiopiperazinediones [448] are applicable to the total synthesis of dehydrogliotoxin [449], sporidesmin A[450], sporidesmin B[451], gliotoxin [452], and hyalodendrin [453]. *cis*-Dialkylation or *cis*-monoalkylation products of bis(methoxymethylthio)-2,5-dioxopiperazines **265** are readily converted to symmetrically and unsymmetrically disubstituted dioxopiperazines **266** [454]. The first total synthesis of d,l-dehydrogliotoxin **268** utilizes dioxopiperazine **267** as starting material [449]. Gliotoxin analogue **273** inhibits reverse transcriptase, the RNA-dependent DNA polymerase of RNA tumor viruses, at concentrations of $3.9 \times 10^{-4} - 10^{-5}$ M, the same order of magnitude as that for gliotoxin. A practical synthesis for **273** [455,456] involves conversion of indolenium-2-carboxamide **269** to chloro-amide **270**, formation of the dioxopiperazine **271**, substitution with thiol functions affording **272**, and oxidation to produce **273**. $NaBH_4$ reduction of epidithio compound **273** regenerates **272**. A formal total synthesis for sporidesmin A (**276**) involves use of acetylthio-substituted dioxopiperazine **274**, which serves as precursor to diacetate-thioacetal **275**, subsequently convertible to target **276** in three steps [450].

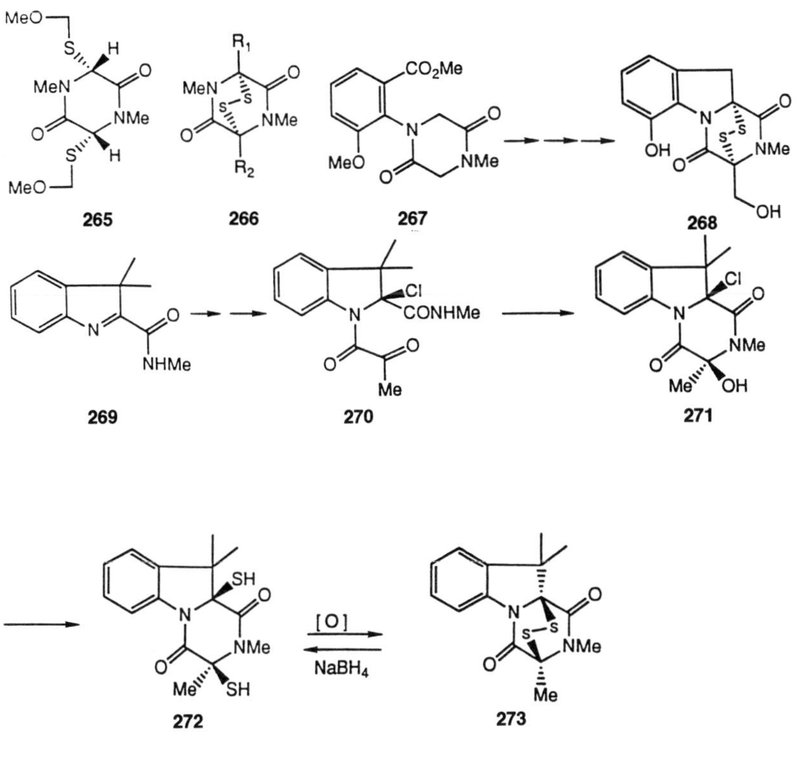

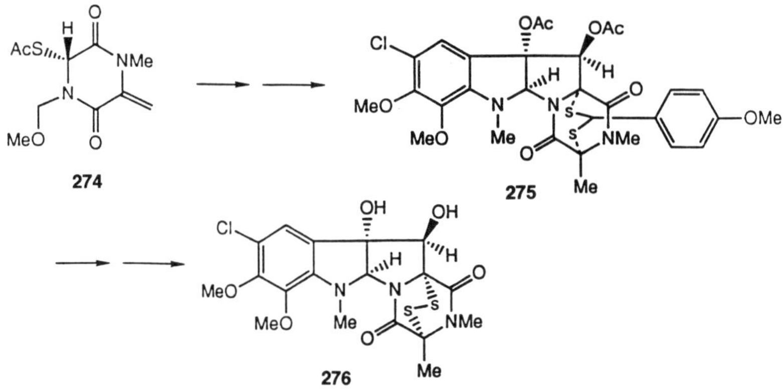

274 275 276

In addition to optically active gliotoxin **277** and racemic dehydrogliotoxin **268**, which are prepared by multi-step syntheses from thioacetal **278** [457], racemic hyaloclendrin **280** is synthesized from bis(methoxymethylthiol) **279** [457]. Such syntheses are particularly significant in light of the important biological properties of these compounds. Dehydrogliotoxin and the sporidesmins have the *R* configuration at the bridgehead carbons and exhibit selective antiviral properties, whereas the antipodal chaetocin does not have this activity [458]. Synthetic racemic dehydrogliotoxin, whose activity is similar to gliotoxin, also inhibits reverse transcriptase [459]. The chemical resolution of **281** may be performed via conversion to diastereomers **282** and **283** as well as via kinetically controlled transformation, but the individual enantiomers of **281** have the same anti-reverse transcriptase activity [459]. The X-ray structure for **278**-related monoethyl dithioketal **284** is available [460].

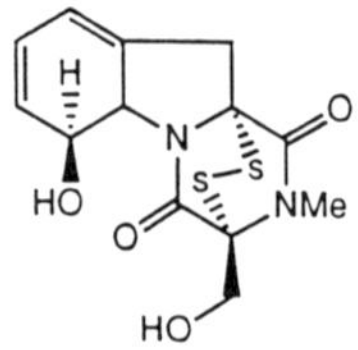

277

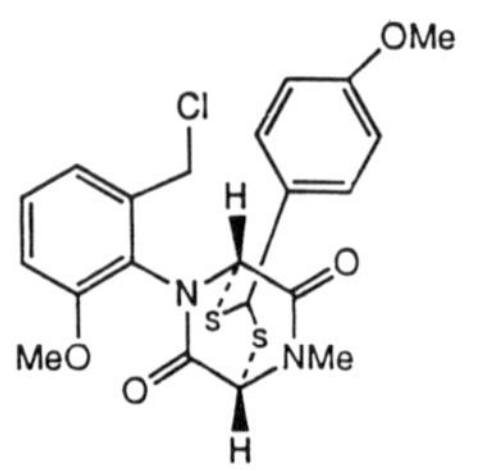

278

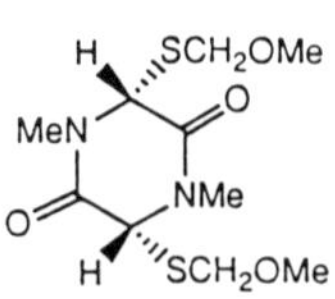

279

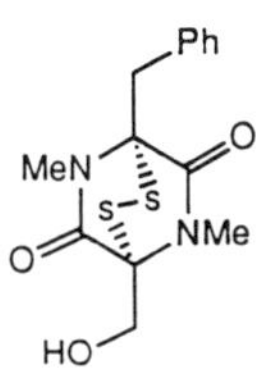

280

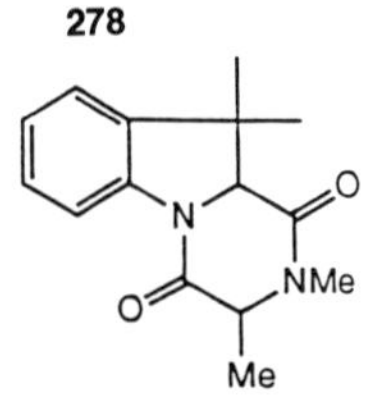

281

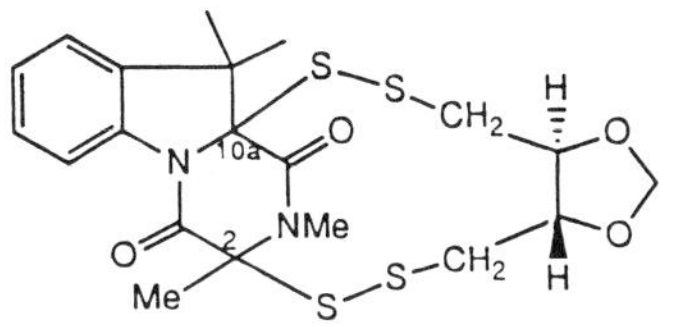

282; S_2S_{10a}
283; R_2S_{10a}

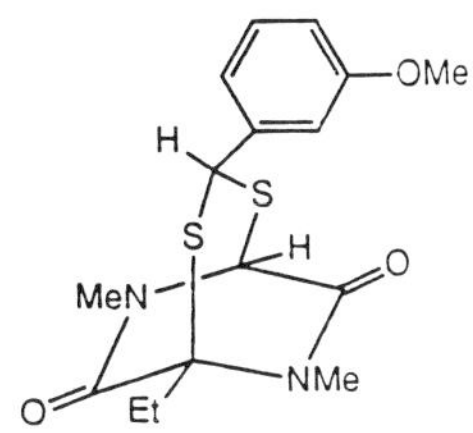

284

Gliotoxin and related sulfur-bridged dioxopiperazines are proposed [461] to be biosynthesized from 2,5-dioxopiperazine precursors **285** via di-N-hydroxy compounds **286** for which are available relatively efficient syntheses [462]. These species are thought to lose water to afford reactive bis(acylimmonium) ions **287**, which undergo reaction with sulfur nucleophiles to produce the bicyclic system **288** [461]. In fact, *cyclo*(L-Phe-L-Ser) is an efficient biosynthetic precursor for gliotoxin [463–465]. Cystine serves as the sulfur atom source [466] and methionine the source of the N-methyl function [467]. Also, biosynthetic introduction of sulfur in gliotoxin isomers may proceed in *Gliocladium deliquescens* without prior N-methylation or oxidative cyclization of the phenyl group [468]. A related biomimetic synthetic sequence involves conversion of N-hydroxy- **289** and N-benzyloxypiperazine **290** to bridged species **291** and **292** in seven to eight steps [371].

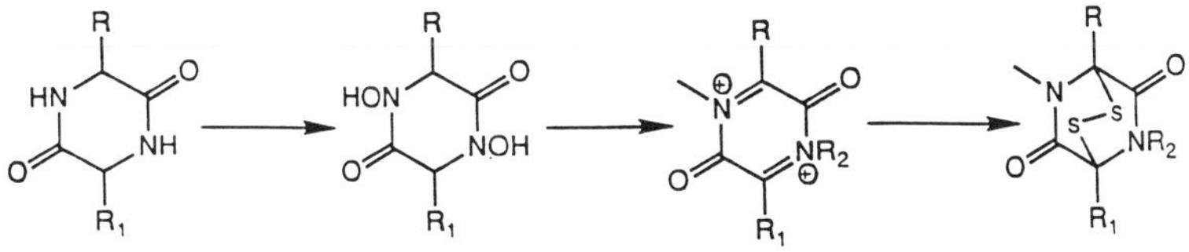

285 **286** **287** **288**

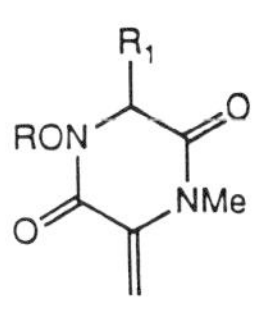

289; R = OH, R_1 = H, Me, or CH_2Ph
290; R = OCH_2Ph, R_1 = H, Me, or CH_2Ph

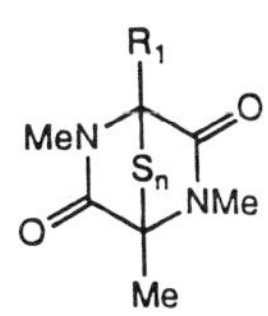

291; n = 1
292; n = 2

The episulfied **293** is convertible to *cis*-3,6-dibenzyl-3,6-bis(methylthio) piperazine-2,5-dione **294** which serves as substrate for bisdethiobis-(methylthio)acetylarantoin **156** in *Aspergillus terreus* [370]. [^{3}H]- and [^{35}S]-labeled epidisulfides **293** are incorporated into the aranotin analogue **156** in *Aspergillus terrus,* but isotope ratio changes for precursor (4.86) and product (3.82) differ [469]. Mycelianamide **106** is biosynthesized in *Penicillium griseofulvum* from tyrosine **295** with retention of the *pro-R* and complete loss of the *pro-S* β-methylene hydrogen atom [470,471]. Also, stereospecifically labeled L-tryptophane **296** is incorporated into cryptoechinuline A **297** via removal of a *pro-S* hydrogen atom from the β-methylene position [472].

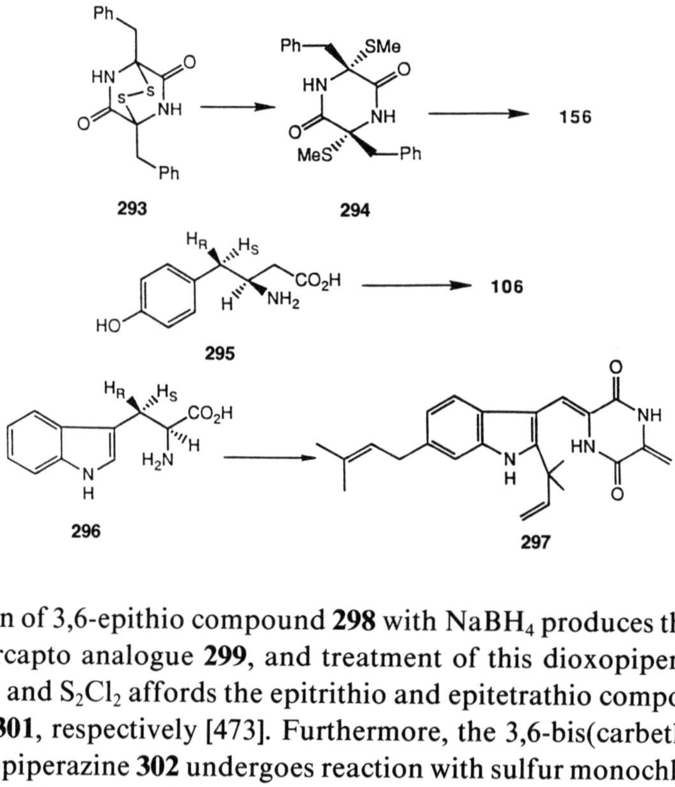

Reduction of 3,6-epithio compound **298** with $NaBH_4$ produces the *cis*-3,6-dimercapto analogue **299**, and treatment of this dioxopiperazine with SCl_2 and S_2Cl_2 affords the epitrithio and epitetrathio compounds **300** and **301**, respectively [473]. Furthermore, the 3,6-bis(carbethoxy) 2,5-dioxopiperazine **302** undergoes reaction with sulfur monochloride (S_2Cl_2) to afford 3,6-epipolythio derivatives **304** [474]. Heating the dipotassium salt **303** with S_2Cl_2 in dioxane provides the epidithio analogue **305**. The *cis*-3,6-bis(ethylthio) isomer of *trans*-**306**, accompanied by traces of **306** and ethylthio derivative **307** are obtained upon reduction ($NaBH_4$) and alkylation (EtI) of the epidithiodioxopiperazine

305 [475]. Reaction of **303** with ethanesulfenyl (and similarly with benzenesulfenyl) chloride gives trans ethylthio ethers of **306** and **307** and hydroxylated materials **308–310** of mixed geometries and in low yield.

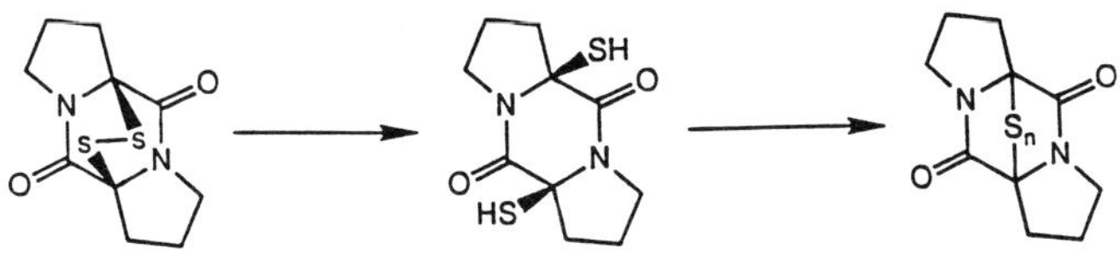

298 **299** **300**; n = 3 **301**; n = 4

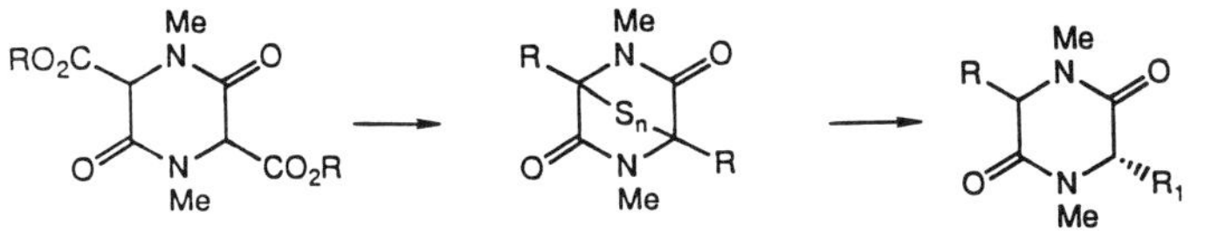

302; R = Et
303; R – K

304; R = CO_2Et, n = 1-4
305; R = H, n = 2

306; R = R_1 = EtS
307; R = EtS, R_1 = H
308; R = OH, R_1 = EtS
309; R = OH, R_1 = H
310; R = R_1 = OH

3.6.7 *Cyclo*(Pro)$_2$ chemistry

Cyclo(Pro)$_2$ analogues have unexplored medicinal chemical potential. *Cyclo*(L-Pro-L-Pro) (**311**) undergoes metalation to form the anion **312**, which when reprotonated affords **311** without loss of optical activity [476]. Alkylation produces the optically active 3-alkyl derivative **313**. Further alkylation leads to optically inactive 3,6-dialkyl compound **315**. The anion **312** also serves as precursor to optically active 3-ethylthio compound **314**. Optically inactive *trans*-3,6-bis(ethylthio) species **317** is prepared from thio anion **316**.

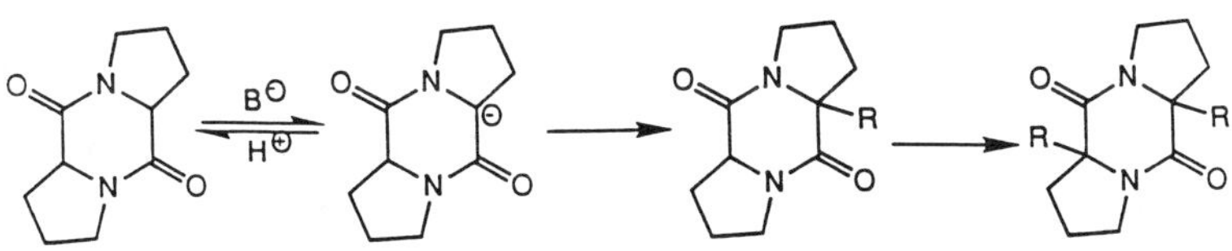

311 **312** **313**; R = Et **314**; R = EtS **315**; R = Et

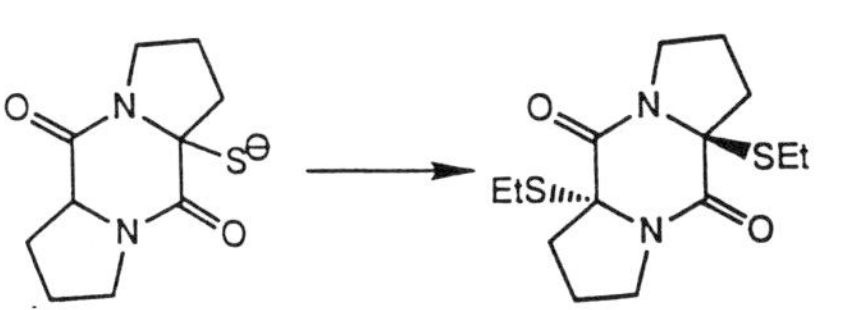

316 **317**

Dehydro analogues **319** and **322** are prepared by elimination of methanthiol from dioxopiperazine **318** and *cis*-**320** or *trans*-**321**, respectively [477]. Oxidation of -SMe functions in **318**, **320** or **321** produce the corresponding sulfones (SO_2Me). Replacement of sulfides by peroxide (-OOH) occurs upon reaction with H_2O_2. Reduction of peroxides with Na_2SO_3 yields the corresponding alcohols [477]. Nucleophilic reactions involving displacement at one asymmetric center proceed with retention of configuration, whereas reaction at both asymmetric carbons provides the thermodynamically more stable isomer (LD or LL + DD) [477].

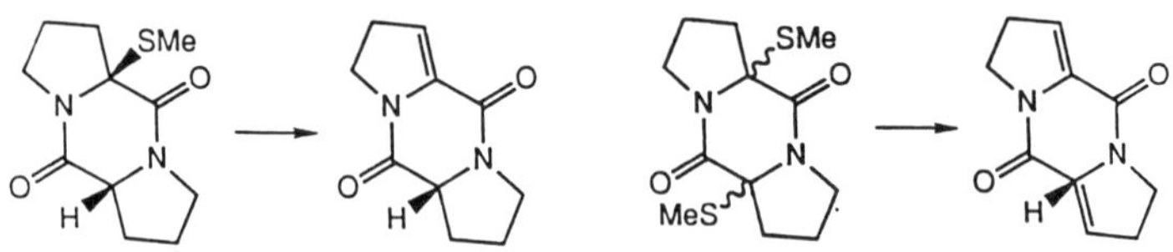

318 **319** **320**; cis **321**; trans **322**

4 2,6-Dioxopiperazines (3)

2,6-Dioxopiperazines are imides of open chain di(carboxymethyl)amines. Unlike the 2,3- and 2,5-dioxo regioisomers, there is a virtual absence of the chemistry and biology of compounds containing a single 2,6-dioxopiperazine heterocycle. Single rings are found as parts of side chains in pipericillin-related semisynthetic antibiotics synthesized for structure-activity comparisons with other oxopiperazines including those of the 2,3-dioxo series discussed previously. The significance of 2,6-dioxopiperazine heterocyclic chemistry and biology evolved as a consequence of exploitation of ethylenediaminetetraacetic acid (EDTA; **323**) pharmacology; cyclized imide derivatives of EDTA and related tetra acids are bis(2,6-dioxopiperazines) wherein the amino N(4) nitrogens of each ring are connected by a central alkene side chain. A review into the early 1980's of their pharmacology [478], and a concise summary [479] of this area are available.

4.1 Early antitumor studies

The bis(2,6-dioxopiperazines) first synthesized in the late 1950's, were thought to have potential for use as textile leveling agents or as pharmaceuticals, but no clear indication for their use was documented [480]. Later, workers at the Imperial Cancer Research Fund (ICRF) introduced rationales for their use as antineoplastic agents [481–484] providing the impetus for a worldwide interest in exploring bis(2,6-dioxopiperazines) as anticancer drugs [479]. Ultimately, the synergistic properties of these heterocycles in combination with clinically efficacious antitumor agents became known, and their ability to ameliorate the toxicity of potent drugs such as the anthracycline antibiotics stimulated considerable research activity.

Whereas many antitumor drugs are chelators of metal ions [485], EDTA (**323**) or its methyl (or ethyl) esters fail to demonstrate any significant antitumor activity [486,487]. Reaction of EDTA with formamide does not yield tetraamide **324**, but generates diimide **325** (ICRF-154) having weak antitumor properties in mice [482,487].

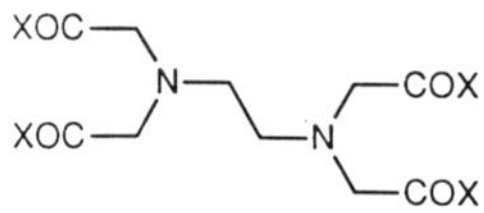

323; X = OH
324; X = NH_2

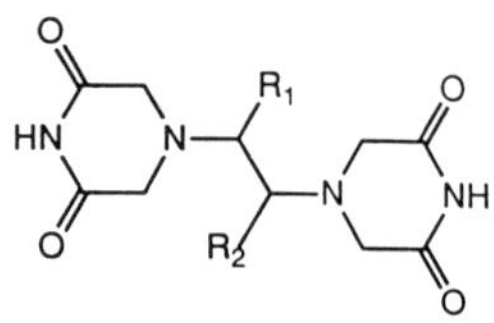

325; $R_1 = R_2$ = H (ICRF-154)
326; R_1 = Me, R_2 = H (ICRF-159)
327; (+) isomer of ICRF-159 (ICRF-187)
328; (–) isomer of ICRF-159 (ICRF-186; also known as ADR-888)
329; R_1 = Et, R_2 = H (ICRF-192)
330; $R_1 = R_2$ = Me (meso isomer ICRF-193)
331; $R_1 = R_2$ = Me (dl isomer ICRF-196)
332; R_1 = Me, R_2 = n-Pr
333; R_1 = Me, R_2 = i-Pr
334; R_1 = Et, R_2 = n-Pr
335; R_1 = Me, R_2 = CH_2OMe (dl-erythro)
336; R_1 = Me, R_2 = CH_2OMe (dl-threo)

ICRF-154 (**325**) produces a 91% inhibition (5 doses at 30 mg/kg) in the growth of Sarcoma 180 (S180) tumor and a 67% (9 doses at 30 mg/kg) increase in survival time of L1210-inoculated mice [482,487]. The related propylene analogue ICRF-159 (razoxane, **326**) effects a 93% (5 doses at 30 mg/kg) inhibition of S180 tumors in mice and a 137% (13 doses at 30 mg/kg) increase in survival time in the L1210 leukemia model [487]. The (+) and (−) enantiomers of razoxane are known as ICRF-187 (**327**) and ICRF-186 (**328**), respectively. ID_{50} inhibition values for mouse L cell colony formation are for ICRF-154 (**325**) 7.3 μM, razoxane (**326**) 3.0 μM, ICRF-192 (**329**) 720 μM, meso isomer ICRF-193 (**330**) 0.09 μM, and racemic ICRF-196 (**331**) 150 μM [488]. Although meso bis(imide) **330** is more potent than razoxane in this assay, its therapeutic index is lower than that of razoxane in vivo [487,488]. The calculated ratio of maximum tolerated dose to the dose required for 90 % inhibition of S180 tumor growth for ICRF-193 = 6.7; for razoxane this ratio = 9.8.

Inhibition of [^{3}H]thyimidine incorporation into monolayers of mouse-embryo fibroblasts reveals IC_{50} values for ICRF-154 = 2.0 μg/ml, razoxane = 0.5 μg/ml, meso ICRF-193 = 0.035 μg/ml, dl-ICRF-196 (inactive), and a compound having a *trans*-cyclobutanediyl linkage (ICRF-197) = 14 μg/ml, or a 1,3-propanediyl linkage (ICRF-161) (inactive) [482]. Other dl-erythro (**332–335**) and dl-threo (**336**) analogues of increased lipophilicity also have considerably decreased activity [482,489].

Only minor modifications of the bis(imide) system are allowed in order to preserve biological activity [490]. Two intact 2,6-dioxopiperazine moities are essential. Replacement of 2,6-dioxopiperazine rings with other heterocycles, or hydrolysis to produce diamide diacids affords inactive materials [482], but these hydrolysis products are now thought to have biological significance (see section 4.15). Generally, linages between the two heterocyclic rings having five or fewer carbons are more effective in vitro.

4.2 Conformationally constrained analogues

Conformational mobility of the linking group leads to arrangements of dioxopiperazine rings, one or more of which likely is important for various biological activities. Crystal structure analysis of racemic razoxane (**326**) reveals a cis face-to-face relationship of dioxopiperazine rings; the (+) isomer ICRF-187 has the extended trans conformation with a parallel arrangement of ring planes in the solid state [491]. These conformations are mimicked by insertion of the 1,2-cyclopropanediyl spacer in place of the 1,2-propanediyl group of razoxane or its optical isomers [492]. The dioxopiperazine rings in these molecules, are held trans as in **337** or cis as in **338**. Biological properties of *trans*-**337** and *cis*-**338** differ markedly. This cis isomer inhibits, and the trans isomer stimulates development of metastases in two different animal models [493,494]. In the Syrian hamster lung adenocarcinoma (LG1002) model, ip. administration of razoxane and *cis*-**338** (15 mg/kg) inhibits lung metastasis, but *trans*-**337** (15 mg/kg) increases the number of metastatic nodules in the lung [493]. Also, using B16-F10 melanoma, pretreatment of cells with either razoxane or *cis*-**338** decreases the yield of lung colonies following iv. injection of tumor cells, whereas iv. injection of *trans*-**337**-pretreated cells produces an increase in lung colonies [494]. Solid state structures of *trans* and *cis*-4,4′-(cyclopropanediyl)bis(2,6-dioxopiperazines) (**337, 338**) confirm that the conformation of cis analogue closely resembles the cis conformation of razoxane in the solid state [495]. The trans isomer and ICRF-187 are somewhat different providing a torsion angle of 138° for *trans*-**337**, but 177° for the trans conformation of ICRF-187 [495]. Dioxopiperazine rings exist as slightly bowed half-chairs [491]. Cytostatic activity seems to reside in the cis conformation [493–495], a conformation likely important to biometal chelation mechanisms discussed in section 4.15, but solubility differences between the *cis*- and *trans*-1,2-cyclopropanediyl analogues may also account for differences in observed activities [494].

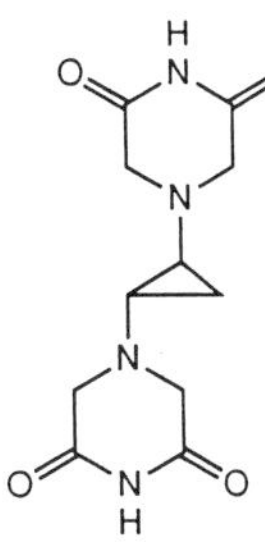

337

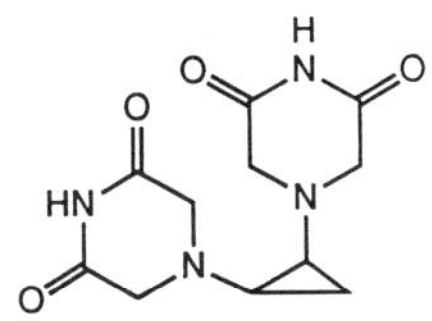

338

The compound *cis*-**338** is prepared from *cis*-1,2-cyclopropanediyl acid (**339**) convertible to diamine **340** which undergoes tetra-N-alkylation to produce intermediate **341** [493]. *Trans*-**337** is similarly synthesized from the trans isomer of **339** [492]. Evaluation of *cis* 1,2-cyclobutanediyl spacers are not reported since attempted preparation of the precursor cis diamines by Curtius rearrangement of the bis(acylchlorides) produces a violent explosion [496]; caution also should be exercised during the synthesis of cis diamine **340** [497].

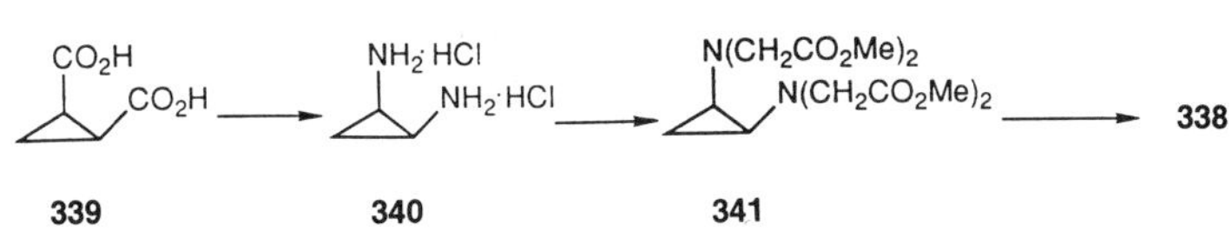

Razoxane has cytotoxic activity against V-79A fibroblasts in tissue culture [492], but *trans*-**337** is inactive. The decrease in activity of *trans*-**337** in this system, and the reported inactivities of **331** and the *trans*-1,2-cyclobutanediyl bis(imide) in S180, leukemia L1210 and [^{3}H]-thymidine assays may be related to similarities in their preferred geometries. Analogues having *cis*, not *trans* geometries, are anticipated to be preferred for most biological activities of bis(dioxopiperazines).

Bis(imides) **342** and **343** are conformationally constrained analogues of ICRF-154 (**325**) and differ from **325** by two hydrogen atoms in their molecular weight [496]. These geometric isomers are also related to *cis*-**338**; removal of the cyclopropyl methylene (CH_2) function of **338** with concomitant bond connection between two C(3) carbons of the dioxopiperazine rings provides isomers *trans*-**342** or *cis*-**343**. These tricycles are tetraazaperhydrophenanthrenes wherein dioxopiperazine rings maintain a cisoid relationship. Formal bond disconnection of one carbonyl group from the central piperazine ring of **342** (or **343**) and rebonding on the opposite carbon of the central ring provides diastereomers *trans*-**344** and *cis*-**345** belonging to the tetraazaperhydroanthracene series. The dioxopiperazine rings of geometric isomers **344** and **345** have an extended rather than cisoid geometry [498].

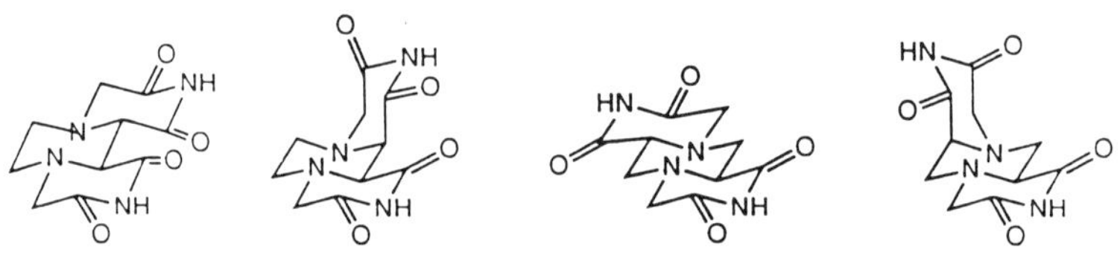

Tricyclic bis(2,6-dioxopiperazines) **342** and **343** are prepared from pyrazine-2,3-dicarboxamide (**346**) convertible to *cis*-**347** (R = Me or Et) via reduction and N-alkylation [496]. Cyclization of **347** in NaOMe/MeOH produces the trans isomer **342**, but in NaOEt/EtOH *cis*-**343** is the exclusive geometric isomer; product geometry is independent of starting esters and likely reflects solubility differences in MeOH vs. EtOH. In MeOH epimerization takes place affording the thermodynamically most stable trans isomer.

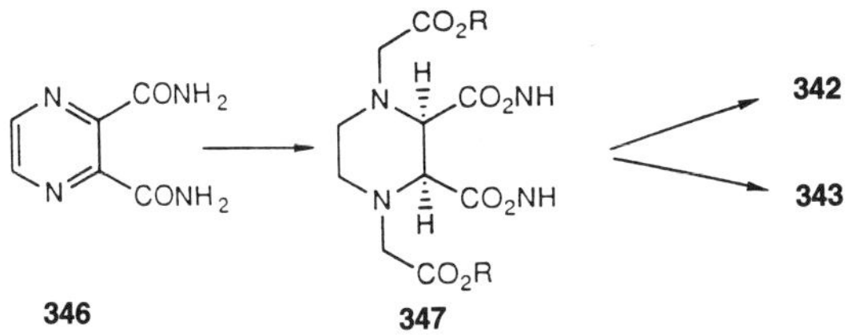

Bis(imide) isomers **344** and **345** are synthesized following similar reaction sequences using 2,5-dimethylpyrazine (**348**) as starting material [498]. Oxidation of methyl groups, ring reduction, and N-alkylation yield respective diastereomers **349** and **350** which are convertible to targets **344** and **345**.

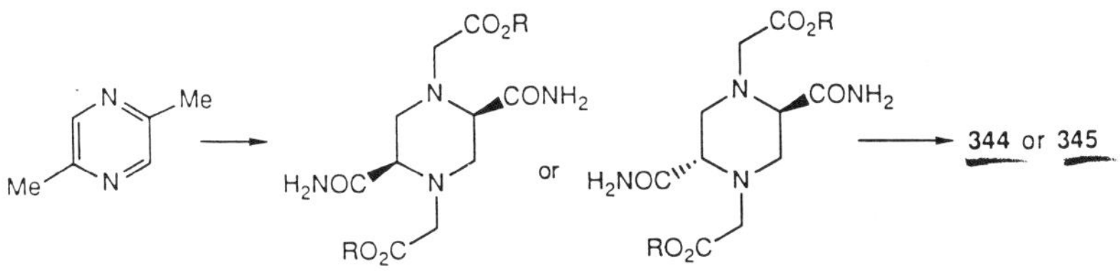

Pretreatment of B16-F10 melanoma cells with *trans*-**342** significantly reduces the metastasis to the lungs of mice at all dose levels (2, 20 and 100 μM). The isomer *cis*-**343** is ineffective in this model [496]. Regioisomeric bis(imides) **344** and **345** have no antimetastatic effects against implanted Lewis lung LL carcinoma in mice [498], and these data again emphasize the need for a cisoid relationship of dioxopiperazine rings.

4.3 N,N′-Bis(morpholinomethyl) and related derivatives

A major problem with bis(2,6-dioxopiperazine) drugs is their poor water solubility. Analogues are administered dissolved in oil or in suspension with carboxymethyl-cellulose. Razoxane has a solubility of only 3 mg/ml in water at 25°C [499]. Use of cosolvents, complexation, or crystalline modification to overcome low water solubility problems has not been successful. Interestingly, resolution provides enantiomers of razoxane, namely (+) ICRF-187 and (−) ICRF-186, which possess significantly greater water solubility than the racemic material [499]. Additionally, bis(morpholinomethyl) derivatives such as bimolane (**351**), a derivative of ICRF-154, has increased water solubility

[500]. However, insertion of morpholinomethyl functions into dioxopiperazines does not always afford water soluble materials. Probimane (**352**), the morpholinomethyl analogue of razoxane [501], is more water soluble than bimolane. Morpholinomethyl derivatives of tricyclic analogues **344** and **345** are insoluble [498].

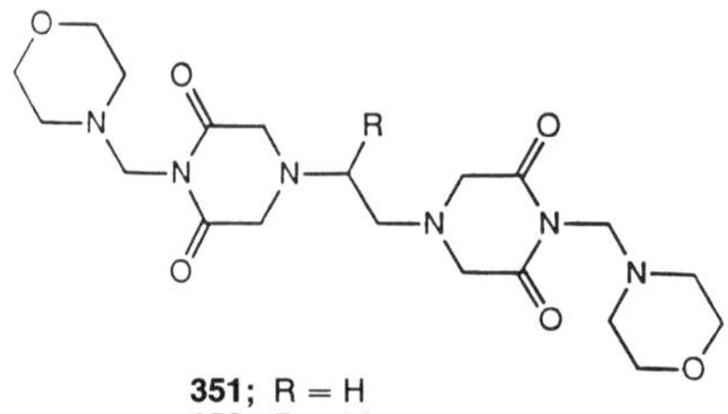

351; R = H
352; R = Me

Whereas few alterations of ICRF-154 (**325**) can be made without loss of antimitotic activity [488], bimolane (**351**) not only exhibits antitumor activity, but also provides a better chemotherapeutic index than the parent drug **325** against S37 [502,503]. Bimolane possesses marked inhibitory effects against S180 and S37 mouse tumors [503]. At doses of 25 mg/kg ip. or 50 mg/kg po., growth of these tumors is nearly completely inhibited. The $ID_{50's}$ against S37 are 1.9 mg/kg ip. and 6.6 mg/kg po. This drug also is effective against L615 leukemia, and inhibits solid mouse tumors such as Ehrlich carcinoma, hepatoma, and brain tumor B22. Toxicity studies in mice reveal that the LD_{50} is 373 mg/kg from a single ip. injection, whereas $LD_{50's}$ determined in mice dosed ip. or orally for eight days with observation for ten days shows values of 89 mg/kg and 244/mg/kg, respectively. Chemotherapeutic indices (LD_{50}/ID_{50}) of bimolane by ip. and oral administration are 48 and 37, respectively. Administration of bimolane to dogs (25 and 50 mg/kg/day im. for 10 days) does not cause significant changes in liver and kidney function tests, or in blood platelet count. Bimolane depresses the humoral-mediated immunity (25 mg/kg ip.) in both the hemolysin and graft versus host reaction tests, but the depressant effect appears to be less than that of ICRF-154 and razoxane. Cell-mediated immunity is not depressed by bimolane [503].

Compared to razoxane, bimolane is less effective at multiple low doses against S180 and leukemia L1210 [T/C = 0.64 and 0.47, respectively, at 5 mg/kg/day (6 doses); T/C = 0.64 and 0.35, respectively, at 6 mg/kg/day (5 doses)] [504]. However, one single high-dose (100

mg/kg) treatment with bimolane is more effective than razoxane against leukemia L1210 (T/C = 131 and 121, respectively). Bimolane inhibits the growth of the LL primary tumors, and significantly reduces the number of pulmonary metastases (T/C = 0.82 for primary tumor growth and 0.35 for formation of lung metastases). Bimolane also has a slight inhibitory effect on tumor B16 primary growth, but it has no effect on metastases to the lung [504].
Other available derivatives (**353–355**) result from substitution of imide nitrogens with maleimide, a sulfhydro reagent, maleic hydrazide, a plant growth inhibitor having weak antitumor activity, and 2-amino-1,3,4-thiadiazole, a weakly antineoplastic heterocycle [505]. When compared for anti-tumor activity against leukemia L1210, S180, LL carcinoma, and Ridgway osteosarcoma, generally, bimolane and the maleic hydrazido analogue **354** have somewhat better properties [505].

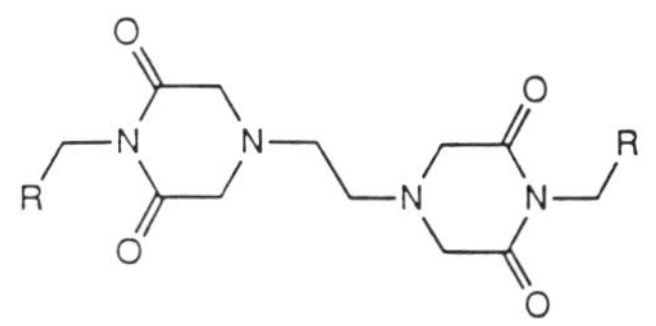

354 (SK 31,342); R =

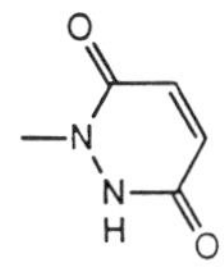

353 (SK 31,318); R =

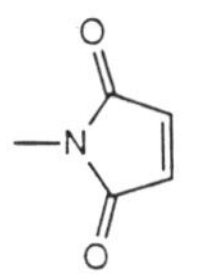

355 (SK 31343); R =

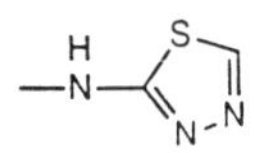

The most potent of many (**352, 356–362**) bimolane analogues prepared [506–508] is probimane (**352**), the bis(morpholinomethyl) derivative of razoxane. Razoxane is sparingly soluble in water and, therefore, unlike derivative **352** cannot be readily administered parenterally [509]. The $ID_{50's}$ of probimane against S37 are 2.8 mg/kg ip. and 4.0 mg/kg ig. per day × 8 days [510]. At doses of 12.5 mg/kg, doses insufficient to inhibit the primary lesion of LL implants in mice, probimane decreases the number of the metastatic lesions to the lungs [510]. Probimane is more active in mice than bimolane against LL carcinoma, B16 melanoma, colon adenocarcinoma, and hepatoma 22 [510]. Moreover, probimane with a therapeutic index of 57 against S37 in mice potentiates irradiation-induced activity against both S37 and S180 [506]. In rats and dogs probimane shows no significant effect on liver, kidney, and lung functional tests, but causes some gastrointestinal disturbance and mild bone marrow depression [506]. Probimane is not mutagenic in the Ames, DNA damage-induction, or sister chromatid exchange tests. Nor does this derivative exhibit a significant effect on cell-mediated immunity in mice [506]. Cytokinetic studies indicate that probimane prolongs G_1 and S phases, and has little effect on the G_2 phase of Ehrlich ascites carcinoma (ECA) cells [511].

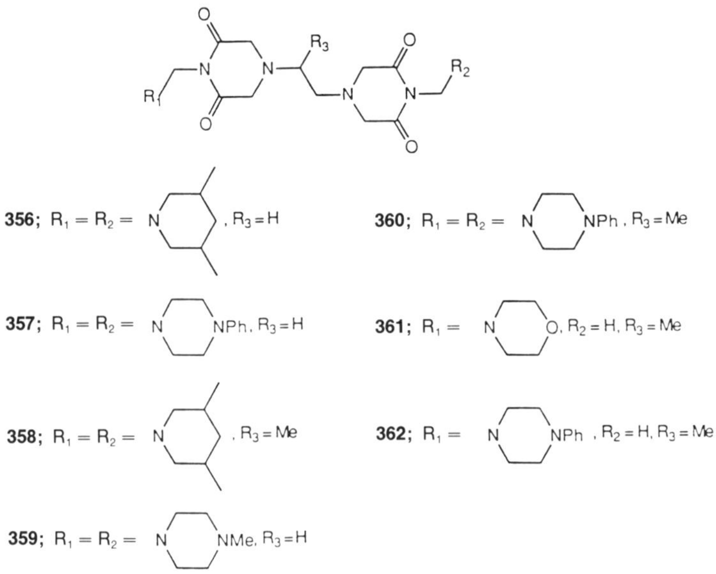

Bimolane und razoxane, when tested against several drug-resistant L1210 sublines [505], have moderate to full activity as compared to the parent L1210 line. The weakest effects of both drugs are on the vincristine-, doxorubicin- and daunorubicin-resistant sublines. In bimolane- and razoxane-resistant L1210 sublines, both drugs show complete cross-resistance, but sensitivity of both resistant lines to many structurally unrelated clinically useful antitumor drugs is similar to the parent L1210 line [505]. Bimolane and probimane inhibit incorporation of [^{3}H]TdR, [^{3}H]UR and [^{3}H]Leu into ECA cells, showing that they block DNA, RNA and protein synthesis [511]. Apparently, cells recover faster following removal of bimolane than after removal of probimane, but no quantification of these observations is available [511].

Bimolane is unstable, and the antitumor properties exhibited by this compound may reflect chemical hydrolysis or metabolism to ICRF-154 in biological systems [512]. However, bis(morpholinomethyl) derivatives of tricyclic bis(imides) **340–345** have activities which differ from the parent drugs. This suggests that bis(morpholinomethyl) species may have intrinsic stereoselective antineoplasic and antimetastatic properties. Perhaps, solubility and transport differences for the various derivatized stereoisomers may be responsible for stereoselective differences in biological activities [478].

Morpholinomethyl derivatives are easily prepared from the parent bis(dioxopiperazines) by reaction with morpholine and formaldehyde, and are of interest because of the important clinical results observed for bimolane. Comparative analysis of bis(morpholinomethyl) derivatives of tricycles **342** and **343** with their respective parent bis(imides) for antitumor effects in mice using a postamputation schedule LL model shows the bis(morpholinomethyl) derivative of *cis*-**343** to be the most effective inhibitor of metastasis [513]. Although bis(morpholinomethyl)-**343** provides the greatest number of 50-day survivors, the bis(morpholinomethyl) derivative of *trans*-**342** also exhibits considerable antimetastatic activity and is more effective than the parent trans bis(imide) **342**. The cis dioxopiperazine **343** is a better inhibitor of metastasis than *trans*-**342** in this assay. However, in the B16-F10 model, *trans*-**342** has greater antimetastatic activity than *cis*-**343** [496]. Examination of regioisomeric bis(imides) **344** and **345** and their corresponding bis(morpholinomethyl) derivatives in the LL model reveals that only the morpholinomethyl derivative of *cis*-**345** inhibits metastasis [498]. Since the parent bis(imide) *cis*-**345** does not exhibit any appreciable antimetastatic effect in this assay, it seems unlikely that the effect of bis(morpholino)-**345** is due to its hydrolysis to *cis*-**345**. Possibly, but not conclusively, morpholinomethyl derivatives of bis(dioxopiperazines) possess intrinsic antitumor properties independent of their obvious instability in aqueous systems and their potential as prodrugs.

Many N-acyloxymethyl derivatives **363–373** of ICRF-154 are available [514]. Antitumor activities depend upon the nature of the acyl moiety. Compound **372** (MST-16) is the most promising antitumor drug among these derivatives, and is more effective than ICRF-154 and razoxane against P388, L1210 and B16 tumors [514]. Compound **372** has low toxicity in both acute and subacute tests, but exhibits myelosuppression [514]. This drug prolongs cell cycle time and duration of G_2 and S phases in leukemia L1210 cells [507], and blocks G_2 to M phase transitions [507].

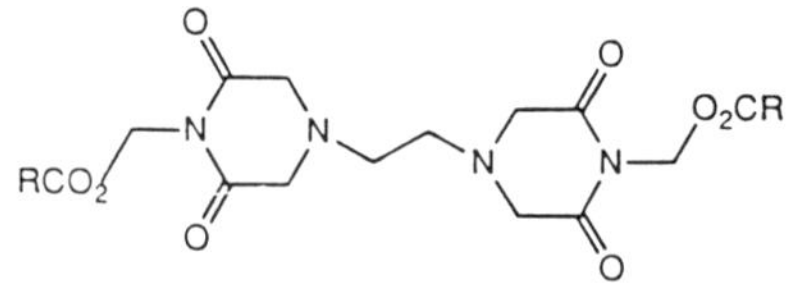

363; R = Me

364; R = 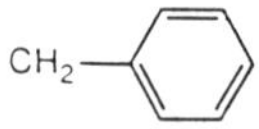

365; R = 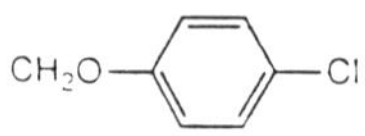

366; R = $CO(CH_2)_2CO_2H$

367; R = $CH_2NH_3^+CF_3COO$

368; R = 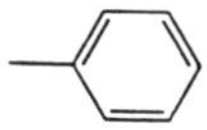

369; R = 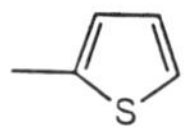

370; R = 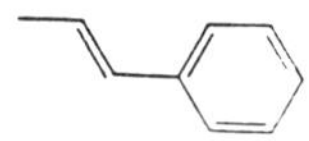

371; R = OCH_3

372; R = $OCH_2CH(CH_3)_2$

373; R = 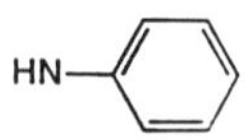

4.4 Cytotoxic and cytokinetic effects of razoxane, its (+) enantiomorph ICRF-187, and the bis(morpholinomethyl) derivative bimolane

Even though there are major solubility differences, the cytotoxic and cytokinetic effects of racemic razoxane and its (+) enantiomorph ICRF-187 in vitro are virtually identical [515]. A 28-hour exposure to 10 μg/ml is required for maximum cytotoxicity. When cytokinetic effects are determined via [^{3}H]thymidine incorporation and mitotic index, these drugs seem to act during the G_2 phase of the cell cycle. No detectable delay in cell passage through the G_1/S boundary or through S phase is observed, and inhibition of DNA synthesis occurs only after G_2 blockage prevents subsequent entry of cells into S phase. Depending upon drug concentration, some cells undergo DNA synthesis without cell division, and this results in a tetraploid cell population. Razoxane and ICRF-187 seem to be G_2-phase-specific agents [515].

Chinese hamster ovary (CHO) cell colony formation is inhibited by 50% when exponentially growing cells are exposed 8.5 hours at 10 μg/ml or for 24 hours at 3μg/ml of ICRF-187 [516]. Early G_1 and G_2 phase cells are more sensitive than are late G_1- or early and mid-S-phase cells. Cells in G_0 are unaffected by the drug at concentrations up to 500 μg/ml. Incubation of human lymphocytes with up to 50 μg/ml

razoxane has no effect on either cell viability or upon phytohemagglutinin stimulation. L1210 cell cycle progression through G_1 and S phases in the presence of 1–50 μg/ml of drug are slowed by 50–100 %. Accumulation of L1210 cells in G_2 phase is most pronounced and occurs almost immediately following addition of drug. ICRF-187 has a cytotoxicity range and cytokinetic effects similar to that of racemic razoxane [516].
Following pulse labeling of S180 cells with [^{3}H]TdR, one ip. dose of bimolane (50 mg/kg) causes the labeling index to drop from 31 % to 15 % in 24 hours [517]. After pulse labeling with [^{3}H]TdR and blocking the S180 cells in metaphase with colchicine, ip. injection of bimolane (50 mg/kg) significantly decreases the percentage of labeling mitosis (PLM). Presumably, bimolane damages S phase cells and prevents cell progression from S to M phase [517]. Incorporation of [^{3}H]TdR into DNA is inhibited significantly after 6 hours of administration of bimolane [517]. The fraction of cells in G_0 phase is approximately 50 % after 14 hours. Little damage occurs in G_0 phase cells [517]. In DBA/$2XC_{57}BL$ rodents the bis(morpholinomethyl) analogue (50 or 25 mg/kg ip.) retards the progression of G_2 to M phase in implanted L1210 cells; this leads to an accumulation of cells in G_2 phase [518, 519].

4.5 Mutagenic effects

Razoxane induces cell polyploidy [520]. Under certain conditions, this bis(dioxopiperazine) prevents cell division in EMT6 mouse tumor cells, but allows DNA synthesis to continue at almost the normal rate [521]. DNA content of cells exposed to 10 μg/ml of razoxane is 4 times that of normal G_1 cells after a 24-hour exposure, and the DNA content continues to increase with drug-exposure time; if drug is removed after 24 hours, the cells divide and begin to proliferate [520]. For the 10 % of surviving EMT6 cells, 95 % are polyploid, having double the usual complement of DNA present in untreated cells [520]. Cells surviving bis(dioxopiperazine)-treatment form colonies and all cell populations arising from these colonies are able to grow as a solid tumor following inoculation into BALB/c mice [520].
The specific chromosomal changes induced by razoxane are restricted to eukaryotic organisms [522]. Razoxane (up to 5 mg/plate) does not cause an increase in the mutation frequency in five strains of *Salmon-*

ella typhimurium (TA 1535, 1537, 1538, 98, 100) with or without the liver S9-mix [522]. Razoxane (200 and 400 mg/kg ip.) is cytotoxic to bone marrow cells, and causes 5-fold increases in micronucleated polychromatic erythrocytes when administered to male and female $CCBF_1$ mice. In Chinese hamsters, razoxane (up to 500 mg/kg orally) induces abnormal chromosome condensation along with 7-fold increases in structural chromosome aberrations. The number of polyploid cells is increased 8-fold in the Chinese hamster metaphase assay [522].

4.6 Immunosuppressive properties

Cytostatic and cytotoxic antineoplastic drugs are expected to have embryopathic [523] and immunosuppressive properties [524]. ICRF-154 and razoxane produce little or no delay of allograft rejection in mice even at near lethal doses of the drugs [525]. Low doses of the drugs (7.5 and 15 mg/kg, respectively) inhibit the graft vs. host reaction in mice; stronger inhibition of the response is observed with single large doses (145 mg/kg) of razoxane in rats. Response differences may be a function of differences in proliferative activity of immunocompetent cells. Small numbers of cells surviving drug treatment may be responsible for allograft rejection [525].

Bimolane suppresses the plaque forming cell response and decreases the number of specific rosette forming cells in mice previously sensitized with sheep red blood cells [526]. Bimolane (6.25 mg/kg/day for 7 days, ip.) inhibits the immediate swelling response measured 3 hours after initiation of delayed hypersensitivity in mice caused by a challenge to heat aggregated bovine serum albumin. At higher doses (50 mg/kg/day for 7 days, ip.) this bis(morpholinomethyl) derivative also suppresses the delayed swelling response determined 24 hours after challenge [526]. Apparently, bimolane inhibits both humoral or cellular immunity in normal mice, with the inhibitory action on humoral immunity being most pronounced [526].

4.7 Treatment of psoriasis

Razoxane is effective in the treatment of severe cutaneous and arthropatic psoriasis. The rational for use of the 2,6-dioxopiperazine in such situations is based upon the similarity of the microvasculature in solid malignant tumors and the microvasculature associated with rapid epidermal proliferation in psoriasis [527]. Razoxane normalizes the microvasculature in the primary tumor [527]. The effectiveness of this drug in psoriatic arthritis might be related to the drug's ability to inhibit collagenase via stimulation of a specific tissue metalloproteinase (TIMP) [528]. In monolayer cultures of rabbit chondrocytes stimulated to produce collagenase with conditioned medium from human peripheral blood mononuclear cells, collagenase production is inhibited and TIMP increased in a dose-dependent manner when cells are treated with the dioxopiperazine daily for 3 days. During this period the effect of razoxane is progressive; 50 μg/ml or less have no effect at day 1, but 50 μg/ml is effective by day 3. Upon drug removal stimulated chondrocytes recover their ability to produce collagenase, and TIMP production returns to near normal [528].

4.8 Embryogenic effects

Razoxane has mainly embryolethal effects in pregnant animals (BALB/c Mice, Sprague-Dawley and Wistar rats, New Zealand White Rabbits) [529]. The margin between no apparent activity and total embryolethality is narrow, but only a small fraction (4/63) of mice fetuses are malformed with some exhibiting retarded development following drug treatment. Similar results are observed in rats, and no malformations are observed in rabbits. The time of maximum sensitivity to the drug is early in gestation, between days 6–10 in mice, 6–8 in rats and 7–10 in rabbits [529].

4.9 Synergism with antitumor drugs

Razoxane is active in L1210 and S180 tumor models at doses of 30 mg/kg for 13 days and 30 mg/kg for 5 days [487], respectively, but these doses are high when compared to many clinically important antineoplastic drugs [479]. Nonetheless, treatment of early L1210 leukemia with a combination of razoxane and cis-platin increases the life span of treated mice 2- to 6-fold over the median survival time (MST) found when either drug is administered alone [530]. In advanced systemic disease, the increase in life span observed with the combination is approximately twice that found when cis-platin is given alone. The optimum combination dose of cis-platin is one-half the amount necessary when employed as a single agent. Razoxane plus daunorubicin also demonstrates enhanced therapeutic responses [531]. The optimum combination (6 mg/kg daunorubicin plus 200 mg/kg of razoxane) increases the MST 2-fold over the optimum dose of either drug individually. In fact, many antitumor drugs including daunorubicin, doxorubicin, camptothecin, cis-platin, and cytosine arabinoside, employed in combination with razoxane, have increased efficacy against leukemia L1210 [532]. Combinations of razoxane with many other antitumor drugs (harringtonine, coralyne, anguidin, chromomycin A_3, 5-azacytidine, 5-HP, camptothecin, methotrexate, cyclophosphamide, ellipticine) produce only a slightly greater effect than that found when either agent is used alone [478,532].

Synergism between ICRF-187 and doxorubicin is observed in the S180 cell line [533]. Incubation for 1 hour with a clinically achievable concentration of either doxorubicin (0.1 μg/ml) or ICRF-187 (10 μg/ml) diminishes S180 colony formation by 20 % and 44 %, respectively. When cell suspensions are incubated simultaneously with both drugs for 1 hour, an additive effect is observed wherein colony formation decreases 15 % over that with ICRF-187 alone. However, combinations of the dioxopiperazine (0.1 μg/ml) and doxorubicin (0.01 μg/ml) produce a 64 % reduction in S180 colony formation following 24-hour incubations. At these concentrations neither drug is effective alone [533].

The activity of either razoxane or its (−) enantiomorph ICRF-186 on HeLa Cell colony inhibition in vitro in combination with daunorubicin or doxorubicin are similar [534]. Both dioxopiperazines increase the cytoxicity of daunorubicin, but not of doxorubixin, a compound which differs from daunorubicin by addition of one side chain hydroxyl group. Concentrations (0.003 or 0.006 μg/ml) of danurobicin provide 26–65 % inhibition of cloning efficiency. The ID_{50} for daunorubicin alone is 0.009 μg/ml, but in combination treatments with ICRF-186 (0.07–0.3 μg/ml; ID_{50} > 1.2 μg/ml) the ID_{50} is reduced to 0.015 μg/ml. No interaction with doxorubicin cytotoxicity is found at low concentrations of razoxane, but at high concentrations (0.62 or 2.5 μg/ml) there is a synergistic cytotoxic interaction between the two drugs. The (−) enantiomorph of razoxane (ICRF-186) has no synergistic interaction with doxorubicin at any dose tested (0.07, 0.15, 0.5, 0.6 and 1.2 μg/ml). Differences in activity observed for these 2,6-dioxopiperazines in combination with either daunuorubicin or doxorubi-

cin cannot be explained by a difference in uptake of the two anthracyclines by the HeLa cells [534].
Twenty-four-hour incubations of S180 cells with either doxorubicin or the (+) enantiomorph of razoxane (ICRF-187) produces a dose-dependent partial G_2M arrest [535]. At high doses, the dioxopiperazine also produces an inhibition of cell division with continued DNA synthesis at a higher ploidy. This results in a second G_2M arrest at 8n population. Co-treatment with the dioxopiperazine and doxorubicin results in enhanced cell cycle blockade at G_2M greater than that of doxorubicin alone. The combination also produces enhanced lethality, a property not associated with enhanced [^{14}C]doxorubicin accumulation by S180 cells. The synergistic interaction between the two drugs seems to involve site-specific enhancement of the anthracycline lethal and cytokinetic effects by the dioxopiperazine at the genome level [535].

4.10 Synergism with radiation on tumor growth

In addition to the synergistic effects of bis(2,6-dioxopiperazines) with other antineoplastic drugs, these substances potentiate the effects of radiation without simultaneously increasing toxicity [536–538]. Combinations of razoxane and radiation exert more than an additive effect on inhibition of S180 tumor growth in mice [536]. A particularly effective combination employs 5 treatments of 1.25 mg/kg of razoxane and 1.25 Gy (100 rads = 1 Gy). This combination inhibits tumor growth to the same extent as 5 treatments of 2.5 Gy alone [536]. Razoxane administered at the same time as radiation seems to be the most effective combination schedule.
Razoxane enhancement of radiation response is also observed using the LL tumor in BDF_1 mice [539,540]. Razoxane (25 mg/kg) administered at 3-hour intervals × 4 before irradiation significantly slows tumor growth relative to a single 100 mg/kg treatment prior to irradiation [540]. Both single and multiple drug treatments produce a transient inhibition of cell proliferation. Protracted drug administration prior to irradiation markedly decreases the postirradiation proliferative recovery of the tumor without significantly altering intestinal recovery [539]. Single acute doses of razoxane fail to alter the growth of LL, and when combined with radiation fail to enhance the radiation effect, but fractionation of the drug results in immediate synergistic

tumor growth inhibitory effects [540]. Thus, low doses of razoxane for extended periods of time are more effective in enhancing radiotherapy than a single high dose [539,540].

Using EMT6 tumor cells in vitro, razoxane potentiates the effects of radiation only when cells are exposed to the drug for more than 10 hours prior to irradiation [538]. No potentiation is seen if the drug is administered after irradiation. The dioxopiperazine produces radiosensitization of log phase cells, but not of plateau phase cells. The cytotoxic effect in EMT6 mouse tumor monolayer cultures is proliferation-dependent [521]. Razoxane prevents cell division while allowing DNA synthesis to continue. This anti-mitotic action is closely related to the cytotoxic effect of the drug. Increases in sensitivity to X-radiation are observed in cultures pretreated for 24 hours with 200 μg/ml of the dioxopiperazine [521].

The interaction of cis-platin and irradiation is a promising combined modality treatment of malignant disease [541]. The effect of razoxane (25 mg/kg every 3 hours $\times$ 4) pretreatment on the antitumor activity of this combination is of interest [541]. These three agents, given alone as single treatments, provide for a transient interruption of LL proliferative activity and only slight to moderate alterations of tumor growth. However, multidose razoxane treatment, provides both low cytoxicity and a partial cytokinetic redistribution of tumor cells, thus enhancing radio- and chemotherapy with cis-platin. For cis-platin-radiation combinations, wherein cis-platin precedes irradiation, increasing the interval between these two treatments from 6 to 24 hours results in synergistic interactions and increased treatment efficacy [541]. When cis-platin-radiation combination follows fractionated administration of razoxane, tumor regression is greatly enhanced. This suggests that relatively non-cytotoxic dioxopiperazines may be used to sensitize low growth fraction-resistant tumors to combination radio- and chemotherapy [541].

The synergistic effect of bimolane (25 mg/kg) and radiation (2.5 Gy/day) in S180 and S37 infected mice is similar to that of a radiation dose of 7.5 Gy/day [542]. Potentiation of the antitumor effect of 25 mg/kg bimolane per day is roughly equivalent to that of a radiation dose of 5 Gy/day [542]. In conjunction with 30 Gy γ-ray irradiation, bimolane (10–100 μg/ml) causes approximately 30% of the cellular DNA to crosslink with protein and induces equal amounts of DNA single-strand breaks [543].

4.11 Antimetastatic properties

A major problem in cancer therapy involves formation of distant metastatic tumors. Development of selective antimetastatic drugs is of considerable significance [544]. Agents classified as antimetastatic are expected to prevent metastasis by a mechanism(s) unrelated to a direct tumoricidal effect [545].
Razoxane inhibits metastasis in a LL carcinoma model [546], and the conformationally constrained cyclopropanediyl analogues **337** and **338** exhibit stereoselective pro- and antimetastatic effects, respectively [493,494]. Tricyclic derivatives also exhibit interesting stereoselective antimetastatic effects [496,498,513]. Razoxane inhibits pulmonary metastases at doses which have little or no effect on the primary tumor. In such studies [546], all control mice show multiple metastatic lung nodules 21 days after tumor implantation, but only five out of the twenty-seven animals treated with the bis(2,6-dioxopiperazine) (30 mg/kg/day for 10 days) have a few small nodules; no significant effect is observed on the weight of the primary tumor. Razoxane may be involved in normalizing the developing tumor blood vessels (angiometamorphic effect) and therefore preventing the release of malignant cells into the circulation, but antimetastatic effects seen with pretreated cells cannot be attributed to normalization of tumor vasculature by the drug [527,547,548]. Significant reductions in metastases are observed at a dose of 12.5 mg/kg of razoxane for 7 days [549].
Studies in vitro with two different lines, C108 and BC215, derived from LL metastases reveal razoxane (200 μg/ml) to be more active against the more metastatic (C108) tumor line [550]. After 8 hours of exposure the surviving fraction is 3.2% for the C108 line and 15% for the BC215 line. In C57BL/6 mice, no effect on the primary tumors is observed after razoxane treatment (30 mg/kg/day $\times$ 5 days, ip.), but an antimetastatic effect is obtained for both the parent C108 line (87%) and its more metastatic subline M1087 (83%) as a function of metastasis reduction. The less metastatic subline BM21548 exhibits a decreased antimetastatic activity (57%) as a result of dioxopiperazine treatment. Thus, the antimetastatic effect of razoxane does not appear to depend entirely on an angiometamorphic effect [550], a conclusion also reached previously based upon studies employing geometric isomers of bis(dioxopiperazines) [494,496].

A better response of LL-derived lines to the dioxopiperazine may be achieved by use of a proper fractionation regimen taking into account the delayed metastasis [551]. Secondary tumors from poorly responsive lines are greatly affected by the fractionation regimen. It seems that a better secondary response is observed after repeated doses of razoxane against moderately advanced tumors than that obtained as a result of surgical therapy [551].

In other studies employing LL lines having different metastatic potentials, a significant prolongation of the survival time of mice is achieved with razoxane (57 mg/kg/day) only after presurgical treatment (7–11 days) [552]. The therapeutic effect is greater using the tumor line with low metastatic ability. The preoperative treatment of the animals with razoxane markedly increases the responses of cyclophosphamide to the low responding M1087 line [552].

Treatment of R3327 MAT-LyLu, a lymphotrophic metastatic prostate-derived rat tumor, with razoxane (20 mg/kg/day) inhibits both the primary tumor growth and the development of metastases [553]. The time required for 50% of the tumors to reach an average diameter of 1.0 cm is increased from 10 (control) to 14 days following drug treatment. Final tumor wet weights are also significantly lower (23%) than that found in the control group [553]. Razoxane suppresses growth of the primary tumor, but is most effective in preventing metastases. For the dioxopiperazine-treated animals, 68% appear to be free of lymphatic metastases and 78% free of observable lung metastases [553].

Razoxane decreases the incidence of metastases following local irradiation of KHT sarcomas in mice [554]. Irradiation (60 Gy) delivered in fractions over 12 days reduce tumor size 55%, but 83% of the mice die from metastases. When razoxane is used to partially synchronize the tumor growth fraction in a radiosensitive state of the growth cycle and promote normalization of the tumor vasculature, combinations of drug with radiation effectively increase local control and decrease the incidence of metastasis [only 1/18 of treated mice have metastatic lesions] [554].

When used at equitoxic doses in LL-infected BD2F1 mice, razoxane (25 mg/kg/day), 4-(3,3-dimethyl-1-triazeno)benzoic acid potassium salt (DM-COOK; 50 mg/kg/day), and 5-(3,3-dimethyl-1-triazeno)imidazole-4-carboxamide (DTIC; 60 mg/kg/day) reduce the number of circulating tumor cells [555]. The most pronounced effects are observed with DM-COOK, which causes the complete disappearance of tumor cells by days 8 and 12. These investigations, employing flow cytometry measurements, corroborate previous studies involving razoxane-induced reduction of circulating tumor cells as assessed by microscopic examinations of blood smears [556].

When comparing the antitumor and antimetastatic effects of razoxane in mice-bearing P388 or L1210 leukemias, and TLX5 lymphoma with cytotoxic (cyclophosphamide, CCNU, GANU) and antiinvasive (vincristine, vinblastine) drugs, or with antimetastatic DM-COOK, different responses are observed for the three tumor lines [557]. Razoxane exerts the greatest response against L1210 leukemia, but is inactive against TLX5 lymphoma and P388 leukemia. DM-COOK shows the greatest response against TLX5 lymphoma, and vincristine against P388 leukemia. Leukemic infiltration of the brain after ip. tumor implantation is reduced by treatment with all drugs except razoxane [557]. Lack of antimetastatic effects observed for razoxane under these experimental conditions may indicate that the angiometamorphic action can only be expressed in a solid tumor [557]. Since razoxane increases the survival time of leukemic mice in ways unrelated to tumor cytotoxicity, the mechanism by which the dioxopiperazine prolongs survival requires additional study [557].

Examination of sections of LL primary tumors derived from the M1087 line show that treatment with razoxane and DTIC, but not DM-COOK, causes cytoplasmic vacuolization and nuclear pycnosis. Razoxane and DTIC reduce markedly the number and size of lung metastases in treated mice; with DM-COOK no lung metastases are detected [558]. Following treatment with these drugs, the histological appearance of the line with high metastatic potential (M1087) becomes similar to that of the line with low metastatic potential (BM21548). It seems that metastatic potential is related to parenchymal organization of the primary tumors; large hemorrhagic areas with detached tumor cells and no endothelialized capillaries are observed in the high metastatic potential line [558].

In a transplantable slowly growing osteosarcoma rat model, characterized by osteoid formation and spontaneous metastases to lungs, kidneys and lymph nodes, razoxane (ip.) administered from 2 days before to 14 days after tumor transplantation (30 mg/kg or 10 mg/kg per day) causes a dose-dependent prolongation of median survival time (83 or 48 days, respectively vs. 38 days for the control group) [559]. No

effect by the dioxopiperazine on primary tumor growth is observed in this model [559]. Razoxane treatment from day –2 to + 14 at 30 mg/kg ip. shows a greater inhibition of lung metastases than treatment from day + 14 to + 28 after transplantation using the same dose. About 60% of the total sectional area of the lungs in the control animals is covered by osteosarcoma metastases in these studies, but this is reduced to about 3 or 26%, respectively, in the early and late treatment groups [559]. The major side-effect is reversible diffuse alopecia [559].
Razoxane blocks invasiveness (42% at 12.5 μg/ml) of the human melanoma cell line A375M into a basement-membrane-like structure composed of polycarbonate filters coated with Matrigel [560]. In this system invasive potential correlates with experimental metastatic potential in mice. However, the more water-soluble (+) enantiomorph of razoxane (i.e. ICRF-187) at 1 μg/ml does not inhibit invasiveness [560]. Unfortunately no data for the (+) enantiomer is available at 12.5 μg/ml [560].
Agents that inhibit platelet aggregation often prevent metastasis [561]. Razoxane markedly inhibits the platelet aggregation in blood vessels, on the surface of rat brain, and in the hamster cheek pouch [561]. However, ICRF-158 (**374**), an N-methylated analogue of ICRF-154, not having antimetastatic properties, inhibits thrombus formation in vivo to the same extent as razoxane. Thus, it appears that the antimetastatic effect of bis(dioxopiperazines) is not related to their anti-aggregatory activities [561].

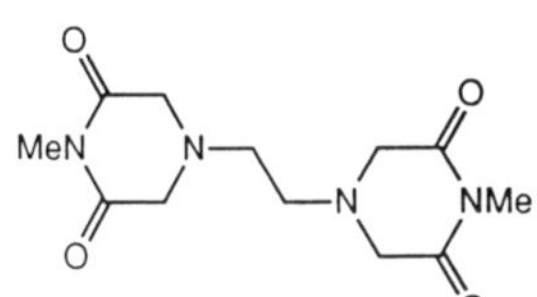

374

Razoxane and its *cis*-**338** cyclopropanediyl analogue reduce the metastatic growth of bronchogenic adenocarcinoma LG1002, but are without effect on the growth of the primary tumor [493]. In contrast, *trans*-**337** stimulates both lung metastasis and primary tumor growth [493]. Similar stereoselective effects of isomers **337** and **338** are observed on lung metastasis of B16–F10 melanoma in mice [494]. Pretreatment of B16–F10 cell lines for 24 hours with *trans*-**337** (2 μM and 20 μM concentrations) results in an increase in lung melanoma colony formation. Such pretreatment with *cis*-**338** reduces metastasis to the lung. Razoxane, like *cis*-**338**, also is antimetastatic at these doses [494]. Results in culture are similar to those in vivo for razoxane and *trans*-**337**. Isomer *cis*-**338**, however, stimulates colony formation in vitro. Results of injecting pretreated cells into mice indicate that the antimetastatic effects of razoxane and *cis*-**338** can be independent of an angiometamorphic mechanism. Intraperitoneal injection (30 mg/kg) of either *trans*-**337** or *cis*-**338** into tumor-bearing mice accelerates tumor growth. Such tumors are detected six days after implantation and grow to a size of 100–250 mm^2 after 22 days. In animals treated with a similar dose of razoxane, tumors do not appear until 17 days and are only 30 mm^2 in size after 22 days. In control animals, tumors appear 12–15 days after implantation and reach a size of 60–80 mm^2 by day 22. These results correlate with stereoselective stimulation of colony formation in vitro.

Conformationally constrained bis(imides) **342–345** also exhibit stereoselective antimetastatic properties. Whereas *trans*- and *cis*-tetraazaperhydrophenanthrenes **342** and **343** have a cisoid relationship of 2,6-dioxopiperazine functions, *trans*- and *cis*-tetraazaanthracens **344** and **345** have extended dioxopiperazine spatial orientations. Pretreatment of B16–F10 melanoma cells with *trans*-**342** significantly reduces metastasis to the lungs of mice at all dose levels. The cis isomer **343** is ineffective in this model [496]. In the LL model, the bis(morpholinomethyl) derivative of cis-syn-trans **343** is more active than the bis(morpholinomethyl) derivative of trans-anti-trans **342**, with parent imide **343** followed by parent imide **342** being least antimetastatic in the postamputation protocol [513]. Thus, a reverse in stereoselectivity is observed between the B16–F10 and LL models. Regioisomeric bis(imides) **344** and **345** with extended conformations are inactive, but the bis(morpholinomethyl) analogue *cis*-**345** is antimetastatic [498]. Possibly, activities observed for morpholinomethyl derivatives reflect differences in solubility and delivery (prodrug), or an intrinsic antitumor activity of the alkylating morpholinomethyl-N functionality [496], not unlike antineoplastic activities attributable to hydroxymethyl-N-metabolites (carbinolamines) of useful drugs [562].

Bimolane (**351**), the bis(morpholinomethyl) derivative of ICRF-154 (**325**), inhibits tumor growth slightly, but markedly inhibits the spontaneous pulmonary metastases of LL in C57BL/6 mice [508]. Inhibition of lung metastases is 93%, and the decrease in primary tumor weight is 32% at 25 mg/kg [508]. Bimolane moderately elevates cAMP levels in primary LL tumor tissue, but does not effect the cAMP levels in plasma of C57BL/6 mice bearing the LL tumor [508].

4.12 Amelioration of anthracyclin-induced toxicity in animals

In addition to antimetastatic and synergistic antitumor properties, bis (2,6-dioxopiperazines) effectively reduce anthracycline-induced toxicity, a toxicity characterized by diffuse myocardial injury leading to irreversible cardiomyopathy [563]. The cardiotoxicity may be acute, subacute, or chronic [564]. Acute cardiotoxicity is characterized by hypotension and electrocardigraphic changes, likely a function of drug-induced histamine release. Subacute toxicity is rare and in the clinic manifests itself in the form of transient myocarditis and pericarditis. Chronic toxicity is particularly important constisting of an often fatal dilated cardiomyopathy which may be of delayed onset. This usually occurs when the drug is given in cumulative doses exceeding 450 mg/m^2 [564]. Anthracyclines also induce increases in perfused dog heart coronary perfusion pressure, and pretreatment with either EDTA (**323**) or razoxane attenuate this acute toxic effect [565]. Daunorubicin (50 mg/kg) administered iv. to Syrian golden hamsters causes 100 % lethality within 2 to 3 days, but pretreatment with razoxane (100 mg/kg ip.) prior to daunorubicin achieves a 50 % survival over 21 days [566].

In mice, the magnitude of protection is altered by varying the timing of razoxane treatment in relation to daunorubicin administration [567]. Treatment simultaneously or 24 hours prior to daunorubicin provides the best protection [567]. In hamsters, the (+) enantiomer ICRF-187 [(100 mg/kg) given 48 hours before daunorubicin (25 mg/kg)] increases the survival time to over 8 weeks compared to the 1–3 weeks found in animals given the anthracycline alone. The number of animals surviving increases from 10 % to 45 % [568]. The survival rate is only 20 % with a 24-hour dioxopiperazine pretreatment. Pretreatment 3 hours before or 3 hours after daunorubicin administration provides the highest survival rate (50–80 %). Unlike mice, no hamsters survive when the bis(dioxopiperazine) is given 12 or more hours after daunorubicin [568].

Amelioration of daunorubicin-induced toxicities in mice by razoxane is not mediated through an alteration in daunorubicin metabolism or its complexation with DNA [569]. Daunorubicinol and the deoxyaglycone are two major metabolites of daunorubicin. Pretreatment with razoxane (100 mg/kg/day × 3 for mice and 80 mg/kg/day × 3 for rats) does not alter such metabolic routes [569]. In vitro, daunorubicin reacts with calf thymus DNA reaching a steady state within two minutes. Preincubation of DNA with the dioxopiperazine does not alter the capacity of DNA to interact with daunorubicin [569].

BDF_1 mice, pretreated with razoxane (200 mg/kg) survive three daunorubicin injections (6 mg/kg) without exhibiting any serious toxicity [570]. The ultrastructure changes in the myocardium induced by high doses of daunorubicin in mice consists of sarcoplasmic translucency, vacuolation of membrane-limited components, and degeneration of mitochondria and lysosomal aggregates [570,571]. Mice pretreated with the dioxopiperazine either fail to show or exhibit minimum ultrastructural alterations in the myocardium [570,571].

Doxorubicin (2 mg/kg ip.) -induced nephropathy in rats, characterized by proteinuria, hyperlipidemia, and glomerular and tubular changes, and degenerative changes of peripheral nerves, is significantly decreased by razoxane administration [572]. In contrast, razoxane enhances the depressant effects of doxorubicin on bone marrow function. Doxorubicin reduces body weight gain, causes ascites, decreases heart and thymus weight, and increases liver and kidney weight; such changes are also inhibited or attenuated by razoxane-pretreatment.

Razoxane (100 mg/kg ip.) does not increase the survival time of doxorubicin-treated (10 mg/kg iv.) rats, but histological findings show a significant prevention of doxorubicin-induced cardiomyopathy by the dioxopiperazine [573]. The total myocardial calcium content of animals treated with doxorubicin is significantly higher, but no difference is seen in animals pretreated with the dioxopiperazine as compared with controls [573]. EDTA (**323**) (124 mg/kg) reduces myocardial calcium levels in doxorubicin-treated animals to near normal values, but the histological cardiac alterations are not prevented [573]. It seems that myocardial calcium metabolism does not have a primary role in the pathogenesis of doxorubicin-induced cardiomyopathy, and that the effect of razoxane on the myocardium is due to other mechanisms [573].

Protection against acute daunorubicin-induced toxicity is also observed in hamsters pretreated with the more water soluble ICRF-187 dioxopiperazine [574]. Differences in protection with varying pretreatment times is observed in the hamster model of daunorubicin toxicity. By 3 weeks only 10% of the hamsters treated with daunorubicin alone (25 mg/kg) are alive, whereas 80% of the animals pretreated with ICRF-187 (100 mg/kg) survive. The highest number of animals surviving daunorubicin administration occurs in those groups where ICRF-187 is given from 3 hours prior to or up to 3 hours after daunorubicin administration [575]. To determine whether ICRF-187 attenuates chronic daunorubicin cardiotoxicity, rabbits were given 3.2 mg/kg ICRF-187 daunorubicin alone or 30 minutes after 12.25 or 25 mg/kg ICRF-187 at 3-week intervals. Administration of daunorubicin to rabbits produces cardiac lesions consisting of cytoplasmic vacuolization and myofibrillar loss. Pretreatment with ICRF-187 eliminates these abnormalities in four out of twelve rabbits and reduces the severity of the lesions in the other eight animals [576]. ICRF-187 also protects against doxorubicin cardiotoxicity. Myocardial alterations such as cytoplasmic vacuolization and myofibrillar loss are absent in four out of six beagle dogs treated with ICRF-187 (12.5 mg/kg) 30 minutes before doxorubicin (1.0 mg/kg) administration [567,577]. Minimal cardiac lesions are found in the remaining two animals. All dogs given doxorubicin alone have severe myocardial lesions.

A long-term protective effect against chronic daunorubicin cardiotoxicity can be demonstrated in rabbits dosed with daunorubicin (3.2 mg/kg once every 3 weeks × 6) and ICRF-187 (25 mg/kg). In contrast to previous studies this experiment was terminated 3 months after the last dosing. All rabbits given daunorubicin alone show evidence of myocardial alterations ranging from minimal (two animals) to mild (five animals) [578–580]. Pretreatment with the dioxopiperazine causes a significant reduction in both the incidence and the severity of cardiac lesions. Hearts from five of seven of the animals given the combination of dioxopiperazine and daunorubicin are normal; myocardial alterations are minimal in the remaining rabbits treated with the protecting drug. Thus, ICRF-187 provides true protection against anthracycline-induced cardiomyopathy, as opposed to only causing a delay in the onset of the toxicity [578]. Heart lesions in rabbits, dogs and miniature pigs associated with anthracycline administration consists mainly of cytoplasmic vacuolization and myofibrillar loss [577,581]. In a variety of animals, pretreatment with ICRF-187 reduces both the incidence and severity of myocardial lesions. Anthracycline treatment is associated with decreases in red and white blood cells, hemoglobin, and serum iron concentrations. ICRF-187 does not prevent or enhance this activity. Thus, concurrent administration of ICRF-187 may offer a means of reducing chronic anthracycline-induced cardiotoxicity without enhancing alterations in other tissues [577,581]. ICRF-187, however, does not affect doxorubicin-reduced histamine responsiveness of right atria in vitro. Thus, this dioxopiperazine may not be specific for all the cardiac affects of doxorubicin [582].

The cardiotoxicity of anthracyclines may be mediated through toxic active oxygen species generated by reactions catalyzed by the antitumor drug [583,584]. Free radical scavengers such as vitamin E, cysteamine, and N-acetylcysteine (NAC) exhibit protective effects against the acute toxicity in mice produced by single large doses of the anthracyclines [584–586]. In studies using miniature swine and beagle dogs, ICRF-187, but not vitamin E or NAC, protects against doxorubicin-induced cardiomyopathy [587–589]. The cardiac alterations observed are similar to those seen in other studies using pigs, dogs and humans treated with doxorubicin [577,590,591]. On a lesion grading scale of 0–4+ with 4 being very severe, miniature swine given only doxorubicin (1.6 mg/kg at 3-week intervals × 6) display grade 1–3 lesions. All seven of the miniature swine receiving the combination of vitamin E (2500 IU twice daily for 4 days, 500 IU twice daily for 4 days and 500 IU twice daily for 17 days) and doxorubicin (1.6 mg/kg once every 3 weeks × 6) demonstrate lesions, but the severity is less in six of these animals [588,589]. All miniature swine receiving doxorubicin (2.4 mg/kg every 3 weeks × 6) develop myocardial lesions of grade 2–3 severity. In contrast, a significant decrease in the severity of myocardial lesions are found in the hearts of all seven miniature swine given this dose of doxorubicin with ICRF-187 (five = lesion score of 0.5–1, and two = lesion score of 0) [588,589]. Severe alterations (grade 3) are found in the hearts of five of six dogs given doxorubicin (1.75 mg/kg × 7) alone [587,589]. Lesions of similar severity are found in animals treated with NAC (200 mg/kg ip.) prior to doxorubicin. In contrast, no abnormalities are observed in the hearts of three of six dogs treated with ICRF-187 alone, or in four of six dogs receiving a combination of ICRF-187 and NAC. The remaining hearts from both these groups show minimal alterations (lesion score = 1) [587–589]. Thus, ICRF-187 is clearly superior to either vitamin E or NAC as a means of preventing doxorubicin-induced cardiotoxicity.

ICRF-187 allows a marked increase in total cumulative doses of doxorubicin tolerated by beagle dogs [592,593]. Animals given doxorubicin alone (1.75 mg/kg) every 3–4 weeks) die or are in poor condition after cumulative doses of 12.25–14 mg/kg (7–8 dosings). All of these animals show severe myocardial lesions (lesion score = 3). In these studies, four of eight animals receiving ICRF-187 (25 mg/kg) and doxorubicin die; two after a cumulative dose of 35 mg/kg (20 dosings), one after 43.75 mg/kg (25 dosings), and one after 52.5 mg/kg (30 dosings) [592]. Two animals survive 30 dosings (52.5 mg/kg). Even at these high cumulative doses three animals show no myocardial lesions (lesion score = 0), three demonstrate minimal lesions (lesion score = 1), and three exhibit lesions of mild severity (lesions score = 2). Thus, ICRF-187 provides significant protection when administered with doxorubicin over 90 weeks and makes it possible to give doses of doxorubicin that would otherwise be lethal [592,593].

Pretreatment of either spontaneously hypertensive rats (SHR) or normotensive rats (WKY) with ICRF-187 attenuates doxorubicin-induced loss in body weight in both types of animals [594]. Myocardial alterations develop in all SHR and most WKY rats treated with doxorubicin (1 mg/kg/week × 12) alone. Myocardial lesions are more severe in SHR than WKY rats. Pretreatment with ICRF-187 significantly attenuates the severity of the heart lesions in both SHR and WKY. ICRF-187 also reduces the severity of renal lesions involving glomerular vacuolization and tubular dilatation found in both strains of rats [594].

Epirubicin is the 4′-hydroxyl epimer of doxorubicin. Morphologic studies show cardiomyopathy and nephropathy in SHR given epirubicin alone (1.5 mg/kg, 12 weekly injections; total cumulative dose 18 mg/kg) is similar to toxic manifestations illicited by doxorubicin in these tissues [595]. Animals receiving ICRF-187 (50 mg/kg) 30 minutes before epirubicin show a marked reduction in the severity of cardiomyopathy and a moderate reduction in nephropathy. Again, it would seem that the dioxopiperazine should be useful for preventing cardiomyopathy by this anthracycline, particularly in situations in which epirubicin may have to be given either in large doses or to patients at high risk of developing drug-related cardiotoxicity [595].

In acute structure-activity studies involving daunorubicin-induced toxicity in hamsters, 70% of the animals pretreated with ICRF-187 (50–200 mg/kg ip.) survive 8 weeks, whereas 90% of the animals die by

4 weeks when daunorubicin (25 mg/kg iv.) is given alone [596]. Both optical isomers of razoxane, (−) ICRF-186 and (+) ICRF-187, are equally protective. This indicates the mechanism of the protective activity is a non-enantioselective process [596]. The ethyl analogue ICRF-192 (**329**) (50–200 mg/kg) only provides 20% survival after 8 weeks. Other analogues and their maximum protective effects (%) under these conditions include ICRF-154 (**325**) (4–15%), ICRF-239 (**375**) (10–20%), ICRF-197 (**376**) (10%), ICRF-161 (**377**) (10%). The *trans*-**337** and *cis*-**338** cyclopropanediyl analogues are not protective. At 122 mg/kg analogue *trans*-**337** affords 20% mortality in hamsters not receiving the anthracycline. The open chain trans cyclopropanediyl tetraacid analogue (**378**) of EDTA (**323**) and its tetramethyl ester (**379**) also are inactive. However, the open chain cis cyclopropandiyl tetraacid **380** and the trans cyclobutanediyl tetraacid (**381**) exhibit 20% protection at 22 and 21 mg/kg, respectively. Bimolane (**351**), which may serve as a prodrug for razoxane in these studies, exhibits 100% protection at 100 mg/kg, 90% at 50 mg/kg, and 20% at either 25 or 12.5 mg/kg [596]. These data show that deviation from the razoxane structure is generally undesirable, and that lipophilicity as well as intrinsic water solubility can markedly effect the ability of bis(2,6-dioxopiperazines) to serve as protectors against anthracycline-induced toxicities. Such observations should be taken into consideration in the design of future drugs.

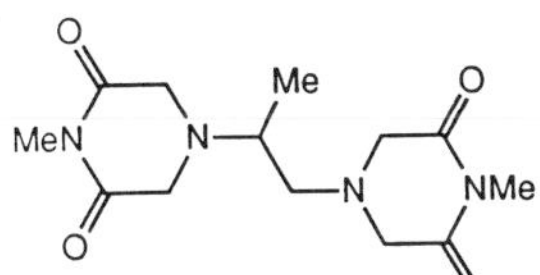

375; (–) isomer

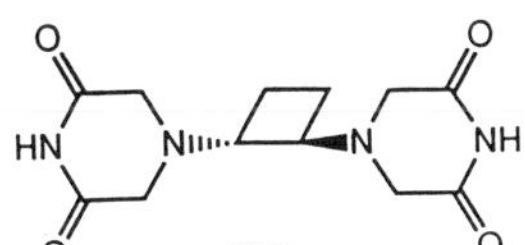

376

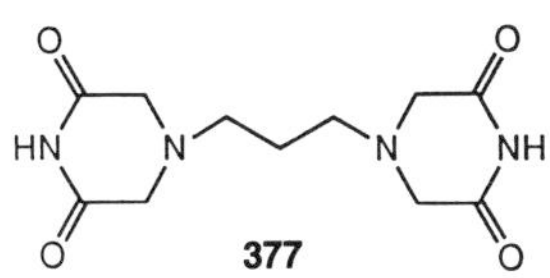

377

378; R = CH_2CO_2H
379; R = CH_2CO_2Me

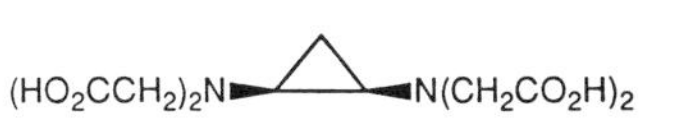

380

381

Severe cardiac lesions (lesion score = 3) are present in all eight beagle dogs given only doxorubicin (1.75 mg/kg at 3-week intervals × 9), but either no (N = 3) or minor abnormalities (four = lesion score of 1, one = lesion score of 2) are found in the hearts of dogs given the more water-soluble probimane (**352**) and doxorubicin [509]. Doxorubicin reduces the erythrocyte count, hemoglobin, and hematocrit when administered alone, but not in the combination. Probimane and ICRF-187 are equally effective in reducing chronic doxorubicin cardiotoxicity in beagle dogs, but it remains to be determined whether or not there are other significantly advantageous antineoplasic properties for the bis(morpholinomethyl) derivative [509].

4.13 Amelioration of acetaminophen-induced toxicities in Syrian golden hamsters

ICRF-187 (300 mg/kg), given 1 hour before acetaminophen (300 mg/kg doses × 2) attenuates the acetominophen-induced increases in enzyme activities, and reduces the incidence and severity of acetaminophen-induced hepatocellular injury in Syrian golden hamsters [597]. Enzyme activities increased by acetaminophen and reduced by the dioxopiperazine are serum glutamic-pyruvic transaminase and alkaline phophatase.

4.14 Protetion against alloxan diabetes in animals

ICRF-187 also protects against experimental induction of alloxan diabetes [598], a condition in which pancreatic beta cell damage in animals is thought to be caused by iron catalyzed generation of activated oxygen species such as superoxide anion and hydrogen peroxide [599]. Blood glucose concentrations are markedly elevated in these animals (CD-1 mice) 48 hours after iv. administration of alloxan (75 mg/kg). Treatment with ICRF-187 (96–345 mg/kg) 60 minutes prior to alloxan significantly attenuates the increases in blood glucose. In alloxan-treated animals, beta cell granules are absent, but only a mild to moderate reduction in the degree of granulation is observed in bis(dioxopiperazine)-protected mice [598].

The dioxopiperazine seems not to interact with superoxide anions [600]. There is no evidence, nor is it to be anticipated on chemical grounds, that any of the known biological effects of ICRF-187 are mediated through free radical scavenging. ICRF-187 also does not alter hexobarbital sleeping time, and therefore likely protects by mechanisms other than microsomal enzyme inhibition. Chelating agents such as EDTA (**323**) protect against alloxan diabetes, and ICRF-187 or its hydrolytic products may also function as chelators [600]. Chelation of iron by ICRF-187 or its hydrolysis products could decrease formation of reactive oxygen radicals in vivo, and provide a means for protecting against alloxan diabetes [600].

4.15 Chelation of biometals and mechanism of action

Stimulation of hydroxyl radical formation in doxorubicin-perfused rat hearts is observed by electron spin resonance spectroscopy [601]. Hydroxyl radical generation is inhibited by superoxide dismutase and catalase, thus providing evidence for the involvement of superoxide radical anions and hydrogen peroxide in anthracycline-induced cardiotoxicity. The bis(dioxopiperazine) ICRF-187 effectively inhibits hydroxyl radical formation without affecting heart rate [601], and this inhibition likely involves the ability of these substances to chelate biometals.

Razoxane reacts with Cu^{2+} [488]. In studies of urinary excretion of copper by the rat, ICRF-158 (**374**) and razoxane increase daily urinary copper excretion (65 mmol/24 hours) to 151 and 87 mmol/24 hours, respectively [602]. It is difficult to understand why bis(N-methylated) derivative (**374**) should cause enhanced copper excretion unless it undergoes more rapid hydrolysis to active chelators. Treatment with ICRF-158 (**374**) for 30 days (0.08 mmol/kg) also produces increases in the mobilization and elimination of lead (67 μg in controls, 184 μg in animals treated with ICRF-158) [603]. Lead-induced inhibition of hemosynthesis is prevented by the 2,6-dioxopiperazine. Furthermore, this analogue does not cause renal damage in excess to lead-induced tubular nephrosis [603]. Administration of ICRF-187 to humans results in an increase in urinary excretion of both iron and zinc [604). Differential binding affinity is observed between each of the two bis(N,N′-dicarboxymethyl-dicarboxamidomethyl) hydrolysis products [ICRF-198 (**382**) and ICRF-226 (**383**)] of cytotoxic razoxane and its inactive 1,2-butanediyl homologue ICRF-192 (**329**), respectively, with Zn^{2+} [605]. The 1:1 complex between Zn^{2+} and the hydrolysis product ICRF-226 of biologically inactive ICRF-192 (**329**) is favored by about two orders of magnitude over that of ICRF-198 (**382**). Other metal ions (Ca^{2+}, Cu^{2+}, Fe^{2+}, Mg^{2+}, Mn^{2+}) exhibit no differences in chelation formation constants with these two hydrolysis products. How such differences in binding with Zn^{2+} relates to differences in the biological activities of these compounds remains unclear. It is of interest that inhibition of nucleic acid synthesis by EDTA in cell culture is reversed by Zn^{2+} [606], but other physicochemical properties unrelated to Zn^{2+} chelation by razoxane also seems to be very important to the cytotoxic effect of dioxopiperazines.

Doxorubicin inhibits the activity of cardiac (Na^+ + K^+)-ATPase at very low concentrations (10^{-15}M) [607]. This inhibition likely is related to chelation by the anthracycline; inhibition of (Na^+ + K^+)-ATPase and ion transport by doxorubicin is counteracted by Ca^{2+} [608], and the complex Fe^{3+}(doxorubicin)$_3$ does not inhibit the ATPase [609]. Doxorubicin and razoxane also inhibit by 50 and 25%, respectively, the capping of surface immunoglobulins at 10^{-15}M concentrations [610]. Possibly, the inhibition of capping at very low concentrations (femtomolar concentrations) can involve single drug molecules (10^{-20}M) wherein the effect is propagated by membrane microfilaments within each cell and from cell to cell [610]. The effect on (Na^+ + K^+)-ATPase and ion transport is proposed to be mediated through a system of microfilaments [610]. More studies are necessary to substantiate these effects at such low concentrations.

Comparative investigations of Cu^{2+} complexing properties of razoxane, ICRF-192 (**329**), ICRF-193 (**330**), ICRF-196 (**331**) and their respective hydrolysis products **382–385**, do not show a simple correlation between metal binding and biological activity [611], but access to binding regions in vivo may be dependent upon lipophilicity and structure-dependent enzymatic hydrolysis. Furthermore, once at the site such binding may be modified by the presence of other ligands and metal ions [611]. The complexing order for Cu^{2+} ions appears to be ICRF-198 (**382**) > ICRF-226 (**383**) > ICRF-236 (**384**) > ICRF-243 (**385**), whereas the biological activity reported for the parent bis(2,6-dioxopiperazines) is ICRF-193 (**330**) > razoxane > ICRF-196 (**331**) > ICRF-192 (**329**) [611]. There also is no obvious correlation between metal ion mobilization in blood plasma and the cytotoxicity of the parent bis(dioxopiperazine). Likely, metal ion complexation in combination with other physico-chemical properties including lipophilicity reflect the biological potency order of these compounds.

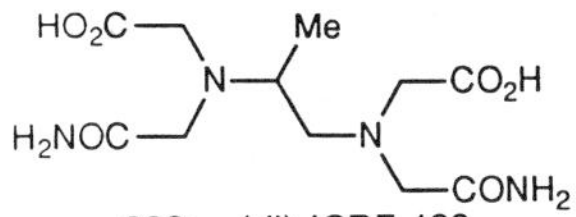

382; (dl)-ICRF-198
382a; (s) ADR-925

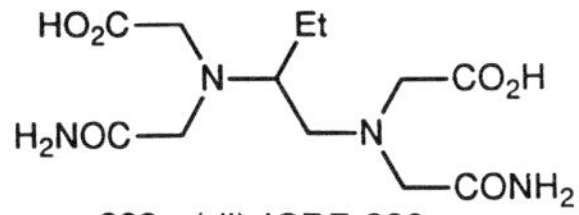

383; (dl)-ICRF-226

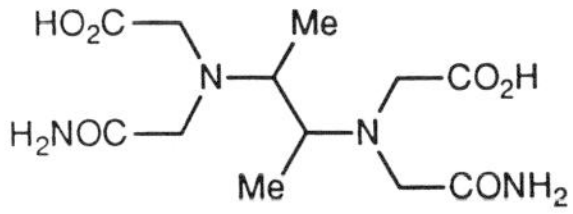

384; (meso)-ICRF-236

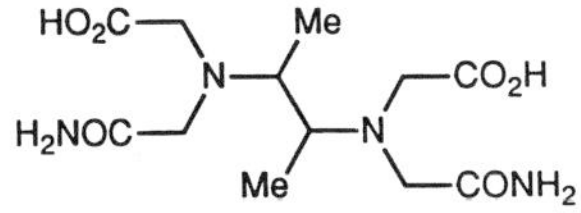

385; (dl)-ICRF-243

Doxorubicin and Fe^{3+} forms a complex that causes destruction of erythrocyte ghost membranes in the presence of glutathione [612]. This destruction of membranes is mediated by superoxide radicals and hydrogen peroxide. Possibly, ICRF-187 inhibits doxorubicin- or daunorubicin-induced toxicity by competing with the anthracycline for chelation of Fe^{3+} cations [600]. This dioxopiperazine provides the most consistent cardioprotective activity of any of the agents that have been examined in the various animal models of anthracycline-induced toxicity, including N-actylcysteine [587,589] and Vitamin E [613]. It seems that anthracyclines induce Fenton-type free radical oxidative stress on the cardiac muscle [614,615]. Much of the lipid peroxidation and damage to cellular components caused by doxorubicin in vitro is dependent upon the presence of iron [612,615–620] or copper [614,617]. ICRF-187 passively diffuses into cultured cells [621], but it is possible that the biological hydrolysis product, (*S*)-ADR–925 (**382a**), may be too polar to cross cell membranes. Uptake and subsequent hydrolysis of ICRF-187 in the myocardium may serve to protect cellular components by reducing the amount of available iron and copper [622], and in turn this could result in a decrease in oxygen radical generating metal ion-anthracycline complexes [622].
In vitro, the kinetics of hydrolysis of ICRF-187 may be followed spectrophotometrically at wave lengths from 205–210 nm, attributable to ICRF-187 in neutral solutions [622]. When the pH is increased this peak is gradually replaced by a new peak at 227 nm which is due to the ionization of the acidic imide hydrogen. The appearance of this new peak corresponds to the ionization of the acidic imide hydrogen (pK_a 10.0 at 25 °C and 9.6 at 37 °C). Hydrolysis and ionization of the imide rings occur independently of each other [622]. At 37 °C the k_{obs} is 7×10^{-4} min^{-1} at pH 7.4, and this is more than 10-fold faster than that observed at pH 2.93. Absorbance changes due to hydrolysis of ICRF-187

occur with a half-life of 8.2 hours for the first ring opening [622]. This compares with a previously determined half-life of 16.3 hours at 37 °C (pH 7.4 Tris buffer) which gives a value of 7.6 hours for razoxane [621]. Several possible explanations for these discrepancies are available [622]. Hydrolysis involves both a hydroxide-catalyzed pathway, and a pH-independent pathway similar to those of other imides. At neutral pH k_w for phthalimide equals 5.4×10^{-4} min^{-1} at 100 °C; k_w for ethosuximide equals 1.1×10^{-2} min^{-1} at 80 °C, and k_w for ICRF-187 equals 2.2×10^{-4} min^{-1} at 30 °C. Under base-catalyzed hydrolysis k_{OH} equals 230 $M^{-1}min^{-1}$, 204 M^{-1} min^{-1} and 45 M^{-1} min^{-1} for ICRF-187, succinimide and diacetylamide at 25 °C, respectively, and 62 M^{-1} min^{-1} for ethosuximide at 80 °C.

Cu^{2+}(doxorubicin)$_2$ is a 2 to 1 complex of anthracycline and Cu^{2+} possessing an absorbance at 600 nm [623]. A decrease in absorbance at 600 nm occurs when a chelating agent such as the ICRF-187 or the razoxane hydrolysis product ICRF-198 (**382**) binds Cu^{2+} in competition with doxorubicin [614,623]. Strong chelators remove Cu^{2+} from Cu^{2+}(doxorubicin)$_2$ complexes. Additionally, the hydrolysis of bis(2,6-dioxopiperazines) is promoted by Cu^{2+} and Fe^{3+} complexes of doxorubicin [605,622,624], and the preparation of ICRF-198 (**382**) is facilitated by stirring and refluxing razoxane in the presence of cupric acetate. Interestingly, divalent and trivalent metal ion complexes also promote the hydrolysis of peptide bonds [625].

Both Fe^{3+}(doxorubicin)$_3$ and Cu^{2+}(doxorubicin)$_2$ react with ICRF-187 to promote dioxopiperazine hydrolysis with concomitant abstraction of the metal ion from the anthracycline complex [626]. Thus, ICRF-187 might be thought of as a suicide protective agent [626]. Metal ion complex-promoted hydrolysis is preceded by mixed ligand complex formation. First-order rates of reaction for loss of metal ion from the anthracycline complex exhibits saturation behavior at high bis(dioxopiperazine) concentrations. When increasing amounts of ICRF-187 are added to a solution containing Cu^{2+} a new absorbance peak forms at 730 nm; isosbestic points are seen at 460 and 645 nm. Addition of varying amounts of the bis(dioxopiperazine) to the Cu^{2+}(doxorubicin)$_2$ complex results in a fast ($\leq$ 30 seconds) drop in absorbance at 600 nm, consistent with the partial replacement of the most weakly complexed anthracycline by the dioxopiperazine [626]. This replacement reaction is one-half complete at 340 μM, a concentration which is less than 1400 μM ICRF-187 required for one-half formation of the $Cu^{2+}(ICRF\text{-}187)_2$ complex. Spectral changes that occur upon the addition of higher concentrations of ICRF-187 to the Cu^{2+}-doxorubicin solution also are consistent with the formation of a Cu^{2+}(daunorubicin) (ICRF-187) mixed ligand complex. The new absorbance and shoulder, not seen in the spectra of the $Cu^{2+}(doxorubicin)_2$ complex, are most likely due to a mixed ligand complex [626]. The $Fe^{3+}(doxorubicin)_3$ complex also exhibits a fast initial drop in absorbance at 600 nm, likely due to an equilibrium displacement of the most weakly bound anthracycline [626]. Subsequent slower spectral changes correspond to the removal of the most tightly bound doxorubicin. The k_{obs} for both of the metal ion-doxorubicin complexes show saturation behavior similar to that expected for complex formation preceding a rate-determining step. The k_{obs} for $Cu^{2+}(doxorubicin)_2$ is one-half its maximal value at about 500 μM dioxopiperazine, while that for $Fe^{3+}(doxorubicin)_3$ occurs at 170 μM bis(dioxopiperazine), suggesting that ICRF-187 interacts more strongly with the $Fe^{3+}(doxorubicin)_3$ complex [626]. This metal ion-doxorubicin complex-promoted hydrolysis likely takes place according to:

(1) $Fe^{3+}(doxorubicin)_3$ + ICRF-187 $\rightleftarrows$ $Fe^{3+}(doxorubicin)_2$(ICRF-187) + doxorubicin, and

(2) $Fe^{3+}(doxorubicin)_2$(ICRF-187) $\rightarrow$ Fe^{3+}(ADR-925) + 2 doxorubicin, wherein the first step of the reaction involves equilibrium displacement of the most weakly bound anthracycline (stability constant $10^{4.4}$) to form the mixed ligand complex [627]. The overall reaction 2 includes the rate-determining step. Following metal ion complex-promoted hydrolysis of the first dioxopiperazine ring, molecular model building shows hydrolysis of the second ring could occur within a strongly bound one-ring opened-metal ion complex intermediate [626]. This second step might occur faster than the first since formation of such an intermediate complex would be strongly favored [626]. Doxorubicin forms very stable complexes with Fe^{3+} (stability constants of 10^{18}, 10^{11} and $10^{4.4}$) [627] and Cu^{2+} (stability constant estimated to be $10^{16.7}$) [623], but, nonetheless, ICRF-198 (**382**) reacts rapidly with such anthracycline-metal species [624]. A 1:1 complex with Fe^{3+} + chelator is produced with $t^{1/2}$ = 1.8 minutes upon treatment of $Fe^{3+}(doxorubicin)_3$ with ICRF-198 in pH 7.4 Tris/KCl buffer [624]. ICRF-198 reacts even more rapidly with Cu^{2+} $(daunorubicin)_2$ than it does with $Fe^{3+}(doxorubicin)_3$ [624]. Expectedly, ICRF-187 reacts with $Fe^{3+}(doxorubicin)_3$ more slowly than does ICRF-198. Scavenging

Fe^{3+} and Cu^{2+} from their potentially damaging complexes with doxorubicin by either the bis(dioxopiperazine) or its hydrolysis product is both kinetically and thermodynamically favorable [624].
The damage to cardiac mitrochondria, a characteristic observation in doxorubicin-induced cardiotoxicity [614], is accompanied by a loss of respiratory enzyme activity [628]. Fe^{3+}(doxorubicin)$_3$ is known to cause a progressive loss of cytochrome c oxidase activity [618,619,628] of both beef heart submitochondrial particles [617,618] and the purified enzyme [629]. ICRF-198 (**382**) concentrations of 50 μM are virtually 100% effective in protecting against loss of cytochrome c oxidase activity [624]. ICRF-187 also is effective in protecting against the loss of cytochrome c oxidase activity, but this bis(dioxopiperazine) is not as effective as its hydrolysis product ICRF-198. Additionally, Fe^{3+} doxorubicin complex-stimulated lipid peroxidation in platelet membranes, as assessed by malondialdehyde formation, is inhibited by ICRF-198 [630]. Thus, the protection offered by these chelators is easily explained on the basis that Fe^{3+} is removed from doxorubicin complexes due to the higher affinity of ICRF-198 for the metal [624]. When ICRF-198 binds Fe^{3+} the Fe^{3+}(doxorubicin)$_3$ complex is no longer available to catalyze redox cycling and tissue-damaging reactions [624].

4.16 Uptake and pharmacokinetics

[^{14}C]Razoxane, [^{14}C]1,2-propanediyldiaminotetracetic acid, and inactive 1,3-propanediyl isomer **377** is taken up by mouse embryo fibrolast cells [631]. The net amount of labeled razoxane within or attached to the cells decreases after 1.5 hours. This is followed by a rise after 5 hours. The highly polar tetraacid is absorbed very slowly, but after 24 hours the concentration in cells is greater than the concentration found for razoxane. [^{14}C]Razoxane has a plasma half-life in rats of about 30 minutes [631]. The clearance of the radioactivity in rats is biexponential consisting of a rapid phase with a half-life of 25 minutes, and a slow phase probably representing the clearance of a metabolite [632]. Bioassay confirms that the rapid exponential phase represents clearance of dioxopiperazine (or active metabolites). By GLC-mass spectrometry, useful for detection of the bis(N-methylated-imide) derivative, the plasma half-life of razoxane is 40–45 minutes in rats and 85 minutes in rabbits. The area under the plasma concentration-time curve after oral administration is significantly higher than that after iv. administration indicating a low oral bioavailability (<50%) [633].

In mice the plasma half-life of the (+) enantiomorph of razoxane (ICRF-187) is about 26 minutes [634]. The bis(dioxopiperazine) is maximally concentrated in the kidney and liver [634].

The plasma concentration-time curves for humans given oral razoxane are similar in magnitude to those obtained after oral doses given to rats and rabbits [633]. In twelve patients administered various doses of [^{14}C]razoxane, analysis of urinary and fecal drug concentrations suggests limited absorption at high dose concentrations [635]. Urinary clearance of this drug is independent of the administration schedule given to phase I patients [636]. In man, steady state serum drug concentrations of ICRF-187 are achieved by infusions lasting longer than 8 hours [637]. Binding to serum proteins is insignificant, and pharmacokinetic factors seem not to account for the schedule dependent toxicity of this drug [637]. Urine and serum specimens collected during and following iv. infusion of drug (1.0 g/m^2) to patients using three different schedules (30 minutes, 8 hours, and 48 hours) show total urinary drug recovery (48%) to be schedule-independent. Serum drug kinetics follow a two-compartment open model wherein mean peak serum levels for the three schedules are 75.3, 22.8 and 2.9 μg/ml, respectively. The mean central compartment volume of distribution is 0.193 L/kg, or slightly greater than the expected extracellular space. Infusion of the dioxopiperazine over a period greater than 8 hours is necessary for achievement of true steady-state conditions [637].

In adult rhesus monkeys [^{14}C]ICRF-187 enters the cerebrospinal fluid to the extent of approximately 10% of the plasma levels observed near the end of a 2-hour infusion (300 mg/m^2) [638]. Assuming no degradation, peak plasma levels of the dioxopiperazine at the end of the infusion are about 9μg/ml. Twenty-four hours after the infusion, radioactivity in the cerebrospinal fluid is approximately 50% that of the plasma; it is unlikely that such drug levels are cytotoxic to leukemia cells, or that this drug could reach significant levels in the cerebrospinal fluid of man [638].

ICRF-187 produces a transient elevation of liver enzymes, and a prolongation of prothrombin and partial thromboplastin time in children at nonmyelosuppressive doses, doses which are 2.8 times higher than those which induce severe myelosuppression in adults [639]. Since no adult leukemia patients have been treated with ICRF-187, it is not known whether hepatotoxicity might surface in such individuals given higher doses of the dioxopiperazine. For ICRF-187 the ratio of the maximum tolerated dose (MTD) reached in children and the MTD in adults with solid tumors is 2.80 (MTD = 3500 mg/m^2 and 1258 mg/m^2 for children and adults, respectively), wherein the MTD is the highest tolerable dose that can be achieved without producing irreversible toxicity [639]. Many factors including intrinsic age-related physiologic differences in drug metabolism, the extent of prior treatment, or increases in bone marrow reserve associated with youth may be responsible for the increased tolerance of the drug to children [640].

ICRF-187 disappears from human plasma biphasically with a terminal half-life of 3.2 ± 0.9 hours [641]. The total clearance is 288.7 ±85.0 ml/hours/kg and the volume of distribution is 1.3 ± 0.4 L/kg. Pharmacokinetic parameters are not dose-dependent from 3.8 − 7.4 g/m^2; no differences are found in patients during the first and second treatments [641].

4.17 Toxicity in phase 1 clinical studies

In phase 1 studies, leukopenia and thrombocytopenia are observed to be the major dose-limiting toxic effects of ICRF-187 [604]. At the lowest dose (25 mg/kg or 500 mg/m^2), no decrease in either white blood cell or platelet counts are observed. There is no indication of hepatotoxicity and the serum enzyme concentrations return to normal by day 17 after drug administration. Other studies indicate that at doses of 800 mg/m^2 and above, leukopenia, not thrombocytopenia, is the dose-limiting toxicity [642]. Hematologic recovery usually occurs by 3 weeks. Dose-limiting myelosuppression is noted and more severe in patients previously treated with nitrosoureas [604,642]. Nonmyelosuppressive toxicity includes reversible increases in liver enzymes such as SGOT and SGPT, mild nausea and vomiting, low-grade fever, alopecia, and pain at the injection site [604,642]. Doses of 800 and 1250 mg/m^2 iv. daily for 5 days are recommended for heavily and nonheavily nitrosourea-pretreated patients, respectively [642]. Recommended doses for phase II studies are 125 mg/m^2/day × 3 and 750 mg/m^2/day × 3 in patients with prior nitrosourea therapy and in patients who have not received nitrosoureas, respectively [604].

During iv. infusion, ICRF-187 causes a dose-limiting granulocytopenia in patients [643]. Granulocyte recovery is noted by 3 weeks, and no evidence of cumulative bone marrow toxicity is observed. Other toxicities again include mild nausea, malaise, and alopecia. The recommended dose for phase II trials is 1000 mg/m^2 given over 48 hours as a continuous iv. infusion every 22 days [643].

Children have a larger volume of distribution per kilogram of body weight in the central compartment, and a more rapid total-body clearance than adults. When ICRF-187 is given as a 2-hour infusion daily for 3 days to patients under 21 years of age, the maximum tolerated dose is about 3500 $mg/m^2/day \times 3$ based upon changes in hepatic function and coagulation abnormalities encountered when larger dosages are administered [644]. Pharmacokinetic parameters explain only part of the increased tolerance of children to the dioxopiperazine [644].

Dose-limiting hematological toxicity is seen in heavily pretreated patients with solid tumors given ICRF-187 infusions of 3.8 $g/m^2/week$, weekly [641]. All patients develop reversible SGOT elevations. Some patients develop anemia at virtually all dose levels tested (0.85 g/m^2 to 7.42 $g/m^2/4$ weeks) [641]. The recommended dose for phase II trials is 3.8 $g/m^2/week \times 4$ for heavily pretreated and 7.42 $g/m^2/week$ for good risk patients [641]. ICRF-187 is not active as a single agent in patients with advanced non-small cell lung cancer when administered on a 1250 mg/m^2 daily $\times$ 3 schedule every 3 weeks [645]. Also, ICRF-187 has no detectable activity against renal adenocarcinoma administered in a schedule of 3.8 $g/m^2/week$ iv. $\times$ 4 [646]. Some response is observed in patients with advanced squamous cell cancer of the head and neck when ICRF-187 is given on a schedule of 1.0 $g/m^2/48$ hours by continuous infusion every 3 weeks; one out of eight patients who did not receive prior chemotherapy responds to this regime [647]. Orally administered razoxane seems to be producing similar response rates in phase II trials [648].

ICRF-154 (**325**) is ineffective in acute lymphoblastic, myeloid, or monocytic leukemia [649], but introduction of a single methyl group on the spacer chain converts this compound to razoxane which is active in patients having acute leukemia and lymphosarcoma [649], non-Hodgkins lymphoma [650], Kaposi's sarcoma [651], and psoriasis [652]. This drug is not particularly useful in the treatment of non-small cell lung cancer [653], malignant melanoma [654], carcinoma of the breast [655,656], squamous cell carcinoma of the head and neck [657], advanced gastric cancer [658], or pediatric solid tumors [659].

4.18 Combination with radiation therapy in humans

Approximately 75% of all patients treated with razoxane plus irradiation experience tumor inhibition [660–663]. Regression of tumors not normally sensitive to irradiation, i.e. melanomas, is observed with these combined modalities [660]. Patients with squamous cell carcinoma of the head and neck respond favorably to radiation alone, but analysis reveals no difference between the combined razoxane- and radiation-treated groups [664]. Other studies also indicate no statistical difference in survival between those patients treated with radiotherapy alone and those who received both radiotherapy and razoxane (125 mg $\times$ 2/day) [665].

4.19 Protection against doxorubicin-induced cardiotoxicity in humans

Protective effects of ICRF-187 against doxorubicin-induced cardiotoxicity is of considerable interest [563,666]. The potential protective effects of ICRF-187 against doxorubicin cardiotoxicity can be demonstrated in patients with breast cancer receiving FDC (5-fluorouracil, doxorubicin, cyclophosphamide) therapy [563]. Eleven patients on the FDC + ICRF-187 arm received cumulative doxorubicin doses at or above 600 mg/m^2. Whereas antitumor response rates and the incidence of fever, infections, alopecia, nausea and vomiting or death due to toxicity are not different between the groups, myelosuppression is slightly greater in the FDC + dioxopiperazine group. Clinical congestive heart failure is found in eleven patients of the FDC group (total of forty-five patients) and in two patients of the FDC + ICRF-187 group (total of forty-seven patients); the mean decrease in the left ventricular ejection fraction is 7 (FDC) vs. 1% (FDC + ICRF-187) when the cumulative dose of doxorubicin is 250 to 399 mg/m^2, 16 vs. 1% at 400 to 499 mg/m^2, and 16 vs. 3% at 500 to 599 mg/m^2. A Billingham miocardial biopsy lesion score is 2 or more in five of thirteen patients in the FDC group whereas lesions are not observed in any of the FDC + ICRF-187 biopsy group [563]. Some patients, having received cumulative doses of doxorubicin as large as 1000 mg/m^2, show no evi-

dence of clinical toxicity [667]. It seems that ICRF-187 offers significant protection against cardiotoxicity caused by doxorubicin without affecting the antitumor properties of doxorubicin or the incidence of noncardotoxic reactions [563].
For breast cancer, standard therapy consists of cyclophosphamide, doxorubicin, and 5-fluorouracil; for small cell lung cancer, standard therapy employs cyclophosphamide, doxorubicin, and vincristine. Current double-blind studies are comparing standard therapy with and without ICRF-187 [667]. If a cardioprotective effect against doxorubicin toxicity is confirmed, the dioxopiperazine should be approved fairly rapidly; the FDA has given ICRF-187 (ADR-529) a 1-A designation [667].

4.20 Clinical studies with morpholinomethyl derivatives

Clinical studies of orally administered bimolane (**351**), the bis(morpholinomethyl) derivative of ICRF-154 (**325**), indicate > 50% regression of tumors in 24 of 155 patients [668]. These effects are more pronounced in certain subgroups; regression is observed in patients with Hodgkin's disease, non-Hodgkin's Lymphoma, and cancer of the vulva. As an adjuvant to radiotherapy in patients with squamous cell carcinoma, lung tumors, and sarcomas of soft tissue, complete or partial responses are observed in over 50% of the cases. Many patients with lung tumors also respond favorably [668]. Patients with psoriasis experience complete or nearly complete responses. Side effects observed in all patients include mild gastrointestinal toxicity, and a slight decrease in leukocyte count [668].

References

1 R. Isaksson, T. Liljefors and J. Sandstrom: J. Chem. Res., 43 (1981).
2 R. Isaksson and I. Liljefors: J. Chem. Soc. Perkin II, 1815–1820 (1980).
3 J.L. Meeks, H.J. Maria, P. Brint and S.P. McGlynn: Chem. Rev. *75*, 603–610 (1975).
4 J.L. Meeks, J.F. Arnett, D.B. Larson and S.P. McGlynn: J. Am. Chem. Soc. *97*, 3905–3908 (1975).
5 J.L. Meeks and S.P. McGlynn: J. Am. Chem. Soc. *97*, 5079–5083 (1975).
6 D. Dougherty, P. Brint and S.P. McGlynn: J. Am. Chem. Soc. *100*, 5597–5603 (1978).
7 D.O. Cowan, R. Gleiter, J.A. Hashmall, E. Heilbronner and V. Hornung: Angew. Chem. *83*, 405–406 (1971).
8 P. Schang, R. Gleiter and A. Rieker: Ber. Bunsengesellschaft Phys. Chem. *82*, 629–633 (1978).
9 J.F. Arnett, G. Newkome, W.L. Mattice and S.P. McGlynn: J. Am. Chem. Soc. *96*, 4385–4392 (1974).
10 D.B. Larson and S.P. McGlynn: J. Mol. Spectroscopy *47*, 469–490 (1973).
11 R.E. Carter and J. Sandström: J. Phys. Chem. *76*, 642–647 (1972).
12 R.N. Jones and A.L. Barry: Rev. Infect Dis. *5*, (Suppl.), S 108 – S 126 (1983).
13 B. Holmes, D.M. Richards, R.N. Brogden and R.C. Heel: Drugs *28*, 375–425 (1984).
14 M. Roach: Br. J. Pharm. Pract. *4*, 32, 34–35, 37, 39, 41 (1982).
15 H.C. Neu: Annu. Rev. Pharmacol. Tox. *22*, 599–642 (1982).
16 T.D. Marsh: Infect. Control *6*, 78–83 (1985).
17 P. Garzone, J. Lyon and V.L. Yu: Drug Intell. Clin. Pharm. *17*, 615–622 (1983).
18 E.A. Funk and L.J. Strausbaugh: Pharmacotherapy *2*, 185–196 (1982).
19 J.M. Brogard and F. Comte: Antibiot. Chemother. *31*, 145–210 (1982).
20 M. Arpi: Chemotherapy (Basel) *34*, 393–400 (1988).
21 A. Matsuura, T. Nagayama and T. Kitagawa: Chemotherapie (Basel) *34*, 345–353 (1988).
22 W.H. Traub, M. Spohr and D. Bauer: Chemotherapy (Basel) *34*, 284–297 (1988).
23 K. Fuursted: Chemotherapy (Basel) *34*, 229–234 (1988).
24 J. Etienne, M. Coulet, Y. Brun, J.-F. Blanchon, F. Demoux and J. Fleurette: Chemotherapy (Basel) *34*, 113–116 (1988).
25 S. Alvarez, M. Jones, S. Holtselaw-Berk and S.L. Berk: Chemotherapy (Basel) *34*, 107–112 (1988).
26 R.H.K. Eng, R. Liu, S.M. Smith, E.S. Johnson and C.E. Cherubin: Chemotherapy (Basel) *34*, 18–26 (1988).
27 R.D. Colucci, L.E. Cobuzzi and N.A. Halpern: Am. J. Hosp. Pharm. *45*, 1357–1358 (1988).
28 D.F. Thompson, M.L. Stiles, L.V. Allen, Jr. and Y.-H. Tu: Am. J. Hosp. Pharm. *45*, 142–145 (1988).
29 R.D. Colucci, L.E. Cobuzzi and N.A. Halpern: Am. J. Hosp. Pharm. *45*, 630–632 (1988).
30 F.V. Warren, Jr., B.A. Bidlingmeyer and M.F. Delaney: Anal. Chem. *59*, 1897–1907 (1987).
31 M. Basilio: Anal. Lett. *21*, 759–772 (1988).
32 L.-Z. Cai and M.-L. Wu: Analyst (London) *113*, 21–22 (1988).
33 P. Calhoun, K.B. Brown, R. Strunk, D.A. Krusch, W.M. Scheld and J.B. Hanks: Ann. Surgery *205*, 420–427 (1987).
34 C.C. Sanders, M.L. Gates and W.E. Sanders, Jr.: Antimicrob. Agents Chemother. *32*, 1893–1895 (1988).

35 A. Yotsuji, J. Mitsuyama, R. Hori, T. Yasuda, I. Saikawa, M. Inoue and S. Mitsuhashi: Antimicrob. Agents Chemother. *32*, 1848–1853 (1988).
36 S.J. Wadsworth and B. Suh: Antimicrob. Agents Chemother. *32*, 1571–1575 (1988).
37 Y. Ikeda and T. Nishino: Antimicrob. Agents Chemother. *32*, 1073–1077 (1988).
38 L.E. Adinolfi and P.F. Bonventre: Antimicrob. Agents Chemother. *32*, 1012–1018 (1988).
39 A.W. Chow, J. Wong and K.H. Bartlett: Antimicrob. Agents Chemother. *32*, 782–784 (1988).
40 T. Hayashi, Y. Watanabe, K. Kumano, R. Kitayama, T. Yasuda, J. Saikawa, J. Katahira, T. Kumada and K. Shimizu: Antimicrob. Agents Chemother. *32*, 912–918 (1988).
41 H.M. Wexler and S.M. Finegold: Antimicrob. Agents Chemother. *32*, 403–406 (1988).
42 Y.W.F. Lam, M.H. Duroux, J.D. Gambertoglio, S.L. Barriere and B.J. Guglielmo: Antimicrob. Agents Chemother. *32*, 298–302 (1988).
43 D.P. Reitberg, T.J. Whall, M. Chung, D. Blickens, H. Swarz and J. Arnold: Antimicrob. Agents Chemother. *32*, 42–46 (1988).
44 J. Russo, Jr. and M.E. Russo: Clin. Pharm. *1*, 207–216 (1982).
45 S.L. Barriere and J.F. Flaherty: Clin. Pharm. *3*, 351–373 (1984).
46 T.A. Tartaglione, L. Nye, N. Vishniavsky, W. Poynor and R.E. Polk: Clin. Pharm. *5*, 911–916 (1986).
47 G.E. Schumacher: Clin. Pharm. *6*, 59–68 (1987).
48 R.A. Hamilton, S.F. Kowalsky, E.M. McCormick and R.M. Echols: Clin. Pharm. *6*, 567–569 (1987).
49 H.M. Wexler and S.M. Finegold: Clin. Ther. *10*, (Suppl. A), 12–18 (1987).
50 R.J. Zabransky and R.J. Birk: Clin. Ther. *10*, 107–125 (1987).
51 J.C.E. Goldstein and D.M. Citron: Clin. Ther. *10*, 487–515 (1988).
52 R. Labia, A. Morand, K. Tiwari, J.S. Pitton, D. Sirot, J. Sirot, H. Ben Yaghlane and A. Boujenah: Drugs Exptl. Clin. Res. *14*, 335–339 (1988).
53 P. van der Auwera, M. Husson and G. Bulliard: Drugs Exptl. Clin. Res. *12*, 871–883 (1986).
54 M. Bonadio, O. Marino, B. Catania and P. Giannetti: Drugs Exptl. Clin. Res. *12*, 895–897 (1986).
55 R.N. Brogden, A. Carmine, R.C. Heel, P.A. Morley, T.M. Speight and G.S. Avery: Drugs *22*, 423–460 (1981).
56 R.J. Fass, J. Barnishan and V.L. Helsel: Experientia *43*, 360–361 (1987).
57 L. Fricke, H. Iven, J. Steinhoff, E. Schulz and K. Sack: Fortschr. Antimikrob. Antineoplast. Chemother. *5*, 941–947 (1986).
58 F. Soergel, U. Stephan, E. Lang, M. Geldmacher, Von Mallinckrodt, R. Seelmann and H.G. Wiesemann: Fortschr. Antimikrob. Antineoplast. Chemother. *5*, 921–936 (1986).
59 K. Uno: Jpn. J. Genet. *63*, 471–493 (1988).
60 T. Fujii, K. Sato, E. Yokota, T. Maejima, M. Inoue and S. Mitsuhashi: J. Antibiot. *41*, 81–85 (1988).
61 O.S. Treskina, L.M. Slepchenko, V.N. Nasonov and E.N. Dutova: Antibiot. Med. Biotekhnol *32*, 855–859 (1987).
62 A.L. Costa: Gaz. Med. It.-Arch. Sc. Med. *146*, 249–253 (1987).
63 S. Yan and J. Li: Clin. Med. J. *100*, 940–944 (1987).
64 Y. Hirat: Hirosaki Igaku *38*, 405–418 (1986).
65 Z. Takase: Biol. Res. Pregnancy Perinatol. *5*, 48–56 (1984).
66 S. Matsuda: Biol. Res. Pregnancy Perinatol. *5*, 57–60 (1984).
67 M. Zajac and A. Molenda: Acta Pol. Pharm. *43*, 428–435 (1986).
68 S. Ripa, F. LaRosa, B. Dainelli, U. Ecari and M.P. Ruffilli: Int. J. Clin. Pharmacol. Res. *6*, 171–175 (1986).

69 P. Boussard and J. Dony: Int. J. Pharm. *46*, 45–48 (1988).
70 C. Yoshida, K. Tanaka, J. Nakano, Y. Todo, T. Yamafuji, R. Hattori, Y. Fukuoka and I. Saikawa: J. Antibiot. *39*, 76–89 (1986).
71 C.L. Branch, M.J. Basker and M.J. Pearson: J. Antibiot. *39*, 1792–1795 (1986).
72 M.J. Basker, C.L. Branch, S.C. Finch, A.W. Guest, P.H. Milner, M.J. Pearson, R.J. Ponsford and T.C. Smale: J. Antibiot. *39*, 1788–1791 (1986).
73 T. Hayashi, Y. Watanabe, K. Kumano, R. Kitayama, T. Yasuda, I. Saikawa, K. Totsuka, T. Kumada and K. Shimizu: J. Antibiot. *39*, 699–712 (1986).
74 M.E. Grace, F.J. Gregory, P.P. Hung and K.P. Fu: J. Antibiot. *39*, 938–942 (1986).
75 L. Dall'Asta, A. Comini, E. Garagnani, D. Alberti and G. Coppi: J. Antibiot. *39*, 1004–1006 (1986).
76 J. Gil, C. Rubio-Clavo, A. Rezusta, M.A. Vitoria and M.S. Salvo. In: Mycobacteria of Clinical Interest Excerpta Medica, International Congress Series 697, M. Casal, Ed., pp. 195–198, Elsevier Science Pub. B.V., Amsterdam 1986.
77 H. Kiyota, S. Onodera, K. Kishimoto, H. Goto and T. Machida: Nippon Hinyokika Gakkai Zasshi *79*, 910–917 (1988).
78 M. Rahman: Curr. Med. Res. Opin. *9*, 71–77 (1984).
79 C. Cantoni, C. Traldi and G. Soncini: Ind. Aliment (Pinerolo, Italy) *25*, 981–984 (1986).
80 D. Jastalska, W. Lupicki, G. Pajchel, M. Zapasnik and W. Chojnowski: Pol. Tyg. Lek. *41*, 1071–1075 (1987).
81 M. Cutten, S. Roberts and S. Lang: N.Z.J. Med. Lab. Technol. *41*, 124–125 (1987).
82 H.C. Neu, N.X. Chin and P. Labthavikul: Chemotherapia *4*, 415–423 (1985).
83 M. Venditti, C. Brandimorte, C. Santini, B. Valente, P. Martino and P. Serra: Chemotherapia *3*, 235–241 (1984).
84 C. Dos Santos, V. Monzillo, P. Marone, E. Concia, P. Grossi and S. Dughetti: Boll. Ist. Sieroter. (Milan) *65*, 374–379 (1986).
85 K.S. Lee and Y.E. Park: Inkan Kwahak 11, 574–581 (1987).
86 K. Gatschow, E.R. Weissenbacher, I. Wachter, D. Adam and A. Schneider: Fortschr. Antimikrob. Antineoplast. Chemother. *6*, 237–247 (1987).
87 I. Wachter, E.R. Weissenbacher, A. Zischler, R. Drexel, D. Adam, A. Schneider and K. Gutschow: Fortschr. Antimikrob. Antineoplast. Chemother. *6*, 249–256 (1987).
88 F. Sorgel, V. Stephan, I. Bruning, C. Jensen and E. Lang: Fortschr. Astimikrob. Antineoplast. Chemother. *5*, 977–982 (1986).
89 K. Matsuo, Y. Tsuzaki, S. Furukawa and T. Uete: Jpn. J. Antibiot. *39*, 2579–2586 (1986).
90 H. Nakahata, Y. Hirai, Y. Kumasaka, T. Miyazawa, T. Nakamura, K. Imamura, I. Makino and K. Takebe: Jpn. J. Antibiot. *39*, 693–700 (1986).
91 K. Saito, K. Yamashita and R. Okachi: Jpn. J. Antibiot. *39*, 1487–1493 (1986).
92 I. Saikawa, A. Takai, Y. Nakashima, T. Ikegami, H. Hayakawa, M. Nogughi, H. Yamauchi and H. Shimizu: Jpn. J. Antibiot. *35*, 2163–2173 (1982).
93 C.D.C. Christie and R.S. Baltimore: West Indian Med. J. *37*, 87–91 (1988).
94 D. Ionescu: Med. Welt. *38*, 318–321 (1987).
95 Y. Yang and D.M. Livermore: J. Antimicrob. Chemother. *22*, 299–306 (1988).

96 A. Boisivon, C. Guiomar and L. Gutmann: J. Antimicrob. Chemother. *22*, 651–657 (1988).
97 P.G.R. Godwin and M.R. Millar: J. Antimicrob. Chemother. *22*, 899–904 (1988).
98 J.A. Smith and D.A. Henry: J. Antimicrob. Chemother. *22*, 849–855 (1988).
99 G.J. Cuchural, S. Hurlbut, M.H. Malamy and F.P. Tally: J. Antimicrob. Chemother. *22*, 785–790 (1988).
100 A.J. de Neeling, A. Rutgers, C.S. Schot and E.E. Stobberingh: J. Antimicrob. Chemother. *22*, 613–621 (1988).
101 D.W. McEniry, S.H. Gillespie and D. Flemingham: J. Antimicrob. Chemother. *21*, 171–175 (1988).
102 R. Labia, A. Morand, K. Tiwar, J.S. Pitton, D. Sirot and J. Sirot: J. Antimicrob. Chemother. *21*, 301–307 (1988).
103 E.G. Raizes and J.R. Cantey: J. Antimicrob. Chemother. *21*, 177–181 (1988).
104 P. van der Auwera, M. Husson and J. Klastersky: J. Antimicrob. Chemother. *21*, 49–55 (1988).
105 C.C. Shepard, R.M. van Landingham, L.L. Walker and R.C. Good: Int. J. Lepr. Other Mycobact. Dis. *55*, 322–327 (1987).
106 A. Yotsuji, Y. Ito, T. Yasuda, I. Saikawa, S. Ohgaku, S. Yano and Y. Ueda: Chemotherapy (Tokyo) *36*, 294–303 (1988).
107 J.-I. Kameyama, M. Tsukamoto, H. Seuda, K. Sikine, W. Iwatani and T. Asano: Chemotherapy (Tokyo) *36*, 549–553 (1988).
108 J. Zhi, C.H. Nightingale and R. Quintiliani: J. Pharma. Sci. *77*, 367–369 (1988).
109 J. Zhi, C.H. Niththingale and R. Quintiliani: J. Pharma. Sci. *77*, 991–992 (1988).
110 I. Komiya, M. Nishio, S. Murata, F. Chiba, T. Sakural, S. Shinkai and M. Fujita: J. Pharmacobio-Dyn. *7*, 545–555 (1984).
111 J. Moravek, O. Schuck, M. Hatala and J. Priborsky: J. Pharmacokinet. Biopharm., *15*, 15–23 (1987).
112 H. Mett, S. Rosa, B. Schucher and R. Frei: Rev. Infect. Dis. *10*, 765–769 (1988).
113 V. Jarlier, M.-H. Nicolas, G. Fournier and A. Philippon: Rev. Infect. Dis. *10*, 867–878 (1988).
114 T. Sawai, A. Yamaguchi and R. Hiruma: Rev. Infect. Dis. *10*, 761–764 (1988).
115 R. Labia, A. Morand, K. Tiwari, J. Sirot, D. Sirot and A. Petit: Rev. Infect. Dis. *10*, 885–891 (1988).
116 G.L. Drusano, P.A. Ryan, H.C. Standiford, M.R. Moody and S.C. Schimpff: Rev. Infect. Dis. *6*, 357–363 (1984).
117 G.L. Drusano, S.C. Schimpff and W.L. Hewitt: Rev. Infect. Dis. *6*, 13–32 (1984).
118 R.J.Z. Anders, A. Lau, R. Sharifi and M. Lee: Ther. Drug Monit. *9*, 472–477 (1987).
119 H. Sadaki, H. Imaizumi, T. Inaba, T. Hirakawa, Y. Murotani, Y. Watanabe, S. Minami and I. Saikawa: Yakugaku Zasshi *106*, 123–128 (1986).
120 H. Sadaki, H. Imaizumi, T. Inaba, T. Hirakawa, Y. Murotani, Y. Watanabe, S. Minami and I. Saikawa: Yakugaku Zasshi *106*, 147–153 (1986).
121 H. Sadaki, H. Imaizumi, T. Inaba, T. Hirakawa, Y. Murotani, Y. Watanabe, S. Minami and I. Saikawa: Yakugaku Zasshi *106*, 129–146 (1986).
122 H. Sadaki, H. Imaizumi, T. Inaba, T. Hirakawa, Y. Murotani and I. Saikawa: Yakugaku Zasshi *106*, 117–122 (1986).
123 P.F. Wheat, J.G.M. Hastings and R.C. Spencer: J. Med. Microbiol. *25*, 95–99 (1988).

124 S.C. Aronoff and D.M. Shlaes: J. Infect. Dis. *155*, 936–941 (1987).
125 R.J. Wallace, Jr., D.R. Nash, W.K. Johnson, L.C. Steele and V.A. Steingrube: J. Infect. Dis. *156*, 959–966 (1987).
126 B. Vogelman, S. Gudmundsson, J. Turnidge, J. Leggett and W.A. Craig: J. Infect. Dis. *157*, 287–298 (1988).
127 M.G. Bergeron, P. Simard and P. Provenchen: J. Clin. Microbiol. *25*, 650–655 (1987).
128 R.F. D'Amato, H.D. Isenberg, G.A. McKinley, E.J. Baron, R. Tepper and M. Shulman: J. Clin. Microbiol. *26*, 1492–1495 (1988).
129 F.S. Nolte, K.K. Krishner, L.A. Beltran, N.P. Christianson and G.E. Sheridan: J. Clin. Microbiol. *26*, 1079–1084 (1988).
130 A.L. Barry, R.N. Jones and The Collaborative Antimicrobial Susceptibility Testing Group: J. Clin. Microbiol. *26*, 13–17 (1988).
131 E.J. Goldstein and D.M. Citron: J. Clin. Microbiol. *26*, 2361–2366 (1988).
132 J.T. Downs, V.T. Andriole and J.L. Ryan: Yale J. Biol. Med. *59*, 11–16 (1986).
133 S.A. Berger, Y. Levy, A. Halevy, A. Gorea and R. Orda: World J. Surg, *12*, 641–644 (1988).
134 N. Thomas, H. Toutain, C. Dadoun, J.P. Morin and J.P. Fillastre: Presse Med. *15*, 2362–2367 (1986).
135 J.B. Fourtillan, W. Couet, A. Mignot, M.A. Lefebvre, Y. Mainguy and P. Friocourt: Presse Med. *15*, 2321–2323 (1986).
136 E. Bergogne-Berezin: Presse Med. *15*, 2324–2327 (1986).
137 B. Ferre and J.F. Acar: Presse Med. *15*, 2317–2320 (1986).
138 M. Derriennic, A. Reynaud, C. Launay and A.L. Courtieu: Pathol. Biol. *35*, 572–576 (1987).
139 E. Ronco, F. Vacheron and A. Orabona: Pathol. Biol. *35*, 558–562 (1987).
140 A.E. Reynaud, B. Coude du Foresto, E.P. Espaze and A.L. Courtieu: Pathol. Biol. *35*, 1023–1026 (1987).
141 A. Sedallian and T. Roudon: Pathol. Biol. *36*, 678–681 (1988).
142 A. Sedallian, J. Debrueres, T. Roudon and M. Deronzier: Pathol. Biol. *36*, 682–684 (1988).
143 Y. Buisson, R. Bercion, P. Mauclere, L. Hugard and H. Schill: Pathol. Biol. *36*, 671–674 (1988).
144 C.J. Soussy, R. Leclercq and J. Duval: Pathol. Biol. *36*, 357–360 (1988).
145 D.A. Bremnmer: Pathology *19*, 274–276 (1987).
146 P.J. Collignon, R. Munro and G. Morris: Pathology *20*, 48–52 (1988).
147 M.K. Aravind, J.N. Miceli and R.E. Kauffman: J. Chromatogr. *233*, 423–426 (1982).
148 F. Jehl, P. Birckel and H. Monteil: J. Chromatogr. *413*, 109–119 (1987).
149 M.A. Riegel and P.P. Ellis: J. Chromatogr. *424*, 177–181 (1988).
150 J. Knöller, W. König, W. Schönfeld, K.D. Bremm and M. Köller: J. Chromatogr. *427*, 257–267 (1988).
151 T. Marunaka, M. Maniwa, E. Matsushima and Y. Minami: J. Chromatogr. *431*, 87–101 (1988).
152 J. Martin, R. Mendez, A. Negro: J. Liq. Chromatogr. *11*, 1707–1716 (1988).
153 D. Jung and N.K. Mahajan: Clin. Chem. *30*, 122–124 (1984).
154 T. Annesley, K. Wilkerson, K. Matz and D. Giacherio: Clin. Chem. *30*, 908–910 (1984).
155 F. Kees, K.G. Naber, B. Bartoschik-Wich, P. Stockmann, G.P. Meyers and H. Grobecker: Arzneim.-Forsch. *35*, 1099–1104 (1985).
156 A.M. Brisson and J.B. Fourtillan: J. Chromatogr. *223*, 393–399 (1981).
157 J. Dokladalova, G.T. Quercia and J.P. Stankewich: J. Chromatogr. *276*, 129–137 (1983).

158 J. Haginaka, J. Wakai, H. Yasuda and T. Uno: J. Chromatogr. *341*, 115–122 (1985).
159 G. LaFollette, S. Kaubisch, J.G. Gambertoglio and E.T. Lin: J. Liq. Chromatogr. *11*, 683–700 (1988).
160 S.H. Schacter, M. Spino, A. Tesoro and S.M. MacLeod: J. Lig. Chromatogr. *7*, 2421–2429 (1984).
161 G. Nygard and S.K. Wahba Khali: J. Liq. Chromatogr. *7*, 1461–1475 (1984).
162 C.M. Selavka, I.S. Krull and K. Bratin: J. Pharm. Biomed. Anal. *4*, 83–93 (1986).
163 C.Y. Chan, K. Chan and G.L. French: J. Antimicrob. Chemother. *18*, 537–545 (1986).
164 S. Ting: J. Assoc. Off. Anal. Chem. *71*, 1123–1130 (1988).
165 V. Das Gupta, C. Bethea and M. dela Torre: J. Clin. Pharm. Ther. *13*, 199–205 (1988).
166 J. Zhi, C.H. Nightingale and R. Quintiliani, J. Pharm. Sci. *75*, 1063–1067 (1986).
167 I. Saikawa. In: Beta-Lactam Antibiotics, pp. 99–106, S. Mitsuhashi, Ed., Japan Sci. Soc. Press, Tokyo, Japan, 1981.
168 I. Saikawa, S. Takano, C. Yoshida, O. Takashima, K. Momonoi, T. Yasuda, K. Kasuya and M. Komatsu: Yakugaku Zasshi *97*, 980–986 (1977).
169 I. Saikawa, N. Matsubara, S. Minami and S. Mitsuhashi. In: [Beta]-Lactam Antibiot.: Mode Action, New Dev., Future Prospects, [Proc. Symp.], Meeting Date 1980, pp. 353–360, R.J.M. Salton, G.D. Stockman Eds. Academic Press, New York, 1981.
170 K. Ueo, Y. Fukuoka, T. Hayashi, T. Yasuda, H. Taki, M. Tai, Y. Watanabe, I. Saikawa and S. Mitsuhashi: Antimicrob. Agents Chemother. *12*, 455–460 (1977).
171 M. Merslavic: Vestn. Slov. Kem. Drus. *32*, 31–40 (1985).
172 K.D. Hardy, F.P. Harrington and A.V. Stachulski: J. Chem. Soc. Perkin Trans *1*, 1227–1235 (1984).
173 A.W. Guest and P.H. Milner: Tetrahedron Lett. *25*, 4845–4848 (1984).
174 A.W. Guest: Tetrahedron Lett. *27*, 3049–3052 (1986).
175 C.L. Branch, M.J. Basker, S.C. Finch, A.W. Guest, F.P. Harrington, A.C. Kaura, S.J. Knott, P.H. Milner and M.J. Pearson: J. Antibiot. *40*, 646–651 (1987).
176 S. Takano, I. Takakura, H. Ochiai, K. Momonoi, K. Tanaka, Y. Todo, T. Yasuda, M. Tai, Y. Fukuoka, H. Taki and I. Saikawa: Yakugaku Zasshi *102*, 629–645 (1982).
177 S. Takano, I. Takakura, H. Ochiai, K. Momonoi, K. Tanaka, H. Shibata, M. Komatsu, T. Yasuda, M. Tai, Y. Fukuoka and I. Saikawa: Yakugaku Zasshi *102*, 617–628 (1982).
178 A.W. Guest, C.L. Branch, S.C. Finch, A.C. Kaura, P.H. Milner, M.J. Pearson, R.J. Ponsford and T.C. Smale: J. Chem. Soc. Perkin Trans *1*, 45–55 (1987).
179 G. Burton, M.J. Basker, P.H. Bentley, D.J. Best, R.A. Dixon, F.P. Harrington, R.F. Kenyon, A.G. Lashford and A.W. Taylor: J. Antibiot. *38*, 721–739 (1985).
180 A.W. Guest, F.P. Harrington, P.H. Milner, R.J. Ponsford, T.C. Smale, A.V. Stachulski, M.J. Basker and B. Slocombe: J. Antibiot. *39*, 1498–1501 (1986).
181 S. Uyeo and H. Ona: Chem. Pharm. Bull. *28*, 1563–1577 (1980).
182 T. Matsuo, T. Sugawara, H. Masuya, Y. Kawano, N. Noguchi and M. Ochiai: Chem. Pharm. Bull. *31*, 1874–1884 (1983).
183 C. Yoshida, T. Hori, K. Momonoi, K. Tanaka, S. Kishimoto and I. Saikawa: Agric. Biol. Chem. *50*, 839–846 (1986).

184 Y. Kawano, T. Watanabe, J. Sakai, H. Watanabe, M. Nagano, T. Nishimura and T. Miyadera: Chem. Pharm. Bull. *28*, 70–79 (1980).
185 P.M. Allison, L.L. Mummah-Schendel, C.G. Kindberg, C.S. Harms, N.U. Bang and J.W. Suttie: J. Lab. Clin. Med. *110*, 180–188 (1987).
186 K. Kawamoto, A. Touchi, K. Sugeno and T. Matsubara: Jpn. J. Pharmacol. *47*, 169–178 (1988).
187 M. Matsuura, S. Satoh, K. Takano, T. Harauchi, T. Yoshizaki, F. Kobayashi, T. Matsubara and K. Uchida: Jpn. J. Pharmacol. *46*, 303–310 (1988).
188 M.J. Shearer, H. Bechtold, K. Andrassy, J. Koderisch, P.T. McCarthy, D. Trenk, E. Jähnchen and E. Ritz: J. Clin. Pharmacol. *28*, 88–95 (1988).
189 T. Matsubara, S. Otsubo, A. Ogawa, K. Kawamoto, J. Okamoto, K. Sugeno, Y. Toschino, T. Yoshida and E. Hirai: Jpn. J. Pharmacol. *45*, 303–315 (1987).
190 T. Matsubara, S. Otsubo, A. Ogawa and H. Nakao: Jpn. J. Pharmacol. *45*, 115–119 (1987).
191 C. Kamei, Y. Sugimoto, N. Muroi and K. Tasaka: J. Pharm. Pharmacol. *38*, 823–828 (1986).
192 J. Mitsuyama, M. Takahata, T. Yasuda and I. Saikawa: J. Antibiot. *40*, 868–872 (1987).
193 I. Saikawa, T. Yasuda, H. Taki, M. Tai, Y. Watanabe, H. Sakai, S. Takano, C. Yoshida and K. Kasuya: Yakagaku Zasshi *97*, 987–994 (1977).
194 A. Yamaguchi, N. Tomiyama, R. Hiruma and T. Sawai: FEBS Lett. *181*, 143–148 (1985).
195 F. Yoshimura and H. Nikaido: Antimicrob. Agents Chemother. *27*, 84–92 (1985).
196 M.C. Nacucchio, M.J.G. Bellora, D.O. Sordelli and M. D'aquino: Antimicrob. Agents Chemother. *27*, 137–139 (1985).
197 M.J. Driver, P.H. Bently, R.A. Dixon, R.A. Edmondson, A.C. Kaura and A.W. Taylor: J. Antibiot. *37*, 297–299 (1984).
198 M.J. Driver, J. Lowther and A.W. Taylor: J. Antibiot. *38*, 550–553 (1985).
199 A. Meulemans, P. Vicart, J. Mohler and M. Vulpillat: Chemotherapy (Basel) *34*, 90–95 (1988).
200 J.P. Akbaraly, J.J. Leng, G. Bozler and J.K. Seydel. In: QSAR Strategies in the Design of Bioactive Compounds, Proc. Fifth Eur. Symp. Quant. Struct.-Act. Relationships (1984), J.K. Seydel, Ed., pp. 313–317, VCH, Weinheim, Fed. Rep. Ger., 1985.
201 S. Okuno, I. Maezawa, Y. Sakuma, T. Matsushita and T. Yamaguchi: J. Antibiot. *41*, 239–246 (1988).
202 N.R. Pattinson, K.E. Willis and B.A. Chapman: Dig. Dis. Sci. *32*, 615–619 (1988).
203 R. Williamson, R. Hakenbeck and A. Tomasz: Antimicrob. Agents Chemother. *18*, 629–637 (1980).
204 N. Matsubara, S. Minami, M. Matsuhashi, M. Takaoka and S. Mitsuhashi: Antimicrob. Agents Chemother. *18*, 195–199 (1980).
205 M. Leyh-Bouille, M. Nguyen-Distèchè, S. Pirlot, A. Veithen, C. Bourguignon and J.-M. Ghuysen: Biochem. J. *235*, 177–182 (1986).
206 N.H. Georgopapadakou and F.Y. Liu: Antimicrob. Agents Chemother. *18*, 834–836 (1980).
207 N.H. Georgopapadakou, S.A. Smith, C.M. Cimarusti and R.B. Sykes: Antimicrob. Agents Chemother. *23*, 98–104 (1983).
208 I. Saikawa, Y. Nakashima, H. Sakai, K. Momonoi, T. Ikegami, H. Minami, H. Hayakawa, H. Ochiai and Y. Todo: Yakugaku Zasshi *100*, 625–640 (1980).
209 I. Saikawa, Y. Nakashima, M. Tai, H. Sakai, I. Takakura, T. Ikegami, H. Hayakawa, Y. Shugimoto, T. Yamafuji, T. Miyaura and M. Noguchi: Yakugaku Zasshi *103*, 414–425 (1983).

210 W. Kramer, F. Girbig, E. Petzoldt and I. Leipe: Biochim. Biophys. Acta *943*, 288–296 (1988).

211 K.-I. Nishimura, Y. Nozaki, A. Yoshimi, S. Nakamura, M. Kitagawa, N. Kakeya and K. Kitao: Chem. Pharm. Bull. *33*, 282–291 (1985).

212 S.K. Branch, A.F. Casy and E.M.A. Ominde: J. Pharm. Biomed. Anal. *5*, 73–103 (1987).

213 A.E. Bird, K.R. Jennings and A.C. Marshall: J. Pharm. Pharmacol. *38*, 913–917 (1986).

214 Z. Plotkowiak: Pharmazie *42*, 449–451 (1987).

215 K. Goldstein, H. Colding and G.E. Andersen: APMIS *96*, 329–332 (1988).

216 I. Saikawa, S. Takamo, C. Yoshida, E. Saito, T. Yasuda, H. Sakai and Y. Takashima: Yakugaku Zasshi *97*, 995–1001 (1980).

217 T. Yamana, Y. Mizukami, F. Ichimura, K. Shimazaki, M. Ishioka and M. Suda: Int. J. Pharm. *11*, 71–80 (1982).

218 P.H. Milner, A.W. Guest, F.P. Harrington, R.J. Ponsford, T.C. Smale and A.V. Stachulski: J. Chem. Soc., Chem. Commun., 1335–1336 (1984).

219 E.A. Cutmore, A.W. Guest, J.D.I. Hatto, T.C. Smale, A.V. Stachulski and J.W. Tyler: J. Chem. Soc., Chem. Commun., 21–23 (1987).

220 C. Fletcher, C. Pearson, S.C. Choi, R.J. Duma, H.J. Evans and G.D. Qureshi: J. Lab. Clin. Med. *108*, 217–223 (1986).

221 L.O. Gentry, J.G. Jemsek and E.A. Natelson: Antimicrob. Agents Chemother. *19*, 532–533 (1981).

222 M.R. Weitckamp, G.M. Caputo, H.A.B. Al-Mondhiry and R.C. Aber: J. Antimicrob. Chemother. *16*, 95–101 (1985).

223 J.C. Petit, G.-L. Daguet, G. Richard and B. Burghoffer: J. Antimicrob. Chemother. *20*, 615–617 (1987).

224 W. Roszkowski, H.L. Ko, K. Roszkowski, J. Jeljaszewicz and G. Pulverer. In: Chemotherapy and Immunity, Zentralblatt für Bakteriologie, Mikrobiologie und Hygiene, I. Abteilung, Suppl. 13, pp. 59–72, G. Pulverer and J. Jeljaszewica, Eds., Gustav Fischer Verlag, Stuttgart, 1985.

225 Y. Ueda, A. Saito, Y. Fukuoka, Y. Yamashiro, Y. Ikeda, B.H. Taki, T. Yasuda and I. Saiwaka: Antimicrob. Agents Chemother. *23*, 374–378 (1983).

226 Y. Yamashiro, Y. Fukuoka, A. Yotsuji, T. Yasuda, I. Saikawa and Y. Ueda: J. Antimicrob. Chemother. *18*, 703–708 (1986).

227 M. Takahata, Y. Yamashiro, A. Yotsuji, Y. Fukuoka, T. Yasuda, I. Saikawa and Y. Ueda: J. Antibiot. *39*, 1167–1171 (1986).

228 K. Roszkowski: Pneum. Pol. *54*, 236–244 (1986).

229 J. Vekemans, C. Pollers-Wieërs and G. Hoornaert: J. Heterocycl. Chem. *20*, 919–923 (1983).

230 G.W.H. Cheeseman and E.S.G. Werstiuk: Adv. Heterocycl. Chem. *14*, 99–209 (1972).

231 M. Bobek, A. Block, P. Berkowitz and T.J. Bardos: J. Med. Chem. *20*, 458–460 (1977).

232 M. Mori, M. Ishikura, T. Ikeda and Y. Ban: Heterocycles *16*, 1491–1494 (1981).

233 M. Ishikura, M. Mori, T. Ikeda, M. Terashima and Y. Ban: J. Org. Chem. *47*, 2456–2461 (1982).

234 S. Iimura and J. Okumura: U.S. US 4316024 A, 16 Feb. (1982) 9pp.

235 Sumitomo Chemical Co.: Jpn. Kokai Tokkyo koho Jp 59/164781 A2 [84/164781]. 17 Sept. (1984) 5 pp.

236 T. Hori, C. Yoshida, Y. Kiba, R. Takeno, J. Nakano, J. Nitta, S. Kishimoto, S. Murakami, H. Tsuda and I. Saikawa: Fr. Demand FR2499569 A1, 13 Aug. (1982), 26 pp.

237 T. Hori, C. Yoshida, Y. Kiba, R. Takeno, J. Nakano, J. Nitta, S. Kishi-

moto, S. Murakami, H. Tsuda and I. Saikawa: U.S. Pat. 4459407 A, 10 July (1984), 13pp. Cont.-in-part of U.S. Ser. No. 169457.
238 Toyama Chemical Co.: Ger. Offen. DE 3027106, 19 Feb. (1981), 86pp.
239 T. Hori, C. Yoshida, S. Murakami, Y. Takeno, Y. Kiba, H. Tsuda, S. Kishimoto and I. Saikawa: Chem. Pharm. Bull. *29*, 1594–1605 (1981).
240 I. Saikawa, T. Hori, C. Yoshida, Y. Kiba, S. Murakami, S. Miyabara, H. Tsuda, M. Sugita, S. Kishimoto and K. Hijikata: Jpn. Kokai Tokkyo Koho JP 54/106481 [79/106481], 21 Aug. (1979) 13pp.
241 T. Hori, C. Yoshida, S. Murakami, R. Takeno, J. Nakano, J. Nitta, H. Tsuda, S. Kishimoto and I. Saikawa: Chem. Pharm. Bull. *29*, 684–698 (1981).
242 T. Hori, K. Momononi, Y. Kiba, C. Yoshida, H. Sakai, R. Takeno, T. Ohashi, S. Kishimoto and I. Saikawa: Yakugaku Zasshi *99*, 730–737 (1979).
243 H. Fischer, H. Möller, M. Budnowski, G. Atassi, P. Dumont, J. Venditti and O.C. Yoder: Arzneim.-Forsch. *34*, 663–668 (1984).
244 G. Atassi, P. Dumont, U. Fischer, M. Zeidler and M. Budnowski: Beitr. Onkol. *18*, 221–234 (1984).
245 T. Hori, C. Yoshida, S. Murakami, Y. Kiba, R. Takeno, J. Nakano, H. Tsuda and I. Saikawa: Chem. Pharm. Bull. *29*, 386–389 (1981).
246 T. Hori, C. Yoshida, Y. Kiba, J. Nakano, S. Kishimoto, S. Murakami, S. Miyahara, H. Tsuda and I. Saikawa: Yakugaku Zasshi *100*, 262–271 (1980).
247 C.D. Lunsford and A.D. Cale, Jr.: Ger. Offen. DE 2120367, 11 Nov. (1971), 28pp.
248 C.D. Lunsford and A.D. Cale, Jr.: U.S. US 3891646, 24 Jun. (1975), 5pp.
249 K. Keppler and E. DeClercq: Arch. Pharm. (Weinheim) 320, 271–274 (1987).
250 F.M.E. Abdel-Megeid, M.A.F. El-kaschef and A.A.G. Ghattas: Egypt. J. Chem. *20*, 427–439 (1980).
251 G.G. Dvoryantseva, K.P. Tetenchuk, V.I. Polshakov and A.S. Elina: Khim.-Farm. Zh. *20*, 399–408 (1986).
252 M. Krishnamurthy, K.A. Iyer and S.K. Dogra: J. Photochem. *38*, 277–287 (1987).
253 D.K. Hess: US 4659713 A, 21 Apr. 1987, 5pp.
254 J.G. Page, H.R. Sullivan and S.L. Due: Xenobiotica *6*, 713–723 (1976).
255 R.L. St. Clair and T.D. Thibault: US 3992378, 16 Nov. 1976, 10 pp.
256 S.B. Kaden and R. Sarges: Eur. Pat. Appl. EP 107455 A1, 2 May 1984, 66pp.
257 I. Saito and T. Matsuura, Biochemistry *6*, 3602–3608 (1967).
258 P. Jacobsen, F.E. Nielsen and T. Honoré: Eur. Pat. Appl. EP 283,959, 28 Sept. 1988, 21 pp.
259 T. Honoré, S.N. Davies, J. Drejer, E.J. Fletcher, P. Jacobsen, D. Lodge and F.E. Nielsen: Science *241*, 701–703 (1988).
260 J.C. Walkins and R.H. Evans: Annu. Rev. Pharmacol. Toxicol. *21*, 165–204 (1981).
261 H. McLennan: Prog. Neurobiol. (NY) *20*, 251-, *271* (1983).
262 D. Lodge, Ed. In: Excitatory Amino Acids in Health and Disease, Wiley, New York, 1988.
263 T.J. Biscoe, J. Davies, A. Dray, R.H. Evans, A.A. Francis, M.R. Martin and J.C. Watkins: Eur. J. Pharmacol. *45*, 315–316 (1977).
264 H. Wamhoff and W. Kleimann: J. Chem. Soc., Chem. Commun., 743–744 (1981).
265 U.V. Venkataram and T.C. Bruice: J. Chem. Soc., Chem. Commun., 899–900 (1984).
266 R.L. Willer and D.W. Moore: J. Org. Chem. *50*, 5123–5127 (1985).

267 W.M. zu Reckendorf, R. Weber and H. Hehenberger: Chem. Ber. *114*, 1306–1717 (1981).
268 H. Aoyama, M. Ohnota, M. Sakamoto and Y. Omote: J. Org. Chem. *51*, 247–249 (1986).
269 J.H. Hoare and P. Yates: Heterocycles *15*, 281–284 (1981).
270 P.G. Sammes: Fortschr. Chem. Org. Naturst. *32*, 51–118 (1975).
271 M. Gdaniec, B. Liberek, A.S. Kolodziejczyk, R. Jankowska and J. Ciarkowski: Int. J. Peptide Protein Res. *30*, 79–92 (1987).
272 A. Kolodziejczyk and J. Ciarkowski: J. Biopolymers *25*, 771–786 (1986).
273 R. Jankowski and J. Ciarkowski: Int. J. Peptide Protein Res. 61–78 (1987).
274 H. Bielinski and J. Ciarkowski: Biopolymers *25*, 795–809 (1986).
275 W. Radding, B. Donzel, N. Ueyama and M. Goodman: J. Am. Chem. Soc. *102*, 5999–6005 (1980).
276 E. Benedetti, R.E. Marsh and M. Goodman: J. Am. Chem. Soc. *98*, 6676–6684 (1976).
277 G. Snatzke. In: Optical Activity and Chiral Discrimination, Proceedings of the Nato Advances Study Institute, held at the University of Sussex, Falmer, England, Sept. 10–22, 1978. S.F. Mason, Ed., pp. 25–41, D. Reidel Publishing Co., Dordrecht, The Netherlands, 1979.
278 H.R. Kricheldorf: Org. Mag. Resonance *13*, 52–58 (1980).
279 G.P. Slater and L.R. Hogge: Can. J. Chem. *57*, 2037–2051 (1979).
280 G.P. Slater and L.R. Hogge: Can. J. Chem. *57*, 2052–2057 (1979).
281 A.B. Mauger: J. Chromatogr. *37*, 315–317 (1968).
282 R.M. Wenger: Helv. Chim. Acta *66*, 2672–2702 (1983).
283 M.J. Sofia, P.K. Chakravarty and J.A. Katzenellenbogen: J. Org. Chem. *48*, 3318–3325 (1983).
284 F.-J. Marner, R.E. Moore, K. Hirotsu and J. Clardy: J. Org. Chem. *42*, 2815–2819 (1977).
285 J.A. Grina and F.R. Stermitz: Tetrahedron Lett. *22*, 5257–5258 (1981).
286 J.A. Grina, M.R. Ratcliff and R.F. Sternitz: J. Org. Chem. *47*, 2648–2651 (1982).
287 S.M. Marcuccio and J.A. Elix: Aust. J. Chem. *37*, 2397–2401 (1984).
288 A.B. Mauger: J. Chem. Soc. Perkin *1*, 1320–1323 (1975).
289 D. Scott. In: Aspartam: Properties, use, present status. Proc. IV. International Congress of Science and Technology, Madrid, pp. 468–473, Vol. 1, 1974.
290 M. Prudel and E. Davidkova: Die Nahrung *25*, 193–199 (1981).
291 M. Prudel and E. Davidkova: Die Nahrung *29*, 381–389 (1985).
292 Federal Register, Vol. 39, No. 145, July 26, 1974.
293 R. Schwyzer, J.P. Carrion, B. Gorup, H. Nolting and A. Tun-kyi: Helv. Chim. Acta *47*, 441–464 (1964).
294 M. Rothe, M. Fähnle and S. Wermuth. In: Pept. Proc. Europ. Pept., Symp., 18th, pp. 573–576, Ragnarsson, Ulf, Almquist and W.K. Sell, Ed. Stockholm, Sweden, 1984.
295 M. Rothe, K. Seffen and I. Rothe: Angew. Chem. *77*, 347–348 (1965).
296 W. Mästle and M. Rothe. In: Peptides, 1978, p. 179–182, G. Kupryszewski, Ed., Wroclaw Univ. Press, Poland, 1979.
297 M. Rothe, M. Fähnle and W. Mästle. In: Peptides, p. 89–92, D.H. Rich and E. Gross, Ed., Pierce Chem. Co., Rockford Ill., 1981.
298 B.H. Arison and J.L. Beck: Tetrahedron *29*, 2743–2751 (1973).
299 G.R. Pettit, R.B. von Dreele, D.L. Herald, M.T. Edgar and H.B. Wood, Jr.: J. Am. Chem. Soc. *98*, 6742–6743 (1976).
300 J.A. Gottlieb, E.T. Freireich, G.P. Bodey, V. Rodriguez, K.B. Meeredie and J.U. Gutterman: Proc. Am. Assoc. Cancer Res. *16*, 86 (1975).
301 S.E. Jones, W.G. Tucker, A. Haut, B.L. Tranum, C. Vaughn, E.M. Chase and B.G.M. Durie: Cancer Treatment Reports *61*, 1617–1621 (1977).

302 B.T. Golding and A.T. Smith: J. Chem. Soc., Chem. Commun., 702–703 (1980).
303 P.S. Steyn: Tetrahedron *29*, 107–120 (1973).
304 P.S. Steyn and R. Vleggaar: Phytochemistry *15*, 355–356 (1976).
305 S. Sekita, K. Yoshihira, S. Natori, S. Udagowa, T. Muroi, Y. Sugiyama, H. Kurata and M. Umeda: Can. J. Microbiol. *27*, 766–772 (1981).
306 S. Udagawa, T. Muroi, H. Kurata, S. Sekita, K. Yoshihira, S. Natori and M. Umeda: Can. J. Microbiol. *25*, 170–177 (1979).
307 D. Brewer and A. Taylor: Can. J. Microbiol. *24*, 1082–1086 (1978).
308 D. Brewer, A.G. McInnes, D.G. Smith, A. Taylor, J.A. Walter, H.R. Loosli and Z.L. Kis: J. Chem. Soc. Perkin *I*, 1248–1251 (1978).
309 D. Brewer, F.W. Calder, T.M. MacIntyre and A. Taylor: J. Agric. Sci. *76*, 465–477 (1971).
310 A.G. McInnes, A. Taylor and J.A. Walter: J. Am. Chem. Soc. *98*, 6741–6742 (1976).
311 W.-C. Jen and G.A. Jones: Can. J. Microbiol. *29*, 1399–1404 (1983).
312 D. Brewer and A. Taylor: Proc. N.S. Inst. Sci. *30*, 101–133 (1980).
313 T. Nagasawa, N. Mori, Y. Tani, K. Ogata and H. Irie: J. Antibiotics *29*, 526–531 (1976).
314 A.G. Sandoz, Europ. Pat. Appl. EP 3286, 8 Aug. 1979, 17pp.
315 D.E. Portlock, A.C. Ghosh, W.E. Schwarzel, R.R. Kurtz, E. Boger, H.C. Dalzell and R.K. Razdan: J. Heterocyclic Chem. *13*, 781–788 (1976).
316 H.G. Floss, J.M. Cassady and J.E. Robbers: J. Pharm. Sci. *62*, 699–715 (1973).
317 S.K. Quadri, J.L. Clark and J. Meites: Proc. Soc. Exptl. Biol. Med. *142*, 22–26 (1973).
318 H. Nagasawa, C.-L. Chen and J. Meites: Proc. Soc. Exptl. Biol. Med. *142*, 625–627 (1973).
319 J.M. Cassady, G.S. Li, E.B. Spitzner and H.G. Floss: J. Med. Chem. *17*, 300–307 (1974).
320 M.J. Sweeney, G.A. Poore, E.C. Kornfield, N.G. Bach, N.V. Owen and J.A. Clemens: Cancer Res. *35*, 106–109 (1975).
321 N. Shimazaki, I. Shima, K. Hemmi and M. Hashimoto: J. Med. Chem. *30*, 1706–1709 (1987).
322 N. Shimazaki, I. Shima, K. Hemmi, Y. Tesurumi and M. Hashimoto: Chem. Pharm. Bull. *35*, 3527–3530 (1987).
323 M. Okamoto, K. Yoshida, M. Nishikawa, T. Ando, M. Iwami, M. Kohsaka and H. Aoki: J. Antibiot. Tokyo *39*, 198–204 (1986).
324 M. Okamoto, K. Yoshida, I. Uchida, M. Nishikawa, M. Kohsaka and H. Aoki: Chem. Pharm. Bull. *34*, 340–344 (1986).
325 K. Yoshida, M. Okamoto, N. Shimazaki and K. Hemmi. In: Progr. Biochem. Pharmacol. Vol. 22, P. Brqquet, H.K. Margold and B.B. Vargaftig Eds., Karger, Basel, 1988, pp. 66–80.
326 M. Okamoto, K. Yoshida, I. Uchida, M. Kohsaka and H. Aoki: Chem. Pharm. Bull. *34*, 345–348 (1986).
327 S. Takase, N. Shigematsu, I. Shima, I. Uchida and M. Hashimoto: J. Org. Chem. *52*, 3485–3487 (1987).
328 N. Shimazaki, K. Hemmi, O. Nakaguti, Y. Miyazaki and M. Hashimoto, Eur. Pat. Appl. EP 243,122; 28, Oct., 1987, 53pp.
329 C. Prasad and A. Peterkofsky: J. Biol. Chem. *251*, 3229–3234 (1976).
330 C. Prasad, T. Matsui and A. Peterkofsky: Nature *268*, 142–144 (1977).
331 T. Matsui, C. Prasad and A. Peterkofsky: J. Biol. Chem. *254*, 2439–2445 (1979).
332 F. Battaini and A. Peterkofsky: Biochem. Biophys. Res. Commun. *94*, 240–247 (1980).

333 A. Enjalbert, M. Ruberg, S. Arancibia, M. Prian, K. Bauer and C. Kordan: Europ. J. Pharmacol. *58*, 97–98 (1979).
334 C. Prasad, T. Matsui, J. Williams and A. Peterkofsky: Biochem. Biophys. Res. Commun. *85*, 1582–1587 (1978).
335 T. Yanagisawa, C. Prasad, J. Williams and A. Peterkofsky: Biochem. Biophys. Res. Commun. *86*, 1146–1153 (1979).
336 R.M. McLeod and J.E. Lehmeyer: Endocrinology *94*, 1077–1085 (1974).
337 L.C. Contley, Jr., L. Josephson, R. Warner, M. Yunagisawa, C. Lechene and G. Guidotti: J. Biol. Chem. *252*, 7421–7423 (1977).
338 R. Tomatis, S. Salvadori and G.P. Sarto: Eur. J. Med. Chem. *16*, 229–232 (1981).
339 H. Ito, M. Yasumatsu and Y. Usui: Fukuoka Igaku Zasshi *76*, 441–450 (1985).
340 Y. Oka, K. Nishikawa and A. Miyake, Japan Pat. Appl., 57–179141, 1982.
341 T. Ogihara, M. Nakamuru, J. Higaki, Y. Kumahara, Y. Hamano, T. Minamino and N. Nakamura: Jpn. Pharmacol. Ther. *11*, 4663–4681 (1983).
342 J.J. Kamanski, M.M. Huycke, S.H. Selk, N. Bodor and T. Higuchi: J. Pharm. Sci. *65*, 1737–1742 (1976).
343 S.D. Worley, S.H. Gerson, N. Bodor and J.J. Kaminski: Chem. Phys. Lett. *60*, 104–107 (1978).
344 K. Hisamitsu, T. Takemoto and T. Yugawa: Jpn. Kokai Tokkyo Koho JP 61/277696 [86/277696], 08 Dec. 1986.
345 E. Kasafireh, J. Krepelka, J. Stuchlik and L. Cvak: Czech. CS, 232,319, 15 Jul. 1986.
346 I. Saikawa, T. Hori, K. Momoi, Y. Kobane, T. Yasuda, M. Miyahara, S. Tai, H. Tsuda, M. Doi and S. Kishimoto: Jpn. Kokai Tokkyo Koho, JP 54/30190 [79/30190], 6 Mar. (1979), 10pp.
347 I. Saikawa, S. Takano, O. Takashima, C. Yoshida, S. Kuroda, M. Komatsu, K. Momonoi, T. Yasuda and Y. Kodama: Japan. Kokai JP 50/15189 [75/151891], 6 Dec. 1975, 16pp.
348 D.E. Nitecki, B. Halpern and J.W. Westley: J. Org. Chem. *33*, 864–866 (1968).
349 J.P. Greenstein and M. Wintz. In: Chemistry of the Amino Acids, pp. 782–804 John Wiley and Sons, Inc., New York, 1961.
350 Y. Yamamoto, H. Yatagai and K. Maruyama: J. Chem. Soc., Chem. Commun., 835–836 (1980).
351 K.D. Kopple and H.G. Ghazanian: J. Org. Chem. *33*, 862–864 (1968).
352 T. Okawara, T. Matsuda and M. Furukawa: Chem. Pharm. Bull. *30*, 1225–1233 (1982).
353 K. Titlestad. In: Chemistry and Biology of Peptides, pp. 59–65, J. Meienhofer, Ed., Ann Arbor Science Publishers, 1972.
354 J. Meienhofer, Y. Sano and R.P. Patel. In: Peptides, Chemistry and Biochemistry, pp. 419–434, B. Weinstein and S. Lande, Eds., Marcel Dekker, Inc., New York, 1968.
355 R. Nagaraj and P. Balaram: Heterocycles 7, 885–890 (1977).
356 J.T. Gerig and R.S. Mcleod: J. Org. Chem. *41*, 1653–1655 (1976).
357 S. Jerumanis and A. Lemieux: J. Heterocyclic Chem. *17*, 39–43 (1980).
358 G. Zoltano, D. Zoltan and B. Khim. Rezso: Geterotsikl. Soedin. *9*, 1211–1220 (1979).
359 N.M. Golyshin, L.G. Fedoseenko, N.S. Ukrainets, N.S. Koltsov, Y.S. Mishchenko, E.M. Storozhenko and V.I. Pustovoitova, U.S.S.R. 267,245 (Cl. A Oln), 17 Sep. 1973. C.A. *81*, 146894 p, 1974.
360 J. Hausler and U. Schmidt: Chem. Bev. *107*, 2804–2815 (1974).
361 T. Fukuyama, B.D. Robins and R.A. Sachlenben: Tetrahedron Lett. *42*, 4155–4158 (1981).
362 S.M. Marcuccio and J.A. Elix: Aust. J. Chem. *38*, 343–346 (1985).

363 A. Ohta, F. Yamamoto, Y. Arimura and T. Watanabe: J. Heterocyclic Chem. *19*, 781–784 (1982).
364 T. Kanmera, S. Lee, H. Aoyagi and N. Izumiya: Tetrahedron Lett., 4483–4486 (1979).
365 N. Izumiya, S. Lee, T. Kanmera and H. Aoyagi: J. Am. Chem. Soc. *99*, 8346–8348 (1977).
366 R. Kuzlauskas, P.T. Murphy and R.J. Wells: Tetrahedron Lett., 4945–4948 (1978).
367 W. Hofheinz and W.E. Oberhansli: Helv. Chim. Acta *60*, 660–669 (1977).
368 R. Kazlauskas, R.O. Lidgard, R.J. Wells and W. Vetter: Tetrahedron Lett., 3183–3186 (1977).
369 C.M. Maes, P.S. Steyn, P.H. van Rooyen and C.J. Rabie: J. Chem. Soc. Chem. Commun., 350–351 (1982).
370 M.I. Pita Boente, G.W. Kirby and D.J. Robins: J. Chem. Soc., Chem. Commun., 619–621 (1981).
371 J.D.M. Herscheid, R.J.F. Nivard, M.W. Tijhuis, H.P.H. Scholten and H.C.J. Ottenheijm: J. Org. Chem. *45*, 1880–1885 (1980).
372 N. Shinmon, M.P. Cava and R.F.C. Brown: J. Chem. Soc., Chem. Commun., 1020–1021 (1980).
373 A.K. Gökturk, A.A.E. Porter and P.G. Sammes, J. Chem. Soc. Perkin *I*, 953–960 (1982).
374 S.M. Marcuccio and J.A. Elix: Aust. J. Chem. *38*, 1785–1796 (1985).
375 S.M. Marcuccio and J.A. Elix: Tetrahedron Lett. *24*, 1445–1448 (1983).
376 M. Barbier: Helv. Chim. Acta *69*, 152–155 (1986).
377 S. Inoue, N. Takamatsu and Y. Kishi: Yakugaku Zasshi *97*, 558–563 (1977).
378 S. Nakatsuka, H. Miyazaki and T. Goto: Tetrahedron Lett. *21*, 2817–2820 (1980).
379 Y. Oikatua, T. Yoshioka and O. Yonemitsu: 12th Temen Yuki Kagobutsu Toronkai (Sapporo), August, 1978 (abstracts pp 22–27).
380 S. Inoue, N. Takamatsu and Y. Kishi: Yakugaku Zasshi *97*, 564–568 (1977).
381 S. Inoue, J. Murata, N. Takamatsu, H. Nagano and Y. Kishi: Yakugaku Zasshi *97*, 576–581 (1977).
382 H.C.J. Ottenheijm, R. Plate, J.H. Noordik, J.D.M. Herscheid: J. Org. Chem. *47*, 2154–2157 (1982).
383 T. Kametani, N. Kanaya and M. Ihara: J. Chem. Soc. Perkin *I*, 959–963 (1981).
384 U. Nagel, H. Menzel, P.W. Lednor, W. Beck, A. Guyot and M. Bartholin: Z. Naturforsh. B. Anorg. Chem., Org. Chem. *36B*, 578–584 (1981).
385 D.W. Henry: J. Heterocycl. Chem. *3*, 503–511 (1966).
386 A.A. Akimova and N.I. Gavrilov: J. Gen. Chem. USSR (Engl. Transl.) *31*, 41–44 (1961).
387 M.E. Jung and J.C. Rohloff: J. Org. Chem. *50*, 4909–4913 (1985).
388 P.A. Sturm, D.W. Henry, P.E. Thompson, J.B. Zeigler and J.W. McCall: J. Med. Chem. *17*, 481–487 (1974).
389 S. Tamai, M. Kaneda and S. Nakamura: J. Antibiot. *35*, 1130–1136 (1982).
390 M. Kaneda, S. Tamai, S. Nakamura, T. Hirata, Y. Kushi and T. Suga: J. Antibiot. *35*, 1137–1140 (1982).
391 S. Nishiyama, K. Nakamura, Y. Suzuki and S. Yamamura: Tetrahedron Lett. *27*, 4481–4484 (1986).
392 A. Kubo, N. Saito, M. Nakamura, K. Ogata and S. Sakai: Heterocycles *26*, 1765–1770 (1987).
393 A. Kubo, N. Saito, H. Yamato and Y. Kawakami: Chem. Pharm. Bull. *35*, 2525–2532 (1987).

394 T. Arai and A. Kubo. In: The Alkaloids, Vol. XXI, pp 55–100, A. Brossi, Ed., Academic Press, Inc., New York, 1983.
395 Y. Ikeda, H. Matsuki, T. Ogawa and T. Munakata: J. Antibiot. *36*, 1284–1289 (1983).
396 J.M. Frincke and D.J. Faulkner: J. Am. Chem. Soc. *104*, 265–269 (1982).
397 errata: ibid., 104, 5004 (1982).
398 A. Kubo, N. Saito, R. Yamauchi and S. Sakai: Chem. Pharm. Bull. *35*, 2158–2161 (1987).
399 T. Kamiya, S. Maeno, M. Hashimoto and Y. Mine: J. Antibiot. *25*, 576–581 (1972).
400 Y. Tokyma, S. Koda, T. Miyoshi and Y. Morimoto: Bull. Chem. Soc. Jpn. *47*, 18–23 (1974).
401 H. Maag, J.F. Blount, D.L. Coffen, T.V. Steppe and F. Wong: J. Am. Chem. Soc. *100*, 6786–6788 (1978).
402 T. Miyoshi, N. Miyairi, H. Aoki, M. Kohsaka, H. Sakai and H. Imanaka: J. Antibiot. *25*, 569–575 (1972).
403 S. Miyamura, N. Ogasawara, H. Otsuka, S. Niwayama, H. Tanaka, T. Take, T. Uchiyama, H. Ochiai, K. Abe, K. Koizumi, K. Asao, K. Mitsuki and T. Hoshino: J. Antibiot. *25*, 610–612 (1972).
404 S. Miyamura, N. Ogaswara, H. Otsuka, S. Niwayama, H. Tanaka, T. Take, T. Uchiyama and H. Ochiai: J. Antibiot. *26*, 479–484 (1973).
405 M. Nishida, Y. Mine, T. Matsubara, S. Goto and S. Kuwahara: J. Antibiot. *25*, 582–593 (1972).
406 T. Miyoshi, M. Iseki, T. Konomi and H. Imanaka: J. Antibiot. *33*, 480–487 (1980).
407 M. Iseki, T. Miyoshi, T. Konomi and H. Imanaka: J. Antibiot. *33*, 488–500 (1980).
408 T. Miyoshi, N. Miyairi, H. Aoki, M. Kohsaka, H. Sukai and H. Imanaka: J. Antibiot. *25*, 559–575 (1972).
409 M. Nishida, Y. Mine, T. Matsubara, S. Goto and S. Kuwahara: J. Antibiot. *25*, 582–593 (1972).
410 M. Nishida, Y. Mine, T. Matsubara, S. Goto and S. Kuwahara: J. Antibiot. *25*, 594–601 (1972).
411 N. Tanaka, M. Iseki, T. Miyoshi, H. Aoki and H. Imanaka: J. Antibiot. *29*, 155–168 (1976).
412 A. Someya, M. Iseki and N. Tanaka: J. Antibiot. *31*, 712–718 (1978).
413 S. Nakatsuka, Y. Yoshida and T. Goto: Tetrahedron Lett. *22*, 2009–2012 (1981).
414 S. Nakatsuka, K. Yoshida and T. Goto: Tetrahedron Lett. *22*, 4973–4976 (1981).
415 S. Nakatsuka, K. Yamada, K. Yoshida, O. Asano, Y. Murakami and T. Goto: Tetrahedron Lett. *24*, 5627–5630 (1983).
416 S. Nakatsuka and T. Goto: Heterocycles *21*, 61–72 (1984).
417 C. Shin, Y. Sato, S. Honda and J. Yoshimura: Bull. Chem. Soc. Jpn. *56*, 2652–2656 (1983).
418 R.M. Williams, O.P. Anderson, R.W. Armstrong, J. Josey, H. Meyers and C. Eriksson: J. Am. Chem. Soc. *104*, 6092–6099 (1982).
419 R.M. Williams: Tetrahedron Lett. *25*, 2341–2344 (1981).
420 R.M. Williams, J.S. Dung, J. Josey, R.W. Armstrong and H. Meyers: J. Am. Chem. Soc. *105*, 3214–3220 (1983).
421 R.M. Williams, R.W. Armstrong and J.S. Dung: J. Am. Chem. Soc. *106*, 5748–5750 (1984).
422 L.V. Dunkerton and R.M. Ahmed: Tetrahedron Lett. *21*, 1803–1806 (1980).
423 R.M. Williams and L.K. Maruyama: J. Org. Chem. *52*, 4044–4047 (1987).

424 M. Yamaura, T. Suzuki, H. Hashimoto, J. Yoshimura and C. Shin: Bull. Chem. Soc. Jpn. *58*, 2812–2820 (1985).
425 Y. Sato, C. Shin, S. Sumiya, Y. Nakajima and J. Yoshimura: Bull. Chem. Soc. Jpn. *57*, 1265–1270 (1984).
426 C. Shin, Y. Sato and J. Yoshimura: Tetrahedron Lett. *22*, 2401–2404 (1981).
427 P. Yates and J. H. Hoare: Can. J. Chem. *61*, 519–528 (1983).
428 P. Yates and J. H. Hoare: Can. J. Chem. *61*, 1397–1404 (1983).
429 A. Taylor. In: Biochemistry of Some Foodborn Microbial Toxins, Pap. Symp., New York, Meeting Date 1966, pp 69–107, R. I. Mateles, Ed., M.I.T. Press, Cambridge, Mass., 1967.
430 C. Leigh and A. Taylor: Adv. Chem. Ser. *149*, 228–275 (1976).
431 M. A. Stillwell, L. P. Magasi and G. M. Stunz: Can. J. Microbiol. *20*, 759–764 (1974).
432 K. H. Michel, M. O. Chaney, N. D. Jones, M. M. Hoehn and R. Nagarajan: J. Antibiot. *27*, 57–64 (1974).
433 R. M. Williams and W. H. Rastetter: J. Org. Chem. *45*, 2625–2631 (1980).
434 A. Srinivasan, A. J. Kolar and R. K. Olsen: J. Heterocyclic Chem. *18*, 1545–1548 (1981).
435 B. Ganem: Tetrahedron *34*, 3353–3383 (1978).
436 B. R. Davis and I. Bernal: Proc. Natl. Acad. Sci. *70*, 279–283 (1973).
437 M. Przybylska and E. M. Gopalakrishna: Acta Cryst. *B30*, 597–605 (1974).
438 G. M. Strunz, M. Kakushima and M. A. Stillwell: Can. J. Chem. *53*, 295–297 (1975).
439 F. R. Ahmed and M. Przybylska: Acta Cryst. *B33*, 168–170 (1977).
440 P. G. Deffieux, M. Gadret, J. M. Leger and A. Carpy: Acta Cryst. *B35*, 2358–2363 (1979).
441 P. G. Deffieux, M. Gadret and J. M. Leger: Acta Cryst. *B33*, 1474–1478 (1977).
442 R. Baute, G. Deffieux, M.-A. Baute, M.-J. Filleau and A. Neveu: Tetrahedron Lett. *44*, 3943–3944 (1976).
443 K. Yokose, N. Nakayama, C. Miyamoto, T. Furumai, H. B. Maruyama, R. D. Stipanovic and C. R. Howell: J. Antibiotics *37*, 667–669 (1984).
444 C. R. Howell and R. D. Stipanovic: Can. J. Microbiol. 321–324 (1983).
445 R. D. Stipanovic and C. R. Howell: J. Antibiotics *35*, 1326–1330 (1982).
446 C. Miyamoto, K. Yokose, T. Furumai and H. B. Maruyama: J. Antibiotics *35*, 374–377 (1982).
447 R. L. DeVault and W. Rosenbrook, Jr.: J. Antibiotics *26*, 532–534 (1973).
448 Y. Kishi, T. Fukuyama and S. Nakatsuka: J. Am. Chem. Soc. *95*, 6490–6492 (1973).
449 Y. Kishi, T. Fukuyama and S. Nakatsuka: J. Am. Chem. Soc. *95*, 6492–6493 (1973).
450 Y. Kishi, S. Nakatsuka, T. Fukuyama and M. Havel: J. Am. Chem. Soc. *95*, 6493–6495 (1973).
451 S. Nakatsuka, T. Fukuyama and Y. Kishi: Tetrahedron Lett., 1549–1556 (1974).
452 T. Fukuyama and Y. Kishi: J. Am. Chem. Soc. *98*, 6723–6724 (1976).
453 G. M. Strunz and M. Kakushima: Experientia *30*, 719–720 (1974).
454 T. Fukuyama, S. Nakatsuka and Y. Kishi: Tetrahedron Lett., 3393–3396 (1976).
455 H. C. J. Ottenheijm, G. P. C. Kerkhoff and J. W. H. A. Bijen: J. Chem. Soc., Chem. Commun., 768–769 (1975).
456 H. C. J. Ottenheijm, J. D. M. Herscheid, G. P. C. Kerkhoff and T. F. Spande: J. Org. Chem. *41*, 3433–3438 (1976).
457 T. Fukuyama, S.-I. Nakatsuka and Y. Kishi: Tetrahedron *37*, 2045–2078 (1981).

458 H.C.J. Ottenheijm, J.D.M. Herscheid and R.J.F. Nivard: J. Org. Chem., 925–926 (1977).
459 D. Hauser, H.P. Weber and H.P. Sigg, Heh. Chim. Acta *53*, 1061–1073 (1970).
460 K. Sasaki, T. Fukuyama, S. Nakatsuka and Y. Kishi: J. Chem. Soc., Chem. Commun., 542–543 (1975).
461 J.D.M. Herscheid, R.J.F. Nivard, M.W. Tijhuis and H.C.J. Ottenheijm: J. Org. Chem. *45*, 1885–1888 (1980).
462 J.D.M. Herscheid, J.H. Colstee and H.C.J. Ottenheijm: J. Org. Chem. *46*, 3346–3348 (1981).
463 J.D. Bu'Lock and C. Leigh: J. Chem. Soc., Chem. Commun., 628–629 (1975).
464 G.W. Kirby, G.L. Patrick and D.J. Robins: J. Chem. Soc. Perkin Trans *1*, 1336–1338 (1978).
465 G.W. Kirby: Pure Appl. Chem. *51*, 705–708 (1979).
466 N. Neuss, L.D. Boeck, D.R. Brannon, J.C. Cline, D.C. DeLong, M. Gorman, L.L. Huckstep, D.H. Lively, J. Maybe, M.M. Marsh, B.B. Molloy, R. Nagarajan, J.D. Nelson and W.M. Stark: Antimicrob. Agents Chemother. 213–219 (1968).
467 J.A. Winstead and R.J. Suhadolnik: J. Am. Chem. Soc. *82*, 1644–1647 (1960).
468 G.W. Kirby, W. Lösel, P.S. Rao, D.J. Robbins, M.A. Sefton and R.R. Talekar: J. Chem. Soc., Chem. Commun., 810–812 (1983).
469 G.W. Kirby, D.J. Robins and W.M. Stark: J. Chem. Soc., Chem. Commun., 812–813 (1983).
470 G.W. Kirby and S. Narayanaswami: J. Chem. Soc. Perkin *I*, 1564–1567 (1976).
471 G.W. Kirby and S. Narayanaswami: J. Chem. Soc., Chem. Commun., 322–323 (1973).
472 R. Cardillo, C. Fuganti, D. Chringhelli and P. Grasselli: J. Chem. Soc., Chem. Comm., 778–779 (1975).
473 E. Ohler, F. Tataruch and U. Schmidt: Chem. Ber. *105*, 3658–3661 (1972).
474 T. Hino and T. Sato: Chem. Pharm. Bull. (Tokyo) *22*, 2866–2874 (1974).
475 T. Sato and T. Hino: Chem. Pharm. Bull. (Tokyo) *24*, 285–293 (1976).
476 H. Poisel and U. Schmidt: Chem. Ber. *105*, 625–634 (1972).
477 E. Ohler, F. Tataruch and U. Schmidt: Chem. Ber. *106*, 165–176 (1973).
478 E.H. Herman, D.T. Witiak, K. Hellmann and V.S. Waravdekar: Adv. Pharmacol. Chemother. *19*, 249–290 (1982).
479 R.V. Nair and D.T. Witiak: J. Chem. Ed. *65*, 534–538 (1988).
480 J.R. Geigy, A.-G. Brit 961,065, June 17 (1964), 4pp.
481 A.M. Creighton and G.D. Birnie: Biochem. J. *114*, 58P (1969).
482 A.M. Creighton: Progr. Antimicrob. Anticancer Chemotherap. *1*, 167–169 (1970).
483 A.M. Creighton: GB 1234935, 9 June (1971).
484 A.M. Creighton: US 3,941,790, Mar. 2 (1976).
485 A. Furst. In: Chemistry of Chelation in Cancer, pp. 46–68, I.N. Kugelmass Ed., Charles C. Thomas, Springfield, IL 1963.
486 J. Leiter, I. Wodinsky and A.R. Bourke: Cancer Res. *19*, Part 2 (Suppl) 309–396 (1959).
487 A.M. Creighton, K. Hellmann and S. Whitecross: Nature (London) *222*, 384–385 (1969).
488 A.M. Creighton, W.A. Jeffery and J. Long. In: Proceedings of the 6th International Symposium on Medicinal Chemistry, pp. 281–288, U.K. Brighton and A. Simkin, Eds., September 4–7, 1978.
489 A.M. Creighton: Ger. Offen. 2,163,601, 22 June (1972), 34 pp.
490 A.M. Creighton and G.D. Birnie: Int. J. Cancer *5*, 47–54 (1970).

491 A. Hempel, N. Camerman and A. Camerman: J. Am. Chem. Soc. *104*, 3453–3456 (1982).
492 D.T. Witiak, H.J. Lee, R.W. Hart and R.E. Gibson: J. Med. Chem. *20*, 630–635 (1977).
493 D.T. Witiak, H.J. Lee, H.D. Goldman and B.S. Zwilling: J. Med. Chem. *21*, 1194–1197 (1978).
494 B.S. Zwilling, L.B. Campolito, N.A. Reiches, T. George and D.T. Witiak: Br. J. Cancer *44*, 578–583 (1981).
495 A. Hempel, N. Camerman and A. Camerman: J. Am. Chem. Soc. *104*, 3456–3458 (1982).
496 D.T. Witiak, B.K. Trivedi, L.B. Campolito, B.S. Zwilling and N. Reiches: J. Med. Chem. *24*, 1329–1332 (1981).
497 J.A. Landgrebe: Chem. Eng. News *59*, 47 (April 27, 1981).
498 D.T. Witiak, R.V. Nair and F.A. Schmid: J. Med. Chem. *28*, 1228–1234 (1985).
499 A.J. Repta, M.J. Baltezor and P.C. Bansal: J. Pharm. Sci. *65*, 238–242 (1976).
500 Y.F. Ren, H.L. Shu, T.M. Zhang, Z.Y. Chen and C. Lin: Kuexue Tongbao *25*, 189–190 (1980).
501 Y.-F. Ren, Eur. Pat. Appl. Ep 125475 A1, 21 Nov. (1984).
502 T. Zhang, Z. Chen and C. Lin: Henan Yixueyuan Xuebeo *15*, 145–150 (1980).
503 T.M. Zhang, Z.Y. Chen and C. Lin: Yao Hsueh Hsueh Pao *15*, 577–583 (1980).
504 X.-T. Li, G.E. Hutchinson, B.A. Pym, M. Finch and K. Hellmann: Eur. J. Cancer Clin. Oncol. *19*, 283–287 (1983).
505 Y.-F. Ren, G.M. Otter and F.A. Schmid: Eur. J. Cancer Clin. Oncol. *21*, 493–497 (1985).
506 Y.-F. Ren, H.L. Shu, J.C. Cai, B. Xu, R.M. Zhang, T. Narita, N. Kiriki and T. Komatsu: 14th International Congress of Chemotherapy, June 23–28, 1985, Kyoto, Japan.
507 H. He, M.-Y. Wang and T.-M. Zhang: Zhongguo Yaoli Xuebao *9*, 369–373 (1988).
508 K.-W. Huang, N.-W. Fu and Y.-F. Ren: Zhongguo Yaoli Xuebao *5*, 69–71 (1984).
509 E.H. Herman, V.J. Ferrans, H.B. Bhat and D.T. Witiak: Cancer Chemother. Pharmacol. *19*, 227–281 (1987).
510 T.-M. Zhang, M.-Y. Wang and Q.-D. Wang: Zhongguo Yaoli Xuebao *8*, 369–374 (1987).
511 M.-Y. Wang, T.-X. Liu, G.-D. Li and T.-M. Zhang: Zhongguo Yaoli Xuebao *9*, 367–369 (1988).
512 N. Camerman, A. Hempel and A. Camerman: Science *225*, 1165–1166 (1984).
513 D.T. Witiak, B.K. Trivedi and F.A. Schmid: J. Med. Chem. *28*, 1111–1113 (1985).
514 J.C. Cai, H.L. Shu, C.F. Tang, T. Narita, T. Komatsu, N. Kiriki, E. Ohki and M. Takase: 14th International Congress of Chemotherapy, June 23–28, 1985, Kyoto, Japan.
515 R.H. Wheeler, D.J. Clauw, R.B. Natale and R.W. Ruddon: Invest. New Drugs *1*, 283–295 (1983).
516 F. Traganos, Z. Darzynkiewicz and M.R. Melamed: Cancer Res. *41*, 4566–4576 (1981).
517 C. Lin and T.-M. Zhang: Yaoxue Xuebao *17*, 654–657 (1982).
518 G. Yu and T.-M. Zhang: Yaoxue Xuebao *19*, 481–485 (1984).
519 T. Zhang, C. Lin and M. Wang: Kexue Tongbao (Foreign Lang. Ed) *31*, 1285–1290 (1986).

520 I.W. Taylor and N.M. Bleehen: Br. J. Cancer *38*, 143–147 (1978).
521 I.W. Taylor and N.M. Bleehen: Br. J. Cancer *35*, 587–594 (1977).
522 R. Albanese and R.A. Watkins: Br. J. Cancer *52*, 725–731 (1985).
523 S. Chaube and M.L. Murphy. In: Advances in Teretolog, Vol. 3, pp. 181–237, D.H.M. Woollem Ed., Academic Press, New York/London, 1968.
524 M.C. Berenbaum: Br. Med. Bull *21*, 140–146 (1965).
525 D.I. Duke, E.O. Field, M. Finch, K. Hellmann, A.S.B. Tucker and D.F. Tucker: Biomedicine *18*, 199–205 (1973).
526 N. Dan and T.-M. Zhang: Zhongguo Yaoli Xuebao *9*, 471–474 (1988).
527 A.W. LeServe and K. Hellmann: Br. Med. J. *1*, 597–601 (1972).
528 S.J. Duncan and J.J. Reynolds: Biochem. Pharmacol. *32*, 3853–3858 (1983).
529 D.I. Duke: Teratology 11, 119–126 (1975).
530 R.J. Woodman: Cancer Chemother. Rep. (Part 2) *4*, 45–52 (1974).
531 R.J. Woodman, R.L. Cysyk, I. Kline, M. Gang and J.M. Venditti: Cancer Chemother. Rep. (Part 1) *59*, 689–695 (1975).
532 I. Kline: Cancer Chemother. Rep. (Part 2) *4*, 33–43 (1974).
533 S. Wadler, M.D. Green and F.M. Muggia: Cancer Res. *46*, 1176–1181 (1986).
534 R. Supino: Tumori *70*, 121–126 (1984).
535 S. Wadler, M.D. Green, R. Basch and F.M. Muggia: Biochem. Pharmacol. *36*, 1495–1501 (1987).
536 K. Hellmann and G. Murkin: Cancer *34*, 1033–1039 (1974).
537 L.J. Peters: Br. J. Radiol. *49*, 708–715 (1976).
538 I.W. Taylor and N.M. Bleehen: Br. J. Cancer *36*, 493–500 (1977).
539 C.J. Kovacs, M.J. Evans, D.R. Burholt and L.L. Schenken: Br. J. Cancer *39*, 524–530 (1979).
540 C.J. Kovacs, M.J. Evans, L.L. Schenken and D.R. Burholt: Br. J. Cancer *39*, 516–523 (1979).
541 C.J. Kovacs, L.L. Schenken and D.R. Burholt: Int. J. Radiat. Oncol. Biol. Phys. *5*, 1361–1364 (1979).
542 Z.-Y. Chen, G.-R. Wang and T.-M. Zhang: Tianjin Yiyao Zhongliuxue Fukan *16*, 183–185 (1983).
543 Y.-P. Zhang, Y.-N. Chen, G.-C. Fan and S.-X. Xia: Zhongguo Yaoli Xuebao *8*, 177–180 (1987).
544 T. Giraldi and G. Sava: Anticancer Res. *1*, 163–174 (1981).
545 S.K. Carter and W.T. Soper: Cancer Treat. Rev. *1*, 1–14 (1974).
546 K. Hellmann and K. Burrage: Nature (London) *224*, 273–275 (1969).
547 K. Burrage, K. Hellmann and A.J. Salsbury: Br. J. Pharmacol. *39*, 205P–206P (1970).
548 A.J. Salsbury, K. Burrage and K. Hellmann: Br. Med. J. *4*, 344–346 (1970).
549 S.E. James and A.J. Salsbury: Cancer Res. *34*, 839–842 (1974).
550 A. Sacchi, M.A. Balduzzi, A. Corsi, C. Greco and G. Zupi. In: Curr. Chemother. Immunother. Proc. Int. Congr. Chemother., 12th, Meeting Date 1981, Vol. 2, 1361–1362, P. Periti, Ed., Gialdroni Grassi, Giuliana, Am. Soc. Microbiol., Washington, D.C., 1982.
551 G. Zupi, A. Cavallari, C. Greco and A. Sacchi: Anticancer Res. *4*, 91–95 (1984).
552 G. Sava, S. Zorzet and L. Perissin: Anticancer Res. *5*, 511–515 (1985).
553 W.D.W. Heston, D. Kadmon and W.R. Fair: Cancer Lett. (Shannon, Irel) *13*, 139–145 (1981).
554 D. Baker, W. Constable, D. Elkon and L. Rinehart: Cancer (Philadelphia) *48*, 2179–2183 (1981).

555 T. Giraldi, G. Sava, R. Cherubino, G. Bottiroli and G. Mazzini: Clin. Exp. Metastasis *2*, 151–159 (1984).
556 A.J. Salsbury, K. Burrage and K. Hellmann: Cancer Res. *34*, 843–849 (1974).
557 G. Sava, T. Giraldi, L. Perissin, S. Zorzet and G. Decorti: Clin. Exp. Metastasis *5*, 27–34 (1987).
558 V. Grill, F. Mallardi, S. Zorzet, L. Perissin and T. Giraldi: Clin. Exp. Metastasis *5*, 233–244 (1987).
559 F. Wingen, H. Spring and D. Schmahl: Clin. Exp. Metastasis *5*, 9–16 (1987).
560 D.R. Welch, T.J. Lobl, E.A. Seftor, P.J. Wack, P.A. Aeed, K.H. Yohem, R.E.B. Seftor and M.J.C. Hendrix: Int. J. Cancer *43*, 449–457 (1989).
561 A. Atherton, D. Bushfield and K. Hellmann: Cancer Res. *35*, 953–957 (1975).
562 A.H. Soloway, R.J. Brumbaugh and D.T. Witiak: J. Theor. Biol. *102*, 361–373 (1983).
563 J.L. Speyer, M.D. Green, E. Kramer, M. Rey, J. Sanger, C. Ward, N. Dubin, V. Ferrans, P. Stecy, A. Zeleniuch-Jacquotte, J. Wernz, F. Feit, W. Slater, R. Blum and F. Muggia: N. Engl. J. Med. *319*, 745–752 (1988).
564 V.J. Ferrans, E.H. Herman and R.L. Hamlin: Dev. Oncol. 53 (Organ Directed Toxic. Anticancer Drugs), 56–63 (1988).
565 E.H. Herman, R.M. Mhatre, I.P. Lee and V.S. Waravdekar: Proc. Soc. Exp. Biol. Med. *140*, 234–239 (1972).
566 E.H. Herman, R.M. Mhatre and D.P. Chadwick: Toxicol. Appl. Pharmacol. *27*, 517–526 (1974).
567 G. Wang, M.D. Finch, D.Trevan and K.Hellmann: Br. J. Cancer *43*, 871–877 (1981).
568 E. Herman, A. El-Hage and D.T. Witiak: Fed. Proc. *41*, 1477 (1982).
569 G.M. Wang and M. Finch: Drug Chem. Toxicol. *3*, 213–225 (1980).
570 V.W. Fischer, L.S. LaRose and G.M. Wang: Drug Chem. Toxicol. *5*, 155–164 (1982).
571 V.W. Fischer, L.S. LaRose and G.M. Wang: J. Am. Coll. Toxicol. *3*, 175–187 (1984).
572 S.T. Hu, E. Brändle and G. Zbinden: Pharmacology *26*, 210–220 (1983).
573 G. Decorti, F. Bartoliklugmann, F. Mallardi, S. Klugmann, B. Benussi, V. Grill and L. Baldini: Cancer Lett. (Shannon, Irel.) *19*, 77–83 (1983).
574 E.H. Herman, B. Ardalan, C. Bier, V. Waravdekar and S. Krop: Cancer Treat. Rep. *63*, 89–92 (1979).
575 E.H. Herman, A.N. El-Hage, V.J. Ferrans and D.T. Witiak: Res. Commun. Chem. Pathol. Pharmacol. *40*, 217–231 (1983).
576 E.H. Herman and V.J. Ferrans: Chemioter. Oncol. *5*, 37–41 (1981).
577 E.H. Herman and V.J. Ferrans: Cancer Res. *41*, 3436–3440 (1981).
578 E. Herman and V. Ferrans: Recent Adv. Chemother., Proc. Int. Congr. Chemother., 14th Volume, Anticancer Sect. 2, pp. 1332–1333, J. Ishigami, Ed., Univ. Tokyo Press, Tokyo, Japan, 1985.
579 E.H. Herman and V.J. Ferrans: Cancer Chemother. Pharmacol. *16*, 102–106 (1986).
580 E.H. Herman, V.J. Ferrans, W. Jordan and B. Ardalan: Res. Commun. Chem. Pathol. Pharmacol. *31*, 85–97 (1981).
581 E.H. Herman and V.J. Ferrans: Drugs. Exp. Clin. Res. *9*, 483–490 (1983).
582 W.E. Perkins, R.L. Schroeder, R.A. Carrano and A.R. Imondi: Br. J. Cancer *46*, 662–667 (1982).
583 N.R. Bachur, S.L. Gorden and M.V. Gee: Mol. Pharmacol. *13*, 901–910 (1977).
584 C.E. Myers, W.P. McGuire, R.H. Liss, I. Ifrim, K. Grotzinger and R.C. Young: Science *197*, 165–167 (1976).

585 J.H. Doroshow, G.Y. Locker, I. Ifrim and C.E. Myers: J. Clin. Invest. *68*, 1053–1064 (1981).
586 R.W. Freeman, J.S. MacDonald, R.D. Olsen, R.C. Boerth, J.A. Oates and R.D. Harbison: Toxicol. Appl. Pharmacol. *54*, 168–175 (1980).
587 E.H. Herman, V.J. Ferrans, C.E. Myers and J.F. Van Vleet: Cancer Res. *45*, 276–281 (1985).
588 E.H. Herman and V.J. Ferrans. In: Proc. Int. Congr. Chemother., 13th Volume 11, 231/16–231/23, K.H. Spitzy, K. Karrer and H.E. Verlag, Eds., Vienna, Austria, 1983.
589 E.H. Herman and V.J. Ferrans: Cancer Treat. Rev. *14*, 225–229 (1987).
590 J.F. Van Vleet, L.A. Greenwood and V.J. Ferrans: Am. J. Vet. Res. *40*, 1537–1552 (1979).
591 M.R. Bristow, M.E. Billingham, J.W. Mason and J.R. Daniels: Cancer Treat. Rep. *62*, 873–879 (1978).
592 E.H. Herman, V.J. Ferrans, R.S.K. Young and R.L. Hamlin: Cancer Res. *48*, 6918–6925 (1988).
593 E.H. Herman, V.J. Ferrans, R.S.K. Young and R.L. Hamlin: Drugs Exp. Clin. Res. *14*, 563–570 (1988).
594 E.H. Herman, A. El-Hage and V.J. Ferrans: Toxicol. Appl. Pharmacol. *92*, 42–53 (1988).
595 M. Dardir, E.H. Herman and V.J. Ferrans: Cancer Chemother. Pharmacol. *23*, 269–275 (1989).
596 E.H. Herman, A.N. El-Hage, A.M. Creighton, D.T. Witiak and V.J. Ferrans: Res. Commun. Chem. Pathol. Pharmacol. *48*, 39–55 (1985).
597 A.N. El-Hage, E.H. Herman and V.J. Ferrans: Toxicology *28*, 295–303 (1983).
598 A.N. El-Hage, E.H. Herman and V.J. Ferrans: Res. Commun. Chem. Pathol. Pharmacol. *33*, 509–523 (1981).
599 L.J. Fisher and S.A. Hamburger: Diabetes *29*, 213–216 (1980).
600 A.N. El-Hage, E.H. Herman, G.C. Yang, R.K. Crouch and V.J. Ferrans: Res. Commun. Chem. Pathol. Pharmacol. *52*, 341–360 (1986).
601 S. Rajagopalan, P.M. Politi, B.K. Sinha and C.E. Myers: Cancer Res. *48*, 4766–4769 (1988).
602 K. Gibbs and J.M. Walshe: Clin. Sci. Mol. Med. *53*, 317–320 (1977).
603 U. Witting and E. Hultsch: Int. Arch. Occup. Environ. Health *48*, 89–98 (1981).
604 D.D. Von Hoff, D. Howser, B.J. Lewis, J. Holcenberg, R.B. Weiss and R.C. Young: Cancer Treat. Rep. *65*, 249–252 (1981).
605 Z.-X. Huang, P.M. May, K.M. Quinlan, D.R. Williams and A.M. Creighton: Agents and Actions *12*, 536–542 (1982).
606 H. Rubin: Proc. Nat. Acad. Sci. USA *69*, 712–716 (1972).
607 M. Gosàlvez and M. Blanco: Int. Biophys. Congr., 5th, Abstr. P. 4 (1975).
608 G.D.V. VanRossum and M. Gosàlvez: Fed. Proc., Fed. Am. Soc. Exp. Biol. *3*, Abstr. 3201 (1976).
609 M. Gosàlvez: Int. Congr. Biochem., 10th, Abstr. 11-4-387 (1976).
610 M. Gosàlvez, L. Pezzi and C. Vivero: Biochem. Soc. Trans. *6*, 659–661 (1978).
611 Z.-X. Huang, A.M. Creighton and D.R. Williams: Inorganica Chimica Acta *107*, L29–L32 (1985).
612 C.E. Myers, L. Gianni, C.B. Simone, R. Klecker and R. Greene: Biochemistry *21*, 1707–1713 (1982).
613 E.H. Herman and V.J. Ferrans: Lab. Invest. *49*, 69–77 (1983).
614 L. Gianni, B.J. Corden and C.E. Myers. In: Reviews in Biochemical Toxicology, Vol. 5, pp. 1–82, E. Hodgson, J.R. Bend and R.M. Philpot Eds., Elsevier Biomedical, New York, 1983.

615 B. Halliwell and J.M.C. Gutteridge. In: Free Radicals in Biology and Medicine, Clarendon, Oxford, 1985.
616 E.G. Mimnaugh, M.A. Trush, M. Bhatnagar and T.E. Gram: Biochem. Pharmacol. *34*, 847–856 (1985).
617 J.M.C. Gutteridge: Biochem. Pharmacol. *33*, 1725–1728 (1984).
618 E.J.F. Demant and P.K. Jensen: Eur. J. Biochem. *132*, 551–556 (1983).
619 E.J.F. Demant: Eur. J. Biochem. *137*, 113–118 (1983).
620 J.H. Doroshaw and K.J. A. Davies: J. Biol. Chem. *261*, 3068–3074 (1986).
621 K.M. Dawson: Biochem. Pharmacol. *24*, 2249–2253 (1975).
622 B.B. Hasinoff: Drug Metab. Disp., in press (1990).
623 H. Beraldo, A. Garnier-Suillerot and L. Tosi: Inorg. Chem. *22*, 4117–4124 (1982).
624 B.B. Hasinoff: Agents and Actions *26*, 378–385 (1989).
625 R.W. Hay and P.J. Morris. In: Metal Ions in Biological Systems 5, pp. 173–243, H. Sigel Ed., Marcel Dekker, New York, 1976.
626 B.B. Hasinoff: Agents and Actions, *29*, 374–381 (1990).
627 P.M. May, G.K. Williams and D.R. Williams: Inorg. Chim. Acta *46*, 221-228 (1980).
628 E. Arena, M. Arico, F. Biondo, N. D'Alessandro, L. Dusonchet, N. Gebbia, F. Garbasi, R. Sanguedolce and L. Rausa. In: Adriamycin Review, Part II, pp. 160–174, M. Staquet, Ed., European Press, Ghent, Belgium 1975.
629 B.B. Hasinoff and J.P. Davey: Biochem. J. *250*, 827–834 (1988).
630 L. Gianni, L. Vigano, C. Lanzi, M. Niggeler and V. Malatesta: J. Natl. Cancer Inst. *80*, 1104–1111 (1988).
631 D.C. Livinston, A.M. Creighton and S.W. Fisher. In: Advances in Antimicrobiol and Antineoplastic Chemotherapy, Vol. 2, pp. 109–110, M. Hejzlar, M. Semonsky and S. Masak, Eds., University Park Press, Baltimore, 1972.
632 E.O. Field, F. Mauro and K. Hellmann: Cancer Chemother. Rep. (Part 1) *55*, 527–530 (1971).
633 W. Sadee, J. Staroscik, C. Finn and J. Cohen: J. Pharm. Sci. *64*, 998–1001 (1975).
634 R.M. Mhatre, A. Rahman, S. Raschild and P.S. Schein: Proc. Am. Assn. Cancer Res. *23*, 212 (1982).
635 P.J. Creaven, L.M. Allen and D.A. Alford: J. Pharm. Pharmacol. *27*, 914–918 (1975).
636 R.H. Earhart, K. Tutsch, J.M. Koeller, H.I. Robins, H.L. Davis and D.C. Tormey: Proc. Am. Assn. Cancer Res. *23*, 128 (1982).
637 R.H. Earhart, K.D. Tutsch, J.M. Koeller, R. Rodriguez, H.I. Robins, C.L. Vogel, H.L. Davis and D.C. Tormey: Cancer Res. *42*, 5255–5261 (1982).
638 D.D. Von Hoff, N. Soares, P. Gormley and D.G. Poplack: Cancer Treat. Rep. *64*, 734–736 (1980).
639 S. Marsoni, R.S. Ungerleider, S.B. Hurson, R.M. Simon and L.D. Hammershaimb: Cancer Treat. Rep. *69*, 1263–1269 (1985).
640 D.L. Glaubiger, D.D. vonHoff, J.S. Holcenberg, B. Kamen, C. Pratt and R.S. Ungerleider. In: Frontiers of Radiation Therapy and Oncology, Vol. 16, pp. 42–49, J.M. Vaeth Ed., S. Karger, Basel 1982.
641 C.L. Vogel, E. Gorowski, E. Davila, M. Eisenberger, J. Kosinski, R.P. Agarwal and N. Savaraj: Invest. New Drugs *5*, 187–198 (1987).
642 J. Liesmann, R. Belt, C. Haas and B. Hoogstraten: Cancer *47*, 1959–1962 (1981).
643 J.M. Koeller, R.H. Earhart and H.L. Davis: Cancer Treat. Rep. *65*, 459–463 (1981).
644 J.S. Holcenberg, K.D. Tutsch, R.H. Earhart, R.S. Ungerleider, B.A. Ka-

men, C.B. Pratt, T.J. Gribble and D.L. Glaubiger: Cancer Treat. Rep. *70*, 703–709 (1986).
645 R.B. Natale, R.H. Wheeler, M.K. Liepman, A. Sauder and L. Bricker: Cancer Treat. Rep. *67*, 311–313 (1983).
646 L.H. Brubaker, C.L. Vogel, L.H. Einhorn and R. Birch: Cancer Treat. Rep. *70*, 915–916 (1986).
647 R.H. Wheeler, L.J. Bricker, R.B. Natale and S.R. Baker: Cancer Treat. Rep. *68*, 427–428 (1984).
648 Lancet *2*, 721 (1987).
649 K. Hellmann, K.A. Newton, D.N. Whitemore, I.W.F. Hanham and J.V. Bond: Br. Med. J. *1*, 822–824 (1969).
650 E.P. Flannery, M.P. Corder, W.W. Sheehan, T.F. Pajak and J.R. Bateman: Cancer Treat. Rep. *62*, 465–467 (1978).
651 C.L.M. Olweny, W. Sikyewunda and D. Otim: Oncology *37*, 174–176 (1980).
652 D.J. Atherton, R.S. Wells, M.R. Laurent and Y.F. Williams: Br. J. Dermatol. *102*, 307–317 (1980).
653 R.T. Eagan, D.T. Carr, D.T. Coles, J. Rubin and S. Frytak: Cancer Treat. Rep. *60*, 947–948 (1976).
654 D.L. Ahmann, J.H. Edmonson, S. Frytak, L.K. Kvols, H.F. Bisel and J. Rubin: Cancer Treat. Rep. *62*, 151–153 (1978).
655 D.L. Ahmann, M.J. O'Connell, H.F. Bisel, J.H. Edmonson, R.G. Hahn and S. Frytak: Cancer Treat. Rep. *61*, 81–82 (1977).
656 R.H. Creech, P.F. Engstrom, D.T. Harris, R.B. Catalano, R.F. Bellet: Cancer Treat. Rep. *63*, 111-114 (1979).
657 M.K. Shah, P.F. Engstrom, R.B. Catalano, A.R. Paul, R.E. Bellet and R.H. Creech: Cancer Treat. Rep. *66*, 557–558 (1982).
658 H.W. Bruckner, J.J. Lokich, D.M. Stablein: Cancer Treat. Rep. *66*, 1713–1717 (1982).
659 P.G. Dyment, K.A. Starling, V.J. Land, A. Cangir, D.M. Komp and C.L. Sexauer: Cancer Treat. Rep. *63*, 1397–1398 (1979).
660 W.U. Rhomberg: Int. J. Radiat. Oncol. Biol. Phys. *4*, 121–126 (1978).
661 T. Bates: Int. J. Radiat. Oncol. Biol. Phys. *4*, 127–131 (1978).
662 R.D.H. Ryall: Int. J. Radiat. Oncol. Biol. Phys. *4*, 133–134 (1978).
663 K. Hellmann, M.B. Grimshaw and G.E. Hutchinson: Int. J. Radiat. Oncol. Biol. Phys. *4*, 109–113 (1978).
664 M.T. Bakowski, E. MacDonald, R.F. Mould, P. Cawte, J. Sloggen, A. Barret, V. Dalley, K.A. Newton, G. Westbury, S.E. James and K. Hellmann: Int. J. Radiat. Oncol. Biol. Phys. *4*, 115–119 (1978).
665 M.F. Spittle, H. Bush, S.E. James and K. Hellmann: Int. J. Radiat. Oncol. Biol. Phys. *5*, 1649–1651 (1979).
666 M.D. Green, J.L. Speyer, P. Stecy, M. Rey, E. Kramer, J. Sanger, F. Feit, R.H. Blum, J.C. Wernz, C. Ward, C. London, N. Dubin and F.M. Muggia: Proc. Am. Soc. Clin. Oncol. *6*, 28 (1987).
667 Cancer Invest. *7*, 101 (1989).
668 Science *213*, 1239 (1981).

Vector-borne diseases*

By Satyavan Sharma
Medicinal Chemistry Division, Central Drug Research Institute, Lucknow 226001, India

1	Introduction	367
2	The vectors of diseases	367
3	Malaria	376
3.1	Geographical distribution	377
3.2	Epidemiology	377
3.3	Clinical characteristics	379
3.4	Diagnosis	380
3.5	Treatment of malaria	381
3.5.1	Natural products	381
3.5.2	Antibiotics	384
3.5.3	9-Aminoacridines	385
3.5.4	4-Aminoquinolines	386
3.5.5	8-Aminoquinolines	388
3.5.6	Arylmethanols	390
3.5.7	Naphthoquinones	393
3.5.8	Antifolates	394
3.5.9	Drug combinations	399
4	Leishmaniasis	400
4.1	Geographical distribution	400
4.2	Epidemiology	401
4.3	Forms of leishmaniasis	402
4.4	Diagnosis of leishmaniasis	406
4.5	Treatment of leishmaniasis	407
4.5.1	Organoantimonials	407
4.5.2	Aromatic diamidine	411
4.5.3	Nitrogen heterocycles	414
4.5.4	Natural products	419
4.5.5	Miscellaneous compounds	420
4.5.6	Local treatment and topical formulations	421
4.5.7	Liposome-drug preparations	422
5	African trypanosomiasis	422
5.1	Geographical distribution	422
5.2	Epidemiology	424
5.3	Clinical features	424
5.4	Diagnosis	425
5.5	Treatment of African trypanosomiasis	425

* Communication No. 4467 from Central Drug Research Institute, Lucknow 226001, India.

5.5.1	Drugs effective in the early stage of sleeping sickness	425
5.5.2	Drugs effective in the late stage of sleeping sickness	427
6	American trypanosomiasis	431
6.1	Geographical distribution	431
6.2	Epidemiology	432
6.3	Clinical characteristics	433
6.4	Diagnosis	433
6.5	Treatment of Chagas' disease	434
7	Filariasis	436
7.1	Geographical distribution	436
7.2	Epidemiology	437
7.3	Clinical characteristics	437
7.4	Diagnosis	440
7.5	Treatment of filariasis	441
7.5.1	Organoantimonials and arsenicals	441
7.5.2	Organophosphorus compounds	443
7.5.3	Suramine	443
7.5.4	Piperazine derivatives	444
7.5.5	Limitations of DEC	446
7.5.6	Structural analogues of DEC	447
7.5.7	Levamisole	448
7.5.8	Benzimidazoles	448
7.5.9	Benzthiazoles	450
7.5.10	Amoscanate and CGP-6140	450
7.5.11	Clofazimine	451
7.5.12	Ivermectin	451
8	Guinea worm infection	452
8.1	Geographical distribution	453
8.2	Epidemiology	453
8.3	Clinical characteristics	454
8.4	Diagnosis	455
8.5	Treatment of dracunculiasis	455
9	Schistosomiasis	456
9.1	Geographical distribution	457
9.2	Epidemiology	457
9.3	Clinical characteristics	459
9.4	Diagnosis	460
9.5	Treatment of schistosomiasis	460
9.5.1	Organometallic compounds	461
9.5.2	Miracils	461
9.5.3	Nitro compounds	462
9.5.4	Quinolines and isoquinolines	464
9.5.5	Miscellaneous compounds	465
10	Flukes other than schistosomes	465
10.1	Treatment of liver and lung fluke infections	467
11	Vector control	468
12	Concluding remarks	469
	References	470
	Addendum	485

1 Introduction

The quest for better health and a happier society has always been one of the prime objectives of man. This has persuaded almost all the advanced countries of the world to conquer most of the metabolic, CNS-CVS, viral, microbial and parasitic diseases. However, a large section of the world population, especially those living in the tropical and sub-tropical zones still continue to reel under the agony and distress of different tropical diseases which are caused by invasion of the human body by a wide variety of viruses, microbes, protozoans and helminths. The tropical diseases thus, pose a serious threat to human as well as animal health and consequently exert detrimental effects on the general well-being, quality of life and socio-economic development of the masses living in the endemic areas [1, 2]. The geographical distribution and rate of incidence of a large number of diseases, prevalent in the tropics, may depend upon several factors, the most important being the abundance of certain biological vectors which not only carry and transmit the pathogens of the diseases but can also serve as the intermediate hosts in the life cycle of the causative agents [3].

2 The vectors of diseases

A large variety of viral, bacterial, protozoal and helminth diseases are transmitted to man through a number of vectors belonging to the phylum arthropoda (such as mites, ticks, lice, bugs, fleas, flies, mosquitoes and crustaceans) and mollusca (mollusks or snails) of which the arthropods play a major role in propagating various tropical diseases in man (Table 1). Depending upon the involvement in propagation, transmission and development of the pathogens, Huff's classification [4] has been used to divide the arthropod vectors broadly into five groups [5].

Group 1: Vectors whose mouth parts become temporarily contaminated with the pathogens (mechanical vectors). Vectors of this class are passive carriers of the disease agents, picking them from the body or excretions of man or animal and depositing them on eatables. Tabanids, for example, carry pathogens of *Francisella tularensis* in USA and several species of *Trypanosoma* in South America and Africa.

Group 2: Vectors in which the pathogens proliferate in gut lumen or gut epithelium without cyclical development. Thus, these vectors cause limited infections as they do not transmit the disease by bite. The arthropods of this class pick up the pathogens from the body of a patient and soon inoculate them directly into wounds, skin sores and blood of another host. The body louse, *Pediculus humans humans,* is responsible for transmitting trench fever, louse-borne epidemic typhus and relapsing fever. Similarly, some fleas carry the pathogens of bubonic plague and murine typhus.

Group 3: Vectors in which the pathogen undergo no cyclical developments but they propagate in salivary glands (propagative vectors). The vectors of this class are responsible for generalized infections as they are transmitted by bite. A large variety of ticks, mosquitoes and a few species of mites and lice are known to transmit various bacterial and viral diseases, the most important being Rickettsial pox, scrub typhus, spotted fever, Q fever, yellow fever, dengue, louping ill, Colorado tick fever and sand fly fever, etc.

Group 4: Vectors in which the infective organisms undergo cyclical changes but do not propagate (cyclodevelopmental vectors). The vectors of this group serve as the intermediate hosts in the life cycle of the pathogen and generally belong to the phylum helminthes. The disease is transmitted through the bites of arthropods. A large number of helminth diseases such as lymphatic filariasis, river blindness, loasis and dipetalonemiasis are transmitted by the bites of various mosquitoes, flies and ticks.

Group 5: Vectors in which the pathogens undergo cyclical developmental changes and also proliferate (cyclopropagative vectors). The development usually takes place in the gut of the vector and transmission of the disease occurs through bites. Such vectors (mosquitoes, flies and bugs etc.) also serve as the intermediate hosts in the development of all protozoal parasites and, therefore, cause diseases like malaria, leishmaniasis, sleeping sickness and Chagas' disease.

Several helminth diseases are also transmitted by mollusks and are described in Table 1. It is also evident from Table 1 that the diseases resulting due to various vectors are numerous and, therefore, it is not possible to deal with the biological and therapeutic aspects of each

Table 1
Vector-borne diseases in humans

Disease	Causative agent	Site of parasite in man	Reservior host(s)	Geographical distribution	Clinical characteristics
1. Mite-borne diseases					
Rickettsial pox	*Rickettsia akari*	Intracellular	Man, mouse, wild rodents	Africa, USA, Korea, USSR	Fever, sweating, chills, headache, backache, lassitude, photophobia
Scrub typhus	*Rickettsia tsutsugamushi*	Intracellular	Mites, field rats	China, Japan, Taiwan, Tibet, Philippines and other parts of Asia	High fever, chills, headache, malaise, anorexia, stupor, restlessness, convulsions, coma with high mortality
2. Tick-borne diseases					
Relapsing fever	*Borrelia* spp.	Peripheral blood	Ticks, rodents, mammals	Africa, Asia, Europe, America	Recurring fever, vomiting, enlarged liver, deafness
Fievre boutonneuse	*Rickettsia conorii*	Lymphatics and endothelial cells	Ticks, dogs (?), wild animals (?)	Africa, India, Mediterranean regions	Fever, body pain, malaise, macular or maculopapular rashes
Queensland tick-typhus	*Rickettsia australis*	Blood	Man	Queensland (Australia)	Headache, malaise, skin rash
Q. fever	*Coxiella* (Rickettsia) *burnetti*	Intracellular	Ticks, sheep, goats, cattle	Worldwide except Sweden, Denmark and Finland	Remittent fever, chills, headache, muscle pain, malaise, anorexia
Encephalitis	Arbovirus B	CNS	Wild birds	Asia, Central Europe, USSR	Fever with CNS involvement

Disease	Causative agent	Site of parasite in man	Reservior host(s)	Geographical distribution	Clinical characteristics
Spotted fever	*Rickettsia rickettsii*	Intracellular	Man	USA, South America	Fever, headache, chills, body pain, rash, necrosis and death
Kyasanur forest disease	Arbovirus B	Blood	Wild mammals and birds	India	Fever, headache, conjunctival inflammation, malaise, nasal haemorrhage
Omsk haemorrhagic fever	Arbovirus B	Blood	Wild mammals and birds	Western Siberia	As above
Louping ill	Arbovirus B	CNS	Wild mammals, sheep, birds	Scotland, U. K., Ireland	Fever with CNS involvement
Powassan	Arbovirus B	CNS	Wild mammals	Northern USA, Canada	As above
3. Bug-borne diseases					
Chagas' disease	*Trypanosoma cruzi*	Peripheral blood, heart, muscles	Dogs, cats, armadillo, squirrel	South and Central America, Mexico	Fever, vomiting, diarrhoea, anorexia, cardiac failure, dialation of tubular organs (megacolon, megaoesophagus)
4. Louse-borne diseases					
Relapsing fever	*Borrelia recurrentis*	Peripheral blood	Man	Some parts of Asia, Europe, Africa and Brazil	Recurring fever, toxaemia
Epidemic typhus (Brill-Zinsser disease)	*Rickettsia prowazekii*	Intracellular	Man	Worldwide except Australia	Fever, bronchitis, bronchopneumonia, nephritis

Disease	Causative agent	Site of parasite in man	Reservior host(s)	Geographical distribution	Clinical characteristics
Trench fever	*Rochalimaea quintana*	Blood, urine	Man	Europe, Egypt, Algeria, Ethiopia, China, Mexico	Fever, headache, vertigo, pain in back and legs
5. Flea-borne diseases					
Bubonic plague	*Yersinia pestis*	Blood, lymph nodes	Rats, wild rodents	Worldwide	High fever, nausea, vomiting, bubo in groin
Murine typhus	*Rickettsia typhi*	Intracellular	Rats	Worldwide (warmer climates)	Fever, rash, macules
Dog hookworm infection	*Dipylidium caninum*	Intestine	Man, dogs	Asia, Europe, USA, Africa, Chile	Abdominal pain, diarrhoea, GI disturbances
6. Fly-borne diseases					
Sand fly fever	Arbovirus	Blood	Man, sand flies	North Africa, Mediterranean area, Middle East	Frontal headache, backache, joint pains, stiffness of neck and back, anorexia, nausea, vomiting, fever
Kala-azar	*Leishmania donovani*	Spleen, liver, bone-marrow and RI system	Man, dogs, jackals	India, China, South and Central America, Mediterranean area	Fever, chills, malaise, diarrhoea, vomiting, anaemia, wasting, enlargement of spleen and liver
Cutaneous leishmaniasis	*L. tropica, L. mexicana*	Skin	Man, dogs	Mediterranean area, Middle East, India, North Africa, South America	Skin ulcerations

Disease	Causative agent	Site of parasite in man	Reservior host(s)	Geographical distribution	Clinical characteristics
Mucocutaneous leishmaniasis	*L. brasiliensis*	Skin, mucous membrane	Forest rodents	Brazil (Amazon forest)	Disfiguration and destruction of tissues of nasal septum, lips and larynx
Sleeping sickness	*Trypanosoma gambiense, T. rhodesiense*	Peripheral blood, spinal fluid	Man, sheep, goats, antelops	Africa	Fever, headache, malaise, CNS discorders, coma, death
Loasis	*Loa loa*	Subcutaneous tissues[a]	Man	Western and Central Africa	Painful calabar swellings
River blindness	*Onchocerca volvulus*	Nodules in subcutaneous and connective tissues[a]	Man	Africa, Mexico, Central America	Ocular problems, blindness
Ozzardi filariasis	*Mansonella ozzardi*	Body cavities[a]	Man	South America	Usually asymptomatic, can cause hydrocele
Dipetalonemiasis	*Dipetalonema perstans*	Peritoneal cavities, pleural cavities, pericardium, mesentries[a]	Man	Africa, South America	Non-pathogenic, can produce oedema of limbs, scrotum and eosinophilia
7. Mosquito-borne diseases					
Malaria	*Plasmodium* spp.	Blood, liver	Man	Tropical countries	Recurring fever, chills, death in some cases
Lymphatic filariasis	*Wuchereria bancrofti, Brugia malayi*	Lymphatics and connective tissues[a]	Man	Asia, Africa, South America	Fever, enlargement of legs, arms, breasts and scrotum
Yellow fever	Arbovirus B	Blood, liver, kidney	Man, monkeys	Central and South America, Africa	High fever, vomiting, haemorrhagic signs, jaundice

Disease	Causative agent	Site of parasite in man	Reservior host(s)	Geographical distribution	Clinical characteristics
Encephalitis[b]	Arbovirus A and B	CNS	Wild birds	Africa, Asia, USA, Canada, Mexico, Panama	CNS involvement, coma, paralysis, death
Dengue fever	Arbovirus B	Blood	Man (?)	Tropical and sub-tropical world	Fever with headache, rash, body pain
8. Crustacean-borne diseases					
Creeping eruption (?)	*Gnasthostoma spinigerum*	Skin	Dogs, cats	Africa, South America, Asia, USA	Dermal problems
Guinea worm infection	*Dracunculus medinensis*	Connective tissues	Man	Africa, Middle East, USSR, India	Erythema, pruritis, giddiness, dyspnea
Lung fluke infection[c]	*Paragonimus westermani*	Primary lungs	Man, carnivores	Africa, Central and South America, Far East, Asia	Chronic bronchitis, abdominal discomfort
Sparganosis	*Spirometra mansonoides*	Subcutaneous tisues	Man, frogs, birds, mammals	Asia, some part of Europe and Africa	Dermal problems
Fish tapeworm infection	*Diphyllobothrium latum*	Intestine	Man, cats, dogs	Asia, Northern Europe, USSR, USA	Abdominal pain, nausea, diarrhoea, loss of appetite
9. Mollusc-borne diseases					
Schistosomiasis	*Schistosoma* spp.	Blood	Man, primates	Africa, Asia, South America	Fever, abdominal discomfort, urinary problems
Fasciolopsiasis	*Fasciolopsis buski*	Intestine	Man, pig	China, Taiwan, Thailand and other parts of Eastern Asia	Abdominal pain, nausea, vomiting
Heterophyiasis	*Heterophyes heterophyes*	Intestine	Man	Japan, China, Korea, Taiwan	Abdominal pain with mucoid diarrhoea

Disease	Causative agent	Site of parasite in man	Reservior host(s)	Geographical distribution	Clinical characteristics
Metagonimiasis	*Metagonimus kogawai*	Intestine	Man	China, Japan	Abdominal discomfort
Clonorchiasis	*Clonorchis sinensis*	Biliary passages	Man, dogs, cats	China, Japan, Korea, Vietnam	Liver cirrhosis, jaundice, cachexia
Fascioliasis	*Fasiola hepatica*	Bile duct, liver parenchyma	Man, sheep	Cosmopolitan	Abdominal pain, vomiting, jaundice, fever
Opisthorchiasis	*Opisthorchis felineus, O. viverrini*	Bile duct	Man, mammals	Southeastern Europe, USSR, Japan, Far-East	Jaundice, liver cirrhosis

[a] Shows the seat of adult filarial worms; the microfilariae live in blood.
[b] Includes Western equine encephalitis, Eastern equine encephalitis, Venezuelan equine encephalitis, Japanese encephalitis, Murray valley encephalitis, St. Louis encephalitis and California encephalitis.
[c] Also uses snails as the first vector.

vector-borne disease in the limited space of this article. However, it has been envisaged to provide a comprehensive coverage of those diseases which are of major public health importance. In this context, the author feels that the parasitic diseases recognized by the World Health Organization and a few other diseases need greater attention. Consequently, the biological and chemotherapeutic aspects of the vector-borne protozoal and helminth diseases (Table 2) have been dealt with in this review as they are expected to generate newer concepts in the treatment and control of such diseases.

Table 2
Vector-borne parasitic diseases of man

Disease	Vectors Common name	Biological name
Protozoal diseases		
1. Malaria	Female mosquitoes	*Anopheles* (nearly 60 species)
2. Leishmaniasis	Female sand flies	*Phlebotomus argentipes, P. chinensis, P. perniciosus, P. orientalis, P. martini, Lutzomyia longipalpis*
3. Sleeping sickness	Tsetse flies	*Glossina palpalis, G. morsitans, G. pallidipes, G. swynnertoni*
4. Chagas' disease	Kissing bugs	Species of the genera *Triatoma, Rhodnius, Panstrongylus, Eutriatoma, Eratyrus, Mepraia* and *Psammolestes* etc.
Helminth diseases		
5. Filariasis		
a) Lymphatic	Female mosquitoes	Several species of the genera *Culex, Anopheles, Aedes, Mansonia*
b) River blindness	Black flies	*Simulium damnosum, S. neavei, Eusimulium avidum, E. mooseri, E. ochraceum*
c) Loasis	Tabanid flies	*Chrysops dimidiata, C. silacea*
d) Dipetalonemiasis	Midge	*Culicoides austeni, C. grahami*
e) Filariasis ozzardi	Midge	*Culicoides furens*
6. Guinea worm infection	Copepods	*Cyclops* spp., *Eucyclops* spp., *Thermocyclops* spp., etc.
7. Schistosomiasis	Freshwater snails	*Bulinus truncatus, Biomphalaria glabratus, Oncomelania nosophora* etc.
8. Lung fluke infection	Snails, crabs, crayfish	*Semisulcospira libertina, S. amurensis* etc.
9. Intestinal fluke infection	Snails	*Segmentina* spp., *Hippeutis cantori* etc.
10. Liver fluke infection	Snails	*Bulimus* spp., *Lymnaea* spp., *Alocinma longicornis, Hua ningpoensis* etc.

3 Malaria

Malaria is one of the most debilitating parasitic diseases of the tropics caused by the protozoans of the genus *Plasmodium*. The disease is transmitted from person to person by female mosquitoes belonging to the genus *Anopheles*. The main species of the *Plasmodium* infecting man are – *Plasmodium falciparum*, *P. vivax*, *P. malariae* and *P. ovale*. Of these, the first two parasites, *P. falciparum* and *P. vivax* account for 95 % of the malaria infections throughout the world. A number of *Plasmodium* species are also known to infect apes, monkeys, rodents, birds and lizards. Some important parasites of animals are *P. knowlesi* and *P. cynomolgi* (monkeys) and *P. berghei* and *P. vinckei* (rodents).

In man, malaria may be of four types:

(a) *Falciparum malaria:* This is caused by *P. falciparum* and is also known as *malignant tertian malaria, tropical malaria* or *pernicious malaria*. It usually produces fulminating infection in non-immune patients which may prove to be fatal if not promptly treated. However, the infection responds well to antimalarial drugs and relapse does not occur. Sometimes the corpuscles infected with *P. falciparum* tend to clump together and adhere to the walls of blood capillaries, which leads to blockage. When this phenomenon occurs in the brain, the condition is known as *cerebral malaria*.

(b) *Vivax malaria:* This form of malaria, caused by *P. vivax*, is called *benign tertian malaria*, or *simple tertian malaria*. It produces clinical attacks milder than those of *P. falciparum* and has low mortality rates even in untreated adult patients. Relapse may occur within two years of primary infection.

(c) *Malariae malaria:* This is also known as *quartan malaria* and is caused by *P. malariae*. Like falciparum malaria, this form of malaria also responds well to chemotherapy if treated early. Although relapse is rare, the disease is known for its chronicity.

(d) *Ovale malaria:* It is a mild disease caused by *P. ovale*. The infection is susceptible to chemotherapy. The parasites of *P. ovale* are morphologically similar to *P. malariae* but the clinical symptoms of the former are similar to that of *P. vivax*.

3.1 Geographical distribution

Malaria infection is widely distributed throughout the tropical zones of the world, especially those falling between 45° north and 40° south latitudes. The disease is endemic in several parts of South America, Africa and Asia. In all the malarious regions, vivax malaria, is more widely distributed than the other forms. However *P. falciparum* tend to predominate in varying degrees of intensity throughout the tropical world. The emergence of chloroquine-resistant *P. falciparum* has become a problem in different parts of the world, especially in East Africa, Central and South America and South-East Asia including India, Bangladesh, Burma, China, Indonesia, Kampuchea, Laos, Vietnam, Thailand, Sri Lanka and Singapore [6, 7].
In 1955, it was estimated that there were 250 million cases of malaria of which 2.5 million patients died annually. The worldwide effort to eradicate malaria since 1956 has caused a substantial decline in the incidence of malaria. In 1972, it was estimated that of the 1840 million people originally living in malarious regions, 1346 million people belonged to areas where a malaria eradication programme had either been completed or was in progress. However 494 million people were living in areas where a malaria eradication programme was not in operation [8]. At the end of 1979, it was reported that nearly 2350 million people were living in malaria prone areas [7]. It is now estimated that nearly 800 million suffer from different forms of malaria with an annual death rate of 1.5 million patients [9a]. There are 215 million people, mostly from Africa, chronically affected and an estimated 150 million new cases occurring every year [9b].

3.2 Epidemiology

Plasmodium is an intracellular protozoal parasite that needs two hosts to complete its life cycle. The first host is a vertebrate (man and animals) where the parasite completes its asexual development (schizogony). The second host is a female *Anopheles* mosquito in which the sexual development (sporogony) of the *Plasmodium* takes place.
The life cycle (Fig. 1) of *Plasmodium* starts when the infected mosquito bites man and injects sporozoites into his/her bloodstream while taking its blood meal. Within less than an hour, the sporozoites

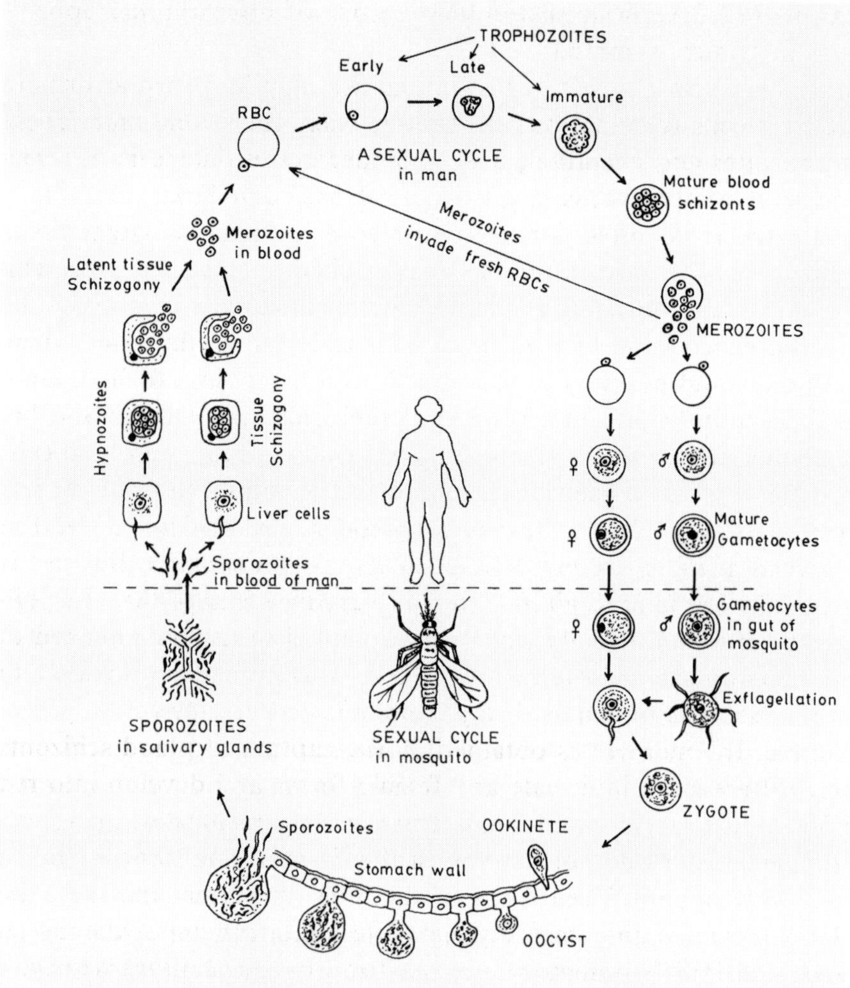

Figure 1
Life cycle of Plasmodium

disappear from the bloodstream and localize in the parenchymal cells of the liver where they grow and undergo asexual division to form merozoites. This constitutes the pre-erythrocytic stage (primary tissue schizogony) of the infection during which the subject remains symptom-free. The duration of this pre-erythrocytic stage is 5.5–7 days (*P. falciparum*), 6–8 days (*P. vivax*), 9 days (*P. ovale*) and 13–16 days (*P. malariae*). This determines the length of the incuba-

tion period, i. e. the time between the day of infection and appearance of clinical symptoms.
In *P. vivax* and *P. ovale,* some sporozoites may form hypnozoites in the liver cells forming the latent tissue stage for a long time. These hypnozoites may eventually also form merozoites (latent tissue schizogony) in the liver cells leading to a relapse of malaria.
When the merozoites, formed either from primary or latent tissue schizogony, increase in number, the packed liver cells (tissue schizonts) rupture, releasing a large number of tissue merozoites into the bloodstream. These merozoites soon invade the erythrocytes where they grow and multiply to form trophozoites that eventually mature to form blood schizonts. The blood schizonts may rupture and liberate merozoites which are capable of invading fresh red blood cells and starting a new asexual cycle in man. Abnormalities and clinical symptoms occur due to the asexual blood stages of *Plasmodium.* The destruction of red blood cells gives rise to fever and chills and is caused by foreign protein set free by decaying erythrocytes. The periodicity in bouts of chills and fever is linked to the time needed to complete an asexual cycle in the blood. This probably led people to term malaria as *tertian* for 3-day cycles and *quartan* for 4-day cycles.
Some of the merozoites obtained by the rupture of blood schizonts may differentiate into male and female forms and develop into red blood cells to form mature male and female gametocytes. These mature gametocytes do not develop further in man until they are taken up by a mosquito. When the mosquito bites the patient and sucks the blood, the male and female gametocytes enter the gut of the insect. Here, the female gametocyte is fertilized by the male gametocytes through a process of exflagellation in the insect's stomach. The resulting zygote attaches itself to the stomach wall in the form of an ookinete. This penetrates the stomach wall and reaches the outer layer. Soon the ookinete is transformed into an oocyst in which large numbers of sporozoites are produced leading to the rupture of the oocyst wall. The liberated sporozoites then reach the salivary glands of the mosquito ready to infect a new vertebrate host.

3.3 Clinical characteristics

The main clinical features of malaria are periodic fever with chills, splenomegaly, anaemia and leukopenia. The enlargement of also liver may be observed. The clinical complications produced by *P. falciparum* are usually dangerous and fatal and, therefore, need immediate medical attendance. The other three forms are free from fatal complications but cause severe illness.

Cerebral malaria, caused by *P. falciparum*, is usually fatal if not treated promptly. The parasitized red blood cells tend to agglutinate and adhere to the capillary endothelium leading to the gradual blockage of blood circulation. The patient complains of progressively increasing headache with little or no fever. In a later stage, the patient falls into a coma. Hyperpyrexia and convulsive seizures are also common.

3.4 Diagnosis

The clinical diagnosis of malaria is sometimes difficult. The disease may be easily confused with influenza, typhoid fever, relapsing fever, yellow fever, kala-azar, amoebic liver abscess and pyelitis. The definite diagnosis of malaria can be carried out by examining the blood film of the patient for the presence of *Plasmodium* spp. [10].

A number of serological tests have been developed to diagnose human malaria. Thus, malaria antibodies may be detected in the sera of infected persons by indirect fluorescent antibody (IFA) [11] and indirect haemagglutination (IH) [12] tests. Recently IFA and ELISA were used by Ferreira and Sanchez [13] to detect antibodies of *P. falciparum* and *P. vivax* in man. Both these tests exhibited low sensitivity in new infections but were highly sensitive in long-standing infection. Franzén and her colleagues [14] have suggested diagnosis of malaria by DNA probe containing cloned repetitive sequences from *P. falciparum*. Although the method could detect parasitaemia levels up to 0.00 1 % in 50 µl of blood, the clinical usefulness of the method is yet to be established [15]. Reibnegger *et al.* [16] have shown elevation of urinary neopterin excretion levels in malaria patients while Bein and Olcen [17] have described a solid-phase radioimmunoassay to detect malaria antigens in the urine. It has been suggested that this method may be used in diagnosing the disease, epidemiological surveys and follow-up treatment studies.

3.5 Treatment of malaria

The chemotherapy of malaria has been reviewed by several researchers [18–22]. The present section will discuss only the salient features of various drugs used as tissue schizontocides, blood schizontocides, gametocytocidal and sprontocidal agents (Table 3).

3.5.1 Natural products

Febrifugine (I) is one of the oldest antimalarials isolated from roots of *Dichroa febrifuga* or leaves of hydrangea. Although a crude extract of the above plants has been used in malaria treatment since 200 BC, their active constituent febrifugine has limited success against falciparum and vivax malaria [19, 23].

A better remedy was prepared from the bark of cinchona trees (*Cinchona ledgeriana*), the most active constituents of which were quinine (II) and quinidine (III). Quinine is effective against asexual blood stages (blood schizontocide) of all the four species of *Plasmodium* infecting man. For the emergency treatment of severe malaria, 600 mg of quinine is injected intravenously in 300 ml of normal saline over a period of 2–4 h. The dose may be repeated with a maximum of 2 g in 24 h. Once the emergency is over, quinine or chloroquine may be given orally [24, 25]. The usual oral dose for adults is 200–300 mg of quinine hydrochloride or quinine sulphate given 4 times daily for 5–10 days [6]. The use of quinine may lead to frequent cinchonism such as headache, nausea, abdominal pain, tinnitus and visual disturbances. The drug may also occasionally cause haemolytic anaemia, hypoglycemia, hypotension and arrhythmias.

Quinidine (III), used as an antiarrhythmic drug (cardiac depressant), shows better activity than quinine against falciparum malaria. Recently, Bunnag *et al.* [26, 27] have used a combination of quinine, quinidine and cinconine in equal ratio (LA-40221) at a dose of 12 mg/kg given every 8 h for 22 doses to cure chloroquine-resistant falciparum malaria in man with good success.

The Chinese herb quinghaosu (*Artemisia annua*) has been used in Chinese traditional medicine for over 1000 years for the treatment of recurring pyrexia, the most notable clinical symptom of malaria. In 1972, Chinese scientists isolated the active principle of this herb which is a sesquiterpene peroxide called artemisinine or quinghaosu

Table 3
Classification of antimalarial drugs

Class	Description	Drugs
1. Primary tissue schizontocides	Drugs which act on the pre-erythrocytic stage of all the species of human *Plasmodium* and, therefore, prevent erythrocytic infection. Drugs of this class are known as 'Causal prophylactics'.	Primaquine, sulfonamides, proguanil and pyrimethamine
2. Secondary tissue schizontocides	Drugs of this class act on the secondary exoerythrocytic stage (hypnozoites) of *P. vivax*, *P. malariae* and *P. ovale* and bring about a radical cure of these infections. They are also called 'anti-relapse drugs' or 'radical curative drugs'.	Primaquine
3. Blood schizontocides[a]	These drugs act on the asexual erythrocytic stage of *Plasmodium*. They are used to achieve a clinical cure against vivax malaria and ovale malaria, and a radical cure against falciparum malaria.	Quinine, chloroquine, mefloquine, quinghaosu
4. Gametocytocidals[b]	Drugs of this class destroy the male and female gametocytes of *P. vivax*, *P. malariae* and *P. ovale*. Thus, such drugs prevent the transmission of infection from man to mosquito.	Primaquine
5. Sprontocidals	These drugs reach the mosquito with its blood meal and check the formation of sporozoites.	Primaquine, sulfonamides, proguanil, pyrimethamine
6. Prophylactics	These drugs are recommended for preventing malaria infection especially for people travelling to malarious regions.	Chloroquine alone or in combination with Fansidar, proguanil or doxycycline

[a] Primaquine, sulfonamides, proguanil and pyrimethamine have also been used as blood schizontocides but their action is too slow for rapid clearance of infection.
[b] Quinine is also gametocytocidal but only for *P. vivax* and *P. malariae* infections.

(IV). Quinghaosu is a fast-acting blood schizontocidal drug but is poorly soluble in water and oils. In order to improve its solubility and absorption profile, a large number of the structural analogues of quinghaosu were prepared. The most active analogues of quinghaosu were found to be artemether (V) and artesunate (VI).

Quinghaosu and is derivatives (IV–VI) present a novel class of compounds possessing a high order of blood schizontocidal activity against chloroquine-resistant *P. falciparum* both in experimental animals and man. Clinical trials carried out in 2000 patients showed

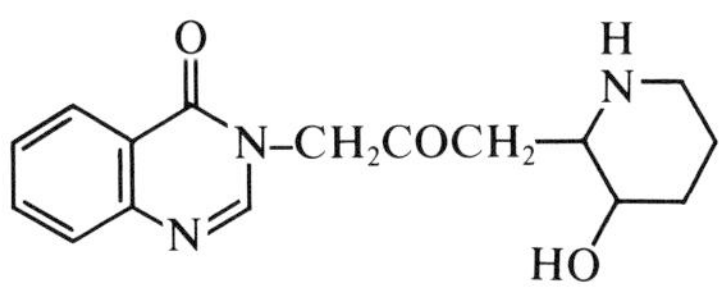

I (Febrifugine)

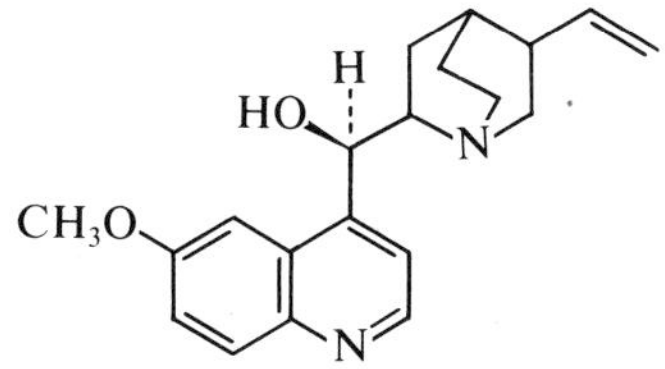

II (Quinine)

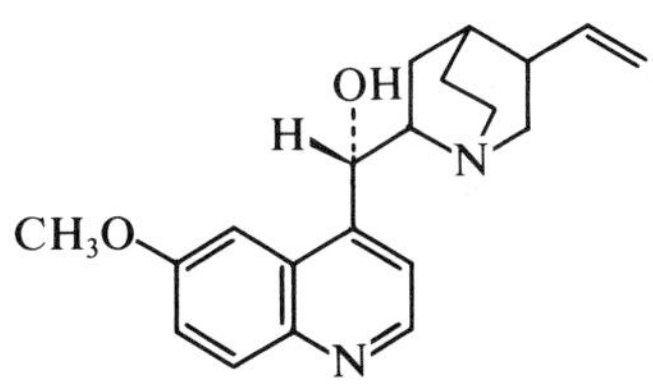

III (Quinidine)

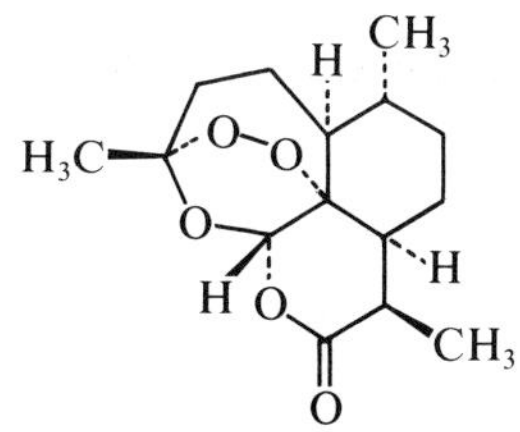

IV (Quinghaosu)

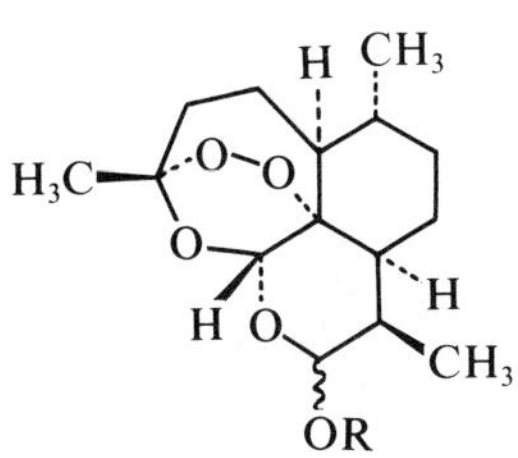

V, $R = CH_3$, Et[Arte(me/e)ther]
VI, $R = COCH_2CH_2COONa$
(Artesunate)

that quinghaosu is effective against vivax and falciparum malaria with 10 % recrudescence occurring after 15–30 days of treatment [22].

Cerebral malaria has been treated by quinghaosu and its analogues. Thus quinghaosu was given in an oily suspension intramuscularly at a dose of 1.5 g for 2 days or 1.2 g as an aqueous suspension over 3 days. Similarly artemether was given in a total dose of 640 mg intravenously or intramuscularly over 3 days. Cure rates were above 90 % [22]. Recently, artemether was used to treat cerebral malaria due to *P. falciparum* at a dose of 200 mg given initially followed by 100 mg every 12 h (total dose 600 mg) intramuscularly when all the patients were cured [28]. It has been suggested that artemether is a useful drug for treating patients with complicated falciparum malaria. The drug also produces more rapid clinical and parasitological resolution of falciparum malaria than quinine [28, 29].

3.5.2 Antibiotics

The antimalarial action of antibiotics has been known for more than three decades. The tetracycline antibiotics such as tetracycline (VII), doxycycline (VIII) and minocycline (IX) as also clindamycin (X)

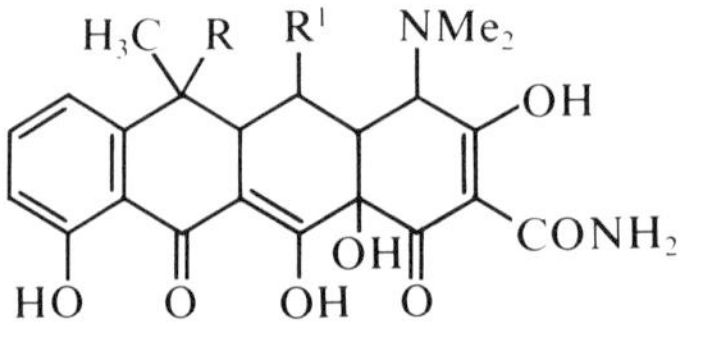

VII, R = OH, R^1 = H (Tetracycline)
VIII, R = H, R^1 = OH (Doxycycline)

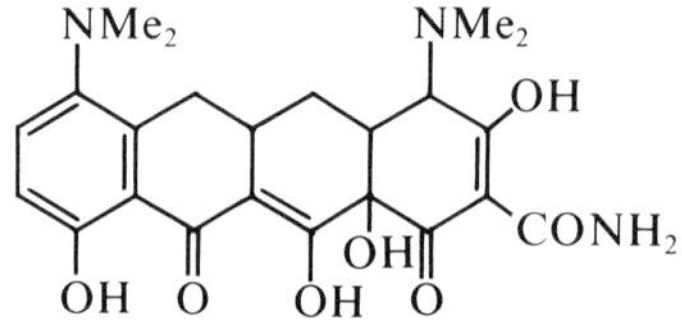

IX (Minocycline)

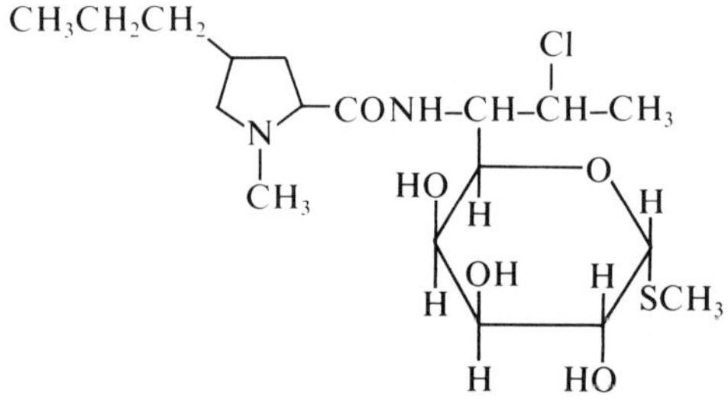

X (Clindamycin)

have been shown to possess marked antimalarial activities in man. However, their use is limited due to their slower clearance of parasitaemia as compared to other clinically used antimalarials. The usefulness of these antibiotics has been reappraised because of their increasing importance in treating multidrug-resistant infections of *P. falciparum* [30, 31].

The tetracyclines inhibit the primary exoerythrocytic stages of *P. falciparum*. They are also active against the asexual erythrocytic stages of *P. falciparum* and *P. vivax*. Clindamycin resembles tetracyclines in its antimalarial activity. Despite the above spectrum of activity, the use of these antibiotics should be restricted only in the treatment of acute attacks of multidrug-resistant falciparum malaria. In patients where parasitaemia is high or symptoms are severe, quinine therapy is an absolute requirement and, therefore, these antibiotics cannot replace this drug initially. The antibiotics have not been recommended for routine treatment of malaria [21].

The dose schedule of antibiotics is as follows – quinine should be given first at a dose of 600 mg three times a day for 5–7 days. Then tetracycline is given at a dose of 250 mg, 4 times a day for 7 days or doxycline (200 mg followed by 100 mg daily for 6 days) or monocycline (200 mg followed by 100 mg twice daily for 7 days). Clindamycin should be given alone at a dose of 5 mg/kg, twice daily for 5 days or may follow a 3-day course of quinine. The above schedule of tetracyclines may also be used alone [21].

The side effects of antibiotics are numerous. Tetracyclines cause teeth discoloration in growing children and can also interfere with enamel and bone development. The drugs also cause frequently nausea, vomiting, diarrhoea, abdominal pain and epigastric distress. Higher doses may cause hepatotoxicity especially in pregnant women and patients with renal impairment.

Clindamycin gives rise to severe diarrhoea and pseudomembranous colitis. Other common side effects of the drug are nausea, vomiting, skin rash and fever.

3.5.3 9-Aminoacridines

Quinacrine (mepacrine, XI) was the first member of this class to have been used clinically in the treatment of malaria until it was replaced by chloroquine and other antimalarials. Although quinacrine provides distinct advantages over quinine, it is largely an obsolete drug for several reasons. It dyes the skin and conjunctiva yellow and may lead to mental disturbances. Vomiting, diarrhoea, headache and dizziness also occur frequently. It must be given daily or at least twice a week for complete suppression of parasitaemia.

Floxacrine (XII), a structural analogue of quinacrine, has strong blood schizontocidal and causal prophylactic activities. Although XII showed marked activity against *P. berghei* in mice, *P. cynomolgi* in rhesus monkeys and *P. falciparum* in owl monkeys [32, 33], it was found to exhibit several toxic effects and, therefore, was not recommended for clinical use.

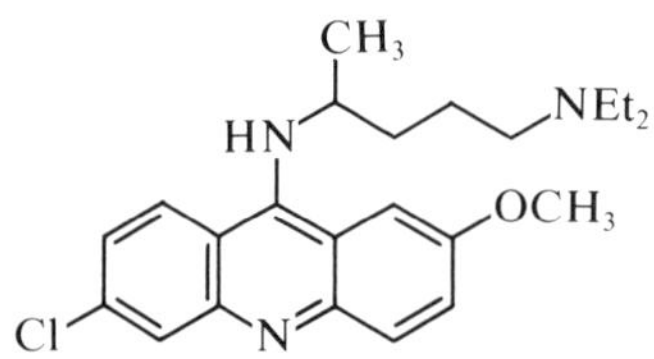

XI (Quinacrine)

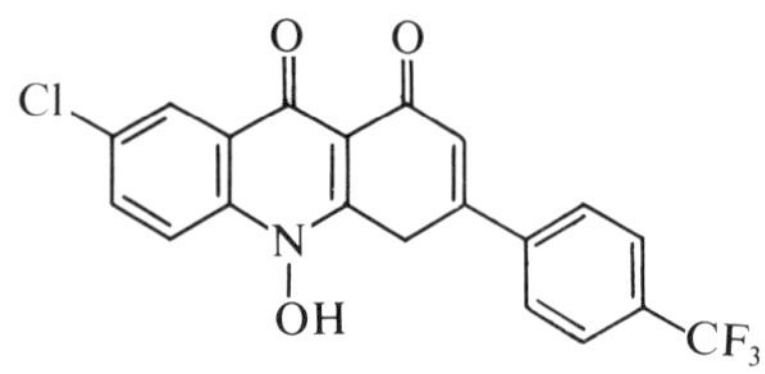

XII (Floxacrine)

3.5.4 4-Aminoquinolines

This is one of the most effective groups of compounds responsible for a series of highly active antimalarials, the most important being chloroquine (XII), hydroxychloroquine (XIV), amodiaquine (XV), amopyroquine (XVI) and cycloquine (XVII). All these drugs are inactive against the sporozoites and primary exoerythrocytic stages but are highly effective against asexual blood stages of all species of human malarial parasites except chloroquine-resistant *P. falciparum.* The four drugs also show high gametocytocidal activity against *P. vivax, P. ovale* and *P. malariae* and immature forms of *P. falciparum.* Thus, these drugs provide a clinical cure for all types of human malaria and a radical cure for falciparum and quartan malaria. Consequently, chloroquine, amodiaquine and cycloquine are used as the drugs of choice for treating acute malaria and for prophylaxis [7, 22].

The recommended adult dose of chloroquine is 600 mg (base) given orally, followed by 300 mg (base) after 6 hours and then a single daily dose of 300 mg for the next 2–3 days. This treatment is effective against all the *Plasmodium* spp. except chloroquine-resistant *P. falciparum* [7, 25]. Children receive the drug at a dose of 10 mg/kg (base) with a maximum of 600 mg, then 5 mg/kg (base) after 6 hours and finally 5 mg/kg (base) daily for 2 days. Amodiaquine is also given at an oral dose of 600 mg (base) followed by 200 mg (base) after 6 hours, then a daily oral dose of 500 mg (base) daily for 2–4 days [7]. Recently, chloroquine (25 mg/kg × 3 days) and quinine (10 mg/kg, three times daily for 5 days) have been found to produce complete clearance of asexual parasitaemia in 20 adult patients with falciparum malaria [34].

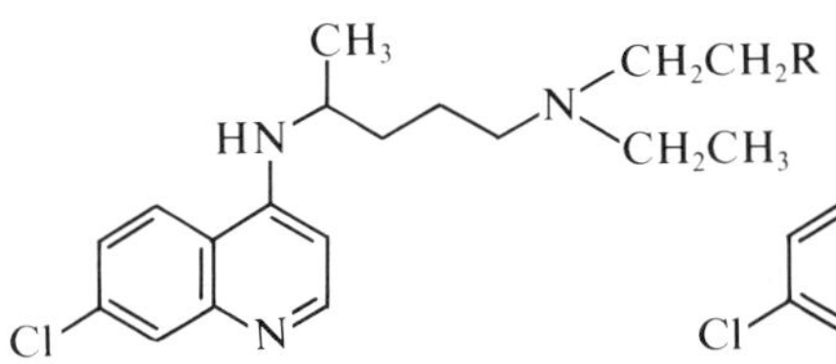

XIII R = H (Chloroquine)
XIV R = OH (Hydroxychloroquine)

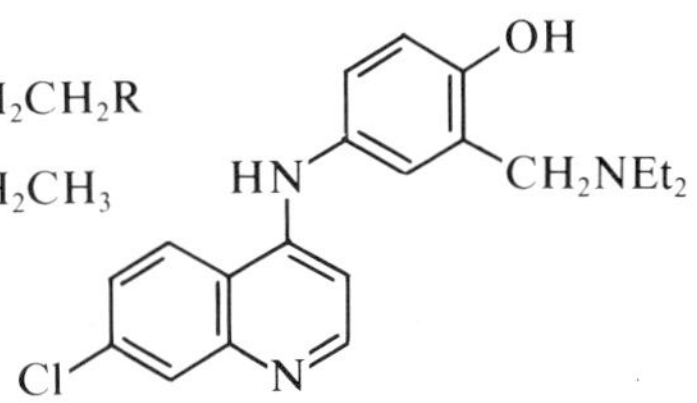

XV (Amodiaquine)

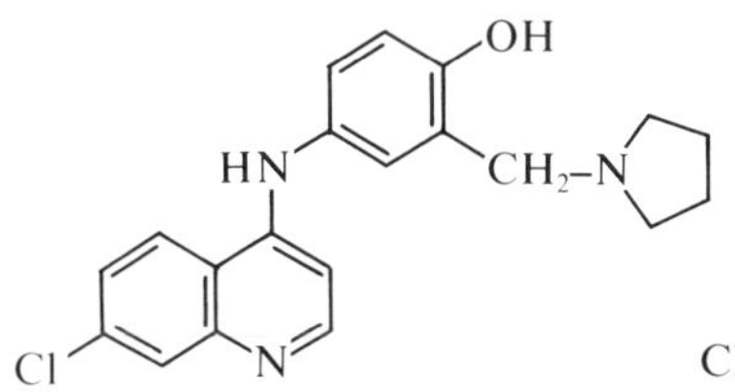

XVI (Amopyroquine)

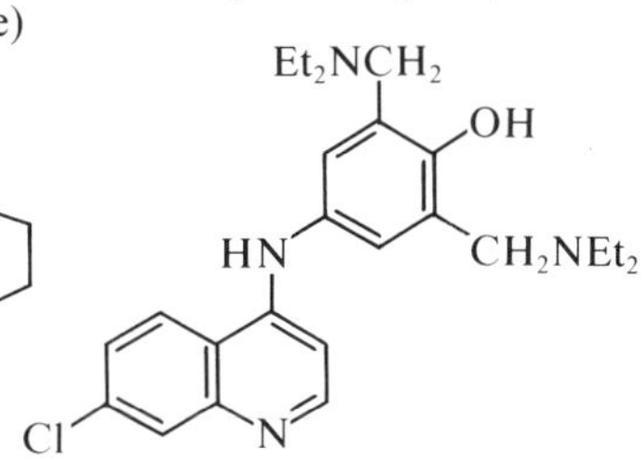

XVII (Cycloquine)

Amodiaquine has also been used to treat falciparum malaria at an oral dose of 25 mg/kg; 90 % of the patients were cured [35]. Most of the 4-aminoquinoline drugs have further been recommended for prophylaxis of malaria. The prophylactic dose of chloroquine is 300 mg (base) given orally two weeks prior to exposure or a single oral dose of 600 mg (base). The recommended doses of amodiaquine (XV) and hydroxychloroquine (XIV) are 400 mg (base)/week. The prophylactic dose schedule for amopyroquine (XVI) is not well defined; 600 mg/week has been found to be well tolerated and effective [21]. Chloroquine has been found to be well tolerated in most cases. However, occasionally, the drug may produce headache, vomiting, confusion, weight loss, corneal opacity and skin problems such as pruritus, eczema and dermal eruption.

3.5.5 8-Aminoquinolines

The pioneering work of German scientist, Schulemann, and his coworkers during the 1920s showing that replacement of one of the methyl groups in methylene blue by dialkylaminoalkyl chain enhances the activity of the parent compound against avian malaria [7], led Roehl [36] to demonstrate marked activity associated with XVIII against *P. relictum*. This was the first 8-aminoquinoline antimalarial compound which prompted an extensive synthetic research programme resulting in the discovery of pamaquine (plasmoquine, XIX) in 1926 [37]. During the beginning of World War II, pamaquine was the only antirelapse drug known, but soon this drug fell into disrepute due to its toxicity. Later a number of other 6-methoxy-8-aminoquinoline drugs were introduced into malaria chemotherapy of which the most important are pentaquine (XX), isopentaquine (XXI), SN-3883 (XXII) and primaquine (XXIII). In 1952, an isoster of primaquine was synthesized in the USSR and used under the name quinocide (XXIV). This compound possesses therapeutic properties and toxicity similar to that of primaquine but its therapeutic index is rather lower.

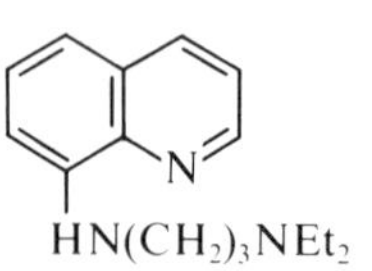

XVIII

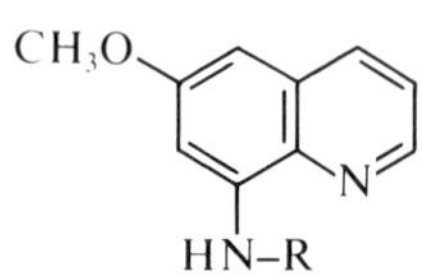

XIX, R = $CH(CH_3)(CH_2)_3NEt_2$ (Pamaquine)
XX, R = $(CH_2)_5$–NH–$CH(CH_3)_2$ (Pentaquine)
XXI, R = $CH(CH_3)(CH_2)_3NHCH(CH_3)_2$ (Isopentaquine)
XXII, R = $(CH_2)_4$–NH_2 (SN-3883)
XXIII, R = $CH(CH_3)(CH_2)_3NH_2$ (Primaquine)
XXIV, R = $(CH_2)_3CH(CH_3)$–NH_2 (Quinocide)

Today, 8-aminoquinolines present one of the most extensively studied groups of compounds in malaria chemotherapy. Accordingly, the structure-activity relationship of this class of compounds has been the subject of much interest and has recently been reviewed by Bhat and co-workers [38].

Primaquine is the most effective antimalarial drug possessing the broadest spectrum of activity against practically all stages of malarial parasites. However, its action on the exoerythrocytic stages and gametocytes is the most pronounced. Although primaquine is known to possess causal prophylactic, antirelapse, blood schizontocidal gametocytocidal and sprontocidal activities, its use is restricted to four conditions: (a) as a prophylactic agent against all four species of malarial parasites because it kills exoerythrocytic stages thereby checking the progression of the infection to blood stages, (b) as an antirelapse (kills hypnozoites) and radical curative (kills tissue schizonts) drug for vivax and ovale malaria, (c) as a gametocytocidal agent to prevent transmission of infection from man to mosquito, (d) in combination with other drugs to treat multidrug-resistant falciparum malaria.

The recommended adult dose of primaquine for prevention of relapses due to *P. vivax* and *P. ovale* is 15 mg (base) given orally for 14 days or 45 mg (base) given per week for 8 doses. Children are given a dose of 0.3 mg (base)/kg daily for 14 days [25]. For radical cure of *P. vivax* or *P. malariae,* a suppressive treatment using 1.5 g of chloroquine (base) given in 3 days has been recommended. Then 15 mg (base) of primaquine is given daily for 14 days [6]. A single oral dose of 45 mg of primaquine kills gametocytes of *P. falciparum* and effectively prevents further transmission of the infection from man to mosquitoes.

Primaquine is usually nontoxic in most of the cases if given in small doses. The ratio between the maximum tolerated dose (MTD) and minimum effective dose (MED) in man is only 10. The primary side effects of the drug are dose-dependent gastrointestinal disturbances (epigastric distress, abdominal pain, anorexia, nausea, etc.) and headache. Higher doses may cause cyanosis and methaemoglobinaemia [39]. All these toxic effects disappear when the drug is withdrawn.

The most important toxicity of primaquine is haemolytic anaemia which occurs in patients whose red cells are genetically deficient in

glucose-6-phosphate dehydrogenase (G6PD) [40–42]. This deficiency is most common in Blacks, Orientals, and Mediterranean people. Since, primaquine causes haemolytic anaemia, it should not be given to subjects known to be deficient of G6PD enzyme. The drug should also not be given in the first trimester of pregnancy. In such cases chloroquine should be administered for a suppressive cure of *P. vivax* and primaquine should be reserved for a radical cure of malaria at a later stage of pregnancy or after delivery.
The mode of action of primaquine is not clear. It is presumed that the drug inhibits the mitochondrial respiration in the parasites through electron transport. Since the half-life of primaquine is only 4–6 hours, the antimalarial response is the function of the total amount of the drug administered rather than the dose [43].

3.5.6 Arylmethanols

The extensive work carried out at the Walter Reed Army Institute of Research, USA, in search of safe and effective antimalarial agents has culminated in the synthesis and antimalarial evaluation of a series of arylmethanols such as 9-phenanthrenemethanols, 4-quinolinemethanols and 4-pyridinemethanols of which a number of compounds have been found to possess marked activity against various stages of *Plasmodium* infecting man and animals [7].

(a) *9-Phenanthrenemethanols:* Among a large number of 9-phenanthrenemethanols prepared and tested for their antimalarial activities [44–55], 6-bromo-α-(diheptylaminomethyl)-9-phenanthrenemethanol (WR-33063, XXV), 3,6-di-(trifluoromethyl)-α-(2-piperidyl)-9-phenanthrenemethanol (WR-122455, XXVI) and 1,3-dichloro-α-[(2-dibutylamino)ethyl]-6-trifluoromethyl-9-phenanthrenemethanol (WR-171669, XXVII) exhibited potent antimalarial activities in mice, owl monkeys and man [7].
WR-33063 has been tested against naturally acquired drug-resistant *P. falciparum* in Vietnam and Thailand. The drug was found to possess rapid action and was superior to any single antimalarial known at that time for resistant strains. An oral dose of 600–1600 mg/day of the hydrochloride of WR-33063 given for 6 days cured patients with multi-drug-resistant *P. falciparum* [56] but could not prevent the disease at a dose of 800 mg/week [57].

The antimalarial potential of WR-122455 has also been studied in detail [58]; however, the drug did not show any promise in clinical practice. WR-171669 (XXVII, halofantrine) is the most active member of this series possessing potent blood schizontocidal activity [59]. A dose of 1 g/day given for 3 days was found to cure 6 out of 6 cases with chloroquine-resistant and 3 out of 3 cases with chloroquine-sensitive infections [60]. The curative dose in man is 3 g/adult [7]. The drug shows no phototoxicity or gastrointestinal blood loss and, therefore, may prove to be a useful alternative to mefloquine [45].

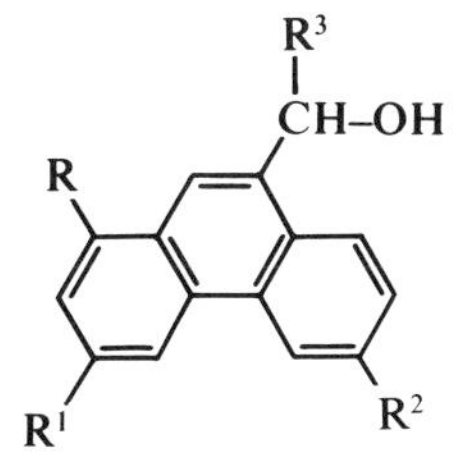

XXV, R = R¹ = H, R² = Br, R³ = $CH_2N(C_7H_{15})_2$
XXVI, R = H, R¹ = R² = CF_3, R³ = 2–piperidinyl
XXVII, R = R¹ = Cl, R² = CF_3, R³ = $(CH_2)_2N(C_6H_9)_2$

(b) *4-Quinolinemethanols:* A large number of 4-quinolinemethanols with structures closely related to quinine have been prepared and evaluated for their antimalarial activity in man. The first member of this series was SN-10275 (XXVIII) which showed good blood schizontocidal activity, however it could not be used clinically due to its serious phototoxicity. Later, a few other 4-quinolinemethanols such as WR-30090 (XXIX), WR-184806 (XXX), WR-226253 (XXXI) and WR-142490 (XXXII) were found to display a wide spectrum of antimalarial activities in man and animals. The most successful outcome

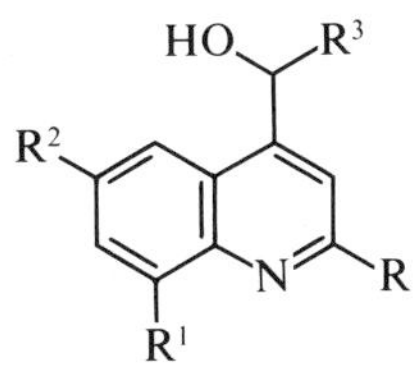

XXVIII, R = Ph, R¹ = Cl, R² = Cl, R³ = 2–piperidinyl
XXIX, R = 3,4–Dichlorophenyl, R¹ = R² = Cl, R³ = $CH_2N(C_4H_9)_2$
XXX, R = CF_3, R¹ = CF_3, R² = H, R³ = $(CH_2)_2NHC(CH_3)_3$
XXXI, R = CF_3, R¹ = R² = Cl, R³ = 2–piperidinyl
XXXII, R = R¹ = CF_3, R² = H, R³ = 2–piperidinyl (Mefloquine)

of this work was the evolution of WR-142490 (XXXII, mefloquine) [61]. Mefloquine is a highly effective human antimalarial drug whose preclinical [62, 63] and clinical [64–69] data have been summarized. Mefloquine is a potent blood schizontocide with a half-life of 20–30 days. The long half-life of this drug makes it suitable for treatment and suppression of malaria. The drug has no effect on the sporozoites and exoerythrocytic stages but is extremely active in the asexual erythrocytic stages. The gametocytocidal action of mefloquine is similar to that of quinine. Although this drug has a marked antimalarial activity, its use in suppressive treatment is restricted due to the possible danger of development of resistant strains to this drug. Consequently, mefloquine is reserved for the treatment of multi-drug-resistant cases of falciparum malaria in combination with Fansidar (a combination of pyrimethamine and sulfadoxine).

Mefloquine (XXXII) is available in Europe in combination with Fansidar under the name Fansimef. A tablet of Fansimef contains 250 mg mefloquine, 25 mg pyrimethamine and 500 mg sulfadoxine. The usual adult dose of the drug is 500 mg mefloquine, 50 mg pyrimethamine and 100 mg sulfadoxine given for 2 days. Mefloquine has also been used alone to treat chloroquine-resistant *P. falciparum* at a single oral dose of 1–1.5 g whereby 100 % cures were obtained. For prophylactic or suppressive treatment, a single oral dose of 250 mg weekly or 500 mg fortnightly of mefloquine has been suggested [6]. Detailed clinical trials carried out with mefloquine in Zimbabwe, Thailand, Burma, Brazil and Europe have established that mefloquine is a successful drug for the prophylactic treatment of vivax and falciparum malaria [64–69].

Mefloquine is a well-tolerated drug. Doses up to 0.5 g per week given for one year have been found to produce no toxic symptoms. However, some patients may experience slight nausea, vomiting, diarrhoea and dizziness. The drug rarely causes psychiatric symptoms.

(c) *Pyridinemethanols:* Of the several pyridinemethanols prepared as possible antimalarial agents none has emerged as a drug for treating malaria in man. However two pyridinemethanols, WR-172435 (XXXIII) and WR-180409 (XXXIV), have been studied in detail. Both these compounds showed marked activity against the multi-drug-resistant Smith strain of *P. falciparum* and the pyrimethamine-resistant Palo Alto strain of *P. vivax* in owl monkeys and, therefore, deserved clinical trials in man. A total dose of 750 mg of XXXIV given in divided doses, appears to be curative but a single dose of 500 mg is not curative [22].

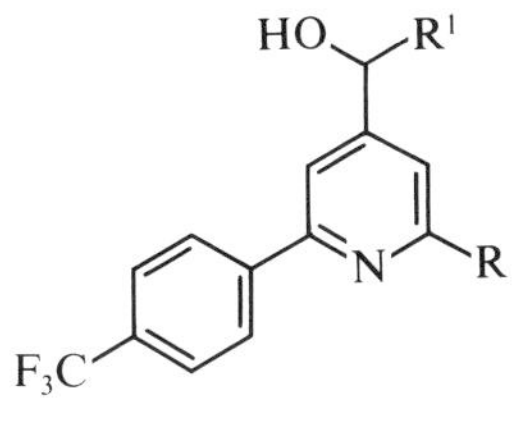

XXXIII, R = C_6H_4–CF_3(p), R^1 = $(CH_2)_2N(C_4H_9)_2$
XXXIV, R = CF_3, R^1 = 2–piperidinyl

3.5.7 Naphthoquinones

The extensive structure-activity relationship studies carried out by Fieser and co-workers [70–74] on naphthoquinone have resulted in the discovery of two promising compounds, lapinone (XXXV) and menoctone (XXXVI). The former was discovered quite early and was found to be effective in treating *P. vivax* infections in man [75]. However, the therapeutic dose of this drug was high and, therefore, it did not find much use in clinical practice.
Menoctone (XXXVI) shows high blood schizontocidal and causal prophylactic activities against *P. berghei* in rodents [76, 77]. However, when tested in man infected with Malayan (Camp) or Uganda I strains of *P. falciparum,* it did not show gametocytocidal, sprontocidal or causal prophylactic activities up to an oral dose of 0.4–0.5 g per day given for 3 days. The inactivity of menoctone in man has been attributed to the drug's poor absorption through the gastrointestinal tract and its strong ability to bind with plasma proteins [22, 78].

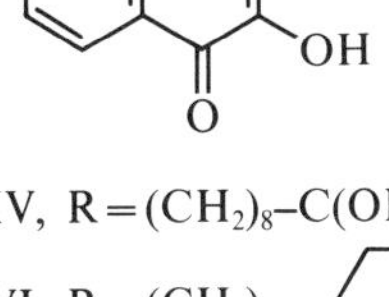

XXXV, R = $(CH_2)_8$–C(OH)$(C_5H_{11})_2$

XXXVI, R = $(CH_2)_8$–

3.5.8 Antifolates

Folic acid (FA) is an important biochemical requirement of all living things. Both mammals and plasmodia utilize FA by converting it successively into dihydrofolic acid (FAH_2) and tetrahydrofolic acid (FAH_4) by enzymes. FAH_4 is then converted into folate co-factors. Mammals can utilize FA which they obtain through their diet. However plasmodia can not utilize FA effectively to meet the required demand of FAH_2. Therefore, they synthesize FAH_2 *de novo*. The first step in the *de novo* synthesis of FAH_2 involves the condensation of pteridine with p-aminobenzoic acid (PABA) in the presence of catalase and synthatase to give dihydropteroic acid (PAH_2). The latter reacts with glutamic acid to form FAH_2.

The striking difference in the synthesis of FAH_2 in plasmodia and mammals (Figure 2) has been exploited in chemotherapy. Drugs that block the synthesis of PAH_2 are known as dihydropteroate synthatase inhibitors (PABA antagonists) and those that block the reduction of FAH_2 into FAH_4 are known as dihydrofolate reductase (DHFR) inhibitors. Collectively these drugs are known as antifolates.

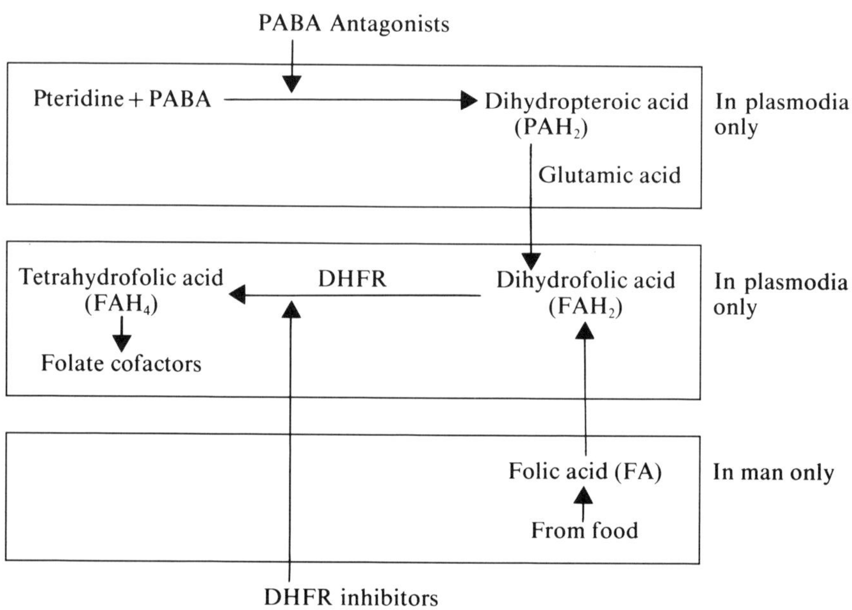

Figure 2
Biosynthesis of tetrahydrofolic acids and sites of antifolate attack.

(a) *PABA antagonists* (Sulfonamides and Sulfones): The drugs of this class were initially developed as antibacterial agents.

Later their use in the chemotherapy of malaria was also established. The PABA antagonists have no action on the sporozoites and exo-erythrocytic stages of the parasites. They are more active against the asexual erythrocytic stages of *P. falciparum* than those of the other species. However the onset of action is slow and resistance develops easily when the drugs are used alone. These compounds show synergistic effects when given in combination with DHFR inhibitors.
Among the large number of sulfonamides prepared [79, 80], only compounds like sulfadiazine (XXXVII), sulfadimethoxine (XXXVIII), sulfamethoxypyridazine (XXXIX), sulfadoxine (XL) and sulfalene (XLI) are of particular interest in malaria chemotherapy because of their long half-life (60–200 hours) in blood.
All the above sulfonamides (XXXVII–XLI) have been evaluated against human malaria including chloroquine-resistant *P. falciparum;* sulfadoxine and sulfalene were found to be better than the other sulfonamides. Although these drugs are effective against falciparum malaria when given alone, such chemotherapy is not advis-

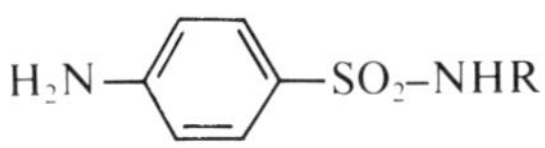

XXXVII, R= 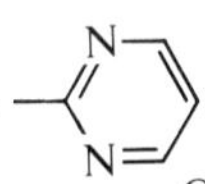 (Sulfadiazine)

XXXVIII, R= 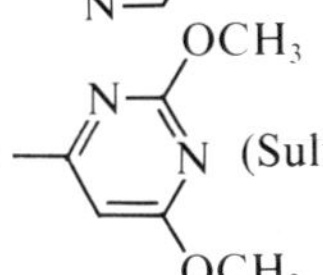 (Sulfadimethoxine)

XXXIX, R= 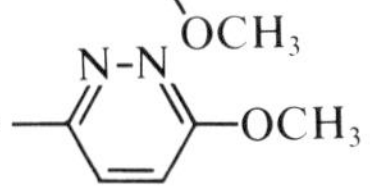 (Sulfamethoxypyridazine)

XL, R= 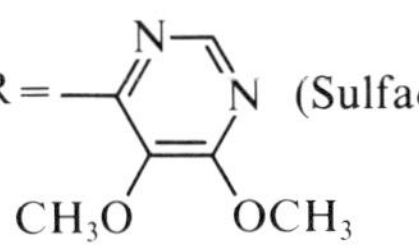 (Sulfadoxine)

XLI, R= 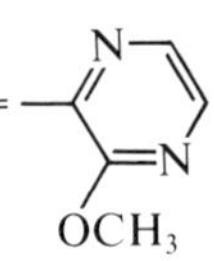 (Sulfalene)

able due to the easy possibility of drug resistance. Thus, the use of sulfonamide drugs is restricted to various combinations with DHFR inhibitors and for the prophylaxis and treatment of falciparum malaria. The clinical profile of various drug combinations of sulfonamides and DHFR inhibitors is discussed in Section 3.5.9.

Following the discovery of antimalarial activity in the antileprotic drug, 4,4'-diaminodiphenylsulfone (dapsone, DDS, XLII) by Coggeshall and co-workers [81], a large number of diphenylsulfones were synthesized [82–84] but none exhibited better antimalarial activity than the parent drug. However, diformyldapsone (DFDDS, XLIII) and acedapsone (DADDS, XLIV) exhibit repository activity and hold some promise in malarial chemotherapy.

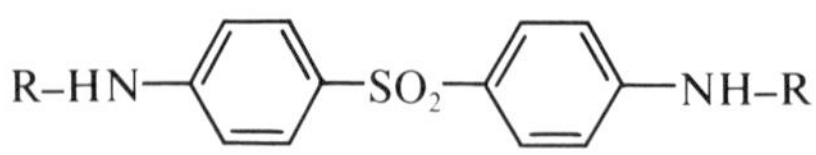

XLII, R = H (DDS)
XLIII, R = CHO(DFDDS)
XLIV, R = COCH$_3$(DADDS)

The curative activity of DFDDS and DADDS has been attributed to their metabolic conversion into DDS [85, 86]. Consequently, among the sulfones only DDS is used clinically. Its curative and suppressive antimalarial profile has been recently reviewed [87].

The toxic effects of sulfonamides may include skin reactions such as utricaria and gastrointestinal disturbances. Higher doses may result in the Stevens-Johnson syndrome. Some individuals may experience hypersensitivity [88].

For dapsone, the side effects are dose related. At a standard daily dose of 25 mg of dapsone, side effects like nausea, vomiting, headaches, blurred vision and insomnia may be observed. Haemolysis may occur in G6PD-deficient patients. Overdoses (›200 mg/day) may cause anaemia while higher doses (100 mg/day) can give rise to dermatitis and hepatitis [21].

(b) *DHFR inhibitors:* A number of DHFR inhibitors have been developed from diaminopyrimidines (trimethoprime and pyrimethamine), biguanides (proguanil and chlorproguanil) and triazines (cycloguanil) which will be discussed under the following heads –

1) *Diaminopyrimidines:* Two compounds of this class, pyrimethamine

(XLV) and trimethoprime (XLVI) have been found to be of significant value in the treatment of human malaria. Of these, pyrimethamine was developed as an antimalarial. Its antimalarial action is due to its ability to block the synthesis of FAH_4 by inhibiting DHFR enzyme in the parasite. Trimethoprim is commonly used as an antibacterial agent. It shows a high degree of inhibition of the bacterial DHFR enzyme. In fact, if pyrimethamine is available commercially, it is the drug of choice in antimalarial chemotherapy [89].
Pyrimethamine (XLV) is very effective against asexual erythrocytic stages of all the four *Plasmodium* spp. and against the erythrocytic stages of *P. falciparum*. The drug brings on a clinical cure in all of the four types of malaria and a radical cure in falciparum malaria. Pyrimethamine has no action against sporozoites, but can cause sterilization of gametocytes and prevents sporogony in mosquitoes. The drug is generally used in combination with PABA antagonists such as sulfadoxine, sulfalene and dapsone (see Sect. 3.5.9). These combinations are more effective and less susceptible to resistance development than pyrimethamine alone and, therefore, have been recommended for use as prophylactic drug combinations in chloroquine-resistant areas.

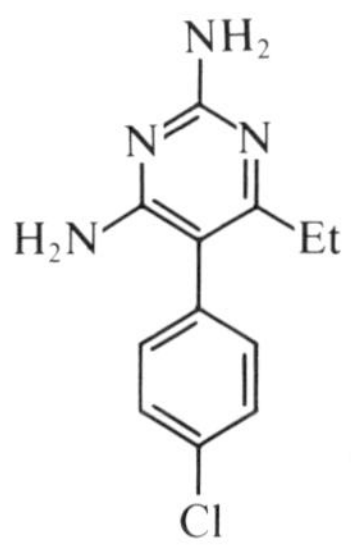

XLV (Pyrimethamine)

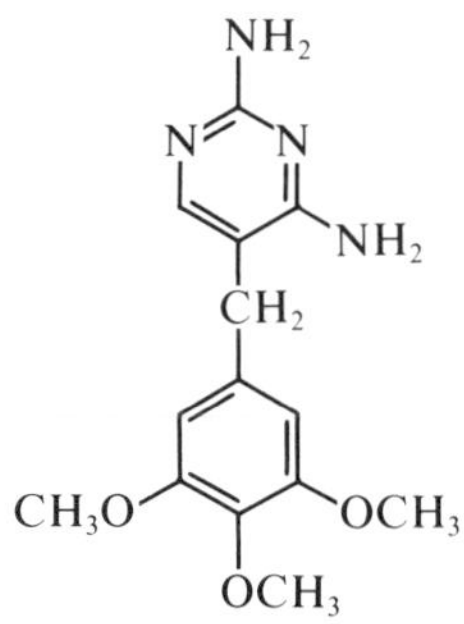

XLVI (Trimethoprime)

The recommended prophylactic dose of pyrimethamine is 25 mg given orally every week. At this dose schedule no adverse effects are observed. Higher doses or daily administration of prophylactic doses (25 mg) may cause megaloblastic anaemia which may be controlled by giving folinic acid at a dose of 10 mg/kg. Pyrimethamine has been found to be teratogenic in some animal studies and, therefore, is not recommended during pregnancy [21].

2) *Biguanides and triazines:* The compounds of this class are potent DHFR inhibitors. Their mode of action is similar to that of diaminopyrimidines [89]. The important guanidines used in malaria chemotherapy are proguanil (XLVII) and chlorproguanil (XLVIII). As such, proguanil is inactive; however, when administered, it is metabolically converted in the liver into its active metabolite, cycloguanil (XLIX), which is also available commercially.

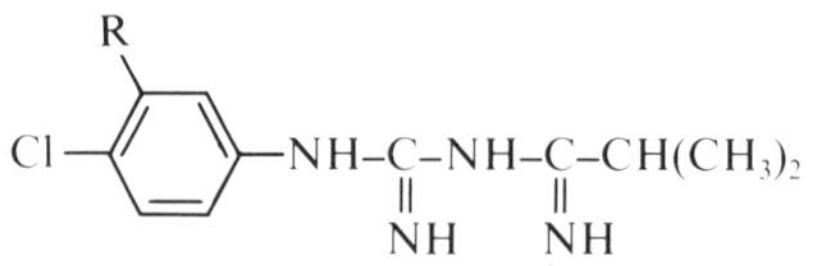

XLVII, R = H (Proguanil)
XLVIII, R = Cl (Chlorproguanil)

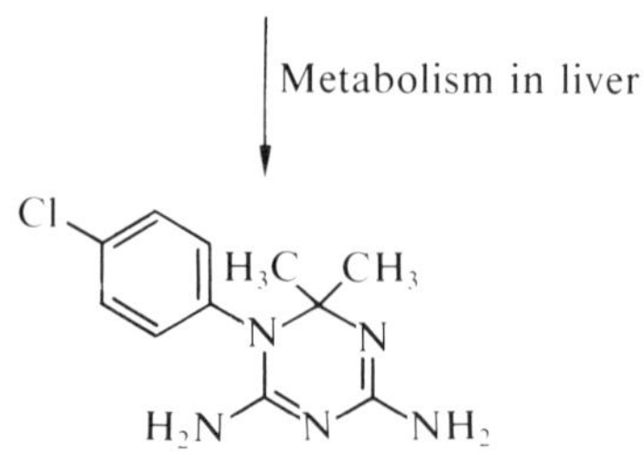

XLIX (Cycloguanil)

Both proguanil and chlorproguanil possess high activity against the exoerythrocytic stages of *P. falciparum* and hence, are used as causal prophylactics in falciparum malaria. They are remarkably active against the asexual erythrocytic stages of all the *Plasmodium* spp. (blood schizontocidals) and also show sprontocidal activity with virtually no toxicity. Further, proguanil is recommended as the best drug for non-immune travellers to malarious regions [90].

The recommended adult dose of proguanil is 100–200 mg/day given as hydrochloride salt. Chlorproguanil is given at a dose of 20 mg at weekly intervals. For semi-immune cases, the usual dose of proguanil is 200 mg given twice a week or 300 mg once a week. Acute attacks of falciparum malaria may need an initial dose of 300–600 mg of proguanil followed by 300 mg daily as long as medication is needed.

Cycloguanil is administered intramuscularly as its pamoate salt at a dose of 5 mg/kg in adult patients. The volume of the injection should be 2–2,5 ml injected preferably in the gluteal muscles. More

than 80 % of the drug remains at the injection site even after 2 weeks and the drug may be detected there up to one year afterward.
Both proguanil and chlorproguanil, as also cycloguanil, are remarkably well tolerated and are amongst the safest antimalarials available today. However, mild utricaria, rash, itching, nausea, anorexia and abdominal discomfort may be observed in a few treated cases [21].

3.5.9 Drug combinations

One of the major problems of malarial chemotherapy is development of resistance to most of the commonly used antimalarials except for primaquine and to some extent quinine. To alleviate this problem various drug combinations have been evaluated in clinical therapy because such additive or synergistic combinations are expected to reduce the risk of development of resistance by *Plasmodium* spp. [91, 92].
Quinine has been used in combination with tetracycline, clindamycine or a PABA antagonist. The recommended dose schedule for the combination of quinine with tetracycline is: quinine 600 mg 3 times a day for 5–7 days followed by 250 mg of tetracycline given 4 times a day for 7 days. The dose regimen for the quinine-clindamycine combination is: quinine 600 mg thrice daily for 3 days, thereafter clindamycine is given at a dose of 5 mg/kg, twice daily for 5 days [21].
A large number of different drug combinations have been evaluated in the treatment of human malaria [22], the most important being the combination of DHFR inhibitors with PABA antagonists. A few such combinations are as follows:

(a) Fansidar	pyrimethamine (25 mg) and sulfadoxine (500 mg)
(b) Co-trimoxazole	trimethoprim (80 mg) and sulfamethoxazole (400 mg)
(c) Maloprim	pyrimethamine (12.5 mg) and dapsone (100 mg)
(d) Metakelfin	pyrimethamine (25 mg) plus sulfalene (500 mg)
(e) Daraclor	pyrimethamine (25 mg) plus chloroquine sulphate (150 mg, base)
(f) Camoprim	amodiaquin (150 mg, base) and primaquine (15 mg, base)
(g) Lapaquine	chloroquine (150 mg) and cycloguanil (20 mg)

All the above drug combinations have been used for prophylaxis or treatment of different forms of malaria in man [7]. However Fansidar has been studied in greater detail showing the combination to be highly successful in treating falciparum malaria [35, 93–98]. Huang and co-workers [99] have found that a combined dose of pyronaridine (500 mg), sulfadoxine (1000–1500 mg) and pyrimethamine (50–75 mg) provides 100 % cures against asexual parasites of *P. falciparum* in man. Similarly cotrimoxazole has been found to be highly effective in treating vivax, falciparum and ovale malaria in man [100].

Following reports of toxicity exhibited by Fansidar, the US Government has recommended certain precautions while using a sulfadoxine-pyrimethamine combination for the prophylaxis of malaria [101]. Bradley and co-workers [102] in England have reommended that Fansidar should not be given concurrently with chloroquine for malaria prophylaxis.

4 Leishmaniasis

Leishmaniasis is a group of several diseases caused by the invasion of the human body by intracellular protozoan parasites of the genus *Leishmania*. The disease is transmitted to man by the bites of over 50 species of small, hairy female sand flies of the genus *Phlebotomus, Sergentomyia* and *Lutzomyia*. The *Leishmania* parasites have been classified [103, 104] into four groups – *Leishmania donovani* complex, *Leishmania tropica* complex, *Leishmania mexicana* complex and *Leishmania brasiliensis* complex of which about ten infect humans. Depending upon the nature of the infection, leishmaniasis can occur in three forms – visceral leishmaniasis, cutaneous leishmaniasis and mucocutaneous leishmaniasis. All these forms of leishmaniasis are basically disfiguring diseases associated with a wide range of morbidity. If not treated, only visceral leishmaniasis is fatal.

4.1 Geographical distribution

Leishmaniases occur in most parts of the world; however, the endemic foci of the disease are the Mediterranean basin, the Middle East, Indian subcontinent, East and North Africa and Central and South America. It is estimated that the worldwide prevalence of leishmaniasis is nearly 12 million cases with an incidence rate of 400 000 new cases every year [105, 106]. It is also believed that about one-third of the world population (some 1600 million people) are exposed to the risk of acquiring leishmaniasis [106]. An early figure shows that nearly 5000 deaths occurred due to leishmaniasis [107]. However according to a recent report, the worldwide deaths due to different forms of leishmaniasis is nearly 1000 [9a].

In India, the history of leishmania epidemics is quite old. The disease has ravaged the Indian population several times with high mortality rates. At the end of the last century, a devastating epidemic of visceral leishmaniasis with nearly 95 % mortality rates swept Bengal, Assam and the Brahmaputra valley, a next epidemic followed in 1918 [108]. Thereafter the incidence of visceral leishmaniasis declined significantly. In 1937, the incidence of kala-azar was 1 in 1000 inhabitants with a death rate of 6–15 % [109]. The kala-azar epidemic again broke in Bihar in 1976–1977 with an estimated 70 000 to 100 000 cases and 4000 deaths [110]. In 1981, the resurgence of the disease was observed in West Bengal causing serious concern to the Indian Health Administration [108].

4.2 Epidemiology

The leishmania parasites occur in two distinct morphological forms: the flagellated promastigotes in the midgut of the sand fly and the unflagellated amastigotes in the mammalian host. The life cycle of the parasite begins when the sand fly injects promastigotes into the skin of man while taking its blood meal. After the inoculation of promastigotes, these flagellated organelles are rapidly taken up by phagocytic cells where they lose their flagella, thus resembling an amastigote (amastigote-type promastigotes). Slowly the promastigotes develop into amastigotes, spread in the blood and multiply through binary fission inside the macrophages of the liver, spleen, bone-marrow, lymph glands, mucosa and the reticuloendothelial system (mononuclear phagocytic network). Finally the macrophage ruptures releasing intracellular and free amastigotes which either invade new macrophages or enter the gut of the sand fly when it sucks the blood an infected person.

72 hours after reaching the vector, the amastigotes become flagellated promastigotes in the midgut. The promastigotes then multiply,

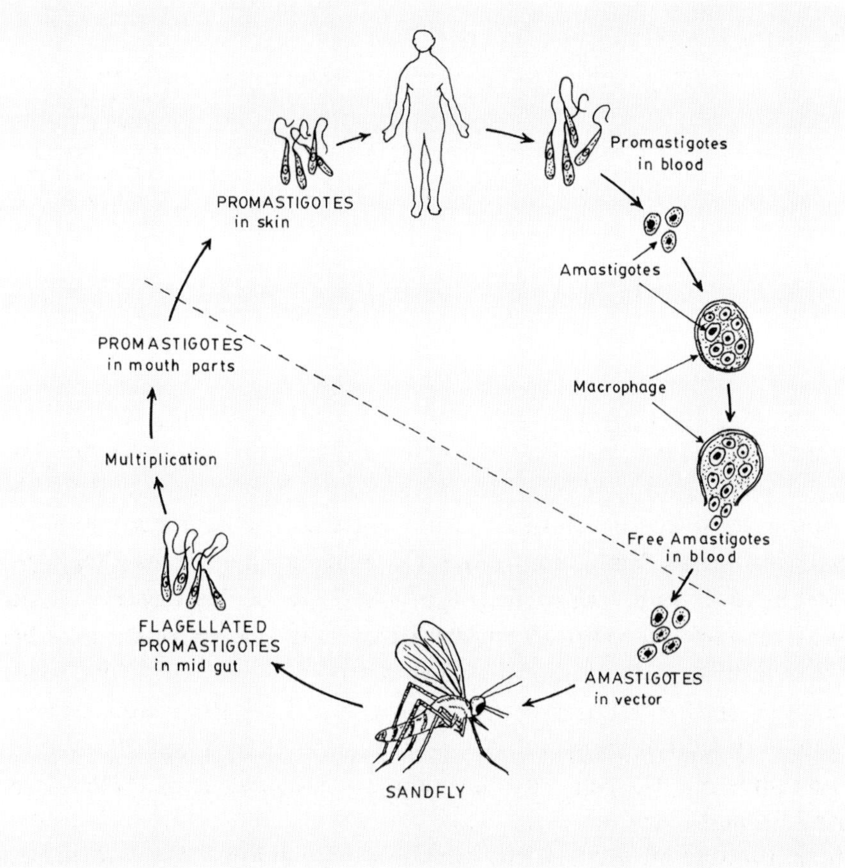

Figure 3
Life cycle of Leishmania

fill up the lumen of the gut of the sand fly and proceed towards the head and mouth parts after 4–18 days. The sand fly is now ready to inoculate promastigotes on a new host (Figure 3).

4.3 Forms of leishmaniasis

(a) *Visceral leishmaniasis* (kala-azar): This is caused by *L. donovani* complex *(L. donovani, L. infantum, L. chagasi)* and is distributed in Southwest and Central Asia, Indian subcontinent, the Middle East, North China, Africa (Kenya, Sudan, Ethiopia, Somalia) and Central and South America (Table 4). The incubation period of visceral

Table 4
The biology of *Leishmania donovani* complex

Parasites	Vector	Reservoir host(s)	Disease caused	Geographical distribution
1. *L donovani donovani*	*Phlebotomus* sp.	Man	Visceral leishmania and post kalazar dermal leishmaniasis	Indian subcontinent and Southwest Asia
2. *L. donovani infantum*	–do–	Dogs, foxes, jackals	Infantile type visceral leishmaniasis	Middle East, Central Asia, North China and North Africa
3. *L. donovani chagasi*	*Lutzomyia* sp.	Dogs, foxes, rarely man	–do–	Central and South America

leishmaniasis is 6 weeks to 6 months but may also spread between 2 weeks to 9 years. The early stage of the disease is marked by mild and irregular fever that slowly becomes a daily occurrence together with chills and sweating. Cough, diarrhoea, vomiting, dizziness, weight loss, loss of appetite, malaise, bleeding gums and pain in limbs may also be observed. Clinical manifestations of kala-azar include anaemia, wasting, skin darkening, enlargement of spleen, liver and lymphatic nodes.

The pathology of visceral leishmaniasis results due to the destruction of the reticuloendothelial system with marked effects on spleen, liver, lymph nodes, bone-marrow and intestine. The spleen is enlarged and its cells are packed with amastigotes. The liver is also grossly enlarged showing fibrosis, cirrhosis and parasitosis in its Kupffer cells. The bone-marrow becomes dark-reddish coloured with impaired production of blood cells. This makes the patient anaemic, leucopenic and thrombocytopenic. The kidneys and lungs are less parasitized but the villi of the jejunum and duodenum can be seen packed with macrophages infected with amastigotes. Lymph nodes are also enlarged and the amastigotes may be found frequently in nasal and pharyngeal secretions.

(b) *Post kala-azar dermal leishmaniasis:* This form of leishmaniasis, also known as post kala-azar disease, is the sequel to a previous infection with *L. donovani.* Sometimes a cutaneous form of leishmaniasis may occur after 1–2 years in patients who have recovered from visceral leishmaniasis or in whom visceral leishmaniasis has subsided. This condition is usually referred to as post kala-azar occurring in nearly 20 % of patients in India and occasionally in East Africa. Hypopigmented and erythematous patches can be seen on face, body trunk and limbs of the patients. Some patients may also develop lepromatous leprosy type nodules.

(c) *Cutaneous leishmaniasis:* This is caused by *L. tropica* complex and *L. mexicana* complex (Tables 5, 6). The disease caused by *L. tropica* complex *(L. tropica* or *L. tropica* minor, *L. major* or *L. tropica* major and *L. aethiopica)* is known as Old World cutaneous leishmaniasis or oriental sore. The infection due to *L. mexicana* complex *(L. mexicana mexicana, L. mexicana amazonensis, L. mexicana pifanoi* and *L. mexicana venezuelensis)* is known as New World cutaneous leishmaniasis. The Old World cutaneous leishmaniasis is prevalent in the Mediterranean countries, Middle East, India, Afghanistan and North Africa including Kenya and Ethiopia. The New World cutaneous leishmaniasis is endemic in Mexico, Brazil, Venezuela, Guatemala and Trinidad.

Table 5
The biology of *Leishmania tropica* complex

Parasites	Vector	Reservoir host(s)	Disease caused	Geographical distribution
1. *L. tropica*[a]	*Phlebotomus* sp.	Man, dog	Cutaneous leishmaniasis and leishmaniasis recidivans	Middle East, Eastern Mediterranean, India, Afghanistan, North Africa
2. *L. major*[b]	–do–	Rats, Gerbils	Zoonotic cutaneous leishmaniasis	Middle East, India, Africa, North Africa, Sub-Saharan Africa
3. *L. aethiopica*	–do–	Hyraxes	Cutaneous and diffused cutaneous leishmaniasis	Ethiopia, Kenya

[a] formerly known as *L. tropica* minor; [b] formerly known as *L. tropica* major.

Table 6
The biology of *Leishmania mexicana* complex

Parasites	Vector	Reservoir host(s)	Disease caused	Geographical distribution
L. mexicana mexicana	*Lutzomyia* Spp.	Rodents	Destructive chronic ear lesions, self-healing body lesions	Mexico, Guatemala
L. mexicana amazonensis	-do-	Rodents, wild animals	Cutaneous and diffused cutaneous leishmaniasis	Brazil, Venezuela, Trinidad
L. mexicana pifanoi	-do-	Rodents (?)	Diffused cutaneous leishmaniasis	Venezuela
L. mexicana venezuelensis	-do-	?	Cutaneous leishmaniasis	Riverine forests of Venezuela

The clinical manifestations of both forms of the above-mentioned leishmaniasis is somewhat similar and relates to skin lesions and ulcerations. Usually a small lesion appears several weeks after the bite of the sand fly. This lesion ulcerates and becomes crusted. The ulcer heals after a few months leaving behind a typical scar. Thus, the oriental sore is self-healing after 1–2 years and gives the patients lifelong immunity to reinfection by the same parasite. Sometimes, multiple unhealing lesions with a thickening of the skin may develop on the face which greatly resemble lepromatous leprosy or tuberculosis of skin. This condition is referred to as leishmaniasis recidivans (LR) and is believed to be an allergic state due to over-developed cell-mediated immunity.

(d) *Mucocutaneous leishmaniasis:* This is a severe and destructive type of cutaneous leishmaniasis caused by *L. brasiliensis* and is restricted to the Amazon forests in Brazil (Table 7). Initially lesions appear on the skin which resemble those of cutaneous leishmaniasis. However, these lesions are large and long-lasting. The parasite may migrate to the tissues of the nasopharyngeal region. After several years, metastatic spread may occur in oronasal and pharyngeal mucosa leading to disfiguration or destruction of the tissues of nasal septum, lips and larynx.

Table 7
The biology of *Leishmania braziliensis* complex

Parasites	Vector	Reservoir host(s)	Disease caused	Geographical distribution
1. *L. brasiliensis brasiliensis*	*Lutzomyia* Sp.	Rodents, rarely man	Mucocutaneous and cutaneous leishmaniasis	Brazil (Amazon Forest)
2. *L. brasiliensis guyanensis*	–do–	Sloths, rarely man	Cutaneous leishmaniasis with multiple sores	Brazil, Surinam, Guyanas
3. *L. brasiliensis panamensis*	–do–	Sloths	Cutaneous leishmaniasis with some leisions	Panama, Costa Rica, Colombia
4. *L. brasiliensis peruviana*	–do–	Dogs	Self-healing cutaneous leishmaniasis	Peru, Argentina

4.4 Diagnosis of leishmaniasis

The diagnosis of visceral leishmaniasis is carried out by showing the presence of amastigotes in (a) the material aspirated from spleen, bone-marrow and enlarged lymph nodes, (b) peripheral blood monocytes. The aspirates and blood may also be cultured for promastigotes. Visceral leishmaniasis can also be diagnosed by testing serum for leishmanial antibodies. Other laboratory diagnostic tests for this disease are (a) formal gel test, a non-specific test for marked increase in IgG, (b) measurement of serum or plasma which is based on the fact that serum albumin is decreased when liver cells are damaged and (c) haematological tests [haemoglobin, total and differential leucocyte count]. (TLC, DLC), platelet count and erythrocyte sedimentation rate (ESR)].
Serological diagnosis of visceral leishmaniasis utilizes the detection of non-specific IgG and IgM which are produced during active visceral leishmaniasis. These specific antileishmanial antibodies may be detected in the patient's sera by various techniques such as indirect fluorescent antibody test (IFAT), counter immunoelectrophoresis (CIEF) and enzyme-linked immunosorbent assay (ELISA).
The diagnosis of cutaneous and mucocutaneous leishmaniases may be done by (a) detecting amastigotes in the smears taken from infected nodules, ulcers or lesions and (b) culturing the material taken

from ulcers and examining the presence of promastigotes. The serological tests discussed above do not work well in the case of cutaneous and mucocutaneous leishmaniases because of poor antibody response in such cases.

4.5 Treatment of leishmaniasis

The chemotherapy of leishmaniasis has been reviewed by several researchers [105, 108–115]. The significant developments in the area are discussed below.

4.5.1 Organoantimonials

The pentavalent antimonials are amongst the earliest known drugs for treating human leishmaniasis and are still used in the clinical practice despite the serious side effects associated with them. The important antimonial drugs are tartar emetic (L), sodium stibogluconate (LI), meglumine antimoniate (LII) and derivatives of stibanilic acid (Stibamine, LIII) such as stibacetin (LIV), ethyl stibamine and urea stibamine (LV). Chemically all the antimonials may be divided in two classes – (a) compounds in which the antimony (Sb^{v}) atom is linked to a carbon atom through an oxygen bridge (e.g. sodium stib-

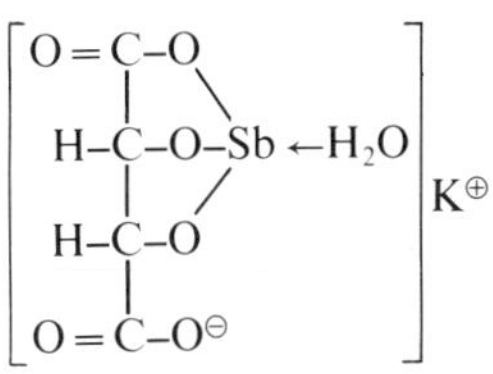

L (Tartar emetic)

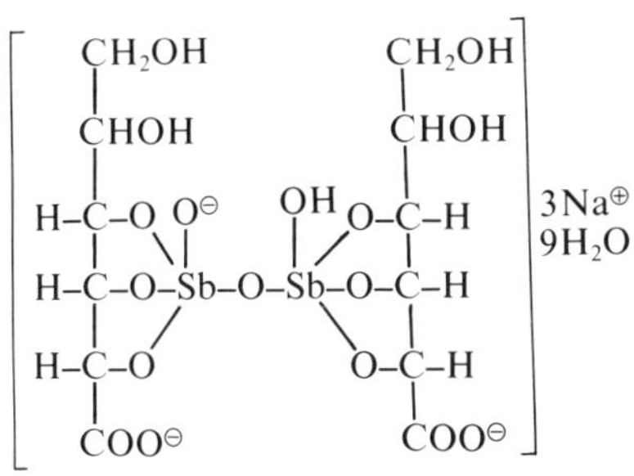

LI (Sodium stibogluconate)

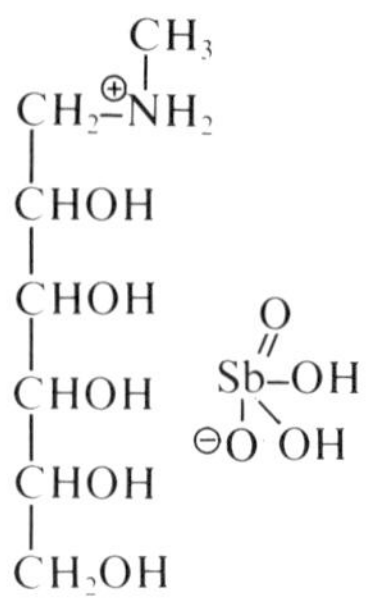

LII (Meglumine antimoniate)

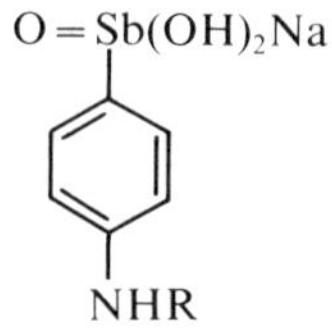

LIII, R = H (Stibamine)
LIV, R = $COCH_3$ (Stibacetin)

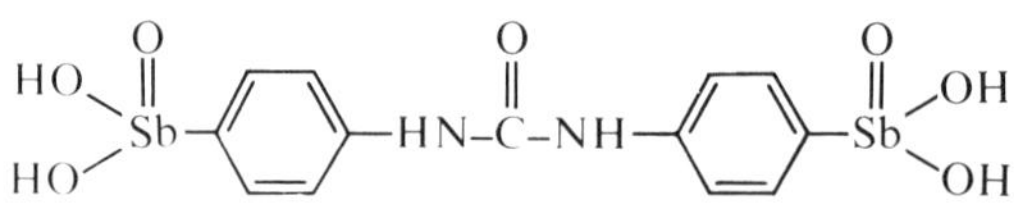

LV (Urea stibamine)

ogluconate) and (b) compounds in which the antimony atom is directly linked to a carbon atom in the benzene ring (e.g. urea stibamine). The precise structures of these drugs are not fully understood. The quality of the drugs is maintained through chemical analysis or physico-chemical measurement of pentavalent antimony (Sb^{v}). After administration, these pentavalent antimonials are reduced to the trivalent form [116] and exhibit activity/toxicity.

Tartar emetic (L): This drug was used earlier to treat visceral leishmaniasis in children [16] and adults [118, 119]; however it is not currently used due to its high toxicity. The use of tartar emetic has also been demonstrated in the control of oriental sores [120].

Sodium stibogluconate (Pentostam, LI): This drug is amongst the safest and most potent pentavalent antimonial drugs. It has been used more widely to treat visceral leishmaniasis than cutaneous or mucocutaneous leishmaniasis [121–124]. The recommended adult and pediatric dose of sodium stibogluconate for treating visceral leishmaniasis is 20 mg/kg given daily (max. 800 mg/day) intramuscularly or intravenously for 20 days. The treatment may be repeated, if necessary, after a gap of 10 days [109, 125]. The drug is available in a 33 % solution (equivalent to 10 % Sb^{v}); 6-ml injections which should be stored at below 25° C [109]. Recent work carried out by Anabwani and coworkers [126] in Kenya has shown that the use of the drug leads to 100 % cures in adults and children suffering from visceral leishmaniasis when given at a dose of 10 mg/kg for 30 days. A 60 % cure was obtained at a lower dose of 2 mg/kg × 30 days.

It may be noted that inadequate drug administration may lead to relapse visceral leishmaniasis following the apparent cure. The risk of relapse visceral leishmaniasis may be greatly reduced by giving a full course of treatment whenever possible. This was clearly demonstrated by Thakur [127] who used 20 mg/kg of the drug to treat an epidemic in Bihar (India) and found that only 0.5 % of 603 patients relapsed after initial cure.

For cutaneous leishmaniasis, sodium stibogluconate is given at a dose of 10 mg/kg daily (max. 600 mg/day) by way of intramuscular or intravenous injections for 6–10 days. A 33 % solution of the drug may be infiltrated around the edges of the lesions at a dose not more than 2 ml at any one time [109, 125]. Thus a 600 mg Sb^{V}/day/person (about 8 mg/kg) given for 10 days cured 10 of 12 patients suffering either with *L. chagasi, L. mexicana* spp., *L. mexicana amazonensis* or other species of *L. mexicana* complex [128]. For simple, cutaneous lesions due to *L. aethopica,* a dose of 18–20 mg/kg Sb^{V} of pentostam given twice daily for 30 days was found to be successful [129]. However such a treatment fails in cases of diffuse cutaneous leishmaniasis caused by *L. aethopica*. In such a situation pentamidine is a preferred drug [130].

Children and adult patients with mucocutaneous leishmaniasis may be treated by sodium stibogluconate at a dose of 20 mg/kg given daily (max. 800 mg/day) intramuscularly or intravenously for 20 days [125]. For more resistant cases, the treatment should be repeated 2 or 3 times with a gap of 10 days [131, 132].

Meglumine antimoniate (Glucantime, LII): This is the preferred drug for treating leishmaniasis [133]. It is more preferred for treating cutaneous and mucocutaneous leishmaniasis [134–136] than visceral leishmaniasis [137, 138]. The drug has also been used to cure post kala-azar dermal leishmaniasis [139].

Recent clinical studies carried out with Glucantime show that the subtoxic levels of the drug (28 mg Sb^{V}/kg per day) given for 10 days is needed to cure patients with cutaneous and mucocutaneous leishmaniases due to *L. braziliensis braziliensis* [140–142]. In an effort to optimize the dose regimen of Glucantime, Sampaio [143] studied three schedules in patients with mucocutaneous leishmaniasis – (a) 28 mg/kg daily × 10 days repeated thrice 15 days apart, (b) 10 mg/kg daily × 30 days and (c) 20 mg/kg daily × 30 days. He found the latter dose schedule to be the most effective curing 11 of 12 patients. Later Marsden and co-workers [144] found that the drug may be given at a dose of 20 mg/kg daily for as long as 85 days with successful results.

Stibanilic acid (Stibamine) and its derivatives: The sodium salt of stibamine (LIII) has been evaluated for its efficacy in human leishmaniasis but soon it was dropped due to its high toxicity and poor stability [145]. The acetyl derivative of stibamine (LIV, stibacetin) was found to be a more effective and better tolerated leishmanicide [145]. It showed high activity against infantile kala-azar [146, 147] but was also not free from toxic effects [148–150].

Ethyl stibamine is a complex structure obtained by mixing stibamine (LIII), stibacetin (LIV), antimonic acid and diethylamine in a molar ratio of 1:2:1:3. It contains 41–44% pentavalent antimony (Sb^{V}). Ethyl stibamine has been shown to be highly effective in treating kala-azar in man [151–154]. However, the effective dose of the drug is close to the maximum tolerated dose. This makes the drug unsuitable for use in clinical medicine.

Urea stibamine (LV): The pioneering work of Upendra Nath Brahmachari led to the discovery of urea stibamine that cured thousands of people suffering from kala-azar [155–157]. In Assam alone, urea stibamine saved nearly 325 000 lives during 1933–1936 [108]. The effectivity of urea stibamine in curing visceral leishmaniasis has also been demonstrated by other workers [158–161]. The drug is usually effective when given intravenously to adults at a dose of 170 mg daily for 6 days [162]. In view of its high toxicity, it has been recommended that the initial dose of 50 mg/adult should be given. A second dose of 100 mg, a third of 150 mg, and fourth and subsequent at a dose of 200 mg/adult may be given only intravenously and dissolved in 5–10 ml of distilled water [163].

Limitations and side-effects of antimonial drugs. Although some pentavalent antimonials like sodium stibogluconate and meglumine antimoniate are used as the front line drugs in the chemotherapy of human leishmaniasis, they are far from being ideal drugs. They are not active orally and long courses of parenteral administrations are needed for a complete cure of the disease. A variable proportion of the patients do not respond to antimony therapy while others may suffer a relapse which becomes difficult to cure. Since most of the

antimonials have ill-defined structures, their efficacy and toxicity may vary from batch to batch; this is possibly due to the difference in concentration of active pentavalent antimony (Sb^{v}). It is also difficult to standardize the manufacturing process or dose schedule in terms of antimony percentage. Furthermore, the activity of the antimonials is known to vary in different regions. For example, kala-azar in India and China may be cured by a course of one injection given daily for 10 days. The same disease in the East African and Mediterranean regions may require three such courses for complete eradication of the infection. The kala-azar patients in Sudan respond poorly to antimonials while South American mucocutaneous leishmaniasis requires three courses of 10 injections with only 50 % cure rates [108]. Despite the fact that the half-life of Sb^{v} in Pentostam and Glucantime in blood ist only 2 hours [164, 165] and 81–97 % of Sb^{v} given as Pentostam is excreted within the first 6–8 hours [166], the accumulation of Sb^{v} in the tissues is not rare. Antimony may be accumulated in cardiac conducting tissues giving rise to changes in ECG and cardiac rhythm. The ECG changes are dose-dependant and, therefore, medical attention is needed during antimony therapy [167]. Sudden death rarely results from severe dysrhythmias such as ventricular fibrillations [312].
The usual side effects of the antimony drugs are vomiting, diarrhoea, abdominal pain, anorexia, malaise, weakness and muscle pain. Liver and renal damage, shock, itching, fever, dizziness, heartburn and headache are also rarely observed [312].
The antimonials are contra-indicated in pneumonia, myocarditis, hepatitis, nephritis, pregnant women and breast-feeding mothers.

4.5.2 Aromatic diamidines

During the search for an alternative to antimony chemotherapy, some aromatic amidines such as stilbamidine (LVI), hydroxystilbamidine (LVIII) and penamidine (LIX) known to have trypanocidal activity [168, 169], were evaluated against leishmaniasis in man and animals. This ultimately culminated in the discovery of pentamidine as the second line of drugs used in the treatment of human leishmaniasis [170–172].
Stilbamidine (LVI): Following the demonstration of activity of stilbamidine against experimental infection of *L. donovani* [173], the

drug was evaluated in one case of kala-azar [174] with favourable response. Subsequently it was given extensive clinical trials in patients with visceral leishmaniasis resistant to antimonial drugs [170, 172, 175–178]. However the drug produced severe side effects such as a marked fall in blood pressure, respiratory distress and neurotoxicity. Later, it was shown that the toxicity of stilbamidine (LVI) was due to its facile dimerisation into *cis, trans, cis*-tetrakis (4-guanylphenyl)-cyclobutane (LVII) in presence of light [179].

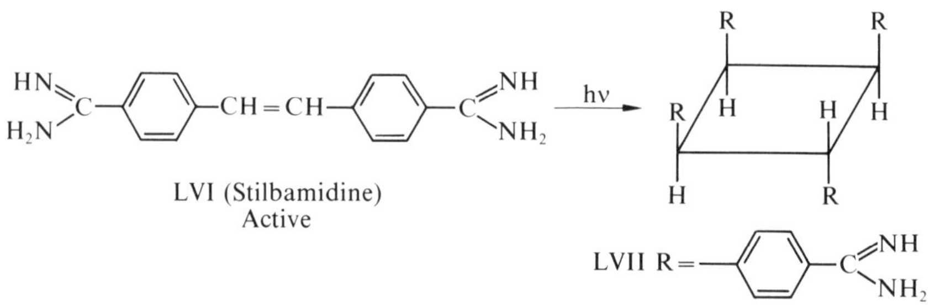

LVI (Stilbamidine)
Active

(Toxic)

2-Hydroxystilbamidine (LVIII): This compound is stable to light and, therefore, is less toxic than stilbamidine. It has been successfully used to treat Indian visceral leishmaniasis [180–182] and American mucocutaneous leishmaniasis [183]. The drug is given intravenously in doses not higher than 0.25 g daily for 10 days to treat kala-azar. However, three such courses may be needed to cure African kala-azar [163].

Pentamidine (LIX): This is the preferred drug in the treatment of visceral leishmaniasis in patients who have failed to respond to antimony-therapy [184, 185]. In East Africa, better clinical results were obtained in kala-azar patients who were first treated with antimonials and then with pentamidine [162]. Pentamidine is less toxic than stilbamidine and can be administered intramuscularly. It is more effective against visceral leishmaniasis [186–189] than cutaneous or mucocutaneous leishmaniases [190–192].

Pentamidine is given in the form of two salts – pentamidine isethionate (pentamidine) and pentamidine dimethane sulfonate (Lomidine). The recommended adult and pediatric doses of the drug are 2–4 mg/kg given intramuscularly (in 3 ml water) for 15 days. The drug has no neurotoxicity like stilbamidine but causes disturbances

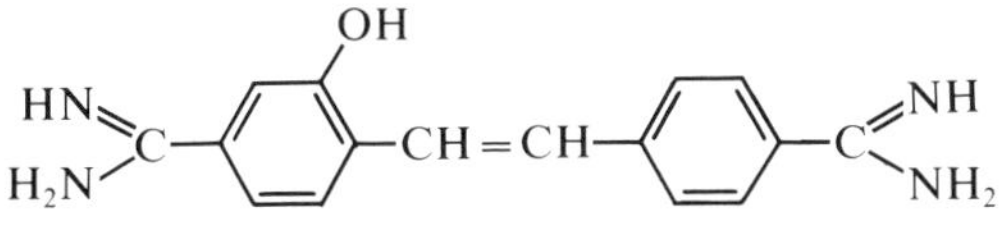

LVIII (Hydroxystilbamidine)

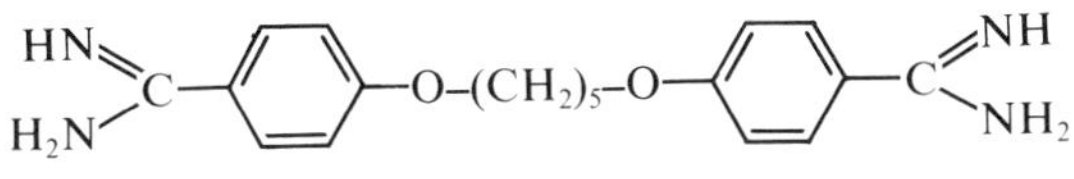

LIX (Pentamidine)

in sugar metabolism and kidney function. It is excreted slowly and thus may be retained in renal and hepatic tissues leading to cumulative toxicity. The frequent sideeffects of pentamidine are gastrointestinal disturbances, vomiting, pain at injection site, hypoglycemia and sudden fall in blood pressure (hypotension). The drug can also give rise to cardiotoxicity, diabetes mellitus, shock and hypocalcemia.

SAR in arylamidines: Diminazene (LX) and HOE-668 (LXI) have been synthesized as structural analogs of stilbamidine. Both compounds exhibited marked activity against *L. donovani* in hamsters [193, 194] but none could be pursued further although LXI was found to be better than antimonials and pentamidine.

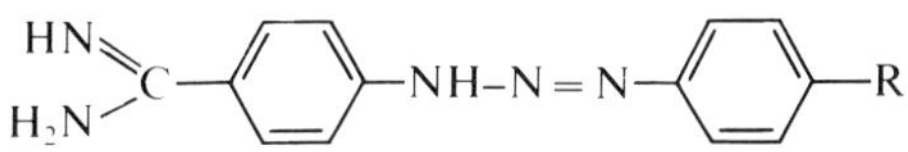

LX, $R = C(NH_2) = NH \cdot (CH_3CONHCH_2COOH)_2$ (Diminazene)

LXI, R = 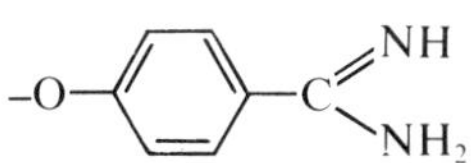 (HOE–668)

The antiprotozoal activity of aryldiamidines is due to their ability to bind with DNA [195, 196]. The mechanism of binding between DNA and diamidines involves an electrostatic interaction between the negative phosphate groups on the DNA helix and the positive centre of the amidines [197]. The examination of the molecular models of diaryldiamidines showed that such compounds may be divided into two classes – one class has the two amidine groups separated by 12 Å and the other by approximately 20 Å. These values correspond well with the distances across the minor and major grooves of DNA [198]. On the basis of the above findings, several aromatic diamidines (LXII–LXV) were synthesized [199–201], all of which showed 71–93 % inhibition of the amastigotes in the macrophages of spleen of hamsters infected with *L. donovani* at an intraperitoneal dose of 2.5 mg/kg given for 5 days. Several other diamidines have also been prepared [202, 203] with no appreciable leishmanicidal activity especially in the compounds of the type LXVI.

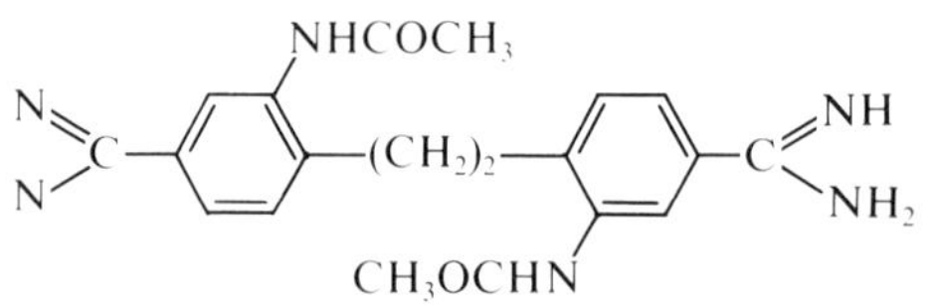

LXII

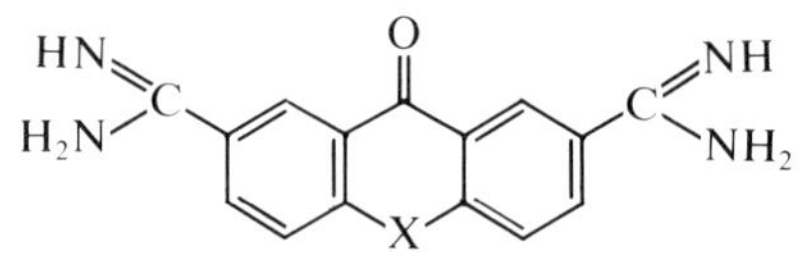

LXIII, X=O,S

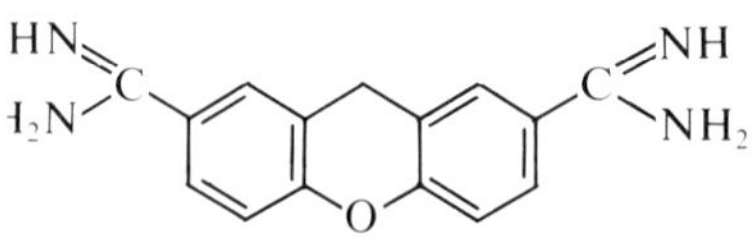

LXIV

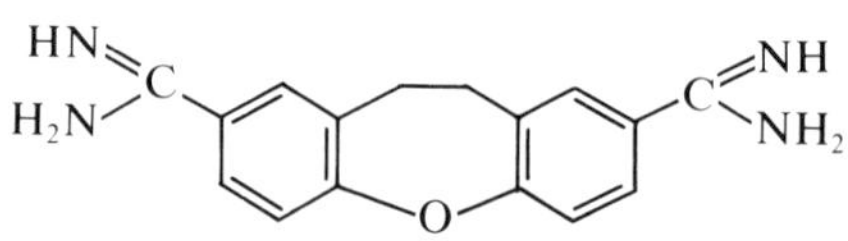

LXV

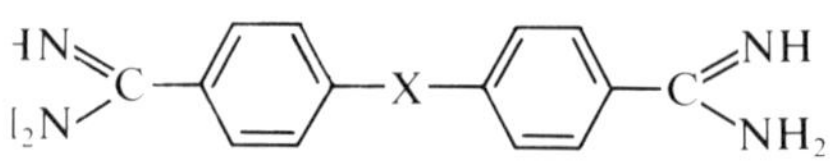

LXVI, X=2,5–Furyl, 2,5–Thienyl etc.

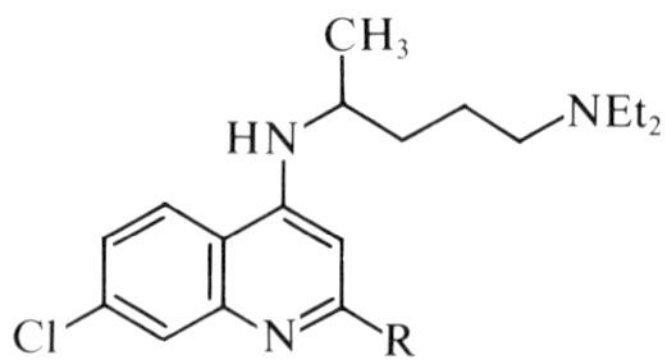

LXVII, R=H (Chloroquine)

LXVIII, R= −CH=CH− 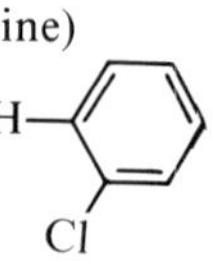

4.5.3 Nitrogen heterocycles

(a) *Quinolines:* The antimalarial drug, chloroquine (LXVII), has been shown to give favourable response in cases of cutaneous leishmaniasis [204, 205] and mucocutaneous leishmaniasis [206]. The chlorostyryl derivative (LXVIII) of chloroquine also shows activity against *L. tropica* in experimental animals [207].

A number of 8-aminoquinoline antimalarials have been evaluated against experimental and clinical leishmaniases [208]. The fact that compounds of this class attain characteristic concentrations in acidic intracellular compartments of liver cells, may provide impetus to explore their leishmanicidal potential [209]. Thus, primaquine (LXIX or XXIII) has been found to possess moderate activity against *L. donovani* in hamsters [210, 211]. This led the scientists of the Walter Reed Army Institute, USA, to explore the potential of 8-aminoquinolines in the chemotherapy of leishmaniasis [212]. Consequently, a number of 8-aminoquinolines (LXX-LXXIV) [213–218] were found to exhibit marked antileishmanial activity; some of the compounds were better than antimonial drugs or pentamidine.

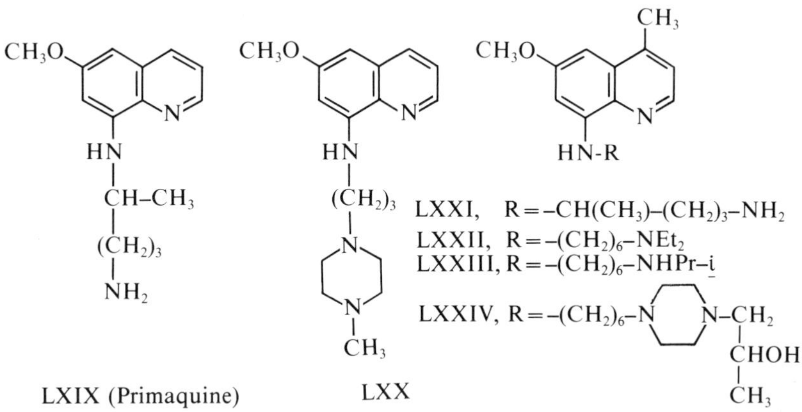

LXIX (Primaquine) LXX

Among the 8-aminoquinolines found to display a high order of leishmanicidal activity, compound LXXII (WR-6026) was picked up as the most effective leishmanicide. WR-6026 shows a 400–700-fold higher activity than meglumine antimoniate against *L. donovani* in hamsters, even when administered orally [212]. The compound shows high efficacy against visceral leishmaniasis in rodents and dogs but is less active against cutaneous leishmaniasis in experimental animals [219]. The metabolites of WR-6026 have been isolated. Like other 8-aminoquinoline drugs, WR-6026 has been found to catalyze haemoglobin oxidation and some amounts of reversible methaemoglobinaemia in dogs [220].

WR-6026 has been found to tolerate up to a single dose of 60 mg and is currently in phase I of clinical trials. Liver appears to be the main site of uptake and metabolism of the drug [221].

(b) *Imidazoles and related compounds:* Metronidazole (LXXV) has been found to be fairly active in treating oriental sores at oral doses of 0.5–1 g daily for 10–70 days [222–226]. In visceral leishmaniasis, metronidazole was given in combination with co-trimoxazole with good results [227]. The schistosomicide niridazole (LXXVI) also shows activity against *L. tropica* and *L. brasiliensis* infections [228, 229]. The antimalarial drugs pyrimethamine (XLV) and cycloguanil (XLIX) have also been used in the treatment of leishmaniasis. At higher doses, pyrimethamine showed beneficial effects in cases of cutaneous and mucocutaneous leishmaniasis [230], while cycloguanil cured 50–80 % of the cases of dermal lesions produced by *L. braziliensis* in Iraq and Ethiopia at an intraperitoneal dose of 350 mg [231–233].

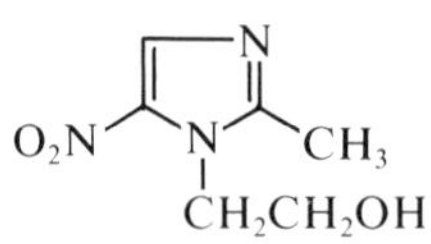

LXXV (Metronidazole)

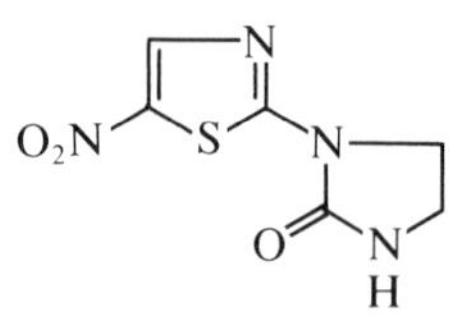

LXXVI (Niridazole)

Ketoconazole (LXXVII) is another imidazole leishmanicide which has emerged as a result of biochemical differences in the mammalian and parasitic cell membrane. The mammalian cell membrane uses cholesterol while the cell membrane of *Leishmania* utilizes ergosterol. Ketoconazole interferes with the biosynthesis of C_{28} sterols in *Leishmania* and inhibits the cytochrome P-450 dependent C-14 demethylation of lanosterol which is the precursor of ergosterol [234].

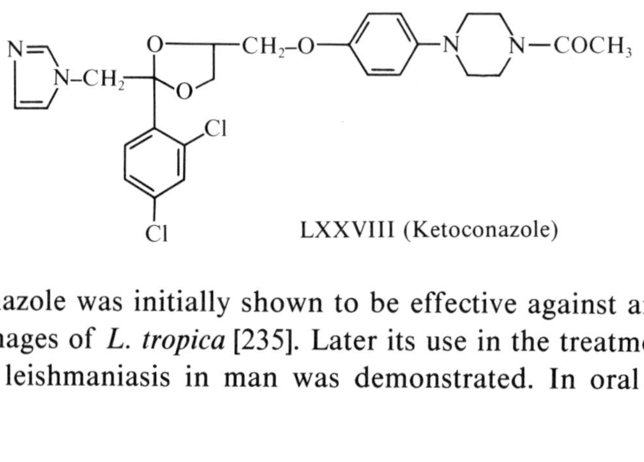

LXXVIII (Ketoconazole)

Ketoconazole was initially shown to be effective against amastigote macrophages of *L. tropica* [235]. Later its use in the treatment of cutaneous leishmaniasis in man was demonstrated. In oral doses of

200–800 mg/day for 4–18 weeks, ketoconazole has been used to treat various forms of leishmaniasis with varying degrees of success [105, 236]. However, Viallet and coworkers [237] have reported the drug to be active in treating cutaneous leishmaniasis at a dose of 400 mg daily for 4–8 weeks.

Levamisole (LXXVIII), an immunomodulator, has been shown to be useful in treating visceral and mucocutaneous leishmaniasis [238]. The major metabolite of levamisole in mammals, 1-β-mercaptoethyl-4(5)-phenylimidazolidin-2-one (LXXIX), has been found to have a fourfold higher immunostimulatory activity than the parent drug [239]. 1-β-Isopropylthioethyl-4(5)-phenylimidazolidin-2-one (LXXX), an structural analog of LXXIX, has been recently shown to cause 100 % inhibition of promastigotes of *L. donovani in vitro* at the MIC of 800 μg/ml [240].

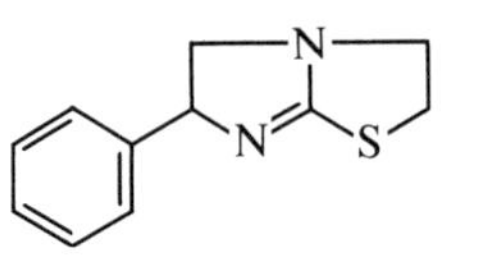

LXXVIII (Levamisole)

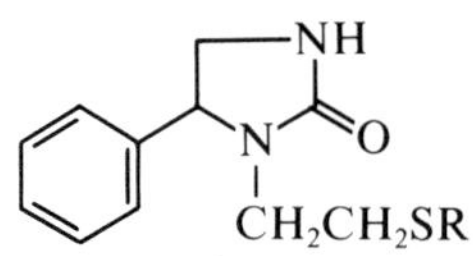

LXXIX, R = H
LXXX, R = –CH(CH_3)$_2$

Cimetidine (LXXXI), an antagonist to histamine H_2 receptor and the drug used in the treatment of peptic ulcers, has been shown to possess high activity against *L. mexicana* in Balb/C mice at a dose of 20 mg/kg daily for 10 days [241].

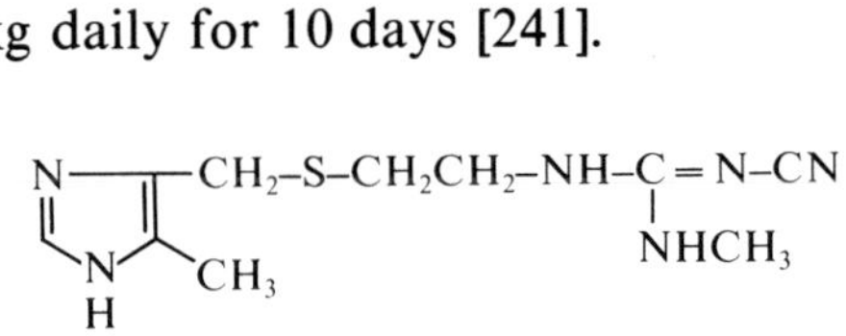

LXXXI (Cimetidine)

(c) *Allopurinol and related compounds:* The antileishmanial activity of the anti-gout drug, allopurinol (LXXXIII), was first demonstrated in 1974. Since then the drug has become a matter of detail study in leishmanial chemotherapy. Apparently allopurinol competes with the natural purine base, hypoxanthine (LXXXII), in the unique salvage pathway of the parasite, leading to the synthesis of abnormal RNA. The parasite finally dies from the inhibition of protein synthesis [242, 243] (Figure 4). This happens because *Leishmania,* like other

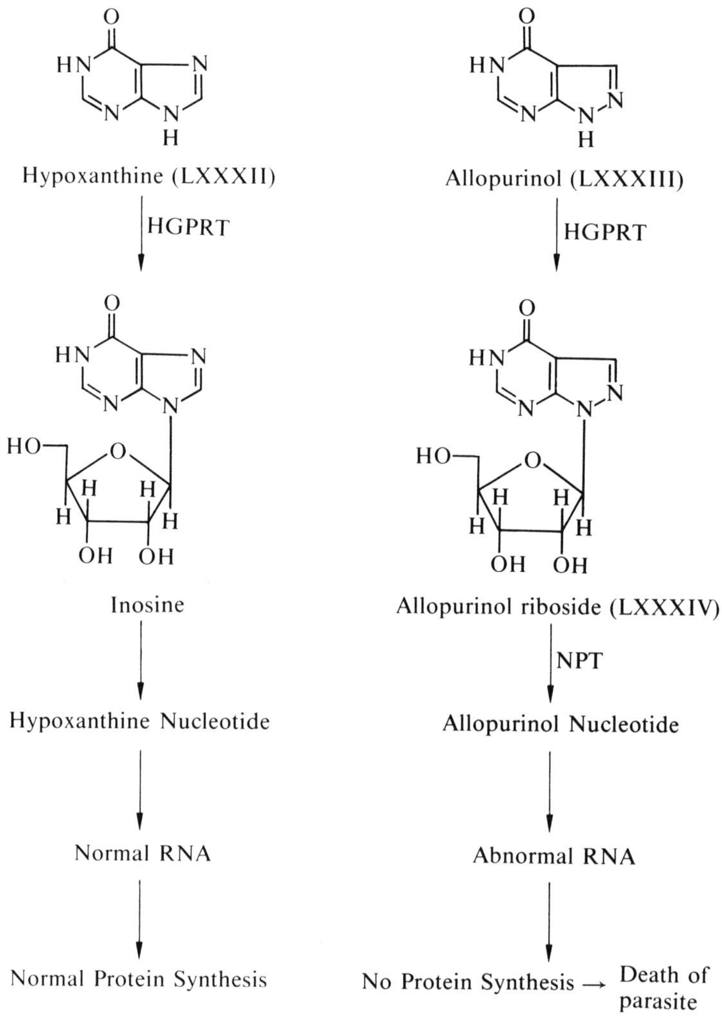

HGPRT = Hypoxanthine-guanine phosphoribosyl transferase
NPT = Nucleoside phosphotransferase

Figure 4
Metabolic pathway of hypoxanthine and allopurinol in *Leishmania.*

protozoan parasites, are unable to synthesize purine base and consequently have developed a salvage pathway to utilize preformed purine bases.

Since allopurinol is orally active and less toxic than antimonials, it has been tested clinically against leishmaniasis. Promising results were obtained in the treatment of kala-azar [244] but results were not

encouraging in the case of leishmaniasis caused by *L. braziliensis* [105]. The poor activity of allopurinol is attributed due to its rapid metabolism into oxipurinol by xanthine oxidase and its faster rate of excretion. Better results have been obtained in treating antimonial-resistant patients using allopurinol (20 mg/kg daily) in combination with sodium stibogluconate [245].
Allopurinol riboside (LXXXIV) has also been evaluated against cutaneous leishmaniasis in man at a dose of 1250 mg four times a day for 28 day when 50 % of the patients were cured. When the drug was used with 500 mg of probenicide, which reduces the excretion rate and increases plasma half-life of purine analog, nearly 66 % of the patients showed clinical improvement [246].

4.5.4 Natural products

Amphotericin B (LXXXV) is an effective drug for treating mucocutaneous leishmaniasis [247]. It also responds very well in mucocutaneous cases resistant to antimonials or pentamidine [247, 248]. The drug treatment is started with a small dose (0.25 mg in 500 ml of glucose) given intravenously every second or third day for 6–16 weeks. The dose of the drug is increased up to 25 mg per infusion. A total of 717.5 mg of amphotericin B given over a period of 3 months has been found to be well tolerated [249]. This schedule of the drug has been used to cure post kala-azar dermal leishmaniasis also [249]. In the case of leishmania recidivans, amophotericin B is injected in a 2 % procain solution at the lesion site with good results [250]. However, the drug is nephrotoxic and cardiotoxic, and also gives rise to nausea, vomiting and fever. It should, therefore, be used under medical supervision only [108].

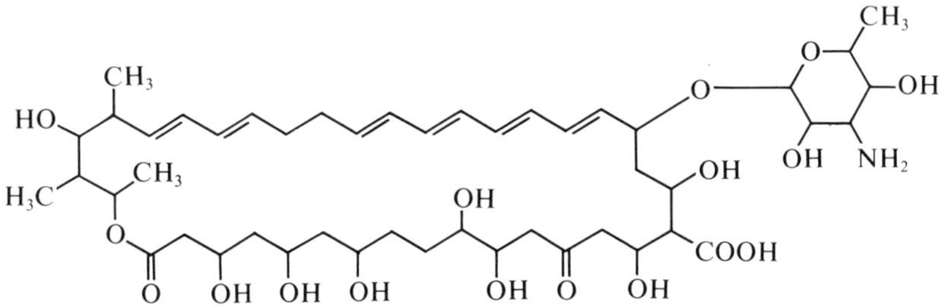

LXXXVI (Nystatin)

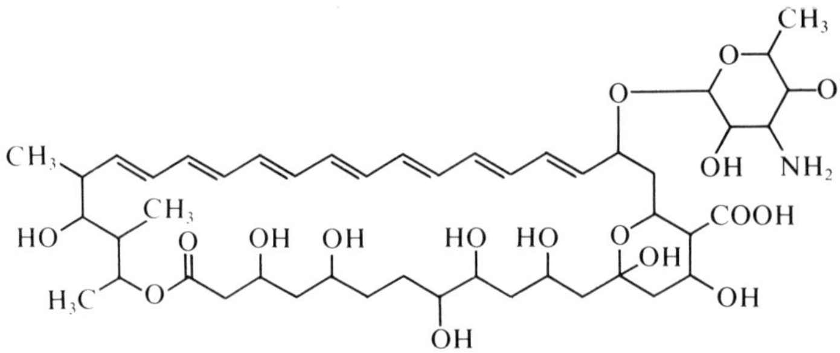

LXXXV (Amphotericin B)

Nystatin (LXXXVI), a closely related antibiotic to amphotericin B, has also been found to be useful in the treatment of kala-azar [251] and post kala-azar dermal leishmaniasis [252]. The toxicity of drug is similar to that of amphotericin B.

Rifampicin (LXXXVII), an orally active antibiotic, cures cutaneous leishmaniasis at higher doses [253–255]. In USSR, monomycin has been used to treat cutaneous leishmaniasis with good results at an intramuscular adult dose of 250,000 units given three times a day for 5–7 days [256, 257]. Some alkaloids such as berberine, emetine and dehydroemetine have also been shown to possess varying degrees of leishmanicidal activity but none of them are used in clinical medicine [108].

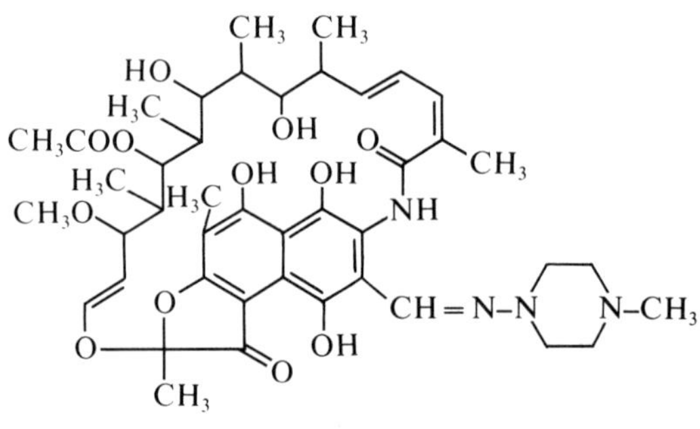

LXXXVII (Rifampicin)

4.5.5 Miscellaneous compounds

A number of 2-substituted-5-nitrofurans have been clinically evaluated against leishmaniasis. Nifurtimox (LXXXVIII, lampit) was shown to be effective in curing a cutaneous form of the disease [258].

In a follow-up study carried out for 1 year in patients with *L. brasiliensis* infection, the drug was found to give 46 % cures in cutaneous cases and only 15 % cures in cases with mucocutaneous leishmaniasis [259].
5-Fluorocytosine (LXXXIX, flucytosine) has been used to cure multiple chronic cutaneous lesions. The latter healed rapidly and the drug was also well tolerated [260].

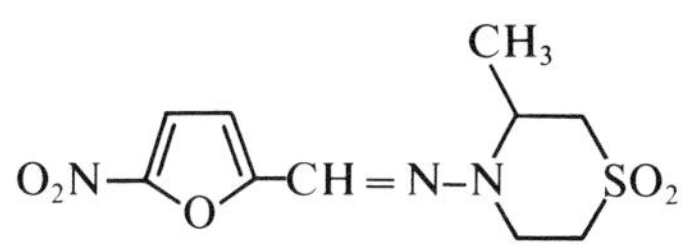

LXXXVIII (Nifurtimox)

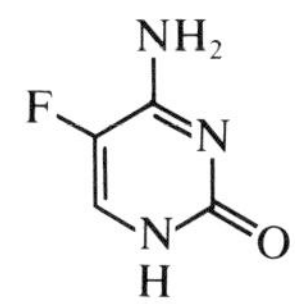

LXXXIX (Flucytosine)

4.5.6 Local treatment and topical formulations

In the Old World where cutaneous leishmaniasis is caused by *L. tropica* and *L. major,* it is usually advisable to give local treatment as it helps to heal the lesions spontaneously without metastasis. Accordingly a variety of physical treatments such as surgical excision [261], cryotherapy [262] and heat [263] have been used to heal the lesions. Local medical treatment using berberine, glucantime, tartar emetic and quinacrine also accelerates the healing [264].
In the case of New World cutaneous leishmaniasis caused by *L. brasiliensis,* local treatments are not advisable as this may make the parasites disseminate and cause mucosal disease. However, in such a situation combination of a local and systemic therapy may prove to be beneficial [265].
El-On and his colleagues [266, 267] have developed a paromomycin-based topical preparation to cure leishmaniasis. The topical application of paromomycin with methyl benzethonium chloride (MBCI) cured *L. major* and *L. mexicana* lesions in mice. In an uncontrolled clinical trial, application of an ointment consisting of 15 % paromomycin + 12 % MBCI for 10–30 days successfully cured over 100 patients infected with *L. tropica, L. major* and *L. aethiopica.* Similarly, a cream formed by 15 % paromomycin and 1 % MBCI has been used to cure cutaneous leishmaniasis [105]. Recently cutaneous leishmaniasis has also been treated using an ointment of ketoconazole and dimethyl-sulfoxide [268].

4.5.7 Liposome-drug preparations

The ability of liposomes, the phospholipid vesicles, to encapsulate drugs and deliver them to specific sites in the mammalian body, has been explored to enhance the activity and reduce the toxicity of antileishmanial drugs [269]. In the initial studies, antimonial drugs encapsulated in multilamellar visicles (MLVs) were shown to be 200–700-fold more active than the free drugs against *L. donovani* in rodents with 35–40 times improvement in the therapeutic index [270]. Similarly amphotericin B and WR-6026 showed 2–4.5 [271] and 11.4 [270] times more enhanced activity than the free drugs when delivered in liposomes. Although liposome-entrapped drugs have shown great promise in experimental *Leishmania* infections, the use of such preparations in clinical medicine has not been fully established.

5 African trypanosomiasis

Sleeping sickness or African trypanosomiasis is an acute and chronic protozoal disease caused by the invasion of the central nervous system by haemoflagellates of the genus *Trypanosoma (Trypanozoon).* The disease is transmitted to man by the bites of various species of tsetse flies belonging to the genus *Glossina.*

Two species of *Trypanosoma* are known to infest man. The infection caused by *T. brucei gambiense* is called Gambian sleeping sickness. The vector of this disease is *Glossina palpalis.* Trypanosomiasis caused by *T. brucei rhodesiense* is known as the Rhodesian sleeping sickness and is transmitted by the flies, *Glossina morsitans.* Both trypanosomes, *T. b. gambiense* and *T. b. rhodesiense*, are morphologically identical with *T. brucei* which is a parasite found in mice and usually does not infect man. The principal host of both the trypanosomes is man while antelops, sheep and goats serve as the reservoirs of the infection.

5.1 Geographical distribution

African trypanosomiasis is endemic throughout tropical Africa especially in areas heavily populated with different species of tsetse flies. The endemic foci of Gambian sleeping sickness are Zaire, Central

African Republic, Congo, Chad, Ghana, Ivory Coast, Mali, Upper Volta and Nigeria etc., while Rhodesian sleeping sickness is endemic in Tanzania, Mozambique, Rhodesia, Zambia and Uganda. It is estimated that 35 million people are at risk of contracting trypanosomiasis [272], while 25 million people are in fact infected with the trypanosomes and 65 000 patients die per year [9a].

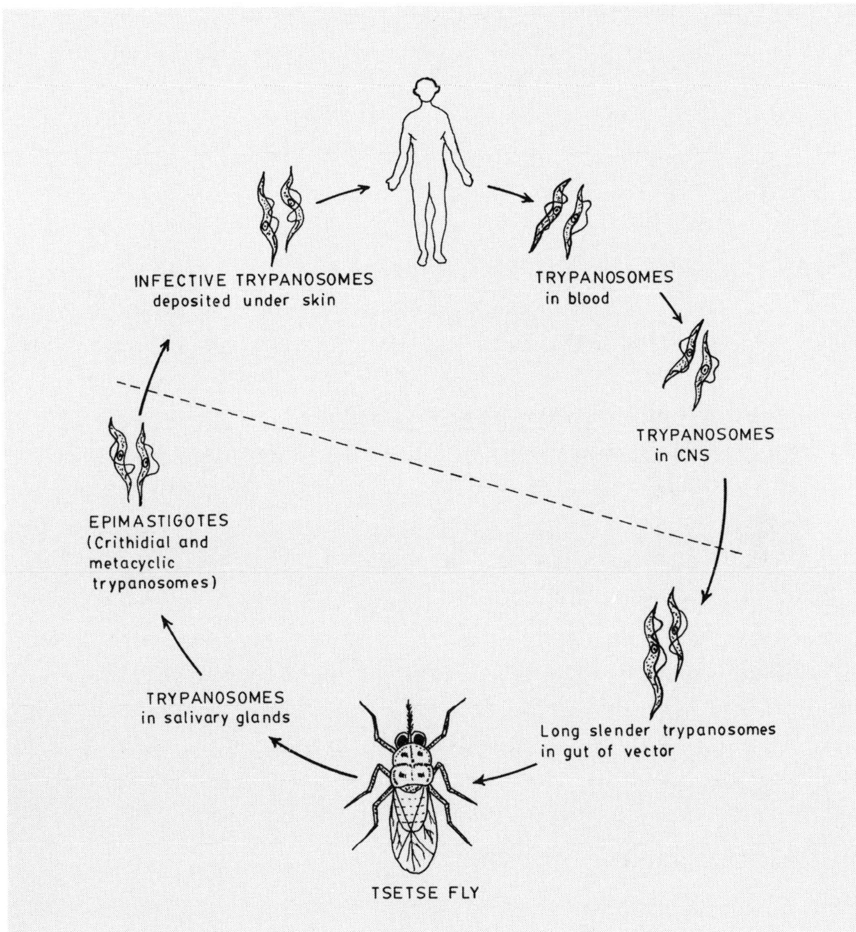

Figure 5
Life cycle of African trypanosomes

5.2 Epidemiology

The life cycle of *Trypanosoma* spp. starts when the infected tsetse fly bites man. The infective metacyclic parasites are deposited subcutaneously causing local irritation. Within 10 days a 3–4-cm sore (chancre) develops at this site. The trypanosomes then invade the blood circulation and multiply. This is the first stage of the disease, and the symptoms are fever, malaise and headache. The parasites eventually invade the CNS bringing on the secondary symptoms of the disease.

When the tsetse fly bites an infected person or the reservoir host, it takes up trypanosomes along with the blood meal. The trypanosomes picked up from man are usually long, slender and motile but some may be short and stumpy. The latter move to the midgut of the fly and multiply. The parasites then migrate to the salivary glands where the epimastigote or crithidial forms develop. The infective metacyclic forms of the trypanosomes are similar to that of stumpy trypanosomes observed in the patient's blood during heavy infections. Now the tsetse is ready to infect a fresh host (Fig. 5).

5.3 Clinical features

Both Gambian and Rhodesian sleeping sickness are characterized by 100 % mortality in untreated patients. The clinical characteristics of the disease may be divided into two stages. The early stage when the blood is invaded by trypanosomes, is marked by fever, headache, malaise and anaemia. The second stage of the disease in which the central nervous system is invaded by trypanosomes, is characterized by meningoencephalitis, meningomyelitis with cachexia. The patient becomes emaciated. In severe cases, the last phase of the disease is coma and death. It may be noted that the second stage involving CNS may not develop for months or even years in the case of Gambian sleeping sickness. However, in the case of Rhodesian sleeping sickness, the disease process is rapid and the patient may usually die before the last stage is observed.

5.4 Diagnosis

An early diagnosis of African trypanosomiasis is rather difficult especially in the mild Gambian form. The disease is easily mistaken for flu, fever or malaria. The conclusive diagnosis of trypanosomiasis may be done by microscopic identification of trypanosomes in blood samples, fluid aspirated from enlarged lymph nodes and cerebrospinal fluid (CSF). The presence of trypanosomes in CSF is indicative of the disease with CNS involvement.
Serological tests, based on detecting trypanosomal antibodies, have also been used to diagnose trypanosomiases. This may include complement fixation, immunofluorescence and ELISA methods.

5.5 Treatment of African trypanosomiases

The chemotherapy of African trypanosomiasis (sleeping sickness) has not made significant progress as the time-tested drugs still find use in clinical medicine [273, 274]. There are two types of drugs available for the treatment of sleeping sickness: (a) those active during the early stage of the disease (haemolymphatic stage) such as suramine and pentamidine, and (b) those active in the late stage of the disease involving CNS, e. g. arsenical drugs, tryparsamide and melarsoprol. The activity of the arsenicals on CNS is due to their ability to cross the blood-brain barrier. However arsenicals should only be used after assessing the stage of the infection by examining the cerebrospinal fluid (CSF) before treatment. If the CSF contains more than 5 cells/ml or shows an elevated protein level then treatment may be carried out using arsenical drugs.

5.5.1 Drugs effective in the early stage of sleeping sickness

(a) *Suramine* (XC): This is a highly active drug during the early stages of sleeping sickness caused by *T. b. gambiense* and *T. b. rhodesiense*. However, it has no effect on the later stage of the disease wherein the trypanosomes invade the CNS. This is due to the fact the suramine is unable to cross the blood-brain barrier.
Suramine shows low *in vitro* trypanocidal activity necessitating a high drug concentration (4 mM) and longer time (> 6 h). It also exhibits delayed *in vivo* trypanocidal action. The drug is known to bind

with blood proteins which can be detected for a long time. Thus, in man, after an intravenous dose of 4 × 1 g given for 3 days, the values of 150–340 mg/l and 5 mg/l of drug were detected after day 1 and 200, respectively [275]. The recommended adult dose of the drug is 1 g/adult given intravenously on days 1, 3, 7, 14 and 21. It is also possible to give the drug at a dose of 1 g, intravenously, each week for 4–6 weeks. However, before starting the treatment, a test dose of 100–200 mg must be given 24–48 h prior to first 1 g dose to test for its idiosyncrasy. The recommended pediatric dose of suramine is 20 mg/kg on days 1, 3, 7, 14 and 21 [25].

Suramine is not a very safe drug. Idiosyncrastic reactions occur in 0,1–0,3 % of the treated cases and include symptoms like nausea, vomiting, circulatory collapse and loss of consciousness [276]. Such reactions are more often found in patients with onchocerciasis [277]. A number of other side effects such as optic atrophy, blindness, nephrotoxicity, haemolytic anaemia, rash, pruritus and granulocytosis may also occur [276, 278, 279]. Since suramine is potentially nephrotoxic, it is advisable to check the urine of the patient at weekly intervals during drug treatment to check any possibility of renal damage.

(b) *Pentamidine* (LIX): The trypanocidal activity of this drug was demonstrated by Lourie and York [280]. The drug is less toxic than suramine but its activity and excretion profile is similar to that of suramine. At a concentration of 2.5 mg/ml, pentamidine shows high *in vitro* trypanocidal activity [281]. The usual adult and pediatric dose of pentamidine to treat the early stage of sleeping sickness is 4 mg/kg given intramuscularly daily for 10 days [25]. The drug also shows prophylactic action against *T. b. gambiense* in man for up to 6 months. The recommended prophylactic dose of pentamidine is 300 mg, given intramuscularly every 3–6 weeks [274]. Like suramine, pentamidine can also be detected in blood. The drug has been detected over months in the liver and kidney in mice and rats [282].

The frequently occurring side effects of pentamidine are gastrointestinal disturbances, vomiting and pain at the injection site. Nephrotoxicity [283], hypoglycemia and hyperglycemia [276] have also been reported. The use of pentamidine may also give rise to hypotension, blood dyscrasias, diabetes mellitus, liver damage, cardiotoxicity, acute pancreatitis and Herxheimer-type reaction [25].

(c) *Berenil* (XCI or LX): In Uganda, berenil has been used successfully to treat early stages of African trypanosomiasis. The drug was used at a dose of 5 mg/kg, given intramuscularly for 3 doses 2 days apart [284].

(d) *Difluoromethylornithine* (DFMO, XCII): α-Difluoromethyl-ornithine (DFMO), a potent inhibitor of polyamine biosynthesis, has been shown to be active against *T. b. brucei* and *T. b. rhodesiensis* in mice [285, 286]. In man, the drug is highly effective against both CNS and non-CNS infections with *T. b. gambiense*. The drug-resistant cases of *T. b. gambiense* infections have also been treated successfully using DMFO which is currently undergoing detailed clinical testing [287–289].

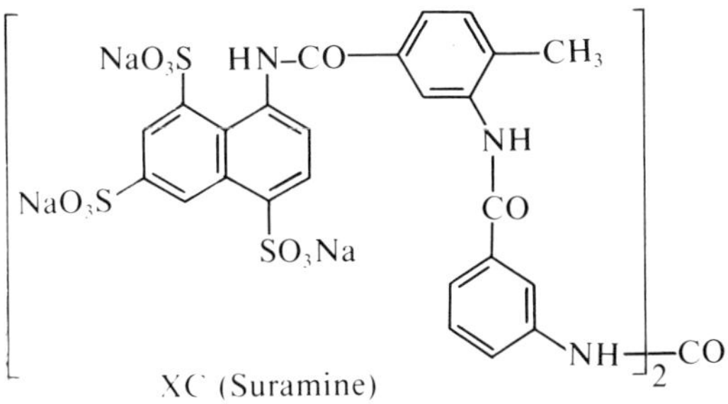

XC (Suramine)

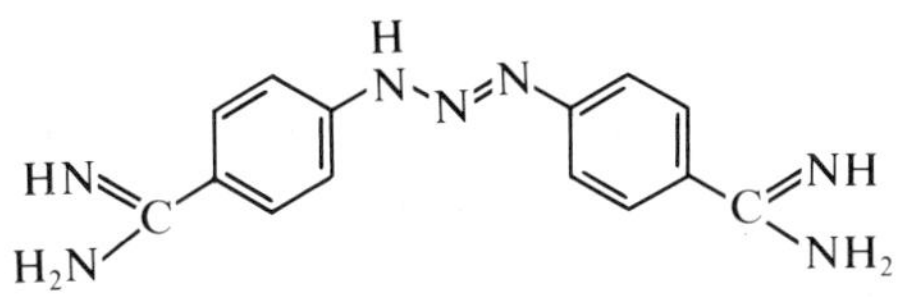

$\cdot (CH_3CONHCH_2COOH)_2$

XCI (Berenil)

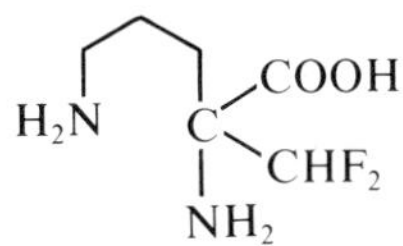

XCII (DFMO)

5.5.2 Drugs effective in the late stage of sleeping sickness

(a) *Tryparsamide* (XCIIa): This is a pentavalent arsenical first described by Jacobs and Heidelberger [290]. The drug shows no activity against *T. brucei in vitro* but is active *in vivo* at a dose of 200 mg/kg given subcutaneously or intravenously to mice. The trypanosomes disappear from blood only after 1–6 hours which is probably due to the fact that the drug requires some time to get converted into its active trivalent form. Delayed action of the drug has also been observed in rabbits. Following an intravenous injection of tryparsamide, 88–95 % of the circulating trypanosomes disappeared from blood within 1 hour and 99 % of the parasites were cleared in 3–4 hours [291].

In man, tryparsamide is used as a second line drug for treating African sleeping sickness. For treating late *T. b. gambiense* infections, the drug is given intravenously at a dose of 2–3 g/patient/week in 10–12 doses. However in the case of advanced *T. b. rhodesiense* infection involving the CNS, tryparsamide is given in combination with suramine [274]. Tryparsamide has also been recommended at a dose of 30 mg/kg given intravenously every 5 days for 12 injections. If required the regimen may be repeated after a month [25].

The use of tryparsamide is associated with frequent nausea and vomiting. Occasionally, the drug may also give rise to optic atrophy, blindness, loss of appetite, diarrhoea, renal damage, exfoliative dermatitis, allergic reaction and death [25, 274].

(b) *Melarsoprol* (Mel B, XCIII): Friedheim [292] introduced this trivalent arsenical in 1949. Unlike tryparsamide, melarsoprol is active *in vitro* and is also less toxic than the former drug. Since melarsoprol crosses the blood-brain barrier, it is the drug of choice for treating the meningoencephalitis stage of Rhodesian or Gambian trypanosomiasis.

The recommended adult dose of Mel B is 2–3.6 mg/kg given daily intravenously for 3 days. After one week a dose of 3.6 mg/kg/day, i.v. × 3 days is repeated. The dose may again be repeated after 10–21 days. For children, an initial dose of 0.36 mg/kg is given intravenously and then gradually increased to a maximum of 3.6 mg/kg at intervals of 1–5 days for a total of 9–10 injections. The total dose in a month should not exceed 18–25 mg/kg [25].

Although Mel B is less toxic than tryparsamide, the side effects of this drug may be serious. The drug is usually associated with myocardial damage, albuminuria, hypertension and colic. Jaundice, diarrhoea and conjunctival infections may also occur [25, 274]. According to Hawking [281], about 1 % of the Mel B treated patients may develop encephalopathy. Headache, tremor, fever, convulsions, coma and 5–10 % deaths may also be observed [277].

(c) *Melarsonyl potassium* (Mel W, XCIV): It is the water soluble derivative of Mel B which can be administered subcutaneously or intramuscularly. Although Mel W is less active and more toxic than

Mel B [293], it may be useful, particularly in children, where intravenous administration is not possible [294]. The usual dose of the drug is 3–4 mg/kg (with a maximum of 200 mg) given 3–4 days consecutively and repeated after 2 weeks [284].

(d) *Melarsen sodium* (XCV): This is a pentavalent arsenical which is almost as toxic as tryparsamide. The drug has been administered intravenously at a dose of 20 mg/kg given 5–7 days apart for 8–12 injections [284].

(e) *Nitrofurazone* (XCVI): The drug is particularly useful in treating patients with *T. b. gambiense* [295] and *T. b. rhodesiense* [296, 297] who have become resistant to arsenical therapy. Nitrofurazone is given orally at a dose of 0.5 g/adult 3–4 times a day for 5–7 days. The treatment may be repeated once or twice after an interval of 1 week [284].
Although nitrofurazone is currently the drug of choice for treating arsenical-resistant cases of sleeping sickness, its unpredictable efficacy and toxicity limits its use in clinical medicine [298]. The drug is associated with polyneuropathy, gastrointestinal disturbances, headache, weight loss, skin reaction and haemolytic anaemia linked to G6PD deficiency [24, 25, 298].

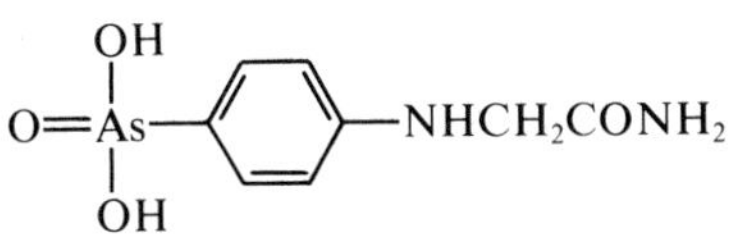

XCII a (Tryparsamide)

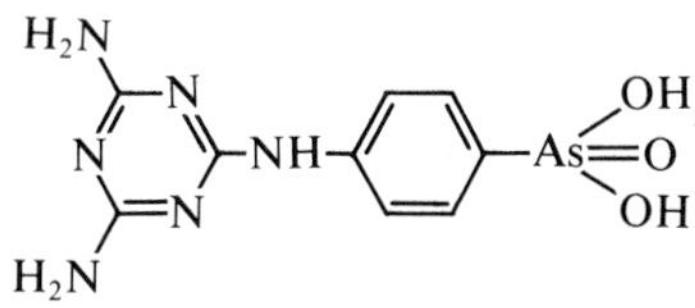

XCV (Melarsen sodium)

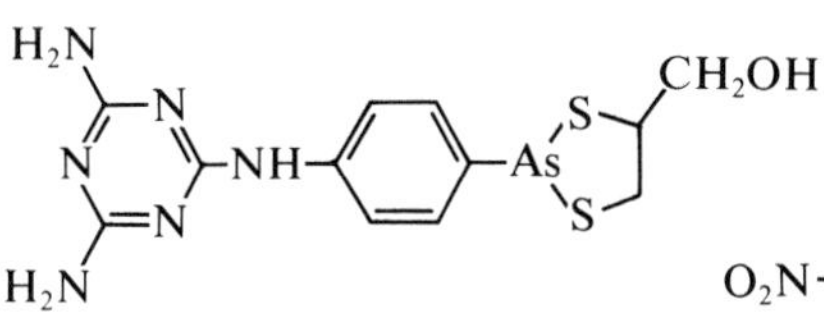

XCIII (Mel-B)

XCVI (Nitrofurazone)

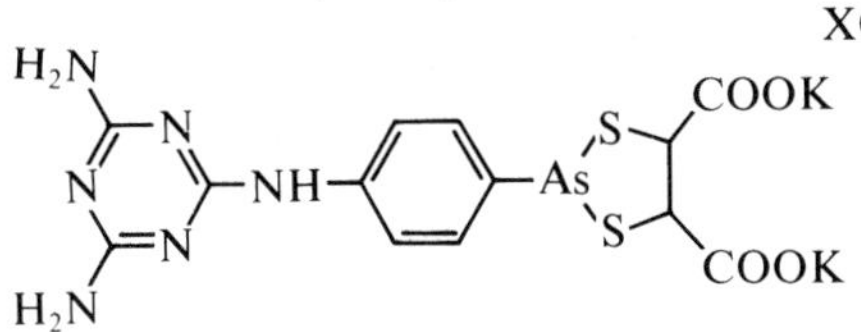

XCIV (Mel-W)

(f) *Drug combinations:* A number of synergistic drug combinations are currently being investigated to cure the late stage of African trypanosomiasis with CNS involvement. Thus, a combination of DFMO and bleomycin, an antibiotic with a glycopeptide backbone, has been found to be highly effective against *T. b. brucei* in mice with CNS involvement. The dose of bleomycin was 3.5–7 mg/kg daily for 3 days and for DFMO it was 2–4 g/kg/day for 14 days. It is interesting to note that neither bleomycin (dose: 7 mg/kg × 7 days) nor DFMO (dose: 4 g/kg × 14 days) was active when given singly to mice infected with *T. b. brucei* with CNS involvement [299].
A combination of DFMO and suramine at a dose of 4 g/kg daily for 7–14 days in water (DFMO) and 20 mg/kg, i.v. (suramine) has been found to be 90–100 % effective against *T. b. brucei* in mice with CNS involvement. Here also, neither of the drugs was effective when given alone [300].

(g) *Miscellaneous compounds:* Dann and coworkers [301–303] have carried out detailed structure-activity relationship studies in diamidines of some benzo-heterocycles. Among the several heterocycle amidines prepared, 2-(4-amidinophenyl)-6-amidinobenzothiophene (XCVIIa) and the corresponding indene analog (XCVIIb) were found to be more active than pentamidine or berenil against *T. b. gambiense*. Similarly compounds XCVIIc and XCVIII exhibited a high order of activity against *T. b. congoliense* and *T. b. rhodesiense*, respectively.
Das and coworkers [196, 304, 305] have also prepared a series of aromatic diamidines wherein the two 4-amidinophenyl moieties are linked through a heterocyclic bridge such as furan, thiophene, pyrrole, oxazole and thiadiazole. The most effective compounds of this series were 2,5-bis(4-amidinophenyl)-1,3-oxazole and 1,3,4-thiadiazole (XCIXa, b) showing a high order of activity against *T. b. rhodesiense* in mice at a dose of 3 mg/kg with cure rates comparable to stilbamidine, hydroxystilbamidine and pentamidine.

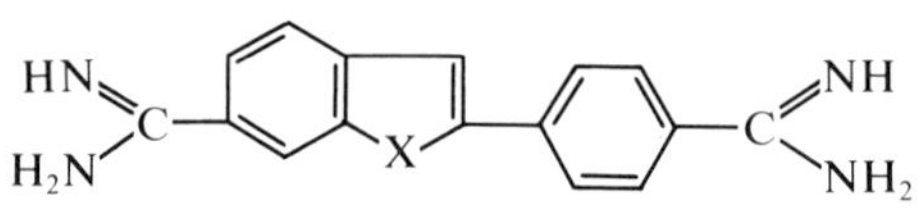

XCVIIa, X = S
b, X = CH_2
c, X = NH

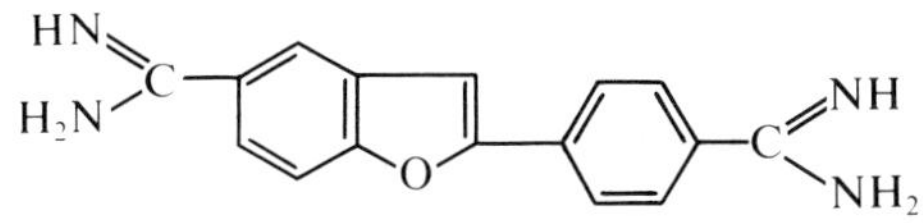

XCVIII

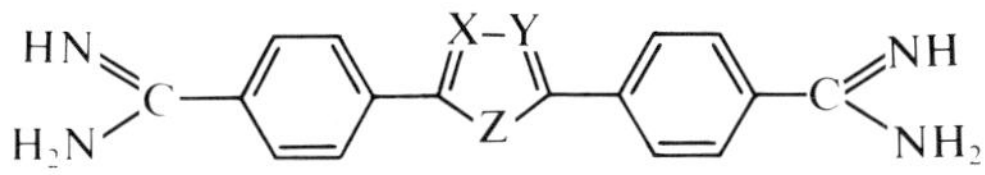

XCIXa, X = CH, Y = N, Z = O
b, X = Y = N, Z = S

Niridazole (LXXVI) has also been found to give permanent cures against *T. brucei* involving CNS in mice at a dose of 200 mg/kg given four times a day for 3 days [306].

6 American trypanosomiasis (Chagas' disease)

Chagas' disease is an acute and chronic disease caused by invasion of blood and tissues by the trypanosomes, *Trypanosoma cruzi*. The trypanosomes inhabit the blood while their leishmanial forms live in the tissues. The disease may occur in people of all ages; however children and young adults are the main victims. Transmission of Chagas' disease takes place through arthropod vectors, triatomine bugs or kissing bugs (*Triatoma gerstaeckeri, T. dimidiata* etc.).

6.1 Geographical distribution

Chagas' disease occurs in several parts of South America especially in Brazil, Argentina, Chile and Venezuela. The disease has also been reported in Central America. An early estimate by the World Health Organisation indicates that 7 million people were infected with *T. cruzi* while 30 million were exposed to the risk of acquiring the infection [307]. In Venezuela alone, there were 2.8 million cases of Chagas' disease [308]. It is now estimated that 35 million people are at risk of contracting this disease and 10 million carry the infection [272].

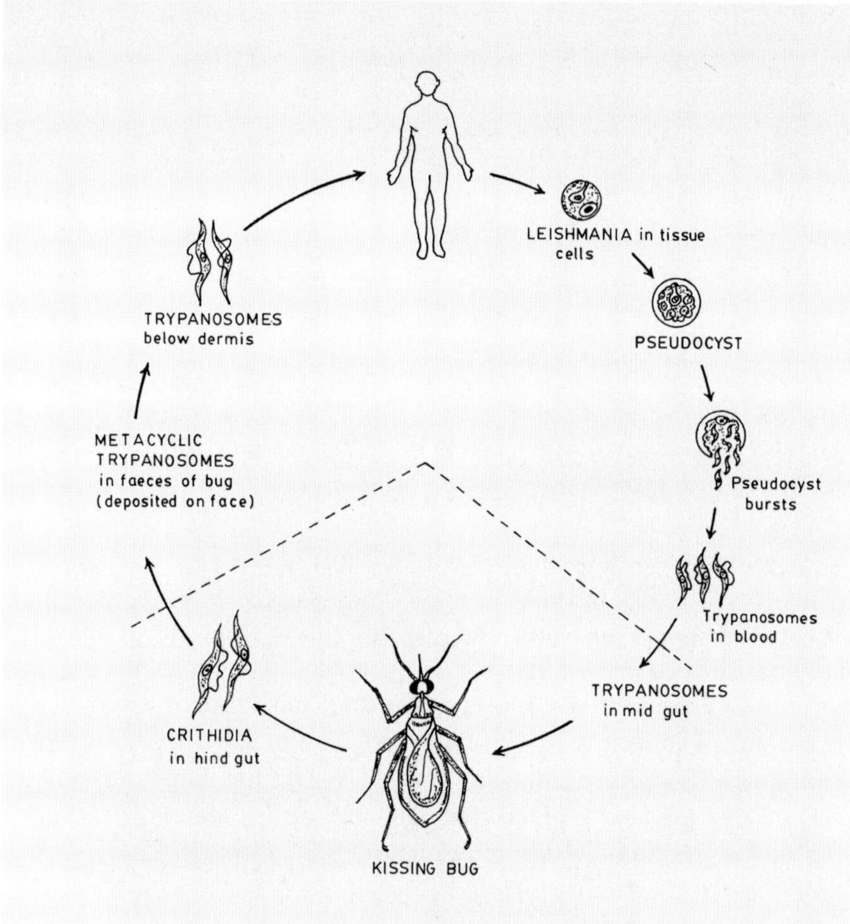

Figure 6
Life cycle of American trypanosomes

6.2 Epidemiology

The transmission of the *T. cruzi* infection is different from that of *T. b. gambiense* or *T. b. rhodesiense*. Here, the mode of transmission of the trypanosomes from vector to host is not 'salivarian' (as in African trypanosomiasis) but 'stercorarian'. The life cycle of *T. cruzi* starts when the infected kissing bug attacks the face of sleeping children at night. The bug then defecates releasing the metacyclic trypanosomes in the faeces. The trypanosomes may enter the body

via skin punctures, through bite wounds, mucous membranes of conjunctiva, skin abrasions or lips.
On entering the body of the host, the trypanosomes invade tissue cells, assume a leishmanial form, become nonmotile and multiply through binary fission till the cell is full of parasites (pseudocyst). The pseudocyst then bursts releasing motile trypanosomes. These circulate in the blood without further development but they may invade another tissue cell and the cycle is repeated. When the bug sucks blood from the patient, trypanosomes enter its midgut from where they move onto the hindgut and get transformed into crithidia and metacyclic trypanosomes that are capable of causing infection in man (Fig. 6).

6.3 Clinical characteristics

The acute phase of the disease, which is known to last 4–8 weeks in children, is associated with high fever, anorexia, vomiting, diarrhoea, enlargement of spleen and liver and ECG changes. Frequently there is unilateral conjunctivitis and oedema of the eyelids (oculoglandular complex). The infection slowly becomes sub-acute or chronic. Its symptoms vary depending on where the parasite is: in the heart, liver, spleen and other organs. The tissue damage continues for years till the disease reaches the chronic stage. The heart is the most common organ to be affected. In such a case, the heart is enlarged and the patient dies of heart failure. A common clinical complication of Chagas' disease is the dilation of tubular organs giving rise to spectacular distension of oesophagus (mega-oesophagus) and colon (megacolon) in the endemic areas [309, 310].

6.4 Diagnosis

Chagas' disease may be diagnosed by detecting trypanosomes in blood samples. Xenodiagnosis, in which a trypanosoma-free bug is allowed to feed on a supposedly infected patient and the faeces of the arthropod is then (2 weeks later) examined for crithidia or metacyclic trypanosomes, can also be used to detect the disease. Several immunodiagnostic tests such as IFA, Machado-Guerreiro test (a complement-fixation test using *T. cruzi* antigen) and intradermal tests have also been suggested. They all vary in degree of sensitivity.

6.5 Treatment of Chagas' disease

The chemotherapy of American trypanosomiasis poses several problems. Since the trypanosomes *T. cruzi* are intracellular parasites living in a variety of tissues, the activity of the circulating drug may depend on its ability to cross the vascular endothelmium as also the cell membranes. Furthermore, the radical cure of the disease would depend on the removal of the circulating trypomastigotes and the tissue amastigotes. The different strains of *T. cruzi* found in different parts of the world vary in tissue tropism, response to chemotherapy and biochemical parameters. This provides yet another bottleneck in evolving new chemotherapeutic agents. The status of chemotherapy and chemoprophylaxis of human trypanosomiasis has been reviewed by some authors [298, 311, 312].

(a) *Nifurtimox* (LXXXVIII or C): This is the drug of choice for treating Chagas' disease in man. The recommended adult dose of nifurtimox is 8–10 mg/kg given orally daily in 4 divided doses [25, 313]. Although drug treatment may be extended for 120 days, a 60-day course can also cure the disease [314]. For children, 1–10 years of age, the drug is given at a dose of 15–20 mg/kg orally in 4 divided doses daily for 90 days. Older children (11–16 years) receive the drug at a dose of 12.5–15 mg/kg/day in 4 divided doses for 90 days [25]. Nifurtimox is readily absorbed through the gastrointestinal tract and metabolized quickly. This leads to a peak plasma concentration within 1–3 hours which declines to zero in 24 hours [312].
Nifurtimox shows marked differences in its efficacy in different geographical areas. Although the drug appears to be active against both the acute and chronic stages of Chagas' disease, reports from Argentina, Chile and El Salvador [315–320] indicate that it is more active and better tolerated by patients in these areas than in Brazil [321–323].
Toxic reactions to nifurtimox occur frequently. Children appear to tolerate the drug better than adults. The frequent side effects of the drug are nausea, vomiting, diarrhoea, weight loss, loss of memory, sleep disorders, weakness and polyneuritis. Convulsions, fever, and joint pains may also be observed occasionally. Higher doses (15–20 mg/kg daily) may cause peripheral neuritis and psychosis. Haemolytic anaemia can also be seen in nifurtimox-treated patients espe-

cially in those with low levels of glucose 6-phosphate dehydrogenase (G6PD) [25, 312].

(b) *Benznidazole* (Cl): Following the observation of Pizzi [324] that metronidazole (LXXV) has suppressive action on *T. cruzi*, benznidazole was developed as an alternative drug in the treatment of *T. cruzi* infections in man. The recommended adult dose of the drug is 5–7 mg/kg given orally daily for 30–120 days [313]. Like nifurtimox, this drug is also rapidly absorbed through the gastrointestinal tract, distributed throughout the body tissues and excreted in urine.

Benznidazole produces several side effects such as gastrointestinal disturbances, weight loss, skin rashes and peripheral neuritis. The drug also causes light-sensitive erythematous skin rashes in 50 % of the treated patients [325]. It may also cause marked thrombocytopenia in man and can depress thymus-dependent immune functions in rabbits [326]. Both drugs, nifurtimox and benznidazole, have also been found to be carcinogenic in animals [312].

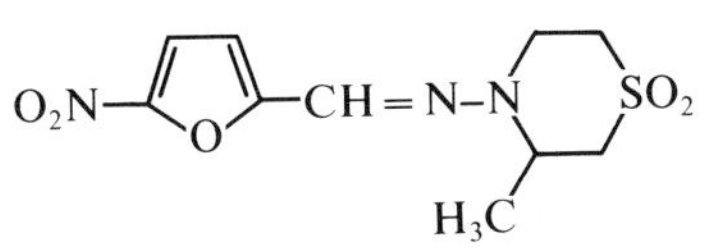

C (Nifurtimox)

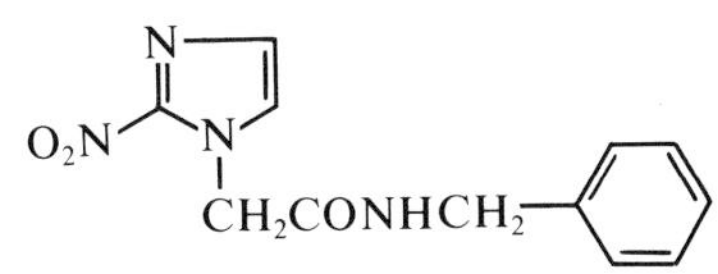

CI (Benznidazole)

(c) *New compounds:* Recently, 2-amino-5-(1-methyl-5-nitro-2-imidazolyl)-1,3,4-thiadiazole (C11) has been found to possess high activity against *T. cruzi* in mice. The drug cures 89 % of the cases in a single dose. Some strains resistant to nifurtimox have also been treated with this agent [327]. It eliminates all the trypomastigotes from blood within 6 hours and the amastigotes are destroyed within 18–36 hours of treatment [328].

Andrade and co-workers [329] have shown that MK-436 (2,5-dinitroimidazole, CIII) can bring about 100 % parasitological cures in mice infected with *T. cruzi* at a dose of 250 mg/kg, given twice daily for 10 days or for 60 days.

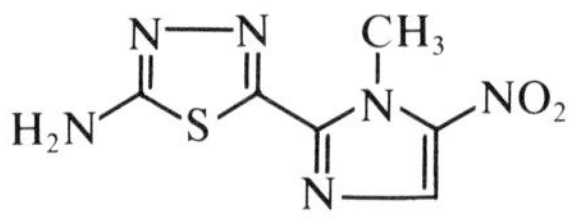

CII

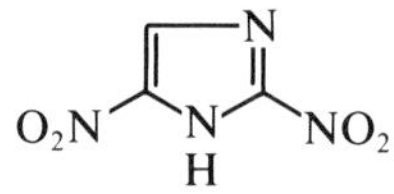

CIII

7 Filariasis

Filariasis, a major communicable disease of the tropics, is caused by the invasion of the human body by a number of elongated and thread-like nematodes called filarial worms. The filarial worms infecting man are *Wuchereria bancrofti, Brugia malayi, B. timori, Onchocerca volvulus, Loa loa, Dipetalonema perstans, Dipetalonema streptocerca* and *Mansonella ozzardi* while those parasitizing animals include *Litomosoides carinii* (cotton rats), *Brugia pahangi* (jirds, dogs), *Brugia malayi (Mastomys,* cats), *Dipetalonema (Acanthocheilonema) viteae* (rodents), *D. repens* (dogs), *Dirofilaria immitis* and *D. repens* (dogs), *Onchocerca reticulata* (horses), *O. gibsoni* (cattle) and *O. cervicalis* (ponies, horses) which are used as primary and secondary test models in experimental chemotherapy of filariasis [330, 331].

The adult filarial worms live in the connective and subcutaneous tissues, lymphatic vessels, lymph nodes, serous cavities and other organs of the host's body, while the microfilariae, produced by the adult female worms, live in the bloodstream and subcutaneous tissues. The filiarial infection is transmitted to man by haematophagus arthropods, usually mosquitoes and flies which also serve as intermediate hosts in the life cycle of the parasites.

7.1 Geographical distribution

Filariasis is one of the most widespread helminth diseases affecting nearly 400 million people in the tropical and subtropical regions of the world [332]. The infection caused by *W. bancrofti* is the most widely distributed form of filariasis. It is prevalent in several countries of Asia, Africa and South America. In India alone, nearly 307 million people are exposed to the risk of *W. bancrofti* (major) and *B. malayi* (minor) infections. There were 22 million microfilariaemic cases reported and nearly 16 million people showed clinical symptoms [333]. *B. malayi* infections have been also reported from other parts of Asia while *O. volvulus* is distributed in several areas of Africa and South America. *Loa loa* infection is prevalent only in the rain forest areas of Western and Central Africa extending roughly from latitude 10° N to latitude 5° S below the equator. Dipetalonemiasis caused by *D. streptocerca* and *D. perstans* has been reported to occur in various parts of Africa and South America. The *M. ozzardi* infections are only prevalent in Central and South America.

7.2 Epidemiology

The life cycle of the human filarial worms involves two hosts. A number of animals and man serve as the definite hosts in which the adult worms and microfilarial live. The intermediate hosts of the filariids are various blood sucking insects such as mosquitoes and flies which transmit the disease while feeding on the blood of man.
The adult female worms produce a large number of microfilariae (larval stage 1 or L_1 stage) which are either sheathed *(W. bancrofti, B. malayi, L. loa)* or unsheathed *(O. volvulus, D. perstans, M. ozzardi)*. After being liberated by the female worms, these L_1 microfilariae either remain in the lymphatics, body cavities and mesenteries, or reach the bloodstream or subcutaneous tissues depending upon the kind of filarial parasites. The L_1 microfilariae are ingested by a suitable insect vector as it feeds on the blood of the infected person. On reaching the body of the insect, the sheathed larvae get exsheathed and then penetrate the stomach wall and move to the thoracic muscles where they undergo morphological changes, enter into a third larval stage (L_3), and become infective to man. These infective larvae live in the buccal parts of the insects and are passed in to the human body when the insect bites. After entering the cutaneous tissues, the larvae make their way to the lymphatics where they slowly mature into adult male and female worms (Fig. 7).

7.3 Clinical characteristics

The pathological changes and clinical manifestations depend greatly on the nature of the disease.
Lymphatic filariasis: The pathological changes resulting from *W. bancrofti* and *B. malayi* infections happen several months after

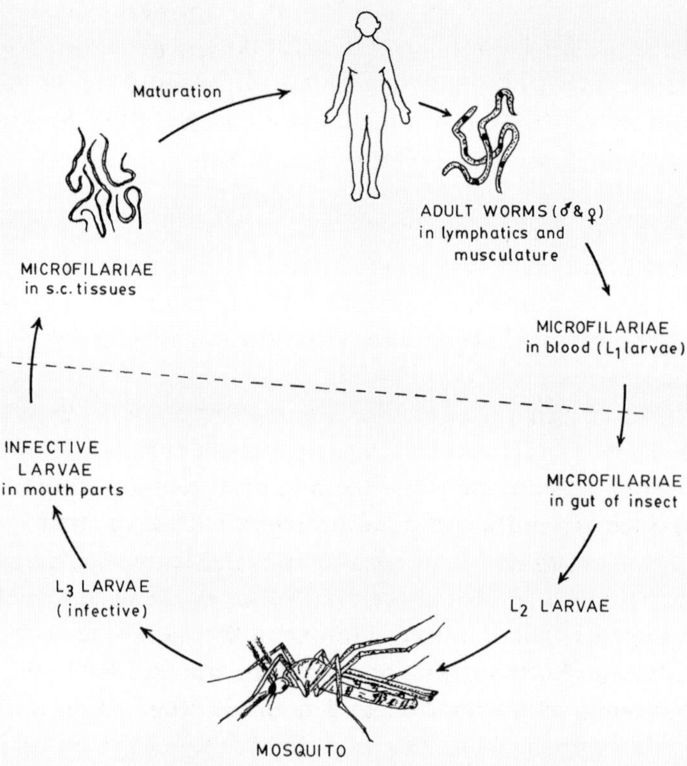

Figure 7
Life cycle of filarial worm

the insects have bitten. Disease usually develops in the lymphatic vessels, glands and tissues around the developing worms. In the acute stage, the area around the filarial worms is inflamed and marked by the presence of polymorphonuclear cells (with eosinophils predominating) as well as oedema and proliferation of reticuloendothelial cells. Later the lymphatic vessels and nodes become enlarged. This in turn leads to fibrosis and progressive obstruction of the lymphatics in different parts of the body such as abdominal cavity, external genitals, mammary glands, arms and legs.

Lymphatic filariasis has three phases – (a) inflammatory phase: high fever with chills resembling malaria, pain and swelling in testes, enlargement of lymph nodes and thickening of spermatic cord; (b) obstructive phase: blockade of lymph circulation. This leads to the enlargement of the scrotum (hydrocele) and the appearance of fatty chyle in urine (chyluria) and (c) elephantiasis: massive enlargement of legs, arms, scrotum and breasts.

River blindness: The bite of *Simulium* (black gnat) injects *O. volvulus* into the skin which may be either asymptomatic or symptomatic depending upon the host's response to the infection. The early clinical symptoms of the infection include pruriginous or oedematous growth on the skin followed by rash and fever. The main clinical features of onchocerciasis may be classified as (a) onchodermatitis, characterized by skin rash, pruritis, scratching, thickening or depigmentation of skin; (b) onchocercoma, marked by the appearance of a number of nodules at the pressure points of different parts of the body; (c) onchophthalmia, a condition which leads to the loss of vision and even blindness in a large numer of patients infected with *O. volvulus;* this is usually called river blindness or ocular onchocerciasis. In the early stage of river blindness, the patient feels the presence of a foreign body in his/her eyes with pain, photophobia, lacrimation, oedema of the eyelids and then chronic conjunctivitis. A series of ocular problems follow and finally the loss of vision occurs due to the presence of filariids in the eyeball; (d) onchoadenitis, characterized by painless growth of lymphatic glands on limbs and groin parts.

Loasis: The main pathological change produced by *Loa loa* infection is the inflammation of subcutaneous tissues due to the migration of adult worms through the groin, breasts, penis, eyelids and eyeballs. The clinical manifestations of loasis are the presence of painful swellings (Calabar swellings) on face, limbs, hands and wrists, fever, eye problems including blindness and nervous disorders caused by the migration of adult worms in specific parts of the body.

M. ozzardi and *Dipetalonema filariasis:* These filariids are usually nonpathogenic, however some pathological changes may be observed. *M. ozzardi* may occasionally be associated with hydrocele and the enlargement of lymph nodes while *D. perstans* may cause vertigo, oedema of the scrotum, abdominal and limb pain, itching, enlargement of the spleen and liver, fever and high eosinophil counts. Infections due to *D. streptocerca* may cause pruritis and skin rashes.

7.4 Diagnosis

(a) *Lymphatic filariasis:* Ways of diagnosing lymphatic filariasis have been reviewed [334]. The classical method involves a parasitological test which enables the detection of microfilariae in peripheral blood. Since microfilaraemia is nocturnal in most of the lympathic filariasis, the blood needed for the examination should be taken at midnight. Usually 20 mm^3 of blood is taken and examined for microfilariae by preparing a thick blood smear. It is also possible to provoke a transitory emergence of microfilariae in peripheral blood during the daytime by giving 100 mg of diethylcarbamazine to the patient [335]. The blood is taken after an hour and examined for microfilariae. However, this test cannot be used in the case of onchocerciasis because it could induce severe reactions (Mazzoti reaction).
Other parasitological tests used in diagnosing lymphatic filariasis are the counting chamber method [336a] and the membrane filtration technique [336b] which have been found to be more reliable than the blood smear or DEC provokation methods.
The fact that most of these parasitological methods suffer from several limitations, such as poor sensitivity and unsuitability in field trials, has led scientists to develop various immunological tests with varying degrees of success. All the immunodiagnostic tests involve the detection of filarial antibodies or antigens in body fluids such as serum and urine.
For antibody assay, complement fixation [337a] and gel diffusion [337b] tests have been used with a low order of reactivity. Other immunodiagnostic tests based on antibody assay are indirect haemagglutination (IHA), indirect immunofluorescent (IIF), radioimmunoassay (RIA), enzyme-linked immunosorbent assay (ELISA) and immediate hypersensitivity reaction (intradermal test or filarial skin test known as FST) [334]. Of these, ELISA [338a] and FST [338b] use filarial antigens and show promise in the diagnosis of lymphatic filariasis.
The detection of antigens in body fluids has recently been recognized to provide a reliable method for the diagnosis of active infections [339]. However the test may pose some complications because

the antigens are generally complexed with antibodies or the antigenic determinant sites are masked. This may lead to poor sensitivity. However, use of monoclonal antibodies are expected to give a high order of sensitivity.

(b) *Onchocerciasis:* The parasitological method involves demonstration of microfilariae in the skin or nodules [340]. This is a reliable method and can be done by skin biopsy (skin-snip examination or observation of aspirates from subcutaneous nodules) or examination of the eyes (direct examination of anterior eye chamber with a slit lamp or conjunctival biopsy) when the eyes are involved.

Several immunodiagnostic tests (intradermal or skin sensitivity tests) have been used to diagnose onchocerciasis using lipid-free antigens prepared from different filariae such as *Dirofilaria immitis* [341]. Among these tests are the complement fixation test, the soluble antigen fluorescent antibody test, the Prausnitz-Kustner test and the Mazzoti test (administer 50 mg of DEC and observe pruritic skin reaction within 24 hours).

Researchers have also tried to assay the circulating antigens in the hope of developing an ideal immunodiagnostic method to detect onchocerciasis [342].

(c) *Loasis:* This disease can be recognized by (1) the presence of calabar swellings; (2) an edematous outline on skin showing the presence of a worm under the skin; (3) high eosinophil counts and (4) demonstration of microfilariae in the peripheral blood by the blood smear method or concentration methods [1].

7.5 Treatment of filariasis

The chemotherapy of filariasis has been recently reviewed [343, 344]. This section will, therefore, describe the advances made in the treatment of lymphatic filariasis, onchocerciasis and loasis in man.

7.5.1 Organoantimonials and arsenicals

Following the demonstration of antifilarial activity exhibited by tartar emetic (L) against *W. bancrofti* in man [345], a large number of organoantimonials were evaluated for their filaricidal efficacy. Some of the compounds were found to kill both the microfilariae and adult worms of a few filarial parasites in different hosts including man

[346–355]. The most effective antimonials of this class are neostibosan (a complex mixture of 4-aminobenzenestibonic acid and 4-acetylaminobenzenestibonic acid), stibocaptate (CIV), stibophen (CV) and MSbB (CVI). However, none of them have found use in clinical practice due to their high toxicity, low therapeutic indices and various side effects.

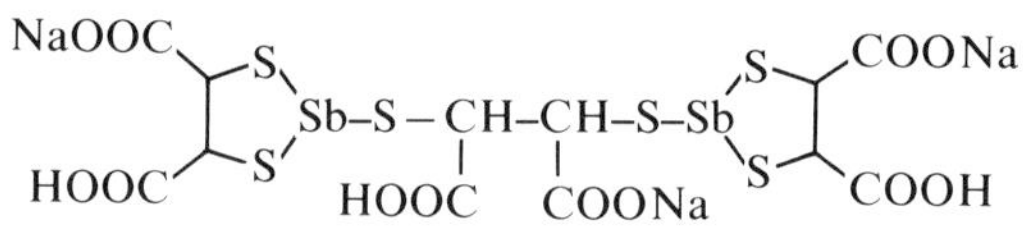

CIV (Stibocaptate)

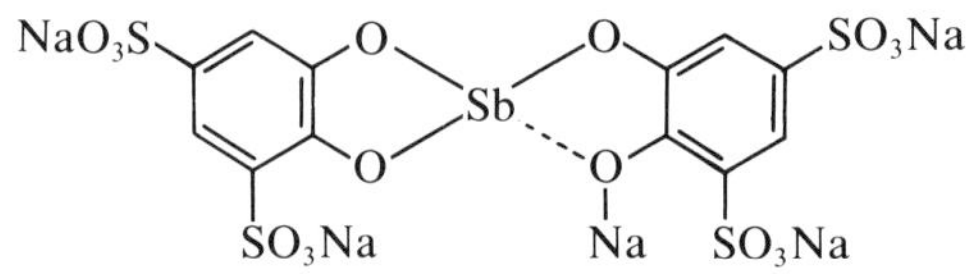

CV (Stibophen)

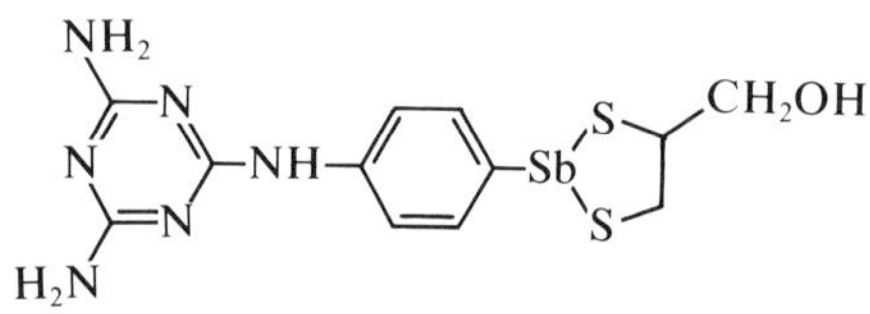

CVI (MSb)

Phenylarsenoxide (Ph–As=O) was the first member of the latter class that was shown to possess *in vitro* and *in vivo* antifilarial activities [350]. This prompted the search for more effective arsenicals which led to the discovery of arsenamide (CVII) and melarsen derivatives Mel-W (XCIV) and F-157 (CVIII).

Arsenamide kills both the microfilariae and adult worms of *W. bancrofti* at an intravenous dose of 1 mg/kg given for 15 days [356, 357] and is also effective against *O. volvulus* [358] and *D. perstans* [356] in man.

Mel-W shows high activity against the microfilariae and adult worms of *W. bancrofti* at an intramuscular dose of 10 mg/kg [359]. The drug is less toxic and, therefore, it has been used to control human filariasis in the Pacific and New Guinea [359, 360]. Mel-W is also effective against river blindness in man [359]. Although the drug exhibits high antifilarial activity in man with the least toxicity, it is not used in modern clinical practice due to the occurrence of encephalopathy in a certain percentage of treated cases [361].

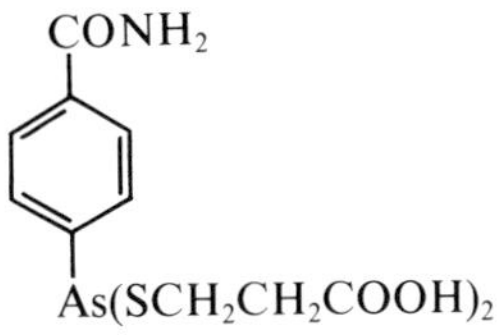

CVII (Arsenamide)

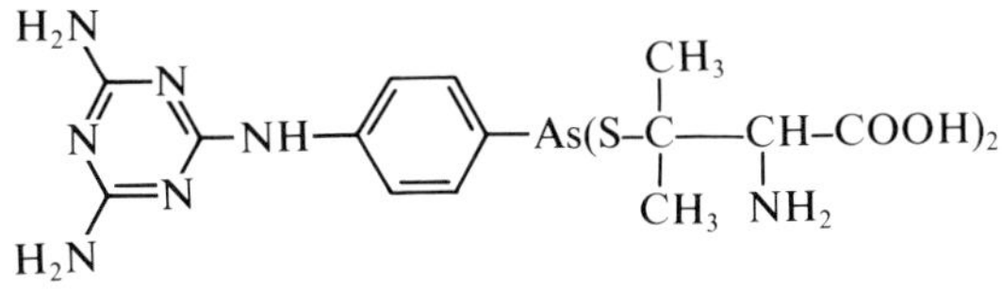

CVIII (F–157)

7.5.2 Organophosphorus compounds

Two compounds of this class, metrifonate (CIX) and fenthione (CX), have been studies in detail in the chemotherapy of filariasis. Of these, metrifonate shows the most promise in the treatment of onchocerciasis in man.

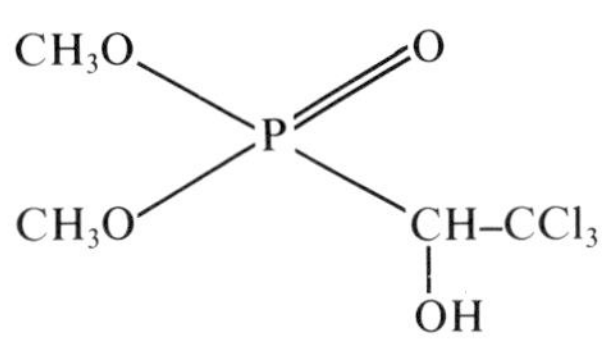

CIX (Metrifonate)

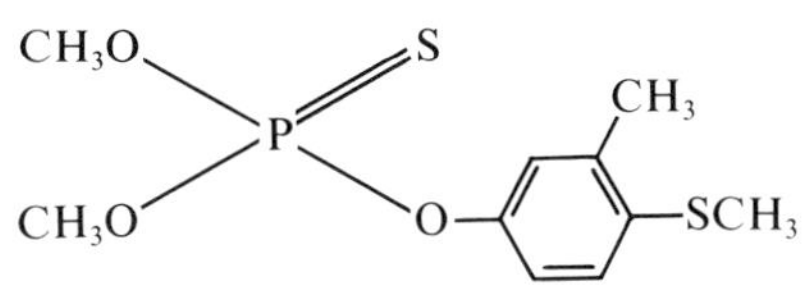

CX (Fenthione)

A dose of 7.5–15 mg/kg of CIX has been found to cure *O. volvulus* infection in man [362–365]. However, metrifonate produces a less intense Mazzoti reaction than does diethylcarbamazine, as the latter destroys the microfilariae [366, 367].

7.5.3 Suramine (XC)

This drug was originally developed as a human trypanosomicidal agent by Bayer laboratories in 1920. Later it was found that suramine also possesses activity against *O. volvulus* in man [368]. Since then it has been extensively used to treat river blindness in man [358, 369–373]. The drug kills the adult worms of *O. volvulus*. However its action on the female adult worms is specific. The female worms die after 4–5 weeks of treatment, the microfilariae and adult male worms are killed rather slowly. The drug has limited or no action on *W. bancrofti, Loa loa* and *D. perstans* in man.

The usual adult dose of suramine is 1 g/patient given weekly by intravenous injections for 5–10 weeks. Before starting the treatment, a test dose of 100–200 mg should be administered intravenously to avoid unpredictable toxicity and deaths associated with suramine treatment. Although the drug is highly effective against *O. volvulus* in man, it should be used only under medical supervision and only in situations where other drugs (such as ivermectin, diethylcarbamazine) have failed to respond to the disease. The side effects of suramine are joint pains, fever, acute pain in the knees and feet, as well as erythematous and pruritic eruptions due to allergic reactions caused by proteins liberated by decaying worms. A later phase of treatment may result in kidney damage and occasional exfoliative dermatitis.

7.5.4 Piperazine derivatives

The Lederle laboratories of American Cyanamid Co. have carried out the synthesis of a large number of 1,4-disubstituted piperazines possessing marked antifilarial activity against *Litomosoides carinii* in cotton rats [374]. The most active compounds of the series were 1-carbethoxy-4-methylpiperazine (CXI) [375], 1-dimethylcarbamoyl-4-methylpiperazine (CXII) and 1-diethylcarbamoyl-4-methylpiperazine (CXIII, DEC, diethylcarbamazine) [376–378] of which the last one is used as the drug of choice for treating filariasis in man [379].

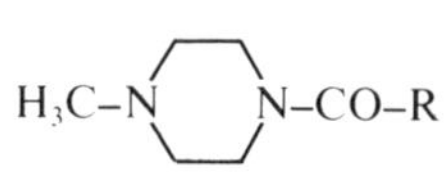

CXI, R = OEt
CXII, R = NMe_2
CXIII, R = NEt_2 (Diethylcarbamazine)

DEC (CXIII) has been used as the drug of choice in the treatment of lymphatic filariasis in man for nearly four decades. The antifilarial profile of this drug may be described as follows:

(a) *Lymphatic filariasis:* DEC is used in the treatment of lymphatic filariasis caused by *W. bancrofti* and *B. malayi* in man. The drug is given at a dose of 4–6 mg/kg (citrate) or 2–3 mg/kg (base) orally, three times a day, preferably after meals for 2–3 weeks. This treatment kills both the microfilariae and adult worms of *W. bancrofti* and *B. malayi*. However, for complete eradication of the microfilariae and adult worms of *W. bancrofti,* a total dose of 72 mg/kg should be given as these worms are less susceptible to DEC treatment than *B. malayi* [380]. If required, the treatment (2–3 mg/kg, t. i. d. × 10–21 days) may be repeated [380, 381, 382].

DEC has been used with promising results in the mass treatment of lymphatic filariasis [383–386]. The drug has been used to control bancroftian filariasis in Malaysia [386] as well as the South Pacific and French Oceania [387]. The use of the drug mixed with table salt has also been suggested for mass treatment programmes [388–391].

There has been a controversy about the ability of DEC to kill adult worms of *W. bancrofti* or *B. malayi*. Ottesen [392] has reviewed this matter and concludes that a full-course treatment with DEC is definitely macrofilaricidal.

(b) *River blindness:* DEC shows rapid action on the microfilariae but has no action on the adult worms of *O. volvulus*. The destruction of microfilariae greatly increases skin pathology. This observation was first made by Mazzoti [393] and, hence it is now known as the Mazzoti reaction. The Mazzoti reaction is so intense and specific that it has been used to diagnose onchocerciasis in those subjects in which the disease is suspected but no microfilariae detected.

The recommended adult dose of DEC citrate is 50 mg on the first day and thereafter 100 mg twice daily for 5 days [394].

More recently the following dose schedule has been recommended: 25 mg/day for 3 days, then 50 mg/day for 5 days and thereafter 100 mg daily for 3 days and then 150 mg daily for 12 days [25]. The risk of side effects may be reduced effectively by starting the treatment with 25 mg of DEC citrate, as indicated above, and then increasing the dose slowly up to 200 mg given for several days [395].

Further work on the clinical efficacy of DEC has established the microfilaricidal action of the drug against *O. volvulus* in man [396–398]. DEC has also been used in some topical preparations and shown to provide considerable relief to patients suffering from ocular onchocerciasis [399].

The use of DEC in onchocerca patient may induce acute host-tissue response due to the destruction of microfilariae. This is manifested by severe itching, skin rashes, fever, headache, joint pains, lymphadenopathy and malaise. There is a chance of collapse and even death [400, 401].

(c) *Loasis:* DEC is the drug of choice in the treatment of *L. loa* infection in man. The drug kills both the microfilariae and adult worms [402]. The recommended dose of DEC is 2–5 mg/kg given orally for 14–21 days [371, 403]. However, care must be taken especially when treating patients with severe microfilaremia, as meningoencephalitis may occur in addition to other side effects. The side effects may be reduced by treating the patient simultaneously with corticosteroids during the first 7–10 days [371]. DEC has also been used as a chemoprophylactic agent at a dose of 5 mg/kg given orally in 2 divided doses for 3 days each month [404].

(d) *Other filarial infections:* DEC has been found to be ineffective against *M. ozzardi* in man [405]. The drug is also less active against the microfilariae and adult worms of *D. perstans* than *D. streptocerca* which responds well to a dose of 2.9 mg/kg daily [1, 406].

7.5.5 Limitations of DEC

The use of DEC is associated with several side effects. In the case of *W. bancrofti* and *B. malayi* infections, fever is the most common reaction. Severe allergic or febrile reactions due to the rapid death of microfilariae, and gastrointestinal disturbances may also occur. Some patients may experience localized swelling and inflammation. Patients infected with *O. volvulus* may show fever as high as 39 °C, headache, itching, skin rashes, desquamation, joint pains, lymphadenopathy, malaise and respiratory distress. Care must be taken while treating patients with heavy infections of *L. loa* as the drug may provoke ocular problems or encephalopathy [25].
Several cases of collapse and death of patients infected with *O. volvulus* [407, 408] and *W. bancrofti* (one case, reason obscure) [409] have been reported during DEC treatment. A good number of encephalitis cases in *L. loa* patients have also been reported [410]. This is due to the presence of microfilariae in the nervous system and cere-

brospinal fluid. The patient first falls into a coma and then dies. A loss of vision may also be observed in *O. volvulus* patients being treated with DEC [411]. These limitations may make DEC often unsuitable for mass treatment of not only onchocerciasis [412] but other filarial infections as well.

7.5.6 Structural analogues of DEC

An array of systematic structural modifications have been carried out in the DEC molecule leading to the synthesis of a large variety of 1,4-disubstituted piperazines, piperazine-heterocycles and related compounds of which only a few exhibit better antifilarial activity than the parent drug [374]. The most effective structural congeners of DEC are 1-methyl-4-piperidin-1-ylcarbonylpiperazine (HOE-28637, CXIV) and 1-methyl-4-morpholin-1-ylcarbonylpiperazine (HOE-26961, CXV) prepared by Hoechst laboratories [413, 414] and 3-ethyl-8-methyl-1,3,8-triazabicyclo[4,4,0]decan-2-one (centperazine, CXVI) [415, 416], 1-isobutoxycarbonyl-4-methylpiperazine (CXVII) [417] and 1-methyl-4-pyrrolidin-1-ylcarbonylpiperazine (CDRI 72–70, CXVIII) [418] prepared at the author's laboratory.

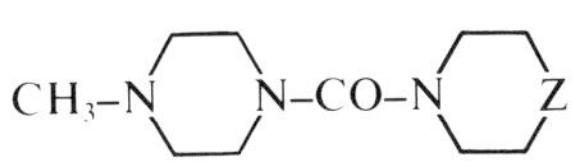

CXIV, Z=CH2
CXV, Z=O

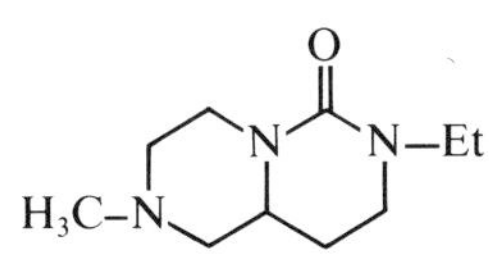

CXVI

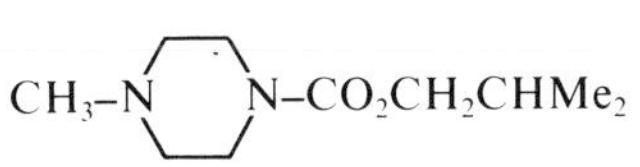

CXVII

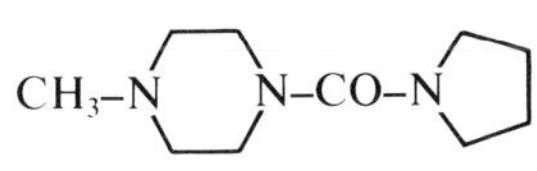

CXVIII

Both the compounds CXVII and CXVIII cause more than 90 % elimination of microfilariae of *L. carinii* in cotton rats at an intraperitoneal dose of 3 mg (base)/kg given for 5 days [417, 418]. Compound CXVII also kills 100 % of the microfilariae and adult worms of *D. viteae* in *Mastomys natalensis* at a subcutaneous dose of 50 mg/kg × 5 days [417]. Centperazine (CXVI) shows more than 90 % clearance of microfilaremia at an intraperitoneal dose of 1 mg/kg given to cotton rats infected with *L. carinii* [416]. The drug is also effective against the microfilariae of *B. malayi* in gerbils [419] and *W. bancrofti* in man [420]. Currently, centperazine is in phase II clinical trials in this institute.

7.5.7 Levamisole (LXXVIII)

Levamisole, a preferred drug for the treatment of various intestinal nematode infections in man and animals [421], has also been found to display antifilarial activity in different hosts including man. The drug has been used to treat *W. bancrofti* infection in man at a dose of 3–6 mg/kg given for 3–8 days [422–424]. In Rhodesia, four patients, infected with *D. perstans* like filarial infections, were given mebendazole (CXX) (100 mg, three times daily after meals) and levamisole (100 mg twice daily) orally for 10 days. The microfilariae disappeared from the bloodstream of three patients [425].

7.5.8 Benzimidazoles

Of a large number of 2,5-disubstituted benzimidazoles showing marked activity against intestinal nematode and cestode infections in man and domestic animals [426, 427], some have been evaluated in the chemotherapy of filariasis. Thiabendazole (CXIX) has been found to be inactive against *O. volvulus, L. loa* and *D. perstans* infections [428]; however, mebendazole (CXX), fenbendazole (CXXI) and flubendazole (CXXII) exhibit potent filaricidal activity.
Mebendazole has been evaluated against *B. pahangi, O. volvulus, B. malayi, W. bancrofti* and *D. perstans* infestations in different hosts with varying degrees of activity [426]. It has been found to be active against *W. bancrofti* and *D. perstans* in man when given in combination with levamisole [425, 429, 430]. This drug combination also

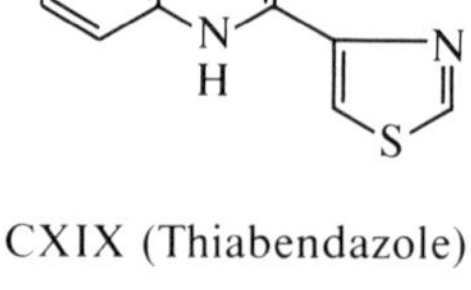

CXIX (Thiabendazole)

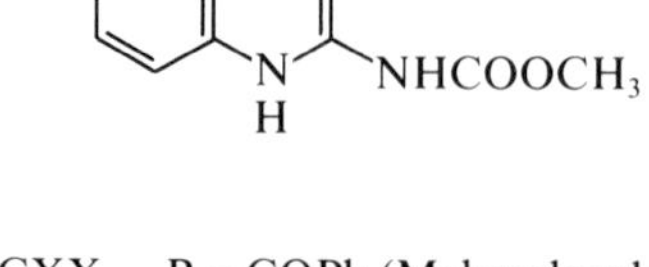

CXX R = COPh (Mebendazole)
CXXI R = SPh (Fenbendazole)
CXXII R = COC_6H_4F (p) (Flubendazole)

shows embryostatic effects on the female worms of *O. volvulus in vivo* and *in vitro* [431]. When given alone, mebendazole did not cause any significant improvement in the microfilarial counts against *O. volvulus* in man up to a daily dose of 200–300 mg given for 1–4 weeks [432, 433].

Fenbendazole kills nearly 100 % of the adult worms of *B. pahangi* located in the peritoneal cavity of jirds at a subcutaneous dose of 2.5 mg/kg [434]. Flubendazole (CXXIII) also kills 83 % of the male and 99 % of the female adult worms of *B. pahangi* in *Mastomys natalensis* when given in feed at a concentration of 250–500 parts/10^6 [435].

In an attempt to develop ideal antifilarials capable of eliminating both the microfilariae and adult worms with no toxic effect on the host, a large variety of 2,5-disubstituted benzimidazoles have been synthesized at this institute. The effective antifilarials to come out of this study were methyl 5(6)-(4-substituted piperazin-1-yl)benzimidazole-2-carbamates (CXXIII a, b) and 2,2′-dicarbomethoxyamino-5,5′-dibenzimidazolyl ketone (CXXIV). A dose of 30 mg/kg of CXXIII a, b given intraperitoneally for 5 days was found to kill 90–95 % of the microfilariae and 90–100 % of the adult worms of *L. carinii* in cotton rats [436]. Compound CXXIV (CDRI 82–437) has been shown to kill 100 % of the microfilariae and adult worms of *L. carinii, A. viteae* and *B. malayi* in experimental animals both by oral (50–200 mg/kg × 5 days) or parenteral (10–50 mg/kg × 5 days) routes of administration [437, 438].

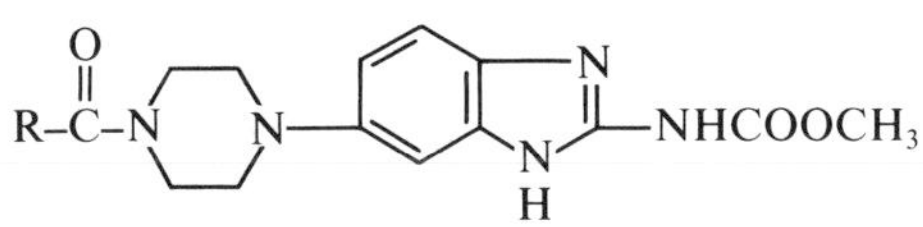

CXXIIIa, R = NEt_2
b, R = 2–Furyl

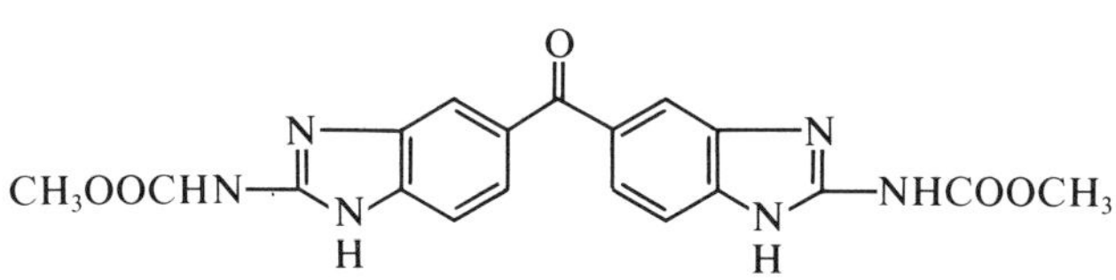

CXXIV

7.5.9 Benzthiazoles

Ciba-Geigy laboratories in Basel have synthesized a large number of substituted benzthiazoles of which three compounds, CGP–20308 (CXXV), CGP–20376 (CXXVI) and CGI–17658 (CXXVII), were found to possess promising activity against several filariids in experimental animals [439–441].

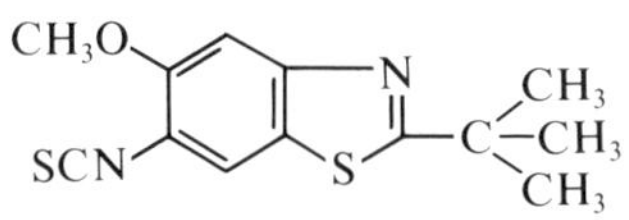

CXXV (CGP–20308)

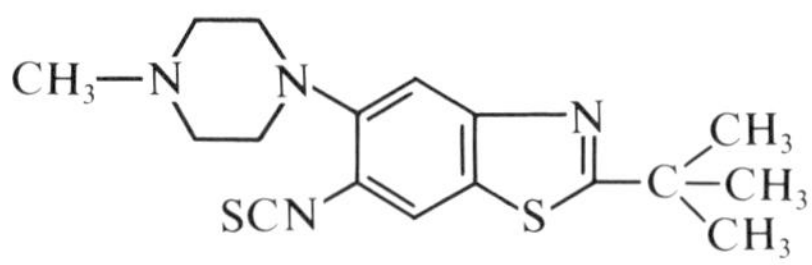

CXXVII (CGI–17658)

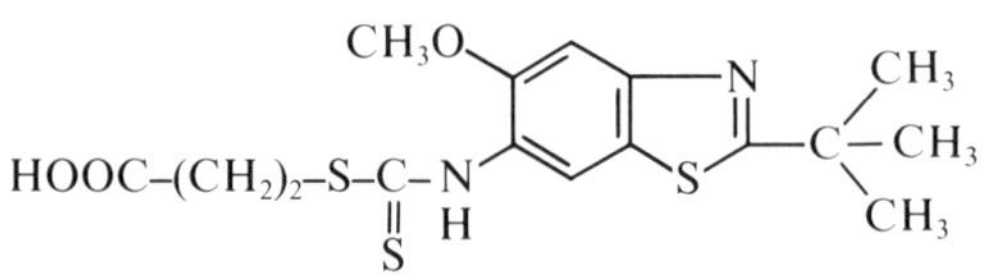

CXXVI (CGP–20376)

The Ciba-Geigy compounds CGP–20308 and CGP–20376 eliminated all the microfilariae of *L. carinii, A. viteae, B. pahangi* and *B. malayi* from *Mastomys natalensis* at an oral dose of 6.25–25 mg/kg given for 5 days [440]. Both compounds were also 100 % microfilaricidal at a single oral dose of 12.5–50 mg/kg. When tested against *O. lienalis* in mice, CGP–20376 (dose 5×25 mg/kg, s. c.) and CGI–17658 (dose 5×12.5 mg/kg, p. o.) caused 100 % clearance of the microfilariae from the blood [441]. In therapeutic efficacy CGI–17658 was next to ivermectin.

7.5.10 Amoscanate and CGP–6140

Amoscanate (CXXVIII), a broad spectrum anthelmintic developed by Ciba-Geigy laboratories, possesses marked activity against different roundworms, tapeworms and fluke. It is also a potent antifilarial [442]. A dose of 100 mg/kg of the drug given for 5 days killed the microfilariae and adult worms of *L. carinii,* while a dose of 600 mg/kg $\times$ 2 days was required to destroy both these stages of *D. viteae* in jirds and *Mastomys,* respectively [442]. Amoscanate is also effective against the microfilariae and adult worms of *B. pahangi* in jirds and *B. malayi* in *Mastomys* at higher doses [443, 444]. Moderate activity of this drug (dose 5 × 100 mg/kg, p. o.) against *O. lienalis* in mice has also been reported [441].

CGP–6140 (CXXIX), an N-methylpiperazinyl derivative of amoscanate, shows significant reduction of microfilariae of *O. lienalis* in mice at an oral dose of 5 × 75 mg/kg [441].

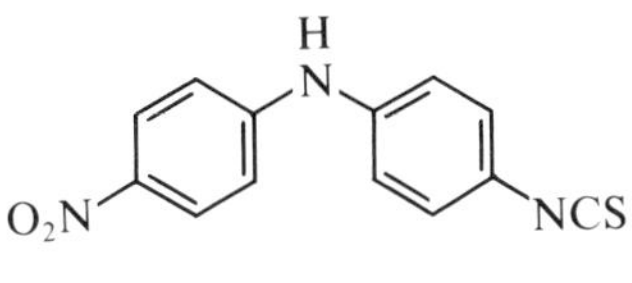

CXXVIII (Amoscanate)

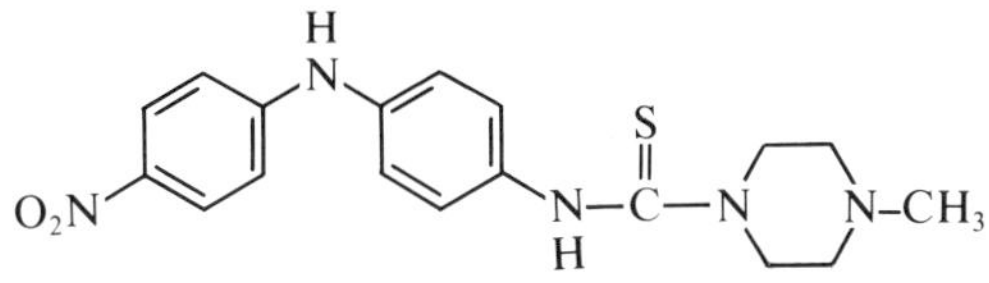

CXXIX (CGP–6140)

7.5.11 Clofazimine (CXXX)

This is a human antileprosy drug which has been found to possess both *in vitro* and *in vivo* antifilarial activity. The drug shows both micro- and macrofilaricidal activity against *L. carinii, B. malayi* and *D. viteae* in rodents at an oral dose of 200 mg/kg given for 6 days. Clofazimine is also effective against the third stage larvae of *D. viteae* in jirds at an oral dose of 200 mg/kg × 6 days [445].

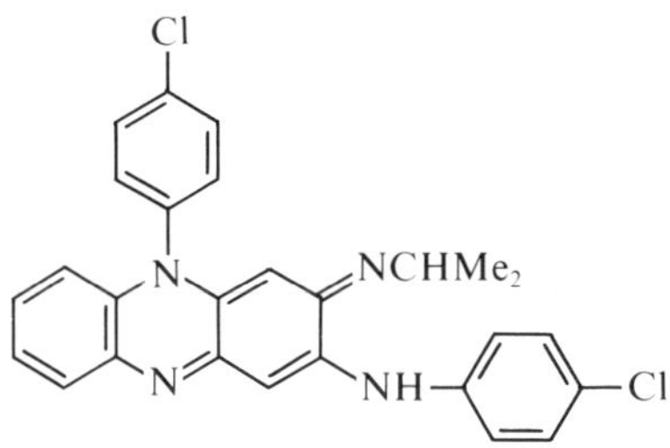

CXXX (Clofazimine)

7.5.12 Ivermectin

Ivermectin (CXXXI), a new class of macrolide lactone antibiotics produced by the actinomycete *Streptomyces avermitilis,* was introduced as an antiparasitic agent in 1981 by Merck Laboratories. The above actinomycete produces a number of such antibiotics which are known collectively as avermectins.

Eight such avermectins have been isolated and characterized. Ivermectin is a 8:2 mixture of 22,23-dihydroavermectin B_{1a} and corresponding B_{1b} homologue [446, 447].

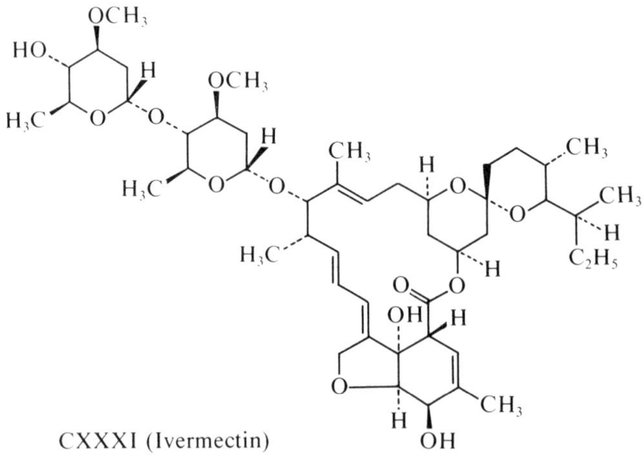

CXXXI (Ivermectin)

Ivermectin possesses very high activity against a variety of intestinal nematodes in cattle, sheep, swine, horses and dogs, and also against the filariids infecting man and animals, by both oral and parenteral routes of administration at doses below 1 mg/kg [447]. The drug is highly effective against the microfilariae of *L. carinii, D. viteae, Setaria equina* and *O. cervicalis* in animals at a dose of 0.2–0.5 mg/kg [448–450].
Clinical trials carried out with ivermectin in human filariases have shown the drug to be effective in the treatment of lymphatic filariasis due to *W. bancrofti* [451], *M. ozzardi* [452], onchocerciasis [453–455] and loasis [456] at a dose of 100–200 µg/kg.

8 Guinea worm infection

Guinea worm infection or dracunculiasis is caused by the guinea worm, *Dracunculus medinensis*. The disease is acquired when man ingests infected *Cyclops* with water. The adult guinea worms are elongated, cylindrical and thread-like structures living in the deep connective and subcutaneous tissues of man. The males measure 1–4 cm while the females may attain a length up to 1 meter or more.

8.1 Geographical distribution

Dracunculiasis is a widespread helminth disease in some parts of Asia and Africa affecting nearly 100 million people [332]. The endemic regions of Africa are Cameroon, Lake Chad, Sudan and Uganda. The disease has also been reported in Mauritania, Senegal, Ghana, Nigeria, Togo Republic, the Republic of Upper Volta and other parts of Africa. In India, guinea worm infection is endemic in Rajasthan as well as a few other parts of the country. Dracunculiasis has also been reported to occur in the Nile Valley, Iraq, Iran, Pakistan, Afghanistan, Russia, Indonesia, South America, Caribbean Islands and Arabia (along the Red Sea).

8.2 Epidemiology

The life cycle of a guinea worm requires two hosts; man as the final host and *Cyclops* as the intermediate host. The fertilized female worms, which live in the connective and subcutaneous tissues, migrate and arrive close to the skin where they discharge their embryos. The worm then secretes a toxin which causes a blister on the skin. The blister begins to itch causing the patient to scratch and eventually the blister bursts. The blister may also burst when it comes in contact with water while the patient works in the muddy fields or takes a bath in a pond or lake.
When the blister bursts, large numbers of motile, rhabditiform larvae are set free in the water. These larvae swim and search their intermediate host, the *Cyclops,* to complete their life cycle. The jerky motion of the larvae attracts the *Cyclops* which ingests it and becomes infected. When man drinks water contaminated by these infected *Cyclops,* the guinea worm infection takes place. On reaching the intestine, the *Cyclops* gets digested by the gastric juices and releases free larvae which invade the intestinal mucosa and reach the connective and subcutaneous tissues where they develop into adult worms in about one year (Fig. 8).

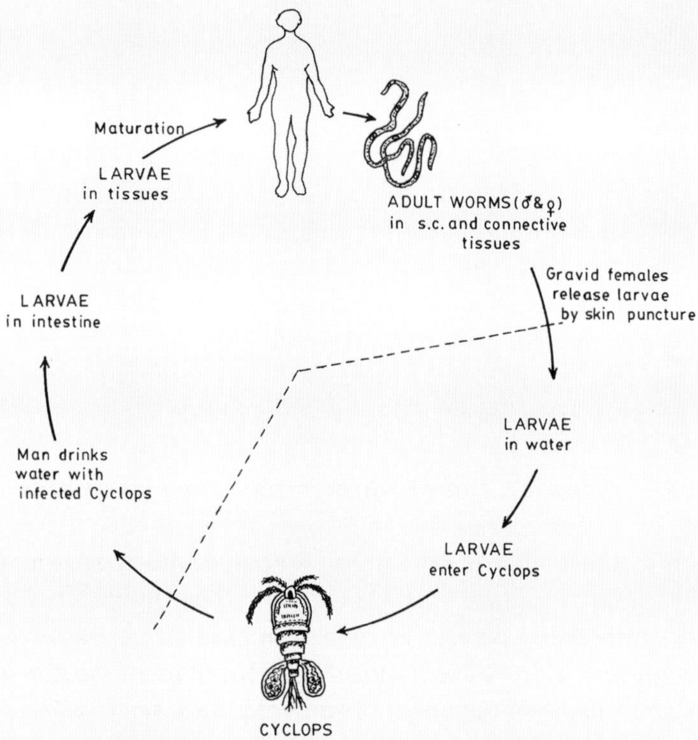

Figure 8
Life cycle of Guinea worm

8.3 Clinical characteristics

No pathological signs are observed immediately after infection. The worms require a 8–12 month incubation period before they are sexually mature. Symptoms appear only when the gravid female worm is ready to discharge the larvae. At this time, reddish papular lesions appear in different parts of the body, especially those, such as legs and arms, which frequently come in contact with water. This may cause itching, fever, giddiness, utricaria, asthma and other allergic reactions. Repeated contact with water or scratching causes the blisters to rupture and a milky fluid to build around the blister opening. The adult may also migrate to other parts of the body leading to neurological damage, joint pains as well as swelling and arthritis.

8.4 Diagnosis

The diagnosis of dracunculiasis is possible only after the cutaneous lesions have appeared or adult worms have migrated to just below the skin. Immunodiagnostic tests, such as the intradermal and complement fixation tests and also the indirect immunofluorescence (I F) test, have been used to diagnose this disease [1]. Larvae may also be detected in the washings of the ulcers (blisters) where the gravid females discharge their embryos. Sometimes larvae may be seen in the fluid aspirated from joint infections.

8.5 Treatment of dracunculiasis

One of the oldest methods of removing adult guinea worms is to wind them on a stick. This practice is still employed today along with chemotherapy. The use of a drug makes it easy to wind the dead worms and also prevents the hosts from secondary infections and allergic reactions resulting from decaying worms [457].

The chemotherapy of guinea worm infection is not satisfactory, although a number of drugs have been shown to possess varying degrees of activity [458]. The drugs which have been evaluated in the clinical therapy of dracunculiasis are metronidazole (LXXV), niridazole (LXXVI), thiabendazole (CXIX), mebendazole (CXX) and diethylcarbamazine (CXIII).

Metronidazole, a preferred drug in the treatment of amoebiasis, giardiasis and trichomoniasis, is a front line drug possessing high activity against guinea worm infection in man. The recommended adult dose of the drug is 250 mg three times a day, given orally for 10 days [25]. A higher dose up to 800 mg/adult given orally thrice daily for 5–7 days has also been used with good clinical response [459–461]. Metronidazole is usually well tolerated causing the expulsion or lysis of the worms. However, the patient may experience nausea, headache, a dry mouth and metallic taste. Occasionally, vomiting, diarrhoea, weakness, insomnia, vertigo, stomatitis, rashes and urethral burning may also occur.

Niridazole (LXXVI), a human schistosomicidal drug, has been widely used to treat guinea worm infection in man. The drug is given at an oral dose of 25 mg/kg in single or divided doses for 10–14 days and has a 91–100 % cure rate [462–465]. Higher doses of 30 or 50 mg/kg of niridazole have also been given for 10 days [466, 467]. In general, the worms were either extracted, or they receded and/or disappeared within 3 weeks. In cases where the worms got broken or did not come out, no toxic symptoms were observed [462]. Niridazole has been found to be well tolerated by the majority of patients; however nausea, vomiting, abdominal pain, diarrhoea and headaches occurred in a few cases which were mild and of short duration. The dead worms were removed by winding them on a stick and side effects controlled by administration of antihistaminics, analgesics or anticholinergics [463, 466].

Both mebendazole (CXX) [468, 469] and thiabendazole (CXIX) [470] have been found to be useful in treating *D. medinensis* infection in man. The usual dose of mebendazole is 200 mg daily (given orally in one or two doses) for 6–7 days [468, 469] while the recommended adult dose of thiabendazole is 25–37.5 mg/kg/day given twice daily for 3 days. If required, the dose may be reduced to minimize toxic effects of the drug [25].

DEC (CXIII) kills the adult worms of *D. medinensis* at higher doses. It is better used as a chemoprophylactic agent as it kills immature worms [471].

9 Schistosomiasis

Schistosomiasis is a major tropical disease caused by the invasion of the circulatory system by four species of blood flukes, *Schistosoma masoni, S. haematobium, S. japonicum* and *S. intercalatum*. A number of schistosomes also infect animals, the most important being *Schistosoma bovis, S. mattheei, S. nasalis, S. indicum* and *S. spindale* which parasitize sheep, goats, cattle and buffalo in India, China and Africa. The adult worms of human schistosomes have separate sexes but they are dioecious (existing together as males and females). The male worm has a groove (gynecophoral canal) along its ventral side in which it carries the female worm during most of its life.

The adult *S. haematobium* worms, responsible for urinary schistosomiasis (bilharziasis), live in the portal system, pelvic veins, particularly in the vesical and pelvic plexus and occasionally in the veins of the colon and rectum. The adult females excrete eggs in the urine of man but rarely in faeces. The other three schistosomes *(S. mansoni, S. japonicum* and *S. intercalatum)* cause intestinal bilharziasis and are found in the portal blood system, mesentric veins and haemorrhoidal plexus. Unlike *S. haematobium* worms, the females of these three blood flukes pass their eggs in the faeces and rarely in urine.

9.1 Geographical distribution

Schistosomiasis has a wide geographical distribution; it has been reported in areas between 36 °N and 34 °S where conditions are ideal (water temperature 20–30 °C) for the breeding of snails, the intermediate hosts of the parasites. Thus, schistosomiasis is prevalent in rural and farmer communities, particularly in areas near slow moving water, the banks of irrigation canals, dams and rivers. The disease is endemic in different parts of Asia, Africa and South America. *S. haematobium* and *S. mansoni* are widely distributed in the Middle East and Africa while *S. mansoni* infections are prevalent in Brazil, Venezuela, Surinam and the Caribbean. *S. japonicum* is widely distributed in China, the Far East, the Philippines, Indonesia and Malaya. It is estimated that nearly 200 million people suffer from schistosomiasis around the world. 20 million of them are in Egypt alone [472, 473]. It has also been reported that nearly 70 million people are infected with *S. mansoni* worldwide [474]. The progress made in assessing the morbidity by *s. haematobium,* has been published by Chen and Mott [475]; it has been estimated that nearly 750 000 people die of schistosomiasis every year [9a].

9.2 Epidemiology

The life cycle of schistosomes requires two hosts; man as the definitive host and snails as intermediate hosts. Humans acquire the infection while working in rice fields, lakes, ponds, canals and streams where the cercariae, present in water, penetrate the skin and enter the bloodstream. The drinking of water containing cercariae may also lead to an infection. After entering the bloodstream, the cercariae undergo some changes and eventually mature into adult worms while feeding on the portal blood. When the female worms get fertilized, they lay their eggs on the walls of the urinary bladder *(S. haematobium)* or intestine *(S. mansoni, S. japonicum, S. intercalatum);* these eggs are later expelled with the urine and faeces, respectively.

When the eggs come in contact with water, they set free swimming embryos called miracidia which seek an intermediate host, usually a freshwater or amphibious snail. If the miracidia do not meet with a snail, they soon die. After finding the mollusc, the miracidia attach themselves to its body, then penetrate the musculature and reach different organs where they produce mother sporocysts. The mother sporocysts then produce several daughter sporocysts which later change into fork-tailed infective larvae called cercariae. These cercariae then leave the snail's body and swim freely in search of their definitive host, man, to complete their life cycle (Fig. 9).

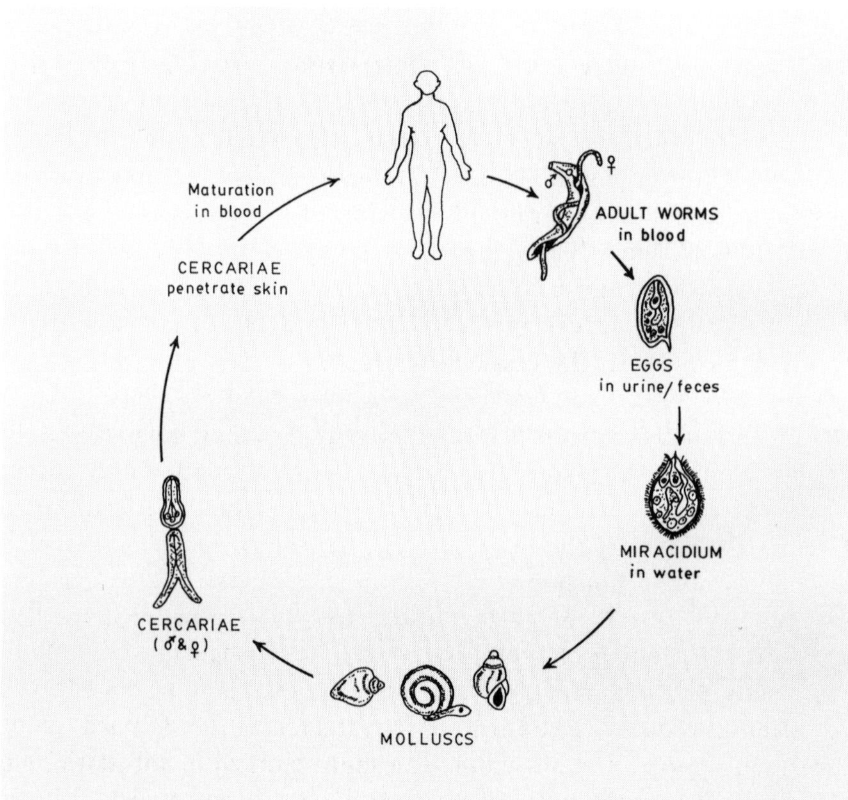

Figure 9
Life cycle of schistosomes

9.3 Clinical characteristics

The pathological changes and clinical characteristics of schistosomiasis depend on the nature and site of infection in man and may be summarized as follows:

(a) *Schistosoma dermatitis:* The penetration of skin by cercariae produces papular rashes which cause intense itching, a condition similar to "swimmer's itch" induced by the cercariae of some bird schistosomes.

(b) *Katayama fever:* After 20–60 days of exposure to *S. japonicum* or *S. mansoni* (less common) infection, the patient experiences fever, chills, malaise, dry cough, utricaria and epigastric discomfort which may last for several weeks but subside gradually.

(c) *Urinogenital schistosomiasis:* This is the most common form of schistosomiasis caused by *S. haematobium*. The disease is endemic in the Nile Valley in Egypt and involves different parts of the male and female urinogenital system. The main clinical features of urinogenital schistosomiasis are the formation of kidney stones as well as fistula in the urinary bladder, penis and vagina, renal failure, haematuria, renal and ureteric colic, the generation of ulcers leading to the formation of pus tracks in the scrotum and penis, the swelling of testes and cancer of the bladder [476].

(d) *Hepatosplenic schistosomiasis:* The early pathological symptoms of this form of schistosomiasis include the inflammation and fibrosis of the spleen. The liver also shows periportal cellular infiltration and fibrosis followed by the wasting and swelling of its cells. The main clinical symptoms of hepatosplenic schistosomiasis are weight loss, weakness, diarrhoea, epigastric discomfort, enlargement of the liver and spleen, swelling of limbs and clubbing of fingers.

(e) *Intestinal schistosomiasis:* Schistosomes can also produce lesions in different parts of the alimentary canal such as the stomach, colon, intestine, appendix and rectum. The main clinical manifestations of this form of schistosomiasis are peptic-like ulceration, fistula in the rectum and colon, rectal prolapse, polyposis (leading to intestinal obstruction due to large groups of polyps), bouts of diarrhoea and other gastrointestinal disturbances.

(f) *Cardiopulmonary schistosomiasis:* This is caused by the presence of ova and worms of schistosomes in the lungs. The penetration of worms into lungs may lead to fibrosis and obstruction of pulmonary blood circulation. The main clinical symptoms of cardiopulmonary schistosomiasis are fatigue, cardiac pain, palpitation giddiness and pulmonary hypertension.

9.4 Diagnosis

Urinogenital schistosomiasis caused by *S. haematobium* can be diagnosed by detecting terminally spined eggs of schistosomes in urine under a low power microscope. The eggs are distributed throughout the urinary tract and are not concentrated in the last few drops of urine at the end of micturination [477]. Viable eggs may also be found in the cervical smears [478], in stool and seminal fluid samples. Sometimes, rectal or sigmoid biopsies are carried out to find *S. haematobium* eggs.

Intestinal schistosomiasis can be recognized by the characteristic eggs in stool samples or rectal biopsy specimens. A *S. mansoni* egg has a lateral spine while a *S. japonicum* egg is like an *Ascaris* egg in shape and size except for the fact that it has a thin wall and is usually a fully developed miracidium. The eggs of *S. intercalatum* are often bent and terminally spined.

Attempts to develop immunodiagnostic tests using the identification of schistosomal antibodies in serum by parasite-derived antigens have been reported [479]. Enzyme-linked immunosorbent assay (ELISA) has also been used to diagnose schistosomiasis with varying degrees of success [480, 481]. The effect of antigen cross-linking on the sensitivity of ELISA has been studied [482, 483]. The reports indicate that the method may not be very helpful in general use [483].

9.5 Treatment of schistosomiasis

The chemotherapy of schistosomiasis has been recently reviewed [472, 484–486]. In this section, emphasis will be laid on the drug treatment of schistosomiasis in man.

9.5.1 Organometallic compounds

A number of organoantimonial compounds such as tartar emetic (L), stibocaptate (CIV), stibophen (CV) and other antimonials have been used to treat different forms of schistosomiasis with varying degrees of success. However none of them find use in clinical practice today because of their high toxicity, low margin of safety, inactivity when administered orally and difficulty in administration by parenteral routes. Further, the advent of new, orally effective and safer drugs has also helped eliminate these organoantimonials from current clinical therapy [486].

The most effective organometallic drug is metrifonate (triclorfon, CIX) which is given orally to treat *S. haematobium* and *S. mansoni* infections. The recommended dose of metrifonate for treating *S. haematobium* in man is 7.5–10 mg/kg, given three times with a 2-week interval, whereby 81–91 % cure rates were achieved [487–490]. Recently, a large number of school children infected with *S. haematobium* were treated in Egypt, Zimbabwe and Kenya using metrifonate at a single oral dose of 10 mg/kg. A 75–90 % reduction in urine egg counts were observed [491–493].

Metrifonate is a quite safe drug. It produces only mild gastrointestinal disturbances and may bring down cholinesterase levels in blood and erythrocytes [485].

9.5.2 Miracils

Introduced in 1946 as a human schistosomicidal agent, Miracil D (lucanthone, CXXXII) was the first metal-free and orally effective drug [494]. Since then it has been used to treat schistosomiasis in man [495, 496]. The drug is most effective against *S. haematobium*, less against *S. mansoni* and inactive against *S. japonicum*. The usual adult dose of lucanthone is 100 mg/kg spread over 5–20 days [497]. A total dose of 60 mg/kg, divided into two doses, given daily for 3 days has also been recommended [498, 499].

Although lucanthone was shown to be valuable in treating *S. haematobium* infection in man, its use in mass therapy was marred due to its numerous side effects like nausea, vomiting, diarrhoea, anorexia, epigastric pain, dizziness, tremor, mental confusion, insomnia and occasional haematemesis in treated patients [497].

Hycanthone (CXXXIII), the active metabolite of lucanthone, shows better activity than the parent drug against schistosomiasis in man. The drug is highly effective against *S. haematobium* and *S. mansoni* in man at a single intramuscular injection of 3 ± 0.5 mg/kg [500–502]. The cure rates may range from 36–90 % depending on the severity of the infection. The drug shows poor activity in patients with a heavy worm burden [503]. Against *S. mansoni,* hycanthone achieves more than 90 % cure rates at a dose of 1.5–3.5 mg/kg given for 3 days [500, 504, 505].

Hycanthone treatment of schistosomiasis in man entails several side effects such as nausea, vomiting, anorexia, vertigo, and abdominal pain which are generally mild and do not last long; however, their frequency is higher at oral doses [500]. In animals, hycanthone has also been found to be possibly mutagenic, teratogenic [506], carcinogenic [507] and with a tendency to develop resistance to *S. mansoni* at higher doses [508].

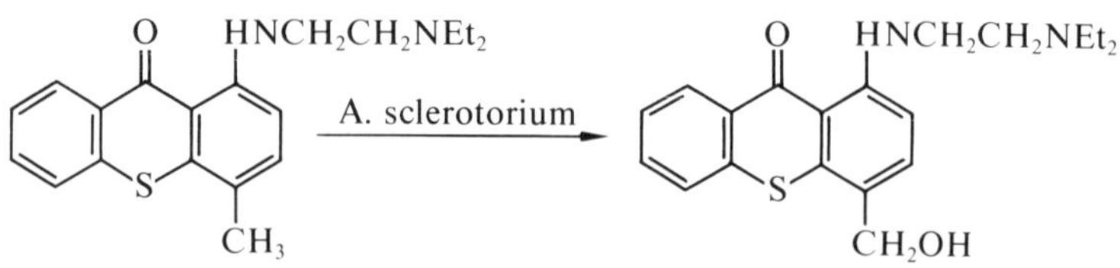

CXXXII (Lucanthone) CXXXIII (Hycanthone)

9.5.3 Nitro compounds

Two compounds of this class, niridazole (LXXVI) and amoscanate (CXXXVIII) have shown potent schistosomicidal activity and, therefore, have been widely evaluated in clinical therapy.

Niridazole is highly effective against *S. haematobium* and *S. mansoni* but less effective against *S. japonicum* and *S. intercalatum* infections. Thus, in an oral dose of 25 mg/kg given for 5–7 days [509–511] or a dose of 30 mg/kg × 4 days [512, 513], it produced 80–100 % cure rates against *S. haematobium* both in adults and children and was also well tolerated. A dose of 50 mg/kg of niridazole, given thrice daily, has also been found to be highly effective in treating urinary schistosomiasis in man [514].

The biological profile of niridazole against *S. mansoni* has been given by Katz [474]. The usual adult dose of the drug is 25 mg/kg given orally for 5–7 days producing 40–100 % cure rates. A dose of 25–30 mg/kg of niridazole given for 5 days gives 40–90 % cure rates [510]. Better results (cure rates 80–100 %) were obtained when the drug was administered at a dose of 25–30 mg/kg × 7 days [515, 516]. Lower doses of 12.5–15 mg/kg have also been used to lessen the side effects while treating *S. mansoni* infection in man [517].

When a dose of 15–25 mg/kg of niridazole was given orally for 5–10 days to treat *S. japonicum* infection in man 0–47 % cures were obtained [515, 518]. However, some workers have achieved a high reduction in egg counts with 48–85 % cure rates by treating *S. japonicum* patients with niridazole [519, 520].

Niridazole shows poor activity against *S. intercalatum* in man at an oral dose of 25 mg/kg given daily for 7 days. Even a second course of treatment given to some patients did not cure the disease [521, 522].

The toxicity and side effects of niridazole are numerous and depend largely on the dose and duration of treatment. The side effects of the drug are nausea, vomiting, anorexia and abdominal pain which are usually mild and of short duration, occurring more frequently in adults than children [511]. The drug also produces CNS disorders such as headaches, muscular spasms, anxiety, confusion, excitation and hallucinations. Occasionally neuropsychic disorders may also be noticed [510, 523]; they can be reduced by simultaneous use of barbiturates. Frequent changes in ECG [523] and some allergic manifestations (utricaria and skin rashes) can also be seen in niridazole-treated cases.

In addition to the above side effects, the use of niridazole has also been reported to be fatal in a few cases [524] and, therefore, its use should be restricted to hospitalized patients only [525]. However, some workers have recommended niridazole for mass treatment of schistosomiasis [513, 516].

Amoscanate, a broad-spectrum anthelmintic possessing high activity against gastrointestinal nematodes, cestodes and filariids, also shows activity against *S. haematobium, S. mansoni* and *S. japonicum* in rodents, birds, dogs and monkeys at a single oral, intramuscular or subcutaneous dose of 30–120 mg/kg [526, 527]. The drug has been widely evaluated against human schistosomiasis in China in a dose of 7 mg/kg given for 3 days whereby a high activity was observed [485].

9.5.4 Quinolines and isoquinolines

The important compound with quinoline skeleton is oxamniquine (CXXXIV) while praziquantel (CXXXV) and dehydroemetine (CXXXVI) belong to the isoquinoline class of heterocycles.

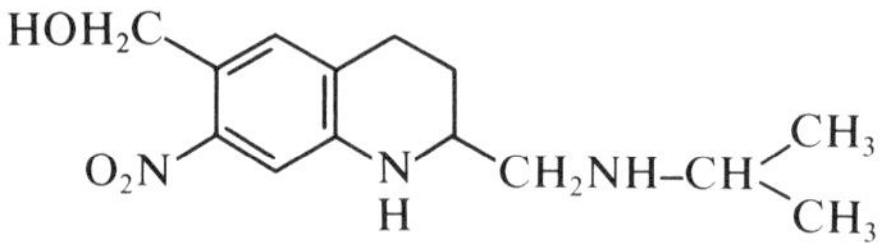

CXXXIV (Oxamniquine)

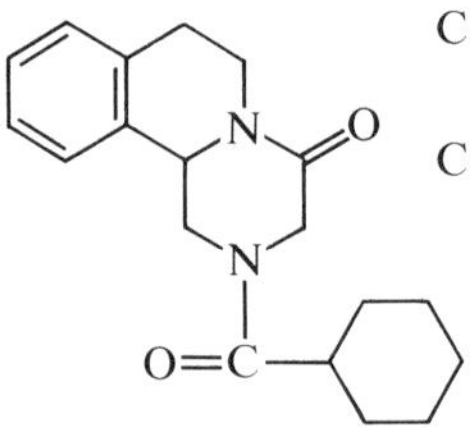

CXXXV (Praziquantel)

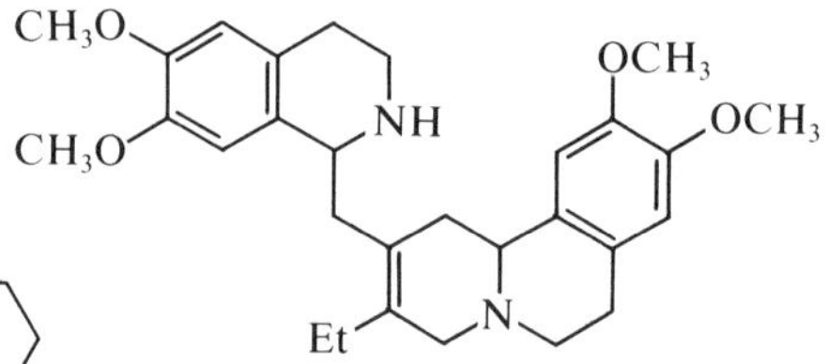

CXXXVI (Dehydroemetine)

Oxamniquine is a highly effective drug for treating *S. mansoni* infection in man but shows little action against *S. haematobium* and is of no value for the treatment of *S. japonicum*. The recommended adult dose of oxamniquine is a single oral dose of 15 mg/kg which produces 70–100 % cure rates against *S. mansoni* infection [25, 528–530]. Children receive the drug at a dose of 10 mg/kg, given two times for one day [25]. In the case of East African patients, the dose should be increased to 30 mg/kg for one day while for Egyptian and South African patients the dose should be 30 mg/kg/day × 2 days. Neuropsychiatric disturbances and seizures may be seen in a few cases [25, 531, 532].

Oxamniquine is usually well tolerated, however side effects like nausea, vomiting, dizziness and generalized seizures may be observed following oral administration of the drug [532, 533].

Praziquantel (EMBAY–8440), a broad-spectrum cestodicidal and flukicidal agent developed jointly by E. Merck, Darmstadt and Bayer, Wuppertal (FRG), is the drug of choice for treating cestode and trematode infections in man and animals [534–539]. In man, praziquantel shows very high activity against *S. haematobium*, *S. man-*

soni, S. japonicum, S. intercalatum and *S. mattheii.* The usual dose of the drug for complete destruction of the schistosomes is 40 mg/kg given in a single dose [537, 540, 541]. A dose of 20 mg/kg, t. i. d. × 1 day has also been recommended [25].
The use of praziquantel is frequently associated with headaches, malaise and dizziness. Abdominal pain, nausea, diarrhoea, rashes, sweating and fever may also occur.
Dehydroemetine has been found to possess 69 % efficacy against *S. haematobium* in man at a dose of 1.5 mg/kg given daily for 10 days [542]. Better results were obtained against the above infection when the drug was given either at a dose of 3–6 mg/kg for 2–3 weeks [543] or a total dose of 60–90 mg/adult given for 10–15 days [544]. Dehydroemetine is also effective against *S. mansoni* infection in man [542, 545].

9.5.5 Miscellaneous compounds

Oltipraz (CXXXVII) shows nearly 90 % cure rates against *S. mansoni, S. haematobium* and *S. intercalatum* infections in man at a total dose of 25–35 mg/kg given in divided doses for 1–3 days [546, 547]. The frequent side effects of the drug are vomiting and mild abdominal pain [547].
Mebendazole (CXX) has been found to cure two patients infected with *S. haematobium* at an oral dose of 1 g/day given in two doses. No eggs were detected in the patients' urine after 10 days [548].

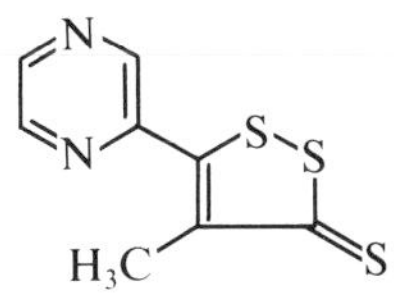

CXXXVII (Oltipraz)

10 **Fluke infections other than schistosomes**

A number of liver flukes *(Fasciola hepatica, F. gigantica, Clonorchis sinensis, Opisthorchis felineus* and *Dicrocoelium dendriticum)* and lung flukes *(Paragonimus westermani)* are known to infect man and domestic animals. The liver flukes utilize snails as intermediate hosts while the lung flukes complete their life cycle through two intermediate hosts, snails and crabs or crayfishes. Man serves as the definitive host for all the above flukes.

The liver flukes live in the liver and biliary passage; the lung flukes are found in lungs. Man acquires the infection by eating raw fish and crabs or by drinking unpurified water. The distribution of liver fluke and lung fluke infections is, therefore, confined to place where proper snail hosts are present. The disease has been reported in China, Japan, Korea, Thailand, Hong-Kong, Africa and different other parts of the world.

The life cycle of liver and lung flukes is simple (Fig. 10). The clinical manifestations of liver fluke infection are related to the liver and in chronic cases may lead to jaundice, cirrhosis of the liver and biliary duct, ascites and cachexia. The lung fluke infections lead to cough-

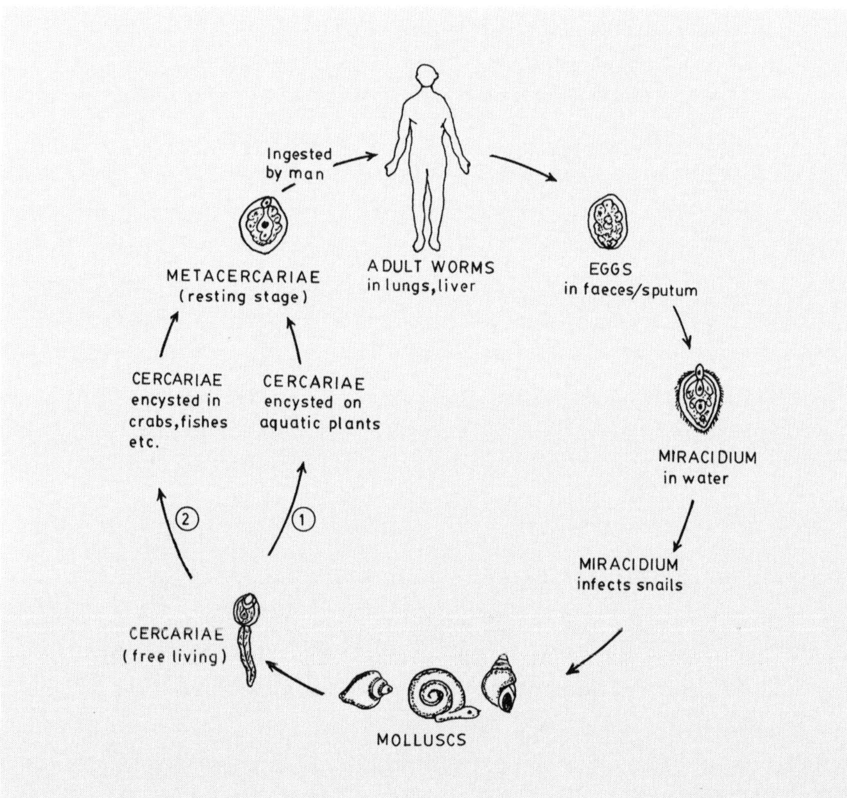

Figure 10
Life cycle of ① liver flukes and ② lung flukes

ing (with gelatinous blood), stained sputum and chest discomfort. Mild anaemia, fever, body pain, abdominal discomfort and diarrhoea may also occur. The parasite may migrate to the brain resulting in Jacksonian-type epilepsy.
The diagnosis of liver fluke infection depends on the detection of characteristic eggs in the faeces or material obtained from the biliary passage or duodenum. Similarly, diagnosis of lung fluke infection may be carried out by detecting characteristic eggs in aspirated pleural effusion, in sputum or faeces.

10.1 Treatment of liver and lung fluke infections

(a) *Fasciola* spp.: Bithionol (CXXXVIII) is the drug of choice for treating sheep liver fluke infection. The recommended adult dose of the drug is 30–50 mg/kg given on alternate days for 10–15 doses [25]. Even a dose of 30 mg/kg given on alternate days for 5 doses cured 100 % of *Fasciola* spp. infection in man with no toxicity [549].

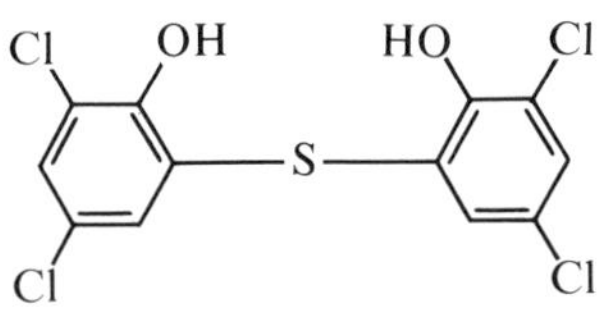

CXXXVIII (Bithionol)

(b) *Clonorchis* spp.: The drug of choice for treating Chinese liver fluke is praziquantel. The recommended adult dose of the drug is 25 mg/kg, t. i. d. given for 2 days [25]. For mass treatment, a single dose of 40 mg/kg has been recommended [550].

(c) *Opisthorchis* spp.: Mebendazole, at a dose of 30 mg/kg, has been found to be active [485] but praziquantel is the drug of choice. The recommended adult dose of praziquantel is 25 mg/kg, t. i. d. for one day [25]. For mass therapy, a single dose of 40 mg/kg may be satisfactory [551]

(d) *Paragonimus* sp.: Both praziquantel (dose 25 mg/kg, t. i. d. ×2 days) and bithionol (dose 30–35 mg/kg on alternate days for 10–15 doses) may be used to treat lung fluke infections in man [25]. The experimental chemotherapy of paragouimiasis has been reviewed recently by Yokogawa [552].

11 Vector control

The control of the vectors of diseases plays an important role in the iradication of the vector-borne diseases. The arthropod and mollusc vectors may be controlled by three methods – mechanical or cultural, biological and chemical [553].

(a) *Mechanical methods:* This method uses elimination of the breeding places and larval habitats of the arthropods such as mosquitoes and flies. Thus the filling or cleaning of ditches and water reservoirs, the disposal of sewage, garbage or trash piles and the improvement of drainage system greatly help in keeping the breeding of the arthropods under control. Improvement in housing will reduce the hiding places of mosquitoes and flies. The mechanical method is the first step in controlling the vector population.

(b) *Biological methods:* This method is based on the development of entomopathogenic organisms such as viruses, bacteria, fungi, protozoans and nematodes that exclusively attack the insects [554, 555]. For example, the introduction of top minnows, e.g. *Gambusia* spp., that feed on larvae and pupae of mosquito, is an efficient way of controlling *Anopheles* in ponds and lakes. Similarly, there are several species of fish which feed on mosquito larvae. The toxins produced by some bacteria like *Bacillus thuringensis* has been found to possess high insecticidal activity.

(c) *Chemical methods:* A large variety of inorganic, organic and organometallic insecticides are now available which are in use throughout the world to control pests affecting vegetation and also to kill the vectors of various viral, protozoal and helminth diseases. Although a number of pesticides may have accumulated in the environment causing damage through the food-chain system, the use of chemical agents to destroy arthropods has not declined except, perhaps, for a more judicious use of such agents.
The modern insecticides are derived from inorganic compounds (such as arsenic oxide, lime sulfur, sodium arsenite etc), chlorinated

hydrocarbons (like DDT, lindane, chlordan, bromocylene, endosulfan etc.), organophosphates (like bromophos, coumophos, malathione, dicaptone, dichlorovos, fenthione, trichlorfon etc), carbamates (such as carbaryl, promacyl etc.) and pyrethroids, as well as related organic compounds. The use of repellents (such as dect, dimethylphthalate etc.) and the synthesis of compounds with juvenile hormone activity that inhibit the development of arthropods in their immature stages, have greatly helped to reduce vector-borne diseases in man. A comprehensive survey of chemical agents used to control arthropod vectors is available [556].

Several chemical molluscicidal agents such as niclosamide, dinitro-*o*-cyclohexylphenol, frescon (N-triphenylmethylmorpholine) and yurimine (3,5-dibromo-4′-nitrobenzene) are also used to control snail population in different parts of the world [557].

12 Concluding remarks

The present review indicates the measures available to treat, control and eradicate various vector-borne diseases. More work is required to develop ideal drugs for the parasitic diseases propagated or transmitted by arthropods and molluscs. Closer attention should be paid to the design of better drugs against malaria (radical curatives and management of cerebral malaria), leishmaniasis, sleeping sickness, Chagas' disease and filariasis (river blindness and lymphatic filariasis). This is only possible by exploring newer biochemical leads for better chemotherapeutics against the various parasitoses responsible for causing vector-borne diseases.

Development of never, safe, effective and economically viable methods to control vector population will also help in eradicating vector-borne diseases. The author has attempted to provide a comprehensive look at the subject by giving factual information as far as possible; however, there are possibilities of omissions and errors. The present review, therefore, should not be treated as a *definitive* source of information. It is suggested that independant verification of the drug dosages and side effects should be made from the original literature published by the respective drug manufacturer.

References

1 G. W. Hunter, J. C. Swartzwelder and D. F. Clyde (eds.): Tropical Medicine. 5th ed., W. B. Saunders Company, Philadelphia (1976).
2 A. W. Woodruff (ed.): Medicine in the Tropics, Churchill Livingstone. London (1974).
3 J. J. McKelvey, Jr., B. E. Eldridge and K. Maramorosch (eds.): Vectors of Disease Agents. Praeger Publishers, New York (1981).
4 C. G. Huff: Science, *74,* 456 (1931).
5 W. Burgdorfer: in Vectors of Disease Agents, eds. J. J. McKelvey, Jr., B. E. Eldridge and K. Maramorosch, Praeger Publishers, New York, p. 44 (1981).
6 P. Nickel: Pharm. Internat. *4,* 37 (1983).
7 L. J. Bruce-Chwatt (ed.): in Chemotherapy of Malaria, II ed., World Health Organisation, Geneva (1981).
8 M. D. Young: in ref. 1, p. 353.
9 (a) C. J. Schofield: Parasitology Today, *1,* 2 (1985); (b) G. Davidson: New Scientist, *96,* 731 (1982).
10 J. F. Trape: Trans. Roy. Soc. Trop. Med. Hyg. *79,* 181 (1985).
11 W. E. Collins and J. C. Skinner: Amer. J. Trop. Med. Hyg. *21,* 690 (1972).
12 I. G. Kagan: Amer. J. Trop. Med. Hyg. *21,* 683 (1972).
13 A. W. Ferreira and M. C. A. Sanchez: Rev. Inst. Med. Trop. São Paulo. *30,* 137 (1988).
14 L. Franzen *et al.*: Lancet *1,* 525 (1984).
15 J. Haworth: Trop. Dis. Bull. *84,* R1 (1987).
16 G. Reibnegger et al: Trans. Roy. Soc. Trop. Med. Hyg. *78,* 545 (1984).
17 K. Bein and P. Olcen: Ethiopian Med. J. *22,* 119 (1984).
18 R. S. Rozman and C. J. Canfield: Adv. Pharmacol. Chemother. *16,* 1 (1979).
19 T. R. Sweeny and R. E. Strube: in Burger's Medicinal Chemistry, Part II, 4th ed., M. E. Wolff (ed.), John-Wiley & Sons, New York, p. 333 (1979).
20 W. Peters and W. H. G. Richards (eds.): Antimalarial Drugs Vol. I and II (Handbook of Experimental Pharmacology, Vol. 68), Springer Verlag, Heidelberg (1984).
21 T. G. Geary and J. B. Jensen: in Chemotherapy of Parasitic Diseases eds. W. C. Campbell and R. S. Rew, Plenum Press, New York, 87 (1986).
22 A. K. Saxena and M. Saxena: Prog. Drug Res. *30,* 221 (1986).
23 R. N. Chaudhuri, B. N. Dutta and N. K. Chakaravarty: Indian Med Gazz. *89,* 660 (1954).
24 M. J. Miller: Prog. Drug Res. *20,* 433 (1976).
25 M. Abramowicz (ed.): The Medical Letter (Medical Letter Inc., New Rochelle, USA), *30,* 15 (1988).
26 D. Bunnag et al.: Trans Roy Soc. Trop. Med. Hyg. *83,* 66 (1989).
27 D. Bunnag and T. Harinasuta: Internat. J. Parasitol. *17,* 169 (1987).
28 P. T. Myint, T. Shwe, L. Soe and Y. Htut: Trans. Roy. Soc. Trop. Med. Hyg. *83,* 72 (1989).
29 P. T. Myint and T. Shwe: Trans. Roy. Soc. Trop. Med. Hyg. *81,* 559 (1987).
30 S. K. Puri and G. P. Dutta: Prog. Drug Res. *26,* 167 (1982).
31 K. H. Riekmann: Ann. Rev. Med. *34,* 321 (1983).
32 W. Raether and E. Fink: Ann. Trop. Med. Parasit. *73,* 505 (1979); *76,* 507 (1982).
33 L. H. Schmidt: Antimicrob. Agents Chemother. *16,* 475 (1979).
34 G. Watt et al.: Trans. Roy. Soc. Trop. Med. Hyg. *82,* 205 (1988).
35 D. L. Heymann, C. O. Khoromana, J. J. Wirima and C. C. Campbell: Trans. Roy. Soc. Trop. Med. Hyg. *81,* 722 (1987).

36 W. Roehl: Naturwissenschaften, *14,* 1156 (1926).
37 P. Mühlens: Naturwissenschaften, *14,* 1162 (1926).
38 B. Bhat, M. Seth and A. P. Bhaduri: Prog. Drug Res.: *28,* 197 (1984).
39 A. S. Alving, C. F. Johnson, A. R. Tarlow, G. J. Brewer, R. W. Kellermeyer and P. E. Carson: Bull. WHD *22,* 621 (1960).
40 P. E. Carson: Fed. Proc. *19,* 995 (1960).
41 P. E. Carson, C. L. Flanagan, C. E. Ickes and A. S. Alving: Science *124,* 484 (1956).
42 A. S. Alving, R. D. Powell, G. J. Brewer and J. D. Arnold: Drugs, Parasites, Hosts, Symp. Middlesex, Hosp. Med. School, p. 83 (1962).
43 L. H. Schmidt, R. Fradkin, D. Vaughan and J. Rasco: Amer. J. Trop. Med. Hyg. *26,* 1116 (1977).
44 E. A. Nodiff, K. Tanabe, C. Seyfried, S. Matsura, Y. Kando, E. H. Chen and M. P. Tyagi: J. Med. Chem. *14,* 921 (1971).
45 T. M. Cosgriff, E. F. Boudreau, C. L. Pamplin, E. B. Doberstyn, R. E. Desjardins and C. J. Canfield. Amer. J. Trop. Med. Hyg. *31,* 1075 (1982).
46 C. W. Muth, B. Bhattacharya, R. L. Mahaffey and H. L. Minigh: J. Med. Chem. *16,* 303 (1973).
47 E. A. Nodiff, A. J. Saggimo, K. Tanabe, E. H. Chen, M. Sinbo, M. P. Tyagi, A. Kozuka, H. Otomasu and B. L. Verma: J. Med. Chem. *18,* 1011 (1975).
48 A. Markovac and M. P. LaMontagne: J. Med. Chem. *19,* 978 (1976).
49 L. C. Washburn and D. E. Pearson: J. Med. Chem. *17,* 676 (1974).
50 R. E. Harmon, T. S. Lin and S. K. Gupta: J. Med. Chem. *16,* 940 (1973).
51 P-Lu Chien, D. J. McCaustland, W. H. Burton and C. C. Cheng: J. Med. Chem. *15,* 28 (1972).
52 M. Loy and M. M. Joullie: J. Med. Chem. *16,* 549 (1973).
53 P-Lu Chien and C. C. Cheng: J. Med. Chem. *16,* 1093 (1973).
54 A. S. Dey and J. L. Neumeyer: J. Med. Chem. *17,* 1095 (1974).
55 G. E. Childs, C. Lambros, J. D. Notsch, C. L. Pamplin, D. E. Davidson Jr. and A. Ager: Ann. Trop. Med. Parasit. *78,* 13 (1984).
56 C. J. Canfield, A. P. Hall, B. S. MacDonald, D. A. Neuman and J. A. Shaw: Antimicrob. Agents Chemother. *3,* 224 (1973).
57 D. F. Clyde, V. C. McCarthy, C. C. Rebert and R. M. Miller: Antimicrob. Agents Chemother. *3,* 220 (1973).
58 M. Porter and W. Peters: Ann. Trop. Med. Parasit. *70,* 271 (1976).
59 L. H. Schmidt, R. Crosby, J. Rasco and D. Vaughan: Antimicrob. Agents Chemother. *14,* 292 (1978).
60 J. Rinehart, J. Arnold and C. J. Canfield: Amer. J. Trop. Med. Hyg. *25,* 769 (1976).
61 W. Peters, R. E. Howells, J. Portus, B. L. Robinson, S. Thomas and D. C. Warhurst: Ann. Trop. Med. Parasit. *71,* 407 (1977).
62 T. R. Sweeney: Med. Res. Rev. *1,* 281 (1981).
63 World Health Organisation: Bull. WHO. *61,* 169 (1981).
64 J. M. K. Ekue et al.: Bull. WHO *61,* 713 (1983).
65 T. Harinasuta et al.: Bull. WHO *61,* 299 (1983).
66 T. Harinasuta et al.: Lancet *1,* 885 (1985).
67 F. Tin et al.: Bull. WHO *60,* 913 (1982).
68 J. P. Couland et al.: Ann. Soc. Belge. Mod. Trop. *63,* 247 (1983).
69 J. M. De Souza: Bull WHO *61,* 815 (1983).
70 L. F. Fieser and H. Heymann: J. Biol. Chem. *176,* 1363 (1948).
71 L. F. Fieser and A. P. Richardson: J. Amer. Chem. Soc. *70,* 3156 (1948).
72 L. F. Fieser et al.: J. Amer. Chem. Soc. *70,* 3151 (1948).
73 L. F. Fieser, J. P. Shirmer, S. Archer, R. R. Lorenz and P. I. Pfaffenbach: J. Med. Chem. *10,* 513 (1967).
74 L. F. Fieser et al.: J. Med. Chem. *10,* 517 (1967).

75 G. Fawaz and F. S. Haddad: Amer. J. Trop. Med. Hyg. *31,* 569 (1951).
76 D. A. Berberian and R. G. Slighter: J. Parasitol. *54,* 999 (1968).
77 D. A. Berberian, R. G. Slighter and H. W. Freele: J. Parasitol. *54,* 1181 (1968).
78 World Health Organisation, Tech. Rep. Ser. No. 529, 62 (1973).
79 J. Hill: Experimental Chemotherapy, Vol. I, p. 513, eds. R. J. Schnitzer and F. Hawking, Academic Press, New York (1963).
80 J. Hill: Experimental Chemotherapy, Vol. IV, p. 448, eds. R. J. Schnitzer and F. Hawking, Academic Press, New York (1966).
81 L. T. Coggeshall, J. Mair and C. A. Best: J. Amer. Med. Assoc. *117,* 1077 (1941).
82 H. Bader et al.: J. Med. Chem. *12,* 709 and 1108 (1969).
83 I. C. Popoff et al.: J. Med. chem. *11,* 631 (1968); *14,* 540 and 1166 (1971).
84 E. F. Elslager et al.: J. Med. Chem. *12,* 363 and 519 (1969).
85 C. Y. Chiou: Biochem. Pharmac. *20,* 2401 (1971).
86 A. C. Sonntag, V. G. Stenger, B. P. Vogh and T. H. Maran: J. Pharmac. Exp. Ther. *182,* 34 (1972).
87 H. J. Scholer, R. Leimer and R. R. Richle: Antimalarial drugs, Part II, p. 123–221, eds. W. Peters and W. H. G. Richards, Springer Verlag, Heidelberg (1984).
88 G. L. Mandell and M. A. Sande: The pharmacological basis of therapeutics, 6th ed. p. 1106–1125 eds. A. G. Gilman, L. S. Goodman and A. Gilman, Macmillan Publishers, New York (1980).
89 R. Ferone: Antimalarial drugs Part II, p. 207–221, eds. W. Peters and W. H. G. Richards, Springer Verlag, Heidelberg (1984).
90 V. V. Olsen: Lancet *1,* 649 (1983).
91 W. Peters, J. H. Portus and B. L. Robinson: Ann. Trop. Med. Parasit. *67,* 143 (1973).
92 W. Peters: Bull WHO *51,* 379 (1974).
93 T. Harinasuta, C. Varavan and P. Buranasin: Trans. Roy. Soc. Trop. Med. Hyg. *82,* 694 (1988); P. Delton, J. D. Sexton, L. Buglimitura and C. Sezibera: Amer. J. Trop. Med. Hyg. *38,* 244 (1988).
94 H. Stemberger et al.: Acta Tropica *41,* 391 (1984).
95 B. Darlow et al.: Amer. J. Trop. Med. Hyg. *31,* 1 and 10 (1982).
96 T. Chongsuphajaisiddhi et al.: Southeast Asian J. Trop. Med. Pub. Health. *14,* 357 (1983).
97 T. J. Ponnampalan and H. A. Frank: Singapore Med. J. *24,* 104 (1983).
98 F. Tin and N. Hlaing: Southeast Asian J. Trop. Med. Pub. Health *15,* 238 (1984).
99 Z. S. Huang et al.: Chin J. Parasit. & Parasitic Dis. *6,* 285 (1988).
100 Y. Yoshida et al.: Jap. J. Trop. Med. Hyg. *15,* 105 (1987).
101 Morbidity and Mortality Weekly Report *33,* 713 (1984) cited in ref. 15.
102 D. J. Bradley et al.: Brit. Med. J. *291,* 136 (1985).
103 R. Lainson and J. J. Shaw: Nature, *273,* 595 (1978).
104 A. Zuckermann and R. Lainson: *Leishmania* in Kreyer Parasitic Protozoa (Academic Press, New York). *1,* 58 (1977).
105 S. L. Croft: Trends in Pharmacol. Sci. *9,* 376 (1988).
106 F. Modabbar: Leishmaniasis in Tropical Diseases Research, a Global partnership. 8th Programme Report of UNDP/World Bank/WHO/Special Programme for Research & Training in Tropical Diseases (1988).
107 J. A. Walsch and K. S. Warren: N. Engl. J. Med. *301,* 967 (1979).
108 R. N. Iyer: Perspective in Parasitology, eds. A. B. Sen, J. C. Katiyar, P. Y. Guru, Print House (India), Lucknow, *1,* 275 (1985).
109 P. F. D'Archy and D. W. G. Harron: Pharm. Internat. 238 (1983).
110 W. Peters: Indian J. Med. Res. *73* (Suppl.) 1 (1981).
111 E. A. Steck: Prog. Drug Res. *18,* 289 (1974).

112 A. D. M. Bryceson: The Leishmaniasis, Vol. II, eds. W. Peters and R. Killick-Kendrick, Academic Press, New York, p. 847 (1987).
113 P. Kern: Antibiot. Chemother. *30,* 203 (1981).
114 Leishmaniasis: Bull WHO 44, 471 (1971).
115 J. D. Berman: Rev. Infect. Dis. *10,* 560 (1988).
116 L. G. Goodwin and J. E. Page: Biochem. J. *37,* 198 (1943).
117 G. de Cristina and G. Caronia: Deut. Arch. Klin. Med. *117,* 263 (1915); Trop. Dis. Bull. *7,* 80 (1916).
118 L. Rogers: Brit Med. J. II, 197 (1915).
119 P. Mackie: Brit. Med. J. II, 745 (1915).
120 G. C. Low: J. Trop. Med. Hyg. *18,* 258 (1915).
121 (a) C. P. Thakur, M. Kumar, P. Kumar, B. N. Mishra and A. K. Pandey: Brit. Med. J. *296,* 1557 (1988); (b) E. Beveridge: Experimental Chemotherapy, Vol. 1, eds. R. Schnitzer and F. Hawking, Academic Press, New York, p. 257 (1963).
122 J. La Pierre: Presse Med. *77,* 134 (1969).
123 H. Schmidt: J. Trop. Med. Hyg. *53,* 95 (1950).
124 M. G. Schultz, K. A. Western and D. Perera: J. Infect. Dis. *119,* 307 (1969).
125 Drugs for Parasitic Infections, Medical Letter on Drugs and Therapeutics *30,* 15 (1988).
126 G. M. Anabwani, J. A. Ngira, G. Dimiti and A. D. M. Bryceson: Lancet I, 210 (1983).
127 C. P. Thakur: Trans. Roy. Soc. Trop. Med. Hyg. *78,* 391 (1984).
128 C. N. Oster et al.: Amer. J. Trop. Med. Hyg. *34,* 856 (1985).
129 J. D. Chulay, E. M. Anzeze, D. K. Koech and A. D. M. Bryceson: Trans. Roy. Soc. Trop. Med. Hyg. *77,* 717 (1983).
130 A. D. M. Bryceson: Trans. Roy. Soc. Trop. Med. Hyg. *64,* 369 (1970).
131 R. E. Adballah, M. Ali, a. J. Wasfi and A. M. E. Hassan: Trans. Roy. Soc. Trop. Med. Hyg. *67,* 549 (1973).
132 B. C. Walton, T. E. Paulson, M. A. Arjona and C. A. Peterson: J. A. M. A. *228,* 1256 (1974).
133 The Leishmaniases, WHO Tech. Rep. Series No. 701 (1984).
134 N. Ansari, C. Mofidi and Chodjazade: Bull. Soc. Path. Exot. *43,* 546 (1950).
135 H. Floch: Bull. Soc. Path. exot. *47,* 509 (1954).
136 H. T. Chaglassian: Derm. Int. *6,* 168 (1967).
137 P. C. Sen Gupta: Indian Med. Gaz. *85,* 291 and 547 (1950).
138 P. J. Carbalho: Hospital, Rio de Janeiro. *54,* 387 (1958).
139 P. C. Sen Gupta, B. Bhattacharya, A. Dass and C. Sanyal: J. Trop. Med. Hyg. *55,* 205 (1952).
140 E. A. Llanos-Cuentas: Masters Thesis, University of Brasilia (1984); cited by P. M. McGreevy and P. D. Marsden, in Chemotherapy of Parasitic Diseases, eds. W. C. Campbell and R. S. Rew, Plenum Press, New York, p. 123 (1986).
141 R. A. A. Rocha, R. N. R. Sampaio, M. Guerra, A. Magalhaes, C. C. Cuba, A. C. Barreto and P. D. Marsden: J. Trop. Med. Hyg. *83,* 131 (1980).
142 R. N. R. Sampaio, R. A. A. Rocha, P. D. Marsden, C. C. Cuba and A. C. Barreto: Ann. Brasil Dermatol. *55,* 69 (1980).
143 R. N. R. Sampaio: Masters Thesis, Universidade Federal de Belo Horizonte (1984); cited by P. B. McGreevy and P. D. Marsden, in Chemotherapy of Parasitic Disease, eds. W. C. Campbell and R. S. Rew, Plenum Press, New York, p. 123 (1986).
144 P. D. Marsden, R. N. R. Sampaio, E. M. Carvalho, J. P. T. Veiga, J. L. M. Costa and E. A. Llanos-Cuentas: Amer. J. Trop. Med. Hyg. *34,* 710 (1985).

145 H. Smidt and F. M. Peter: Advances in the Therapeutics of Anitmony, Thieme Verlag, Leipzig (1938).
146 U. N. Brahmachari: J. Trop. Med. Hyg. *24,* 213 (1921).
147 A. Lignos: Bull. Soc. Path. Exot. *25,* 1043 (1932).
148 F. P. Mackie: Indian Med. Gaz. *56,* 76 (1921).
149 L. E. Napier, Lancet. I, 280 (1923).
150 D. N. Banerji: Calcutta Med. J. *17,* 265 (1923).
151 L. E. Napier: Indian J. Med. Res. *15,* 81 (1927).
152 M. H. Sati: Ann. Trop. Med. Parasit. *36,* 1 (1942).
153 P. C. Sen Gupta: Indian Med. Gaz. *88,* 20 (1953).
154 E. A. Cacchia and F. F. Fench: Trans. Roy. Soc. Trop. Med. Hyg. *58,* 234 (1964).
155 U. N. Brahmachari: Indian J. Med. Res. *10,* 492, 508, 948 (1922); *11,* 393, 405 (1923); *12,* 397 (1924); *13,* 111 (1925).
156 U. N. Brahmachari: J. Trop. Med. Hyg. *44,* 67 (1941).
157 U. N. Brahmachari and J. Das: Indian J. Med. Res. *12,* 423 (1924); *13,* 693 (1925).
158 H. E. Shortt and R. T. Sen: Indian J. Med. Res. *11,* 653 (1923); *12,* 335 (1924).
159 H. E. Shortt: Trans. Roy. Soc. Trop. Med. Hyg. *39,* 13 (1945).
160 E. A. Ho, T. H. Soong and Y. Li: Trans. Roy. Soc. Trop. Med. Hyg. *41,* 573 (1948); *42,* 573 (1949).
161 T. T. Chen: Ann. Trop. Med. Parasit. *43,* 174 (1949).
162 P. E. C. Manson-Bahr and R. B. Heisch: Trans. Roy. Soc. Trop. Med. Hyg. *50,* 465 (1956).
163 F. Biagi: in Tropical Medicine, 5th ed., eds. G. W. Hunter, J. C. Swartzwelder and D. F. Clyde, W. B. Saunders Company, p. 415 (1976).
164 WHO Offset Publication, TDR/CHEMLEISH/VL/82.3, p. 19 (1982); cited by P. M. McGreevy and P. D. Marsden in Chemotherapy of Parasitic Diseases eds. W. C. Campbell and R. S. Rew, Plenum Press, New York p. 119 (1986).
165 J. D. Chulay, S. M. Bhatt, R. Muigai, M. Ho, G. Gachihi, J. B. O. Were, C. Chunge and A. D. M. Bryceson: J. Infect. Dis. *148,* 148 (1983).
166 P. H. Rees, M. I. Keating, P. A. Kager and W. T. Hochmeyer: Lancet. *2,* 226 (1980).
167 J. D. Chulay, H. C. Spencer and M. Mugambi: Amer. J. Trop. Med. Hyg. *34,* 702 (1985).
168 J. N. Ashley, J. Barber, A. J. Ewins, G. S. Newberry and A. D. H. Self: J. Chem. Soc. 103 (1942).
169 J. N. Ashley and H. O. Harris: J. Chem. Soc. 567 (1946).
170 E. B. Schoenbach and E. M. Greenspan: Medicine *27,* 327 (1948).
171 F. N. Fastier: Pharm. Rev. *14,* 37 (1962).
172 A. Kreutzberger: Prog. Drug Res. *11,* 356 (1968).
173 S. Alder and I. Tchernomoretz: Ann. Trop. Med. Parasit. *33,* 313 (1939); *36,* 11 (1942).
174 A. R. D. Adams and W. Yorke: Ann. Trop. Med. Parasit. *33,* 323 (1939); *34,* 173 (1940).
175 S. Alder and M. Rachmilewitz: Ann. Trop. Med. Parasit. *33,* 329 (1939).
176 A. L. Wingfield: Ann. Trop. Med. Parasit. *35,* 55 (1941).
177 P. C. Sen Gupta: Indian Med. Gaz. *78,* 537 (1943); *79,* 50 (1944).
178 H. L. Chung, H. K. Chow, N. Chang and Y. P. Ts'ao: Chinese Med. J. *69,* 171 (1951).
179 J. D. Fulton: Ann. Trop. Med. Parasit. *37,* 48 (1943).
180 P. C. Sengupta: Indian Med. Gaz. *85,* 547 (1950).
181 P. C. Sengupta: Lancet II, 97 (1949).
182 R. C. Morris, D. O'Brien and J. C. Gonick: Amer. J. Med. *30,* 624 (1961).

183 I. Snapper: Amer. J. Med. *13,* 655 (1952).
184 P. C. Sengupta: Indian Med. Gaz. *79,* 49 (1944).
185 P. Kirk: Proc. 6th Internat. Cong. Trop. Med. Mal. Lisbon. *3,* 815 (1958).
186 N. Ansari, C. Mofidi and C. Zade: Bull. Soc. Path. Exot. *43,* 546 (1950).
187 T. M. Lee and C. C. Ling: Chin. Med. J. *69,* 160 (1951).
188 A. N. Hazarika: Indian Med. Gaz. *84,* 140 (1949).
189 P. C. Lopes and M. A. DeAlmeida: Hospital, Rio de Janeiro *73,* 223 (1968).
190 Y. P. Silva: Hospital, Rio de Janeiro, *42,* 261 (1952).
191 A. Correa and G. da R. Brito: Bull. Soc. Path. Exot. *47,* 513 (1954).
192 T. P. Waalkes, C. Denham and V. T. de Vita: Clin. Pharmac. Ther. *11,* 505 (1970).
193 H. Meith: Z. Tropenmed. Parasit. *12,* 320 (1966).
194 W. Raether, H. Seidenath and H. Loewe: Ann. Trop. Med. Parasit. *72,* 543 (1978).
195 B. Festy, J. Sturm and M. Daune: Biochem. Biophysica Acta, *407,* 24 (1975).
196 B. P. Das and D. W. Boykin: J. Med. Chem. *20,* 1219 (1977).
197 M. J. Waring: J. Mol. Biol. *54,* 247 (1970).
198 R. M. Wartell, J. E. Larson and R. D. Wells: J. Biol. Chem. *250,* 2698 (1975).
199 P. M. S. Chauhan, K. V. B. Rao, R. N. Iyer, V. Shankhdhar, P. Y. Guru and A. B. Sen: Indian J. Chem. *26B,* 248 (1987).
200 P. M. S. Chauhan, R. N. Iyer and D. S. Bhakuni: Indian J. Chem. *27B,* 144 (1988).
201 P. M. S. Chauhan and R. N. Iyer: Indian J. Chem. *22B,* 898 (1983).
202 E. A. Steck, K. E. Kinnamon, D. S. Rane and W. L. Hanson: Expt. Parasit. *52,* 404 (1981).
203 B. P. Das, V. B. Zalkow, M. L. Forrester, P. F. Molock and D. W. Boykin: J. Pharm. Sci. *71,* 465 (1982).
204 A. K. Kurban: Hautarzt. *24,* 369 (1973).
205 P. J. Hare: Brit. J. Derm. *74,* 102 (1962).
206 J. Homez: Ann. Derm. Syphilol. *83,* 271 (1956).
207 G. N. Pershin and N. Yu. Moskalenko: Farmac. Toksikol. *26,* 446 (1963).
208 E. Beveridge, G. Goodwin and L. P. Walls: Nature *182,* 316 (1958).
209 J. D. Berman: in The Leishmaniasis, eds. K. P. Chang and R. S. Bray, Elsevier Science Publishers, Amsterdam (1985).
210 E. Beveridge and R. A. Neal: Dermat. Internat. *6,* 163 (1967).
211 W. Peters: Postgrad. Med. J. *49,* 573 (1973), cited by E. A. Steck: Prog. Drug Res. *18,* 324 (1974).
212 R. A. Neal: in The Leishmaniasis Vol. II, eds. W. Peters and R. Killick-Kendrick, Academic Press, New York, pp. 793–845 (1987).
213 P. A. Barrett, A. C. Caldwell and L. P. Walls: J. Chem. Soc. 2404 (1961).
214 K. E. Kinnamon, E. A. Steck, P. S. Loiseaux, W. L. Hanson, W. L. Chapman, Jr. and V. B. Waits: J. Trop. Med. Hyg. *27,* 751 (1978).
215 M. P. La Montagne, P. D. Dagli, M. S. Khan and P. Bhumbergs: J. Med. Chem. *23,* 981 (1980).
216 M. S. Khan and M. P. La Montagne: J. Med. Chem. *22,* 1005 (1979).
217 F. I. Carroll, B. O. Berrang and C. P. Linn: J. Med. Chem. *23,* 581 (1980).
218 J. L. Johnson and L. M. Werbel: J. Med. Chem. *26,* 185 (1983).
219 R. A. Neal, S. L. Croft and D. J. Nelson: Trans. Roy. Soc. Trop. Med. Hyg. *79,* 122 (1985).
220 J. C. Anders. H. Chung and A. D. Theoharides: Fundam. Appl. Toxicol. (in press); cited by S. L. croft: Trends Pharm. Sci. *9,* 376 (1988).
221 A. D. Theoharides, H. Chung and H. Velazquez: Biochem. Pharmacol. *34,* 181 (1985).

222 T. A. Furtado and A. C. Viegas: Trop. Dis. Bull. *65,* 987 (1968).
223 P. I. Long: J. A. M. A. *223,* 1378 (1973).
224 J. K. P. Derson and S. Sawicki: Arch. Dermat. *111,* 1343 (1975).
225 H. F. Beltran, M. Guttierez and F. F. Biagi: Bull. Soc. Path. Exot. *60,* 61 (1967).
226 J. Masramon, J. Aubia, J. Lloveras, J. Llorach and M. Llorce: Lancet *1,* 669 (1981).
227 K. J. Murphy and A. C. W. Bong: Lancet. *1,* 323 (1981).
228 M. C. Baranski: Trop. Dis. Bull. *68,* 1058 (1970).
229 T. A. Furtado, J. Contijo and A. C. Viegas: Hospital Raio de Janeiro. *75,* 75 (1969); Trop. Dis. Bull. *66,* 530 (1969).
230 A. C. Viegas and T. A. Furtado: Anais Bras. Derm. *43,* 163 (1968); Trop. Dis. Bull. *67,* 277 (1970).
231 H. H. Salem, H. M. Elkomy and G. El-Allaf: Trans. Roy. Soc. Trop. Med. Hyg. *63,* 388 (1969).
232 C. M. Johnson: Amer. J. Trop. Med. Hyg. *17,* 819 (1968).
233 A. D. H. Bryceson, W. A. Forster and A. Lenna: Trans. Roy. Soc. Trop. Med. Hyg. *63,* 152 (1969).
234 J. D. Berman, L. J. Goad, D. H. Beach and G. G. Holz: Mol. Biochem. Parasitol. *20,* 85 (1986).
235 J. D. Berman: Amer. J. Trop. Med. Hyg. *30,* 566 (1981).
236 R. Kubba, Y. Al-Gindan, A. M. El-Hassan and A. H. S. Omer: Saudi Med. J. *7,* 596 (1986).
237 J. Viallet et al.: Amer. J. Trop. Med. Hyg. *35,* 491 (1986).
238 P. G. Butler: J. Trop. Med. Hyg. *81,* 221 (1978).
239 R. VanGinckel and M. De Brabander: J. Reticuloendothel. Soc. *25,* 125 (1979).
240 P. M. S. Chauhan, R. N. Iyer and G. P. Dutta: Indian J. Chem. *22B,* 894 (1983).
241 R. E. Coleman, J. D. Edman and L. H. Semprevivo: Trans. Roy. Soc. Trop. Med. Hyg. *82,* 232 (1988).
242 J. J. Marr and R. L. Berens: Mol. Biochem. Parasitol. *7,* 339 (1983).
243 B. Ullman: Pharm. Res. *5,* 194 (1984).
244 P. A. Kager, P. H. Rees, B. T. Wallde, W. U. Hockmeyer and W. H. Lyerly: Trans. Roy. Soc. Trop. Med. Hyg. *75,* 556 (1981).
245 (a) C. N. Chunge et al.: Trans. Roy. Soc. Trop. Med. Hyg. *79,* 715 (1985); (b) S. Martinez et al.: Amer. J. Trop. Med. Hyg. *39,* 250 (1988).
246 R. E. Saenz, H. Paz, C. M. Johnson, M. D. Rogers, D. J. Nelson and K. H. Pattishall: IXth Internat. Congr. Infect. Parasit. Dis., Munchen, July 20–26 (1986).
247 S. A. P. Sampaio, M. Castro, N. L. Dillon and J. E. Costa-Martins: Internat. J. Dermat. *10,* 179 (1971).
248 P. J. T. Mayer: Munch. Med. Wschr. *116,* 1539 (1974).
249 P. Yesudian and A. S. Thambiah: Arch. Dermat. *109,* 720 (1974).
250 S. Ganor: Derm. Internat. *6,* 141 (1967).
251 B. K. Ghose and A. N. Chatterjee: Antiobot. Chemother. *12,* 221 (1962).
252 B. K. Ghose and A. N. Chatterjee: Ann. Biochem. Expt. Med. (Calcutta) *21,* 25 (1961).
253 M. M. El-din Selim and E. Kandil: J. Kuwait Med. Assoc. *75,* 159 (1972).
254 I. O. Iskander: J. Inst. Med. Res. *6,* 280 (1978).
255 N. Even-Paz, L. Weinrauch, R. Livshine, J. El-On and C. L. Greenblatt: Internat. J. Dermat. *21,* 110 (1982).
256 O. I. Kellina, A. V. Inyakhina and R. I. Yastrebova: Trop. Dis. Bull. *63,* 1187 (1966).
257 A. A. Akoubyan and S. M. Mukhamedov: Trop. Dis. Bull. *65,* 987 (1968).
258 M. Restrespo and J. Velaquez: Trans. Roy. Soc. Trop. Med. Hyg. *67,* 616 (1973).

259 P. D. Marsden, C. C. Cuba, A. C. Barreto, R. N. Sampaio and R. A. Rocha: Trans. Roy. Soc. Trop. Med. Hyg. *73,* 391 (1979).
260 G. Gonzalez-Ochoa and M. O. Collado C.: Trop. Dis. Bull. *70,* 732 (1973).
261 A. S. Azab, M. S. Kamal, M. S. El Haggar, B. A. Metawaa and D. S. Hindawy: J. Dermatol. Surg. Oncol. *9,* 1007 (1983).
262 A. Bassiouny, M. El Meshad, M. Talaat, K. Kutty and B. Metawaa: Brit. J. Dermatol. *107,* 467 (1982).
263 F. A. Neva, E. A. Petersen, R. Corsey, D. H. Bogaert and D. Martinez: Amer. J. Trop. Med. Hyg. *33,* 800 (1984).
264 E. A. Steck: The chemotherapy of Protozoal Diseases, Vols. I–IV, Division of Medicinal Chemistry, Watter Reed Army Institute of Research, Washington, D. C. (USA), 1972.
265 E. Solano, H. Hidalgo and R. Zeledon: Med. Cutan. Iber. Lat. Am. *12,* 19 (1984).
266 J. El-On, G. P. Jacobs, E. Witzum and C. L. Greenblatt: Antimicrob. Agents Chemother. *26,* 745 (1984).
267 J. El-On, G. P. Jacobs and L. Weinrauch: Parasitol Today *4,* 76 (1988).
268 W. Dutz, N. Agarval, M. Z. Bashardost, G. Bentley and C. O. Kindmark: Int. J. Dermatol. *26,* 199 (1987).
269 S. L. Croft: Pharm. Internat. *7,* 229 (1986).
270 C. R. Alving: Pharm. Ther. *22,* 407 (1983).
271 R. C. C. New, M. L. Chance and S. Heath: Antimicrob Chemother. *8,* 371 (1981).
272 J. R. Brown: Pharm. Internat. *4,* 61 (1983).
273 African Trypanosomiasis: WHO Tech. Rep. Ser. No. 434 (1969).
274 R. S. Rew: in Chemotherapy of Parasitic Diseases, eds. W. C. Campbell and R. S. Rew, Plenum Press, New York, p. 129 (1986).
275 F. Hawking: Trans. Roy. Soc. Trop. Med. Hyg. *34,* 37 (1940).
276 I. Rollo: in The Pharmacological Basis of Therapeutics eds. A. G. Gillmann, L. S. Goodman and A. Gillman, MacMillan Publishing Co. New York, p. 1070 (1980).
277 The African Trypanosomiasis: WHO Tech. Rep. Ser. No. 635 (1979).
278 B. Thylefors and A. Rolland: Bull. WHO, *57,* 479 (1979).
279 C. Chesterman: Trans. Roy. Soc. Trop. Med. Hyg. *18,* 311 (1924).
280 E. M. Lourie and W. York: Am. Trop. Med. Parasit. *33,* 289 (1939).
281 F. Hawking in Experimental Chemotherapy, Vol. I, eds. R. J. Schnitzer and F. Hawking, Academic Press, New York, p. 129 (1963).
282 L. Launoy, M. Guillot and H. Jonchere: Ann. Pharm. Fr. *18,* 273 (1960).
283 T. P. Waalkes, C. Denham and V. T. DeVita: Clin. Pharmacol. Ther. *11,* 505 (1970).
284 F. Hawking: in Tropical Medicine, 5th ed., eds. G. W. Hunter, J. C. Swartzwelder and D. F. Clyde, W. B. Saunders Company, Philadelphia, p. 437 (1976).
285 C. J. Bacchi, H. C. Nathan, S. H. Hunter, P. P. McCann and A. Sjoerdsma: Science *210,* 332 (1980).
286 P. P. McCann, C. J. Bacchi, S. L. Hanson, G. D. Cain, H. C. Nathan, S. H. Hunter and A. Sjoerdsma: Adv. Polyamine Res. (eds. C. M. Calderera, V. Zuppia and U. Bachrach, Raven Press, New York), *3,* 97 (1981).
287 H. Taelman et al.: Amer. J. Med. *82,* 607 (1987).
288 F. Doua et al.: Amer. J. Trop. Med. Hyg. *37,* 525 (1987).
289 J. Pepin et al.: Lancet *2,* 1431 (1987).
290 W. A. Jacobs and M. Heidelberger: J. Expt. Med. *30,* 411 (1919).
291 L. Launoy and O. Fleury: Bull. Soc. Path. Exot. *30,* 315 (1937).
292 E. A. H. Friedheim: Amer. J. Trop. Med. *29,* 173 (1949).
293 J. O. L. Arroz, R. F. P. Carmo and J. L. Neves: Trop. Dis. Bull. *70,* 27 (1973).

294 H. J. C. Watson: Trans. Roy. Soc. Trop. Med. Hyg. *59,* 113 (1965).
295 F. Evans, K. Niemegreens and A. Packchanian: Amer. J. Trop. Med. Hyg. *6,* 665 (1957).
296 F. I. C. Apted: Trans. Roy. Soc. Trop. Med. Hyg. *54,* 225 (1960).
297 K. Adriaenssens: Ann. Soc. Belge Med. Trop. *1,* 32 (1962).
298 F. I. C. Apted: Pharmacol. Ther. *11,* 391 (1980).
299 A. B. Clarkson Jr., C. J. Bacchi, G. H. Mellow, H. C. Nathan, P. P. McCann and A. Sjoerdsma: Proc. Nat. Acad. Sci. USA *80,* 5729 (1983).
300 A. B. Clarkson Jr., E. J. Bienen, C. J. Bacchi, P. P. McCann, H. C. Nathan, S. H. Hunter and A. Sjoerdsma: Amer. J. Trop. Med. Hyg. *33,* 1073 (1984).
301 O. Dann, E. Hiecke, H. Hahn, H. H. Miserre, G. Lurding and R. Rosser: Justus Liebigs Ann. Chem. *734,* 23 (1970).
302 O. Dann, G. Bergen, E. Demant and G. Volz: Justus Liebigs Ann. Chem. *749,* 68 (1971).
303 O. Dann, G. Volz, E. Demant, W. Pfeifer, G. Bergen, H. Fick and E. Walkenhorst: Justus Liebigs Ann. Chem. 1112 (1973).
304 B. P. Das and D. W. Boykin: J. Med. Chem. *20,* 531 (1977).
305 B. P. Das, R. A. Wallace and D. W. Boykin Jr.: J. Med. Chem. *23,* 578 (1980).
306 F. W. Jennings: Trans. Roy. Soc. Trop. Med. Hyg. *81,* 616 (1987).
307 P. D. Marsden: in Medicine in the Tropics, ed. A. W. Woodruff, Churchill Livingstone, London, p. 91 (1974).
308 A. C. Gomez Nunez, J. C. Gallimore, J. M. Fernandez and A. Gross: Acta Cient. Venez. *13,* 46 (1963).
309 F. Köberle: Bull WHO *42,* 739 (1970).
310 R. J. Earlam: Amer. J. Dig. Dis. *17,* 559 (1972).
311 S. Brener: Pharmacol. Ther. *7,* 71 (1979).
312 P. B. McGreevy and P. D. Marsden: in Chemotherapy of Parasitic Diseases, eds. W. C. Campbell and R. S. Rew, Plenum Press, New York, p. 115 (1986).
313 V. Macedo: in Cecil Textbook of Medicine, 16th ed., Eds. J. B. Wyngaarden and L. H. Smith, W. B. Saunders Company, Philadelphia, p. 1728 (1982).
314 D. M. James and H. M. Gilles: Human Antiparasitic Drugs: Pharmacology and Usage, John-Wiley & Sons, New York, p. 269 (1985).
315 C. L. Bocca, T.: Bol. Chile Parasit. *24,* 24 (1969).
316 M. Rubio and F. Donoso: Bol. Chile Parasit. *24,* 43 (1969).
317 J. J. Fernandez, R. A. Cedillos and G. A. Godoy: Bol. Chile Parasit. *24,* 51 (1969).
318 J. A. Chichero, E. Segura and J. C. Quatrochi: Bol. Chile Parasit. *24,* 59 (1969).
319 H. Schenone, L. Coucha, R. Aranda, A. Rojas, F. Knierim and M. Roja: Bol. Chile Parasit. *27,* 11 (1972).
320 J. R. Cancado and Z. Brener, Terapeutica: in Trypanosoma cruzi: Doenca de Chagas, eds. Z. Brener and Z. Andrade, Guanabra Koogan, Rio de Janeiro, p. 362 (1979).
321 J. R. Concado et al.: Bol. Chile Parasit. *24,* 28 (1969).
322 G. C. Levi and V. A. Amato: Rev. Inst. Med. Trop. São Paulo *13,* 369 (1971).
323 A. Rossi and H. D. Ferreira: Rev. Soc. Bras. Med. Trop. *5,* 235 (1971).
324 T. Pizzi: Bol. Chile Parasit. *16,* 35 (1961).
325 E. Boainain: Rev. Goiana Med. *25,* 1 (1979).
326 A. R. L. Teixeira, E. Jabur, J. C. Cordoba, I. C. S. Maior and E. Solorzano: Rev. Soc. Brasil Med. Trop. *16,* 11 (1983).
327 L. S. Filardi and Z. Brener: Ann. Trop. Med. Parasit. *76,* 293 (1982).

328 T. A. Maria, L. S. Filardi and Z. Brener: Rev. Soc. Brasil. Med. Trop. *17*, 89 (1984).
329 S. G. Andrade, R. C. Silva, C. M. G. Santiago and L. A. R. Freitas: Bull. WHO *65*, 625 (1987).
330 S. Gupta, S. Sharma and J. C. Katiyar: J. Sci. Industr. Res. *47*, 602 (1988).
331 J. C. Katiyar, S. Gupta and S. Sharma: Rev. Infect. Dis. *11*, 638 (1989).
332 P. A. J. Janssen: Prog. Drug. Res. *18*, 191 (1974).
333 C. K. Rao: J. Commun. Dis. *14*, 206 (1982).
334 J. C. Katiyar and P. K. Murthy: J. Sci. Industr. Res. *49*, 112 (1990).
335 F. Partono and J. H. Cross et al.: Southeast Asian J. Trop. Med. Publ. Health *3*, 366 (1972).
336 (a) B. A. Southgate: Trans. Roy. Soc. Trop. Med. Hyg. *68*, 177 (1974); (b) D. T. Dennis, E. McConnell and G. B. White: Amer. J. Trop. Med. Hyg. *25*, 257 (1976).
337 (a) J. Z. Farooqui, R. P. Saxena, A. Tandon, V. K. Srivastava and K. C. Saxena: Indian J. Parasit. *3*, 91 (1979); (b) P. Kaliraj, S. N. Ghirnikar and B. C. Harinath: Curr. Sci. *46*, 603 (1977).
338 (a) A. Tandon, P. K. Murthy, R. P. Saxena, A. B. Sen and K. C. Saxena: Indian J. Med. Res. *87*, 429 (1988); (b) J. C. Katiyar, R. Chandra, P. K. Murthy, K. Tyagi and A. B. Sen: Indian J. Med. Res. *81*, 465 (1985).
339 R. G. Hamilton: Rev. Infect. Dis. *7*, 837 (1985).
340 A. A. Buck (ed.): Onchocerciasis-Symptomatology, pathology, diagnosis. WHO, Geneva, pp. 80 (1974).
341 (a) J. C. Mueller, D. W. Mitchell et al.: Amer. J. Trop. Med. Hyg. *22*, 337 (1973); (b) P. G. Janssen: in ref. 2 p. 239 (1974).
342 M. A. Ouaissi, L E. Kouemeni, I. D. Moutis, A. Haque and A. Capron: Medecine et Armee *10*, 157 (1982).
343 G. Edwards: Trends in Pharmacol. Sci. *5*, 192 (1984).
344 S. Sharma: Prog. Drug Res. *30*, 473 (1986).
345 L. Rogers: Brit. Med. J. *1*, 596 (1920).
346 J. T. Culbertson and H. M. Rose: J. Pharmacol. *81*, 189 (1944).
347 J. T. Culbertson, Trans. Roy. Soc. Trop. Med. Hyg. *41*, 18 (1947).
348 J. T. Culbertson: Ann. N. Y. Acad. Sci. *50*, 73 (1948).
349 P. Sewell and F. Hawking: Brit. J. Pharmacol. *5*, 239 (1950).
350 G. F. Otto and T. H. Maren: Science *106*, 105 (1947); Ann. N. Y. Acad. Sci. *50*, 39 (1948); Amer. J. Hyg. *50*, 92, 94, 121, 135 (1949).
351 S. Kume: Amer. J. Vet. Res. *18*, 912 (1957).
352 W. E. Kershaw and J. Williamson: Ann. Trop. Med. Parasit. *46*, 268 (1952).
353 E. A. H. Friedheim: Trans. Roy. Soc. Trop. Med. Hyg. *56*, 236 (1962).
354 T. B. Panse: Bombay Technologists p. 53 (1951).
355 F. Hawking: Prog. Drug. Res. *9*, 191 (1966).
356 J. A. McFadzean and F. Hawking: Brit. Med. J. *1*, 956 (1954).
357 G. F. Otto et al.: Amer. J. Trop. Med. Hyg. *1*, 470 (1952); *2*, 495 (1953).
358 L. L. Ashburn, T. A. Burch and F. J. Brady: Bol. Ofic. Sanit. Pan-Am. *28*, 1107 (1949).
359 E. A. H. Friedheim: Ann. Trop. Med. Parasit. *56*, 343, 387 (1962).
360 D. D. McCarthy, E. Simpson and P. Robit: N. Z. Med. J. *61*, 143 (1962).
361 B. O. L. Duke: Trans. Roy. Soc. Trop. Med. Hyg. *60*, 691 (1966); Bull. WHO *39*, 157, 179 (1968).
362 H. Fuglsang and J. Anderson: Z. Tropenmed. Parasit. *28*, 439 (1977).
363 D. G. Barranco: Rev. Invest. Salud. Pub. *37*, 37 (1977).
364 D. E. Abaru and J. E. McMohan: Z. Tropenmed. Parasit. *29*, 175 (1978).
365 M. S. Mallen, D. G. Barranco and J. J. Mendozo: Ann. Trop. Med. Parasit. *65*, 393 (1971).

366 M. S. Mallen, D. G. Barranco and H. D. C. Moutes: Rev. Inst. Med. Trop. São Paulo *13*, 363 (1971).
367 K. Awadzi and H. M. Gilles: Ann. Trop. Med. Hyg. *74*, 199 (1980).
368 M. Wanson: Ann. Soc. Belge Med. Trop. *30*, 671 (1950).
369 T. A. Burch and L. L. Ashburn: Amer. J. Trop. Med. *31*, 617 (1951).
370 J. W. R. Sarkies: Trans. Roy. Soc. Trop. Med. Hyg. *46*, 435 (1952).
371 M. J. Miller: Prog. Drug Res. *20*, 433 (1976).
372 F. Hawking: Adv. Pharmacol. Chemother. *15*, 289 (1978).
373 B. O. L. Duke: Bull. WHO *39*, 157 (1968).
374 S. Sharma: Pharm. Res. *3*, 75 (1986).
375 R. I. Hewitt, E. White, S. Kushner, W. S. Wallace, H. W. Stewart and Y. Subbarow: Ann. N. Y. Acad. Sci. *50*, 128 (1948).
376 R. I. Hewitt, W. S. Wallace, E. White and Y. Subbarow: J. Lab. Clin. Med. *32*, 1293 (1947).
377 R. I. Hewitt, E. White, W. S. Wallace, H. W. Stewart, S. Kushner and Y. Subbarow: J. Lab. Clin. Med. *32*, 1304 (1947).
378 R. I. Hewitt, S. Kushner, H. W. Stewart, E. White, W. S. Wallace and Y. Subbarow: J. Lab. Clin. Med. *32*, 1314 (1947).
379 F. Hawking: Adv. Pharmacol. Chemother. *16*, 129 (1979).
380 World Health Organisation: WHO Tech. Rep. Ser. No. 542 (1974); cited in ref. 371.
381 F. Hawking: Prog. Drug Res. *18*, 173 (1974).
382 B. S. Seo and J. W. Lee: Korean J. Parasit. *11*, 1 (1973).
383 F. Ciferri, N. Siliga, G. Long and J. F. Kessel: Amer. J. Trop. Med. Hyg. *18*, 369 (1969).
384 B. S. Seo and K. Whang: Korean J. Parasit. *12*, 21 (1974).
385 R. M. Sundaram, N. K. Rao, S. K. Rao and C. K. Rao: J. Commun. Dis. *6*, 290 (1974).
386 J. L. K. Hii et al.: Amer. J. Trop. Med. Hyg. *38*, 582 (1988).
387 F. Hawking and D. A. Denham: Trop. Dis. Bull. *73*, 348 (1976).
388 F. Hawking and R. J. Marques: Bull. WHO, *37*, 405 (1967).
389 A. Davis and D. R. Bailey: Bull. WHO *41*, 195 (1969).
390 A. B. Sen, R. Chandra, J. C. Katiyar and S. Chandra: Indian J. Med. Res. *62*, 1181 (1974).
391 P. C. Fan, Y. C. Wang, J. C. Liu and M. L. King: Ann. Trop. Med. Parasit. *69*, 515 (1975).
392 E. A. Ottesen: WHO/FIL/84.174 p. 1–21 (1984), cited by D. A. Denham, in Chemotherapy of Parasitic Diseases. Eds. W. C. Campbell and R. S. Rew. Plenum Press, New York, p. 277 (1986).
393 L. Mazotti: Rev. Inst. Sal. Enferm. Trop. *9*, 235 (1948).
394 B. O. L. Duke and J. Anderson: Trop. Doctor *2*, 107 (1972).
395 B. O. L. Duke: Internat. Symp. Onchocerciasis, Washington, p. 46 (1974).
396 O. O. Kale: J. Helminthol. *53*, 169 (1979).
397 K. Awadzi and H. M. Gilles: Ann. Trop. Med. Parasit. *74*, 199 (1980).
398 H. R. Taylor, M. E. Langham, E. M. de Stahl, L. N. Figueroa and F. Beltranena: Z. Tropenmed. Parasit. *31*, 357 (1980).
399 M. E. Langham, Z. D. Traube and R. Richardson: Z. Tropenmed. Parasit. *29*, 156 (1978).
400 A. P. Oomen: Trans. Roy. Soc. Trop. Med. Hyg. *63*, 548 (1969).
401 H. Fuglsand and J. Anderson: Trans. Roy. Soc. Trop. Med. Hyg. *68*, 72 (1974).
402 H. Bonnin and G. F. Moretti: Bull. Soc. Path. Exot. *43*, 279 (1950).
403 D. A. Denham: in Chemotherapy of parasitic diseases. Eds. W. C. Campbell and R. S. Rew. Plenum Press, New York, p. 277 (1986).
404 B. O. L. Duke: Ann. Trop. Med. Parasit. *57*, 82 (1963).

405 C. F. Bartholomew, M. B. Nathan and E. S. Tikasingh: Trans. Roy. Soc. Trop. Med. Hyg. *72,* 423 (1978).
406 P. E. Adolph, I. G. Kagan and R. M. McQuay: Amer. J. Trop. Med. Hyg. *11,* 76 (1962).
407 A. P. Omen: Trans. Roy. Soc. Trop. Med. Hyg. *63,* 548 (1969).
408 H. Fuglsang and J. Anderson: Trans. Roy. Soc. Trop. Med. Hyg. *68,* 72 (1974).
409 B. R. Jones: Trans Ophtalmol. Soc. U. K. *90,* 299 (1970).
410 C. G. B. Downie: J. R. Army Med. Corps. *112,* 46 (1966).
411 A. C. Bird, H. El-Seikh, J. Anderson and H. Fuglsang: Lancet *2,* 46 (1979).
412 H. O. Chlebowski and E. Zielke: Z. Tropenmed. Parasit. *31,* 339 (1980).
413 G. Laemmler, H. Herzog and H. R. Schuetze: Bull. WHO *44,* 757 (1971).
414 G. Laemmler: Pestic Sci. *8,* 563 (1977).
415 R. Saxena, S. Sharma, R. N. Iyer and N. Anand: J. Med. Chem. *14,* 929 (1971).
416 S. Sharma, R. Bindra, R. N. Iyer and N. Anand: J. Med. Chem. *18,* 913 (1975).
417 R. Rastogi, S. Sharma, N. Anand, N. Fatma, A. Dutta, R. K. Chatterjee and A. B. Sen: J. Helminthol. *58,* 251 (1984).
418 S. Sharma, S. K. Dubey, V. K. Agrawal, R. N. Iyer, N. Anand, R. K. Chatterjee, S. Chandra and A. B. Sen: Indian J. Chem. *26B,* 748 (1987).
419 P. E. Thompson, J. B. Zeigler and J. W. McCall: Antimicrob. Agents Chemother. *3,* 693 (1973).
420 V. P. Arya: Drugs of Future *5,* 229 (1980).
421 P. A. J. Janssen: Prog. Drug Res. *20,* 347 (1976).
422 M. Merlin, B. Carme, H. Kaeuffer and J. Laigret: Bull. Soc. Path. Exot. *69,* 257 (1976).
423 M. V. V. L. Narasimham, S. P. Roychowdhury, M. Das and C. K. Rao: Southeast Asian J. Trop. Med. Pub. Health *9,* 571 (1978).
424 J. E. McMahon: Ann. Trop. Med. Parasit. *73,* 465 (1979).
425 H. G. Bernberg, V. de V. Clark and M. Gelfand: Trans. Roy. Soc. Trop. Med. Hyg. *73,* 233 (1979).
426 H. Vanden Bossche, F. Rochette and C. Horig: Adv. Pharmacol. Chemother. *19,* 67 (1982).
427 S. Sharma and Syed Abuzar: Prog. Drug Res. *27,* 85 (1983).
428 W. C. Campbell and A. C. Cuckler: Tex. Rep. Biol. Med. *27* (Suppl. 2), 665 (1969).
429 L. Chantin, M. Merlin and J. Laigret: Bull. Soc. Path. Exot. *68,* 198 (1975).
430 K. Maertens and M. Wery: Trans. Roy. Soc. Trop. Med. Hyg. *69,* 359 (1975).
431 R. Rivas-Alcala, C. D. Mackenzie, E. Gomez-Rojo, B. M. Green and H. R. Taylor: Tropenmed. Parasitol. *35,* 71 (1984).
432 O. Kale: Z. Tropenmed. Parasit. *29,* 63 (1978).
433 A. R. Rivas-Alcala et al.: Lancet *2,* 485 (1981).
434 D. A. Denham, R. R. Suswillo and R. Rogers: Trans. Roy. Soc. Trop. Med. Hyg. *72,* 546 (1978).
435 I. V. Kerckhoven and V. Kumar: Trans. Roy. Soc. Trop. Med. Hyg. *82,* 890 (1988).
436 R. Dubey, S. Abuzar, S. Sharma, R. K. Chatterjee and J. C. Katiyar: J. Med. Chem. *28,* 1748 (1985).
437 S. Abuzar, S. Sharma, N. Fatma, S. Gupta, P. K. Murthy, J. C. Katiyar, R. K. Chatterjee and A. B. Sen: J. Med. Chem. *29,* 1296 (1986).
438 R. K. Chatterjee, N. Fatma, S. Abuzar and S. Sharma: Acta Trop. *36* (1989) (in press).

439 A. B. Reddy and D. Subrahmanyam: J. Commun. Dis. *18,* 310 (1986).
440 H. Zahner, H. P. Striebel, H. R. Schuetze, I. Sänger, H.-A. Müller and K. Schultheiss: Trop. Med. Parasit. *39,* 14 (1988).
441 S. Townson, A. Dobinson, C. Connelly and R. Muller: J. Helminth. *62,* 181 (1988).
442 H. P. Striebel: Experientia. *32,* 457 (1976).
443 H. J. Saz, G. A. Dunbar and E. Bueding: Amer. J. Trop. Med. Hyg. *26,* 574 (1977).
444 I. Sänger, G. Lämmler and P. H. Wegenhof: Acta Tropica *27,* 262 (1980).
445 J. J. Jaffe and J. W. McCall: 4th Internat. Congr. Parasit. Warshaw, p. 31 (1978).
446 W. C. Campbell, M. H. Fischer, E. O. Stapley, G. Allersschönberg and T. A. Jacobs: Science *221,* 823 (1983).
447 W. C. Campbell: Parasit. Today *1,* 10 (1985).
448 L. S. Blair and W. C. C. Campbell: Amer. J. Vet. Res. *40,* 1031 (1979); J. Parasit. *64,* 1032 (1978).
449 J. W. McCall and W. C. Campbell: ASB Bull. *26,* 43 (1979).
450 T. R. Klei, B. J. Torbert and R. Ochoa: J. Parasit. *66,* 43 (1980).
451 S. Diallo et al.: Lancet *1,* 1030 (1987).
452 T. B. Nutman et al.: J. Infect. Dis. *156,* 662 (1987).
453 P. Vingtain et al.: Bull. Soc. Path. Exot. Fil. *81,* 260 (1988).
454 J. A. G. Whiteworth, D. Morgan, G. H. Mande and D. W. Taylor: Lancet *2,* 97 (1988).
455 H. R. Taylor et al.: Amer. J. Trop. Med. Hyg. *40,* 494 (1989).
456 D. Richard-Lenoble et al.: Amer. J. Trop. Med. Hyg. *39,* 480 (1988).
457 R. Muller: Adv. Parasit. *9,* 73 (1971).
458 R. Muller: Bull. WHO *57,* 683 (1979).
459 S. C. Sastry, Y. R. K. Rao and D. L. Narayana: Indian J. Med. Assoc. *62,* 412 (1974).
460 B. B. Gaitonde, D. M. Renapurkar, G. S. Ramchandani and M. V. Sant: Bull. Haffkin Inst. *2,* 90 (1974).
461 V. P. Sharma, H. S. Rathore and M. M. Sharma: Amer. J. Trop. Med. Hyg. *28,* 658 (1979).
462 M. L. Kothari, D. S. Pardanani, L. Mehta, D. L. Kothari and M. P. Anand: Trans. Roy. Soc. Trop. Med. Hyg. *63,* 608 (1969).
463 G. Raffier: Ann. N. Y. Acad. Sci. *160,* 720 (1969).
464 T. A. Morsy and M. S. Soliman: J. Kuwait Med. Assoc. *9,* 165 (1975).
465 O. O. Kale: Ann. Trop. Med. Parasit. *68,* 91 (1974).
466 S. O. Oduntan, A. O. Lucas and H. M. Gilles: Lancet *2,* 73 (1967).
467 M. L. Kothari, D. S. Pardanani and M. P. Anand: Amer. J. Trop. Med. *17,* 864 (1968).
468 A. Z. Shafei: J. Trop. Med. Hyg. *79,* 197 (1976).
469 B. B. Gaitonde, D. M. Renapurkar and V. H. Apte: Bull. Haffkin Inst. *7,* 20 (1979).
470 S. C. Sastry, J. K. Kumar and V. Lakshminarayan: J. Trop. Med. Hyg. *81,* 32 (1978).
471 F. Hawking: Trop. Dis. Bull. *53,* 829 (1956).
472 P. F. D'Arcy and D. W. G. Harron: Pharm. Internat. *4,* 16 (1983).
473 T. P. Pesign: World Med. J. *14,* 18 (1967).
474 N. Katz: Adv. Pharmacol. Chemother. *14,* 1 (1977).
475 M. G. Chen and K. E. Mott: Trop. Dis. Bull. *86,* Rl (1989).
476 M. A. Ghoneim, A. Ashmalla, M. A. Gaballa and E. I. Ibrahim: Brit. J. Urol. *57,* 303 (1985).
477 R. B. Gove: Trans. Roy. Soc. Trop. Med. Hyg. *64,* 431 (1970).
478 S. R. Smithers: Brit. Med. Bull. *28,* 49 (1972).
479 K. E. Mott and H. Dixon: Bull. WHO *60,* 729 (1982).

480 L. S. Qui, H. C. Xue, Y. X. Tao, Y. H. Zhang and Z. X. Zhu: J. Parasitol. Parasitic Dis. *1*, 144 (1983).
481 J. P. Rotmans: Acta Leidensia *49*, 63 (1982).
482 J. P. Rotmans and H. R. Delwel: J. Immun. Methods *57*, 87 (1983).
483 J. P. Rotmans and B. A. A. Scheven: J. Immun. Methods *70*, 53 (1984).
484 J. K. Bennett and J. W. Depenbusch: in Parasitic Diseases Vol. 2: The Chemotherapy. Ed. J. M. Mansfield, Dekker, New York, pp. 73–131 (1984).
485 W. C. Campbell and E. G. Garcia: in Chemotherapie of Parasitic Diseases, Eds. W. C. Campbell and R. S. Rew. Plenum Press, New York, p. 385 (1986).
486 S. Sharma: Prog. Drug Res. *31*, 9 (1987).
487 A. Davis and D. R. Bailey: Bull. WHO *41*, 209 (1969).
488 M. Saif, M. Koura and F. Abdel-Fattah: J. Egypt. Med. Assoc. *56*, 527 (1973).
489 A. H. S. Omer and C. H. Teesdale: Ann. Trop. Med. Parasit. *12*, 145 (1978).
490 K. Sato, M. Shimada et al.: Amer. J. Trop. Med. Hyg. *38*, 81 (1988).
491 A. Elkholy, S. Boutros, F. Tamara, K. S. Warren and A. A. F. Mahmond: Amer. J. Trop. Med. Hyg. *33*, 1170 (1984).
492 P. R. Manson and S. A. Tswana: Amer. J. Trop. Med. Hyg. *33*, 599 (1984).
493 C. H. King et al.: Amer. J. Trop. Med. Hyg. *39*, 295 (1988).
494 W. Kikuth, R. Goennert and H. Mauss: Naturwissenschaften *33*, 253 (1946).
495 R. Goennert: Bull WHO *25*, 702 (1961).
496 R. E. M. Lees: Trans. Roy. Soc. Trop. Med. Hyg. *60*, 233 (1966).
497 R. Cavier and F. Hawking (eds): Chemotherapy of Helminthiasis, Vol. 1, Pergamon Press, Oxford, U. K., 1973.
498 W. Alves: Bull. WHO *18*, 1109 (1958).
499 D. M. Blair: Bull WHO *18*, 989 (1958).
500 V. de V. Clarke, D. M. Blair and M. C. Weber: Cent. Afr. J. Med. *15*, 1 (1969).
501 J. C. Maritz: South Afr. Med. J. *44*, 126 (1970).
502 J. A. Cook, L. Woodstock and P. Jordan: Amer. J. Trop. Med. Hyg. *23*, 910 (1974).
503 V. L. Ongom: Egypt. J. Bilh. *1*, 209 (1974).
504 A. D. Coutinho and V. S. Barreto: Rev. Inst. Med. Trop. São Paulo. *13*, 57 (1971).
505 A. S. Cunha, D. G. Carvalho, J. N. S. Cambraia and J. R. Concado: Rev. Inst. Med. Trop. São Paulo. *13*, 131 (1971).
506 E. Bueding: J. Toxicol. Envir. Health *1*, 329 (1975).
507 A. Yarinsky, H. P. Drobeck, H. Freele, J. Wiland and K. I. Gumaer: Toxic. Appl. Pharmac. *27*, 169 (1974).
508 S. H. Rogers and E. Bueding: Science *172*, 1057 (1971).
509 G. Raffier: Ann. N. Y. Acad. Sci. *160*, 696 (1969).
510 A. Abdallah and M. Saif: Ann. N. Y. Acad. Sci. *160*, 686 (1969).
511 H. H. Salem, Z. G. Hayaty, A. M. Awaness and G. Al-Allaf: J. Trop. Med. Hyg. *72*, 137 (1969).
512 A. Ruas: Ann. N. Y. Acad. Sci. *160*, 930 (1969).
513 F. Arfaa, I. Farahmandian and M. Soleimani: Trans. Roy. Soc. Trop. Med. Hyg. *64*, 130 (1970).
514 R. A. Gulieve: Nauch. Issl. Inst. Med., Med. Kirova *56* (1982).
515 Acta Tropica (various authors) *9* (suppl.) (1966).
516 H. E. Shazali, J. Kusel and A. Dirdery: J. Trop. Med. Hyg. *73*, 229 (1970).
517 M. E. Kilpatrick, N. A. El-Masry, S. Bassily and Z. Farid: Amer. J. Trop. Med. Hyg. *31*, 1164 (1982).

518 M. Yokogawa et al.: Ann. N. Y. Acad. Sci. *160,* 933 (1969).
519 A. T. Santos, B. L. Blas, J. S. Nosenas and G. P. Portillo: J. Philipp. Med. Assoc. *47,* 203 (1971).
520 M. Yokogawa, M. Tsuji, K. Araki, T. Iijima, Y. Ito and T. Sasaki: Japan J. Parasit. *17,* 175 (1968).
521 P. G. Janssens et al.: Ann. Soc. Belge Med. Trop. *47,* 455 (1967).
522 M. Barbier: Bull. Soc. Path. exot. *62,* 874 (1969).
523 A. Abdallah, M. Saif, M. G. El-Mawla, F. Abdel-Fattah and I. M. Aly: J. Egypt. Med. Assoc. *51,* 923 (1968).
524 N. G. Nicholson and J. E. McMahon: Brit. Med. J. *2,* 1261 (1966).
525 S. R. Kanani, R. Knight and A. W. Woodruff: J. Trop. Med. Hyg. *73,* 162 (1970).
526 H. P. Striebel: Experientia *42,* 457 (1976).
527 H. P. Striebel in: Advanced Pharmacology and Therapy, Ed. M. Adolphe, Pergamon Press, Oxford, p. 17 (1979).
528 L. C. Silva et al.: Rev. Inst. Med. Trop. São Paulo. *16,* 103 (1974).
529 A. L. C. Dominiques and O. Continho: Rev. Inst. Med. Trop. São Paulo. *17,* 164 (1975).
530 K. Taddese and Z. A. Zein: Ann. Trop. Med. Parasit. *82,* 175 (1988).
531 C. H. Teesdale, L. Chitsulo and R. N. H. Pugh: East Afr. Med. J. *64,* 40 (1984).
532 H. Stokvis et al.: Amer. J. Trop. Med. Hyg. *35,* 330 (1986).
533 G. I. Higashi and Z. Farid: Brit. Med. J. *2,* 830 (1979).
534 U. Kranz: Drugs made in Germany *24,* 39 (1981).
535 H. Thomas and P. Andrews: Pestic. Sci. *8,* 556 (1977).
536 P. Andrews, H. Thomas, R. Pohlke and J. Seubert: Med. Res. Rev. *3,* 147 (1984).
537 Supplementary issue on praziquantel: Arzneim.-Forsch. *31* (3a), 535 (1981).
538 E. Groll: Adv. Pharmac. Chemother. *20,* 219 (1984).
539 J. C. Katiyar and S. Gupta: J. Sci. Industr. Res. *43,* 219 (1984).
540 B. Gryseels, L. Nkulikyinka and M. H. Coosemans: Trans. Roy. Soc. Trop. Med. Hyg. *81,* 641 (1987).
541 M. M. A. Homeida et al.: Amer. J. Trop. Med. Hyg. *38,* 496 (1988).
542 M. Gelfand, E. T. Reid and B. E. Simpson: Trans. Roy. Soc. Trop. Med. Hyg. *56,* 77 (1962).
543 H. Abd-Rabbo and M. Montasir: J. Trop. Med. Hyg. *70,* 117 (1967).
544 H. H. Salem: Trans. Roy. Soc. Trop. Med. Hyg. *59,* 307 (1965).
545 O. F. de Gouveia and D. Teixeira: Hospital Rio de Janeiro. *63,* 639 (1963).
546 R. Pieron, B. Lesobre, Y. Mafart and F. Lancastre: Med. Trop. *40,* 302 (1980).
547 H. Bella et al.: Amer. J. trop. Med. Hyg. *31,* 775 (1982).
548 N. S. Al-Waili: Trans. Roy. Soc. Trop. Med. Hyg. *81,* 781 (1987).
549 H. F. Farag, A. Salem, S. A. El-Hifini and M. Kandil: J. Trop. Med. Hyg. *91,* 240 (1988).
550 S. H. Lee: Arzneim.-Forsch. *34,* 1227 (1984).
551 S. Sorumani et al.: Arzneim.-Forsch. *34,* 1231 (1984).
552 M. Yokogawa: Arzneim.-Forsch. *34,* 1193 (1984).
553 R. H. Grothaus and D. E. Weidhaas: in Tropical Medicine Eds. G. W. Hunter, J. C. Swartzwelder and D. F. Clyde, W. B. Saunders Company, Philadelphia, p. 783 (1976).
554 R. F. Smith: Bull. WHO *48,* 685 (1973).
555 Biological control of Vectors: WHO Report (TDR/AR(5)/81, 10-BCV, Geneva. Cited by P. Nickel: Pharm. Internat. *4,* 38 (1983).

556 G. W. Ivine and L. D. Rowe: in Chemotherapy of parasitic diseases. Eds. W. C. Campbell and R. S. Rew. Plenum Press, New York, p. 507 (1986).
557 E. A. Malek: in Tropical Medicine. Eds. G. W. Hunter, J. C. Swartzwelder and D. F. Clyde, W. B. Saunders Company, Philadelphia, p. 672 (1976).
558 S. Ram: Prog. Med. Chem. (Ed. G. P. Ellis and G. B. West, Elsevier Science Publishers, Amsterdam) *25*, 233 (1988).
559 Various authors: Parasitology Today, *6*, 106–136 (1990).
560 W. C. Campbell (ed.): Ivermectin and Abamectin, Springer Verlag, Heidelberg (1989).
561 W. Raether: in Parasitology in Focus. Facts and Trends. Ed. H. Mehlhorn. Springer Verlag, Heidelberg (1988).
562 K. Wessely et al.: Trans. Roy. Soc. Trop. Med. Hyg. *82*, 743 (1988).
563 L. Loutan et al.: Lancet *2*, (8659), 383 (1989).
564 A. Mukherjee, M. Seth and A. P. Bhaduri: Prog. Drug Res. *34*, 447 (1990).

Addendum

Since the article was submitted for publication, a number of interesting articles have appeared which describe drug treatment of Vector-borne diseases. The detailed chemical and biological profile of benzimidazole anthelmintics [558, 559] and ivermectine [560] has appeared. A comprehensive account of the chemotherapy and control measures of parasitic diseases in domestic animals and man was published [561]. Two recent benzimidazole anthelmintics, luxabendazole and triclabendazole have been found to possess high order of flukicidal activity. Luxabendazole shows 95–100% efficacy against *F. hepatica* and *D. dendriticum* in sheep at a dose of 7.5–10 mg/kg [561]. Triclabendazole, on the other hand, has been successfully used to treat fascioliasis in man at a single oral dose of 10–12 mg/kg [562, 563]. A recent article describes the work on biochemical and biological aspects of leishmaniasis [564].

Interaction of cancer chemotherapy agents with the mononuclear phagocyte system

By Alberto Mantovani and Annunciata Vecchi
Istituto di Ricerche Farmacologiche "Mario Negri", Via Eritrea 62, 20157 Milan, Italy

1	Introduction	488
2	Effect of chemotherapeutic agents on mononuclear phagocytes	489
2.1	Glucocorticoids	489
2.1.1	GC and mononuclear phagocytes	489
2.1.2	GC and tumor growth and metastasis	492
2.2	Antimetabolites	494
2.3	Alkylating agents	497
2.4	Intercalating agents	501
2.5	Plant alkaloids	505
3	Modulation of tumor cell susceptibility to macrophage cytotoxicity	507
4	Modulation of mononuclear phagocyte and antitumor efficacy	509
5	Concluding remarks	513
	Acknowledgements	514
	References	514

1 Introduction

Cytotoxic agents used in the medical treatment of neoplasia have profound effects on host immunity. Traditionally the interaction of cancer chemotherapy agents with immunity has been considered in connection with immunodepression. Immunodepression is in fact a prime determinant in the pathogenesis of some of the acute and long-term toxicities associated with cancer chemotherapy, including infections and second malignancies. Interference with immunity has also been considered an intrinsic limiting factor in the efficacy of anticancer agents. In this perspective, chemotherapeutic agents are, in a way, self-defeating. More recently, evidence has accumulated that the interaction of cancer chemotherapeutic agents with immunity is more complex than originally envisaged and that these drugs should be considered from the wider perspective of immunomodulation rather than from the narrower angle of simple immunodepression. In this light, it appears that the relationship between antitumor efficacy and modulation of host resistance is more complicated than expected on the somewhat simplistic assumption that chemotherapeutic agents act as pure depressants of immunity and that no substantial differences exist among antitumor drugs in their interaction with host defence mechanisms.

Cells of the mononuclear phagocyte lineage play a central role in the immunobiology of neoplasia [1–3]. Appropriately activated macrophages can kill extracellularly tumor cells and have been suggested to act as a surveillance mechanism [1–3]. In established malignancy, mononuclear phagocytes represent a major component of the lymphoreticular infiltrate of neoplastic tissues. Thus, being strategically located at the very interface between tumor and host, tumor-associated macrophages are crucial cells in the light of therapeutic intervention.

The purpose of this review is to analyze the interaction of cancer chemotherapy agents with cells of the monocyte-macrophage lineage. Previous reviews on the modulation of specific immunity by chemotherapeutic agents [4, 5] provide a framework for the present effort. We will extend and update previous analysis [6–8] on the effects of cancer chemotherapeutic agents on the mononuclear phagocyte system and discuss available information as to the *in vivo* relevance of modulation at this level.

2 Effect of chemotherapeutic agents on mononuclear phagocytes

2.1 Glucocorticoids (G C)

2.1.1 G C and mononuclear phagocytes

Glucocorticoid hormones are used not only in the treatment of neoplastic diseases, but also to suppress immunity and inflammation under a variety of clinical conditions. Thus the significance of modulation of mononuclear phagocyte function by GC extends beyond the field of cancer chemotherapy. We have recently reviewed the immunosuppressive activity of GC [7]. Here we will concisely summarize available information on the interaction of GC with the mononuclear phagocyte system and emphasize recent results of the mode of action of these agents.

Mononuclear phagocytes are a major target of the immunoregulatory and antiinflammatory effects of glucocorticoid hormones as indicated, for instance, by peripheral blood counts of monocytes that are affected more than those of other leucocyte populations [8a–10]. Cells of the monocyte-macrophage lineage have high-affinity binding sites for GC. Glucocorticoid-binding macromolecules have been demonstrated in monocytes, macrophages and macrophage cell lines [11, 12]. The dissociation constant of the receptors was within physiological ranges and the specificity and affinity correlated with biological function [12].

In vivo administration of GC causes dramatic monocytopenia both in man and mice [10, 13, 14]. Although GC have the capacity to affect monocyte production from bone marrow precursors, this is not the major determinant of the rapid fall of monocyte counts following GC administration. GC are not directly toxic for mature mononuclear phagocytes, interfere with the release of newly formed monocytes from the blood compartment and probably cause a redistribution of cells out of the circulation into unidentified compartments.

The proliferation of myeloid precursors in colony stimulating factor (CSF)-stimulated cultures of bone marrow cells is suppressed by GC *in vitro* [15, 16]. It is of interest that in this *in vitro* system the myelosuppressive effect of GC is prevented by IL-1 [16]. GC also inhibit the proliferative capacity of mature macrophages.

GC alter the differentitation of mononuclear phagocytes. In an *in vitro* system consisting in culturing human monocytes, Rinehart et al.

[17] reported that the transition from monocytes to macrophage-like cells was inhibited by GC. Similarly, differentiation of an established monocytic cell line, was partially blocked by dexamethasone [11]. Along the same line *in vivo* or *in vitro* exposure to GC prevented the spontaneous fusion of rabbit alveolar macrophages to form multinucleated giant cells [18].

Recruitment of mononuclear phagocytes into tissues is a major target of the immunosuppressive and antiinflammatory action of GC [19, 20]. Rinehart et al. [21] reported that hydrocortisone inhibits the responsivencess of human monocytes to lymphocyte-derived chemoattractants. *In vivo*, GC have been shown to inhibit monocyte recruitment at sites of inflammation, infection and neoplastic growth [20–22].

In addition to causing monocytopenia and to inhibiting recruitment of mononuclear phagocytes in tissues, GC interfere with various functions of differentiated mature macrophages, including production of prostaglandins, plasminogen activator and collagenase [18, 23, 24]. Of particular relevance in the perspective of GC as antitumor agents is the effect of these compounds on the cytotoxic activity of mononuclear phagocytes. GC have been shown to reduce the cytostatic and cytotoxic activity of *in vivo* and *in vitro* activated macrophages of rodent or human origin [25–27]. Recent evidence indicates that GC interfere both with priming (IFNγ) and triggering (LPS) signals of macrophage activation for tumor cytotoxicity [27]. Correspondingly *in vivo* treatment with cortisone acetate abolishes the non-specific protection induced by *Corynebacterium parvum* in mice bearing the P 815 mastocytoma [28].

Analysis of the interaction of GC with the production of cytokines by macrophages as well as with their capacity to respond to soluble mediators of immunity acting in autocrine or paracrine loops has considerably improved our understanding of the mode of action of these agents. Snyder and Unanue [29] reported that hydrocortisone inhibited the production of interleukin 1 (IL1) by mouse macrophages exposed to bacterial lipopolysaccharides (LPS). In the same study they also showed that GC blocked the lymphokine-induced expression of Ia antigens by macrophages. Similar results were reported by Smith [30], who studied mouse macrophages triggered by LPS under treatment with dexamethasone. Spontaneous or carrageenan-induced production of IL-1 by rat macrophages was suppressed by hydrocortisone *in vitro* [31]. In apparent contrast with these data in rodents, methylprednisolone did not affect the production of IL-1 by human monocytes exposed to a phorbol ester [32]. However, subsequent studies [33, 34] revealed that methylprednisolone inhibited the production of lymphocyte-activating factor by monocytes cultured with PPD. It was not excluded that under those circumstances the glucocorticoid in fact acted upon contaminating T cells responsible for triggering the release of the monokine by mononuclear phagocytes.

Reconstitution experiments provided further evidence that block at the IL 1 level can be important in glucocorticoid-mediated immunosuppression at least *in vitro.* Exogenous IL 1 has been shown to prevent glucocorticoid inhibition of the murine primary humoral response *in vitro* and to protect helper but not suppressor T cells from these agents [35–38]. However, using human cells Bendtzen et al. [33] failed to reconstitute the production of leukocyte migration inhibitory factor (LIF), inhibited by methylprednisolone, with exogenous IL 1. IL 1 has also been reported to counteract the myelosuppressive effect of dexamethasone *in vitro* in a system of granulocyte/macrophage colony formation in response to CSF [16]. Renelletti et al. [39] studied the role of accessory cells in suppression of mitogen responsiveness in blood lymphocytes and thymus cells. Monocytes or IL 1 reconstituted the dexamethasone-inhibited response to phytohemagglutinin of blood lymphocytes, but mononuclear phagocytes or IL 1 failed to reconstitute the response of thymocytes.

The isolation of cDNA clones encoding cytokines has provided the conditions for more in depth analysis of the mode of action of GC on mononuclear phagocytes. Dexamethasone has been shown to reduce expression of IL 1 mRNA transcripts. Reduced expression of IL 1 mRNA was related to inhibited transcription of the gene. A block in translation into protein of IL 1 mRNA in the presence of GC has also been reported [40, 41]. It is of interest that IL 1 production has been reported to be differentially regulated in macrophages versus vascular endothelium, a cell type that shares many properties with mononuclear phagocytes [42–44]. Dexamethasone was found to inhibit IL 1 production in macrophages but not in endothelial cells. Similarly, dexamethasone did not affect GM-CSF induction in endothelial cells while it strongly suppressed it in macrophages.

In considering the modulatory effect of GC on monokine circuits, pharmacological concentrations of these agents, i.e. those used in chemotherapy protocols, should be considered. It must however be emphasized that under physiological conditions GC play a crucial role as modulators of the action of monokines. For instance, in mice, the acute toxicity of IL 1 is prevented by endogenous production of GC [45]. Moreover GC have a cooperative, permissive influence on the induction of certain acute phase proteins (e. g. fibrinogen) by monokines [7].

Tumor necrosis factor (TNF), a pleiotropic monokine functionally related to IL 1, is also affected by GC. GC inhibit the expression of TNF mRNA in macrophages by interfering with gene transcription [46]. Inhibition of TNF transcript expression by GC results in lower levels of monokine being produced. Since TNF is involved in the cytotoxic action of mononuclear phagocytes, at least against certain tumor targets, (for review see ref. 2), inhibition of production of this monokine could at least partially account for the block of macrophage cytotoxicity by GC discussed above. In addition to interfering with TNF production, dexamethasone has recently been shown to inhibit the tumor necrotizing activity of this cytokine [47]. Dexamethasone blocked TNF-induced tumor necrosis only when the monokine was injected intratumorally. Necrosis of neoplastic tissues induced by local administration of TNF involves recruitment of leukocytes from the blood compartment and participation of vascular endothelium. It is conceivable that GC inhibit TNF-induced necrosis by blocking the endothelial cell responses to this cytokine and leukocyte extravasation.

2.1.2 GC and tumor growth and metastasis

Although it is clear that GC can inhibit growth of primary tumor, metastatic dissemination can be augmented by these agents [22, 48–59]. For instance, in the first (to the best of the author's knowledge) of these observations in a syngeneic model. Agosin et al. [48] found that cortisone inhibited the growth of C3H spontaneous mammary adenocarcinoma but induced the appearance of metastasis. A more widespread distribution of metastases after treatment with glucocorticoid hormones has also been reported in humans with recurrent malignancy [60, 61]. Enhanced metastasis is not caused by increased numbers of tumor cells leaving the primary site [49], although this point has not been extensively investigated. In fact, enhanced metastasis is also observed when tumor cells are injected intravenously, thus excluding a contribution at the level of the primary tumor [22, 58]. Tumors comprise subpopulations heterogenous in various biological properties and metastases can differ both among themselves and from primary tumors in several biological properties [62]. Hence it could be argued that the different effects of GC on primary tumors and metastases may be related to differences in the intrinsic properties of tumor cells that populate these two sites. To test this hypothesis, cells from metastases were transplanted intramuscularly and the effects of hydrocortisone under these conditions were studied [22]. Hydrocortisone had similar effects on tumors originated by transplanting cells from the primary lesion and from metastases, with enhancement of secondaries and inhibition of intramuscular malignancy. Thus the divergent effects of GC on growth of primary tumors and metastases are not related to differences in the intrinsic properties of tumor cell populations at these sites [22].

Metastasis is a multistep event [62], and the augmentation by GC of secondary spread is not indiscriminate for all steps of the process. Release of neoplastic cells from the primary lesions does not seem to be affected [22, 49]. Moreover, once established, secondary deposits are not enhanced by hydrocortisone, but actually tumor growth at these sites is inhibited by GC as found at the primary lesion [22]. Thus GC augment metastasis by affecting some early step of implantation and growth at secondary sites [22].

In addition to augmenting metastasis at sites already involved by the secondary spread under "normal" conditions, GC alter the pattern of metastasis. In fact, glucocorticoid hormones caused the appearance of extrapulmonary metastases in tumors that otherwise colonize only in lungs [22, 50]. It is noteworthy that similar observations have been reported in humans [60, 61].

Host defense mechanisms have been suggested to promote tumor growth at least in some neoplasms and at certain phases of tumor progression [63]. In particular, evidence suggests that the macrophage infiltrate of neoplastic tissues can provide the conditions for optimal tumor growth, possibly by favoring angiogenesis and by providing growth factors for neoplastic cells [64, 65]. In a series of murine tumors, hydrocortisone diminished the macrophage infiltrate of neoplastic tissues and concomitantly the growth of primary tumors [22]. Since there was no correlation between *in vivo* macrophages suppression of tumor growth on the one hand and glucocorticoid receptor content or *in vitro* susceptibility of glucocorticoid cytotoxicity of neoplastic cells on the other, the possibility that interference with some host components was responsible for inhibition of neoplastic growth by hydrocortisone was explored. A role for NK cells or T lymphocytes was excluded on the basis of data obtained in mice with congenital or acquired defects of these reactivities. Hydrocortisone depleted the macrophage content of the neoplastic tissues, and admixture of macrophages to tumor cells reconstituted malignant growth in hydrocortisone-treated mice at least for the mFS6 sarcoma, but not for another sarcoma used in the same study [22]. Thus these and other [66] data strongly suggest that at least in some tumors glucocorticoid may inhibit neoplastic growth by reducing macrophage accumulation in neoplastic tissues, tumor-associated macrophages providing the optimal conditions for neoplastic proliferation by favoring angiogenesis and providing growth factor [67, 68]. This mode of action of GC, formally demonstrated so far only in one tumor model [22], may be of interest in relation to the recently observed antitumor activity of combinations of these agents with heparin, which have been suggested to act by interference with angiogenesis [69].

As discussed in the previous section, glucocorticoid hormones enhance metastasis in experimental animals, and, possibly, in humans. The mechanisms of this effect of GC have not been elucidated. NK cells and macrophages have been suggested to act as a mechanism of restraint of the vascular phase of secondary spread of cancer cells and of the early steps of cell implantation and growth [for review see 65, 70]. It is at this (these) level(s) that GC augment metastasis [22, 49]. Therefore one can speculate that inhibition of NK cells and mononuclear phagocytes contribute to enhancement of metastasis by GC [22]. Effects on the vessel wall could also play a role in augmentation of secondaries by GC [58].

2.2 Antimetabolites

This discussion will focus on thiopurine, 6-mercaptopurine (6-MP) and its nitroimidazole derivative azathioprine (AZA), which, possibly because of the widespread use of the latter in clinical practice as an immunosuppressant, have been extensively studied for their interaction with mononuclear phagocytes. AZA is rapidly transformed *in vivo* into its parent compound, 6-MP [71], and has similar immunosuppressive properties [72, 73]. The available information concerning the effects of other antimetabolites on cells of the monocyte-macrophage series is scanty and fragmentary and will be mentioned only briefly.

Treatment with relatively low doses of thiopurines resulted in a profound monocytopenia in various animal species [74–82]. The mechanisms responsible for the reduction in monocyte counts following treatment with AZA were extensively investigated by Van Furth and co-workers [81, 82]. These investigators found a decreased mitotic activity of promonocytes during AZA treatment in mice; however, the labeling index of promonocytes exposed to AZA increased and a higher percentage of these cells were tetraploid. Therefore, it was concluded that AZA arrests the cell cycle of the promonocytes late in the DNA synthesis phase or in the postsynthesis (G_2) phase, thus preventing mitosis. During acute inflammation, the cell cycle time of bone marrow promonocytes decreased and monocyte production increased: relatively low doses of AZA (3 mg/kg) abolished this response elicited by inflammatory stimuli. Moreover, while in untreated mice 40% of the monocytic cells leaving the circulation entered the site of inflammation, this fraction was reduced to 10% in AZA-treated mice. It is of interest in this connection that Phillips and Zweiman [83] showed that high concentrations of 6-MP directly affected macrophage migration in guinea pigs. Similarly, *in vitro* exposure to AZA or to amethopterin inhibited the response of rabbit macrophages to MIF [84]. Interestingly enough, production of MIF by lymphocytes was not markedly affected by AZA in rabbits and baboons [84, 85]. In this line, we have observed that an other antimetabolite, 5-fluorouracil (5FU), only weakly affected IL-1 production by peritoneal macrophages when cells were tested 24 hours after *in vivo* treatment and the effect was completely recovered 72 hours after treatment (Table 1).

Treatment with AZA profoundly inhibited the accumulation of macrophages within a murine sarcoma [86]. The mechanism of this effect was not fully elucidated but the above-described interference of AZA with monocyte production and impairment of macrophage responsiveness to lymphokine could play a role in the reduction of tumor-associated macrophages. A reduced accumulation of macrophages in AZA-pretreated, tumor-transplanted animals was associated with a reduced growth of this sarcoma. This observation, together with data in X-irradiated mice [87], suggested that tumor-associated macrophages were providing a stimulus for tumor cell proliferation *in vivo*.

Table 1
Effect of chemotherapeutic agents on IL-1 production by murine macrophages

Drug	Dose (mg/kg)	Days after treatment	c.p.m. at supernatant dilution 1/8	1/32
–	–	1	14453 ± 1043	6845 ± 1093
5 FU	200 i.p.	1	9712 ± 635*	4643 ± 1018
–	–	3	17281 ± 3160	5092 ± 983
5 FU	200 i.p.	3	19664 ± 3357	4278 ± 1695
–	–	2	24358 ± 2078	6117 ± 997
Cy	200 i.p.	2	6733 ± 1700**	2708 ± 357**

c.p.m.: counts per minutes. Macrophages (1.5×10^6/ml) from C3H/HeN control and treated mice were exposed to 20 μg/ml LPS for 24 hr. Supernatants were collected, centrifuged, dialyzed and frozen. IL-1 was evaluated as LAF activity (lymphocyte activating factor) on thymocytes (7.5×10^5/well) cultured with PHA (1 μg/ml) and various dilutions of supernatants for 72 hr in a final volume of 0.2 ml/well. Proliferation was evaluated as ^{3}H-thymidine incorporation during the last 6 hr of incubation.
* $p < 0.05$ by Student's t-test; ** $p < 0.01$ by Student's t-test.

The *in vivo* clearance of carbon particles, colloidal gold, BSA, or ^{51}Cr-labeled erythorcytes, was not appreciably affected by AZA treatment unless hepatotoxic doses were used [88–91]. While these reports suggest that thiopurines do not affect phagocytosis, Phillips and Zweiman [83] found a decrease of phagocytosis of ^{14}C-labeled particles and of latex in guinea pigs, as well as an impaired spreading capacity on glass.

Antibody-dependent cellular cytotoxicity (ADCC) mediated by macrophages was not inhibited by AZA in mice [92]: If anything, AZA-treated mice exhibited enhanced macrophage-mediated ADCC. In contrast, in the same series of experiments, Purves [92] reported a selective impairment of ADCC mediated by nonadherent K cells. While the antibody-dependent cytocidal activity of macrophages was not inhibited by AZA, direct antibody-independent tumoricidal activity of peritoneal macrophages was drastically reduced following *in vivo* exposure to AZA [93]. Suppression of direct macrophage-mediated cytotoxicity was observed using both untreated or BCG-stimulated effector cells [93]. Since monocytes or young monocyte-derived macrophages have cytotoxic activity on tumor cells [93–96] and better responsiveness (in terms of cytotoxicity) to lymphokine supernatants [97, 98], it was speculated that AZA inhibited macrophage-mediated tumoricidal activity by interfering with the supply of young bone marrow-derived mononuclear phagocytes, endowed with appreciable cytotoxic capacity [93].

In conclusion, it is well established that thiopurines cause a profound monocytopenia even at low doses by interfering with the cell cycle of promonocytes. As far as mature macrophages are concerned, different effects were reported for different cell functions. Phagocytosis and ADCC generally are not inhibited by these drugs. In contrast, there are indications that entry of monocytes at sites of inflammation and responsiveness of mononuclear phagocytes to MIF are inhibited by exposure to thiopurines. Similarly, *in vivo* treatment with AZA was found to inhibit the direct tumoricidal activity of mouse macrophages. The latter effect might in part be related to an impaired production of lymphokine responsive-monocytes and to a reduced responsiveness to activating lymphokines by analogy with results obtained with MIF.

The subcellular basis for the interaction of thiopurines with the mononuclear phagocyte system is not entirely clear. These agents have antiproliferative activity by blocking *de novo* purine synthesis and interconversion and thus inhibiting DNA, RNA, and protein synthesis [71]. While the antiproliferative activity can account for effects on monocyte production, it remains unclear whether it can represent the sole basis for effects such as the inhibition of monocyte entry at sites of inflammation and inhibition of macrophage responsiveness to MIF.

2.3 *Alkylating agents*

The most extensively studied alkylating agent is cyclophosphamide (Cy). This compound, inactive per se, requires metabolic activation at the level of the liver microsomal enzymes [99, 100]. Cy reduces blood monocyte counts in various animal species [101–103]. The relative contribution to this effect of direct destruction of mononuclear phagocytes and of interference with monocyte precursors has not been elucidated. Since at least large doses of Cy rapidly reduced the long-lived resident peritoneal macrophage count [91, 93, 104], it appears likely that at least high doses of this compound can be cytotoxic on mature mononuclear phagocytes. Therefore in this respect Cy appears to differ from AZA, which only affects monocyte production, without killing mature macrophages [81, 82].

In addition to reducing peritoneal macrophage counts, treatment with high doses of Cy caused morphological alterations in macrophages, with increase in size, heavy vacuolization, and appearance of giant cells [91]. Morphological alterations have also been documented in human monocytes exposed to 4-hydroperoxycyclophosphamide [105].

High doses of various alkylating agents, but not Cy, impaired reticulo-endothelial function as assessed by particle clearance [104]. Following Cy administration, Gadeberg et al. [91] reported impaired *in vitro* phagocytosis of bacteria by peritoneal macrophages and a reduced ability to degrade antigen-antibody complexes. However, following long-term Cy treatment, Buhles and Shifrine [106] noted an increased phagocytosis by murine macrophages. Modulation of arachidonate metabolism has also been reported in peritoneal macrophages from Cy-treated mice [107].

The effect of Cy on ADCC mediated by mononuclear phagocytes from different anatomical sites has been extensively studied by Fauci and co-workers [108–110]. *In vitro* exposure to serum from Cy-treated guinea pigs, presumably containing active metabolites, caused a transient depression of the capacity of blood mononuclear cells to mediate ADCC and mitogen-induced cytotoxicity against erythrocytes [109]. Monocytes can play a role in both reactivities, but their relative contribution was not analyzed in this study [109].

The effect of *in vivo* exposure to Cy on antibody- or mitogen-dependent lysis of red cells was critically affected by the schedule of administration. Using doses causing similar levels of monocytopenia and leukopenia, a single treatment with a large dose of Cy did not affect the antibody- or mitogen-dependent cytotoxicity of blood mononuclear cells or alveolar macrophages [102, 110]. In contrast, repeated daily drug administrations caused a significant impairment of these effector functions in blood or lung alveoli [102, 110].

The effect of Cy on direct macrophage cytotoxicity was studied in mice using cytolysis [93] or cytostasis assays [111, 112]. Unlike AZA and 5-(3,3-dimethyl-1-trazenyl)-1H-imidazole-4-carboxamide (DTIC), Cy did not interfere with the expression of tumoricidal activity by normal or BCG-stimulated mouse peritoneal macrophages [93]. Using a cytostasis assay, Chirigos and coworkers [111, 112] reported enhanced growth inhibitory capacity of Cy-treated macrophages. When the effect of Cy on activation of macrophages by various agents was studied, high concentrations of Cy (300 mg/kg) did not interfere with the augmentation of cytostatic activity by pyran

copolymer or glucan (111). In contrast, Cy given concurrently with BCG decreased macrophage-mediated cytostasis measured 6 days later in one study [111], but this was not confirmed in a subsequent investigation [112].
The apparent discrepancy between the latter finding [111] and the above-mentioned observations in a cytolysis assay [93] is probably related to the different timing of Cy administration relative to the activating stimulus (BCG). Schultz et al. [111] gave Cy on the same days as BCG, and the drug, which causes monocytopenia [103], might have interfered either with the recruitment in the peritoneal cavity of lymphokine-responsive monocytes [96–98] or with the lymphokine production by lymphocytes, this agent inhibiting the production of MIF [108, 109, 113]. We have also obtained different effects on macrophage cytolysis if Cy or Mafosfamide, (Mafo) a direct-acting-derivative of Cy, were given 2 or 8 days after *C. parvum*. As shown in Table 2, inhibition of cytolysis was observed when Cy or Mafo were given 2 days after the activating stimulus, while the drugs were no more active when given on already activated cells (day 8). These and previous data [93] show that activated macrophages are resistant to Cy effects, at least as far as cytolysis B is concerned. The hypothesis, that recruitment of lymphokine-responsive monocytes in the peritoneal cavity is relevant for the subsequent expression of cytotoxicity, is also supported by the data obtained with doxorubicin (DX) that did not affect the expression of cytotoxicity by normal or activated macrophages, but inhibited the induction of cytotoxic activity when administered together with *C. parvum* [114].
McBride et al. [115] recently investigated the effect of *in vivo* or *in vitro* exposure to Cy on cytochine production by human monocytes. Afer *in vivo* administration of the drug, IL-1 producing capacity was decreased, while TNF production was augmented. In contrast *in vitro* exposure to 4-hydroxyperoxy Cy or maphosphamide (Cy active metabolites), at concentrations that were not overtly cytotoxic for monocytes caused a reduction of both IL-1 and TNF release. In *in vivo* studies performed in mice [116], IL 1 production by peritoneal macrophages was markedly reduced 2 days after treatment with 200 mg/kg Cy (Table 1), in agreement with results obtained in humans *in vivo*. The partial dissociation between the *in vivo* and *in vitro* effects of Cy on cytokine production may reflect effects of this agent on non-monocytic cells.

The mode of action of Cy in its interaction with mononuclear phagocytes remains in part to be elucidated. While it appears likely that the cytotoxic activity of this agent can account for the monocytopenia, reversible modulation of cytocidal functions (109, 110) remains to be explained.

A large number of nitrosourea analogs are now available, but the interaction of these compounds with macrophage-mediated host defense mechanisms has received limited attention [117, 118]. Following *in vivo* administration of antitumorally active doses which profoundly reduced T- or B-cell responses in mouse spleen [117], antibody-dependent cytotoxicity of red cells, mediated to a large extent by macrophages, was not appreciably modified by 1,3-bis(2-chloroethyl)-1-nitrosourea (BCNU) or by 1-(2-chloroethyl)-3-(4-methylcyclohexyl)-1-nitrosourea (MeCCNU). In rats, *in vivo* administration of 1-(4-amino-2-methyl-5-pryrimidinyl)methyl-3-(2-chloroethyl)-3-nitrosourea (ACNU) inhibited the spontaneous tumoricidal activity of peritoneal macrophages and this reduction was reversed by BCG [117]. Phagocytosis of human monocytes from normal subjects have been reported to be usually unaffected after *in vitro* exposure to BCNU [119]. One interesting aspect of nitrosoureas is their effect on target cell susceptibility to mononuclear phagocytes. Nathan et al. [120] reported that macrophage-metiated cytolisis was augmented by exposing tumor target cells to BCNU, an effect related to inhibition of the glutathione redox cycle.

DTIC causes a profound and long-lasting suppression of immune responsiveness in mice [121]. This agent also inhibited macrophage-mediated cytotoxicity [93].

Cis-Platinum diamine dichloride (DDP) is an antitumor agent active in humans [122]. Muchmore and coworkers investigated the effect of this agent on the *in vitro* generation of human monocyte cytotoxicity against red blood cells. *In vitro* exposure to DDP or to Doxorubicin (DX) augmented the generation of cytotoxicity by directly affecting killer monocytes [123–125]. Unlike DDP and DX, L-phenylalanine mustard (L-PAM) augmented the generation of monocytes cytotoxic against red cells by interfering with suppressor lymphocytes too [125]. These *in vitro* observations with DDP appeared to have some bearing on *in vivo* conditions in as much as ovarian cancer patients responding to DDP-containing regimens showed enhanced generation of monocytes lytic against erythrocytes [123, 124]. Preincubation of human monocytes with DDP (1 μM for 60 min) selectively inhibited the chemotactic migration of monocytes, while other functions were unaffected [126].

While a relatively large number of studies have dealt with the effect of alkylating agents, Cy in particular, on various aspects of the mononuclear phagocyte system, including cytotoxicity, little is known concerning possible effects of these agents on macrophages at the tumor site. In a series of studies, Evans and coworkers [103, 127, 128] investigated the effect of antitumorally active doses of Cy on tumor-associated macrophages in murine sarcomas. Drug-induced regression of the tumor was associated with an increase in the proportion of mononuclear phagocytes relative to tumor cells. The increased host cell to tumor cell ratio did not prevent recurrence of the sarcomas.

In a series of studies, we investigated the tumoricidal activity of macrophages from ascitic ovarian tumors treated with chemotherapy [93]. Interestingly enough, at the beginning of these experiments we had access to a few patients treated with Cy alone. The direct cytostatic or cytolytic activity of tumor-associated macrophages was not appreciably affected by treatment with Cy, alone or in combination with other antitumor agents. Similarly, the levels of antibody-dependent macrophage-mediated cytostasis were not appreciably modified by the chemotherapy protocols employed in these studies (unpublished data). The limitations inherent to these studies in humans (e.g., lack of time course or dose-response experiments) caution against drawing definite conclusions, but these observations tentatively suggest that in humans as well as in mice, Cy, alone or in combination with DX, DDP, or hexamethylmelamine, does not inhibit the expression of macrophage-mediated tumoricidal activity.

2.4 *Intercalating agents*

The anthracycline antibiotics doxorubicin (DX) and daunomycin (DM) are the intercalators most extensively investigated for their effects on cells of the monocyte-macrophage lineage. No direct information is available concerning the effects of these agents on macrophage precursors. Mature peritoneal macrophages were relatively resistant to the cytotoxic activity of DX, when exposed to the drug *in vitro* or *in vivo* [93, 95, 112, 114, 129, 130–135]. Following *in vivo* administration of the drug, the relative proportion of spleen mononuclear phagocytes was increased, as assessed by counting phagocytic-adherent cells [130] or histologically [135]. The increased percentage of splenic macrophages was attributed to a preferential killing of T and B cells [130]. DM, was both *in vitro* and *in vivo*, somewhat more toxic than DX murine macrophages [129, 136].

Several reports have dealt with effects of anthracycline on macrophage cytolytic or cytostatic activity on tumor cells [93, 112, 114, 129, 132, 133, 137–139]. DX did not inhibit macrophage-mediated cytostatic or cytolytic activity expressed spontaneously or induced by BCG or *C. parvum* [93, 112, 114, 129, 132, 133, 137–139]. In contrast, equitoxic doses of DM caused some inhibition of *C. parvum*-induced cytostasis [129].

The route of administration of DX is an important determinant of the effects of this agent on peritoneal macrophage cytotoxicity. Upon i.v. injection of doses devoid of overt toxicity ($\leq$ LD10), peritoneal macrophages were unaffected [129, 139]. In contrast, i.p. administration of DX induced cytotoxicity in peritoneal macrophages [138, 139, Mantovani, unpublished data]. DX induced cytotoxicity was short-lived, with return to baseline values within 2–4 days.

Circumstantial evidence suggested that macrophages exposed *in vivo* (but not *in vitro*) to DX served as a depot for the drug [138, 139]. A major problem associated with i.p. inoculation of DX not taken into account in these studies, is that this route of administration markedly augments the toxicity of this agent, possibly as a consequence of peritonitis [140]. Direct evidence for the importance of a macrophage-accumulated drug as a mechanism of the cytotoxicity observed after local administration of DX was obtained by Giavazzi et al. [141]. They found that "DX-activated" macrophages, unlike conventionally activated macrophages, were not cytotoxic for DX-resistant tumor target.

DX treatment augmented the *in vivo* radioresistant antitumor response mediated by hemopoietic histocompatibility-like mechanisms [142]. This effect of the drug was abrogated by the antimacrophage agents silica and carrageenan [143]. NK cells, relatively unaffected by pharmacologic doses of DX [137, 144], could play a role in this type of radioresistant activity.

The effect of DX on the *in vivo* induction of macrophage-mediated cytostasis by *C. parvum* was examined in some detail [114]. Administration of DX around the time of *C. parvum* injection inhibited the induction of cytotoxicity measured 6 to 14 days later. This was interpreted as an indication of a cytotoxic effect of the drug at the level of proliferating precursors. In contrast, treatment with DX 3 to 7 days prior to *C. parvum* did not interfere with the subsequent induction of cytostasis. Actually, at least in part of the experiments, mice given DX 5 days before *C. parvum* showed an accelerated appearance of enhanced cytostasis possibly as a consequence of rebound proliferation of precursors [114].

The effect of DX on other aspects of macrophage function, such as ADCC or MIF responsiveness, have been the object of limited attention. As expected on the basis of results discussed so far, DX did not reduce ADCC against chicken red blood cells (Tagliabue and Mantovani, unpublished data] or against tumor cells [131]. Recently, Mace et al. [145] examined the effect of DX or DM encapsulated into liposomes on macrophage function. Drug-treated macrophages could be activated *in vitro* for tumor cytotoxicity and encapsulated DX induced some cytotoxicity on its own. Drug-treated macrophages produced IL-1-like activity.

Little information is available on the effects of anthracyclines on cells of the monocate-macrophage lineage in humans. *In vitro* exposure to DX augmented the generation in culture of monocytes cytotoxic against erythrocytes. This effect was dependent on direct stimulation of the killer monocyte [125]. In our laboratory, we have studied the antibody-dependent or -independent tumoricidal activity of tumor-associated macrophages in patients with ascitic tumors treated with chemotherapy regimes including DX (95). Chemotherapy with drug combinations including DX did not reduce antibody-dependent or -independent cytotoxicity against tumor cells, thus tentatively suggesting a resistance of human mature mononuclear phagocytes to drug toxicity analogous to that observed in mice [146]. The lack of interference with human monocyte macrophage cytotoxicity was recently confirmed [147]. *In vitro* exposure of human monocytes to DX did not affect activation for tumor cytotoxicity by liposome-encapsulated immunomodulators. Similarly, monocytes obtained from patients with osteogenic sarcoma treated with DX were fully responsive *in vitro* to activating signals [147].

The relative sparing of macrophages by DX, or the augmentation of selected mononuclear phagocyte functions by this agent, has important consequences for the effects of DX on other immune parameters that are macrophage-dependent. At least under selected experimental conditions, DX resulted in augmented generation of cytotoxic T lymphocytes and of PHA- or LPS-induced lymphocyte colonies; these effects were possibly macrophage-mediated [131, 135, 148–152].

The pharmacological basis of the effects of anthracycline antibiotics on the immune system has not been elucidated. The kinetics of spleen cell depletion by DX or DM appeared to be related to the different rates of accumulation of the drugs in these two organs [130, 132, 153], and the levels of cellular depletion caused by DX in the peritoneal cavity, spleen and lymph nodes were inversely related to the drug levels in these organs and to the macrophage concentration at these anatomical sites [132].

The mechanisms through which anthracyclines, DM in particular, can kill non-proliferating macrophages *in vitro* are unclear; in addition to inhibiting DNA template function, anthracyclines have been suggested to interact with membranes and this level of action would account for the selectivity of these agents [154, 155].

DX and DM concentrate mainly in the nucleus, but when the cytoplasmic drug concentrations are considered, higher levels of DM than of DX are measured within lysosomes [156]. A preferential accumulation of DM in lysosomes could account for the greater *in vitro* toxicity of this agent, compared to DX, for lysosome-rich mature macrophages [129, 136]. In addition to DNA synthesis, anthracyclines inhibit RNA and protein synthesis [157]. Protein synthesis is important for the expression of tumoricidal activity by murine and human mononuclear phagocytes [158–162], and one might speculate that inhibition of macrophage cytostasis by DM is in part related to an effect of this agent at this level, as observed with actinomycin D [160, 162].

Following the extensive studies with DX summarized above, other anthracycline antibiotics were examined for their interaction with the mononuclear phagocyte system. Mitomycin C, Bisantrene and Aclacinomycin were reported to display a pattern of modulation of mononuclear phagocyte function that, though less extensively studied, is reminiscent of that of DX [163–167]. By and large, these agents do not inhibit mature macrophage function. Under certain conditions augmentation of macrophage activities is observed including tumor cytotoxicity. Evidence was also obtained that macrophage-accumulated drugs were not responsible for the tumor cytotoxicity of treated mononuclear phagocytes [166].

Thus, the peculiar immunopharmacology of DX in relation to mechanisms of natural non-specific resistance including macrophages and NK cells, is not unique to this agent and may serve as a paradigm for other drugs of the same class.

2.5 *Plant alkaloids*

The *Vinca rosea* alkaloids, vinblastine and vincristine, represent a useful tool in the cancer chemotherapy arsenal und are among the most frequently employed drugs for several types of cancer; they belong to a group of mitosis inhibitors that exert their activity by depolymerization of microtubules [168, 169]. Within this group, colchicine is widely investigated as a microtubule-disrupting agent in experimental conditions, whereas the *V. rosea* alkaloids are clinically useful drugs.
As the microtubular system is involved in a variety of macrophage functions, several studies have been performed to clarify the interaction between microtubule-disrupting agents and macrophages. Employing an antiserum against microtubule proteins to permit visualization of these structures by immunofluorescent light microscopy, Frankel [170] showed that microtubules of mouse macrophages radiate from a small region at the cell center. After 1 hr in 0.1 μg/ml of colchicine, all the microtubules had depolymerized but large numbers still remained at the cell center. After 2 hr, most of these had also depolymerized. Within 1 hr after removal of the drug, the cells again showed normal distribution of the tubules, which regrew out from the cytocenter. The microtubule-deprived macrophages generally lose complexity of shape and functional activity. After 48 hr of *in vitro* culture, mouse revealed a shift in cellular locomotion when colchicine (10^{-5}M) or vinblastine (10^{-6}M) were added to the incubation medium [169]. The macrophage movement was changed from a gliding form of locomotion to an induced ameboid form. Since electron microscopy showed that after the disappearance of the microtubules from the cytoplasm of the drug-treated cells, the normal polarization of cytoplasmic organelles was disrupted, Bhisey and Freed [169] concluded that in the absence of cytoskeletal structures, differences in cortical tension may lead to cytoplasmatic streaming and consequent ameboid movement. This finding prompted Pick and Abrahamer [171] to investigate the susceptibility to MIF after treatment with microtubule-disrupting agents. They found that macrophage migration from capillary

tubes was enhanced by colchicine and vinblastine. Moreover, drug-treated macrophages were not susceptible to MIF. Therefore, it was concluded that integrity of the microtubular cytoskeleton of macrophages is a requisite for inhibition of motility by MIF. In a subsequent report Pick and Grunspan-Swirzky [172] showed that D_20, a microtubule-stabilizing agent [173], blocks spontaneous migration and intensifies MIF responsiveness. These results were confirmed by McCarthy et al. [174], who also presented evidence that lumocolchicine, a derivative of colchicine that lacks the antimicrotubular properties of its analog but retains the non-tubulin-associated effects of cholchicine on membrane transport [175], does not affect macrophage responsiveness to MIF. Colchicine was also shown to prevent the capability of MIF to induce macrophage refractoriness to adenylate cyclase stimulators such as β-adrenergic agents or prostaglandins (PG) of the E series [172]. The refractoriness induced by MIF was increased by D_20. Thus, the control of MIF responsiveness via cyclic nucleotides [176] seems to involve the microtubules. Further studies are needed to understand this point better.

To further support the observation that the mitosis inhibitors can affect adenylate cyclase, it was shown in two reports that cholchicine [177] and vinblastine [178] increase the intracellular cAMP levels after the stimulation of macrophages with PGE. Unfortunately, the contradiction between the latter study, where lumocolchicine induced the same effect as colchicine, and the former, where lumocholchicine did not, leaves some doubts about the relationship between microtubules and cyclic nucleotides.

An important feature of macrophages is the induction of lysosomal enzymes and the intracellular degradation of materials endocytosed. Colchicine, but not lumocholchicine, inhibited induction of acid phosphatase resulting from both phagocytic and pinocytic stimuli [179]. Furthermore, collagenase and elastase secretion by macrophages was enhanced by colchicine and vinblastine, whereas lysozyme secretion was inhibited [180].

Effects of microtubule-disrupting agents on certain macrophage membrane reactivities were also shown [181, 182]. Vinblastine enhances the formation of rosettes with sheep red blood cells coated with IgG1-type antibodies [181] and cholchicine promoted Con A capping in alveolar macrophages [182].

Various investigators have studied the effect of microtubule-disrupting agents on macrophage tumorical activity with conflicting results. Keller [158] reported that cholchicine did not modify rat macrophage cytotoxicity. In contrast, Sharma and Piessens [160] and Martin et al. [183] found a reduction of cytotoxicity of rat and guinea pig macrophages by colchicine. Similarly, cytochalasin B, active on microfilaments [184], was reported to inhibit guinea pig macrophage cytotoxicity [160] and to have no effect on rat macrophage tumoricidal activity [183]. Cytochalasin A and B inhibited binding of human monocytes to K 562 cells [184].

Thus, a variety of macrophage functions can be affected by microtubule-disrupting drugs either at the intracellular level or at the membrane level. This class of drugs has proved to be an extremely useful tool in basis studies directed at understanding macrophage biology. So far no information is available on the *in vivo* effect of *V. rosea* alkaloids on macrophage functions.

3 Modulation of tumor cell susceptibility to macrophage cytotoxicity

Evidence suggests that the intrinsic susceptibility of tumor cells to killing by activated effectors can represent a crucial determinant of the final outcome of the interaction between neoplastic cells and immunocompetent cells. Murine tumors vary considerably in their susceptibility to non-specific cytotoxicity mediated by macrophages [96]. Tumor progression has been associated with the emergence of cell variants possessing increased resistance to macrophage killing [185, 186]. Data on human tumors are still limited and ovarian carcinomas are the neoplasms more extensively investigated in this respect. In such tumors it has been found that, in analogy to what has been seen in animals, freshly isolated neoplastic cells from different patients exhibit wide differences in their *in vitro* susceptibility to killing by monocytes or macrophages, obtained from the tumor-bearing patients or from unrelated individuals [187–189]. Similar data, although more limited, have been obtained with respect to other human tumors [190]. Hence, the regulation of target cell resistance or susceptibility to mononuclear phagocytes may have considerable importance in the role actually played by these cells *in vivo*. We have therefore undertaken a series of studies to investigate, using a pharmacologic approach, the regulation of target cell susceptibility to macrophage cytotoxicity. It was found that exposure of selected tumor lines (most notably the WEHI 164) to certain antitumor agents dramatically increased the cytotoxic killing mediated by human or murine mononuclear phagocytes. We have referred to this reactivity as drug-dependent cellular cytotoxicity (DDCC). The main effectors of DDCC are cells of the monocyte-macrophage lineage and killing is mediated via released or membrane-associated TNF [for review see ref. 2]. The DDCC reaction can thus be effected by cells having the capacity to produce TNF. The main effectors in peripheral blood are mononuclear phagocytes but activated B cells can also be involved [see for review ref. 2].

Table 2
Effect of Cy and Mafo on *C. parvum*-induced macrophage cytotoxicity.

Drug	Dose mg/kg	Day of treatment	% Specific cytotoxicity
–	–	2	41.1 ± 2.3
Cy	50	2	39.8 ± 4.5
Cy	100	2	19.2 ± 1.8**
Cy	200	2	14.4 ± 2.7**
Mafo	200	2	17.6 ± 3.1**
–	–	8	45.2 ± 1.8
Cy	50	8	42.3 ± 3.1
Cy	100	8	39.0 ± 4.3
Cy	200	8	46.1 ± 2.2
Mafo	200	8	41.7 ± 3.7

C3H/HeN female mice were given 0.7 mg/mouse i.p. *C. parvum* on day 0. Drugs were given 2 or 8 days later and cytotoxicity expressed by peritoneal macrophages was assessed on day 10. Results reported are for the 40:1 A:T ratio. Cytotoxicity was evaluated in terms of ^{3}H-thymidine release from prelabeled mKSA-TU5 target cells after 48 hr of incubation. ** $p < 0.01$ by Dunnett's test.

Various drugs can sensitize appropriate cell lines to TNF-dependent monocyte killing, including Actinomycin D, DX, emetine, cycloheximide. Inactive agents include nitrosoureas, Cy active metabolites, 5-fluorouracile, methotrexate. Actinomycin D is the agent most extensively investigated for its capacity to sensitize targets to mononuclear phagocyte killing. This agent inhibits DNA-dependent RNA polymerase and, secondarily, protein synthesis. In addition, it directly interacts with membranes and affects membrane microviscosity [191, 192]. It was possible to dissociate changes in membrane fluidity from sensitization to macrophage tumoricidal activity. In contrast, a correlation was observed between inhibition of protein synthesis and induction of susceptibility to macrophages. Thus, chemotherapeutic agents may sensitize tumor cells to the TNF-dependent cytotoxic action of mononuclear phagocytes by interfering with protein synthesis-dependent mechanisms of repair of cellular damage. Inhibition of topoisomerase I and II by various chemotherapeutic agents has been also suggested as a relevant effect for increasing tumor cell sensitivity to natural cell-mediated cytotoxicity [193].

Thus, antitumor chemotherapy can modulate, in the sense of sensitization, the susceptibility of tumor lines to macrophage cytotoxicity. A related question is whether tumor variants resistant to conventional chemotherapeutic agents retain susceptibility to killing by immunocompetent cells. This question is of crucial importance in the perspective of combined use of chemotherapeutic agents and Biological Response Modifiers (BRM). It was found that activated monocytes and LAK cells were able to kill tumor lines with pleiotropic drugs resistance [194, 195]. Along the same line non-specific effector cells were also toxic on cell lines resistant to a variety of individual chemotherapeutic agents [194–196].

These results suggest that activated monocytes and LAK cells have a definite potential to control tumor cell populations with pleiotropic or non-pleiotropic resistance to chemotherapeutic agents. Hence they provide a rationale for immunological approaches designed to eradicate residual tumor cells surviving cytotoxic chemotherapy.

4 Modulation of mononuclear phagocytes and antitumor efficacy

Although host defense mechanisms have frequently been credited with an important role in determining the antineoplastic efficacy of chemotherapy [197], a relative paucity of systematic studies is available in this area. Indications that, for a given drug, the therapeutic activity is the result of cooperation between host resistance and direct tumor cell killing have been obtained through different approaches, by comparing therapeutic efficacy in specifically preimmunized or immunodepressed hosts [133, 198–205] or in tumor sublines of different immunogenicity [133, 206], or by evaluating therapeutic efficacy in tumors transplanted across minor histocompatibility barriers [142].

In some selected murine experimental systems, there is convincing evidence that immune responses may contribute to the antitumor action of selected chemotherapeutic agents. Basically, agents such as

Cy, arabinosyl cytosine, melphalan and DX show reduced efficacy when given to immunosuppressed hosts in selected experimental murine systems [133, 142, 198–205]. While evidence in these particular models is convincing, one should guard against generalization of these findings. In two studies in murine sarcomas, host immunity was a prerequisite for optimal expression of Cy antitumor efficacy [199, 201, 202]. However, in a large series of chemically induced sarcomas, Evans [103, 127, 128] found no correlation between Cy-induced regression and tumor immunogenicity, and immunosuppressed hosts were as responsive as intact mice to chemotherapy. The role of host defense mechanisms in the antitumor activity of DX has been studied in murine solid tumors and leukemias [133, 200, 206]. While in immunogenic neoplasms impairment of host resistance by various procedures discussed in detail below compromised the antineoplastic efficacy of the drug, in poorly immunogenic tumors no evidence for an appreciable role of host immunity was obtained [133, 206].

Therefore, while host resistance appears to contribute to the antitumor activity of selected anticancer agents in some experimental murine tumors, these observations cannot be generalized to other drugs or neoplasms. Even in the experimental models in which indications have been obtained for a role of host immunity in chemotherapeutic efficacy, an analysis of the role played by different populations in the antitumor activity of cancer chemotherapeutic agents is largely lacking. In particular, the relationship, if any, between effects of drugs on various facets of immune responsiveness and participation of different host defence mechanisms to the antitumor action of these drugs remains to be elucidated.

In the last few years, we have studied the interaction of anthracyclines antibiotics, DX in particular, with host defense mechanisms and we have investigated the relationship, if any, between the immunological and the antitumor efficacy of this compound, alone or in combination, with macrophage activators [93, 94, 114, 129–134, 207]. The lymphomas with greater immunogenicity responded better to DX chemotherapy but not to DM [133]. In immunogenic tumor models, the antineoplastic effectiveness of DX was markedly reduced in thymus-deprived mice or by inhibition of host defense mechanisms with silica and carrageenan, or with DTIC, this drug suppressing thymus-dependent responses [121] and macrophage cytotoxicity [93] but not NK cytotoxicity and hemopoietic histocompatibility reactions [144]. Using the L1210Ha leukemia resistant to cytotoxicity of DTIC, suppression with this drug could be applied after tumor transplantation and after DX injection. DX cured up to 80% of L1210Ha-transplanted mice, but suppression by DTIC of host defence mechanisms, including macrophages, as late as 9 days after chemotherapy resulted in progressive tumor growth. Thus, as late as 9 days after "curing" the mice with DX, viable tumor cells persisted in the peritoneal cavity, restrained by host defence mechanisms. This observation could be interpreted as an indication of a "dormancy"-like status induced by chemotherapy and maintained by host resistance, possibly by macrophages that, when recovered from "DX-cured" L121Ha-transplanted mice, showed enhanced cytostatic activity [133].

These series of observations were interpreted as evidence that, although DX can markedly inhibit humoral antibody production and cell-mediated responses [208], specific immunity was a prerequisite for optimal expression of the antineoplastic effectiveness of this agent, and that the residual host immune responsiveness represented a critical determinant of the ultimate curative efficacy of DX in particular immunogenic tumor models. Moreover, the data appeared consistent with the possibility, though did not conclusive proof, that sparing of mononuclear phagocytes, possibly activated as a consequence of immune responses to strongly immunogenic tumors, contributed to the antitumor activity of DX [133]. Similar results were reported by Radov et al. [203] in a murine carcinoma treated with melphalan. Specific immunity to the tumor was required for optimal antitumor efficacy of chemotherapy, and the cytotoxic activity of tumor-associated mononuclear cells was not affected by the drug regimens [203].

The results discussed so far indicate that host defence mechanisms, and macrophages in particular, could play a role in the antitumor activity of selected anticancer agents at least in some experimental models. Compounds capable of activating macrophages, such as BCG and *C. parvum,* have limited but significant antitumor efficacy in rodents [209, 210]. In various transplanted murine tumors, combinations of "immunostimulatory" compounds with cytoreductive chemotherapy have given better therapeutic results than either modality alone [211–213]. In such combined approaches, whereby chemotherapeutic agents are used in conjunction with macrophage activators, it appears reasonable to hypothesize that the interaction of cytoreductive therapy with macrophages has important consequenses for therapeutic results.

Several investigators have observed that the choice of the agent used for chemotherapy is a critical determinant of the antineoplastic effectiveness of chemoimmunotherapy [207, 211–213]. The reasons why, in the various models investigated, chemotherapeutic agents having similar cytoreductive efficacy have a different capacity to synergize with non-specific immunotherapeutic compounds remain largely unclear. With the anthracycline antibiotics DX and DM, better antitumor activity was observed when DX was combined with *C. parvum*, but little therapeutic advantage was obtained when DM was substituted for DX in the chemotherapy arm of the combination [133, 207]. It is important to note that the addition of *C. parvum* really increased the therapeutic index of DX, which was employed up to its maximally nontoxic, chemotherapeutically optimal dose [133]. Too often can we observe experimental chemoimmunotherapy protocols where the effectiveness of the combination with a suboptimal dose of the cytotoxic drugs is not higher than of the drug administered alone at its best dosage. Given the greater *in vitro* and *in vivo* toxicity of DM for mature murine macrophages [129], the therapeutic "synergism" of DX combined with *C. parvum* was attributed to the relative lack of cytotoxicity for mature mononuclear phagocytes of the latter chemotherapeutic drug [112, 129, 133, 135, 136, 207].

The antineoplastic efficacy of chemoimmunotherapy combinations is schedule-dependent and this could in part be related to the time dependency of the effects of chemotherapeutic drugs on mature macrophages or macrophage precursors. As discussed above, DX did not interfere with the induction of macrophage cytotoxicity by *C. parvum*, provided at least 3–5 days had elapsed before injection of the macrophage activator [114]. Interestingly enough, when different schedules of administration of the DX- *C. parvum* combination were tested for antitumor efficacy in murine lymphomas, a 5-day interval between chemotherapy and treatment with the anaerobic coryneform was found optimal [114]. Thus, in these murine models and for this chemoimmunotherapy combination, the possibility is that the time dependency of the antitumor activity of combinations of DX and *C. parvum* is a reflection of the interplay of these two agents with cells of the monocyte-macrophage lineage.

5 Concluding remarks

The studies discussed here indicate that, in rodents, cancer chemotherapeutic drugs can be heterogeneous in terms of their interaction with cells of the monocyte-macrophage lineage. Not only do drugs from different classes have different effects, but even closely structurally related analogs can differ significantly in their effects on mononuclear phagocytes, the anthracycline antibiotics being particularly significant in this respect. The suppressive or at times stimulatory effects of cytotoxic agents can be relatively selective, only or mainly a maturation stage or a cell function being affected. Although it appears that the modulation of mononuclear phagocytes by chemotherapeutic drugs is not solely a consequence of cytotoxicity, the cellular mode of action of these agents at this level remains largely elusive and our understanding remains, with a few exceptions, at a descriptive, phenomenological level.

In selected experimental models, there is evidence that the cells of the monocyte-macrophage lineage contribute to the antitumor activity of drugs. These observations are confined to specific model systems and drugs, and, although little systematic work has been done in this area, it appears that generalizations and extrapolations in terms of drugs, tumors, or species are unwarranted.

In spite of a considerable effort, in human neoplasia therapeutic results of immunotherapy approaches, usually applied after cytoreductive chemotherapy, have been, by and large, disappointing and at best marginal. Limited findings in transplanted animal tumors suggest that the antitumor efficacy of combinations of cytoreductive chemotherapy and nonspecific immunomodulators capable of activating macrophages is critically affected by the interaction of chemotherapeutic drugs with mononuclear phagocytes. Therefore, it appears conceivable that a better understanding of the effects at this level of cytotoxic agents could provide a basis for a less empirical design of chemoimmunotherapy approaches.

In the perspective of clinical relevance and application, the observation that resistant tumor cells are susceptible to activated mononuclear phagocytes is encouraging.

Adoptive transfer or *in vivo* local activation of effector cells with cytotoxic activity may represent a valid complementary strategy with the potential to eliminate residual tumor cells surviving conventional pharmacological treatment.

Acknowledgements

This work was supported by the CNR (National Research Council, Rome, Italy) Contracts No. 88.00733.44 and 88.00914.44.
The generous contribution of the Italian Association for Cancer Research, Milan, Italy, is gratefully acknowledged.

References

1 J.B. Hibbs: The Macrophage in Neoplasia. Academic Press, New York. p. 83, 1976.
2 A. Montovani: Human Monocytes. Academic Press, London, p. 303, 1989.
3 A. Mantovani, B. Bottazzi, P. Allavena and C. Balotto: Immune Responses to Metastases, vol. 1. CRC Press, Boca Raton, p. 106, 1987.
4 P. Sirois and M. Rola-Pleszczynski eds: Immunopharmacology. Elsevier Biomedical Press, Amsterdam 1982.
5 E. Mihich and Y. Sakurai eds: Biological Responses in Cancer, vol. 3. Immunomodulation responses in cancer. Plenum Press, New York 1985.
6 A. Mantovani and A. Tagliabue: The Reticuloendothelial System: A Comprehensive Treatise, vol. 5. Plenum Press, New York, p. 253, 1983.
7 A. Mantovani: Biological Responses in Cancer, vol. 3. Immunomodulation by anticancer drugs. Plenum Press, New York, p. 155, 1985.
8 F. Spreafico, A. Vecchi, F. Colotta and A. Mantovani: Springer Seminars Immunopathology *8,* 361 (1985).
8a H.N. Claman: New Engl. J. Med. *287,* 388 (1972).
9 C.G. Craddock: Ann. intern. Med. *88,* 564 (1978).
10 A.S. Fauci and D.C. Dale: J. clin. Invest. *53,* 240 (1974).
11 F.O. Ranelletti, G. Starace, M. Piantelli, G. Lambertenghi-Deliliers and R.P. Revoltella: J. cell. Physiol. *116,* 329 (1983).
12 Z. Werb, R. Foley and A. Munck: J. exp. Med. *147,* 1684 (1976).
13 J. Thompson and R. Van Furth: J. exp. Med. *131,* 429 (1970).
14 J. Thompson and R. Van Furth: J. exp. Med. *137,* 10 (1973).
15 Y. Ishii, M. Shinoda and M. Shikita: Exp. Haemat. *11,* 178 (1983).
16 R.I. Mishell, D.A. Lee, K.H. Grabstein and L.B. Lachman: J. Immun. *128,* 1614 (1982).
17 J.J. Rinehart, D. Wuest and G.A. Ackerman: J. Immun. *129,* 1436 (1982).
18 S.W. Crawford, A.H. Kang and C.L. Mainardi: Am Rev. resp. Dis. *127,* 46 (1983).
19 W.S. Casey and C.E. McCall: Immunology *21,* 225 (1971).
20 J.E. Balow and A.S. Rosenthal: J. exp. Med. *137,* 1031 (1973).
21 J.J. Rinehart, S.P. Balcerzak, A.L. Sagone and A.F. Lo Buglio: J. clin. Invest. *54,* 1337 (1974).
22 R. Acero, N. Polentarutti, B. Botazzi, S. Alberti, M.R. Ricci, A. Bizzi and A. Mantovani: Int. J. Cancer *33,* 95 (1984).
23 M.A. Bray and D. Gordon: Br. J. Pharmac. *57,* 461 (1976).
24 J. Vassalli, J. Hamilton and E. Reich: Cell *8,* 271 (1976).
25 J.B. Hibbs Jr.: Science *184,* 468 (1974).
26 R. Keller: Immunology *27,* 285 (1974).
27 M.M. Hogan and S.N. Vogel: J. Immun. *140,* 513 (1988).
28 M.T. Scott: J. natn. Cancer Inst. *54,* 789 (1975).
29 D.S. Snyder and E.R. Unanue: J. Immun. *129,* 1803 (1982).
30 K.A. Smith: Immun. Rev. *51,* 337 (1980).

31 S. Stosic-Crujicic and M. M. Simié: Cell. Immun. *69*, 235 (1982).
32 K. Bendtzen and J. Petersen: Immun. Letter *5*, 79 (1982).
33 K. Bendtzen, J. Petersen and B. Soeberg: Acta path. microbiol. immun. scand. section C *91*, 159 (1983).
34 R. Lomnitzer, R. Phillips and A. R. Rabson: Clin. Immun. Immunopathol. *27*, 378 (1983).
35 R. I. Mishell, A. Lucas and B. B. Mishell: J. Immun. *119*, 118 (1977).
36 R. I. Mishell, L. M. Bradley, Y. U. Chen, K. H. Grabstein and S. M. Shiigi: Ann. N.Y. Acad. Sci. *332*, 433 (1979).
37 R. I. Mishell, J. M. Shiigi, B. B. Mishell, K. H. Grabstein and S. M. Shiigi: Immunopharmacology *2*, 233 (1980).
38 L. M. Bradley and R. I. Mishell: Eur. J. Immun. *12*, 91 (1982).
39 F. O. Raneletti, P. Musiani, N. Maggiano, I. Lauriola and M. Piantelli: Cell Immun. *76*, 22 (1983).
40 W. Lew, J. J. Oppenheim and K. Matsushima: J. Immun. *140*, 1895 (1988).
41 A. C. Allison and S. W. Lee: Agents Actions (Suppl.) *24*, 207 (1988).
42 A. Mantovani and E. Dejana: Immun. Today *10*, 370 (1989).
43 W. K. Seelentag, J. J. Mermod, R. Montesano and P. Vassalli: EMBO J. *6*, 2261 (1987).
44 S. H. Zuckerman, J. Shellhaas and L. D. Butler: Eur. J. Immun. *19*, 301 (1989).
45 R. Bertini, M. Bianchi and P. Ghezzi: J. exp. Med. *167*, 1708 (1988).
46 B. Beutler, N. Krochin, I. W. Milsark, C. Luedke and A. Cerami: Science *232*, 977 (1986).
47 H. Rosen, S. Gordon and R. J. North: J. exp. Med. *170*, 27 (1989).
48 M. Agosin, R. Christen, O. Badinez, G. Gasic, A. Neghme, O. Pizzarro and A. Jarpa: Proc. Soc. exp. Biol. Med. *80*, 128 (1952).
49 G. Gasic and T. Gasic: Br. J. Cancer *11*, 88 (1957).
50 T. C. Pomeroy: Cancer Res. *14*, 201 (1954).
51 G. E. Moore and T. Kondo: Durgery *44*, 199 (1958).
52 T. Kondo and K. Tsuki: Gann *47*, 339 (1956).
53 L. Nemeth, K. Lapis and E. Bihari: Acta Un. int. Cancer *16*, 709 (1960).
54 T. A. M. Stoker: Br. J. Surg. *55*, 870 (1968).
55 I. Zeidman: Cancer Res. *22*, 501 (1962).
56 B. Kallum and T. Saldeen: Acta path. microbiol. scand. *70*, 12 (1967).
57 E. V. Sugarbaker, A.M. Cohen and A.S. Ketcham: J. surg. Oncol. *2*, 277 (1970).
58 I.J. Fidler and S. Lieber: Res. Commun. chem. path. Pharmac. *4*, 607 (1972).
59 H.S. Bhakoo, N.S. Paolini, R.J. Milholland, R.E. Lopez and F. Rosen: Cancer Res. *41*, 1695 (1981).
60 H. G. Iversen and G. H. Hjort: Acta path. microbiol. scand. *44*, 205 (1958).
61 P. Sherlock and W. H. Hartman: J. Am. med. Ass. *181*, 313 (1962).
62 G. Poste and I. J. Fidler: Nature *283*, 139 (1980).
63 R. T. Prehn: J. natn. Cancer Inst. U.S.A. *59*, 1043 (1977).
64 R. Evans: Cancer Metastasis Rev. *1*, 227 (1982).
65 A. Mantovani: Progress in Immunology, vol. 5. Academic Press, New York, p. 1001, 1984.
66 M. Nelson, D. S. Nelson and K. E. Hopper: Am. J. Path. *104*, 114 (1981).
67 P. J. Polverini, P. S. Cotran, M. A. Gimbrone Jr. and E. R. Unanue: Nature, Lond. *269*, 804 (1977).
68 A. Mantovani: Int. J. Cancer 22, 741 (1978).
69 J. Folkman, R. Langer, R.J. Linhardt, C. Haudenschild and S. Taylor: Science *221*, 719 (1983).
70 M. Introna and A. Mantovani: Cancer Metastasis Review *2*, 337 (1983).

71 G.B. Elion and G.H. Hutchings: Handbook of Experimental Pharmacology: vol. 38/2. Antineoplastic and Immunosuppressive Agents, Springer-Verlag, Berlin, p. 404, 1975.
72 M.C. Berenbaum: Clin. exp. Immun. *8*, 1 (1971).
73 F. Spreafico, M.G. Donelli, A. Bossi, A. Vecchi, S. Standen and S. Garattini: Transplantation *16*, 269 (1973).
74 E.R. Hurd and M. Ziff: J. exp. Med. *128*, 785 (1978).
75 J.S. Latta and R.P. Gentry: Anat. Rec. *132*, 1 (1958).
76 A.R. Page, R.M. Condie and R.A. Good: Am. J. Path. *40*, 519 (1962).
77 Y. Borel and R. Schwartz: J. Immun. *92*, 754 (1964).
78 M. Ziff, E.R. Hurd, E.M. Lemmel and H.E. Jasin: Mononuclear Phagocytes. Blackwell, Oxford, p. 282, 1970.
79 Zweimann and S.M. Phillips: Science *169*, 284 (1970).
80 H.L. Spielgelberg and P.A. Miescher: J. exp. Med. *118*, 869 (1963).
81 A.E. Gassmann and R. Van Furth: Blood 46, 51 (1975).
82 R. Van Furth, A.E. Gassmann and M.M.C. Diesselhoff-Den Dulk: J. exp. Med. *141*, 531 (1975).
83 S.M. Phillips and B. Zweiman: J. exp. Med. *137*, 1494 (1973).
84 J. Pekarek, J. Svejcar, K. Nouza and J. Johanovsky: Immunology *31*, 773 (1976).
85 T.E. Brown, A. Ahmed, R.S. Filo, R.C. Knudsen and K.W. Sell: Transplantation *21*, 27 (1976).
86 R. Evans: Br. J. Cancer *35*, 557 (1977).
87 R. Evans: Int. J. Cancer *20*, 120 (1977).
88 R.S. Schwartz and J. Andre: Proc. Soc. exp. Biol. Med. *104*, 228 (1960).
89 D.B. Kaufman and R.M. McIntosh: Clin. Res. *19*, 221 (1971).
90 A. Winkelstein, C.G. Craddock and J.S. Lawrence: J. Reticuloendothel. Soc. *9*, 307 (1971).
91 O.V. Gadeberg, J.M. Rhodes and S.O. Larsen: Immunology *28*, 59 (1975).
92 E.C. Purves: Clin. exp. Immun. *22*, 348 (1975).
93 A. Mantovani, W. Luini, G.P. Candiani and F. Spreafico: Int. J. Immunopharmac. *2*, 333 (1980).
94 A. Mantovani, T.R. Jerrells, J.H. Dean and R.B. Herberman: Int. J. Cancer *23*, 18 (1979).
95 A. Mantovani, G. Peri, N. Polentarutti, P. Allavena, C. Bordignon, C. Sessa and C. Mangioni: Natural Cell-Mediated Immunity Against Tumors. Academic Press, New York, p. 1271, 1980.
96 A. Tagliabue, A. Mantovani, M. Kilgallen, R.B. Herberman and J.L. McCoy: J. Immun. *122*, 2363 (1979).
97 L.P. Ruco and M.S. Meltzer: J. Immun. *120*, 1054 (1978).
98 G. Poste and R. Kirsh: Cancer Res. *39*, 2582 (1979).
99 H.J. Hohorst, U. Draeger, G. Peter and G. Voelcker: Cancer Treat. Rep. *60*, 409 (1976).
100 B.E. Domeyer and N.E. Sladek: Cancer Res. *40*, 174 (1980).
101 E. Lemmel, E.R. Hurd and M. Ziff: Clin. exp. Immun. *8*, 355 (1971).
102 G.W. Hunninghake and A.S. Fauci: J. Immun. *117*, 337 (1976).
103 R. Evans, L.D. Madison and D.M. Eidlen: Cancer Res. *40*, 395 (1980).
104 W. Zschiesche: J. Reticuloendothelial Soc. *12*, 16 (1972).
105 M. Jason, B.S. Andrews, M. Colvin and G.J. Friou: Cancer Res. *44*, 3936 (1984).
106 W.C. Buhles and M. Shifrine: J. Reticuloendothel. Soc. *21*, 285 (1977).
107 M. Giordano, J.R. Geffner, A. Prat, M.S. Palermo, G.P. Serebrinsky and M.A. Isturiz: Int. J. Immunopharmac. *10*, 939 (1988).
108 J.E. Balow, D.L. Hurley and A.S. Fauci: J. clin. Invest. *56*, 65 (1975).
109 J.E. Balow, J.E. Parrillo and A.S. Fauci: Immunology *32*, 899 (1977).
110 G.W. Hunninghake and A.S. Fauci: Clin. exp. Immun. *27*, 555 (1977).

111 R. M. Schultz, N. A. Pavlidis, M. A. Chirigos and J. F. Weiss: Cell Immun. *38,* 302 (1978).
112 J. N. Stoychkov, R. M. Schultz, M. A. Chirigos, N. A. Pavlidis and A. Goldin: Cancer Res. *39,* 3014 (1979).
113 A. Winkelstein, H. R. Naukin, J. M. Mitulla: Proceedings of the Seventh Leukocyte Conference. Academic Press, New York, p. 429, 1973.
114 A. Mantovani, A. Tagliabue, W. Luini, T. Facchinetti and F. Spreafico: The Macrophage and Cancer. University of Edinburgh, Edinburgh, p. 203, 1977.
115 W. H. McBride, D. B. Hoon, T. Jung, J. Naungayan, A. Nizze and D. L. Morton: J. Leukocyte Biol. *42,* 659 (1987).
116 F. Fazioli, M. Sironi, A. Vecchi, S. Sozzani and F. Spreafico: Meth. Find. exp. clin. Pharmac. *9,* 595 (1987).
117 A. Ghaffar, W. Lichter, L. L. Welham and M. M. Sigel: J. natn. Cancer Inst. U.S.A. *60,* 1483 (1978).
118 N. Saijo, N. Irimajiri, A. Ozaki, E. Shimizu and H. Niitani: Br. J. Cancer *42,* 162 (1980).
119 W. Pruzanski, S. Saito and G. DeBoer: Cancer Res. *43,* 1420 (1983).
120 C. F. Nathan, B. A. Arrick, H. W. Murray, N. M. DeSantis and W. S. Cohn: J. exp. Med. *153,* 766 (1980).
121 A. Vecchi, M. C. Fioretti, A. Mantovani, A. Barzi and F. Spreafico: Transplantation *22,* 619 (1976).
122 L. H. Einhorn and S. D. Williams: New Engl. J. Med. *300,* 289 (1979).
123 E. S. Kleinerman, L.A. Zwelling, D. Howser, A. Barlock, R. C. Young, J. M. Decker, J. Bull and A. V. Muchmore: Lancet 2, 1102 (1980).
124 E. S. Kleinerman, L. A. Zwelling and A. V. Muchmore: Cancer Res. *40,* 3099 (1980).
125 E. S. Kleinerman and A. V. Muchmore: Proc. Am. Assoc. Cancer Res. *22,* 277 (1981).
126 H. Nielsen: Cancer immun. Immunother. *18,* 223 (1984).
127 R. Evans: Int. J. Cancer *21,* 611 (1978).
128 R. Evans: Am. J. Path. *99,* 667 (1980).
129 A. Mantovani: Cancer Res. *37,* 815 (1977).
130 A. Mantovani, A. Tagliabue, A. Vecchi and F. Spreafico: Eur. J. Cancer *12,* 381 (1976).
131 A. Mantovani, A. Vecchi, A. Tagliabue and F. Spreafico: Eur. J. Cancer *12,* 371 (1976).
132 A. Mantovani, P. Candiani, W. Luini, M. Salmona, F. Spreafico and S. Garattini: Current Trends in Tumor Immunology. Garland STPM Press, New York, p. 139, 1979.
133 A. Mantovani, N. Polentarutti, W. Luini, G. Peri and F. Spreafico: J. natn. Cancer Inst. U.S.A. *63,* 61 (1979).
134 A. Mantovani, A. Vecchi, A. Tagliabue and F. Spreafico: Tumor-Associated Antigens and their Specific Immune Response. Academic Press, New York, p. 271, 1979.
135 F. R. Orsini, Z. Pavelic and E. Mihich: Cancer Res. *37,* 1719 (1977).
136 T. Facchinetti, A. Raz and R. Goldman: Cancer Res. *38,* 3944 (1978).
137 A. Santoni, C. Riccardi, V. Sorci and R. B. Herberman: J. Immun. *124,* 2329 (1980).
138 J. S. Haskill: Cancer Res. *41,* 3582 (1981).
139 F. Martin, A. Caignard, N. O. Olsson, J. F. Jeamin and A. Leclerc: Cancer Res. *42,* 3851 (1982).
140 L. Lenaz and A. Di Marco: Cancer Chemother. Rep. *56,* 431 (1972).
141 R. Giavazzi, C. D. Bucana and I. R. Hart: J. natn. Cancer Inst. U.S.A. *73,* 447 (1984).

142 C. Riccardi, A. Bartocci, P. Puccetti, F. Spreafico, E. Bonmassar and A. Goldin: Eur. J. Cancer *16,* 23 (1979).
143 C. Riccardi, P. Puccetti, A. Santoni, R.B. Herberman and E. Bonmassar: Immunopharmacology *1,* 211 (1979).
144 A. Mantovani, W. Luini, G. Peri, A. Vecchi and F. Spreafico: J. natn. Cancer Inst. U.S.A. *61,* 1255 (1978).
145 K. Mace, E. Mayhew, E. Mihich and M.J. Ehrke: Cancer Res. 48, 130 (1988).
146 A. Mantovani, N. Polentarutti, P. Gritti, G. Bolis, A. Maggioni and F. Spreafico: Eur. J. Cancer *15,* 797 (1979).
147 M.M. Hudson, J.S. Snyder, N. Jaffe and E.S. Kleinerman: Cancer Res. *48,* 5256 (1988).
148 F. Orsini, C. Eppolito, M.J. Ehrke and E. Mihich: Cancer Treat. Rep. *64,* 211 (1980).
149 F. Orsini and E.S. Henderson: Immunopharmacology *1,* 203 (1979).
150 F.R. Orsini and E.S. Henderson: Immunopharmacology *2,* 375 (1980).
151 M.J. Ehrke, V. Tomazic, C.L. Eppolito and E. Mihich. Fedn. Proc. *37,* 1652 (1978).
152 V. Tomazic, M.J. Ehrke and E. Mihich: Cancer Res. *40,* 2748 (1980).
153 D.W. Yesair, E. Schwartzbach, D. Shuck, E.P. Denine and M.A. Asbell: Cancer Res. *32,* 1177 (1972).
154 H.S. Shwartz: Biomedicine *24,* 317 (1976).
155 R. Goldman, T. Facchinetti, D. Bach, A. Raz and M. Shinitzky: Biochim. biophys. Acta *512,* 254 (1978).
156 G. Noel, A. Trouet, A. Zenebergh and P. Tulkens: Adriamycin Review. European Press, Ghent, p. 99 (1974).
157 A. Di Marco: Antimicrob. Agents Chemother. *23,* 216 (1978).
158 R. Keller: Immunology *27,* 285 (1974).
159 R. Keller, R. Keist and R.J. Ivatt: Int. J. Cancer *14,* 675 (1974).
160 S.D. Sharma and W.F. Piessens: Cell Immun. *38,* 264 (1978).
161 D.J. Cameron: Int. J. Immunopharmac. *2,* 135 (1980).
162 D.J. Cameron and W.H. Churchill: Int. J. Immunopharmac. *3,* 77 (1981).
163 H. Shindo, T. Ogura, T. Masuno, S. Hayashi and S. Kishimoto: Cancer Immun. Immunother. *20,* 145 (1985).
164 H. Bodenheimer Jr., C. Charland and J. Leith: J. Leukocyte Biol. *43,* 265 (1988).
165 B.S. Wang, A.L. Lumanglas and F.E. Durr: Cancer Res. *46,* 503 (1986).
166 A.B. Cuellar, D.S. Algara, G. Metzger and S. Orbach-Arbouys: Cancer Res. *47,* 3477 (1987).
167 A. Bravo-Cuellar, G. Balercia, J.P. Levesque, X.H. Liu, F. Osculati and S. Orbach-Arbouys: Cancer Res. *49,* 1578 (1989).
168 S. Inoué and H. Sato: J. cell Physiol. *50,* 259 (1967).
169 A.N. Bhisey and J.J. Freed: Exp. Cell Res. *64,* 419 (1971).
170 F.R. Frankel: Proc. natn. Acad. Sci. U.S.A. *73,* 2798 (1976).
171 E. Pick and H. Abrahamer: Int. Archs. Allergy appl. Immun. 44, 215 (1973).
172 E. Pick and A. Grunspan-Swirzky: Cell Immun. *32,* 340 (1977).
173 D. Marsland, L.G. Tilney and M. Hirshfield: J. cell Physiol. *77,* 187 (1971).
174 P.L. McCarthy, J.E. Shaw and H.G. Remold: Cell Immun. *46,* 409 (1979).
175 S.B. Mizel and L. Wilson: Biochemistry *11,* 2573 (1972).
176 W.J. Koopman, M.H. Gillis and J.R. David: J. Immun. *110,* 1609 (1973).
177 D. Gemsa, L. Steggman, G. Till and K. Resh: J. Immun. *119,* 524 (1977).
178 E. Remold-O'Donnell and H.R. Alpert: Cell Immun. *45,* 221 (1979).
179 E.L. Pesanti and S.G. Axline: J. exp. Med. *141,* 1030 (1975).
180 S. Gordon and Z. Werb: Proc. natn. Acad. Sci. U.S.A. *73,* 872 (1976).

181 G. A. Medgyesi, G. Foris, B. Dezso, J. Gergely and H. Bazin: Immunology *40,* 317 (1980).
182 D. A. Williams, L. A. Boxer, J. M. Oliver and R. L. Baehner: Nature, Lond. *267,* 255 (1977).
183 F. Martin, N. O. Olsson and J. F. Jeamin: Cancer Immun. Immunotherapy *10,* 113 (1981).
184 N. K. Wessells, B. S. Spooner, J. F. Ash, M. A. Bradley, M. A. Luduena, E. L. Taylor, T. Wrenij and K. M. Yamada: Science *171,* 135 (1971).
185 D. Gemsa, W. Kramer, I. Napierski, E. Barlin, G. Till and K. Resch: J. Immun. *126,* 2143 (1981).
186 K. M. Miner and G. L. Nicolson: Cancer Res. *43,* 2063 (1983).
187 S. Haskill, H. Koren, S. Becker, W. Fowler and L. Walton: Br. J. Cancer *45,* 747 (1982).
188 A. Mantovani, N. Polentarutti, G. Peri, Z. Bas-Shavit, A. Vecchi, G. Bolis and C. Mangioni: J. natn. Cancer Inst. U.S.A. *64,* 1307 (1980).
189 G. Peri, N. Polentarutti, C. Sessa, C. B. Mangioni and A. Mantovani: Int. J. Cancer *28,* 143 (1981).
190 B. M. Vose: Cancer Immun. Immunother. *5,* 173 (1978).
191 S. L. Schlager and S. H. Ohanian: J. Immun. *124,* 626 (1980).
192 B. K. Sinha and C. F. Chignell: Biochem. biophys. Res. Commun. *86,* 1051 (1979).
193 T. Utsugi, S. Demuth and N. Hanna: Cancer Res. *49,* 1429 (1989).
194 P. Allavena, M. Grandi, M. D'Incalci, O. Geri, F. C. Giuliani and A. Mantovani: Int. J. Cancer *40,* 104 (1987).
195 C. Gambacorti-Passerini, L. Rivoltini, R. Supino, M. Rodolfo, M. Radrizzani, G. Fossati and G. Parmiani: Cancer Res. *48,* 2372 (1988).
196 P. Allavena, G. Damia, T. Colombo, D. Maggioni, M. D'Incalci and A. Mantovani: Cell. Immun. *120,* 250 (1989).
197 R. S. Schwartz: Cancer Res. *28,* 1452 (1968).
198 E. Mihich: Cancer Res. *29,* 848 (1969).
199 M. Moore and D.E. Williams: Int. J. Cancer *11,* 358 (1973).
200 H. S. Schwarts and G. B. Grindey: Cancer Res. *33,* 1837 (1973).
201 G. Steele and G. E. Pierce: Int. J. Cancer *13,* 572 (1974).
202 G. Steele, H. O. Sjogren and J. Ankerst: Int. J. Cancer *74,* 743 (1974).
203 L. A. Radov, J. S. Haskill and J. H. Korn: Int. J. Cancer *17,* 773 (1976).
204 G.H. Heppner and P. Calabresi: J. natn. Cancer Inst. U.S.A. *48,* 1161 (1972).
205 G. H. Heppner and P. Calabresi: A. Rev. Pharmac. Toxic. *16,* 367 (1976).
206 F. Giuliani, A. M. Casazza, A. Di Marco: Biomedicine *21,* 435 (1974).
207 A. Tagliabue, N. Polentarutti, A. Vecchi, A. Mantovani and F. Spreafico: Eur. J. Cancer *13,* 657 (1977).
208 A. Vecchi, A. Mantovani, A. Tagliabue and F. Spreafico: Cancer Res. *36,* 1222 (1976).
209 P. Alexander: Cancer *40 Suppl. 1,* 467 (1977).
210 W. D. Terry and D. Windhorst: Immunotherapy of Cancer: Present Status of Trials in Man, vol. VI. Raven Press, New York, 1978.
211 B. Fisher, N. Wolmark, H. Rubin and E. Saffer: J. natn. Cancer Inst. U.S.A. *55,* 1147 (1975).
212 B. Fisher, N. Wolmark, E. Saffer and E. Fisher: Cancer *35,* 134 (1975).
213 B. Fisher, H. Rubin, E. Saffer and N. Wolmark: Cancer Res. *36,* 2714 (1976).

Drug molecules of marine origin

By Pushkar N. Kaul
Clark Atlanta University,
James P. Brawley Drive at Fair Str. S. W.,
Atlanta, Georgia 30314, USA

1	Introduction	522
2	Cytotoxic/anticancer molecules	523
2.1	Cembranes	523
2.2	Macrolides	524
2.3	Depsipeptides	525
2.4	Other compounds	528
3	Cardiovascular-active molecules	532
3.1	Cardiotonics	532
3.1.1	Marine peptides	532
3.1.2	Marine glycosides	534
3.2	Hypotensive compounds	535
3.2.1	Marine nucleosides	535
3.2.2	Hypotensive peptides and other compounds	535
4.	Other pharmacologically active molecules	537
4.1	Autonomium	537
4.2	Enzyme inhibitors	537
4.3	Ionophores/antibiotics	539
4.4	Latrunculins	540
4.5	Miscellaneous molecules	542
5	Marine toxins	543
5.1	Palytoxin	543
5.2	Red tide toxins	546
5.2.1	Saxitoxins and Gonyautoxins	547
5.2.2	Brevetoxins	548
5.3	Ciguatera toxins	549
5.3.1	Maitotoxin	549
5.3.2	Ciguatoxin	550
5.4	Conotoxins	551
6	Summary and conclusions	552
	References	552

1 Introduction

The word drug in this review has been used in line with the classical concept, i. e., a substance possessing a pharmacological effect particularly on the mammalian organism. It is only through a planned and careful evaluation of this effect during years of preclinical and successful clinical studies that the effective substance finally becomes a therapeutically useful drug.

Substances of marine origin in various crude forms have been used by man for thousands of years [1]. However, attention to isolation, chemical and pharmacological characterization, and clinical interest of these substances in our times is only five decades old [2]. Several reviews and monographs in the last thirty years have dealt with the biological activities and toxicities of marine-derived substances in relatively crude and impure forms [3–13]. In a few of these reviews, some pure and structurally well-defined molecules have also been included [14–17]. Several excellent reviews on the chemistry of thousands of compounds derived from marine organisms have appeared during the eighties. The most comprehensive of these by Krebs [18] covers 1980 to 1985, giving 1093 structures and citing 760 references. An annual feature in the Natural Product Reports by Faulkner [19] since 1984 reviews the chemistry of marine-derived molecules published during each of the preceding years.

The most recent review on pharmacology of chemically well-defined molecules of marine origin appeared in 1986 [20]. This chapter is intended to highlight the essential evidence in that review but largely to summarize the progress made in the field between 1985 and 1990. Altogether, the evidence keeps on mounting in favor of the belief of this author for over 20 years that there exists a distinct possibility for new drugs to emerge out of the marine environment.

It is imperative that the reader's attention be drawn particularly to marine toxins. Because of very high potency of some of the pharmacologically active molecules of marine origin, it became customary early on to refer to these substances as toxins. However, both pharmaceutical and medical scientists are aware of the philosophy of Paracelsus that clinical usefulness of an active substance is only a matter of the right dose given by the right route. Therefore, like atropine, curare, digitalis and tetrodotoxin [21], it is conceivable that many of the currently labeled marine toxins will some day become useful drugs and/or physiological tools which may help us unravel the mechanisms of cellular processes characteristic of life but not well understood at present. It is only a matter of time and a turning point when chemists and pharmacologists will focus attention to the sea as they did toward plants a century ago and toward soil samples in search of antibiotics 45 years ago following the clinical introduction of penicillin.

2 Cytotoxic/anticancer molecules

In various screening programs sponsored by the U. S. National Cancer Institute (NCI) and the U. S. National Sea Grant Office, perhaps thousands of substances of marine origin in various grades of purity have been found to possess cytotoxic activities in a variety of cell lines and in vivo activity against leukemias and tumors in animal models [20]. However, this review will include only those well-characterized molecules that have shown confirmed and/or potent activities. The primary structural types among pure compounds that have received more attention include the cembranes, macrolides and depsipeptides.

2.1 Cembranes

Scores of 14-membered cyclic diterpenes (cembranoids) from various species of soft corals have been well characterized [22–25]. Most of these compounds contain an exocyclic lactone. Sinularin (**I**) and its dihydro congener obtained from *Sinularia flexibilis* possess anticancer activity as guaged by the NCI screening program [22]. Crassin acetate (**II**) from the Caribbean gorgonian *Pseudoplexaura porosa* was found to be relatively innert to the mammalian systems but highly cytotoxic to mouse fibroblasts and human leukemic as well as HeLa cells in

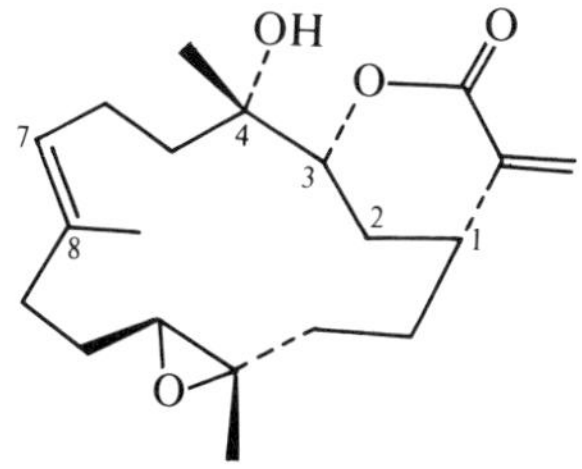

I

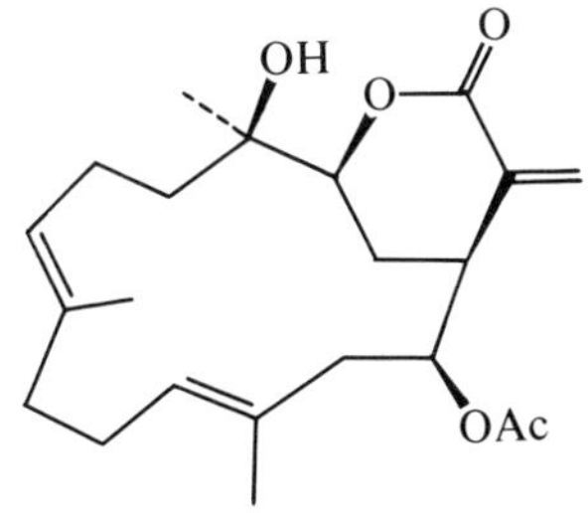

II

vitro; interestingly this activity is lost if the exocyclic lactone of the compound is opened by titration with dilute sodium hydroxide [26]. Hydrolysis of acetate to generate hydroxyl may make this compound more water soluble and useful, provided the lactone is kept intact. Surprisingly, some of the cembranoids at 5 ppm were found to deciliate protozoa [27], suggesting that some of these compounds or their structural modifications may also possess spermicidal and/or parasiticidal activities. These molecules of marine origin clearly need further preclinical evaluation, especially since they are relatively inert to the mammalian systems [26]. Most recent additions to the active cembranoids are 8 new compounds with no exocyclic lactones isolated from the soft coral *Sarcophyton glaucum,* which are capable of inhibiting tumor promoters [28].

2.2 Macrolides

A series of 26-membered macrocyclic lactones have been isolated from the bryozoan *Bugula neritina* and related species. Of these, bryostatin-1 and 2 (**III,IV**) possess antileukemic activity [29]. Like phorbol esters, bryostatins in low concentrations activate protein kinase C (PKC) and thereby inhibit the growth of A549 human lung carcinoma cells [30]. This PKC is the key enzyme in the signal transduction pathway involving activation of phospholipase C which hydrolyzes pho-

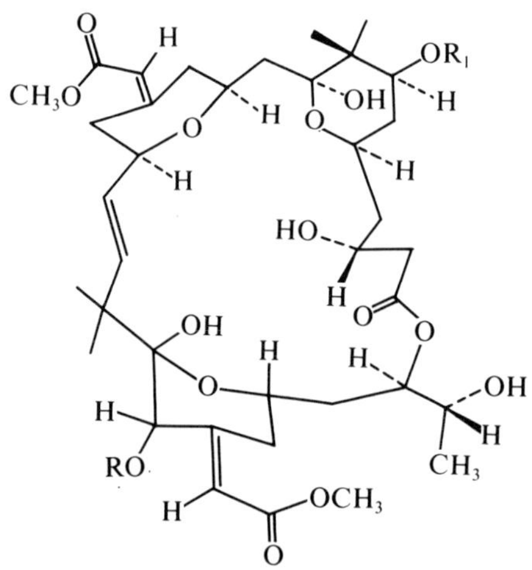

III R = CO–CH = CH–CH = CHC$_3$H$_7$ R$_1$ = COCH$_3$
IV R = CO–CH = CH–CH = CHC$_3$H$_7$ R$_1$ = H

phatidylinositol biphosphate to yield diacylglycerol and inositol triphosphate. The diacylglycerol directly activates PKC which in turn affects both growth and differentiation. The bryostatins possess high affinity for creatinine phosphokinase and at nanomolar concentrations stimulate this enzyme [31], suggesting a possible mechanism of action of these macrolides. Their ability to stimulate human hematopoietic progenator cells has also been demonstrated [32]. Although more than 8 bryostatins with anticancer activity have been isolated, only 2 are currently under detailed preclinical and early clinical evaluations. We should hear more about these compounds during this decade.

2.3 Depsipeptides

A depsipeptide contains amino acid as well as hydroxyacid units and therefore both peptide and lactone features. A number of macrocyclic peptides of varying structural complexities have been isolated from marine organisms during the eighties. Didemnins A, B and C from the Caribbean tunicate *Tridemnum solidum* are potent inhibitors of L1210 leukemic cells in vitro and of P388 leukemia, B-16 melanoma and *Herpes simplex* type-2 virus in vivo [33]. Of these, didemnin B (**V**) is by far the most potent in vitro and in vivo anticancer, antiviral and immunosuppressive compound. Acute as well as chronic toxicitiy studies have been carried out on **V**, and it has been claimed as the first ma-

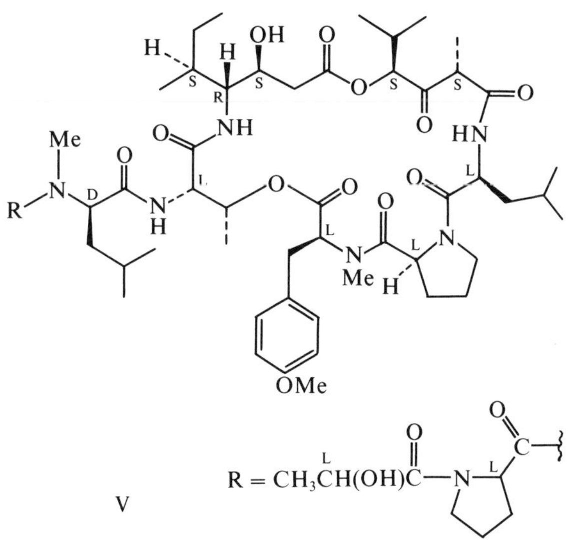

V

rine-derived compound to enter clinical trials as an antineoplastic agent [34] currently in Phase II of the trials [35]. To date, however, it has shown a high general host toxicity accompanying the clinical effectiveness. Perhaps molecular modifications of this molecule will yield a more desirable and clinically acceptable agent of this type.
Antitumor and immunosuppressive effects of purified fractions of the sea squirt (tunicate) *Ecteinascidia turbinata* were extensively reported during the seventies [36–39]. Now only recently, several active ectascidins have been isolated from the tunicate by countercurrent chromatography and tissue culture bioautoradiography guided separations. The most active of the 6 compounds isolated showed ID_{50} of 0.4 ng/ml against L1210 leukemic cells and a very good survival ratio at 3.8 mcg/kg dose in P388 leukemic mice [40].
Several depsipeptides of interesting macrocyclic structures have been recently isolated also from other ascidian species. The hexapeptides bistratamides A (**VI**) and its dehydro congener B (**VII**), isolated from

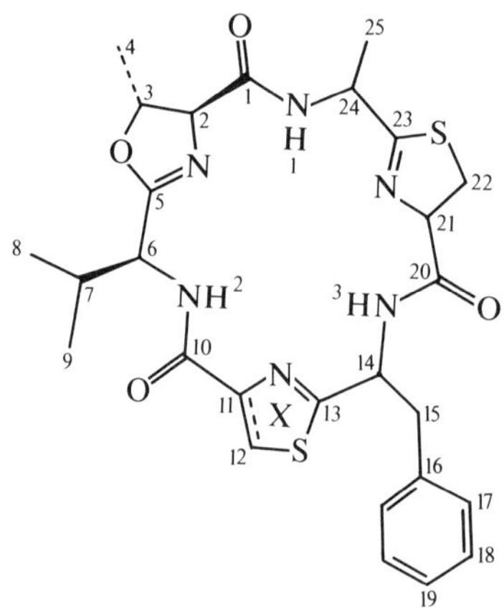

VI X = thiazole
VII X = thiazoline

the aplousobranch ascidian *Lissoclinum bistratum* and structurally characterized by two-dimensional NMR, exhibited cytotoxic activities, but at relatively higher concentrations [41]. The same species also yielded macrocyclic ethers, bistratenes, which, however, were considerably more potent cytotoxic agents comparable to the potency of bryostatins.

Related cyclic peptides patellamide D (**VIII**) and lissoclinamides 4 and 5 (**IX, X**) were isolated from *Lissoclinum patelle* [42]. Of these compounds, **IX** was the most active against fibroblasts and tumor cells in vitro. However, another research group has isolated a novel but very

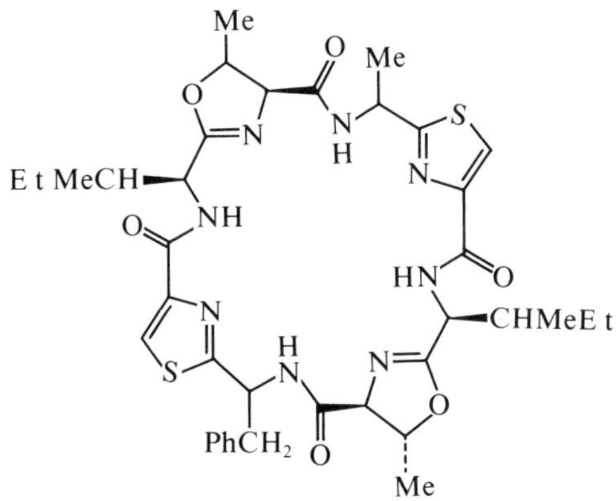

VIII

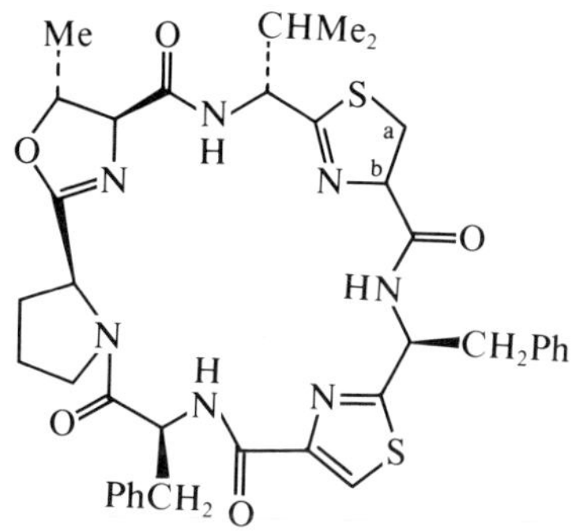

IX a–b = saturated
X a–b = unsaturated

potent thiazole containing macrolide, patellide (**XI**), with a minimum inhibitory concentration (MIC) of 0.3 ng/ml against KB cells in vitro [43]. More recently, cyclodepsipeptide dolastin 13 (**XII**) from *Dolabella auricularia* inhabiting the Indian Ocean was found to inhibit (ED_{50}: 13 ng/ml) P388 lymphocytic leukemic cells, suggesting a potential antineoplastic activity [44]. Clearly, these depsipeptide macromolecules provide complex and fascinating possibilities for investigations. Their clinical potential is already apparent.

2.4 Other compounds

During the eighties, cultured blue-green algae (cyanobacteria) of both marine and terrestrial origin have produced a large number of cytotoxic and antifungal macrolides and nucleosides [43]. Similar compounds isolated from sponges may be attributed to the symbiotic al-

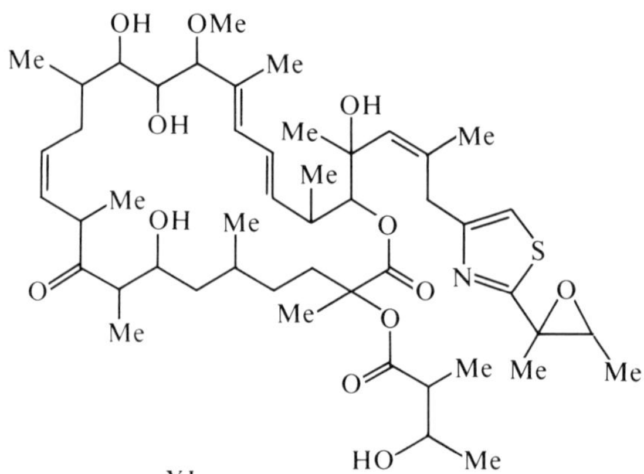

XI

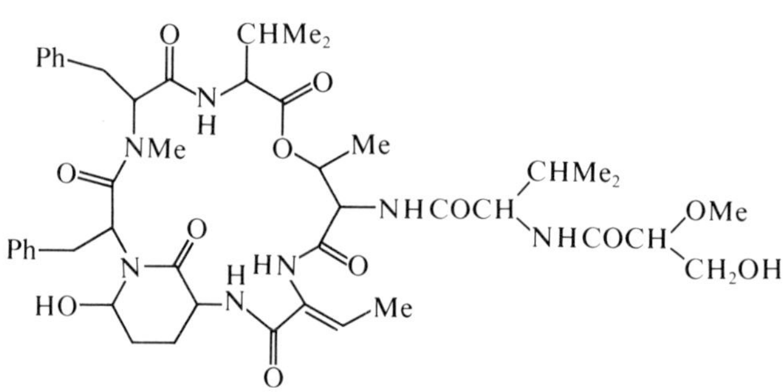

XII

gae found in these sponges. A novel macrolide, laulimalide (**XIII**), which showed an MIC of 15 ng/ml against KB in vitro, was isolated from a sponge, *Hyattella* sp., in the Indonesian waters. An interesting discovery is the presence of 5'-glucosylated nucleosides (**XIV**) in the

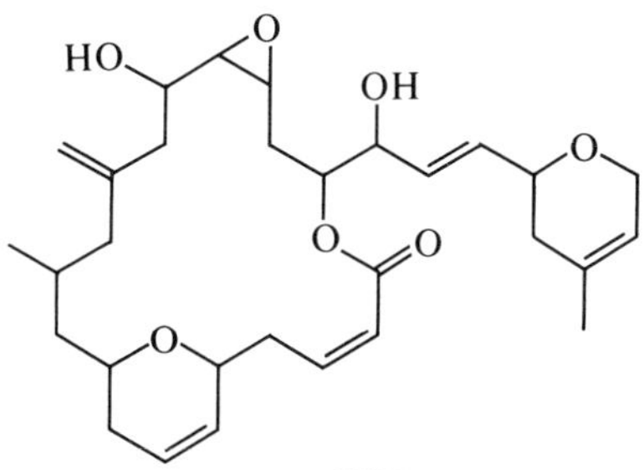

XIII

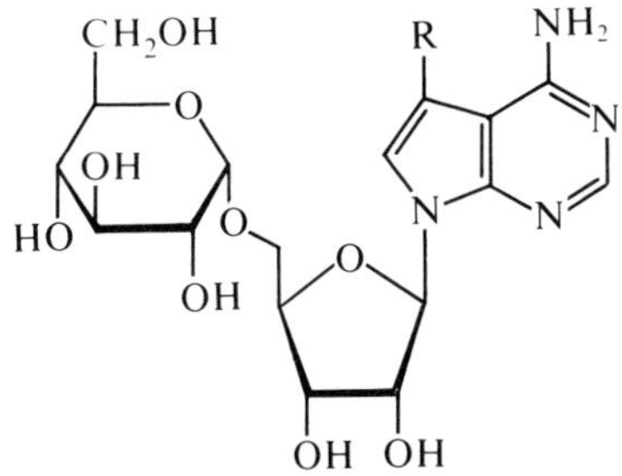

XIV

blue-green algae, which would be far more water soluble than the nucleosides themselves. The low water solubility of some nucleosides has limited their delivery and thereby pharmacologic usefulness. It has been suggested that the naturally occurring glucosyltransferase can be used to glucosylate some of the natural and/or synthetic antiviral nucleosides to enhance their solubility and thus therapeutic effectiveness [43], a possibility worth testing.

Antifungal macrolides mycalamides A and B, isolated from sponges of the *Mycale* spp. from New Zealand and Japanese waters, are novel heterocycles with potent antitumor activities [45]. Their mechanism of action has been attributed largely to their ability to inhibit protein synthesis and not to any apparent effect on DNA.

Ara A, 9-*beta*-D-arabinosyladenine, a well-known antileukemic drug, was originally developed as a semi-synthetic modification of a marine nucleoside, but now it has been found to exist in nature as well [46]. Until recently, these marine nucleosides were believed to be present only in sponges, but now corals (gorgonians) also have been found to contain these compounds. For example, spongouridine possessing antiviral activity was originally isolated from the sponge *Creptotethia crypta,* but now a coral, *Eunicella cavolini,* has been discovered to contain it [46].

More recent studies on sponges collected from depths of 30 to 800 meters in the Caribbean and Mediterranean waters have revealed inter-

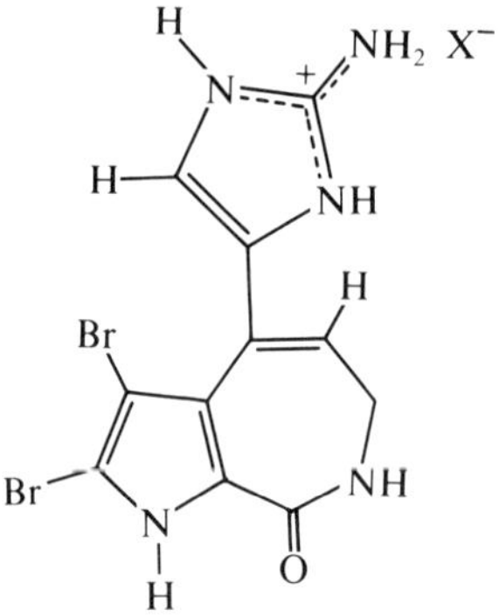

XV

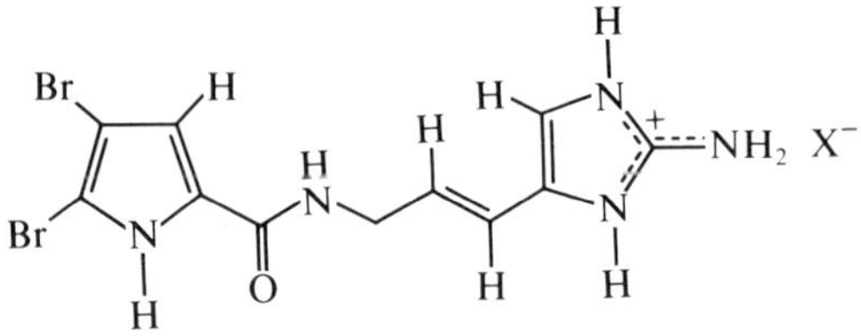

XVI

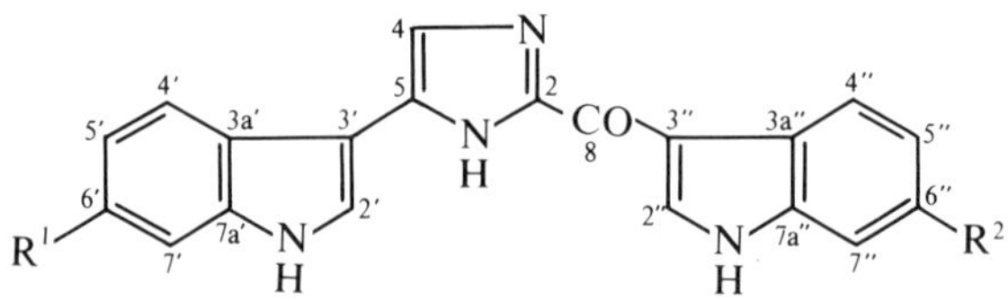

$R^1 = H,\ R^2 = OH$

XVII

esting heterocycles odiline (**XV**), oroidin (**XVI**) and topsentin (**XVII**), all possessing antitumor and cytotoxic activities [40].

A polycyclic orthoquinone, stypoldione (**XVIII**) from the brown seaweed *Stypodium zonale,* is a potent inhibitor of soluble tubulin, and was suggested as an inhibitor of microtubular reassembly [47]. More recent studies by the same investigators, however, have revealed that this compound inhibits cytokinesis at a dose much lower than that inhibiting microtubular tubulin [48]. It appears that **XVIII** uncouples cytokinesis from mitosis at low concentrations perhaps by binding to the sulfhydryl groups of a variety of cytoplasmic molecules other than tubulin. Although molecules such as **XVIII** represent interesting

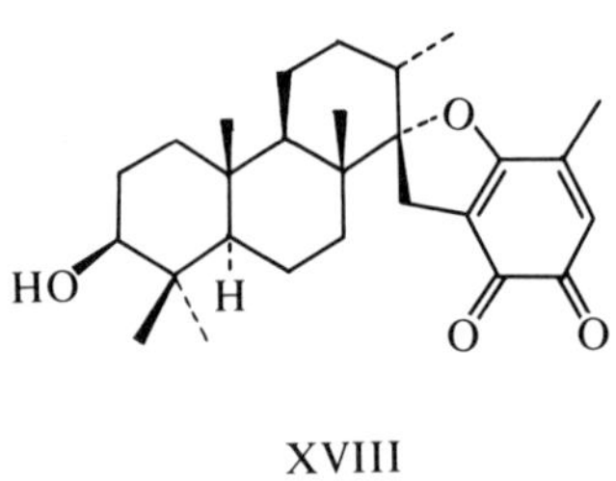

XVIII

nonmetallic inactivators of SH-containing cellular components, lack of their selective toxicity may limit their clinical potential as antineoplastic agents.

A novel steroid, 5,6-epoxylitosterol (**XIX**) from the soft coral *Litophyton viridis,* showed antileukemic activity (IC_{50}: 0.5 mcg/ml) against

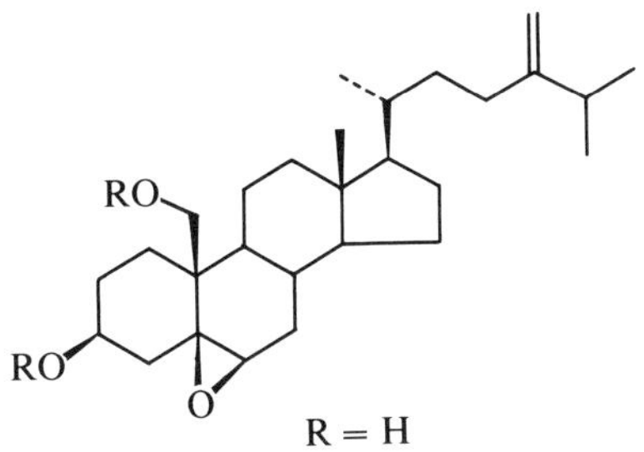

XIX

P388 cells in vitro [49]. One of the two new scarlene-type bishomosesterterpenes, phyllofoliaspongin, isolated from the Okinawa marine sponge *Phyllospongia foliascens,* showed cytotoxic, thrombolytic and vasodilatory activities [50]. A number of polyoxygenated macromolecules possessing antineoplastic/cytostatic activities have been reviewed recently [51], but it is premature to emphasize any specific potential for these compounds.

In this era of AIDS, a specific interest has been stimulated by the isolation of averanol (**XX**) and averone (**XXI**) from *Dysidea avara,* both of which are inhibitors of the human immunedeficiency virus (HIV) in vitro. [52].

Undoubtedly, hundreds of isolated and structurally characterized compounds have been screened for in vitro cytotoxic activites and many have exhibited activities at some dose level. Many of these have

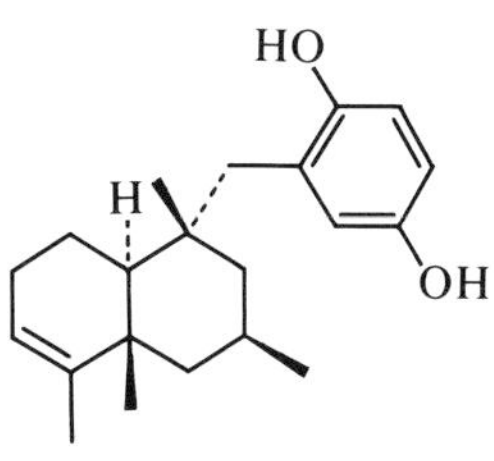

XX

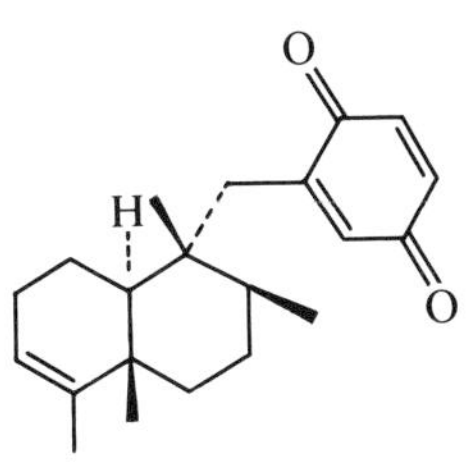

XXI

been designated as "anticancer", "antineoplastic", "antitumor" etc. agents, based on the results of these in vitro screening tests. It is perhaps an improper use of such terminology, and certainly premature to emphasize their activity unless and until sustained investigations confirm their potential for more serious studies. Therefore, though reviewed, these hundreds of compounds have not been included in this chapter. The best source for structural information for such compounds is the annual feature of marine natural products in the Natural Product Reports [19].

3 Cardivascular-active molecules

During cardiovascular screening of thousands of extracts, fractions and pure isolates of marine origin in various laboratories over the past 22 years, the cardiovascular activity in these samples has been detectable almost as commonly as the antimicrobial and cytotoxic activities. However, to date only a limited number of compounds have been studied in any depth and none has reached the stage of clinical trials.

3.1 Cardiotonics

For the purpose of this chapter, the cardiotonic compounds included are those that show an in vitro and/or in vivo positive inotropic activity on part or whole heart.

3.1.1 Marine peptides

A century-old classical concept of cardiotonicity of exogenous molecules belonging primarily to the steroidal skeleton of digitalis or strophanthus aglycones was dispelled when a group of cardiotonic peptides, anthopleurins A, B and C (AP-A, B, C) were isolated from the sea anemone *Anthopleura xanthogrammica* [53]. Of these, AP-A containing 49 amino acid residues and having a molecular weight of 5,195 was reported to be the most potent in its cardiotonic effect on the isolated as well as in situ heart of various species [54–57]. The cardiotonic effect was seen to occur without any changes in the heart rate or the mean arterial pressure and was not mediated via norepinephrine release [54]. Studies on the effect of AP-A on nonischemic and ischemic hearts of the anesthetized dogs, with partially ligated left anterior descending coronary artery to mimic stenosis, revealed a dose-dependent increase in peak dp/dt in both nonischemic and ischemic myocardium [58]. The ability of AP-A to exert a sustained positive initropic effect under ischemic condition gives it a distinct advantage over cardiac glycosides. Apparently, unlike ouabain, AP-A did not produce a major increase in the regional blood flow; only a slight increase was observed which may at best account for only a part of the myocardiac improvement seen in the ischemic area following AP-A administration. The authors of these studies have suggested that AP-A might be particularly useful in patients with a concomitant *beta*-adrenergic blockade [58].

Of at least 7 other cardiotonic peptides, the anemonia toxin II (ATX-II) from the sea anemone *Anemonia sulcata* containing 47 amino acid residues (4,770 daltons) and exhibiting considerable structural resemblance to AP-A [20], was reported to evoke a potent dose-dependant cardiotonic response in various mammalian heart preparations [59]. Apparently the changes in Na+, K+ and Ca++ affected by ATX-II (also called Cardiotoxin-II) are not adequate to explain the observed cardiotonic effect [60]. In its 3 disulfide bridges ATX-II resembles AP-A, but the primary difference between the two peptides seems to be at the residue 38 where lysine in ATX-II is replaced by alanine in AP-A. The differences in their relative potencies have been attributed to this lysine/alanine variance. Both peptides have an aspartate as their 9th residue. Any modification of the acidic carboxyl group of this residue leads to complete inactivation of ATX-II [61]. Perhaps the ionization of this carboxylic and thereby ionic bond formation, or hydrogen bonding of the unionized group, is involved in the interaction of ATX-II with the appropriate membrane receptors on the cardiac muscle cells [20].

Based on the data obtained by application of circular dichroism, IR spectroscopy and Chou-Fasman calculations, the conformation of the two peptides are very similar [62]. This and other similarities, e. g., amino acid sequence, basicity, disulfide bridges etc., may explain why the two peptides show very similar profiles of cardiotonic activity, though differing in their potencies. Expectedly, AP-A given systemically induces an immune response and given orally loses its activity in the acid stomach [63]. As suggested previously [20], a systematic hydrolysis of the peptide coupled to structure-activity studies may reveal a smaller but cardiotonic moiety, analogous to the case of pentagastrin. In fact, a cardiotonic molluscan neurosecretory tetrapeptide, FMRFamide, has already been reported [64]. Also, new neuropeptides of 4–7 residues have been isolated from *A. elegantissima* [65]. Most recently, another group of 3 cardiotonic peptides, tenebrosins A, B and C have been isolated from the Australian sea anemone *Actina tenebrosa* [66]. Although they possess a similar number of residues and N-terminal amino acid sequences with a molecular weight of about 20,000, tenebrosin C appears to be the most potent cardiotonic as well as hemolytic compound. Considering the case of pentagastrin and the recent isolation of smaller active peptides from marine organisms, there clearly remain distinct possibilities for the existence in nature of smaller 4–7 residue cardio-active peptides or for synthetic efforts to obtain the same.

3.1.2 Marine glycosides

Aglycones of the saponins called holothurins, isolated from the echinoderms of class *Holothuroidae*, resemble the stroidal moiety of the aglycones of digitalis glycosides. During the past decade, several scores of hydroxylated stroidal glycosides have been isolated from holothuroids (sea cucumbers), asteroids (star fishes), ophiuroids (brittle stars) and echinoids, mostly at the Soviet marine research laboratories. The chemistry of these molecules has been reviewed regularly over the past 5 years, but access to any pharmacological data on the newer compounds is limited, if in fact such studies have been carried out at all.

Earlier work on holothurins from Belgian and U. S. laboratories has been reviewed [20]. These compounds generally show cardiotoxic activity with a very narrow margin between the effective dose and the toxic dose. Both the sulfated glycosides (astrosaponins from star fishes) and the holothurins (nonsulfated) inhibit Na^+, K^+ as well as Mg^{++}-ATPases, but the holothurins are far more active on the Na^+, K^+-ATPase [64]. Cholesterol appears to prevent this enzyme inhibitory activity, as also their potent hemolytic effect.

3.2 Hypotensive compounds

3.2.1 Marine nucleosides

The Caribbean sponge *Cryptotethia crypta* first made history in marine biomedicine when it yielded an unusual arabinosylnucleoside which on slight structural alteration gave the drug ara C, an antileukemic drug of choice today [67]. The same sponge has yielded several other nucleosides of varying interests. One of these, spongosine, exhibits various coronary effects similar to those of adenosine, e. g., negative inotropy and coronary vasodilation. Spongosine, however, shows not only more pronounced and prolonged effects, but it also is hypotensive at doses at which adenosine is inactive.

Another nucleoside doridosine, N′-methylisoguanosine, isolated from the nudibranch *Anisodoris nobilis,* is the most potent hypotensive of the nucleosides studied [68]. Our preliminary look at the kinetics of enzymatic degradation of these nucleosides revealed that doridosine, the most active and lasting nucleoside, does not get oxidatively deaminated in vitro when incubated with adenosine deaminase, whereas adenosine disappears rapidly. Spongosine, the intermediate acting, disappears rather slowly. Perhaps the duration and intensity of the cardiovascular effects of these marine nucleosides is related to their half-lives in vivo.

A side observation relative to spongosine and doridosine is that when given intracerebroventricularly to guinea pigs both the compounds lower body temperature by several degrees for 4–7 hours [17]. This hypothermic effect does not seem to relate to their systemic actions. Another nucleoside, 5′-deoxy-5-iodobercidin isolated from the red alga *Hypnea valentiae,* has also been found to lower body temperature in mice [69]. This compound is the first 5′-deeoxyribosyl nucleoside found in nature, as also the first iodinated nucleoside known. A potent inhibitor of adenosine kinase, this iodinated nucleoside may become an interesting biochemical research tool.

3.2.2 Hypotensive peptides and other compounds

Urotensins I and II (UI, UII), initially isolated from the caudal neurosecretory system of the teleost *Gillichthys miralilis,* are polypeptides with potent hypotensive activity demonstrated in rat, dog, sheep and squirrel monkey [70]. The hypotensive effect appears to be due to their vasodilatory action. The same peptides have also been isolated from another fish, *Catostomus commersoni* [71, 72]. Several other fresh water as well as marine fishes have been found to contain these vasoactive peptides by detecting the UII-immunoreactive neurons in the caudal neurosecretory systems of these fishes [73]. Although clinical potential of these peptides was originally suggested by its discoverers more than a decade ago, no report on their clinical trials has yet surfaced. More recently a 147-residue peptide (20,000 daltons), equitoxin from the sea anemone *Actinia equina,* was found to cause rapid hypotension and bradycardia in the rat [74].

DL-Laminine (**XXII**), a basic amino acid widely distributed in brown algae of the *Laminariaceae* family, lowered blood pressure in rats ap-

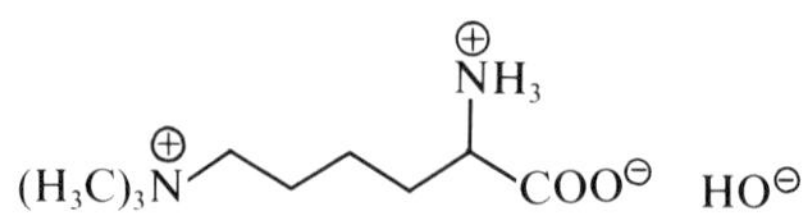

XXII

parently via ganglionic blockade when given in high doses [75]. However at lower doses, only a slight fall in blood pressure was observed, but this fall could be potentiated by propranolol and blocked by atropine, indicative of a muscarinic parasympathomimetic effect of laminine at these doses. The brown algae have been used in oriental medicine for a long time for treating hypertension [76]. Whether or not laminine is solely responsible for this antihypertensive effect of the algae remains to be established, but it certainly deserves closer pharmacological scrutiny.

Aaptamine (**XXIII**) from *Aaptos aaptos* and hymenin (**XXIV**) from *Hymeniacidon aldis* are *alpha*-adrenergic blockers as screened on iso-

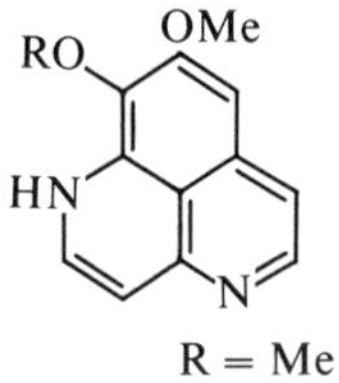

XXIII

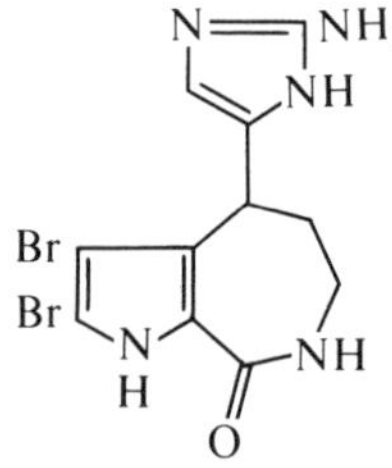

XXIV

lated rabbit aorta [77]. If not rapidly metabolized in the intact mammalian species, these compounds may show a hypotensive effect. Rather interesting heterocycles, the molecular size and structural types make these molecules good candidates for extensive pharmacological studies, should their toxicities be relatively low.

4 Other pharmacologically active molecules

4.1 Autonomium

This compound (**XXV**) from the sponge *Verongia fistularis* is an unusual molecule in that its structure is an isosteric hybrid of epineph-

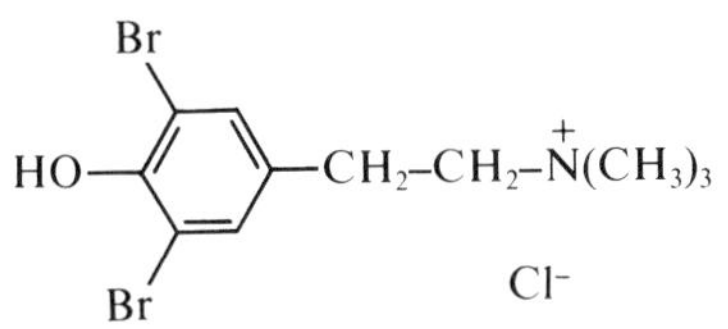

XXV

rine and acetylcholine, possessing elements of both *beta*-phenethylamine and the quaternary ammonium head [78]. The distance between the quaternary nitrogen and the oxygen-bearing carbon in the molecule is 4.2 °A compared to 4.1 °A in acetylcholine. As might be predictable from its structure, autonomium possesses both *alpha* and *beta* adrenergic as well as cholinergic activities. Existence of such amphoteric autonomic compounds in nature is interesting.

4.2 Enzyme inhibitors

Enzyme inhibitors are now being used clinically. For example, inhibitors of angiotensin converting enzyme, the ACE inhibitors, are effective antihypertensive agents in current use. Furthermore, there is an enormous interest in the domain of molecular biology for specific inhibitors of restriction enzymes and polymerases. Many diseases in humans are characterized by overactivity or overproduction of enzymes. Therefore increased attempts are warranted to screen for and identify enzyme inhibitors among abundant marine molecules.

Of a number of halogenated cyclic ethers with exocyclic enine features isolated from an alga and its predator *Aplysia dactylomela* (sea

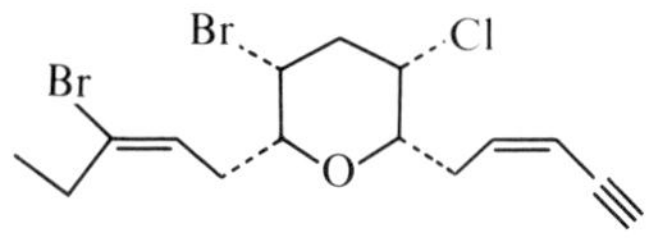

XXVI

hare), dactylyne (**XXVI**) was found to be the most potent inhibitor of cytochrome P_{450} [20]. Recently, extracts of 300 fresh water and 200 marine eukaryotic algae were tested for their ability to inhibit three different glycosidases; 38 were found to be active [79]. Likewise, an aqueous extract of the marine red alga, *Schizymenia pacifica*, was found to inhibit reverse transcriptase of avian and mammalian retrovirus [80]. The active molecule if isolated and identified could be a valuable agent against diseases caused by retroviruses.

Several inhibitors of H^+, K^+-ATPase have been isolated from sponges of *Epipolasis* and *Dysideae* species [81,82]. One of these compounds, a hexaprenylhydroquinone sulfate, also inhibits phospholipase A_2 as well as gastric secretion in the rat [81].

Manoalide (**XXVII**) containing an *alpha, beta*-unsaturated *delta*-lactol function was isolated from *Luffariella variabilis* [83]. It was found to be an antiinflammatory analgesic and a potent inhibitor of phospholipase A_2 [84]. Based on these results, it was proposed that the antiinflammatory activity of **XXVII** may be due to its enzyme-inhibiting

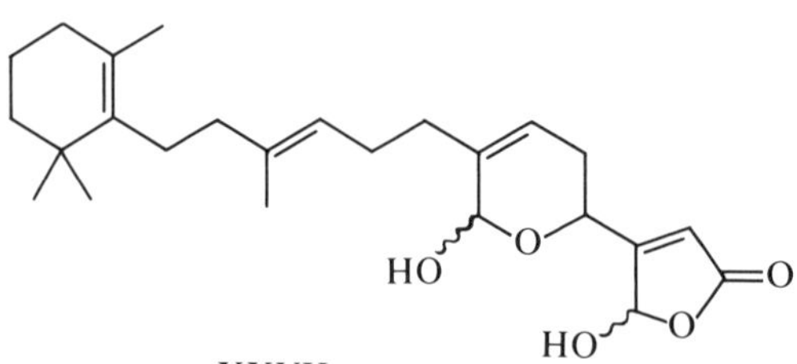

XXVII

property [84]. However, another recent study has shown this compound to be an inhibitor of calcium mobilization including intracellular Ca^{++} release, the hormone-operated plasma membrane Ca^{++} channels, and of the voltage-operated membrane Ca^{++} channels [85]. Because at these low concentrations affecting Ca^{++} mobilization **XXVII** showed no effect on phospholipases, these authors concluded that the antiinflammatory activity of manoalide may be mediated via calcium effects rather than the phospholipase A_2 inhibition.

4.3 Ionophores/antibiotics

Testing for antibiotic activities is perhaps one of the simplest and least expensive undertakings. For this reason, antibiotic activity of marine-derived extracts, fractions and pure isolates have been reported most abundantly [5, 6, 18, 19, 86, 87]. However, despite several thousands of leads over a period of 30 years, no marine-derived compound has emerged as a clinically useful antibiotic. It is probably very difficult to surpass the quality and spectrum of the available antibiotics developed from terrestrial organisms and/or synthetic efforts.

Acanthifolicin (**XXVIII**), a novel ionophore isolated from the sponge *Pandaros acanthifolium,* is an antibacterial and cytotoxic molecule,

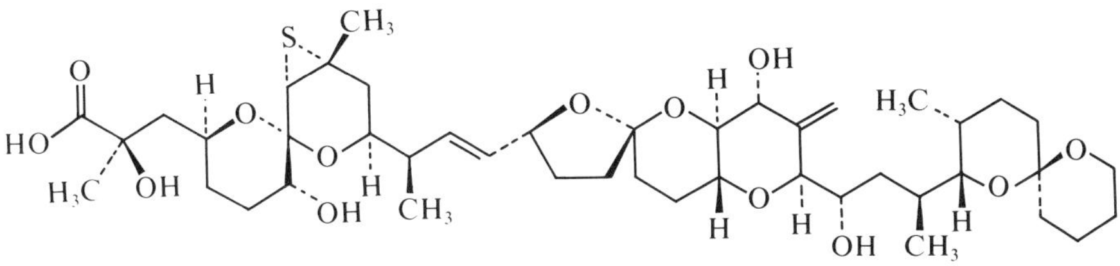

XXVIII

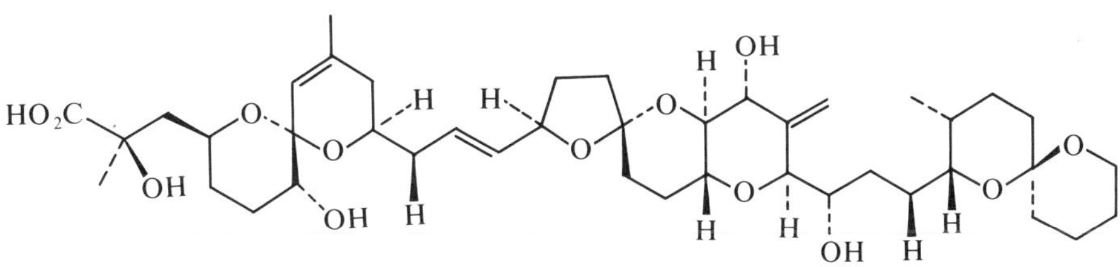

XXIX

but also lethal to mice at a low dose of 0.14 mg/kg i. v. [88]. A similar polyether, okadaic acid (**XXIX**) isolated from two sponges *Halichondria okadai* and *H. melanodocia,* was far more cytotoxic to KB cells and to mice than was **XXVIII** [89]. These ionophore molecules are capable of carrying divalent cations such as Ca^{++} across lipoidal membranes. Okadaic acid produces prolonged contraction of the human umbilical artery and rabbit aorta even in the absence of extracellular Ca^{++}, and does not affect the Na^{+}, K^{+}-ATPase [90]. The total synthesis of okadaic acid has already been accomplished [91].

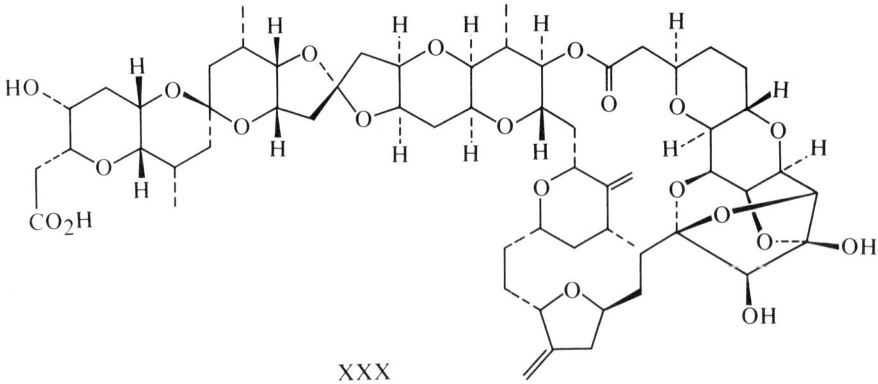

XXX

The antitumor polyether macrolide norhalichondrin A (**XXX**) and several other halichondrin analogs [92, 93] have also been isolated from *H. okadai.* Most of these molecules appear to be capable of exerting multiple biological activities and a high degree of toxicity. Structure-activity correlational studies on their derivatives and structural fragments should be of considerable interest. The other significant polyether macromolecules that are also very toxic to mammals include brevitoxins from dinoflagellates discussed under Section 5 dedicated to marine toxins.

The antibiotics SS-228R (**XXXI**) and SS-228Y (**XXXII**), originally isolated from marine microbes out of the sea mud, have now been syn-

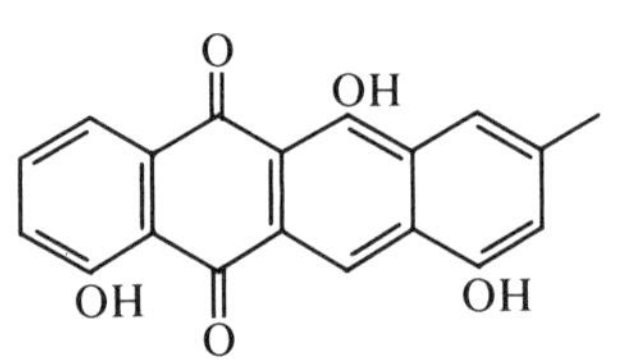

XXXI

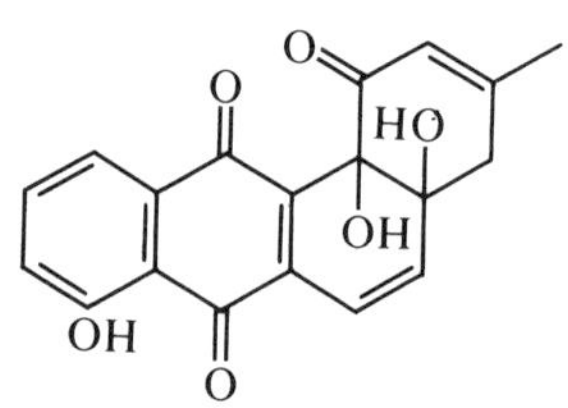

XXXII

thesized [94]. The presence of many other antibacterial or otherwise bioactive molecules from various marine microorganismus has been reviewed elsewhere [95].

4.4 Latrunculins

A large number of 14- and 16- membered macrocyclic diterpene alkaloids have been isolated from corals and sponges, most of which possess some biological activity. Of particular interest are the toxic latrunculins A–D isolated from the magnificent red branching sponge *Latrunculia magnifica* inhabiting the Red Sea [96–98]. At nanomolar concentrations in cultured mouse neuroblastoma and fibroblasts, the latrunculins A (**XXXIII**) and B (**XXXIV**) produce significant disrup-

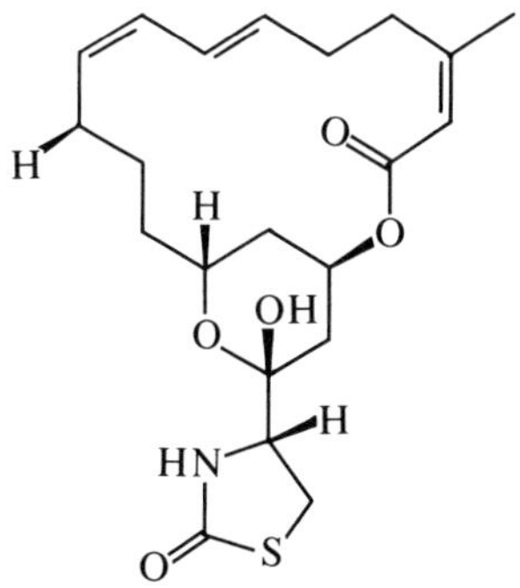

XXXIII

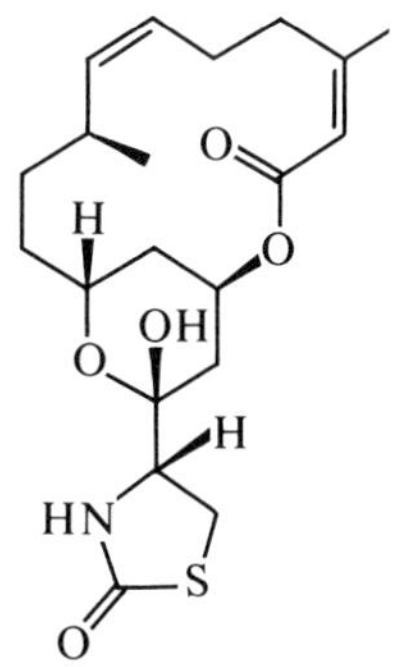

XXXIV

tion of microfilament organization without affecting the microtubules, analogous to what the less active cytochalasins do [99]. However, unlike cytochalasins, the latrunculins do not alter the polymerization rate of active filaments. From another species of *Latrunculia* from New Zealand waters has been isolated the alkaloidal pigment discorhabdin C (**XXXV**), which is a potent cytotoxic agent [100]. A related compound prienosin A (**XXXVI**) from the sponge *Prianos melanos* is a potent antileukemic molecule [101].

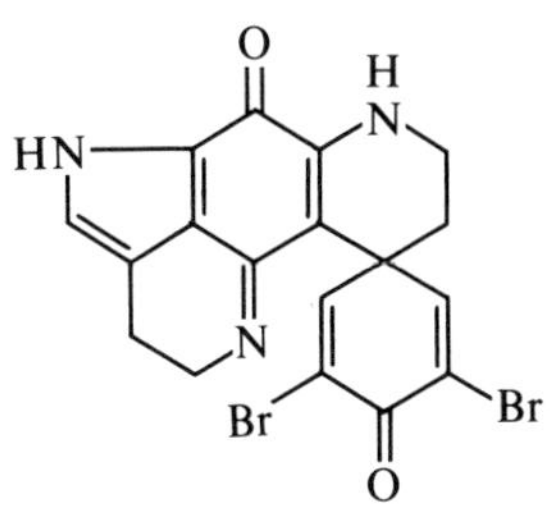

XXXV

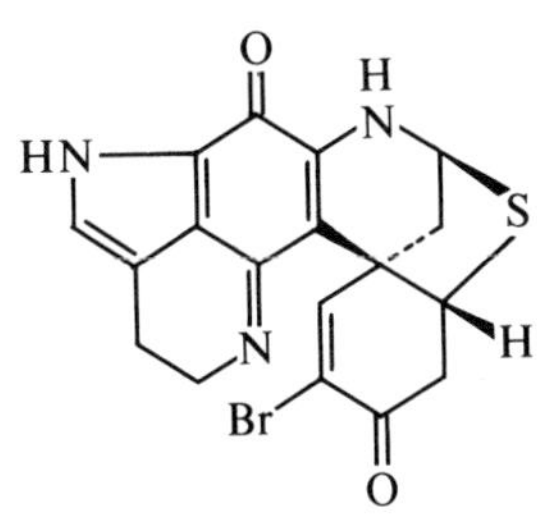

XXXVI

4.5 Miscellaneous molecules

Aplysinopsin (**XXXVII**), originally isolated from the yellow sponge *Verongia spengelii,* has also been isolated from anthozoan *Astroides calicularis* [102]. It is an interesting tryptophan derivative with cytotoxic activity against KB, P388 and L1210 leukemia cell lines, and possesses antidepressant activity similar to that of imipramine [20]. A related indole alkaloid barettin (**XXXVIII**) from the coldwater sponge

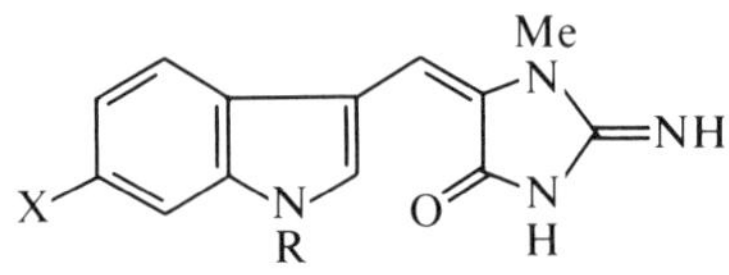

R = X = H

XXXVII

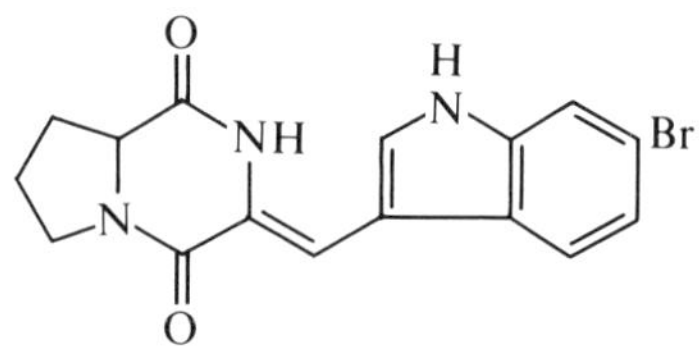

XXXVIII

Geodia beretti is a smooth muscle stimulant [103]. Jaspamide, a cyclic depsipeptide from a species of *Jaspis* sponge possesses antifungal and insecticidal activities [104].

A very interesting finding, if substantiated by the isolation of the pure active component(s), suggests that endogenous opioids may be present in marine invertebrates. In a study, extracts of 11 organisms ranging phyllogenetically from sponges to tunicates were found to contain substances that competed with the opioid peptides when tested in the opioid receptor binding and radioimmunological assay [105].

Recently, two new alkaloids, 28-deoxyzoanthenamine (**XXXIX**) and its 22-epi-isomer were isolated from a *Zoanthus* sp. from the Bay of

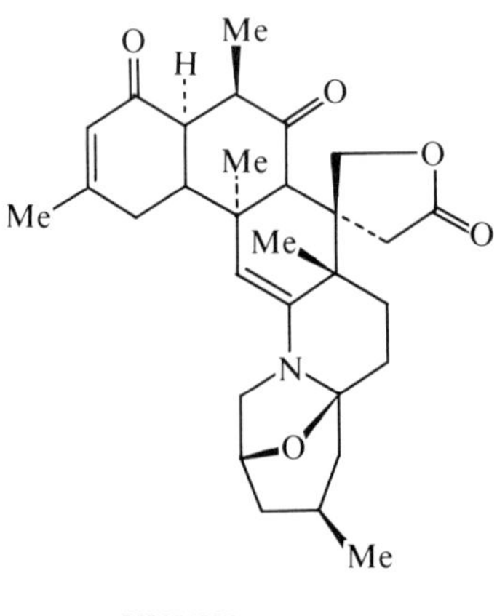

XXXIX

Bengal, the former possessing strong antiinflammatory and analgesic activities [106].

5 Marine toxins

Volumes have been written on the venomous and toxic substances from marine organisms in their crude, semi-pure and other forms [1–8]. The isolation and structure determination of most toxins other than tetrodotoxin and saxitoxin remained untackled for many decades. During the eighties, however, many toxins with some of the most complex structures have been chemically characterized, probably due to rapid advances in the analytical tools and global cooperative efforts of marine chemists and pharmacologists. This review deals with only those toxins that have been either well studied or possess a very potent and/or unusual biological activity.

5.1 Palytoxin

Discovered independently and concurrently by two research groups thousands of miles apart, palytoxin (PTX) from the zoanthid *Palythoa* inhabiting the Caribbean [107] and Pacific [108] oceans is a relatively linear, polycycling, polyether molecule, $C_{129}H_{223}N_3O_{54}$ (mol. wt. 2,677), and one of the most interesting compounds found in nature. Being probably the largest nonpeptide molecule, PTX is the most potent marine toxin [109] and the most potent coronary vasoconstrictor known [16]. A dose of 16 attomoles (10^{-18}moles) of PTX in 0.1 ml saline injected into the inflowing ringer in the isolated perfusing heart of the guinea pig produced a near total constriction of the coronary artery, virtually stopping coronary flow [110]. Considering the dilution factor of the inflowing ringer, the actual effective concentration of PTX reaching vascular cells may be visualized in terms of 10^{-20} moles or less.

The structure of PTX (**XL**) was determined after nearly 10 years of its discovery and isolation in a pure form [111]. After intense effort over a two-year period, this structure and its stereochemistry were confirmed by a remarkable feat indeed; the total synthesis of PTX [112]. More recently, the same laboratory has accomplished total synthesis of a fully protected PTX-carboxyclic acid and PTX-amide [113]. The fully protected acid molecule contained 42 protecting groups of 8 different

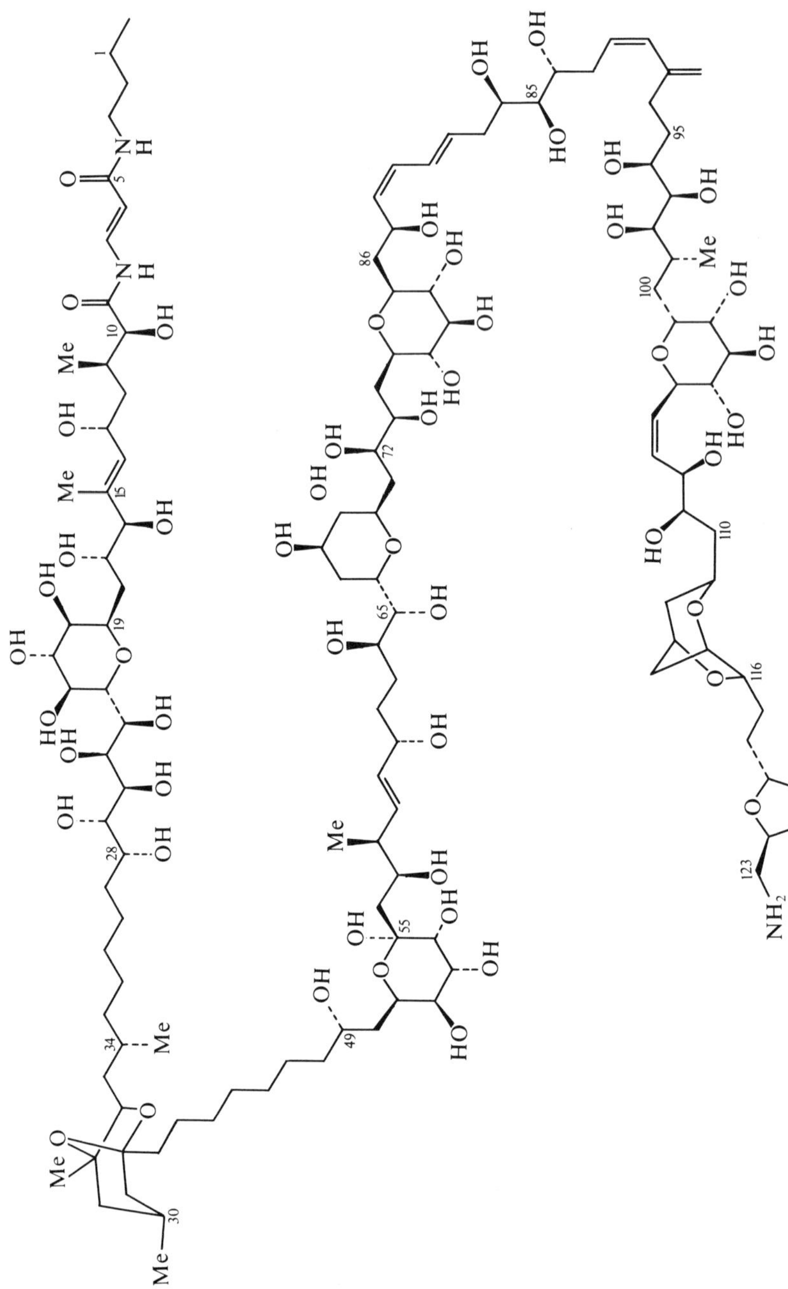

XL

types, all of which could be successfully removed in 5 synthetic and hydrolytic steps. The overall synthetic and conformational aspects of PTX have recently been reviewed by Kishi [114]. It is interesting that among several PTXs isolated from *Palythoa* sp. from Okinawa, one of the new isolates was the PTX-carboxylic acid already synthesized by Kishi et al. [113] prior to this isolation.

Early work in dogs and rats under anesthesia had revealed PTX to (i) act as a strong coronary vasoconstrictor, (ii) elevate the T-wave of the EKG, (iii) increase the force of ventricular contraction, (iv) increase the mean arterial pressure as a result of profound peripheral vasoconstriction, and (v) finally produce death, the cardiac arrythmia followed by ventricular fibrillation always preceding the respiratory arrest [16; Kaul, unpublished data]. Our work had ruled out any central mechanism of action involved in the lethal toxicity of PTX [115]. This was further confirmed or at least supported by Habermann's group [116]. More recently, PTX was named a ciguatera poison cansing respiratory distress, muscular spasm and extremely elevated (up to 200-fold) CPK, LDH and SGOT [117], the enzymes typically associated with coronary ischemia. The most recent and comprehensive review on PTX by Habermann concludes that the toxin acts through Na^+, K^+-ATPase [116].

It is important to note that PTX is not a general cell poison, for it does not necessarily affect all cells and tissues. However, smooth, skeletal and cardiac muscle cells all contract in response to PTX. Depolarization exchanging Na^+ for K^+ has been observed to be the key event in all affected cells [116], followed by various individual events. In the striated muscle, there is a release of intracellular Ca^{++} leading to contraction [116]. This may also be the case with the vascular smooth muscle, for we observed profound contraction of the aorta even in Ca^{++} deprived ringer and to some extent also in the presence of calcium channel blockers [115].

In other organs, the depolarization may open up calcium channels to extracellular Ca^{++}, leading to increased cytoplasmic free Ca^{++} [118–12], which in turn may activate (i) inositol phosphate production [123], (ii) phospholipase A_2 and thereby arachidonic acid metabolism [124, 125], (iii) norepinephrine release [120, 125, 126], and (iv) mast cell degranulation [127]. Based on this evidence and the fact that Na^+-K^+ exchange is instantaneous and massive following exposure to PTX, it has been concluded that this toxin acts primarily by affecting the Na^+, K^+-ATPase [116]. Supporting this, and of particular interest, is that the Na^+-K^+ exchange occurs at very low subpicomole doses of PTX, as do also the cardiovascular effects [110]. All other effects occur at 100 to 1,000-fold higher doses.

Further support of the K^+ efflux as the mechanism of action PTX comes from the experiments with erythrocytes in which only a few molecules of PTX per erythrocyte were adequate to efflux 50% of the intracellular K^+ [116, 128]. All evidence taken together, Habermann [116] has concluded that PTX at high concentrations interacts with the ouabain receptors of the NA^+, K^+-ATPase in the membranes of all vertebrate cells, but a low dose of PTX triggers open the neighboring or ATPase-associated channels to allow K^+ to flow out. Apparently, the overall effect is massive ionic efflux primarily of the intracellular K^+.

Now that the synthesis of PTX and its derivatives has been accomplished, it should be of considerable interest to determine which structural feature or subcomponent is responsible for the unique actions of PTX, or if indeed the entire molecule is essential.

5.2 Red tide toxins

The earliest record of red tide toxicity came from the Gulf of Mexico region and the first mention of the toxin appeared in 1958 [128a]. Now it is well known that dinoflagellates periodically bloom to produce various toxins and the red carotenoid pigment peridinin wherefrom the term "red tide" has emerged. The most important among various species that produce toxins are *Ptychodiscus brevis* (previously called *Gymnodinium breve*), *Protogonyaulax* (formerly *Gonyaulax*) *catenella*, and *P. tamarensis*. Since these organisms have a symbiotic relationship with shellfishes, the toxins get accumulated in these fishes and upon consumption by man elicit a toxic epidemic called "paralytic shellfish poisoning" (PSP). The latest studies on PSP in Japan revealed that it was first confirmed in 1975 in Owase Bay. However, the red tide has since been spreading to wide apart areas [129]. In 1983 PSP incidents were reported in Thailand and Philippines, and in 1986 several incidents were reported in Taiwan and Korea. Table 1 gives some of the dinoflagellate species along with their toxins and reported geographical locations [129].

Table 1
Red tide organisms and their toxins

Organisms	Predominant toxins	Reported site
Protogonyaulax catenella	GTXs and STXs	Pacific USA, Japan, S. America
P. cohorticula	GTXs	Gulf of Thai
P. tamarensis	GTXs and STXs	N. Atlantic coastal regions
Pyrodinium bahamense	STXs and GTXs	Palau Island
Gymnodinium catenatum	GTXs and PX	Japan, Tasmania
GTXs = Gonyautoxins	STXs = Saxitoxins	PX = Protogonyautoxin

5.2.1 Saxitoxins and Gonyautoxins

The first PSP was described in 1957 [130]. The toxin responsible for this poisoning episode was isolated from the Alaskan butter clam *Saxidomus gigantus* and hence the name saxitoxin (**XLI**). However, we

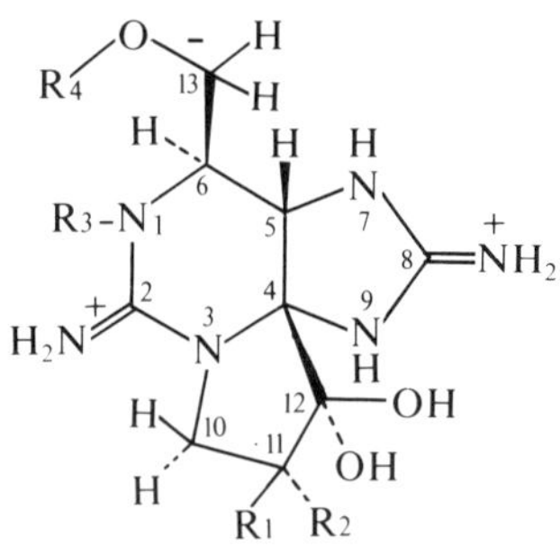

XLI

$R_1 = R_2 = R_3 = H$
$R_4 = CONH_2$

now know that toxins consist of more than 10 closely related derivatives of the basic structural skeleton of STX. Like the well-known puffer fish toxin, tetrodotoxin (TTX), the PSP toxins all contain guanidinium groups. There are at least 8 GTXs and 4 PXs known. Like TTX, these bisguanidinium tricyclic molecules also bind specifically to the Na^+ channels on the outside of excitable membranes, inducing Na^+ influx in exchange for K^+ efflux within a few milliseconds of the accompanying membrane depolarization [130, 131].

5.2.2 Brevetoxins

A number of polycyclic polyether metabolites, brevetoxins, have been isolated from the dinoflagellate *Ptychodiscus brevis* during the late seventies and early eighties. Early on they were given various names by different investigators but this early confusion of their nomenclature was cleared in 1982 [132]. The earlier abbreviation BTX is now called PbTx denoting *Ptychodiscus brevis* toxin. The suffixes of arabic numerals name all the brevetoxins known and yet to be discovered. Thus BTX-A is now PbTx-1 and the old BTX-B is PbTx-2, and so on. The PbTxs in general are unique activators of voltage-sensitive Na^+ channels, with an ability to bind to specific receptor sites on the rat brain synaptosomes [132a]. The chemistry of PbTxs has been reviewed by Nakanishi [133].

These lipid soluble PbTxs have been classified into two major groups based on whether they are hemolytic or neurotoxic, with an overall preponderance of the latter type. The most potent reported is PbTx-1 (**XLII**) whose structure differs from that of other 9 PbTxs. Unlike

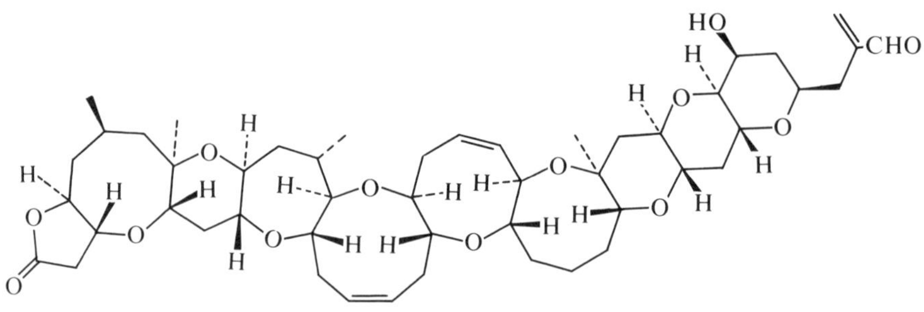

XLII

STXs and GTXs, all of the PbTxs contain no nitrogen. The toxin PbTx-2 is cardiotonic and arrythmogenic in the rat and guinea pig heart; it finally produces an A-V block [134]. It activates Na^+ channels at sites other than those specific for TTX [135], and produces a neuromuscular junction blockade by a persistent depolarization of the nerve terminal [136]. Specific antibodies have been developed to brevetoxins for detection of their traces in shellfishes prior to consumption, but such attempts have failed to prevent PSP incidents [137].

Baden [137a] has recently reviewed the unique polyether toxins from dinoflagellates, especially with reference to their biological activities.

Their excitatory effect on isolated neuromuscular and other cells is mediated by an enhanced Na^+ influx following their specific binding to site 5 on the voltage-sensitive sodium channel, the site to which ciguatoxin also binds.
There should be a similarity in the pharmacological and toxicological profiles of PbTxs and the earlier described polyether ionophores such as the okadaic acid and norhalichondrins isolated from *Halichondria okadai*. Both these types of molecules contain unusual polycyclic, polyether structures with no nitrogen, and yet are highly potent in biological activities.

5.3 Ciguatera toxins

These toxins, though not directly associated with PSP of the red tide, are present in toxic dinoflagellates which then reach the human food chain not through shellfishes but through reef fishes. The initial factor was isolated from the Pacific red snapper, termed ciguatoxin (CTX), and found to be a complex lipoidal molecule [138] originating from *Gambierdiscus toxicus* [139]. The same organism was subsequently found to also produce a more powerful toxin, maitotoxin (MTX).

5.3.1 Maitotoxin

Although not associated with the red tide, MTX has been isolated from the toxic dinoflagellate *Gambierdiscus toxicus* [140] and found to be a Ca^{++} channel activator [141, 142]. It increases Ca^{++} uptake in cultured NG108-15 neuroblastoma X glioma cells by altering voltage dependence of calcium channel activation [143].
Maitotoxin has been erroneously claimed as the most potent marine toxin [144], but a careful comparison of the LD_{50} or MLD of various toxins will clearly reveal that palytoxin is in fact the most potent marine toxin known to date [20, 116]. In the rat myocardiac cells, MTX increases both force and rate of contraction followed by arrythmias [144]. These effects are potentiated by increasing the extracellular calcium, inhibited by verapamil, and abolished by Ca^{++}-free medium or by Ca^{++} antagonists. In reaggregate cultures of striated neurons, MTX induces a dose-dependent release of GABA, which can however be blocked by the calcium blocker D-600 [145]. In other experiments, calcium-activated increases in the release of progesterone and relaxin were observed in the swine luteal tissue treated with MTX [146].

The first direct evidence that MTX activates voltage-independent calcium channels in the cardiac myocytes came from the patch-clamp experiments [147]. In these studies, the MTX-induced current was enhanced by epinephrine and abolished by cadmium. Since the mean open time of the MTX-activated channels was ten times longer than the normal voltage-gated calcium channels, these investigators have suggested that MTX activates a new class of calcium channels.

The latest studies on MTX further confirm its effect on Ca^{++} flux into the cultured myocardial cell, which begins with cardiotonicity and is followed by depletion of ATP as the intracellular Ca^{++} accumulates, and finally ends in cell injury [148]. All these effects can be blocked by a calcium channel blocker such as verapamil.

The structure of MTX has not yet been established, but studies so far reveal it to be a non-peptide with a large molecular weight of 3,425 as determined by the fast atom bombardment mass spectrum [149]. It reacts with Dragendorff's reagent but not with ninhydrin, suggesting that it is alkaloidal in nature. The molecule possesses a large number of hydroxyls, ether oxygens, and two sulfate ester groups. It appears to contain no repeating amino acid or carbohydrate units, no carbonyl groups, no side chains other than a methyl or a vinyl group, and no carbocycles. All these features, excepting the sulfated ester groups, would seem to match the features of PTX. In toxicity also, MTX comes close to PTX, though the latter is clearly much more toxic. It should be interesting to see if MTX is not, in fact, another variant molecule of the PTX family.

5.3.2 Ciguatoxin

Characterized by nausea, diarrhea, headache and hypothermia, ciguatoxin (CTX) poisoning has been known for a long time. The structural studies on this toxin are currently in progress, but the pure homogeneous CTX has been assigned a molecular formula of $C_{53}H_{77}NO_{24}$ and a molecular weight of 1,112 [150].

At low i. v. doses CTX elicits bradycardia and respiratory stimulation in anesthetized cat and rat, while at higher doses both the heart rate and respiration are depressed accompanied by hypertension [151], probably due to reflex compensation of bradycardia. At very low picogram concentrations, CTX is strongly cardiotonic [152]. Comparative studies have suggested that CTX acts at the nerve endings whereas MTX acts on the cell membranes [152].

Purified CTX at 0.1 to 10 ng/ml caused depletion of GABA and dopamine in the rat brain synaptosomes, but this effect was blocked by TTX [153]. In electrophysiological experiments carried out with cultured neuroblastoma cells, CTX caused an increased influx of Na^+ which was abolished by TTX. Based on these data, it has been concluded that CTX acts selectively on voltage-dependent Na^+ channels in nerve and muscle cells as well as the synaptic terminals [153].
Recent studies on the mode of cardiotonic action of CTX revealed a dose-dependent increase in the contractile force of the isolated guineapig left atrium exposed to low doses, which was markedly inhibited by TTX and fully abolished by practolol and reserpine. Higher doses of CTX however were only partially blocked by these agents. The effect of CTX on action potential of atrial strips was also abolished by TTX. These data suggest that CTX belongs to a new class of Na^+ channel activators interacting with a specific site different from the site where other toxins bind [154].

5.4 Conotoxins

A number of toxins isolated from the venom of a fish-eating snail, *Conus geographus,* may be grouped into (i) μ-conotoxins, a group of 7 homologous peptides of 22 residues each [155], and (ii) ω-conotoxins. In the first group, Geographutoxins I and II described by Kobayashi et al. [156] appear to be identical with conotoxins GIIIA and BIIIB [157]. Like guanidinium toxins TTX and STX, the μ-conotoxins block the action potential of skeletal muscle cells, but unlike TTX they do not affect the channels of the nerves [155]. The conotoxin CIIIA has been found to bind to the same site on the Na^+ channel where STX and TTX bind [158].
The ω-conotoxins are a subgroup in a class of peptides that are neurotoxic [159]. Designated *w*-CgTX, they irreversibly inhibit the transmitter release at the frog neuromuscular junction and block the Ca^{++} component of the action potential in the chick dorsal root ganglion neurons [160]. Both these actions appear to be due to Ca^{++} channel blockade [161].
Both ω-CgTX and the related *w*-CmTX from *C. magnus* were shown to inhibit a depolarization-evoked ATP release from synaptosomes of ray electric organ, apparently by blocking the Ca^{++} flux through the voltage-gated calcium channels [162]. A specific high affinity binding

of ω-CgTX VIA to synaptosomal and membrane preparations from chick brain has also been demonstrated, suggesting a new site on the calcium channel for the toxin distinct from both dihydropyridine and verapamil binding sites [163].
A glycoprotein with a molecular weight of 25,000 isolated from *C. striatum,* striatotoxin (StTX), inactivated Na^+ channels in the mouse neuroblastoma cells and rat brain synaptosomes [164]. Earlier, StTX was shown to elicit a long-lasting stimulant effect on the isolated guinea-pig atria [165]. The mechanism of this action studied with the help of various pharmacological, biochemical and electrophysiological techniques revealed that (i) StTX caused a dose-dependent cardiotonic effect blockable by TTX, and (ii) the duration of action potential was prolonged perhaps due to slow inactivation of Na^+ current which would increase the available intracellular calcium for the cardiac myocytes; hence the cardiotonicity observed with StTX [166].

6 Summary and conclusion

Somewhat accelerated developments in the chemistry and pharmacology of marine molecules during the eighties are a clear indication of the biomedical potential of marine organisms for the twenty-first century. Unfortunately, the overall effort toward the field is still insignificant. Both industry and governments are spending only a token share of R&D funds in pursuit of pharmacologically active substances from the sea. A critical appraisal of the literature reveals the existence of fascinating molecules with unusual and potent activities. The challenge of harnessing the clinical potential of these molecules is clearly evident. It is only awaiting the awakening of the academic, industrial, and federal researchers and resources. Only a concerted and a massive effort can shorten the time between now and the first clinical drug from the sea.

References

1 B.W. Halstead: Poisonous and Venomous Marine Animals of the World, vol. 1. US GPO, Washington, DC 1965. pp. 1–29.
2 G.A. Emerson and C.H. Taft: Toxicol. Rep. Biol. Med. *3,* 302 (1945).
3 R.F. Nigrelli: Biochemistry and Pharmacology of Compounds Derived from Marine Organisms, Ann. NY Acad. Sci., New York 1960.
4 H.C. Freudentahl: Drugs From the Sea, Marine Technol. Soc., Washington, DC 1968.

5 A. Der Marderosian: J. Pharm. Sci. *58,* 1 (1969).
6 M.H. Baslow: Marine pharmaceuticals, Williams & Wilkins, Baltimore 1969.
7 H.W. Youngken: Food Drugs from the Sea Proceedings, Marine Technol. Soc., Washington, DC 1969.
8 F.E. Russel: International Encyclopedia of Pharmacology and Therapeutics *2,* 3 (1971).
9 L.R. Worthen: Food Drugs from the Sea Proceedings, Marine Technol. Soc., Washington, DC 1972.
10 D. Martin and G. Padilla: Marine Pharmacognosy, Academic, New York 1973.
11 H.H. Webber and G.D. Ruggieri: Food Drugs from the Sea Proceedings, Marine Technol. Soc., Washington, DC 1976.
12 G.D. Ruggieri: Science *194,* 491 (1976).
13 Y. Hashimoto: Marine Toxins and Other Bioactive Marine Metabolites, Japan Sci., Tokyo 1979.
14 P.N. Kaul and C.J. Sindermann: Drugs and Food from the Sea – Myth or Realitiy? Oklahoma Univ. Printing, Norman 1978.
15 P.N. Kaul: Impact Sci. on Soc. *29,* 123 (1979).
16 P.N. Kaul: Fed. Proc. *40,* 10 (1981).
17 P.N. Kaul: Pure Appl. Chem. *54,* 1963 (1982).
18 M.Chr. Krebs: Fortschr. Chem. Organ. Naturstoffe *48,* 151 (1986).
19 D.J. Faulkner: Nat. Prod. Rep. *1,* 251 and 551 (1984); *3,* 1 (1986); *4,* 539 (1986); *5,* 613 (1988).
20 P.N. Kaul and P. Daftari: Ann. Rev. Pharmacol. Toxicol. *26,* 117 (1986).
21 Y. Ogura: Eur. J. Pharmacol. *3,* 58 (1968).
22 A.J. Weinheimer, J.A. Matson, T.K.B. Karns, M.B. Hossain and D. van der Helm: pp. 117–21 in Ref. 14.
23 Y. Kashman, A. Groweiss, S. Carmely, Z. Kinamoni and D. Czarkie: Pure Appl. Chem. *54,* 1995 (1982).
24 B. Tursch, J.C. Braekman, D. Caloze and M. Kaisin: Marine Natural Products, vol 2. Academic, New York/London 1978. pp. 247–96.
25 M.G. Hadfield and L.S. Ciereszko: pp. 145–50 in Ref. 14.
26 P.N. Kaul: pp. 209–307 in Ref. 9.
27 D.L. Perkins and L.S. Ciereszko: Hydrobiologia *42,* 77 (1973).
28 Y. Kobayashi and K. Osabe: Chem. Pharm. Bull. *37,* 631 (1989).
29 G.R. Pettit, C.L. Herald, D.L. Doubej and D.L. Herald: J. Am. Chem. Soc. *104,* 6846 (1982).
30 I.L. Dale, T.D. Bradshaw, A. Gescher and G.R. Pettit: Cancer Res. *49,* 3242 (1989).
31 H.G. Drexler, S.M. Gignac, R.A. Jones, C.S. Scott, G.R. Pettit and V. Hoffbrand: Blood *74,* 1747 (1989).
32 W.S. May, S.T. Sharkis, E.H. Ahmed, V. Gebbia, A.S. Kraft and G.R. Pettit: Proc. Natl. Acad. Sci. (USA) *84,* 8483 (1987).
33 K.L. Rinehart, J.B. Gloer, G.R. Wilson, R.G. Hughes and H.L. Li: Fed Proc. *42,* 87 (1983).
34 H.G. Chun. B. Davies, D. Hoth, M. Suffness, J. Plowman, K. Flora, C. Grieshaber and L. Jones: Invest. New Drugs, *4,* 279 (1986).
35 K.L. Rinehart, V. Kishore, K.C. Bible, R. Saki, D.W. Sullivan and K.M. Li: J. Nat. Prod. *51,* 1 (1988).
36 W. Lichter, L.L. Wellham, B.A. van der Werf, R.E. Middlebrook and M.M. Sigel: pp. 117–27 in Ref. 9.
37 W. Lichter. A. Ghaffar, L.L. Wellham and M.M. Sigel: pp. 137–44 in Ref. 14.
38 M.M. Sigel, W. Lichter, A. Ghaffar, L.L. Wellham and A.J. Weinheimer: Adv. Exp. Med, Biol. a. *121,* 577 (1979).

39 W.C. Dunn, W.L. Carrier and J.D. Regan: Toxicon *20,* 703 (1982).
40 K.L. Rinehart: Pure Appl. Chem. *61,* 525 (1989).
41 B.M. Degnan, C.J. Hawkins, M.F. Lavon, E.J. McCaffrey, D.L. Parr and D.J. Walters: J. Med. Chem. *32,* 1354 (1989).
42 B.M. Degnan, C.J. Hawkins, M.F. Lavin, E.J. McCaffrey, D.L. Parry, A.L. van der Brenk and D.J. Walters: J. Med. Chem. *32,* 1349 (1989).
43 R.E. Moore, S. Banerjee, V. Bornemann, F.R. Caplan, J.L. Chen. D.C. Corley, L.K. Larsen, B.S. Moore, G.M.L. Patterson, V.J. Paul, J.B. Stewart and D.E. Williams: Pure App. Chem. *61,* 521 (1989).
44 G.R. Pettit, Y. Kamano, C.L. Herald, C. Dufresne, R.L. Cerny, D.L. Herald, J.M. Schmidt and H. Kizu: J. Amer. Chem. Soc. *111,* 5015 (1989).
45 N.S. Burrcs and J.J. Clement: Cancer Res. *49,* 2935 (1989).
46 G. Cimino, S. DeRosa and S. De Stefano: Experientia *40,* 339 (1984).
47 E.T. O'Brien, R.S. Jacobs and L. Wilson: Mol. Pharmacol. *24,* 26 (1983).
48 E.T. O'Brien, D.J. Asai, R.S. Jacobs and L. Wilson: Mol. Pharmacol. *35,* 635 (1989).
49 K. Iguchi, S. Saitoh and Y. Yameda: Chem. Pharm. Bull. *37,* 2553 (1989).
50 I. Kitagawa, M. Kobayashi, N.K. Lee, Y. Oyama and Y. Kyogoku: Chem. Pharm. Bull. *37,* 2078 (1989).
51 Y. Hirata: Pure Appl. Chem. *61,* 293 (1989).
52 P.S. Sarin, D. Sun, A. Thornton and W.E.G. Miller: J. Natl. Cancer Inst., *78,* 663 (1987).
53 T.R. Norton, S. Shibata, M. Kashigawa and J. Bentley: J. Pharm. Sci. *65,* 1368 (1976).
54 S. Shibata, T.I. Zumi, T. Matsuo and S. Katsuki: J. Pharmacol. Exp. Ther. *199,* 298 (1976).
55 S. Shibata, T.I. Zumi, D.G. Seriguchi and T.R. Norton: J. Pharmacol. Exp. Ther. *205,* 683 (1978).
56 A. Scriabine, C.G. van Arman, G. Morgan, A.A. Morris, C.D. Bennett and N.R. Bohidar: J. Cardiovasc. Pharmacol. *1,* 571 (1979).
57 I. Kodama, J. Toyama, S. Shibata and T.R. Norton: J. Cardiovasc. Pharmacol. *3,* 75 (1981).
58 G.J. Gross, D.C. Warltier, M.F. Hardman and S. Shibata: Eur. J. Pharmacol. *110,* 271 (1985).
59 L. Beress: pp. 59–72 in Ref. 14.
60 R. Alsen, T. Peters and E. Scheufler: J. Cardiovasc. Pharmacol. *4,* 63 (1982).
61 J. Barhanin, M. Hughes, H. Scheweitz, J.P. Vincent and M. Lazdunski: Toxicon *20,* 59 (1982).
62 A.A. Nabiullin, S.E. Odinikov, E.P. Kozlovskaya and G.B. Elyakov: FEBS Lett. *141,* 124 (1982).
63 T.R. Norton: Fed. Proc. *40,* 21 (1981).
64 B.A. Gorshkov, V.A. Stonik and G.B. Elyakov: Toxicon *20,* 655 (1982).
65 D. Graff and C.J.P. Grimmelikhuijzen: Brain Research *4442,* 354 (1988).
66 R.S. Norotn, G. Bobek, J.O. Ivanov, M. Thompson, E. Fiala-Beer, R.L. Moritz and R.J. Simpson: Toxicon *28,* 29 (1990).
67 S.S. Cohen: Cancer *40,* 509 (1977).
68 F.A. Fuhrman, G.J. Fuhrman, Y.H. Kim, L.A. Pavelka and H.S. Mosher: Science *207,* 193 (1980).
69 L.P. Davis, D.D. Jamieson, J.A. Baird-Lambert and R. Kazlauskas: Biochem. Pharmacol. *33,* 347 (1984).
70 K. Laderis and K.L. MacCannell: pp. 51–59 in Ref. 14.
71 K. Laderis, A. Letter, D. McMaster, G. Moore and D. Schlesinger: Science *218,* 162 (1982).
72 D. McMaster and K. Laderis: Peptides *4,* 367 (1983).
73 K. Owada, M. Kawata, L. Akagi, A. Takagi, M. Moriga and H. Kobyashi: Cell Tissue Res. *239,* 349 (1985).

74 W. M. Lafranconi, I. Ferlan, F. E. Russell and R. J. Huxtable: Toxicon *22,* 347 (1984).
75 J. P. Girard, C. Marion, M. Liutkus, M. Boucard, E. Recheneq, J. P. Vidal and J. C. Rossi: Planta Medica *54,* 193 (1988).
76 H. A. Hoppe and T. Levring: Marine Algae in Pharmaceutical Sciences. vol. 2. de Gruyter, Berlin 1982.
77 J. Kobayashi, Y. Ohizumi, H. Nakamura, Y. Hirata, W. Wakamatsu and T. Miyazawa: Experientia *42,* 1064 (1986).
78 K. H. Hollenbeak, F. J. Schmitz, P. N. Kaul and S. K. Kulkarni: pp. 81–87 in Ref. 14.
79 R. J. P. Cannell, S. J. Kellam, A. M. Owsianka and J. M. Walker: J. Gen. Microbiol. *133,* 1701 (1987).
80 H. Nakashima, Y. Kido, N. Kobayashi, Y. Motoki, M. Neeshul and N. Yamamoto: J. Cancer Res. Clin. Oncol. *113,* 413 (1987).
81 N. Fusetani, M. Sugano, S. Matsunaga, K. Hashimoto, H. Shikama, A. Ohta and N Nagano: Experientia *43,* 1233 (1987).
82 N. Fusetani, M. Sugano, S. Matsunaga and K. Hashimoto: Experientia *43,* 1234 (1987).
83 E. D. De Siva and P. J. Scheuer: Tetrahedron Lett. *21,* 1611 (1980).
84 K. B. Glaser and R. S. Jacobs: Biochem. Pharmacol. *35,* 449 (1986); *36,* 2079 (1987).
85 L. A. Wheeler, G. Sachs, G. DeVries, D. Goodrum, E. Woldemussie and S. Muallem: J. Biol. Chem. *262,* 6531 (1989).
86 P. J. Scheuer: Chemistry of Marine Natural Products. Academic, New York 1973.
87 J. Baker and V. Murphy: Handbook of Marine Sciences. CRC, Cleveland 1976.
88 F. J. Schmitz, Y. Gopichand, D. Michaud, R. S. Prasad and S. Ramaley: Pure Appl. Chem. *103,* 2469 (1981).
89 K. Tachibana, P. J. Scheuer, Y. Tsukitani, H. Kiruchi and D. van Engen: J. Am. Chem. Soc. *103,* 2469 (1981).
90 S. Shibata, Y. Ishida, H. Kitano, Y. Ohizumi and J. Habon: J. Pharmacol. Exp. Ther. *223,* 135 (1982).
91 M. Isobe, Y. Ichikawa and T. Goto: Tetrahedron Lett. *27,* 963 (1986).
92 D. Uemura, K. Takahashi, T. Yamamoto, C. Katayama, J. Tanaka, Y. Okumura and Y. Hirata: J. Am. Chem. Soc. *107,* 4796 (1985).
93 Y. Hirata and D. Uemura: Pure Appl. Chem. *58,* 201 (1986).
94 D. W. Cameron, G. I. Fentrill, C. L. Gibson and R. W. Reed: Tetrahedron Lett. *26,* 3887 (1985).
95 Y. Okami: Microb. Ecol. *12,* 65 (1986).
96 Y. Kashman, A. Groweiss and U. Shmueli: Tetrahedron Lett. *21,* 3629 (1980).
97 A. Groweiss, U. Shmueli and Y. Kashman: J. Org. Chem. *48,* 3512 (1983).
98 Y. Kashman, A. Groweiss, R. Lidor, D. Blasberger and S. Carmely: Tetrahedron *41,* 1905 (1985).
99 I. Spector, R. R. Schochet, Y. Kashman and A. Groweiss: Science *219,* 493 (1983).
100 N. B. Perry, J. W. Blunt, J. D. McCombsand, M. H. G. Munro: J. Org. Chem. *51,* 5476 (1986).
101 J. Kobayashi, J. F. Cheng, M. Ishibashi, H. Nakamura, Y. Ohizumi. Y. Hirata, T. Sasaki, H. Lu and J. Clardy: Tetrahedron Lett. *28,* 4939 (1987).
102 E. Fattorusso, V. Lanzotti, S. Magno and E. Novellino: J. Nat. Prod. *48,* 924 (1985).
103 G. Lindgren, L. Bohlin and J. Bergman: Tetrahedron Lett. *27,* 3283 (1986).
104 T. M. Zabriskie, J. A. Klocke, C. M. Ireland, A. H. Marcus, T. F. Molinski, D. J. Faulkner, C. F. Xu and J. Clardy: J. Am. Chem. Soc. *108,* 3123 (1986).

105 R. Piccoli, D. Melek, A. Spagnuolo, S. Vescia and L. Zanetti: Comp. Biochem. Physiol. *80C*, 237 (1983).
106 D.V. Rao, V.S.N. Raju, B.W. Sullivan and D.J. Faulkner: Heterocycles *28*, 103 (1989).
107 D.H. Attaway and L.S. Ciereszko: Proc. Soc. Int. Coral Reef Sym. *7*, 497 (1974).
108 R.E. Moore and P.J. Scheuer: Science *172*, 495 (1971).
109 P.N. Kaul: Proc. West. Pharmacol. Soc. *17*, 294 (1974).
110 P.N. Kaul: p. 311 in Ref. 11.
111 R.E. Moore and G. Bartolini: J. Am. Chem. Soc. *103*, 2491 (1981).
112 J.K. Cha, W.J. Christ, J.M. Finin, M. Fujioka, Y. Kishi, L.L. Klein, S.S. Ko, J. Leder, W.W. McWhorter, K.P. Pfaff, M. Yonaga, D. Uemura and Y. Hirata: J. Am. Chem. Soc. *104*, 7369 (1982).
113 R.W. Armstrong, J.M. Beau, S.M. Cheon, W.J. Christ, M. Fujioka, W.M. Ham, L.D. Hawkins, H. Jin, S.H. Kang, Y. Kishi, M.J. Martinelli, W.C. McWhorther, M. Mizuno, M. Nakato, A.E. Stutz, F.X. Talamas, M. Taniguchi, J.A. Tino, K. Ueda, J.I. Uenishi, J.B. White and M. Yonaga: J. Am. Chem. Soc. *111*, 7525 (1989); *111*, 7530 (1989).
114 Y. Kishi: Pure Appl. Chem. *61*, 313(1989).
115 S.K. Kulkarni, W.G. Kirlin and P.N. Kaul: pp. 73–80 in Ref. 14.
116 E. Habermann: Toxicon *27*, 1171 (1989).
117 A.M. Kodama, Y. Hokama, T. Yashimota, M. Fukni, S.J. Manea and N., Sutherland: Toxicon *27*, 1051 (1989).
118 M.D. Rayner, B.J. Sanders, S.M. Harris, Y.C. Lin and B.E. Morton: Res. Commun. Chem. Path. Pharmacol. *11*, 55 (1975).
119 K. Ito, H. Karaki and N. Urakawa: Eur. J. Pharmacol. *46*, 9 (1979).
120 M. Tatsumi, M. Takahashi and Y. Ohizumi: Molec. Pharmacol. *25*, 379 (1984).
121 Y. Ishida, N. Satake, J. Habon, H. Kitano and S. Shibata: J. Pharmacol. Exp. Ther. *232*, 551 (1985).
122 H. Ozaki, H. Nagase, H. Karaki and N. Urakawa: Comp. Biochem. Physiol. *86C*, 387 (1987).
123 E. Habermann and M. Lanx: Naunyn-Schmiedeberg's Arch. Pharmacol. *334*, 1 (1986).
124 L. Levine and H. Frijika: Carcinogenesis *6*, 1631 (1985).
125 H. Nagasi and H. Karaki: J. Pharamcol. Exp. Ther. *242*, 1120 (1987).
126 Y. Ishida, A. Kajiwara, K. Takagi, Y. Ohizumi, and S. Shibata: J. Pharmacol. Exp. Ther. *232*, 551 (1985).
127 G.S. Chhatwaal, G. Ahnert-Hilger, L. Beress and E. Habermann: Intro. Arch. Allergy Appl. Immun. *68*, 97 (1982).
128 E. Habermann, G. Ahnert-Hilger, G.S. Chhatwaal and L. Beress: Biochem. Biophys. Acta. *649*, 481 (1981).
128a T.J. Starr: Texas Rep. Biol. Med. *16*, 813 (1958).
129 K. Hashimoto and K. Noguchi: Pure Appl. Chem. *61*, 7 (1989).
130 T. Narahashi: Physiol. Rev. *54*, 813 (1974).
131 C.Y. Kao: Red. Proc. *40*, 30 (1981).
132 Y. Schmizu: Pure Appl. Chem. *54*, 1973 (1982).
132a M.A. Poli, T.J. Mende and G. Baden: Mol. Pharmacol. *30*, 129 (1986).
133 K. Nakanishi: Toxicon *23*, 473 (1985).
134 R.L. Rodgers, H.N. Chou, K. Temma, T. Akera and Y. Shimizu: Toxicol. Appl. Pharmacol. *76*, 296 (1984).
135 J.M.C. Huang, C.H. Wu and D.G. Baden: J. Pharmacol. Exp. Ther. *229*, 615 (1984).
136 D.G. Baden, G. Bikenazi, S.J. Decker, F.F. Foldes and I. Leung: Toxicon *22*, 75 (1984).
137 D.G. Baden, T.J. Mende, J. Walling and R. Schultz: Toxicon *22*, 783 (1984).

137a D. G. Baden: FASEB J. *3*, 1807 (1989).
138 P. J. Scheuer, W. Takanashi, J. Tsutsumi and T. Yoshida: Science *155*, 1267 (1967).
139 R. Adachi and T. Fukuyo: Bull. Jpn. Soc. Sci. Fish. *45*, 67 (1979).
140 R. W. Dickey, D. M. Miller and D. R. Tindell: ACS Symposium Series *262*, 257 (1984).
141 M. Takahashi, M. Tatsumi and Y. Ohizumi: J. Biol. Chem. *258*, 10944 (1983).
142 T. Miyamoto, Y. Ohizumi, H. Washio and Y. Yasumoto: Pfluegers Arch. *400*, 439 (1984).
143 S. B. Freedman, R. J. Miller, D. M. Miller and D. R. Tindal: Proc. Natl. Acad. Sci. USA *81*, 4582 (1984).
144 M. Kobayashi, G. Miyakado, T. Nakamura and Y. Ohizumi: Eur. J. Pharmacol. *111*, 121 (1985).
145 I. A. Shalaby, S. Kongsamut and R. J. Miller: J. Neurochem. *46*, 1161 (1986).
146 J. D. Veldhuis, K. Yoshida, W. DuBois and M. J. Fields: Amer. J. Physiol. *252E*, 8 (1987).
147 M. Kobayashi, R. Ochi and Y. Ohizumi: Br. J. Pharmacol. *92*, 665 (1987).
148 G. Santostasi, R. Kutty, A. L. Bartorelli, T. Yasumota and G. Krishna: Toxicol. Appl. Pharmacol. *102*, 164 (1990).
149 A. Yokoyama, M. Mureta, Y. Oshima, T. Iwashita and T. Yasumoto: J. Biochem. *104*, 184 (1988).
150 M. Nukina, L. M. Koyanagi and P. J. Scheuer: Toxicon *22*, 169 (1984).
151 A. M. Legrand, M. Galonnier and R. Bagnis: Toxicon *20*, 311 (1982).
152 A. M. Legrand and R. Bagnis: Toxicon *22*, 471 (1984).
153 J. N. Bidard, H. P. M. Vijverberg, C. Frelin, E. Chungue, A. M. Legrand, R. Bagnis and M. Lazdunski: J. Biol. Chem. *259*, 8353 (1984).
154 A. Seino, M. Kobayashi, K. Momose, T. Yasumoto and Y. Ohizumi: Br. J. Pharmacol. *95*, 876 (1988).
155 L. J. Cruz, W. R. Gray, B. M. Olivera, R. D. Zeikus, L. Kerr, D. Yoshikami and F. Moczyllowski: J. Biol. Chem. *260*, 9280 (1985).
156 M. Kobayashi, C. H. W, M. Yoshii, T. Narahashi, H. Nakamura, J. Kobayashi and Y. Ohizumi: Pflugers Arch. *407*, 241 (1986).
157 S. Sato, H. Nakamura, Y. Ohizumi, J. Kobayashi and Y. Hirata: FEBS Lett. *155*, 277 (1983).
158 Y. Yanagawa, T. Abe and M. Satake: J. Neurosci. *7*, 1498 (1987).
159 B. M. Olivera, J. M. McIntosh, R. Zeikus, W. R. Gray, J. Verga, J. Rivier, V. de Santos and L. J. Cruz: Science *230*, 1338 (1985).
160 L. M. Kerr and D. Yoshikami: Nature *308*, 282 (1984).
161 D. H. Feldman, B. M. Olivera and D. Yoshikami: FEBS Lett. *214*, 295 (1987).
162 R. E. Yeager, D. Yoshikami, J. Rivier, L. J. Cruz and G. P. Miljanich: J. Neurosci. *7*, 2390 (1987).
163 L. J. Cruz and B. M. Olivera: J. Biol. Chem. *261*, 6230 (1986).
164 T. Gonoi, Y. Ohizumi, J. Kobayashi, H. Nakamura and W. A. Catterall: Mol. Pharmacol. *32*, 691 (1987).
165 J. Kobayashi, H. Nakamura, Y. Hirata and Y. Ohizumi: Toxicon *20*, 823 (1982).
166 Y. Ohizumi, M. Kobayashi, A. Muroyama, H. Nakamura and J. Kobayashi: Br. J. Pharmacol. *95*, 876 (1988).

Index Vol. 35

The references of the Subject Index are given in the language of the respective contribution.
Die Stichworte des Sachregisters sind in der jeweiligen Sprache der einzelnen Beiträge aufgeführt.
Les termes repris dans la Table des Matières sont données selon la langue dans laquelle l'ouvrage est écrit.

Aaptamine 534
Aaptos aaptos 534
Acanthifolicin 537
Acebutol 158
Acedapsone 396
Acetaminophen 330
Acetylcholine 32, 55
Aclazinomycin 504
Actina equina 534
Actinomycin 271, 504, 508
Adenocarcinoma 309, 492
Adenosine 533
Adenosine kinase 533
Adrenergic blockers 534
Adrenergic receptor 31
Adrenocorticotropin 88
Agitation 141
Agoraphobia 99
Akathisia 140
Alaproclate 87
Alboursin 263
Albumin 216
Allergic disorders 276
Allopurnole 417
Alloxan 330
Amino penicillin 254
Aminoacridines 385
Aminobenzamide 126
Aminobenzoic acid 394
Aminobenzylpenicillin 254
Aminoquinolines 386, 388
Aminoarone 156, 158, 175, 198, 223
Amodiaquine 386
Amopyroquine 386
Amoscanate 450, 463
Amphotericin 419
Ampicillin 260
Amygdala 40
Anemonia elegantissima 532
Anemonia sulcata 531
Angiotensin 34, 535
Anguidine 316
Ansidoris nobilis 533
Anthopleura xanthagrammica 530
Anthracyclin 324
Anthracycline antibiotics 251, 303
Anti-interferon 114
Antibiotics 384
Anticancer drugs 261
Antifolates 394
Antihypertensive agents 62, 535
Antimetastatic drugs 321
Antineoplastic drugs 251, 303
Antinvasive drugs 321
Antitumor drugs 251, 261, 315
Antitumor studies 302
Anxiety 99
Aplysia dactylomela 535
Aplysinopsin 540
Apnea 54
Ara C 533
Aranotins 295
Arbovirus B 369
Arginine-vasopressin 38
Arrhythmia 152
Arsenamide 442
Arsenicals 441
Artemisia annua 381
Arterial blood pressure 27
Arylmethanols 390
Arylpiperazines 90
Arythmogenesis 152
Aspergillus terreus 282, 300
Aspergillus ustris 272
Astroides calicularis 540
Ataxia 142
Atenolol 142, 156, 158
Atrial fibrilation 167
Atrial flutter 167
Atropin-like drugs 181
Atropine 199, 214
Autonomium 535
Averanol 529
Averone 529
Azacytidine 316
Azathioprine 494

Bacillus subtilis 289
Barbiturates 140
Barettin 540
Baroreceptor 30
Benzimidazoles 448
Benzoctamine 93
Benzodiazepine 139, 140
Benthiazoles 450
Benzylpenicillin 289
Berenil 426
Beta blockers 165, 175, 194, 140
Biclomycin 289
Bicuculline 34, 39, 44, 56, 58
Bicyclomycin 291
Biguanides 398
Bilhariasis 457
Bimolane 308, 341
Bipolaris sorokiniana 282
Bisantrene 504
Bistratamide 524
Bleomycin 261
Blocking agents 158
Borellia spp. 369
Borrelia recurrentis 370
Botrytis cinerea 280
Bradyarrhythmia 160
Bradycardia 52, 54
Brain tumor 308
Brevetoxins 546
Brevianamide 284
Bromides 140
Bromo-ergokryptine 275
Bronchoconstriction 276
Brucella abortus 114, 128
Brugia malayi 372
Bryostatin 522
Bufotenin 90
Bugs 367
Bugula neritina 522
Buspirone 91

Calcium channel blockers 120
Calcium ion channel 154
Calmidazolium 120
Calmodulin antagonists 120
Camptothecin 316
Cancer chemotherapy 99, 487
Capsaicin 53
Carbachol 38
Carbamazepine 140
Cardiomyopathy 324
Cardiotonics
Cardiotoxicity 340
Cardiotoxin 531
Catalase 331
Catecholamines 27, 33, 55, 158, 194, 198
Catostomus commersoni 534
Cefoperazone 253, 256, 259, 261
Cembranes 521
Central nervous system 29
Cephalosporin 251, 253
Ceratocytis ulmi 294
Chaetomins 295
Chaetomium 273
Chaetomium cochliodes 273
Chaetoxin 273
Chagas' disease 375, 431
Chetomin 273, 294
Chloramphetamine 88, 96
Chlorimipramine 47, 96
Chlormethylamphetamine 88
Chlorophenylalanine 94
Chloropromazin 120, 143
Chloroquine 385, 386, 414
Chlorproguanil 398
Cholescystokinin 120
Cholesterol 257, 532
Chromomycin 316
Ciguatera toxins 547
Ciguatoxin 548
Cimetidine 170, 175, 210, 417
Cinanserin 93
Cinchona ledgeriana 381
Cinchona trees 381
Cis-platin 316
Cis-platinum anthracycline 501
Citalopram 87
Clafazimine 451
Clindamycin 384
Clinical research 9
Clonazepam 142
Clonidin 32, 37, 42, 55, 62
Clonorchis sinensis 374
Coccidiostats 251, 266
Collagenase 315
Coma 14
Concanavalin 261
Confusion 142
Conotoxins 549
Conus geographus 549
Conus striatum 550
Coralyne 316
Corticosterone 88, 89
Corticotropin 88
Cortisol 99, 125
Cortisone 490
Corynebacterium parvum 490
Coumadin 175, 181, 200
Coxiella burnetti 369
Creatinine 205
Creptotethia crypta 527
Crustaceans 367

Cryptotethia crypta 533
Cyclodipeptides 251, 281
Cycloguanil 398, 416
Cycloheximide 508
Cyclooxygenase inhibitors 124
Cyclophosphamide 316, 346, 497
Cycloquine 386
Cyproheptadine 93
Cytokines 111
Cytosine arabinoside 316
Cytotoxic agents 488

Dactylyne 536
Danitracen 93
Daphnane 125
Dapsone 396
Daunomycin 501
Daunorubicin 310, 316, 324
Decarboxylase 86
Dehydroemetine 465
Dehydrogliotoxin 296
Delirium 141
Deoxyzoanthenamine 540
Depetalonema perstans 372
Depression 99
Depsipeptides 523
Dermatitis 213
Desmethyl encainide 168
Dexamethasone 491
Diacetylamide 334
Diacylglycerol 120, 122
Diamidines 411
Diaminopyridine 38
Diazepam 44, 141
Dichroa febrifuga 381
Didermnins 523
Difluoromethylornithine 427
Diformyldapsone 396
Digitalis 155
Digoxin 165, 170, 175, 200, 216
Dihydropteroic acid 394
Dihydroxytryptamine 95
Diltiazem 156
Dimethyl-sulfoxide 421
Diminazene 413
Dioxopiperazines 349
Diphyllobotrium latum 373
Dipylidium caninum 371
Discorhabdin 539
Disopyramide 156, 158, 166, 171, 182, 213
Disphoria 140
Disulfiram 257
Dolabella auricularia 526
Dolastin 526
Dopamine 87, 88, 92, 146, 278
Doridosine 533
Doxocycline 261, 384
Doxorubicin 261, 310, 316, 340, 501
Dracunculus medinensis 373
Dysidea avara 529
Dysidea herbacea 282
Dysideae spp. 536

Echinulin 284
Ecteinascidia turbinata 524
Ehrlich carcinoma 303
Ellipticine 316
Embryogenic effects 315
Emetine 508
Encainide 158, 166, 177
Endorphin 57
Enkephalin 49
Epicorazine A 295
Epinephrine 30, 54
Epipolasis spp. 536
Epoxylitosterol 529
Ergocornine 275
Ergopeptide-related compounds 275
Escherichia coli 254, 261, 289
Estrogen metabolites 124
Ethmozine 156, 189
Ethosuccimide 334
Ethyl stibamine 407
Eunicella cavolini 527

Fansidar 392
Fansimef 392
Fasciola hepatica 374
Fasciolopsis buski 373
Febrifugine 381, 383
Femoxetine 87
Fenfluramine 88, 89, 95, 96
Fenthione 443
Filariasis 375, 436
Fleas 367
Flecainide 156, 163, 177
Flesinoxan 64
Floxacrine 386
Flubendazole 448
Flucytosine 421
Fluorocytosine 421
Fluorotryptophan 94
Fluorouracil 261, 340
Fluoxetine 55, 87, 89
Fluvoxamine 87
Fusarium culmorum 280

GABA 44
GABA antagonist 34
Galactosidase 115
Gambierdiscus toxicus 547

Geodia beretti 540
Geographutoxins 549
Gepirone 91
Gillichthys miralilis 533
Gliocladium deliquescens 299
Gliocladium virens 296
Gliotoxin 296
Gliotoxins 295
Gliovicin 294
Gliovirin 296
Glucan 128
Glucantime 409
Glucocorticoid hormones 124
Glutamate 34, 39, 45
Gnasthosoma spinigerum 373
Gonyautoxins 545
Guinea worm 375

H*alichondra akadai* 537
Halofantrine 391
Haloperidol 140
Harringtonine 316
Hepatoma 308
Heterophyes heterophyes 373
His-Parkinje tissue 152
Hodgkin's disease 341
Holoturins 532
Hyalodendrin 294, 295, 297
Hyatella spp. 527
Hycanthone 463
Hydralzine 221
Hydrangea 381
Hydrocortisone 490
Hydroxychloroquine 386
Hydroxydopamine 37
Hydroxyindoleacetic acid 96
Hydroxytryptophan 86, 88
Hymeniacidon aldis 534
Hymenin 534
Hyphomycete 294
Hypnea valentiae 533
Hypoprothrombinemia 257
Hypotension 54, 140, 276
Hypothalamus 38
Hypothermia 278
Hypoxanthine 417

Idazoxan 63
Image, public 9
Imidazoles 416
Immunodepression 488
Immunohistochemistry 50
Immunological techniques 26
Immunomodulation 488
Immunosuppressive properties 314
Indalpine 87
Indecainide 158
Indoles 90
Indomethacin 124, 125
Indorenate 90
Inflammation 488
Inflammatory diseases 110
Inflammatory disorders 276
Influenza 265
Infranodal 204
Ingenane 125
Inositol triphosphate 122
Insulin 140
Interferon 112, 131
Interleukin 261, 490
Intestinal fluke infection 375
Ipsapirone 91
Ischemia 154
Isoniazid 221
Isopentaquine 388
Isoproterenol 173
Isoquinolines 465
Ivermectin 451

Jaspamide 540
Jaspis 540

Kainate 53
Kaposi's sarcoma 339
Katalase 394
Ketanserin 42, 55, 98
Ketokonazole 416
Klebsiella falcalis 258
Klebsiella pneumonia 254, 258, 261
Kynurenic acid 52, 58

L-Dopa 42
L-Glutamic acid 52
β-Lactam antibiotics 253
Lamina terminalis 29
Laminine 534
Lapinone 393
Latrunculins 538
Laulimalide 527
Leischmania brasiliensis 372
Leischmania donovani 371
Leischmania tropica 371
Leischmaniasis 375, 400
Leischmanii mexicana 371
Leu-enkephalin 32, 57
Leukemia 321, 339
Leukopenia 338
Levamisole 417, 448
Lidocaine 38, 156, 158, 180, 183
Lipopolysaccharides 490
Liposomes 130
Lissoclinamides 525

Lissoclinum bistratum 524
Lissoclinum patelle 525
Listeria monocytogenes 111, 116, 121
Lithium 140
Litomosoides carinii 286
Litophyton viridis 529
Liver fluke infestion 375
Loa loa 372
Lorazepam 141
LSD 275
Lucanthone 462
Luffariella variabilis 536
Lung cancer 341
Lungfluke infestion 375
Lymphokines 111
Lymphoma 321
Lyngbia majuscula 270
Lysergic acid 275

Macrolides 522
Macrophages 110, 261
Mafosamide 499
Maitotoxin 547
Majusculamide B 270
Malaria 375
Maloney sarcomas 129
Manoalide 536
Mansonella ozzardi 372
Mebendazole 448, 466
Medical research 9
Mefloquine 391
Meglumine antimoniate 407
Melanoma 309
Melanoma cell 118
Melarsen sodium 429
Melarsonyl potassium 428
Melarsoprol 428
Melincacidins 295
Menoctone 393
Mercaptopurine 494
Met-enkephalin 32, 57
Metagonimus kogawai 374
Metastatic properties 251
Metergoline 46, 49, 55, **64, 93**
Methiothepin 93
Methotrexate 316, 508
Methoxamine 55
Methoxytryptamine 90
Methyl-dopa 62
Methylprednisolone 491
Methylserotonin 90, 93
Methysergide 46, 47, 64, 93
Metoprolol 156, 158
Metrifonate 443, 462
Metronidazole 416, 456
Mexiletine 156, 158, 180, 187, 200
Mezlocilline 261
Mianserin 93
Microiontophoresis 26
Midazolam 143
Minocycline 384
Miracils 462
Mites 367
Mitomycin C 261, 504
Moluscs 367
Monoamino oxidase 86
Monocytopenia 490
Morizicine 189
Mosquitos 367
Muramyl dipeptide 116
Muscimol 36, 39, 55, 56
Mycalamides 527
Mycale spp. 527
Mycelianamide 274, 282
Mycobacterium tuberculosis 115
Mycotoxins 273

Nadolal 156, 158
Naloxone 49, 57
Naphthoquinones 393
Narcotics 181
Nausea 142
Neoechinulin 284
Neoplasia 488
Neoplastic cells 110
Neuroleptic drugs 142
Neuroleptics 140
Neuropeptide Y 32
Neurotensin 32
Nicotinamide 32
Nifedipine 120
Nifurtimox 420, 434
Nipecotic acid 56
Niridazole 416, 431, 463
Nitrofurans 420
Nitrofurazone 429
Nitrosurea 500
Noradrenergic cell group 27
Noralichondrin 538
Norepinephrine 33, 39, 54, 87, 88
Nystatin 420

Obex 29
Odiline 528
Okadaic acid 537
Oltipraz 466
Onchocerca volvulus 372
Oomycetes 296
Opistorchis felineus 374
Opistorchis viverrini 374
Organoantimonials 407, 441
Organometallic compounds 462

Organophosphorous compounds 443
Orijzachlorins 295
Osteosarcoma 309
Ouabain 278, 531
Oxamniquine 465
Oxytocin 50

P*alythoa* spp. 543
Palytoxin 541
Pamaquine 388
Pandaros acanthifolium 537
Panic disorder 99
Panuramine 87
Paragonimus westermani 373
Parainfluenza 265
Paroxetine 87
Patellamide 525
Penbutol 94
Penicillium 274
Penicillium griseofulvum 300
Penicillium turbatum 294
Pentagastrin 532
Pentamidine 412, 426
Pentaquine 388
Pentobarbital 30
Pentostam 408
Peptides 530
Phagocyte system 487
Phagocytes 110. 488
Phenobarbital 181, 210
Phenylephrine 61
Phenytoin 171, 210, 216
Phorbol ester 125
Phosphatidylcholine 257
Phosphodiesterase 115
Phospholipase 124
Phthalimide 334
Phyllofoliaspongain 529
Phyllospongia foliascens 529
Physostigmine 38
Phytohemaglutinin 313
Picrocellin 284
Picrotoxin 59
Pindolol 94, 156, 158
Piperacillin 255, 257, 259
Piperazine derivatives 444
Piperazines 90
Piperazinomycin 287
Piperazinones 262
Piperoxane 31, 41, 63
Pirandamine 87
Pirmenol 158
Pirmenol 204
Pizotiphen 93
Plasmodium spp. 372
Platelet-activating factor 276
Polymerases 535
Praziquantel 465
Prazosin 40, 41
Prianos melanos 539
Prienosin 539
Primaquine 388, 415
Probimane 308, 309, 330
Procainamide 156, 158, 175, 182, 200, 219
Proguanil 398
Prolactin 99, 275
Propafenone 156, 158, 173, 200, 223
Propanolol 156
Prostaglandin 124, 130
Protein kinase C 120
Proteinases 124
Proteus mirabidis 255
Proteus vulgaris 254
Prothrombin 175, 200
Protogonyalax catenella 544
Pseudomonas aeruginosa 253
Pseudomonas vulgaris 261
Pseudoplexaura porosa 521
Psoriasis 213, 314, 339
Psychosis 141
Ptychodiscus brevis 544, 546
Puromycin 276
Pyrimethamine 392, 396

Quinacrine 385
Quinghaosu 381
Quinidine 156, 158, 166, 171, 175, 182, 200, 207, 223, 381
Quinine 381
Quinocide 388
Quinolines 414, 465
Quipazine 90
Quisqualate receptors 53, 267

Rauwolscine 64
Razoxane 263, 303, 308, 324
Recainam 156, 191
Renieramycins 288
Renin 50
Reserpine 88, 94
Retroviruses 536
Rheumatoid arthritis 110
Riboflavin 266
Rickettsia akari 369
Rickettsia australis 269
Rickettsia conorii 269
Rickettsia prowazekii 370
Rickettsia tsutsugamushi 269
Rickettsia typhi 371
Rifampicin 261, 420
Rifampin 181, 205, 210, 216

Rocella fuciformis 284
Rochalimaea quintana 371

Saframycins 288
Salmonella typhimurium 314
Sarcoma 303
Sarcophyton glaucum 522
Saxitoxins 545
Schistosoma spp. 373
Schistosomiasis 375, 457
Schizophrenia 99
Serotonin 32, 35, 45, 86
Serotonin receptor agonists 90
Serratia aureus 259
Serratia falcalis 258
Serratia marcescens 254
Sertraline 85
Sinularia flexibilis 521
Sleeping sickness 375
Snails 367
Society 9
Somatostatin 32, 57
Sotalol 156, 158, 175, 194
Spiperone 64
Spirometra mansonoides 373
Spongosine 533
Sporidesmins 295, 296, 298
Staphylococcus asureus 254
State 9
Stibacetin 407
Stibamine 407, 410
Stibocaptate 442
Stibogluconate sodium 407
Stibophen 442
Stilbamidine 411
Streptococcus faecium 262
Streptomyces aizunensis 289
Streptomyces sapporensis 289
Streptoverticillium olivoreticuli 287
Striatotoxin 550
Stypodium zonale 528
Stypoldione 528
Succinimide 334
Sulfadiazine 395
Sulfadimethoxine 395
Sulfadoxine 392, 395
Sulfalene 395
Sulfamethoxypyridazine 395
Sulfonamides 395
Sulfones 395
Sumatriptan 93
Superoxide dismutase 331
Suramine 425, 443
Sympatomimetic activity 158
Synthetase 394

Tetracyclines 384, 385
Tachycardia 63, 160, 167
Tenebrosin 532
Tetrabenazine 88
Terodotoxin 36
Thiopurine 494
Thrombocytopenia 338
Thyrotropin 277
Thyrotropic hormone 48
Thyroxin 199
Ticks 367
Tigliane 125
Timolol 156, 158
Tocainide 156, 158, 180, 183
Transcriptase 298
Trazodone 93
Trehalose dimycolate 115
Triazines 398
Triazoloquinoxaline 266
Triclorfone 462
Tridemnium solidum 523
Trimethoprime 397
Trypanosoma cruzi 118, 370
Trypanosoma gambiense 372
Tryparsamide 427
Tryptophan 46, 86
Tryptophan hydroxylase 86, 94
Tumor cells 110
Tumor necrosis factor 492
Tyramine 37

Urea stibamine 407
Ureidopenicillins 261
Uretensin 533

Vagus 29
Vasoconstriction 51, 162
Vasodilator 162
Vasopressin 36, 50
Vasopressin release 38
Verapamil 120, 156, 159, 170, 210
Verongia fistularis 535
Verongia spengelii 540
Verticillins 295
Vinblastine 321, 505
Vinca rosea 505
Vincristine 310, 321, 505
Vitamin K 256

Wucheria bancrofti 372

Yohimbine 41, 63

Zacopride 94
Zanthoxylium arborescens 270, 286
Zimelidine 87, 97

Index of titles
Verzeichnis der Titel
Index des titres
Vol. 1–35 (1959–1990)

Acetylen-Verbindungen als Arzneistoffe, natürliche und synthetische
14, 387 (1970)
Adenosine receptors: Clinical implications and biochemical mechanisms
32, 195 (1988)
Adipose tissue, the role of in the distribution and storage of drugs
28, 273 (1984)
Steroidogenic capacity in the adrenal cortex and its regulation
34, 359 (1990)
β-Adrenergic blocking agents
20, 27 (1976)
β-Adrenergic blocking agents, pharmacology and structure-activity
10, 46 (1966)
β-Adrenergic blocking drugs, pharmacology
15, 103 (1971)
Adrenergic receptor research, recent developments
33, 151 (1989)
Adverse reactions of sugar polymers in animals and man
23, 27 (1979)
Allergy, pharmacological approach
3, 409 (1961)
Alzheimer's disease, implications of immunomodulant therapy
32, 21 (1988)
Neuroimmune axis as a basis of therapy in Alzheimer's disease
34, 383 (1990)
Amebic disease, pathogenesis of
18, 225 (1974)
Amidinstruktur in der Arzneistofforschung
11, 356 (1968)
Amines, biogenic and drug research
28, 9 (1984)
Amino- und Nitroderivate (aromatische), biologische Oxydation und Reduktion
8, 195 (1965)
Aminonucleosid-Nephrose
7, 341 (1964)
Amoebiasis, chemotherapy
8, 11 (1065)
Amoebiasis, surgical
18, 77 (1974)
Amoebicidal drugs, comparative evaluation of
18, 353 (1974)
Anabolic steroids
2, 71 (1960)
Analgesia and addiction
5, 155 (1963)
Analgesics and their antagonists
22, 149 (1978)
Ancyclostomiasis in children, trial of bitoscanate
19, 2 (1975)
Androgenic-anabolic steroids and glucocorticoids, interactions
14, 139 (1970)
Anthelmintic action, mechanisms of
19, 147 (1975)
Anthelminticaforschung neuere Aspekte
1, 243 (1959)
Anthelmintics, comparative efficacy
19, 166 (1975)
Anthelmintics, laboratory methods in the screening of
19, 48 (1975)
Anthelmintics, structure-activity
3, 75 (1961)
Anthelmintics, human and veterinary
17, 110 (1973)
Antiarrhythmic compounds
12, 292 (1968)
Antiarrhythmic drugs, recent advances in electrophysiology of
17, 34 (1973)

Antiarrhythmic drugs, the oral
35, 151 (1990)
Antibacterial agents of the nalidixic acid type
21, 9 (1977)
Antibiotics, structure and biogenesis
2, 591 (1969)
Antibiotic activities, in vitro models for the study of
31, 349 (1987)
Antibiotika, krebswirksame
3, 451 (1961)
Antibody titres, relationship to resistance to experimental human infection
19, 542 (1975)
Anticancer agents, metabolism of
17, 320 (1973)
Antidiabetika, orale
30, 281 (1986)
Antifertility substances, development
7, 133 (1964)
Antitumor antibiotics, the chemistry of DNA modification
32, 411 (1988)
Anti-filariasis campaign: its history and future prospects
18, 259 (1974)
Antifungal agents
22, 93 (1978)
Antihypertensive agents
4, 295 (1962), *13,* 101 (1969), *20,* 197 (1976)
Antihypertensive agents 1969–1981
25, 9 (1981)
Anti-inflammatory agents, nonsteroid
10, 139 (1966)
Anti-inflammatory drugs, biochemical and pharmacological properties
8, 321 (1965)
Antikoagulantien, orale
11, 226 (1968)
Antimalarials, 8-aminoquinolines
28, 197 (1984)
Antimetabolites, revolution in pharmacology
2, 613 (1960)
Antimicrobials, penetration of human cells
33, 83 (1989)
Antituberculous compounds with special reference to the effect of combined treatment, experimental evaluation of
18, 211 (1974)
Antiviral agents
22, 267 (1978)
Antiviral agents
28, 127 (1984)
Art and science of contemporary drug development
16, 194 (1972)
Arterial pressure by drugs
26, 353 (1982)
Arzneimittel, neue
1, 531 (1959), *2,* 251 (1969), *3,* 369 (1961), *6,* 347 (1963), *10,* 360 (1966)
Arzneimittel, Wert und Bewertung
10, 90 (1966)
Arzneimittelwirkung, Einfluss der Formgebung
10, 204 (1966)
Arzneimittelwirkung, galenische Formgebung
14, 269 (1970)
Aspirin as an antithrombotic agent
33, 43 (1989)
Asthma, drug treatment of
28, 111 (1984)
Atherosclerosis, cholesterol and its relation to
1, 127 (1959)
Axoplasmic transport, pharmacology and toxicology
28, 53 (1984)
Ayurveda
26, 55 (1982)
Ayurvedic medicine
15, 11 (1971)

Bacterial cell surface and antimicrobial resistance
32, 149 (1988)
Bacterial vaccines, approaches to the rational design
32, 375 (1988)
Bacteria and phagocytic cells, surface interaction between
32, 137 (1988)
Barbiturates and the $GABA_A$ receptor complex
34, 261 (1990)
Basic research, in the US pharmaceutical industry
15, 204 (1971)
Benzinmidazole anthelmintics chemistry and biological activity
27, 85 (1983)
Benzodiazepine story
22, 229 (1978)
Benzodiazepine augmentation of the treatment of disruptive psychic behavior
35, 139 (1990)

Beta blockade in myocardial infarction
30, 71 (1986)
Bewertung eines neuen Antibiotikums
22, 327 (1978)
Biliary excretion of drugs and other xenobiotics
25, 361 (1981)
Biochemical acyl hydroxylations
16, 229 (1972)
Biological activity, stereochemical factors
1, 455 (1959)
Biological response quantification in toxicology, pharmacology and pharmacodynamics
21, 105 (1977)
Bitoscanate, a field trial in India
19, 81 (1975)
Bitoscanate, clinical experience
19, 96 (1975)
Bitoscanate, experience in the treatment of adults
19, 90 (1975)

Calmodulin-mediated enzymatic actions, effects of drugs on
33, 353 (1989)
Cancer chemotherapy
8, 431 (1965), *20,* 465 (1976)
25, 275 (1981)
Cancer chemotherapy agents, interaction with the mononuclear phagocyte system
35, 487 (1990)
Cancerostatic drugs
20, 251 (1976)
Systemic cancer therapy: Four decades of progress and some personal perspectives
34, 76 (1990)
Carcinogenecity testing of drugs
29, 155 (1985)
Carcinogens, molecular geometry and mechanism of action
4, 407 (1962)
Cardiovascular drug interactions, clinical importance of
25, 133 (1981)
Cardiovascular drug interactions
29, 10 (1985)
Central dopamine receptors, agents acting on
21, 409 (1977)
Central nervous system drugs, biochemical effects
8, 53 (1965)
Cestode infections, chemotherapy of
24, 217 (1980)
Chemical carcinogens, metabolic activation of
26, 143 (1982)
Chemotherapy of schistosomiasis, recent developments
16, 11 (1972)
Cholera infection (experimental) and local immunity
19, 471 (1975)
Cholera in Hyderabad, epidemiology of
19, 578 (1975)
Cholera in non-endemic regions
19, 594 (1975)
Cholera pandemic, and bacteriology
19, 513 (1975)
Cholera pathophysiology and therapeutics, advances
19, 563 (1975)
Cholera, research in India on the control and treatment of
19, 503 (1975)
Cholera toxin induced fluid, effect of drugs on
19, 519 (1975)
Cholera toxoid research in the United States
19, 602 (1975)
Cholera vaccines in volunteers, antibody response to
19, 554 (1975)
Cholera vibrios, interbiotype conversions by actions of mutagens
19, 466 (1975)
Cholesterol, relation to atherosclerosis
1, 127 (1959)
Cholinergic mechanism-monoamines relation in certain brain structures
6, 334 (1972)
Cholinergic neutrotransmitter system, behavioral correlates of presynaptic events
32, 43 (1988)
Clostridium tetani, growth in vivo
19, 384 (1975)
Communicable diseases, some often neglected factors in the control and prevention of
18, 277 (1974)
Computer-assisted structure – Antileukemic activity analysis of purines and their aza and deaza analogs
34, 319 (1990)
Conformation analysis, molecular graphics
30, 91 (1986)

Contraception
21, 293 (1977)
Contraceptive agents, development of
33, 261 (1989)
Convulsant drugs – relationships between structure and function
24, 57 (1980)
Cooperative effects in drug-DNA interactions
31, 193 (1987)
Cytokines, the role of in macrophage activation
35, 109 (1990)
Cyclopropane compounds
15, 227 (1971)

Deworming of preschool community in national nutrition programmes
19, 136 (1975)
1,2-Diamine functionality, medicinal agents
33, 135 (1989)
Diarrhoea (acute) in children, management of
19, 527 (1975)
Diarrhoeal diseases (acute) in children
19, 570 (1975)
3,4-Dihydroxyphenylalanine and related compounds
9, 223 (1966)
Dioxopiperazines: Chemistry and biology
35, 249 (1990)
Diphtheria, epidemiological observations in Bombay
19, 423 (1975)
Diphtheria, epidemiology of
19, 336 (1975)
Diphtheria in Bombay
19, 277 (1975)
Diphtheria in Bombay, age profile of
19, 417 (1975)
Diphtheria in Bombay, studies on
19, 241 (1975)
Diphtheria, pertussis and tetanus, clinical study
19, 356 (1975)
Diphtheria, pertussis and tetanus vaccines
19, 229 (1975)
Diphtheria toxin production and iron
19, 283 (1975)
Disease control in Asia and Africa, implementation of
18, 43 (1974)
Disease-modifying antirheumatic drugs, recent developments in
24, 101 (1980)
Diuretics
2, 9 (1960)
DNA technology, significance in medicine
33, 397 (1989)
Dopamine agonists, structure-activity relationships
29, 303 (1985)
Drug action and assay by microbial kinetics
15, 271 (1971)
Drug action, basic mechanisms
7, 11 (1964)
Drug combination, reduction of drug action
14, 11 (1970)
Drug discovery, organizing for
32, 329 (1988)
Drug discovery, serendipity and structural research
30, 189 (1986)
Drug in biological cells
20, 261 (1976)
Drug latentiation
4, 221 (1962)
Drug-macromolecular interactions, implications for pharmacological activity
14, 59 (1970)
Drug metabolism
13, 136 (1969)
Drug metabolism (microsomal), enhancement and inhibitions of
17, 12 (1973)
Drugs affecting serotonin neurons
35, 85 (1990)
Drug-metabolizing enzymes, perinatal development of
25, 189 (1981)
Drug potency
15, 123 (1971)
Drug research
10, 11 (1966)
Drug research and development
20, 159 (1976)
Drugs, biliary excretion and enterohepatic circulation
9, 299 (1966)
Drug molecules of marine origin
35, 521 (1990)
Drugs, structures, properties and disposition of
29, 67 (1985)

Eegg-white, reactivity of rat and man
13, 340 (1969)
Endocrinology, twenty years of research
12, 137 (1968)
Endotoxin and the pathogenesis of fever
19, 402 (1975)
Enterobacterial infections, chemotherapy of
12, 370 (1968)
Enzyme inhibitors of the renin-angiotensin system
31, 161 (1987)
Estrogens, oral contraceptives and breast cancer
25, 159 (1981)
Excitation and depression
26, 225 (1982)
Experimental biologist and medical scientist in the pharmaceutical industry
24, 83 (1980)

Fifteen years of structural-modifications in the field of antifungal monocyclic 1-substituted 1 H-azoles
27, 253 (1983)
Filarial infection, immuno-diagnosis
19, 128 (1975)
Filariasis, chemotherapy
9, 191 (1966)
Filariasis in India
18, 173 (1974)
Filariasis, in four villages near Bombay, epidemiological and biochemical studies in
18, 269 (1974)
Filariasis, malaria and leprosy, new perspectives on the chemotherapy of
18, 99 (1974)
Fluor, dérivés organiques d'intérêt pharmacologique
3, 9 (1961)
Fundamental structures in drug research Part I
20, 385 (1976)
Fundamental structures in drug research Part II
22, 27 (1978)
Fungal metabolites and Chinese herbal medicine as immunostimulants
34, 395 (1990)
Further developments in research on the chemistry and pharmacology of synthetic quinuclidine derivatives
27, 9 (1983)

GABA-Drug Interactions
31, 223 (1987)
Galenische Formgebung und Arzneimittelwirkung
10, 204 (1966), *14*, 269 (1970)
Ganglienblocker
Glucocorticoids: anti-inflammatory and immuno-suppressive effects
33, 63 (1989)
Oligosaccharide chains of glycoproteins
34, 163 (1990)

Heilmittel, Entwicklung
10, 33 (1966)
Helminthiasis (intestinal), chemotherapy of
19, 158 (1975)
Helminth infections, progress in the experimental chemotherapy of
17, 241 (1973)
Helminthic infections, immunodiagnosis of
19, 119 (1975)
Helminth parasites, treatment and control
30, 473 (1986)
High resolution nuclear magnetic resonance spectroscopy of biological samples as an aid to drug development
31, 427 (1987)
Homologous series, pharmacology
7, 305 (1964)
Hookworm anaemia and intestinal malabsorption
19, 108 (1975)
Hookworm disease and trichuriasis, experience with bitoscanate
19, 23 (1975)
Hookworm disease, bitoscanate in the treatment of children with
19, 6 (1975)
Hookworm disease, comparative study of drugs
19, 70 (1975)
Hookworm disease, effect on the structure and function of the small bowel
19, 44 (1975)
Hookworm infection, a comparative study of drugs
19, 86 (1975)
Hookworm infections, chemotherapy of
26, 9 (1982)
Human sleep
22, 355 (1978)
Hydatid disease
19, 75 (1975)

Hydrocortisone, effects of structural alteration on the antiinflammatory properties
5, 11 (1963)
5-Hydroxytryptamine and related indolealkylamines
3, 151 (1961)
5-Hydroxytryptamine receptor agonists and antagonists
30, 365 (1986)
Hypertension and brain neurotransmitters
30, 127 (1986)
Hypertension, recent advances in drugs against
29, 215 (1985)
Hypertension: Relating drug therapy to pathogenic mechanisms
32, 175 (1988)
Hypolipidemic agents
13, 217 (1969)

Immune system, the pharmacology of
28, 83 (1984)
Immunization, host factors in the response to
19, 263 (1975)
Immunization of a village, a new approach to herd immunity
19, 252 (1975)
Immunization, progress in
19, 274 (1975)
Immunology
20, 573 (1976)
Immunology in drug research
28, 233 (1984)
Immunostimulation with peptidoglycan or its synthetic derivatives
32, 305 (1988)
Immunopharmacology and brain disorders
30, 345 (1986)
Immunosuppression agents, procedures, speculations and prognosis
16, 67 (1972)
Immunotherapy for leprosy and tuberculosis
33, 415 (1989)
Impact of natural product research on drug discovery
23, 51 (1979)
Impact (the) of state and society on medical research
35, 9 (1990)
Indole compounds
6, 75 (1963)
Indolstruktur, in Medizin und Biologie
2, 227 (1960)
Industrial drug research
20, 143 (1976)
Influenza virus, functional significance of the various components of
18, 253 (1974)
Interaction of drug research
20, 181 (1976)
Intestinal nematodes, chemotherapy of
16, 157 (1972)
Ion and water transport in renal tubular cells
26, 87 (1982)
Ionenaustauscher, Anwendung in Pharmazie und Medizin
1, 11 (1959)
Isotope, Anwendung in der pharmazeutischen Forschung
7, 59 (1964)

Ketoconazole, a new step in the management of fungal disease
27, 63 (1983)

Leishmaniases
18, 289 (1974)
Present status of Leishmaniasis
34, 447 (1990)
Leprosy, some neuropathologic and cellular aspects of
18, 53 (1974)
Leprosy in the Indian context, some practical problems of the epidemiology of
18, 25 (1974)
Leprosy, malaria and filariasis, new perspectives on the chemotherapy of
18, 99 (1974)
Progress in the chemotherapy of leprosy: Status, issues and prospects
34, 421 (1990)
Levamisole
20, 347 (1976)
Light and dark as a “drug”
31, 383 (1987)
Lipophilicity and drug activity
23, 97 (1979)
Lokalanästhetika, Konstitution und Wirksamkeit
4, 353 (1962)
Lysostaphin: model for a specific enzymatic approach to infectious disease
16, 309 (1972)

Malaria, advances in chemotherapy
30, 221 (1986)
Malaria chemotherapy, repository antimalarial drugs
13, 170 (1969)
Malaria chemotherapy, antibiotics in
26, 167 (1982)
Malaria, eradication in India, problems of
18, 245 (1974)
Malaria, filariasis and leprosy, new perspectives on the chemotherapy of
18, 99 (1974)
Mast cell secretion, drug inhibition of
26, 277 (1985)
Mass spectrometry in pharmaceutical research, recent applications of
18, 399 (1974)
Mechanism of action of anxiolytic drugs
31, 315 (1987)
Medical practice and medical pharmaceutical research
20, 491 (1976)
Medicinal chemistry, contribution to medicine
12, 11 (1968)
Medicinal chemistry: A support or a driving force in drug research?
34, 343 (1990)
Medicinal research: Retrospectives and perspectives
29, 97 (1985)
Medicinal science
20, 9 (1976)
Membrane drug receptors
20, 323 (1976)
Mescaline, and related compounds
11, 11 (1968)
Metabolism of drugs, enzymatic mechanisms
6, 11 (1963)
Metabolism (oxidative) of drugs and other foreign compounds
17, 488 (1973)
Metronidazol-therapie, Trichomonasis
9, 361 (1966)
Pharmacology of migraine
34, 209 (1990)
Molecular pharmacology
20, 101 (1976)
Molecular pharmacology, basis for drug design
10, 429 (1966)
Monitoring adverse reactions to drugs
21, 231 (1977)
Monoaminoxydase-Hemmer
2, 417 (1960)
Monoamine oxidase, inhibitors of
30, 205 (1986)
Mycoses, chemotherapy for
33, 317 (1989)

Narcotic antagonists
8, 261 (1965), *20*, 45 (1976)
Natriuretic hormones
34, 231 (1990)
Necator americanus infection, clinical field trial of bitoscanate
19, 64 (1975)
Nematoide infections (intestinal) in Latin America
19, 28 (1975)
Neurotransmitters (Role of) in the central regulation of the cardiovascular system
35, 25 (1990)
Nicotine: an addictive substance or a therapeutic agent
33, 9 (1989)
Nitroimidazoles as chemotherapeutic agents
27, 163 (1983)
Noise analysis and channels at the postsynaptic membrane of skeletal muscle
24, 9 (1980)

Ophthalmic drug preparations, methods for elucidating bioavailability mechanisms of
25, 421 (1981)

Parasitic infections in man, recent advances in the treatment of
18, 191 (1974)
Parasitosis (intestinal), analysis of symptoms and signs
19, 10 (1975)
Bioactive peptide analogs: In vivo and in vitro production
34, 287 (1990)
Pertussis agglutinins and complement fixing antibodies in whooping cough
19, 178 (1975)
Pertussis, diphteria and tetanus, clinical study
19, 356 (1975)
Pertussis, diphteria and tetanus vaccines
19, 229 (1975)
Pertussis, epidemiology of
19, 257 (1975)

Pertussis vaccine
19, 341 (1975)
Pertussis vaccine composition
19, 347 (1975)
Pharmacology of the brain: the hippocampus, learning and seizures
16, 211 (1972)
The pharmacology of caffeine
31, 273 (1987)
Phenothiazine und Azaphenothiazine
5, 269 (1963)
Photochemistry of drugs
11, 48 (1968)
Podophyllotoxin glucoside – etoposide
33, 169 (1989)
Pyrimidinones as biodynamic agents
31, 127 (1987)
Placeboproblem
1, 279 (1959)
Platelets and atherosclerosis
29, 49 (1985)
Progesterone receptor binding of steroidal and nonsteroidal compounds
30, 151 (1986)
Propellants, toxicity of
18, 365 (1974)
Prostaglandins
17, 410 (1973)
Protozoan and helminth parasites
20, 433 (1976)
Psychopharmaka, Anwendung in der psychosomatischen Medizin
10, 530 (1966)
Psychopharmaka, strukturelle Betrachtungen
9, 129 (1966)
Psychosomatische Medizin, Anwendung von Psychopharmaka
10, 530 (1966)
Psychotomimetic agents
15, 68 (1971)
Pyrimidinones as biodynamic agents
31, 127 (1987)
The use of quantum chemical methods to study molecular mechanisms of drug action
34, 9 (1990)

Quaternary ammonium salts, chemical nature und pharmacological actions
2, 135 (1960)
Quaternary ammonium salts – advances in chemistry and pharmacology since 1960
24, 267 (1980)
Quinazoline derivatives
26, 259 (1982)
Quinazolones, biological activity
14, 218 (1970)
Quinolones
31, 243 (1987)
Quinuclidine derivatives, chemical structure and pharmacological activity
13, 293 (1969)

Red blood cell membrane, as a model for targets of drug action
17, 59 (1973)
Renin-angiotensin system
26, 207 (1982); *31,* 161 (1987)
Reproduction in women, pharmacological control
12, 47 (1968)
Research, preparing the ground: importance of data
18, 239 (1974)
Rheumatherapie, Synopsis
12, 165 (1968)
Ribonucleotide reductase inhibtors as anticancer and antiviral agents
31, 101 (1987)
Risk assessment problems in chemical oncogenesis
31, 257 (1987)
RNA virus evolution and the control of viral disease
33, 93 (1989)

Schistosomiasis, recent progress in the chemotherapy of
18, 15 (1974)
Schwefelverbindungen, therapeutisch verwendbare
4, 9 (1962)
Shock, medical interpretation
14, 196 (1970)
Serum electrolyte abnormalities caused by drugs
30, 9 (1986)
Social pharmacology
22, 9 (1978)
Spectrofluorometry, physiochemical methods in pharmaceutical chemistry
6, 151 (1963)
Stereoselective drug metabolism and its significance in drug research
32, 249 (1988)
Stoffwechsel von Arzneimitteln, Ursache von Wirkung, Nebenwirkung und Toxizität
15, 147 (1971)

Strahlenempfindlichkeit von Säugetieren, Beeinflussung durch chemische Substanzen
9, 11 (1966)
Structure-activity relationships
23, 199 (1979)
Substruktur der Proteine, tabellarische Zusammenstellung
16, 364 (1972)
Sulfonamide research
12, 389 (1968)

T-cell factors, antigen-specific and drug research
32, 9 (1988)
Teratogenic hazards, advances in prescreening
29, 121 (1985)
Terpenoids, biological activity
6, 279 (1963), *13,* 11 (1969)
Tetanus and its prevention
19, 391 (1975)
Tetanus, autonomic dysfunction as a problem in the treatment of
19, 245 (1975)
Tetanus, cephalic
19, 443 (1975)
Tetanus, cholinesterase restoring therapy
19, 329 (1975)
Tetanus, diphteria and pertussis, clinical study
19, 356 (1975)
Tetanus, general and pathophysiological aspects
19, 314 (1975)
Tetanus in children
19, 209 (1975)
Tetanus in Punjab and the role of muscle relaxants
19, 288 (1975)
Tetanus, mode of death
19, 439 (1975)
Tetanus neonatorum
19, 189 (1975)
Tetanus, pertussis and diphtheria vaccines
19, 229 (1975)
Tetanus, present data on the pathogenesis of
19, 301 (1975)
Tetanus, role of beta-adrenergic blocking drug propranolol
19, 361 (1975)
Tetanus, situational clinical trials and therapeutics
19, 367 (1975)
Tetanus, therapeutic measurement
19, 323 (1975)
Tetracyclines
17, 210 (1973)
Tetrahydroisoquinolines and β-carbolines
29, 415 (1985)
Thymoleptika, Biochemie und Pharmakologie
11, 121 (1968)
Toxoplasmosis
18, 205 (1975)
Treatment of helminth diseases – challenges and achievements
31, 9 (1987)
Trichomonasis, Metronidazol-Therapie
9, 361 (1966)
Trichuriasis and hookworm disease in Mexico, experience with bitoscanate
19, 23 (1975)
Tropical diseases, chemotherapy of
26, 343 (1982)
Tropical medicine, teaching
18, 35 (1974)
Tuberculosis in rural areas of Maharashtra, profile of
18, 91 (1974)
Tuberkulose, antibakterielle Chemotherapie
7, 193 (1964)
Tumor promoters and antitumor agents
23, 63 (1979)

Unsolved problems with vaccines
23, 9 (1979)

Vaccines, controlled field trials of
19, 481 (1975)
Vector borne-diseases
35, 365 (1990)
Vibrio cholerae, cell-wall antigens of
19, 612 (1975)
Vibrio cholerae, recent studies on genetic recombination
19, 460 (1975)
Vibrio cholerae, virulence-enhancing effect of ferric ammonium citrate on
19, 564 (1975)
Vibrio paraphaemolyticus in Bombay
19, 586 (1975)
Vibrio parahaemolyticus infection in Calcutta
19, 490 (1975)

Wurmkrankheiten, Chemotherapie
1, 159 (1959)

Author and paper index
Autoren- und Artikelindex
Index des auteurs et des articles
Vol. 1–35 (1959–1990)

Pertussis agglutinins and complement fixing antibodies in whooping cough *19*, 178 (1975)	K. C. Agarwal M. Ray N. L. Chitkara Department of Microbiology, Postgraduate Institute of Medical Education and Research, Chandigarh, India
Pharamcology of clinically useful beta-adrenergic blocking drugs *15*, 103 (1971)	R. P. Ahlquist A. M. Karow, Jr. M. W. Riley School of Medicine, Medical College of Georgia, Augusta, Georgia, USA
Adrenergic beta blocking agents *20*, 27 (1976)	R. P. Ahlquist Medical College of Georgia Augusta, Georgia, USA
Trial of a new anthelmintic (bitoscanate) in ankylostomiasis in children *19*, 2 (1975)	S. H. Ahmed S. Vaishnava Department of Paediatrics, Safdarjung Hospital, New Delhi, India
Development of antibacterial agents of the nalidixic acid type *21*, 9 (1977)	R. Albrecht Department of Drug Research, Schering AG, Berlin, FRG
The mode of action of anti-rheumatic drugs. 1. Anti-inflammatory and immunosuppressive effects of glucocorticoids *33*, 63 (1989)	Anthony C. Allison Simon W. Lee Syntex Research, 3401 Hillview Avenue, Palo Alto, CA 94002, USA
Biological activity in the quinazolone series *14*, 218 (1970)	A. H. Amin D. R. Mehta S. S. Samarth Research Division, Alembic Chemical Works Co. Ltd., Alembic Road, Baroda 3, India

The pharmacology of caffeine *31,* 273 (1987)	M. J. Arnaud Nestec-Ltd., Nestlé Research Centre, Vers-chez-les-Blanc, CH-1000 Lausanne 26, Switzerland
Enhancement and inhibition of microsomal drug metabolism *17,* 11 (1973)	M. W. Anders Department of Pharmacology, University of Minnesota, Minneapolis, Minnesota, USA
Reactivity of rat and man to egg-white *13,* 340 (1969)	S. I. Ankier Allen & Hanburys, Ltd., Research Division, Ware, Hertfordshire, England
Enzyme inhibitors of the renin-angiotensin system *31,* 161 (1987)	Michael J. Antonaccio, Ph. D. John J. Wright, Ph. D. Bristol Myers Company, Pharmaceutical Research and Development Division, 5 Research Parkway, Wallingford, CT 06492, USA
Narcotic antagonists *8,* 261 (1965)	S. Archer L. S. Harris Sterling-Winthrop Research Institute, Rensselaer, New York, USA
Recent developments in the chemotherapy of schistosomiasis *16,* 11 (1972)	S. Archer A. Yarinsky Sterling-Winthrop Research Institute, Rensselaer, New York, USA
Recent progress in the chemotherapy of schistosomiasis *18,* 15 (1974	S. Archer School of Science, Department of Chemistry, Rensselaer Polytechnik Institute, Troy, NY 12181, USA
Recent progress in research on narcotic antagonists *20,* 45 (1976)	S. Archer School of Science, Department of Chemistry, Rensselaer Polytechnic Institute, Troy, NY 12181, USA W. F. Michne Sterling-Winthrop Research Institute, Rensselaer, New York, USA
Molecular geometry and mechanism of action of chemical carcinogens *4,* 407 (1962)	J. C. Arcos Department of Medicine and Biochemistry, Tulane University, U.S. Public Health Service, New Orleans, Louisiana, USA

Molecular pharmacology, a basis for drug design *10*, 429 (1966) Reduction of drug action by drug combination *14*, 11 (1970)	E. J. Ariëns Institute of Pharmacology, University of Nijmegen, Nijmegen, The Netherlands
Stereoselectivity and affinity in molecular pharmacology *20*, 101 (1976)	E. J. Ariëns Dr. J. F. Rodrigues de Miranda Pharmacological Institute, University of Nijmegen, Nijmegen, The Netherlands P. A. Lehmann Departamento de Farmacologia y Toxicologia, Centro de Investigación y Estudios Avanzados, Instituto Politécnico Nacional, México, D. F., México
Recent advances in central 5-hydroxytryptamine receptor agonists and antagonists *30*, 365 (1986)	Lars-Erik Arvidsson Uli Hacksell Department of Organic-Pharmaceutical Chemistry, Uppsala Biomedical Center, University of Uppsala, S-75123, Uppsala, Sweden Richard A. Glennon Department of Medicinal Chemistry, Virginia Commonwealth University, Box 581 MCV Station, Richmond, Virginia 23298, USA
Drugs affecting the renin-angiotensin system *26*, 207 (1982)	R. W. Ashworth Pharmaceuticals Division, Ciba-Geigy Corporation, Summit, New Jersey, USA
Tetanus neonatorum *19*, 189 (1975) Tetanus in children *19*, 209 (1975)	V. B. Athavale P. N. Pai A. Fernandez P. N. Patnekar Y. S. Acharya Department of Pediatrics, L. T. M. G. Hospital, Sion Bombay 22, India
Toxicity of propellants *18*, 365 (1974)	D. M. Aviado Department of Pharmacology, School of Medicine, University of Pennsylvania, Philadelphia, USA
Neuere Aspekte der chemischen Anthelminticaforschung *1*, 243 (1959)	J. Bally Sandoz AG, Basel, Switzerland
Problems in preparation, testing and use of diphtheria, pertussis and tetanus vaccines *19*, 229 (1975)	D. D. Banker Glaxo Laboratories (India) Ltd., Bombay 25, India

Recent advances in electrophysiology of antiarrhythmic drugs *17,* 33 (1973)	A. L. Bassett and Dr. A. L. Wit College of Physicians and Surgeons of Columbia University, Department of Pharmacology, New York, NY, USA
Stereochemical factors in biological activity *1,* 455 (1959)	A. H. Beckett Chelsea College of Science and Technology, Chelsea, London, England
Industrial research in the quest for new medicines *20,* 143 (1976) The experimental biologist and the medical scientist in the pharmaceutical industry *24,* 38 (1980)	B. Berde Sandoz Ltd., Basel, Switzerland
Newer diuretics *2,* 9 (1960)	K. H. Beyer, Jr. Sharp and Dohme Research Laboratories, West Point, Pennsylvania, USA J. E. Bear Merck Institute für Therapeutic Research, West Point, Pennsylvania, USA
Recent developments in 8-aminoquinoline antimalarials *28,* 197 (1984)	A. P. Bhaduri B. K. Bhat M. Seth Central Drug Research Institute, Lucknow, 226001 India
Studies on diphtheria in Bombay *19,* 241 (1975)	M. Bhaindarkar Y. S. Nimbkar Haffkine Institute, Parel, Bombay 12, India
Bitoscanate in children with hookworm disease *19,* 6 (1975)	B. Bhandari L. N. Shrimali Department of Child Health, R. N. T. Medical College, Udaipur, India
Recent studies on genetic recombination in *Vibrio cholerae* *19,* 460 (1975)	K. Bhaskaran Central Drug Research Institute, Lucknow, India
Interbiotype conversion of cholera vibrios by action of mutagens *19,* 466 (1975)	P. Bhattacharya S. Ray WHO International Vibrio Reference Centre, Cholera Research Centre, Calcutta 25, India

Experience with bitoscanate in hookworm disease and trichuriasis in Mexico *19*, 23 (1975)	F. Biagi Departamento de Parasitologia, Facultad de Medicina, Universidad Nacional Autónoma de Mexico, Mexico
Analysis of symptoms and signs related with intestinal parasitosis in 5,215 cases *19*, 10 (1975)	F. Biagi R. López J. Viso Departamento de Parasitologia, Facultad de Medicina, Universidad Nacional Autónoma de Mexico, Mexico
Untersuchungen zur Biochemie und Pharmacologie der Thymoleptika *11*, 121 (1968)	M. H. Bickel Medizinisch-Chemisches Institut der Universität Bern Bern, Switzerland
The role of adipose tissue in the distribution and storage of drugs *28*, 273 (1984)	M. H. Bickel Universität Bern, Pharmacologisches Institut 3008 Bern, Switzerland
The β-adrenergic-blocking agents, pharmacology, and structure-activity relationships *10*, 46 (1966)	J. H. Biel Aldrich Chemical Company Inc. Milwaukee, Wisconsin, USA B. K. B. Lum Department of Pharmacology, Marquette University School of Medicine, Milwaukee, Wisconsin, USA
Prostaglandins *17*, 410 (1973)	J. S. Bindra and R. Bindra Medical Research Laboratories, Pfizer Inc., Groton, Connecticut, USA
In vitro models for the study of Antibiotic Activities *31*, 349 (1987)	J. Blaser Medizinische Poliklinik, Departement für Innere Medizin, Universitätsspital, Rämistr. 100, CH-8091 Zurich, Switzerland S. H. Zinner Department of Medicine, Brown University, Roger Williams General Hospital, Providence, Rhode Island 02908, USA
The red blood cell membrane as a model for targets of drug action *17*, 59 (1973)	L. Bolis Università degli Studi di Roma, Istituto di Fisiologia Generale, Rome, Italy

Epidemiology and public health. Importance of intestinal nematode infections in Latin America *19*, 28 (1975)	D. Botero School of Medicine, University of Antioquia, Medellin, Columbia
Clinical importance of cardiovascular drug interactions *25*, 133 (1981)	D. C. Brater Division of Clinical Pharmacology, Departments of Pharmacology and Internal Medicine, The University of Texas, Health Science Center at Dallas, 5323 Harry Hines Boulevard, Dallas, Texas 75235, USA
Update of cardiovascular drug interactions *29*, 9 (1985)	D. Craig Brater Michael R. Vasko Departments of Pharmacology and Internal Medicine, The University of Texas Health Science Center at Dallas and Veterans Administration Medical Center, 4500 Lancaster Road, Dallas, TX 75216, USA
Serum electrolyte abnormalities caused by drugs *30*, 9 (1986)	D. Craig Brater Departments of Pharmacology and Internal Medicine, The University of Texas Health Science Center, 5323 Harry Hines Boulevard, Dallas, Texas 75235, USA
Some practical problems of the epidemiology of leprosy in the Indian context *18*, 25 (1974)	S. G. Browne Leprosy Study Centre, 57a Wimpole Street, London, England
Brain neurotransmitters and the development and maintenance of experimental hypertension *30*, 127 (1986)	Jerry J. Buccafusco, Department of Pharmacology and Toxicology, and Psychiatry, Medical College of Georgia and Veterans Administration Medical Center, Augusta, Georgia 30912, USA Henry E. Brezenoff Department of Pharmacology, University of Medicine and Dentistry of New Jersey, Newark, New Jersey 07103, USA
Die Ionenaustauscher und ihre Anwendung in der Pharmazie und Medizin *1*, 11 (1959) Wert und Bewertung der Arzneimittel *10*, 90 (1966)	J. Büchi Pharmazeutisches Institut der ETH, Zürich, Switzerland

Cyclopropane compounds of biological interest *15,* 227 (1971) The state of medicinal science *20,* 9 (1976)	A. Burger University of Virginia, Charlottesville, Virginia, USA
Human and veterinary anthelmintics (1965–1971) *17,* 108 (1973)	R. B. Burrows Mount Holly, New Jersey, USA
The antibody basis of local immunity to experimental cholera infection in the rabbit ileal loop *19,* 471 (1975)	W. Burrows J. Kaur University of Chicago, P. O. B. 455, Cobden, Illinois, USA
Les dérivés organiques du fluor d'intérêt pharmacologique *3,* 9 (1961)	N. P. Buu-Hoï Centre National de la Recherche Scientifique, Gif-sur-Yvette, France
Teaching tropical medicine *18,* 35 (1974)	K. M. Cahill Tropical Disease Center, 100 East 77th Street, New York City NY 10021, USA
Anabolic steroids *2.* 71 (1960)	B. Camerino G. Sala Department of Clinical Chemistry Farmitalia, Milan, Italy
Immunosuppression agents, procedures, speculations and prognosis *16,* 67 (1972)	G. W. Camiener Research Laboratories, The Upjohn Company, Kalamazoo, Michigan, USA W. J. Wechter The Upjohn Company, Kalamazoo, Michigan, USA
Dopamine agonists: Structure-activity relationships *29,* 303 (1985)	Joseph G. Cannon The University of Iowa, Iowa City, Iowa 52242, USA
Analgesics and their antagonists: recent developments *22,* 149 (1978)	A. F. Casy Norfolk and Norwich Hospital and University of East Anglia, Norwich, Norfolk, England
Chemical nature and pharmacological actions of quaternary ammonium salts *2,* 135 (1960)	C. J. Cavallito School of Pharmacy, University of North Carolina, Chapel Hill, North Carolina, USA A. P. Gray Neisler Laboratories Inc., Decatur, Illinois, USA

Contributions of medicinal chemistry to medicine – from 1935 *12*, 11 (1968) Quaternary ammonium salts – advances in chemistry and pharmacology since 1960 *24*, 267 (1980)	C. J. Cavallito Medicinal Chemistry, School of Pharmacy, University of North Carolina, Chapel Hill, North Carolina, USA
Changing influences on goals and incentives in drug research and development *20*, 159 (1976)	C. J. Cavallito Ayerst Laboratories, Inc., New York, NY, USA
Über Vorkommen und Bedeutung der Indolstruktur in der Medizin und Biologie *2*, 227 (1960)	A. Cerletti Sandoz AG, Basel, Switzerland
Cholesterol and its relation to atherosclerosis *1*, 127 (1959)	K. K. Chen Department of Pharmacology, University School of Medizine, Indianapolis, Indiana, USA Tsung-Min Lin Division of Pharmacologic Research, Lilly Research Laboratories, Indianapolis, Indiana, USA
Effect of hookworm disease on the structure and function of small bowel *19*, 44 (1975)	H. K. Chuttani R. C. Misra Maulana Azad Medical College & Associated Irwin and G. B. Pant Hospitals, New Delhi, India
The psychomimetic agents *15*, 68 (1971)	S. Cohen Division of Narcotic Addiction and Drug Abuse, National Institute of Mental Health, Chevy Chase, Maryland, USA
Implementation of disease control in Asia and Africa *18*, 43 (1974)	M. J. Colbourne Department of Preventive and Social Medicine, University of Hong Kong, Sassoon Road, Hong Kong
Structure-activity relationships in certain anthelmintics *3*, 75 (1961)	J. C. Craig Department of Pharmaceutical Chemistry, University of California, San Francisco, California, USA M. E. Tate University of New South Wales, Department of Organic Chemistry, Kensington, N. S. W., Australia

Contribution of Haffkine to the concept and practice of controlled field trials of vaccines *19*, 481 (1975)	B. Cvjetanovic Bacterial Diseases, Division of Communicable Diseases, WHO, Geneva, Switzerland
Antifungal agents *22*, 93 (1978)	P. F. D'Arcy E. M. Scott Department of Pharmacy, The Queen's University of Belfast, Belfast, Northern Ireland
Some neuropathologic and cellular aspects of leprosy *18*, 53 (1974)	D. K. Dastur Y. Ramamohan A. S. Dabholkar Neuropathology Unit, Grant Medical College and J. J. Group of Hospitals, Bombay 8, India
Autonomic dysfunction as a problem in the treatment of tetanus *19*, 245 (1975)	F. D. Dastur G. J. Bhat K. G. Nair Department of Medicine, Seth G. S. Medical College and K. E. M. Hospital, Bombay 12, India
Studies on *V. parahaemolyticus* infection in Calcutta as compared to cholera infection *19*, 490 (1975)	B. C. Deb Cholera Research Centre, Calcutta, India
Biochemical effects of drugs acting on the central nervous system *8*, 53 (1965)	L. Decsi University Medical School, Pécs, Hungary
Some reflections on the chemotherapy of tropical diseases: Past, present and future *26*, 343 (1982)	E. W. J. de Maar
Drug research – whence and whither *10*, 11 (1966)	R. G. Denkewalter M. Tishler Merck Sharp & Dohme Research Laboratories, Rahway, New Jersey, USA
Serendipity and structured research in drug discovery *30*, 189 (1986)	George de Stevens Drew University, Madison, New Jersey 07940, USA
Hypolipidemic agents *13*, 217 (1969)	G. de Stevens W. L. Bencze CIBA Pharmaceutical Company, Summit, New Jersey, USA R. Hess CIBA Limited, Basel, Switzerland

The interface between drug research, marketing, management, and social political and regulatory forces *20*, 181 (1976)	G. de Stevens Pharmaceuticals Division, CIBA-GEIGY Corporation, Summit, New Jersey, USA
Antihypertensive agents *20*, 197 (1976)	G. de Stevens M. Wilhelm Pharmaceuticals Division CIBA-GEIGY Corporation, Summit, New Jersey, USA
Medicinal research: Retrospectives and Perspectives *29*, 97 (1985)	George de Stevens Department of Chemistry, Drew University, Madison, N. J., USA
Medicinal chemistry: A support or a driving force in drug research? *34*, 343 (1990)	George de Stevens Department of Chemistry and The Charles Dana Research Institute, Drew University, Madison, New Jersey 07940, USA
RNA virus evolution and the control of viral disease *33*, 93 (1989)	Esteban Domingo Department of Biology, University of California San Diego, La Jolla, California 92093, USA
Transport and accumulation in biological cell systems interacting with drugs *20*, 261 (1976)	W. Dorst A. F. Bottse Department of Pharmacology, Vrije Universitiet, Amsterdam, The Netherlands G. M. Willems Biomedical Centre, Medical Faculty, Maastricht, The Netherlands
Immunization of a village, a new approach to herd immunity *19*, 252 (1975)	N. S. Deodhar Department of Preventive and Social Medicine, B. J. Medical College, Poona, India
Surgical amoebiasis *18*, 77 (1974)	A. E. de Sa Bombay Hospital, Bombay, India
Epidemiology of pertussis *19*, 257 (1975)	J. A. Sa Glaxo Laboratories (India) Ltd., Worli, Bombay 25, India

Profiles of tuberculosis in rural areas of Maharashtra *18*, 91 (1974)	M. D. Deshmukh K. G. Kulkarni S. S. Virdi B. B. Yodh Memorial Tuberculosis Reference Laboratory and Research Centre, Bombay, India
The pharmacology of the immune system: Clinical and experimental perspectives *28*, 83 (1984)	Jürgen Drews, Sandoz Ltd., Pharmaceutical Research and Development, CH-4002 Basel, Switzerland
An overview of studies on estrogens, oral contraceptives and breast cancer *25*, 159 (1981)	V. A. Drill Department of Pharmacology, College of Medicine, University of Illinois at the Medical Center, Chicago, ILL 60680, USA
Aminonucleosid-nephrose *7*, 341 (1964)	U. C. Dubach Medizinische Universitäts-Poliklinik Basel Basel, Switzerland
Impact of researches in India on the control and treatment of cholera *19*, 503 (1975)	N. K. Dutta Director, Vaccine Institute, Baroda, India
The perinatal development of drug-metabolizing enzymes: What factors trigger their onset? *25*, 189 (1981)	G. J. Dutton J. E. A. Leakey Department of Biochemistry, The University of Dundee, Dundee, DDI 4 HN, Scotland
Laboratory methods in the screening of anthelmintics *19*, 48 (1975)	D. Düwel Helminthology Department, Farbwerke Hoechst AG, Frankfurt/Main 80, FRG
Progress in immunization *19*, 274 (1975)	G. Edsall Department of Microbiology, London School of Hygiene and Tropical Medicine, London W.C. 1, England

Host factors in the response to immunization *19,* 263 (1975)	G. Edsall Department of Microbiology, London School for Hygiene and Tropical Medicine, London W.C. 1, England M. A. Belsey World Health Organization, Geneva, Switzerland R. Le Blanc Tulan University School of Public Health and Tropical Medicine, New Orleans, LA, USA L. Levine State Laboratory Institute, Boston, MA, USA
Drug-macromolecular interactions: implications for pharmacological activity *14,* 59 (1970)	S. Ehrenpreis Department of Pharmacology, New York Medical College, Fifth Avenue at 106th Street, New York, NY 10029, USA
Betrachtungen zur Entwicklung von Heilmitteln *10,* 33 (1966)	G. Ehrhart Farbwerke Hoechst AG, Frankfurt a. M., FRG
Progress in malaria chemotherapy, Part I. Repository antimalarial drugs *13,* 170 (1969) New perspectives on the chemotherapy of malaria, filariasis and leprosy *18,* 99 (1974)	E. F. Elslager Chemistry Department, Parke, Davis & Company, Ann Arbor, Michigan, USA
Recent research in the field of 5-hydroxytryptamine and related indolealkylamines *3,* 151 (1961)	V. Erspamer Institute of Pharmacology, University of Parma, Parma, Italy
The chemistry of DNA modification by antitumor antibiotics *32,* 411 (1988)	Jed. F. Fisher and Paul A. Aristoff Cardiovascular Diseases Research, The Upjohn Company, Kalamazoo, MI 49001, USA
Drugs affecting serotonin neurons *35,* 85 (1990)	Ray W. Fuller Eli Lilly and Company, Lilly Corporate Center, Indianapolis, IN 46285, USA
Bacteriology at the periphery of the cholera pandemic *19,* 513 (1975)	A. L. Furniss Public Health Laboratory, Maidstone, England
Iron and diphteria toxin production *19,* 283 (1975)	S. V. Gadre S. S. Rao Haffkine Institute, Bombay 12, India

Effect of drugs on cholera toxin induced fluid in adult rabbit ileal loop *19*, 519 (1975)	B. B. Gaitonde P. H. Marker N. R. Rao Haffkine Institute, Bombay 12, India
Drug action and assay by microbial kinetics *15*, 519 (1971) The pharmacokinetic bases of biological response quantification in toxicology, pharmacology and pharmacodynamics *21*, 105 (1977)	E. R. Garrett The J. Hillis Miller Health Center, College of Pharmacy, University of Florida, Gainesville, Florida, USA
The chemotherapy of enterobacterial infections *12*, 370 (1968)	L. P. Garrod Department of Bacteriology, Royal Postgraduate Medical School, Hammersmith Hospital London, England
The use of neutrophils, macrophages and organ cultures to assess the penetration of human cells by antimicrobials *33*, 83 (1989)	Zell A. McGee, Gary L. Gorby and Wanda S. Updike The Center for Infectious Diseases, Diagnostic Mircobiology and Immunology, University of Utah School of Medicine, Salt Lake City, Utah 84132, USA
Progress in the chemotherapy of leprosy: Status, issues and prospects *34*, 421 (1990)	Robert H. Gelber Kuzell Institute for Arthritis and Infectious Diseases, 2200 Webster Street, San Francisco, CA 94115-1896, USA
Metabolism of drugs and other foreign compounds by enzymatic mechanisms *6*, 11 (1963)	J. R. Gillette Laboratory of Chemical Pharmacology, National Heart Institute, Bethesda 14, Maryland, USA
Orale Antidiabetika *30*, 281 (1986)	Heiner Glombik Rudi Weyer Hoechst AG D-6230 Frankfurt (M) 80, FRG
The art and science of contemporary drug development *16*, 194 (1972)	A. J. Gordon S. G. Gilgore Pfizer Pharmaceuticals, 235 East 42nd Street, New York, USA
Basic mechanisms of drug action *7*, 11 (1964) Isolation and characterization of membrane drug receptors *20*, 323 (1976)	D. R. H. Gourley Department of Pharmacology, Eastern Virginia Medical School, Norfolk, Virginia, USA

Zusammenhänge zwischen Konstitution und Wirksamkeit bei Lokalanästhetica *4*, 353 (1962)	H. Grasshof Firma M. Woelm, Eschwege, FRG
Das Placeboproblem *1*, 279 (1959)	H. Haas Knoll AG, Ludwigshafen, FRG H. Fink and G. Härtefelder Forschungslaboratorien der Knoll AG, Ludwigshafen, FRG
Approaches to the rational design of bacterial vaccines *32*, 377 (1988)	Peter Hambleton Stephen D. Prior Andrew Robinson Public Health Laboratory Service, Centre for Applied Microbiology and Research Porton Down, Salisbury, Wilts. SP 4 OJG, U. K.
Clinical field trial of bitoscanate in *Necator americanus* infection, South Thailand *19*, 64 (1975)	T. Harinasuta D. Bunnag Faculty of Tropical Medicine, Mahidol University, Bangkok, Thailand
Pharmacological control of reproduction in women *12*, 47 (1968) Contraception – retrospect and prospect *21*, 293 (1977)	M. J. K. Harper The University of Texas, Health Science Center at San Antonio, San Antonio, Texas, USA
Drug latentiation *4*, 221 (1962)	N. J. Harper University of Aston, Birmingham 4, England
Chemotherapy of filariasis *9*, 191 (1966) Filariasis in India *18*, 173 (1974)	F. Hawking Clinical Research Centre, Watford Road, Harrow, Middlesex, England
Recent studies in the field of indole compounds *6*, 75 (1963)	R. V. Heinzelmann J. Szmuszkovicz The Upjohn Company, Kalamazoo, Michigan, USA
Neuere Entwicklungen auf dem Gebiete therapeutisch verwendbarer organischer Schwefelverbindungen *4*, 9 (1962)	H. Herbst Farbwerke Hoechst, Frankfurt a. M., FRG
The management of acute diarrhea in children: an overview *19*, 527 (1975)	N. Hirschhorn Management Sciences for Health, One Broadway, Cambridge, Mass., USA

The tetracyclines *17,* 210 (1973)	J. J. Hlavka and J. H. Booth Lederle Laboratories, Pearl River, NY, USA
The use of quantum chemical methods to study molecular mechanisms of drug action *34,* 9 (1990)	H.-D. Höltje, M. Hense, S. Marrer and E. Maurhofer Department of Pharmacy, Free University of Berlin, Königin-Luise Strasse 2 + 4, D-1000 Berlin 33, FRG
Chemotherapy for systemic mycoses *33,* 317 (1989)	Paul D. Hoeprich Section of Medical Mycology, Department of Internal Medicine, School of Medicine, University of California, Davis, CA 95616, USA
Relationship of induced antibody titres to resistance to experimental human infection *19,* 542 (1975)	R. B. Hornick R. A. Cash J. P. Libonati The University of Maryland School of Medicine, Division of Infectious Diseases, Baltimore, Maryland, USA
Recent applications of mass spectrometry in pharmaceutical research *18,* 399 (1974)	G. Horváth Research Institute for Pharmaceutical Chemistry, Budapest, Hungary
Risk assessment problems in chemical oncogenesis *31,* 257 (1987)	G. H. Hottendorf Medical University of South Carolina, Charleston, South Carolina 29425, USA
Recent developments in disease-modifying antirheumatic-drugs *24,* 101 (1980)	I. M. Hunneyball Research Department, Boots Co. Ltd., Pennyfoot Street, Nottingham, England
The pharmacology of homologous series *7,* 305 (1964)	H. R. Ing Oxford University, Department of Pharmacology, Oxford, England
Progress in the experimental chemotherapy of helminth infections. Part. I. Trematode and cestode diseases *17,* 241 (1973)	P. J. Islip The Wellcome Research Laboratories, Beckenham, Kent, England
Pharmacology of the brain: the hippocampus, learning and seizures *16,* 211 (1972)	I. Izquierdo A. G. Nasello Departamento de Farmacologia, Facultad de Ciencias Quimicas, Universidad Nacional de Córdoba, Estafeta 32, Córdoba, Argentina

Cholinergic mechanism – monoamines relation in certain brain structures *16*, 334 (1972)	J. A. Izquierdo Department of Experimental Pharmacology, Facultad de Farmacia y Bioquimica Buenos Aires, Argentina
The development of antifertility substances *7*, 133 (1964)	H. Jackson Experimental Chemotherapy, Christie Hospital and Holt Radium Institute, Paterson Laboratories, Manchester 20, England
Agents acting on central dopamine receptors *21*, 409 (1977)	P. C. Jain N. Kumar Medicinal Chemistry Division, Central Drug Research Institute, Lucknow, India
Recent advances in the treatment of parasitic infections in man *18*, 191 (1974) The levamisole story *20*, 347 (1976)	P. A. J. Janssen Janssen Pharmaceutica, Research Laboratories, Beerse, Belgium
Recent developments in cancer chemotherapy *25*, 275 (1981)	K. Jewers Tropical Product Institute, 56/62, Gray's Inn Road, London, WC1X8LU, England
Search for pharmaceutically interesting quinazoline derivatives: Efforts and results (1969–1980) *26*, 259 (1982)	S. Johne Institute of Plant Biochemistry, The Academy of Sciences of the German Democratic Republic, 4010 Halle (Saale), PSF 250, GDR
A review of advances in prescribing for teratogenic hazards *29*, 121 (1985)	E. Marshall Johnson Daniel Baugh Institute, Jefferson College, Thomas Jefferson University, 1020 Locust Street, Philadelphia, PA 19107, USA
A comparative of bitoscanate, bephenium hydroxynaphthoate and tetrachlorethylene in hookworm infection *19*, 70 (1975)	S. Johnson Department of Medicine III, Christian Medical College Hospital, Vellore, Tamilnadu, India
Tetanus in Punjab with particular reference to the role of muscle relaxants in its management *19*, 288 (1975)	S. S. Jolly J. Singh S. M. Singh Department of Medicine, Medical College, Patiala, India

Virulence-enhancin effect of ferric ammonium citrate on *Vibrio cholerae* *19*, 546 (1975)	I. Joó Institute for Serobacteriological Production and Research ‚HUMAN' WHO International Reference Centre for Bacterial Vaccines, Budapest, Hungary
Drug molecules of marine origin *35*, 521 (1990)	Pushkar N. Kaul Clark Atlanta University, James P. Brawley Drive at Fair Str. SW, Atlanta, Georgia 30314, USA
Toxoplasmosis *18*, 205 (1974)	B. H. Kean The New York Hospital – Cornell Medical Center, 525 East 68th Street, New York, NY, USA
Tabellarische Zusammenstellung über die Substruktur der Proteine *16*, 364 (1972)	R. Kleine Physiologisch-Chemisches Institut der Martin-Luther-Universität, 4020 Halle (Saale), GDR
Bioactive peptide analogs: In vivo and in vitro production *34*, 287 (1990)	Horst Kleinkauf and Hans von Doehren Institut für Biochemie und Molekulare Biologie, Technische Universität Berlin, Franklinstrasse 29, D-1000 Berlin 10, FRG
Experimental evaluation of antituberculous compounds, with special reference to the effect of combined treatment *18*, 211 (1974)	F. Kradolfer Biological Research Laboratories, Pharmaceutical Division, Ciba-Geigy Ltd., Basel, Switzerland
The oxidative metabolism of drugs and other foreign compounds *17*, 488 (1973)	F. Kratz Medizinische Kliniken und Polikliniken, Justus-Liebig-Universität, Gießen, FRG
Die Amidinstruktur in der Arzneistofforschung *11*, 356 (1968)	A. Kreutzberger Institut für pharmazeutische Chemie der Westfälischen Wilhelms-Universität Münster, Münster (Westfalen), FRG
Present data on the pathogenesis of tetanus *19*, 301 (1975) Tetanus: general and pathophysiological aspects: achievement, failures, perspectives of elaboration of the problem *19*, 314 (1975)	G. N. Kryzhanovsky Institute of General Pathology and Pathological Physiology, AMS USSR, Moscow, USSR

Lipophilicity and drug activity *23,* 97 (1979)	H. Kubinyi Chemical Research and Development of BASF Pharma Division, Knoll AG, Ludwigshafen/Rhein, FRG
Klinisch-pharmakologische Kriterien in der Bewertung eines neuen Antibiotikums. Grundlagen und methodische Gesichtspunkte *23,* 327 (1978)	H. P. Kuemmerle München/Eppstein, FRG
Adrenergic receptor research: Recent developments *33,* 151 (1989)	George Kunos Laboratory pf Physiologic and Pharmacologic Studies, National Institute on Alcohol Abuse and Alcoholism, National Institutes of Health, Bethesda, MD 20892, USA
Über neue Arzneimittel *1,* 531 (1959), *2,* 251 (1960), *3,* 369 (1961), *6,* 347 (1963), *10,* 360 (1966)	W. Kunz Dr. Schwarz GmbH, Monheim (Rheinland), FRG
Die Anwendung von Psychopharmaka in der psychosomatischen Medizin *10,* 530 (1966)	F. Labhardt Psychiatrische Universitätsklinik, Basel, Switzerland
The bacterial cell surface and antimicrobial resistance *32,* 149 (1988)	Peter A. Lambert Pharmaceutical Sciences Institute, Aston University, Birmingham B4 7ET, U. K.
Therapeutic measurement in tetanus *19,* 323 (1975)	D. R. Laurence Department of Pharmacology, University College, London, and Medical Unit, University College Hospital Medical School, London, England
Physico chemical methods in pharmaceutical chemistry, I. Spectrofluorometry *6,* 151 (1963)	H. G. Leemann K. Stich Margrit Thomas Analytical Department Research Laboratory, Sandoz Ltd., Basel, Switzerland
Biochemical acyl hydroxylations *16,* 229 (1972)	W. Lenk Pharmakologisches Institut der Universität München Nussbaumstrasse 26, FRG

Cholinesterase restoring therapy in tetanus *19*, 329 (1975)	G. Leonardi Department of Medicine, St. Thomas Hospital, Portogruaro, Venice, Italy K. G. Nair F. D. Dastur Department of Medicine, Seth G. S. Medical College and K. E. M. Hospital, Bombay 12, India
Biliary excretion of drugs and other xenobiotics *25*, 361 (1981)	W. G. Levine Department of Molecular Pharmacology, Albert Einstein of Medicine, Yeshiva University, 1300 Morris Park Avenue, Bronx, NY 10461, USA
Structures, properties and disposition of drugs *29*, 67 (1985)	Eric J. Lien Biomedicinal Chemistry, School of Pharmacy, University of Southern California, Los Angeles, CA 90033, USA
Ribonucleotide reductase inhibitors as anticancer and antiviral agents *31*, 101 (1987)	Erich J. Lien Section of Biomedicinal Chemistry, School of Pharmacy, University of Southern California, Los Angeles, CA 90033, USA
Fungal metabolites and Chinese herbal medicine as immunostimulants *34*, 395 (1990)	Erich J. Lien Biomedicinal Chemistry, School of Pharmacy, University of Southern California, 1985 Zonal Avenue, Los Angeles, California 90033, USA
Interactions between androgenic-anabolic steroids and glucocorticoids *14*, 139 (1970)	O. Linèt Sinai Hospital of Detroit, Department of Medicine, 6767 West Outer Drive, Detroit, Michigan 48235
Drug inhibition of mast cell secretion *29*, 277 (1985)	R. Ludowyke D. Lagunoff Department of Pathology, St. Louis University, School of Medicine, 1402 S. Grand Blvd, St. Louis, MO 63104
Reactivity of bentonite flocculation, indirect haemagglutination and casoni tests in hydatid disease *19*, 75 (1975)	R. C. Mahajan N. L. Chitkara Division of Parasitology, Department of Microbiology, Postgraduate Institute of Medical Education and Research, Chandigarh, India

Interaction of cancer chemotherapy agents with the mononuclear phagocyte system *35*, 487 (1990)	Alberto Mantovani Annunciata Vecchi, Istituto di Ricerche Farmacologiche „Mario Negri", Via Eritrea 62, 20157 Milan, Italy
Epidemiology of diphtheria *19*, 336 (1975)	L. G. Marquis Glaxo Laboratories (India) Limited, Worli, Bombay 25, India
Biological activity of the terpenoids and their derivatives *6*, 279 (1963)	M. Martin-Smith University of Strathclyde, Department of Pharmaceutical Chemistry, Glasgow, C. 1, Scotland T. Khatoon The Eden Girls College, Dacca, East Pakistan
Biological activity of the terpenoids and their derivatives – recent advances *13*, 11 (1969)	M. Martin-Smith W. E. Sneader University of Strathclyde, Glasgow, C. 1, Scotland
Antihypertensive agents 1962–1968 *13*, 101 (1969) Fundamental structures in drug research – Part I *20*, 385 (1976) Fundamental structures in drug research – Part II *22*, 27 (1978) Antihypertensive agents 1969–1980 *25*, 9 (1981)	A. Marxer O. Schier Chemical Research Department, Pharmaceuticals Division, Ciba-Geigy Ltd., Basel, Switzerland
Relationships between the chemical structure and pharmacological activity in a series of synthetic quinuclidine derivatives *13*, 293 (1969)	M. D. Mashkovsky L. N. Yakhontov All-Union Chemical Pharmaceutical Research Institute, Moscow, USSR
Further developments in research on the chemistry and pharmacology of synthetic quinuclidine derivatives *27*, 9 (1983)	M. D. Mashkovsky L. N. Yakhontov M. E. Kaminka E. E. Mikhlina S. Ordzhonikidze All-Union, Chemical Pharmaceutical Research Institute, Moscow, USSR
Role of neutrotransmitters in the central regulation of the cardiovascular system *35*, 25 (1990)	Robert B. McGall The Upjohn Company Kalamazoo, MI 49001, USA

On the understanding of drug potency *13*, 123 (1971) The chemotherapy of intestinal nematodes *16*, 157 (1972)	J. W. McFarland Pfizer Medical Research Laboratories, Groton, Connecticut, USA
Zur Beeinflussung der Strahlenempfindlichkeit von Säugetieren durch chemische Substanzen *9*, 11 (1966)	H.-J. Melching C. Streffer Radiologisches Institut der Universität Freiburg i. Br., Freiburg i. Br., FRG
Analgesia and addiction *5*, 155 (1963)	L. B. Mellett University of Michigan Medical School, Ann Arbor, Michigan, USA L. A. Woods Department of Pharmacology, College of Medicine, State University of Iowa, Iowa City, USA
Comparative drug metabolism *13*, 136 (1969)	L. B. Mellett Kettering-Meyer Laboratories, Southern Research Institute, Birmingham, Alabama, USA
Mechanism of action of anxiolytic drugs *31*, 315 (1987)	T. Mennini S. Caccia S. Garattini Istituto di Ricerche Farmacologische „Mario Negri", Via Eritrea 62, 20157 Milan, Italy
Pathogenesis of amebic disease *18*, 225 (1974) Protozoan and helminth parasites – a review of current treatment *20*, 433 (1976)	M. J. Miller Tulane University, Department of Tropical Medicine, New Orleans, Louisiana, USA
Medicinal agents incorporating the 1,2-diamine functionality *33*, 135 (1989)	Erik T. Michalson Jacob Szmuszkovicz Department of Chemistry, University of Notre Dame, Notre Dame, IN 46556, USA
Synopsis der Rheumatherapie *12*, 165 (1968)	W. Moll Basel, Schweiz
On the chemotherapy of cancer *8*, 431 (1965) The relationship of the metabolism of anticancer agents to their activity *17*, 320 (1973) The current status of cancer chemotherapy *20*, 465 (1976)	J. A. Montgomery Kettering-Meyer Laboratory, Southern Research Institute, Birmingham, Alabama, USA

Present status of Leishmaniasis *34,* 447 (1990)	Anita Mukherjee, Manju Seth and A. P. Bhaduri Division of Medicinal Chemistry, Central Drug Research Institute, Lucknow 226 001, India
The significance of DNA technology in medicine *33,* 397 (1989)	Hansjakob Müller Dept of Human Genetics, University Childrens Hospital, CH-4005 Basel, and Lab. of Human Genetics, Dept of Research, Kantonsspital, CH-4031 Basel, Switzerland
Der Einfluß der Formgebung auf die Wirkung eines Arzneimittels *10,* 204 (1966) Galenische Formgebung und Arzneimittelwirkung. Neue Erkenntnisse und Feststellungen *14,* 269 (1970)	K. Münzel F. Hoffmann-La Roche & Co. AG, Basel, Switzerland
A field trial with bitoscanate in India *19,* 81 (1975)	G. S. Mutalik R. B. Gulati A. K. Iqbal Department of Medicine, B. J. Medical College and Sassoon General Hospital, Poona, India
Comparative study of bitoscanate, bephenium hydroxynaphthoate and tetrachlorethylene in hookworm disease *19,* 86 (1975)	G. S. Mutalik R. B. Gulati Department of Medicine, B. J. Medical College and Sassoon General Hospital, Poona, India
Ganglienblocker *2,* 297 (1960)	K. Nádor Chemisches Institut der Tierärztlichen Universität, Budapest, Ungarn
Nitroimidazoles as chemotherapeutic agents *27,* 162 (1983)	M. D. Nair K. Nagarajan Ciba-Geigy Research Centre, Goreagon East, Bombay 40063
Recent advances in cholera pathophysiology and therapeutics *19,* 563 (1975)	D. R. Nalin Johns Hopkins School of Medicine and Public Health, Dacea, Bangladesh
Preparing the ground for research: importance of data *18,* 239 (1974)	A. N. D. Nanavati Department of Virology, Haffkine Institute, Bombay, India

Computer-assisted structure – antileukemic activity analysis of purines and their aza and deaza analogs *34*, 319 (1990)	V. L. Narayanan, Mohamed Nasr and Kenneth D. Paull National Institutes of Health, Bethesda, Maryland, USA
Mechanism of drugs action on ion and water transport in renal tubular cells *26*, 87 (1982)	Yu. V. Natochin I. M. Sechenov Institute of Evolutionary Physiology and Biochemistry, Leningrad, USSR
Progesterone receptor binding of steroidal and nonsteroidal compounds *30*, 151 (1986)	Neelima M. Seth A. P. Bhaduri Division of Medicinal Chemistry, Central Drug Research Institute, Lucknow 226001, India
Recent advances in drugs against hypertension *29*, 215 (1985)	Neelima B. K. Bhat A. P. Bhaduri Central Drug Research Institute, Lucknow 226001, India
High resolution nuclear magnetic resonance spectroscopy of biological samples as an aid to drug development *31*, 427 (1987)	J. K. Nicholson Department of Chemistry, Birkbeck College, University of London, Gordon House, 29, Gordon Square, London WC1 E6BT, England Ian D. Wilson Department of Safety of Medicines, ICI Pharmaceutical Division, Mereside, Alderley Park, Macclesfield, Cheshire SK 10 4TG, England
Antibody response to two cholera vaccines in volunteers *19*, 554 (1975)	Y. S. Nimbkar R. S. Karbhari S. Cherian N. G. Chanderkar R. P. Bhamaria P. S. Ranadive B. B. Gaitondé Haffkine Institute, Parel, Bombay 12, India
Surface interaction between bacteria and phagocytic cells *32*, 137 (1988)	L. Öhman G. Maluszynska K.E. Magnusson and O. Stendahl Department of Medical Microbiology, Linsköping University, S-581 85 Linsköping, Sweden
Die Chemotherapie der Wurmkrankheiten *1*, 159 (1959)	H.-A. Oelkers C. F. Asche & Co., Hamburg-Altona, FRG

GABA-Drug Interactions *31*, 223 (1987)	Richard W. Olsen Department of Pharmacology, School of Medicine, and Brain Research Institute, University of California, Los Angeles, CA 90024, USA
Drug research and human sleep *22*, 355 (1978)	I. Oswald University Department of Psychiatry, Royal Edinburgh Hospital, Edinburgh, Scotland
Effects of drugs on calmodulin-mediated enzymatic actions *33*, 353 (1989)	Judit Ovádi Institute of Enzymology, Biological Research Center, Hungarian Academy of Sciences, P.O.B. 7, H-1502 Budapest, Hungary
An extensive community outbreak of acute diarrhoeal diseases in children *19*, 570 (1975)	S. C. Pal C. Koteswar Rao Cholera Research Centre, Calcutta, India
Drug and its action according to Ayurveda *26*, 55 (1982)	Shri Madhabendra Nath Pal
Oligosaccharide chains of glycoproteins *32*, 163 (1990)	Y.T. Pan Alan D. Elbein Department of Biochemistry, University of Texas Health Science Center, San Antonio, Texas 78284, USA
3,4-Dihydroxyphenylalanine and related compounds *9*, 223 (1966)	A. R. Patel A. Burger Department of Chemistry, University of Virginia, Charlottesville, Virginia, USA
Mescaline and related compounds *11*, 11 (1968)	A. R. Patel Department of Chemistry, University of Virginia, Charlottesville, Virginia, USA
Experience with bitoscanate in adults *19*, 90 (1975)	A. H. Patricia U. Prabakar Rao R. Subramaniam N. Madanagopalan Madras Medical College, Madras, India
The impact of state and society on medical research *35*, 9 (1990)	C. R. Pfaltz University of Basel, Basel, Switzerland

Monoaminoxydase-Hemmer *2*, 417 (1960)	A. Pletscher K. F. Gey P. Zeller F. Hoffmann-La Roche & Co. AG, Basel, Switzerland
The oral antiarrhythmic drugs *35*, 151 (1990)	Lisa Mendes, Scott L. Beau, John S. Wilson Philip J. Podrid Boston University, School of Medicine, Boston, MA 02115, USA
What makes a good pertussis vaccine? *19*, 341 (1975) Vaccine composition in relation to antigenic variation of the microbe: is pertussis unique? *19*, 347 (1975) Some unsolved problems with vaccines *23*, 9 (1979)	N. W. Preston Department of Bacteriology and Virology, University of Manchester, Manchester, England
Antibiotics in the chemotherapy of malaria *26*, 167 (1982)	S. K. Puri G. P. Dutta Division of Microbiology, Central Drug Research Institute, Lucknow 226001, India
Clinical study of diphtheria, pertussis and tetanus *19*, 356 (1975)	V. B. Raju V. R. Parvathi Institute of Child Health and Hospital for Children, Egmore, Madras 8, India
Epidemiology of cholera in Hyderabad *19*, 578 (1975)	K. Rajyalakshmi P. V. Ramana Rao Institute of Preventive Medicine, Hyderabad, Andhra Pradesh, India
Adenosine receptors: Clinical implications and biochemical mechanisms *32*, 195 (1988)	Vickram Ramkumar George Pierson Gary L. Stiles Departments of Medicine and Biochemistry, Duke University Medical Center, Durham, NC 27710, USA
Problems of malaria eradication in India *18*, 245 (1974)	V. N. Rao Health Services (Health), Maharashtra, Bombay, India
Pharmacology of migraine *34*, 209 (1990)	Neil H. Raskin Dept. of Neurology, University of California, School of Medicine, Box 0114, San Francisco, CA 94143, USA

The photochemistry of drugs and related substances *11*, 48 (1968)	S. T. Reid Experimental Pharmacology Division, Institute of Physiology, The University, Glasgow, W2, Scotland
Orale Antikoagulantien *11*, 226 (1968)	E. Renk W. G. Stoll Wissenschaftliche Laboratorien der J. R. Geigy AG, Basel, Switzerland
Mechanism-based inhibitors of monoamine oxidase *30*, 205 (1986)	Lauren E. Richards Alfred Burger Department of Chemistry, University of Virginia, Charlottesville, Virginia 22901, USA
Tetrahydroisoquinolines and β-carbolines: Putative natural substances in plants and animals *29*, 415 (1985)	H. Rommelspacher R. Susilo Department of Neuropsychopharmacology, Free University, Ulmenallee 30, D-1000 Berlin 19, FRG
Functional significance of the various components of the influenza virus *18*, 253 (1974)	R. Rott Institut für Virologie, Justus-Liebig-Universität, Gießen, Deutschland
Behavioral correlates of presynaptic events in the cholinergic neurotransmitter system *32*, 43 (1988)	Roger W. Russell Department of Pharmacology, University of California, Los Angeles, CA 90024-1735, USA
Role of beta-adrenergic blocking drug propranolol in severe tetanus *19*, 361 (1975)	G. S. Sainani B. J. Medical College and Sassoon General Hospitals, Poona, India K. L. Jain V. R. D. Deshpande A. B. Balsara S. A. Iyer Medical College and Hospital, Nagpur, India
Studies on *Vibrio parahaemolyticus* in Bombay *19*, 586 (1975)	F. L. Saldanha A. K. Patil M. V. Sant Haffkine Institute, Parel, Bombay 12, India

Pharmacology and toxicology of axoplasmic transport *28,* 53 (1984)	Fred Samson, Ph. D., Ralph L. Smith J. Alejandro Donoso Research Center, The University of Kansas, Kansas City, Kansas 66103, USA
Clinical experience with bitoscanate *19,* 96 (1975)	M. R. Samuel Medical Division, Hoechst Pharmaceuticals Limited, Bombay, India
Tetanus: Situational clinical trials and therapeutics *19,* 367 (1975)	R. K. M. Sanders M. L. Peacock B. Martyn B. D. Shende The Duncan Hospital, Raxaul, Bihar, India
Epidemiological studies on cholera in non-endemic regions with special reference to the problem of carrier state during epidemic and non-epidemic period *19,* 594 (1975)	M. V. Sant W. N. Gatlewar S. K. Bhindey Haffkine Institute, Parel, Bombay 12, India
Epidemiological and biochemical studies in filariasis in four villages near Bombay *18,* 269 (1974)	M. V. Sant, W. N. Gatlewar and T. U. K. Menon Department of Zoonosis and of Research, Division of Microbiology, Haffkine Institute, Bombay, India
Hookworm anaemia and intestinal malabsorption associated with hookworm infestation *19,* 108 (1975)	A. K. Saraya B. N. Tandon Department of Pathology and Department of Gastroenterology, All India Institute of Medical Sciences, New Delhi, India
The effects of structural alteration on the anti-inflammatory properties of hydrocortisone *5,* 11 (1963)	L. H. Sarett A. A. Patchett Merck Sharp & Dohme Research Laboratories, Rahway, New Jersey, USA S. Steelman Merck Institute for Therapeutic Research, Rahway, New Jersey, USA
The impact of natural product research on drug discovery *23,* 51 (1979)	L. H. Sarett Merck & Co., Inc., Rahway, New Jersey, USA

Anti-filariasis campaign: Its history and future prospects *18*, 259 (1974)	M. Sasa Institute of Medical Science, University of Tokyo, Tokyo, Japan
Barbiturates and the $GABA_A$ receptor complex *34*, 261 (1990)	Paul A. Saunders and I. K. Ho Department of Pharmacology and Toxicology, The University of Mississippi, Medical Center, 2500 North State Street, Jackson, MS 39216–4505, USA
Platelets and atherosclerosis *29*, 49 (1985)	Robert N. Saunders, Sandoz Research Institute, East Hannover, NJ, USA
Pyrimidinones as biodynamic agents *31*, 127 (1987)	Anil K. Saxena Shradha Sinha Division of Medicinal Chemistry, Central Drug Research Institute, Lucknow 226 001, India
Immuno-diagnosis of helminthic infections *19*, 119 (1975)	T. Sawada K. Sato K. Takei Department of Parasitology, School of Medicine, Gunma University, Maebashi, Japan
Immuno-diagnosis in filarial infection *19*, 128 (1975)	T. Sawada K. Sato K. Takei Department of Parasitology, School of Medicine, Gunma University, Maebashi, Japan M. M. Goil Department of Zoology, Bareilly College, Bareilly (U. P.), India
Quantitative structure-activity relationships *23*, 199 (1979)	A. K. Saxena S. Ram Medicinal Chemistry Division, Central Drug Research Institute, Lucknow, India
Advances in chemotherapy of malaria *30*, 221 (1986)	Anil K. Saxena Mridula Saxena Division of Medicinal Chemistry, Central Drug Research Institute, Lucknow 226001, India

Phenothiazine und Azaphenothiazine als Arzneimittel *5,* 269 (1963)	E. Schenker Sandoz AG, Basel, Schweiz H. Herbst Farbwerke Hoechst, Frankfurt a. M., FRG
Antihypertensive agents *4,* 295 (1962)	E. Schlittler CIBA Pharmaceutical Company, Summit, New Jersey, USA J. Druey CIBA Ltd., Basel, Switzerland
Die Anwendung radioaktiver Isotope in der pharmazeutischen Forschung *7,* 59 (1964)	K. E. Schulte Ingeborg Mleinek Institut für Pharmazie und Lebensmittelchemie der Westfälischen Wilhelms-Universität Münster, Münster (Westfalen), FRG
Natürliche und synthetische Acetylen-Verbindungen als Arzneistoffe *14,* 387 (1970)	K. E. Schulte G. Rücker Westfälische Wilhelms-Universität Münster, Münster (Westfalen), FRG
The role of cytokines in macrophage activation *35,* 105 (1990)	Richard M. Schulz Lilly Research Laboratories, Indianapolis, IN 46285, USA
Central control of aterial pressure by drugs *26,* 353 (1982)	A. Scriabine D. G. Taylor Miles Institute for Preclinical Pharmacology, P. O. Box 1956, New Haven, Connecticut 06509, USA E. Hong Instituto Miles de Terepeutica Experimental, A. P. 22026, Mexico 22, D. F.
The structure and biogenesis of certain antibiotics *2,* 591 (1960)	W. A. Sexton Imperial Chemical Industries Ltd., Wilmslow, Cheshire, England
Role of periodic deworming of preschool community in national nutrition programmes *19,* 136 (1975)	P. M. Shah A. E. Junnarkar R. D. Khare Institute of Child Health, J. J. Group of Government Hospitals and Grant Medical College, Bombay, India

Quinolones *31*, 243 (1987)	Pramod M. Shah Zentrum der Inneren Medizin, Klinikum der J.-W.-von-Goethe-Universität, Theodor-Stern-Kai 7, D-6000 Frankfurt/Main 70, FRG
Advances in the treatment and control of tissue-dwelling helminth parasites *30*, 473 (1986)	Satyavan Sharma Cogswell Laboratory, Chemistry Department, Rensselaer Polytechnic Institute, Troy, New York, 12180, USA
Chemotherapy of cestode infections *24*, 217 (1980)	Satyavan Sharma S. K. Dubey R. N. Iyer Medicinal Chemistry Division, Central Drug Research Institute, Lucknow 226001, India
Chemotherapy of hookworm infections *26*, 9 (1982)	Satyavan Sharma Elizabeth S. Charles Medicinal Chemistry Division, Central Drug Research Institute, Lucknow 226001, India
The benzimidazole anthelmitics chemistry and biological activity *27*, 85 (1983)	Satyavan Sharma Medicinal Chemistry Division, Central Drug Research Institute, Lucknow 226001, India
Treatment of helminth diseases challenges and achievements *31*, 9 (1987)	Satyavan Sharma Medicinal Chemistry Division, Central Drug Research Institute, Lucknow 226001, India
Vector-borne diseases *35*, 365 (1990)	Satyavan Sharma Central Drug Research Institute, Lucknow 226001, India
Ayurvedic Medicine – past and present *15*, 11 (1971)	Shiv Sharma ‚Baharestan', Bomanji Petit Road, Cumballa Hill, Bombay, India
Mechanisms of anthelmintic action *19*, 147 (1975)	U. K. Sheth Seth G. S. Medical College and K. E. M. Hospital, Parel, Bombay 12, India
Aspirin as an antithrombotic agent *33*, 43 (1989)	Melvin J. Silver and Giovanni Di Minno Cardeza Foundation of Hematologic Research, Thomas Jefferson University, Philadelphia, PA 19107, USA and Second Medical School, Naples University, Naples, Italy

Immunopharmacological approach to the study of chronic brain disorders *30*, 345 (1986)	Vijendra K. Singh and H. Hugh Fudenberg Department of Basic and Clinical Immunology and Microbiology, Medical University of South Carolina, 171 Ashley Avenue, Charleston, SC 29425, USA
Implications of immunomodulant therapy in Alzheimer's disease *32*, 21 (1988)	Vijendra K. Singh and H. Hugh Feudenberg Department of Microbiology and Immunology, Medical University of South Carolina Charleston, SC 29425, USA
Neuroimmune axis as a basis of therapy in Alzheimer's disease *34*, 383 (1990)	Vijendra K. Singh Utah State University, Biomedical Division, Development Center for Handicapped Persons and Department of Biology, Logan, Utah 84322, USA
Some often neglected factors in the control and prevention of communicable diseases *18*, 277 (1974)	C. E. G. Smith London School of Hygiene and Tropical Medicine, Keppel Street, London, England
Tetanus and its prevention *19*, 391 (1975)	J. W. G. Smith Epidemiological Research Laboratory, Central Public Health Laboratory, London, England
Growth of *Clostridium tetani in vivo* *19*, 384 (1975)	J. W. G. Smith Epidemiological-Research Laboratory, Central Public Health Laboratory, London, England A. G. MacIver Department of Morbid Anatomy, Faculty of Medicine, Southampton University, Southampton, England
The biliary excretion and enterohepatic circulation of drugs and other organic compounds *9*, 299 (1966)	R. L. Smith St. Mary's Hospital Medical School (University of London), Paddington, London W. 2, England
Noninvasive pharmacodynamic and bioelectric methods for elucidating the bioavailability mechanisms of ophthalmic drug preparations *25*, 421 (1981)	V. F. Smolen Pharmacontrol Corp. 661 Palisades Ave, P.O. Box 931, Englewood Cliffs, New Jersey, 07632, USA

On the relation between chemical structure and function in certain tumor promoters and anti-tumor agents *23* 63 (1979) Relationships between structure and function of convulsant drugs *24,* 57 (1980)	J. R. Smythies Department of Psychiatry, University of Alabama in Birmingham Medical Center, Birmingham, Atlabama, USA
Gram-negative bacterial endotoxin and the pathogenesis of fever *19,* 402 (1975)	E. S. Snell Glaxo Laboratories Limited, Greenford, Middlesex, England
Benzodiazepine augmentation of the treatment of disruptive psychotic behavior *35,* 139 (1990)	David A. Solomon Brown University, Providence, Rhode Island, USA Edison Miyawaki Harward Medical School, Boston, MA, USA Carl Salzman Harward Medical School and Massachusetts Mental Health Center, Boston, MA 02115 USA
Emerging concepts towards the development of contraceptive agents *33,* 267 (1989)	Ranjan P. Srivastava and A. P. Bhaduri Division of Medicinal Chemistry, Central Drug Research Institute, Lucknow 226001, India
Strukturelle Betrachtungen der Psychopharmaka: Versuch einer Korrelation von chemischer Konstitution und klinischer Wirkung *9,* 129 (1966)	K. Stach C. F. Boehringer & Söhne GmbH, Mannheim-Waldhof, Deutschland W. Pöldinger Psychiatrische Universitätsklinik Basel, Basel, Switzerland
From podophyllotoxin glucoside to etoposide *33,* 169 (1989)	H. Stähelin and A. von Wartburg Preclinical Research, Sandoz Ltd., Basel, Switzerland
Chemotherapy of intestinal helminthiasis *19,* 158 (1975)	O. D. Standen The Welcome Research Laboratories, Beckenham, Kent, England
Immunotherapy for leprosy and tuberculosis *33,* 415 (1989)	J. L. Stanford Department of Medical Microbiology, University College and Middlesex School of Medicine, London W1P 7LD, England
The leishmaniases *18,* 289 (1974)	E. A. Steck Department of the Army, Walter Reed Army Institute of Research, Division of Medicinal Chemistry, Washington, D.C., USA

The benzodiazepine story *22,* 229 (1978)	L. H. Sternbach Research Department, Hoffmann-La-Roche Inc. Nutley, New Jersey, USA
Immunostimulation with peptidoglycan or its synthetic derivatives *32,* 305 (1988)	Duncan E. S. Stewart-Tull Department of Microbiology, University of Glasgow, Glasgow G 61 1 QH, Scotland
Hypertension: Relating drug therapy to pathogenic machanisms *32,* 175 (1988)	David H. P. Streeten and Gunnar H. Anderson Jr. State University of New York, Health Science Center, Syracuse, NY 13219, USA
Progress in sulfonamide research *12,* 389 (1968) Problems of medical practice and of medical-pharmaceutical research *20,* 491 (1976)	Th. Struller Research Department F. Hoffmann-La-Roche & Co. Ltd., Basel, Switzerland
Antiviral agents *22,* 267 (1978)	D. L. Swallow Pharmaceuticals Division, Imperial Chemical Industries Limited, Alderley Park, Macclesfield, Cheshire, England
Antiviral agents 1978–1983 *28,* 127 (1984)	D. L. Swallow, Imperial Chemical Industries PLC, Pharmaceutical Division, Alderley Park, Macclesfield, Cheshire SK 10 4 TG, England
Ketoconazole, a new step in the management of fungal disease *27,* 63 (1983)	J. Symoens G. Cauwenbergh Janssen Pharmaceutica, B-2340 Beerse, Belgium
Antiarrhythmic compounds *12,* 292 (1968)	L. Szekeres Department of Pharmacology, School of Medicine, University of Szeged, Szeged, Hungary J. G. Papp Senior Lecturer, University Department of Pharmacology, Oxford, England
Practically applicable results of twenty years of research in endocrinology *12,* 137 (1968)	M. Tausk State University of Utrecht, Faculty of Medicine, Utrecht, Netherlands

Stereoselective drug metabolism and its significance in drug research *32*, 249 (1988)	Bernard Testa and Joachim M. Mayer Ecole de Pharmacie, Université de Lausanne, CH-1005, Lausanne, Switzerland
Age profile of diphtheria in Bombay *19*, 412 (1975)	N. S. Tibrewala R. D. Potdar S. B. Talathi M. A. Ramnathkar A. D. Katdare Topiwala National Medical College, BYL Nair Hospital and Kasturba Hospital for Infectious Diseases, Bombay 11, India
On conformation analysis, molecular graphics, fentanyl and its derivatives *30*, 91 (1986)	J. P. Tollenaere H. Moereels M. van Loon Department of Theoretical Medicinal Chemistry, Janssen Pharmaceutica Research Laboratories, 2340 Beerse, Belgium
Antibakterielle Chemotherapie der Tuberkulose *7*, 193 (1964)	F. Trendelenburg Medizinische Universitätskliniken, Homburg, Saar, Deutschland
Diphtheria *19*, 423 (1975)	Dr. P. M. Udani M. M. Kumbhat U. S. Bhat M. S. Nadkarni S. K. Bhave S. G. Ezuthachan B. Kamath The Institute of Child Health, J. J. Group of Hospitals, and Grant Medical College, Bombay 8, India
Biologische Oxydation und Reduktion am Stickstoff aromatischer Amino- und Nitroderivate und ihre Folgen für den Organismus *8*, 195 (1965) Stoffwechsel von Arzneimitteln als Ursache von Wirkungen, Nebenwirkungen und Toxizität *15*, 147 (1971)	H. Uehleke Pharmakologisches Institut der Universität Tübingen, 7400 Tübingen, FRG
Mode of death in tetanus *19*, 439 (1975)	H. Vaishnava C. Bhawal Y. P. Munjal Department of Medicine, Maulana Azad Medical College and Associated Irwing and G. B. Pant Hospitals, New Delhi, India

Comparative evaluation of amoebicidal drugs *18,* 353 (1974) Comparative efficacy of newer anthelmintics *19,* 166 (1975)	B. J. Vakil N. J. Dalal Department of Gastroenterology, Grant Medical College and J. J. Group of Hospitals, Bombay, India
Cephalic tetanus *19,* 443 (1975)	B. J. Vakil B. S. Singhal S. S. Pandya P. F. Irami J. J. Group of Hospitals and Grant Medical College, Bombay, India
The effect and usefulness of early intravenous beta blockade in acute myocardial infarction *30,* 71 (1986)	Anders Vedin, Claes Wilhelmsson, Cardiac Unit and the Department of Medicine, Östra Hospital, University of Göteborg, S-41685 Göteborg, Sweden
Methods of monitoring adverse reactions to drugs *21,* 231 (1977) Aspects of social pharmacology *22,* 9 (1978)	J. Venulet Division of Clinical Pharmacology, Department of Medicine, Hospital Cantonal and University of Geneva, Geneva, Switzerland
The current status of cholera toxoid research in the United States *19,* 602 (1975)	W. F. Verwey J. C. Guckian J. Craig N. Pierce J. Peterson H. Williams, Jr., The University of Texas Medical Branch, Galveston, State University of New York Medical Center (Downstate), and Johns Hopkins University School of Medicine, USA
Systemic cancer therapy: Four decades of progress and some personal perspectives *34,* 76 (1990)	Charles L. Vogel AMI Kendall Cancer Center, 11750 Bird Road, Miami, Florida 33330, USA
Cell-kinetic and pharmacokinetic aspects in the use and further development of cancerostatic drugs *20,* 521 (1976)	M. von Ardenne Forschungsinstitut Manfred von Ardenne, Dresden, GDR
The problem of diphtheria as seen in Bombay *19,* 452 (1975)	M. M. Wagle R. R. Sanzgiri Y. K. Amdekar Institute of Child Health, J. J. Group of Hospitals and Grant Medical College, Bombay 8, India

Nicotine: an addictive substance or a therapeutic agent? *33,* 9 (1989)	David M. Warburton Department of Psychology, University of Reading, Building 3, Earley Gate, Whiteknights, Reading RG6 2AL, United Kingdom
Cell-wall antigens of *V.-cholerae* and their implication in cholera immunity *19,* 612 (1975)	Y. Watanabe R. Ganguly Bacterial Diseases, Division of Communicable Diseases, World Health Organization, Geneva 27, Switzerland
Steroidogenic capacity in the adrenal cortex and its regulation *34,* 359 (1990)	Michael R. Watermann and Evan R. Simpson Departments of Biochemistry and Obstetrics and Gynecology, and The Cecil H. and Ida Green Center of Reproductive Biology Sciences, The University of Texas, Southwestern Medical Center, Dallas, TX 75235, USA
Antigen-specific T-cell factors and drug research *32,* 9 (1988)	David R. Webb Synthex Research, Palo Alto, CA 94303, USA
Where is immunology taking us? *20,* 573 (1976)	W. J. Wechter Barbara E. Loughman Hypersensitivity Diseases Research, The Upjohn Company, Kalamazoo, Michigan 49001, USA
Immunology in drug research *28,* 233 (1984)	W. J. Wechter Barbara E. Loughman, The Upjohn Company, Kalamazoo, Michigan 49001, USA
Natriuretic hormones *34,* 231 (1990)	William J. Wechter and Elaine J. Benaksas Department of Medicine, Loma Linda University School of Medicine, Loma Linda, California 92354, USA
Metabolic activation of chemical carcinogens *26,* 143 (1982)	E. K. Weisburger Division of Cancer Cause and Prevention, National Cancer Institute, Bethesda, Maryland 20205, USA
A pharmacological approach to allergy *3,* 409 (1961)	G. B. West Department of Pharmacology, University of London, London, England

A new approach to the medical interpretation of shock *14*, 196 (1970)	G. B. West The British Industrial Biological Research Association, Woodmansterne Road, Carshalton, Surrey, England M. S. Starr Department of Pharmacology, St. Mary's Hospital Medical School, University of London, London, England
Adverse reactions of sugar polymers in animals and man *23*, 27 (1979)	G. B. West Department of Paramedical Sciences, North-East London Polytechnic, London, England
Biogenic amines and drug research *28*, 9 (1984)	G. B. West Department of Paramedical Sciences, North-East London Polytechnik, England
Some biochemical and pharmacological properties of antiinflammatory drugs *8*, 321 (1965)	M. W. Whitehouse The University of Oxford, Oxford, England
Wirksamkeit und Nebenwirkungen von Metronidazol in der Therapie der Trichomonasis *9*, 361 (1966)	K. Wiesner H. Fink Farbenfabriken Bayer AG, Leverkusen, FRG
Carcinogenicity testing of drugs *29*, 155 (1985)	G. M. Williams, J. H. Weisburger Naylor Dana Institute for Disease Prevention, American Health Foundation, Valhalla, NY 10595, USA
Organizing for drug discovery *32*, 329 (1988)	Michael Williams and Gary L. Neil * Pharmaceutical Division Ciba-Geigy, Summit, NJ, USA * Biotechnology and basic Research Support, The Upjohn Company, Kalamazoo, MI 49001, USA
Drug treatment of asthma *28*, 111 (1984)	Archie F. Wilson University of California, Irvine Medical Center, Orange, CA 92683, USA
Cooperative effects in drug-DNA interactions *31*, 193 (1987)	W. David Wilson Department of Chemistry and Laboratory for Microbial and Biochemistry Sciences, Georgia State University, Atlanta, Georgia 30307, USA

Nonsteroid antiinflammatory agents *10,* 139 (1966)	C. A. Winter Merck Institute for Therapeutic Research, West Point, Pennsylvania, USA
A review of the continuum of drug-induced states of excitation and depression *26,* 225 (1982)	W. D. Winters Departments of Pharmacology and Internal Medicine, School of Medicine, University of California, Davis, California 95616, USA
Basic research in the US pharmaceutical industry *15,* 204 (1971)	O. Wintersteiner The Squibb Institute for Medical Research, New Brunswick, New Jersey, USA
Light and dark as a „drug" *31,* 383 (1987)	Anna Wirz-Justice Psychiatric University Clinic, Wilhelm-Klein-Straße 27, CH-4025 Basel, Switzerland
Dioxopiperazines: Chemistry and biology *35,* 249 (1990)	Donald T. Witiak Yong Wey College of Pharmacy, The Ohio State University, Columbus, Ohio 43210, USA
The chemotherapy of amoebiasis *8,* 11 (1965)	G. Woolfe Research Department at Boots Pure Drug Company Ltd., Nottingham, England
Antimetabolites and their revolution in pharmacology *2,* 613 (1960)	D. W. Woolley The Rockefeller Institute, New York, USA
Noise analysis and channels and the postsynaptic membrane of skeletal muscle *24,* 9 (1980)	D. Wray Lecturer, Pharmacology Department, Royal Free Hospital School of Medicine, Pont Street, London NW3 2QG, England
Krebswirksame Antibiotika aus Actinomyceten *3,* 451 (1961)	Kh. Zepf Farbwerke Hoechst, Frankfurt a. M., FRG Christa Zepf Kelkheim (Taunus), FRG
Fifteen years of structural modifications in the field of antifungal monocyclic 1-substituted 1 H-azoles *27,* 253 (1983)	L. Zirngibl Siegfried AG, Zofingen, Switzerland
Lysostaphin: Model for a specific enzymatic approach to infectious disease *16,* 309 (1972)	W. A. Zygmunt P. A. Tavormina Mead Johnsohn Research Center, Evansville, Indiana, USA